THE EVOLUTION OF RELIGIONS

THE EVOLUTION OF RELIGIONS

A HISTORY OF RELATED TRADITIONS

LANCE GRANDE

Columbia University Press
New York

Columbia University Press
Publishers Since 1893
New York   Chichester, West Sussex

Library of Congress Cataloging-in-Publication Data
Names: Grande, Lance, author.
Title: The evolution of religions : a history of related traditions / Lance Grande.
Description: New York : Columbia University Press, [2024] |
Includes bibliographical references and index.
Identifiers: LCCN 2023010675 (print) | LCCN 2023010676 (ebook) |
ISBN 9780231212304 (hardback) | ISBN 9780231559317 (ebook)
Subjects: LCSH: Religion and culture—History—21st century. |
Multiculturalism—History—21st century.
Classification: LCC BL65.C8 G733 2024 (print) | LCC BL65.C8 (ebook) |
DDC 200.9—dc23/eng/20231016

Cover design: Henry Sene Yee
Cover illustration: Shutterstock

To those who have sought

knowledge over authoritarianism,

compatibility over conquest,

and tolerance over bigotry.

CONTENTS

END MATERIALS

PREFACE

A Modern Evolutionary Approach to History of Religion Studies

Every major religion is a branch of a complex evolutionary tree of ideological diversification. This tree involves many broadly held traditions that have been shaped over time by cultural selection. Countless ideological branches have evolved. Some of them have expanded and continued to diversify, while others have drifted to extinction. Religions have historically been a source of human social cohesion as well as an accelerant of intercultural conflict. A better understanding of how different major religious belief systems are historically interrelated, and the reasons for their diversification, could increase our mutual understanding and respect for each other as members of the same species.

Religions account for only one aspect of human culture. It is one, however, that has had a profound effect on the social development of our species. Religion has sometimes enhanced empathic and altruistic behavior in humans as well as aided in the growth of literacy, writing, and other beneficial aspects of human culture. It has inspired some of humanity's greatest works of art, music, and literature. It was responsible for the early development of higher education and for most of medicine and healing until the twentieth century. However, people have also used religion to inflame deep-seated intercultural conflict and violence. Understanding the broader interconnected histories of religions, and developing a greater tolerance of ideological diversity, may be critical to the survival of our species through the twenty-first century.

The study of religion's complex history has often been contentious and partisan. Here I examine the evolutionary history of major groups of religions in an *agnostic* context. For more detailed discussion of various points touched on briefly in the chapters of this book, there are boldface, parenthetical references to **Notes** throughout the text. These notes are keyed to more extensive discussions (sometimes several pages in length) located in the back of this book for readers who want more depth on one or more of these points. For example, the *agnostic* approach (just mentioned) is discussed in more detail in **Note PrA**. A glossary at the back of this book defines how I use certain terms that may have variable meanings elsewhere. Consistency and clarity in terminology is key to any study as broad as this one.

In the first two chapters, I introduce certain tenets of *classification* and *evolutionary process theory* to make a case for how it can apply to the study of cultural ideologies by social scientists. I support social scientists who are today attempting to reincorporate the use of comparativism and evolutionary perspective back into studies of human culture, after some badly misdirected colonialist attempts in the late nineteenth and early twentieth centuries. In chapters 3 through 11, I explore the broad evolutionary relationships among the ideologies of the world's major Organized Religions (explanation of the use of capitalization to follow). I also present simplified hypotheses about how these ideological variants are historically interrelated. My intended presentational context takes a nonpartisan look at the history and diversification of supernatural belief systems. There is no scientific way to know which such systems, if any, are "correct." Advocating for one particular religion or another, or for atheism, is not my goal. If anything, this book advocates for greater tolerance of diversity among belief systems (pluralism). I use the terms *religion* and *supernaturalism* interchangeably, defining this aspect of culture simply as the traditional belief in a realm or spiritual force beyond the physical world. I define *Organized Religion* as a more formalized version that involves institutionalized doctrine, rituals, or practice. *Religion* is a notoriously difficult and controversial concept to define.[1] It is primarily an anthropological term, not a theological one.[2]

Most branches of what I term *Organized Religion* in this book have been covered in much greater detail by specialists of particular religions. Here I have greatly simplified the characterization of each branch, emphasizing general ideologies (those parts of religion that evolve and diversify) supplemented with contemporaneous written historical records of diversification wherever possible. Collectively, this facilitates the study of broad patterns. I am more interested in the forest than the trees, so to speak. However, there is one type of metaphorical tree I use extensively throughout this book: *phylogenetic trees*. Such trees are simplified diagrams of historical diversification and interrelationships. They are used most often by evolutionary biologists to express historical relationships among different groups. (The use of phylogenetic trees is discussed further in **Note PrB**). The phylogenetic trees in chapters 3 through 11 are hypotheses of how the major named branches of Organized Religion discussed in those chapters are historically interrelated. The tree at the beginning of chapter 12 (the summary chapter) is a hypothesis of how all of the trees in this book fit together. When group names are used formally (e.g., Organized Religion), they are capitalized as proper nouns to distinguish them from their vernacular usage elsewhere.

As an evolutionary biologist and systematist for more than forty years, I have made considerable efforts to emphasize that modern evolutionary studies are no longer subjective searches for ladders of "progress" or universal "improvements." Evolution is not necessarily a qualitatively progressive process, nor does it follow predetermined goals. There is no universal pinnacle for evolution. It is simply the process of change and diversification over time: history in the making. Evolutionary research today focuses on deriving hypotheses of interrelationship among different groups, based on historical records and shared, hypothetically derived characteristics (**Note PrC**). A comparative, modern evolutionary approach provides a more inclusive understanding of diversity and historical relationships. It can provide a broader and deeper scientific perspective to studies of cultural history, much like it does for studies of natural history. I broadly synthesize the work of many historians and scientists here. I agree with E. O. Wilson in his assessment of where we are today in the modern information age: "We are drowning in information, while starving for wisdom. The world henceforth will be run by synthesizers, people able to put together the right information at the right time, think critically about it, and make important choices wisely."[3]

All year dates given in this book conform to the *Common Era* (BCE/CE) system. About forty different calendar systems are used in the world today. Many are specific to particular religions. Just to name a few, year AH 1 (the Islamic calendar) is roughly equivalent to year AD 622 (the Christian Gregorian calendar), year 1166 BE (the Buddhist calendar), and year AM4383 (the Jewish or Hebrew calendar). The Common Era system is used here, because it is currently the world's most widely used calendar system. Although the Common Era system is numerically equivalent to the Christian Gregorian system, it uses the terms *Before Common Era* (BCE) and *Common Era* (CE) to secularize the Christian terminology of *Before Christ* (BC) and *Anno Domini* (AD).

And so, we begin . . .

INTRODUCTORY SECTION

APPLYING EVOLUTIONARY THEORY AND PHILOSOPHY TO ASPECTS OF HUMAN CULTURE

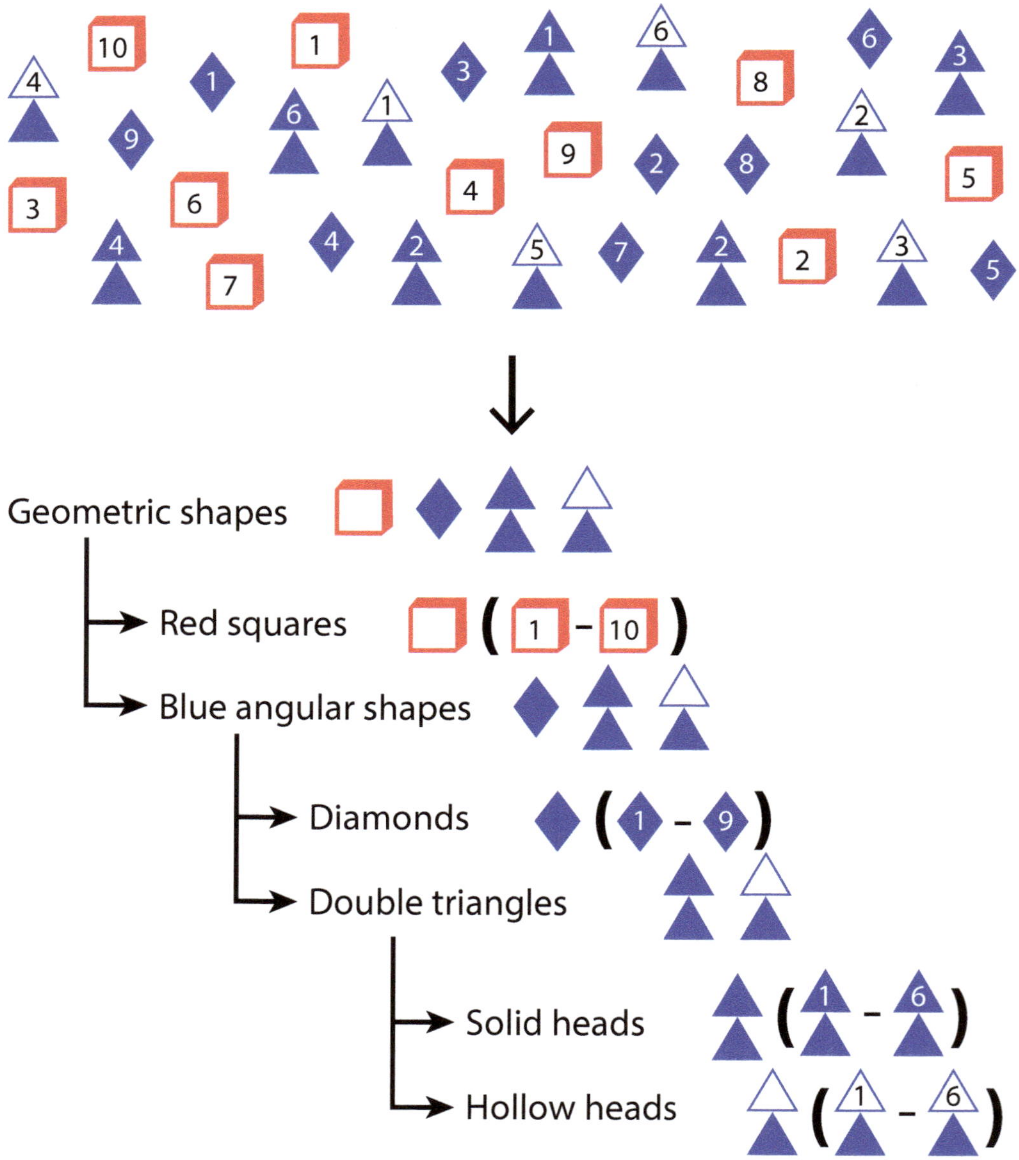

Classification: the identification and systematic organization of different hypothetical groups

1

RELIGIONS, CLASSIFICATION, AND PHYLOGENETIC PATTERN

All real science rests on classification, and only in the case we cannot succeed in classifying the various dialects of faith, shall we have to confess that a science of religion is really an impossibility.

—F. MAX MÜLLER, 1873

Religion is one of humanity's most diverse, yet narrowly understood, social systems. Like major groups of organisms (e.g., animals, plants, or bacteria), or like other major aspects of human culture (e.g., language, art, or governance), religion has evolved into many different groups, kinds, and samples of diversity over time. Although specific differences distinguish one kind of religion from another, certain characteristics and historical records also interconnect and help classify them. The hierarchical organization of a modern classification is not meant as a qualitative ranking or ladder of progress. It is instead simply an efficient organization of nested groups within groups, according to increasing levels of certain shared characteristics (such as the highly simplified example on the opposing page). Many of those shared characteristics can be hypothesized as evidence of historical relationship, especially in the absence of contemporaneous text records documenting otherwise. As a scientist who has studied and applied the philosophy and methods of evolutionary biology for several decades, I also became interested in the evolution of cultural systems. I was intrigued by the fact that many social scientists reject cultural evolutionism, cultural classification, and even the practice of *comparativism*, particularly within History of Religion studies.[1] That rejection drew me into a comparative study of one of human culture's most impactful aspects: religion. Much like lineages of biological organisms, lineages of religion have a constant and unavoidable tendency to diversify over time.[2] Although I am not a social scientist by training, my decades of experience as an evolutionary systematist, who is interested in change and diversification of groups over time, gives me a particular perspective on the study of deep history. It allows me to approach the subject of religious history in an unconventional way. I also use my analytical experience to tie together the works of many social scientists , general historians, and History of Religion specialists.

To understand the history and origin of diversification, we must first identify the groups of comparison (referred to by some social scientists as *comparands*).[3] Then, comparative analysis can help construct hypothetical patterns of organization among those groups based on contemporaneous historical records, and where no such records exist, on shared group characteristics. A highly simplified graphic example of a character-based pattern is the previously mentioned classification of shapes included at the beginning of this chapter. Evolutionary systematists theorize that such statistically strong hierarchical patterns of organization are at least partly a reflection of historical diversification (evolution).[4] Under such theoretical parameters, we can theorize that certain uniquely shared group characteristics are shared because they were inherited from a common (uniquely shared) ancestor.

The first two chapters of this book are largely technical and introductory. They explain why certain methods of comparative analysis and modern evolutionary theory should be applicable to historical studies of culture, much like they are applicable to historical studies of nature. These two chapters also establish the context for the rest of the book, which focuses on the historical diversification (evolution) of most of the world's major religious belief systems. Religions evolve as ideological (conceptual) traditions and practices, not as biological (physical) features. Where the evolution of culture is sometimes confused with the evolution of human physical capabilities or universal qualitative advance,[5] here it is not.

Chapter 1 (this chapter) begins with a brief section on the predilection of humans toward religion (supernatural belief systems). The rest of this chapter focuses on methods of comparativism and classification used to identify hierarchically organized patterns of diversity. The empirical search for systematic *patterns* is an analytical stage of analysis that precedes the application of *evolutionary process theories* to explain how those patterns came to be (discussed further in **Note 1A**). Such patterns can most simply be rendered either as hierarchically nested group arrangements (e.g., the *classification* on page 2) or as character-based branching diagrams (e.g., the *cladogram* on page 20 and the *phylogenetic trees* throughout this book). See **Note 1B** for more on the difference between cladograms and phylogenetic trees.

Chapter 2 focuses on the types of *evolutionary process theories* used by evolutionists to explain the origin of hypothetical group *patterns* in nature or culture. It highlights a number of similarities between biological evolutionary theory and cultural evolutionary theory. It also briefly reviews missteps in the early use of cultural evolutionism by early social scientists during the late nineteenth and early twentieth centuries, who attempted to apply a universal "evolutionary ladder" to human cultures. As stated in the preface, I use the term *evolution* in a modern context as simply "the process of change and diversification over time." I do not use the term in its older connotation as "a ladder of progress."

Chapters 3 through 11 each cover a major branch of Organized Religion, provide simplified hypothesis about the interrelationships of its major subbranches, and explain how major religions have diversified through human history. These chapters also briefly discuss iconic founders, gods, sacred texts, and various other historical details concerning the religions highlighted within those chapters.

Chapter 12 summarizes four hypothetical trends in the history of known Organized Religions regarding the following: (1) how all major branches of Organized Religion in chapters 3–11 appear to be historically interrelated; (2) how there has historically been a symbiotic linkage between the most successful major Organized Religions and imperial or monarchical ruling powers; (3) how there appears to have been a general decline in the role of women in most major Organized Religions; and (4) how religions have had a significant, but changing, impact on the evolutionary fitness of our species over time. Chapter 12 concludes that developing better tolerance and deeper understanding of ideological diversity among religions, as well as among other aspects of culture, is key to the future long-term survival of our species.

THE HUMAN PREDILECTION TOWARD RELIGION: IS IT INSTINCTUAL?

According to the *World Religion Database*, there were around six billion religiously affiliated adults and children around the globe by 2010, representing more than 88 percent of the world's population.[6] Many social anthropologists contend that religion is present in all human societies, even though many of these societies lack a word for it or do not consciously separate it from other aspects of their culture.[7] Many or most branches of what is termed here as *Organized Religion* (formally codified or institutionalized forms of religion) contain certain partisan elements of unquestionable doctrine for believers. Such emotionally entrenched convictions add to the sensitivity and resistance toward the comparative study of religions. It can be hard for many people to accept that their particular religious belief system is but a piece of a complex evolving social network that has unfolded over tens of thousands of years. Many people consider it acceptable to question the function and origin of *some* religions, yet consider it sacrilege, taboo, or otherwise wrong to ask the same questions about their own. They often subjectively distinguish their supernatural beliefs as being "truth" while depicting other supernatural beliefs as myth, misdirected superstition, ignorance, or heresy. Such cultural chauvinism has historically been common in major branches of Organized Religion (as defined here), particularly within Abrahamic Monotheism. Christian theologian Clement of Alexandria, from the second century CE, dismissed Traditional Christianity's Hellenistic (Greek) rivals as creations of fallen angels and therefore not true religions. Islamic fundamentalists classify all religions into three categories: wholly true, partially true, or wholly false, with Islam being the only true religion. There are countless other such examples. To the true believer, the function of religion is not primarily to make someone think logically or enrich his or her knowledge of the physical world. Instead, it is often a guide for how to gain help, or at least approval, from God, the gods, saints, ancestors, or other spiritual entities, to improve one's present life. For some, it can also represent a guide for success in the afterlife (achieving eternal peace in heaven, joining one's departed loved ones, achieving a better fate in the next reincarnation, or achieving nirvana). To yet others, it can be a guide for peaceful existence within a complex world. For many people, a religious assumption about "absolute truth" is comforting to embrace because of its unquestioning simplicity.

Moral and ethical behavior did not originate with religion; it was instead an independent product of evolution. It has been found to be present even in some of the great apes.[8] However, religion, in its most positive form, helped establish certain codified rules of ethical human behavior that often aided in the development of large civilizations (as civilization is defined in **Note 3C**). Some examples of such codification include the *Ten Yamas* of Hinduism, discussed in chapter 4; the *Five Precepts* of Buddhism, discussed in chapter 4; the *Forty-Two Negative Confessions* of Egyptian Polytheism, discussed in chapter 6; and the *Ten Declarations* or *Ten Commandments* of Abrahamic Monotheism, discussed in chapters 6–8. Honesty, altruism, and social cohesion have often been reinforced by such liturgical precepts. On a deeper emotional note, the attraction of religion is powerfully seductive at a personal level. It often provides a clear and simple way to address the most fundamental questions that many people have about their lives: Why am I here? What is the meaning of life? What happens to my spiritual essence after death? Some religions provide definitive answers to these questions in return for unquestioning faith in their doctrine. For many people, faith provides answers without a lot of complex qualifications. If you follow the rules, the rains will come, your side will win, or you will achieve eternal peace. It reaches to the core of human longing. In addition, by placing trust in religious philosophies, religious authorities, or religious institutions, many troubling moral dilemmas are often avoided: What is right? What is wrong? What constitutes the greater good?. Max Weber (1864–1920), one of the cofounders of modern social science and its study of religion, saw Organized Religion as one of the core forces of human society.[9] He thought it represented humanity's effort to impose intellectual order on the chaos of existence and to discover the ultimate meaning of the cosmos.[10] David Emil Durkheim (1859–1917), the other cofounder of modern social science, saw religion as having laid the foundation for modern society and the interactions that governed it.[9] Some evolutionary psychologists, such as Robin Dunbar, contend that as human societies got bigger, they required more elaborate and law-like forms of religion to prevent social breakdown.[2]

Historically, no major society has persisted very long without some element of religion. Perhaps it was a selective advantage in the early overall success of our species. Although several smaller countries are more secular than the United States (e.g., Scandinavian countries and some European countries), twentieth-century experiments to eliminate culturally embedded religions have been unsuccessful. The Soviet Union, for example, tried to extirpate religions but failed. The regime lasted only seventy years, and today, Eastern Orthodox Christianity is back as the official religion of Russia.[11] A 2008 international survey indicated that by 2008, 72 percent of all Russians self-identified as Eastern Orthodox Christians, 5 percent as Islamic, and 5 percent as other Christian (Protestant or Roman Catholic).[12] Only 18 percent self-identified as atheists. Similarly, Mao Zedong (1893–1976) and the Cultural Revolution of the late 1960s tried to eliminate religions from China, deeming them as "anticommunist." This too proved to be unsuccessful. In 1978, the Chinese government backtracked somewhat and released their *Freedom of Religion* act, which along with the 1982 *Constitution of the Peoples Republic of China* allowed "officially sanctioned" religions back into Chinese society (Taoism, Buddhism, Roman Catholicism, Protestantism, and Islam). Forcibly imposed atheism is no better than forcibly imposed religious belief. Religion seems to be an ingrained or instinctual cultural element for most of humanity.

Some scientists have proposed that the human species is genetically predisposed to believe in general concepts of a soul, afterlife, and deities.[11,13] Well-known geneticist Dean Hamer even proposed a specific gene as the source of that predisposition,[14] although this is highly controversial. In any case, the human predilection toward religious or spiritual belief seems embedded in most of human society. However, even if most humans are genetically hardwired to form supernatural beliefs, they are not genetically programmed to believe in a specific god, prophet, practice, or doctrine. There lies the niche for the evolution and diversification of religions.

THE COMPLEX DIVERSIFICATION OF RELIGIONS: HOW MANY ARE THERE?

It is estimated that at least ten thousand distinct Organized Religions are recognized today[6] and that ten times that number may have existed throughout history.[15,16] My phylogenetic trees (branching diagrams of historical diversification) focus mainly on major branches, most of which could be expanded into many detailed subbranches and reticulations by specialists (*evolutionary networks*, discussed in **Note 2E**). The exact number of different religious belief systems at the most detailed level is impossible to know, because nearly every individual person has their own particular perspective deviating in one detail or another from the typified form of the religion with which they self-identify. This is true even when a core doctrinal framework exists (e.g., a Veda, Tanakh, Bible, or Quran). Many people's personal beliefs are unique combinations of principles and practices from more than one religion, and many reject one or more elements of their particular branch. It is uncommon for someone to even have knowledge of every historical detail of the tradition they identify with. Consequently, potentially millions of personal belief variants have existed over tens of thousands of years. Such "individual variation" seems like an insurmountable challenge to the classification and evolutionary analysis of religious belief systems. However, the scale of the problem is no greater than the continued classification and evolutionary analysis of biological organisms, with their billions of genetic variations (genotypes) that have diversified over 3.7 billion years. Also, although an estimated ten thousand recognized kinds of Organized Religion are still being practiced today, an estimated nine million recognized species of organisms are still living today, each of which includes some level of intraspecific variation.

To understand the broader phylogenetic relationships of organisms, evolutionary biologists generally simplify the problem of complex diversity by comparing putative, representative groups (e.g., species, genera, families, or larger groups) rather than the detailed genotype of every individual biological organism. Similarly, cultural evolutionists also approach the problem by comparing putative, representative ideological subgroups of religion (e.g., typified or generalized sects, branches, schools, sampradayas, or denominations) rather than the detailed personal beliefs of every individual. Here, I focus on the broader elements of how each major branch of known Organized Religion has established itself over time, with the realization that uncountable minor variants also exist and that we know very little about the earliest origin of Organized Religion. In that regard, I take the role of generalist for this book. For centuries, specialists and generalists in biology have been collaboratively chipping away at deciphering the broad history of life. As a result, the most comprehensive phylogenetic tree of biological organisms becomes resolved at finer and finer levels of detail over time. I hope that the same will be true for the tree of religious ideologies, as well as for trees of other aspects of culture (**Note 1C**). One of the general principles for the study of historical diversification is that comparative science goes hand in hand with the classification of different general kinds of diversity.

USING CLASSIFICATION AS AN ORGANIZATIONAL TOOL FOR COMPARATIVE STUDIES

A classification is merely an efficiently organized (i.e., systematic) arrangement of perceived groups and characteristics in an attempt to better understand their differences and similarities (e.g., see figure at the beginning of this chapter). Its essential function is to organize descriptive information about different kinds (taxonomic groups) for storage, retrieval, and comparative purposes. It is a primary step in the search for statistically determined *patterns* of diversity.[17] Classification allows us to condense a great deal of information into a single name and to form a simplified hypothesis about the natural order of complex systems. Classification and scientific comparativism are inseparable for the understanding of diversity, even in the social sciences.[18] Together, they serve to clarify differences and specializations while simultaneously recognizing special similarities indicative of historical relationship. Classification began thousands of years ago with the ancient Greek philosophers, such as Plato and Aristotle. They attempted to systematically differentiate and organize "natural kinds" that were diagnosed on the basis of fixed properties (discussed further in **Note 1D**). This resulted in what was interpreted as a "natural order" into which these kinds could be arranged.[17]

Over the centuries, methods of classification continued to develop. By the early eighteenth century, a much more rigorous and formalized system of classification, known as the Linnaean system, developed. This system originated with the Swedish biologist Carolus Linnaeus (1707–1778). It contained official rules for the naming and for the hierarchical organization of groups of organisms (taxa).[19] The Linnaean system is "hierarchical" only in the sense that the smaller groups are embedded in larger and larger groups (i.e., higher taxonomic groups). Increasing group size is based on increasingly widespread occurrence of uniquely shared characters among taxa. I emphasize that in a taxonomic sense, "higher" means more broadly encompassing or inclusive of subgroups, rather than qualitatively better or superior. Groups of species form genera, groups of genera form families, groups of families form orders, groups of orders form classes, groups of classes form phyla, and groups of phyla form kingdoms. "Super" and "sub" groups between each level increase the number of hierarchical levels of organization even further (e.g., super genus and subfamily between genus and family, or superfamily and suborder between family and order). The Linnaean system developed more than a century before Darwin's 1859 publication of *Origin of* Species.[19,20] Linnaeus was a creationist building on earlier ideas of classification, rather than an evolutionist. He explained the origin of natural order in his classification of plants and animals as representing the "propinquity in God's creative scheme."[21]

After Darwin's *Origin of Species* was published, evolutionary process theory began to be used to explain the origin of natural order. It was proposed that the underlying patterns of organization that Linnaeus perceived in nature were the result of evolutionary diversification and *natural selection*. Classifications that incorporated evolutionary parameters, such as time elements and hypotheses of historical interrelationship, eventually became known as *phylogenies*. However, the importance of the Linnaean system of classification as a consistent way to identify specific groups has not diminished. It remains an invaluable tool for comparative studies and it has become the international rule for the official naming of biological groups phylogenetically (the International Codes of Nomenclature discussed further in **Note 1E**).[22] This provides a necessary level of consistency for broad studies of taxonomic diversity by scientists around the world. Such formalized rules for naming do not currently exist in the social sciences, which sometimes leads to confusion over the inconsistent way many authors use certain group names.

Taxa are hypothetical constructs. They are comparands used to identify putative groups whose members share certain characteristics. Taxonomy is a necessary organizational tool for conceptualizing and better understanding historical diversity and diversification. Just as taxonomy is used in studies of evolutionary biology, it can be used in History of Religion studies. In the words of History of Religion scholar, Jonathan Z. Smith: "Scholarly labor is a disciplined exaggeration in the direction of knowledge; taxonomy is a valuable tool in achieving that necessary distortion."[23]

CHARACTERIZING AND CLASSIFYING RELIGIOUS IDEOLOGIES

To compare and classify religious ideologies, we must first characterize (diagnose) the groups to be compared. We can start with the general anthropological concept of religion itself. In a most inclusive and unrestrictive sense (the way it is used here), the ideological nature of religion is simply a system of shared beliefs and practices connecting the physical world to the supernatural. In that context, I refer to religion also as *supernaturalism*. Supernatural belief systems probably began as something more instinctual, such as the general veneration of animistic forces, ancestors, or motherhood (discussed in chapter 3). It is impossible to know how old the roots of those cognitive elements are in humans (discussed further in **Note 1F**). However, as certain supernatural beliefs were seen by later generations as worth preserving, they became institutionalized. They became preserved with *narratives* (written scripture or oral transmissions), *practices* (rituals), and often with *objects of reverence* (iconic objects and structures); and they developed into one of the most influential aspects of human culture: *Organized Religion* (discussed in chapters 4–11). As indicated in the preface, the formally used names of Organized Religions used on phylogenetic trees in this book are treated as proper nouns and are capitalized to distinguish them from the vernacular usage of those and other terms elsewhere. Over thousands of years of history, Organized Religion evolved into the countless subgroups that exist today as well as countless other subgroups that became extinct or unrecognizably altered (**Note 1G**).

Some authors claim that religions "cannot be reduced to belief systems," often using Judaism as an example.[24] That point of view, however, often confuses religion with the broader context of undifferentiated ethnic culture. For example, Jewish culture is much more than simply a religious tradition. Many people even self-identify as nonreligious secular Jews (e.g., the *Hiloni*) and consider Jewishness to be an ethnicity rather than a religious tradition. This is why I specify "supernatural belief systems" or "religious ideologies" as the focus of this book (e.g., only the *religious* aspect of Jewish culture). *Culture*, in its broadest context, is a collective term that includes many different aspects, each with its own evolutionary history. These aspects include language, governance, art, religion, and many other manifestations of human intellect (discussed further in chapter 2). Religion, at its core (and as defined here), involves some sort of supernatural belief system. At the most detailed level, it is a specific combination of ideologies.

Many authors define religion in a more or a less restrictive sense than I do here. For example, some consider religion to be an invention of so-called Western (i.e., European, North American, and Australian) culture,[25,26] or even an invention of Christianity.[27] This is simply a case of subjectively defining the term "religion" more narrowly with a Western bias. Some define religion in a more complex discursive sense than I do here, preferring not to view it as a distinct form of culture at all.[26,28,29] Some also criticize the use of names for religions because the original adherents did not self-identify with those names. There are no official rules for formal names in comparative studies of culture, nor do names have to be tied to the way early members self-identified. Names are retroactively created tools for the purpose of description and comparison. The scientific name for birds (*Aves*) was not invented until the mid-eighteenth century CE, and birds have never self-identified as birds. However, that has not stopped ornithologists from naming and classifying fossil species of Aves that are as old as 150 million years. The focus of comparative and evolutionary analysis is not about the names; it is instead about the groups that those names represent. The name is only a tool of group categorization and organization. As articulated by History of Religion scholar Jonathan Z. Smith (1938–2017): "Religion [in the sense of being a classifiable set of different kinds] is solely the creation of the scholar's study. It is created for the scholar's analytic purposes by his imaginative acts of comparison and generalization."[30] ("Defining religion" is discussed further in **Note 1H**.)

In this book, I organize branches of religion based on my interpretation of written records and previous studies by History of Religion specialists wherever possible. Where no such records exist, I use collective patterns of shared characteristics to hypothesize historical patterns of relationship. I also supplement the phylogenetic trees with time estimates from various sources, although these are highly generalized estimates in most cases. At the very least, I hope this book projects the sense that most Organized Religions are historically interrelated at one level or another. No major Organized Religion today originated without influence (positive and negative) from traditions that came before it.

Friedrich Max Muller (1823–1900) was one of the founders of Comparative Religion as an academic discipline.[10] As quoted in the beginning of this chapter, he summarized the importance of classification in History of Religion studies when he stated: "All real science rests on classification, and only in the case we cannot succeed in classifying the various dialects of faith, shall we have to confess that a science of religion is really an impossibility."[31] In the late nineteenth century, Muller was severely criticized by theologians and officers of the Roman Catholic Church for such "anti-Christian" views, and he was accused of blasphemy.[32] Ironically, in the twenty-first century, some Comparative Religion specialists and social scientists now criticize Muller's work as being overly biased *toward* Christian conceptions of God.[33] The comparative study of religions never fails to attract passionate controversy.

The first requirement for any rigorous classification or comparative scientific analysis is to identify and define the groups of comparison within a consistent contextual framework. We need to know that when a group name is used in different studies, it consistently refers to the same group. Although such a system has been worked out within the biological sciences (as previously mentioned), nomenclatural rules have never been developed within the social sciences. The practice of naming and defining groups within the social sciences is often inconsistent. For example, some group names and definitions used for religions are based on the group's traditions (e.g., Hinduism or Abrahamic Monotheism), iconic individuals (e.g., Buddhism, Christianity, or Lutheranism), the geographic regions they originated or are sampled from (e.g., Romanian Orthodox, Ukrainian Orthodox, Russian Orthodox), or even nonreligious aspects of ethnic culture (e.g., the broadest context of Judaism). (The practice of naming groups for comparative analysis is further discussed in **Note 1I** and **Note 1J**.)

Taxonomic groups in the comparative sciences are never absolute. Each taxon is a hypothetical tool used for comparative study. Science is a process, and not a system of dictated authority. The most common terminal (end or smallest) taxon in biology is usually the "species." However, questions such as "What is a species?" and

"Are species real?" have long been debated within the biological sciences, and these questions have by no means been settled even today.[34,35,36] Sometimes the question is whether there are several closely related but distinct species or a single species that shows great variation. Some biologists recognize subspecies or regional populations as their terminal groups of comparison. Perhaps the safest, most pragmatic way to define biological taxa for evolutionary analysis is simply as *representative, typified samples of biological diversity.*

Similarly, in History of Religion studies, taxa can be thought of simply as *representative, typified samples of religious diversity.* Just as in studies of biological evolution, systematic identification of most typified cultural taxa (i.e., kinds of religion in this case), and the boundaries separating them, are hypothetical and ideologically generalized. However, identification of such groups represents the first step in any comparative study. As stated by History of Religion scholar Oliver Freiberger, "comparativism cannot work without the postulation that the comparands [units of comparison] are both similar and different."[3] To an evolutionist, the *differences* (sometimes enhanced by gaps in our knowledge of the evolutionary continuum) are what allow us to identify comparands as distinctly different kinds or taxa. However, it is the *similarities* at one level of generality or another that help us better understand them, systematically classify (organize) them, and generate hypotheses of historical relationship among them.

Discrete boundaries separating taxonomic groups are sometimes enhanced by missing information about intermediate forms. In rare cases in which we observe the complete history of an unbroken gradual continuum, (e.g., detailed, minute variation over time within biological populations or religious traditions), it can be difficult to define where one kind ends and the next begins (sometimes referred to as "fuzzy" or "blurred" boundaries).[37,38] However, the empirical record of deep historical change is intermittent, rather than continuous, because it is confined to excavation sites (fossil or archaeological) and to rare, sporadic examples of contemporaneous documentation (**Note 1K**). Gaps in the record of evolving lineages effectively enhance the distinctiveness of taxonomic groups (see the figure on page 14).

The use of classification and comparativism in the study of culture (such as the history and interrelationships of religions) is sometimes criticized as a product of Western efforts toward cultural dominance.[39] I contend that unbiased comparative classification is the most reliable way to understand the true diversity of foreign cultures. In fact, to avoid an unbiased comparative look at other cultures is to avoid attempting to understand those other cultures at all. As psychologist and anthropologist Benson Saler argues, comparison is an inescapable aspect of human cognition and reasoning because the unfamiliar can be conceptually assimilated only in terms of what is familiar.[40] In science, comparativism is not an effort to claim

HOW GAPS IN THE HISTORICAL (EVOLUTIONARY) RECORD ENHANCE THE DISTINCTIVENESS OF TAXONOMIC GROUPS

At the right is a character-based tree of relationship for terminal taxonomic groups *1* through **13**, and for higher taxonomic groups **A** through **I**. Terminal groups *1* through *8* (in *light italics*) are known only from fossil or archaeological evidence. Terminal groups **9** through **13** (in **bold face**) are extant forms. The distinctiveness of all taxonomic groups is enhanced by missing intermediate forms, as in the scenario below.

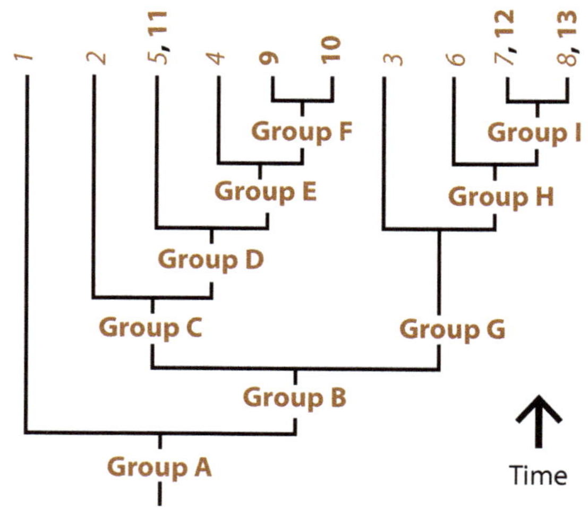

At the right is an evolutionary scenario including terminal groups *1* through **13** from above. The distinctiveness of each of those groups is enhanced by the many missing intermediate stages (e.g., missing stages between extinct group *8* and extant group **13**) whether those stages are biological or cultural.

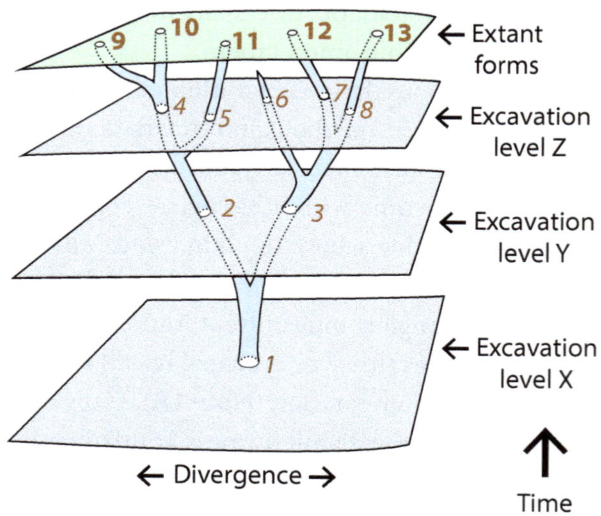

superiority or inferiority. It is, instead, simply an effort to better understand the diversity of things being compared.

Cross-cultural identification and classification of religious diversity is still in a very early stage, with only the most general groups identified. As stated by History of Religion author Stephen Sharot, the comparative study of even popular religions remains "under systematized and under theorized."[41] If we were to clearly identify the detailed evolution of all variations of religious belief systems, there would be many times the number we currently recognize, and the branch tips of my phylogenetic trees would be much more finely divided with reticulations (**Note 2E**). Within early Old World Hard Polytheism, for example, each local polity had its own pantheon and favored gods, and the status and role of particular gods often changed over time. To fully understand the detailed developmental history of religions, we would need to distinguish tens of thousands more kinds than we currently recognize. Figuring out the finer branching elements of any phylogenetic tree is the job of the specialists who are also comparativists. Most specialists have barely scratched the surface in the comparative study and classification of religions. In addition, most "extreme" syncretizations have never been distinguished or comparatively analyzed in detail (e.g., later branches of animistic polytheism in Mesoamerica and Africa showing strong influence from Christianity, or various syncretic combinations of Christianity and Eastern religions as "New Religionists"[6]).

Comparative Biologists have been chipping away at formally classifying species of organisms for centuries, with more than 1.2 million species named and defined so far, and an estimated 7.5 million more remaining to be done.[42] As noted by Jonathan Z. Smith (1938–2017), the famous History of Religion scholar from the University of Chicago, "There is much about method that can be learned [by History of Religion scholars] from colleagues in the Biological Sciences."[43] He further contended, "classification is a necessary prerequisite" to scientific analysis, whether biological or cultural.[43]

Some social scientists are wary of classifying religions in general. Graham Harvey, professor of Religious Studies at the Open University in the United Kingdom, warns against "the dangerous temptation to systematize [religions] because the things academics deal with are often very much alive" and that "classification names can be misleading."[48] Many social anthropologists share his views. I contend, however, that it is precisely *the job* of a scientist to express his or her thoughts as systematically as possible. The job of systematizing biological organisms is not easy either. Like religion, the biota is also "very much alive," and some classification names in biology can be misleading as well. However, because so many biologists have been working at it one way or another for so long, they have made significant progress in deciphering the historical diversification of biological organisms. We cannot hope to develop a broader historical understanding of religions or other aspects of culture without some sort of impartial systematization, no matter how messy or controversial the problem might seem in the beginning.

The comparative systematic analysis of religions is a relatively young practice, with broad studies going back only to the nineteenth century CE. Most early studies were flawed in that they were often biased in favor of touting the benefits of Western Christianity over *paganism* (a derogatory name for non-Abrahamic religions, as discussed further in chapter 2). In contrast, the comparative systematic study of biological organisms goes back at least 2400 years to the ancient Greek philosophers.[17] Methods of comparative biological analysis have been continually revised, even into the late twentieth century, as our understanding of evolutionary theory has continued to develop. (The basic method used for this book is discussed in **Note PrC**). I nevertheless agree with Harvey's sentiment that we should be careful that our classifications of culture do not imply universal value judgments (e.g., intellectual ability, qualitative superiority, or innate universal fitness). Evolutionary fitness is contextual: theoretically dependent on time, place, and associated environmental factors.

USING CHARACTER-BASED ANALYSIS TO DECIPHER PATTERNS OF RELATIONSHIP

Once we have identified groups for comparison, we can then search for hypothetical patterns of relationship among them. We do this by searching for hierarchies of *monophyletic groups* (groups of subgroups whose members show special similarities that are theoretically derived from a uniquely-shared common ancestor) (**Note 1L**). When reliable contemporaneous historical records are lacking, one way such patterns of relationship can be hypothesized is to search for strong statistical congruence of uniquely shared characteristics indicating groups within groups. This method of pattern analysis, called *cladistics*, produces a graphic pattern of relative relationship called a *cladogram*.[4] A cladogram does not require time calibrations or identification of specific ancestors. It is only a branching diagram based on special shared characteristics that have been statistically determined to be nonrandom and theoretically indicative of group relationship. Such characteristics are called *synapomorphies*. For example, the data matrix and cladogram shown in the accompanying figure for taxonomic groups **A**, **B**, and **C**, support a hypothesis that groups **B** and **C** are more closely related to each other than either is to group **A**. I stress here that each synapomorphy is hypothetical and forever subject to further testing and revision, as are the cladistic relationships that they indicate.

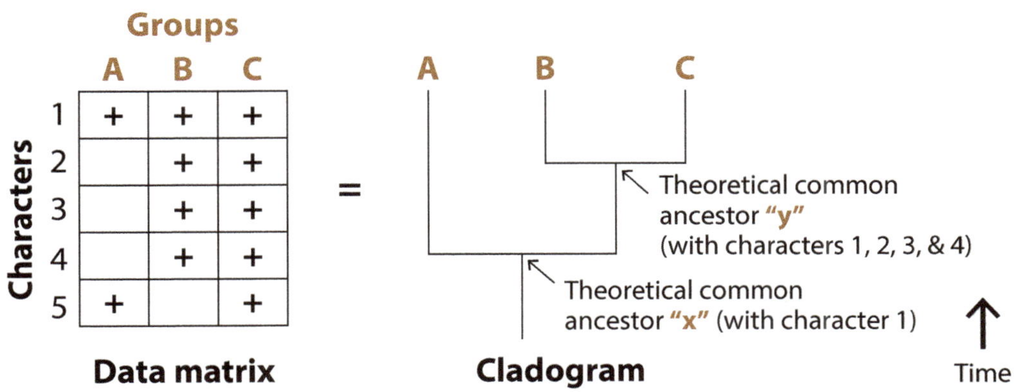

Simplified explanation of how a data matrix of characters 1–5 for groups A–C translates into a hypothetical pattern (cladogram) theorized to be the result of evolutionary history. The hypothetical pattern is determined through the use of statistics and parsimony analysis. Based on parsimony, the incongruent character 5 is theorized to have either been lost in group B or to have been independently derived in groups A and C (*parsimony* explained in **Note 1M**).

The particular pattern of relationship in this example is the result of the statistical preponderance of characters 2, 3, and 4, which are uniquely shared between groups **B** and **C** in the data matrix. This type of character is termed *synapomorphy* (shared, derived characters theorized to have been inherited from a common ancestor). Conflicting character number 5 in the data matrix, which is seemingly shared by groups **A** and **C**, is statistically overpowered by characters 2, 3, and 4. Therefore, the similarity between **A** and **C** is thought to have been either mistakenly identified, secondarily lost in group **B**, or an example of what is called *evolutionary convergence* (similarity not due to inheritance from a common ancestor). Most nonsynapomorphic similarities are collectively known as *homoplasies* in cladistic studies.[17,21]

Many social anthropologists contend that homoplasies are more common with aspects of culture (e.g., religions) than with aspects of nature (e.g., species of animals), although I see this as less of a problem (further explained in **Note 1M**). While a common default position in the social sciences is that different religions develop independently without relationship to other branches unless proven otherwise, this book takes a different approach. Here the default theory is that all Organized Religions are influenced at some level by earlier beliefs (i.e., are historically related to each other at some level) unless there is evidence to believe otherwise.

In terms of *evolutionary process theory* (discussed further in chapter 2), the cladogram implies that theoretically there was a common ancestor to a group containing **B** and **C** (unknown ancestor *y*) that was historically younger than the most recent common ancestor to a group containing **A**, **B**, and **C** (unknown ancestor *x*). This theoretical chronology of ancestors (be they biological or cultural) is a logical deduction, even if we have no idea exactly what those ancestors were. Since the late twentieth century, cladistic methods have been adopted by most evolutionary biologists.

A real-life example from the biological sciences could include three animal species, as in the next branching diagram, where **B** represents a species of sparrow, **C** represents a species of robin, and **A** represents a species of sheep. Based on cladistic analysis of uniquely shared characteristics (e.g., wing structure, flight feathers), the sparrow and robin appear more closely related to each other than either is to the sheep, even without knowledge of the numerous other closely related animal species or "missing links." This diagram also implies that the sparrow and robin belong to a group that has a more recent common ancestor (theoretical common ancestor *y*) than the group containing the sparrow, robin, and sheep (theoretical common ancestor *x*). Cladistic principles can similarly be applied to aspects of culture, such as three subbranches of the group known as *Abrahamic Monotheism*. In the accompanying diagram, **B** could represent Roman Catholicism, **C** could represent Protestantism, and **A** could represent Orthodox Judaism. Based on an analysis of shared characteristics (e.g., various characteristics of Traditional Christianity hypothesized to have been uniquely derived), Roman Catholicism and Protestantism appear to be more closely related to each other than either is to Orthodox Judaism, even without the added evidence from the numerous historical records that exist.

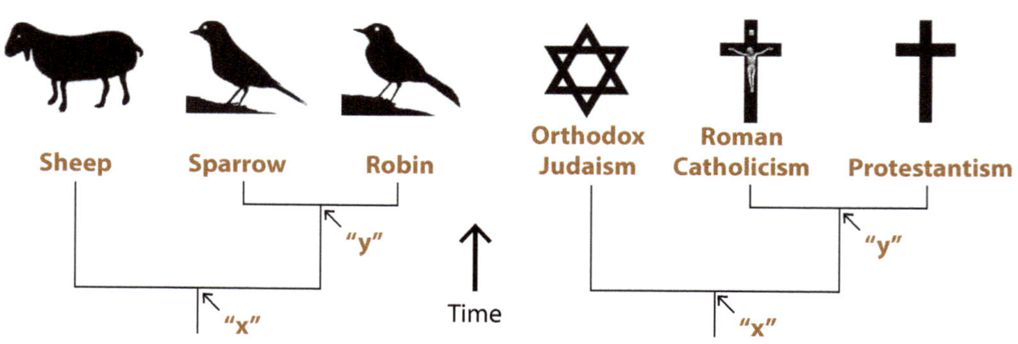

(*Left*) Cladistic relationships for three species of *Vertebrata* (animals with vertebrae). (*Right*) Cladistic relationships for three branches of *Abrahamic Monotheism* (religions revering the God of Abraham). For both examples, available evidence suggests that the most recent common ancestor **y** is younger than the most recent common ancestor **x**.

We can graphically illustrate cladistic methodology in a highly simplified way, and its association with idealized classification, by applying it to the thirty-one different numbered shapes (i.e., thirty-one different "kinds") at the beginning of this chapter (reproduced at the bottom of this page).

Cladistics can express what appears to be a number of special relationships among these different kinds, by organizing them in the most efficient (i.e., parsimonious) way possible. The thirty-one different kinds of shapes at the bottom of this page can be most parsimoniously represented by the cladogram on the next page, using seven hypothetical synapomorphies (shown in italics). (See **Note 1N** for further discussion about *parsimony* and different types of character similarities.)

This example is a highly idealized case. In practice, analyses involving biological or cultural groups are usually less idealized, with some of the branching order incompletely resolved and with some conflicting characters, such as character 5 on page 17. In spite of imperfections in the data matrix, many clear, statistically relevant patterns of relationship reveal themselves through cladistic analysis. Some evolutionary scientists, such as Darwinian scholar Niles Eldredge, have even used such techniques to propose theories of historical interrelationship and development for certain musical instruments (e.g., coronets).[45] Even the technology of musical instrumentation has a history of diversification. It has continually changed in response to new innovations and to the changing requirements of the user community. Most any group of different kinds (comparands) that has had a long history, has left a record of past (including extinct) variants, plus a shared character-based pattern among extinct and extant variants. Collectively, this information suggests a pattern of historical development and diversification (an evolutionary history).

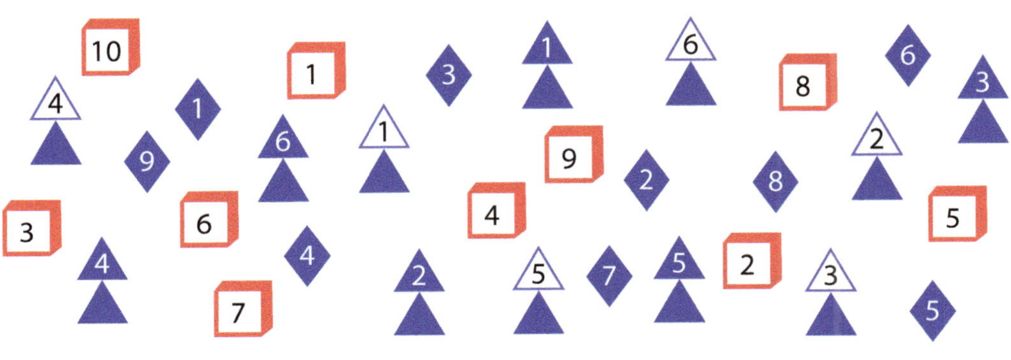

An assortment of different kinds that are cladistically organized in the next figure.

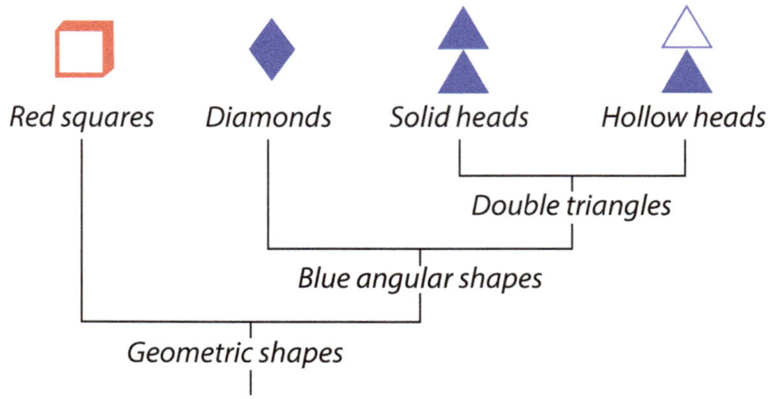

Cladogram of the kinds from the previous figure together with the synapomorphies used to organize them.

In the next chapter, I briefly discuss how certain concepts of *evolutionary process theory* from the biological sciences can be applied to the social sciences, particularly with regard to explaining historical patterns of cultural group diversity over time. I also discuss some early misdirected attempts by sociologists of the late nineteenth and early twentieth century to mistakenly apply evolutionary theory to culture as a ladder of progressive "improvement."

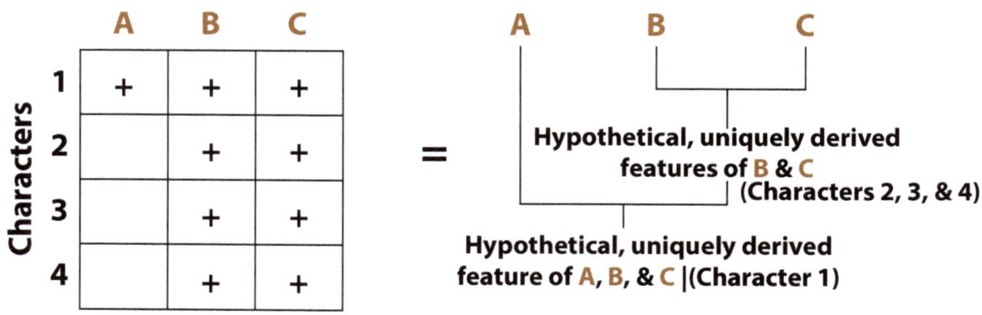

Character-based *pattern hypothesis* for groups A, B, and C

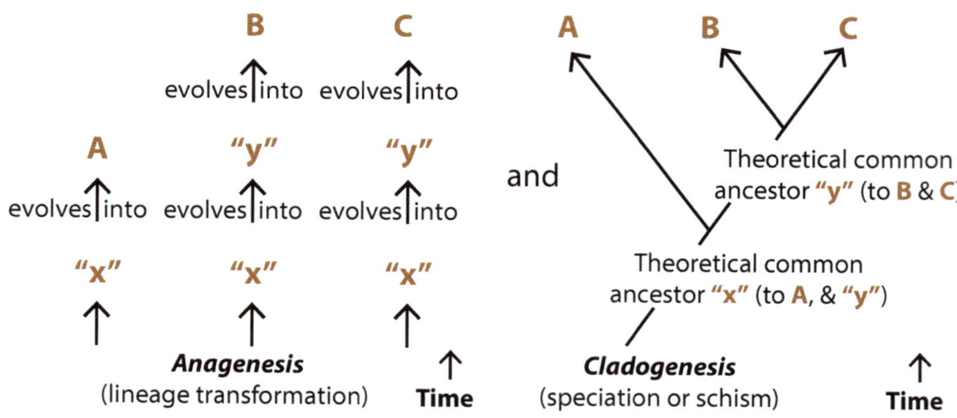

Evolutionary *process theories* to explain the character-based pattern for groups A, B, and C

2

EXPLAINING HYPOTHETICAL PATTERNS WITH EVOLUTIONARY PROCESS THEORY

Evolution is really the same as history if we take it in its objective sense.

—F. MAX MÜLLER, 1889

Hypothetical group patterns of organization derived from comparative studies, such as the classifications and the cladograms described in chapter 1, can be explained with modern evolutionary process theories. Although these theories were originally proposed to explain patterns of biological diversity, they can also be used to explain patterns of cultural diversity (e.g., the diversity of religions). Unfortunately, the application of evolutionary theory to cultural diversity got off to a bad start in the social sciences. Certain biologists and social anthropologists of the late nineteenth and early twentieth centuries misappropriated evolutionism in a culturally biased context (e.g., Spencerian Evolutionism, racial hygiene, and worse; discussed later in this chapter). The subsequent backlash against these early, misdirected efforts resulted in a widespread rejection of cultural evolutionism and comparativism by many social scientists.

After its early rejection in the social sciences, evolutionary theory and methods continued to develop within the biological sciences through the late twentieth and early twenty-first centuries. Modern evolutionary research came to focus on characteristics of *relationship* among groups (as discussed in chapter 1) rather than on differences and qualitative value judgments. Evolution was no longer seen as a ladder of universal "progress" (sometimes referred to as *progressive evolution*[1].) Evolutionary analysis moved toward being an empirical search for character-based patterns of relationship. Such patterns were hypothesized to be retrospective maps of historical diversification. In this revised methodological context, comparativism has been slowly coming back in the social sciences, although some resistance persists there.

THE EVOLUTION OF CULTURE VERSUS THE EVOLUTION OF NATURE

Another reason some social scientists reject evolutionary theory and comparativism in the study of culture is because they claim that cultural is much more complex and more difficult to define than biology.[2] I contend that such analytical challenges are no greater for social scientists than they are for evolutionary biologists. However, this requires that cultural anthropologists focus on one aspect of culture at a time. Cultural history, like natural history, does not evolve as a homogeneous entity. Even though different aspects of culture can influence each other's evolutionary trajectory, each aspect generally diversifies and evolves in its own way. To paraphrase anthropologist Gary Feinman, aspects of culture need to be decoupled from each other in order to decipher their evolutionary histories.[3]

Within cultural history, for example, religions may evolve at a different rate and move in a different direction than other aspects of culture such as language, governance, or art. A culture could capture elements of religion from one neighboring culture, and elements of linguistic or architectural traditions from yet other neighboring cultures by simple imitation. In general, different aspects of culture are not necessarily linked any more than different aspects of natural history are. The evolution of a particular group of bird species, for example, may follow a different pattern than the evolution of the bacteria that resides within the gut of those birds, or the geologic evolution of the earth's continents where those species live. To look at history as a homogenous entity may inevitably obscure the evolutionary patterns of individual aspects as well as possible congruence between some of them. Some social scientists contend that religious belief cannot be viewed as distinct from other aspects of culture.[4] However, if we are interested in the evolutionary history of religion, looking at it as a distinct aspect is a necessary first step in the analytical process.

Even though different aspects of culture can evolve independently, once social scientists have analyzed them independently, they can then look for signs of broader influence between them. Such influence is indicated by a congruence of historical patterns (e.g., between the evolution of *major empires* and the evolution of *major religions*). Similarly, once biologists analyze different aspects of natural history independently, they often look for signs of influence between them (e.g., influence between *continental drift* and the *evolution of certain groups of plants and animals*). The subsequent study of congruent patterns between different aspects of culture or of nature is an important part of investigating deep history.[5,6] However, as previously stated, the congruence of such secondary patterns can be empirically supported only after the different aspects have been analyzed independently.

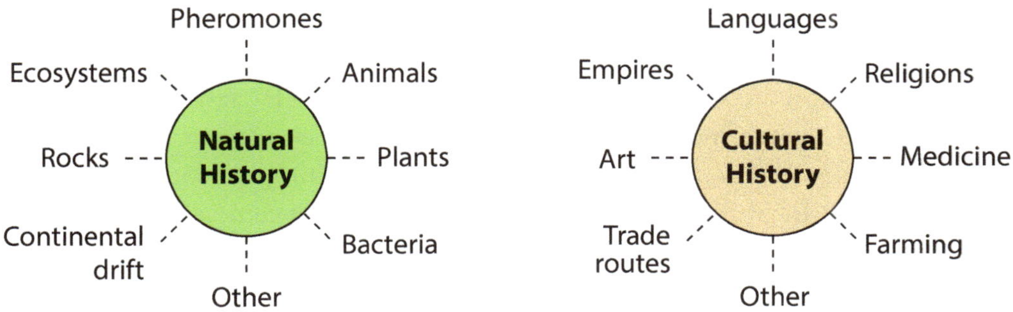

Examples of historical "aspects": (*left*) in studies of natural history and (*right*) in studies of cultural history.

Historical complexity is not the only reason that many social scientists have avoided evolutionary theory in the study of culture. Another reason, as mentioned at the beginning of this chapter, is the unethical way those methods were used in the past by several biologists and anthropologists of the late nineteenth and early twentieth centuries. Such subjective bias often went under the name of *Social Darwinism*.

SOCIAL DARWINISM, EUGENICS, AND SOCIAL ANTHROPOLOGY'S BACKLASH AGAINST CULTURAL EVOLUTIONISM

Several biologists and early anthropologists of the late nineteenth and early twentieth centuries supported a theory called Social Darwinism. This theory mischaracterized cultural evolution as a ladder of progress, or a scale of social improvement, among races and cultures. It ultimately blamed poverty, illiteracy, and crime on genetic inferiority. This eventually led to a number of ethical abuses ranging from eugenics (i.e., racial hygiene) to genocide.[7] Although those early misappropriations of evolutionary theory are discredited today, they left lasting scars within the social sciences.

The name Social Darwinism is a misleading and misappropriated name because it did not originate with Darwin.[8] The term was first used by various social scientists of the late nineteenth and early twentieth centuries who were intent on riding Darwin's coattails and the popularity of his groundbreaking *Origin of Species*.[9] It is true that Darwin was a product of colonial times. Nevertheless, in his *Origin of Species* book, he avoided discussions of the different races of humans. In a later book called *On the Races of Man*, he concluded that all living races of humans belonged to the same species.[10]

The later rise of Social Darwinism envisioned human races as qualitatively different. The tenets of that movement are most often associated with social anthropologist and political theorist Herbert Spencer (1820–1903).[11] He saw social evolution not merely as change and diversification over time, but as a ladder of progression toward a greater level of perfection, typified by Western European society. Spencer's sense of Social Darwinism eventually became adopted by some early paleontologists, biologists, and social scientists. Various politicians used it to justify political conservatism, imperialism, and racism. Some even used it to discourage humanitarian efforts for the disadvantaged, with the justification that those efforts interfered with nature's laws to improve the human species.[8]

In the latter half of the nineteenth century, German embryologist and Social Darwinist Ernst Haeckel (1834–1919) used Social Darwinism to produce a tree with which he identified dozens of distinct human races classified among twelve living *species of men*.[12,13] While Darwin had been cautious not to use his theories to construct any phylogeny of different species of living humans, Haeckel was not. Haeckel's progressive tree is based on biological characteristics, primarily on hair type, but also considering skin color and head shape. The problem with Haeckel's tree was not so much his use of comparativism, but rather with his qualitative assessments and progressive rankings of race-based human species that he identified. Haeckel, like Spencer, held that biological evolution resulted in races that were unequal in intelligence and ability.[12] He promoted the idea that living humans could be subdivided into innately superior and inferior races (evolutionary progressionism applied to race). Many of his ideas collectively fueled what became known as "scientific racism." Haeckel, like other Social Darwinists of the late nineteenth and early twentieth century, believed the Indo-Germanic race to be culturally and biologically superior to races such as "Kaffres" and "Negroes."[12] He advocated for racial purity among the German people. He proposed that many of what he called the "underdeveloped human species" were on a predetermined course for extinction.[12,14,15] Haeckel also believed that the "laws of evolution" conferred the right of some human races to dominate others. His views are thought by many authors to have even contributed to the rise of Nazism.[16] However, Haeckel was not embraced by the Nazis, despite isolated attempts to use his fame to support Nazi ideology.[15] For one thing, he was not an anti-Semite. He ranked Semites (and Jews) on the highest hierarchical level, along with the Germans.[12,15] The racist ideas of Spencer and Haeckel fueled the program of *eugenics* in parts of the United Kingdom and in the United States during the so-called Progressive Era of the late nineteenth and early twentieth centuries.[17]. Haeckel's concept of evolutionary "progress" is discussed further in **Note 2A**.

The name *eugenics* was coined in 1883 by Darwin's half-cousin, Francis Galton (1822–1911), who pioneered the movement.[17] Galton was an anthropologist and statistician who took Darwin and Haeckel's ideas down an even darker pathway, although this time with Semites no longer included in the preferential category. Eugenics sought to preserve and "improve" dominant social groups perceived as superior, and weed out lower forms.[18] The program's adherents believed that through selective breeding of people, the human species could control its own evolution. Eugenics generally held the blond-haired, blue-eyed Nordic race (i.e., Germanic, Slavic, Northern European, and Anglo-Saxon peoples) was a genetically superior "master race." The program advocated forcible sterilization of the poor, disabled, and subjectively "immoral," while it encouraged those it deemed as "superior" races to reproduce.[8]

American conservation biologist and amateur anthropologist Madison Grant (1865–1937) took Galton's ideas and wrote an influential book in 1916 called *The Passing of the Great Race*. It argued for a strong eugenics program, and went through several editions in the United States during the early twentieth century. It was a virulent alarmist warning against the thinning of "superior Aryan blood" (by which Grant meant the Northwestern Europe and Northern European settlers of the original thirteen colonies) through interbreeding with immigrant "inferior races" (by which he meant Poles, Czechs, Italians, Jews, Africans, and Asians). (It should be noted that the term *Aryan* originally referred to aspects of culture rather than race until the mid-nineteenth and early twentieth century when it began to be usurped by some as a racist term.[19]) Grant's book also recommended segregating "unfavorable" races in ghettos. The eventual rise of Adolf Hitler's obsessive attempt to create an Aryan master race in mid-twentieth-century Germany was partly inspired by the earlier eugenics work in the United States. Nazism claimed that the Indo-Germanians were the descendants of the "superior" Aryan Nordic migrants. Even today, white supremacy groups in the United States still often refer to themselves as Aryans in the same bigoted sense.[19] Hitler personally wrote to Grant thanking him for writing the book and referring to it as "my Bible."[18] He emphasized its anti-Semitic element to help justify the Holocaust.

It is little wonder that many historians and social scientists are reluctant to come back to looking at human culture in an evolutionary context. The early pseudoscientific elements of Social Darwinism, associated with its colonialist mindset and racist ideology, gave evolutionism and comparativism a bad reputation in the social sciences for many decades. Humans are capable of appropriating most any ideology for nefarious purposes. Because of Social Darwinism and many similar abuses during the early twentieth century, many social anthropologists and ethnologists moved far away from looking at culture in an evolutionary context.

CHANGING IDEAS ABOUT EVOLUTION AND COMPARATIVISM IN THE LATE TWENTIETH CENTURY

After the early rejection of evolutionism in the social sciences, evolutionary theory continued to develop and change within the biological sciences (**Note PrC**). By the late twentieth century, much of evolutionary science had moved away from the progressive-ladder model of evolutionary process, toward a qualitatively neutral, nonprogressive branching model. Evolutionary science became focused on using shared characteristics among different groups of organisms to construct hypothetical patterns of relationship and historical diversification (cladograms and phylogenetic trees). These patterns were neither implicitly progressive nor predetermined (as described in chapter 1). Biologists no longer portrayed evolution as a universal progression of lineages from inferior "lower" forms to superior "advanced" forms. The term "highly evolved" became a meaningless vernacular rather than a term indicating general superiority. Most modern evolutionary biologists became focused on the search for *shared characteristics* and *relationships* rather than qualitative *differences* or *improvements*.

By the time biological scientists had transformed evolutionary theory away from the ladder-of-progress approach, most social scientists had long abandoned the idea of studying human culture in any sort of evolutionary context. Furthermore, many social scientists even rejected the idea of historical comparativism altogether, favoring what was sometimes referred to as the "particularist" approach (e.g., Franz Boas early in his career, discussed in chapter 3). This discouraged the broad comparative study of culture and effectively threw out the baby with the bathwater. The dirty bathwater in this case was racial or cultural discrimination. The baby was the comparativism necessary for deciphering deep historical relationships among different cultures and recognizing their common interests. This backlash within the social sciences exists to the present day.[20,21,22,23,24] Some History of Religion specialists even in the early twenty-first century have been quoted as saying "comparison is over" within their field.[22] Such a stance impedes the broader understanding of cultures outside of our own. I believe that the philosophy of modern comparative analysis is a necessary component of the study of deep history. The goal of broad analyses is not to erase distinct features of the different "kinds" being compared (as feared by some history of religion specialists). Instead, the goal is to look for special similarities that suggest historical relationship among those clearly different kinds. It should not be an either/or contest between comparativism and specialization. It should instead be a synergistic collaboration between these two methods of analysis.

Although modern evolutionary theory is interested in historical interrelationships among groups, it is not a search for subjective "progress" from universally "inferior" groups to "superior" groups. Paraphrasing Stephen J. Gould (1941–2002), the famous late-twentieth-century evolutionary biologist: *to equate evolution with progress is one of the strongest impediments to understanding what evolution* is.[25] Part of the problem, historically, may have been confusion over interpretation of the word "progress." According to the Merriam-Webster dictionary,[26] it has two quite different meanings: It can mean nonqualitative "forward movement" (simply a change over time). It can also mean "a gradual betterment" (progressive superiority). Conflicting interpretations of language often adds to philosophical conflict.

Social scientists have spent an inordinate amount of time and ink debating comparativism.[27] As stated by History of Religion scholar Wendy Doniger, "The academic world . . . now suffers from a post-post-colonial backlash; in the age of multinationalism to assume that two texts from different cultures are 'the same' in any significant way is regarded as demeaning to the individualism of each, a reflection of the old racist attitude that 'all wogs look alike.'"[24] The effect of cultural anthropology's rejection of comparativism in twentieth century slowed efforts to understand many aspects of human culture, such as religion, in the broadest historical context. This is not to say that the comparativist approach disappeared completely from History of Religion studies. Scholars such as Mircea Eliade (1907–1986) and Jonathan Z. Smith (1938–2017) occasionally supported certain forms of the comparative approach in their study of religions.[20,28,29,30] Other more recent History of Religion scholars also promote a return to comparativism, including Oliver Freiberger,[21] William Paden,[31] and many others.[32] As Freiberger argues, religious studies must be grounded in comparativism to be considered as an academic discipline.[21]

Comparativism is an inescapable element of the scientific method; and scientific method is a critical element in the study of deep history and diversity. Today, evolutionists do not argue about whether or not to use scientific method. They only debate which analytical methods should be used and which specific hypotheses should be supported if conflicting hypotheses result from different scientific studies. Scientific (empirically based) hypotheses are, by definition, meant to be testable and potentially falsifiable.[33] That is a critically important part of science. Comparativism is also necessary to establish interpretive context. Even most opponents of cultural comparativism in the anthropological sciences use some level of it for their descriptions to be intelligible to the rest of the scientific community.[27]

As imperfect as broad comparative studies of culture may be, the alternative of completely avoiding comparativism is worse. Strict particularism can often add to subjective bias, because History of Religion specialists who promote particularism sometimes self-identify with the religion they focus on. As an extreme case, the racist colonialist assessment of non-European culture in the nineteenth century was not based on comparative knowledge; it was based on chauvinistic ignorance of this culture and a strong Western bias. The colonialists had no interest in understanding shared characteristics and historical connections to cultures outside their own. Instead, they were focused on supporting exclusionary *differences* and the preconceived notion of Western superiority. They could then propose that the religious belief systems of the West were not only different but also *better.* As stated by Doniger, "European colonialism was supported by a discourse of difference."[34]

For a deep understanding of the evolution of religions, we must consider the shared similarities and historical origins of different core defining elements and focus on the phylogenetic patterns of relationship that emerge from such studies. To focus only on exclusivist characters of individualism and differences is to ignore the social connections that bind us together as a species and provide clues to our common history. As in biology, and as noted previously, the comparativist (generalist) and the particularist (specialist) should be collaborative, not adversarial. Their combined studies should complement each other in a way that provides a deeper, more widely understandable, historical picture.

USING EVOLUTIONARY PROCESS THEORIES TO EXPLAIN HOW SYSTEMATIC PATTERNS DEVELOPED OVER TIME

Evolutionary process theories used to explain diversification patterns of *biological species* can also be used to explain diversification patterns of *supernatural belief systems.* I discuss only a few general process theories here, because the focus of the remaining chapters in this book is on the patterns of diversification and cultural interrelationships among religious ideologies.

In *biological evolution,* the origin of new groups (e.g., species or genera) can be explained as a product of *biological diversification* (*speciation*). It generally is not possible to identify how long or rapidly a biological speciation event took or to know precise ancestors based on contemporaneous records. Because biologists do not observe the complete process of speciation as a seamless continuum in real time, they observe only samples of it from various points in time (preserved as fossil and living species). They nevertheless use this series of intermittent stages to build a historical sequence of diversification and to construct phylogenetic trees. The process of even a single biological speciation event, from start to finish, is generally thought to be an extremely long one, transcending humanly observed spans of time. However, even though evolutionary biologists cannot document speciation events in their temporal entirety, they can observe short-term biological processes in modern time (e.g., genetic inheritance, competition, extinction) that can be extrapolated into process theories about how speciation has occurred. Evolutionary biologists also continually discover new fossil and living species, which fill in missing "links" and add to the temporal record of biological evolution.

In *cultural evolution*, the origin of new groups (e.g., schools, sects, or ideological traditions) can be explained as a product of *cultural diversification* (the cultural equivalent to *biological speciation*). The study of cultural diversification has one significant advantage over the study of biological speciation. Unlike in biology, where the complete process of diversification has rarely, if ever, been recorded in its entirety, many examples of cultural diversification (e.g., the evolution of new religions) have been humanly observed in their entirety. This is due to the shorter durations and more recent occurrences of religious schisms and transformations. All things considered, a more detailed documentation of evolutionary processes exists for culture than for biology.

TRANSMISSION OF TRAITS OVER TIME AND THE EFFECTS OF NATURAL OR CULTURAL SELECTION

In *biological evolution*, the microevolutionary process explanation includes populations of interbreeding organisms that perpetuate billions of genetic mutations and gene recombinations forming a *genotype*. Normally, the genotype is passed down "vertically" by way of reproductive success, from parent to offspring, from biological generation to generation. Certain compatibilities are necessary for genetic transmission (e.g., animals being of the same species, enabling transmission of genetic traits to succeeding generations). There are also rare cases of "horizontal" gene transfer from one species to another. (Vertical versus horizontal transmission of traits is discussed further in **Note 2B**). Genetic variations eventually result in different physical expressions (phenotypes). The resulting physical variations interact competitively within their ecological environment that is also ever changing. As a result, some lineages survive and diversify, whereas others become extinct.

In *cultural evolution*, the microevolutionary process explanation is similar to that of biological evolution. The evolution of religions, for example, includes countless small variations due to new ideas and meme recombinations forming what is termed here as a *memotype* (memes are discussed further in **Note 2C**). The memotype is passed from one ideological generation to another by way of oral transmission, written documentation, and imitation (i.e., learning[1]). Certain compatibilities are necessary for memetic transmission (e.g., groups having a similar language, enabling transmission of ideas to succeeding generations). Many resulting variations interact competitively within their social environments that are also ever changing. Some lineages survive and diversify, whereas others become extinct. Just as each major biological system (e.g., various anatomical, physiological, and immunological systems) has a different evolutionary history, each major aspect of culture (e.g., religion, language, and governance) has a different evolutionary history. "Horizontal" transfer of traits may be more common in aspects of cultural evolution than it is in biological evolution (e.g., merging of elements from contemporaneous religious ideologies), but examples of ideological syncretization do not diminish the value of looking at culture in an evolutionary (phylogenetic) context. Like biological evolution, cultural evolution is a process of change and diversification over time.

The general process of evolution involves descent with modification in response to the ever-changing and fractionating environment. The nonrandom differential survival of one type over another in response to the environmental factors in which it exists is traditionally termed *selection* by evolutionary biologists. Selection is defined as the result of *fitness* (reproductive success in transmission of traits to succeeding generations). (Fitness is discussed further in **Note Note 2D**). Whether we are dealing with biological species or aspects of culture, some lineages survive and others become extinct. With survival comes inevitable change and diversification over time. Darwin originally coined the term *Natural Selection* as a driver of biological evolution among populations in nature.[9] In much the same way, *Cultural Selection* is seen here as a driver of cultural evolution among social comparands.

Natural Selection proposes that the biological phenotype (the physical expressions of genotype) most adaptable to its natural environment has the best chances for survival (i.e., it is *naturally selected for survival*). The lineages with individuals that are the longest-lived and most reproductively successful are generally the lineages that are most competitive for a given environment. However, when the environment changes, rates of survivability and reproduction (i.e., positive or negative selection pressure) can also change substantially. For example, tree frogs in the genus *Hyla* in the southern United States have two color variants: a gray one and a green one. In heavily wooded areas, the gray variety predominates because its color blends with the gray bark of the trees, hiding it from predators. In that environment, the gray variant survives to reproductive age in greater numbers than the green variety, leading to their eventual dominance over the green variety. However, in wetland areas that become filled with bright green vegetation, the green variety becomes dominate (i.e., it is naturally selected) because its camouflage enables its greater survival and reproductive rate.

Sometimes genes or memes become linked, presenting an even more complex scenario. The net selective advantage or disadvantage of a genetically or memetically linked trait is dependent on the specific environment in which that trait exists. For example, the gene that causes sickle-cell anemia also carries resistance to malaria.[8] Therefore, the net selective advantage of that gene is partly dependent on whether or not the population lives in an environment with exposure to malaria. It may also depend on whether a cure is available for malaria or cycle-cell anemia. Some seemingly detrimental characteristics that persist in evolutionary lineages (e.g., why most moths instinctively fly into a flame) may be vestiges of an earlier time and a different environment when the trait was linked to a selective advantage (e.g., a time when the instinct of flying toward light provided some net advantage for moth species). The same vestige principle may exist for certain cultural traits (e.g., the propensity for competing religions to initiate Holy Wars or forceful proselytization efforts). This may be a vestige of an earlier time and a different environment when it provided a net advantage for group survival (e.g., localized beneficial social cohesion and controlled expansion in a yet uncrowded planet that is lacking in weapons of mass destruction). It is important to consider that while such linked characteristics may have benefited a species or culture in one time and place, it may lead to its extinction in a later time or different place. The cultural environment, like the biological environment, inevitably changes over the course of time.

More than a decade after publishing *The Origin of Species*, Darwin proposed the general concept of *cultural group selection* as a driver for cultural evolution and applied it to language and altruism.[10] There was a backlash in the mid-twentieth century against this concept, largely because it was confused with Spencer's model of Social Darwinism (discussed earlier in this chapter). More recently, there has been a resurgence of support for group selection to help explain the evolution of social behavior.[35,36,37]

THE EXPLANATORY POWER OF PHYLOGENETIC TREES AND THE *LAW OF CULTURAL CONNECTIVITY*

Phylogenetic trees, such as the ones beginning each of the remaining chapters of this book, are simplified, intuitively understandable, visual aids. They summarize infinitely more complex, microevolutionary networks of relationship, and they provide a relative temporal order for diversification events. (Evolutionary networks discussed further in **Note 2E**.) As I mentioned before, these trees are not meant to represent universal progress or improvement. A group at the base of a particular phylogenetic tree is not necessarily a lesser or better form than a group higher up on that diagram. Its basal position simply indicates that it has a deeper and sometimes broader history within the specific context in which it is being portrayed. With religions, for example, the position where a particular ideology falls on a phylogenetic tree is not a qualitative, progressive, or moral ranking of that particular belief system. It merely represents a hypothesis about how each religious group is historically related to other religious groups on that particular tree (a phylogeny).

Change and diversification are primary, unavoidable elements of evolution, whether we are looking at the physical universe, biological organisms, or aspects of human culture. Scientific theory holds that the big bang resulted in the physical diversification of the universe, including the stars, planets, and pretty much everything else. It also proposes that the evolution of life resulted in the biological diversification of genotypes, phenotypes, and species. Similarly, the origin of human cognition eventually resulted in the diversification of countless supernatural belief systems. Evolutionary process begins with mutations or transformations within something to produce variation. This is then followed by environmental selection that either supports or eliminates specific variants over time. The process of diversification (a form of increasing disorder or "entropy") never stops (**Note 2F**).

Systems that have diverged over time can retrospectively be traced back to common sources (ancestral forms) based on contemporaneous written records where they exist (e.g., for much of the postliterate diversification of Organized Religion). Where no such records exist (e.g., for biological speciation or for most preliterate religious diversification), common sources can sometimes be extrapolated from hypothetical patterns of shared characteristics. My conclusion from such patterns is that most Organized Religions are historically connected at one level or another, just as all biological species or all physical aspects of the physical universe are. For aspects of culture, I call this the *law of cultural connectivity* (a retrospective aftereffect of entropy). Today's branches of Organized Religion were all influenced by branches before them that were influenced by branches before them. The continual process of change and diversification over time resulted in tens of thousands of religious belief systems whose interrelationships can be graphically illustrated with phylogenetic trees.

One of the arguments sometimes made against the use of phylogenetic trees in evolutionary studies, is that they do not reflect the detailed complexity of cultural evolution. However, most phylogenetic trees (biological or cultural) are not intended to show microevolutionary process detail. Instead, they are empirically derived hypotheses meant to illustrate broader (macroevolutionary) interrelationships among putative groups. Most modern phylogenetic studies are not focused on a search for specific ancestors. They instead focus on the discovery of sister group relationships (two groups more closely related to each other than to any other group). It is hard to beat the clear explanatory power of a phylogenetic tree for illustrating hypothetical interrelationships among different groups, as well as the relative timing of diversification (speciation events or schisms). That explanatory power is due to so much of human conceptual thought being intrinsically metaphorical.

As mentioned earlier in this chapter, the phylogenetic study of Organized Religions (as well as other aspects of human culture) has one major advantage over the phylogenetic study of biological groups. Many detailed records of humanly observed evolutionary diversification can used to help reconstruct historical detail. Contemporaneous written records for religions include observations of *origins* (ancestral forms, founders, and other iconic inspirational figures), *ideological transformations* (innovations and syncretizations), and *schisms* (major splitting events resulting in diversification), because they happened within the past few thousand years. For example, the split between Christianity and other forms of Abrahamic Monotheism happened within the past two thousand years, as did the hundreds, perhaps thousands, of subsequent schisms within Christianity. Conversely, contemporaneous written records for the diversification of major biological groups existing today (e.g., speciation events resulting in major groups of plants or animals) do not exist because they happened millions of years ago. For example, the theoretical split between the ancestor to birds and the ancestor to the group containing crocodiles is believed to have happened more than 150 million years ago.

Because so much of the diversification of Organized Religion was recorded over the course of human history, I include much of that historical detail, including date estimates, on my trees. This makes the branching diagrams of religions in this book *phylogenetic trees* rather than simple *cladograms*. In addition, I must qualify these trees with two important caveats: First, the relative branch lengths on the trees are not proportionate to time intervals. Second, many or most of the year dates are only rough approximations. They are also only estimates of *minimum* age, because the chance of the very earliest form of most ancient lineages being preserved in the anthropological record is often slim. Nevertheless, these trees provide a broad theoretical picture of historical diversification. These phylogenetic trees are also testable hypotheses of group relationships (phylogenies) that can be revised and expanded with further study and new discoveries. Hypotheses building and testing is the core of any scientific discipline.

DECIPHERING EVOLUTIONARY PATTERNS IN SPITE OF MISSING LINKS

A major challenge for both biological and cultural evolutionary studies is missing data on extinct forms (e.g., "missing links" in the form of missing fossil species of organisms, missing dialects of languages, or missing stages of religion). In general, the further back in time we look in our studies, the less complete and reliable the historical data are. With regard to evolutionary biology, for example, it is estimated that less than 0.05 percent of species that ever lived on this planet were preserved as fossils.[38] The further back in time that we take our studies, in general, the more incomplete and unreliable the fossil record is. It is not only missing species that are a challenge, it is also missing information on the fossil species that we have. Only select parts of organisms ever fossilize (e.g., usually only skeletal parts for animals). Also, we do not always understand the function of fossil parts that have no obvious modern analogs. The less that fossils resemble living species, the more speculative our interpretations of them often are.

Similarly, the further back in time we look in our studies of preliterate religions, in general, the less complete and reliable the records of them become (see chart on next page). It is not only missing sects that are the challenge. We often also do not understand the precise significance of the archaeological objects we do have. Interpreting history from only carvings, cave paintings, and statues is an inexact science at best. If the only evidence that future archaeologists had to interpret our present-day religious belief systems were statues and paintings, including those of the Easter Bunny and Santa Claus, or glorified statues of iconic leaders and military heroes, what would they make of that? The interpretation of what past people believed would no doubt be influenced by whatever traditions were embedded in that future archaeologist's own culture. Furthermore, missing records are not the only challenge to reconstructing the evolutionary pattern of religions. Misrepresented information of existing records also serves to confuse. Many written records were later distorted (intentionally or unintentionally) by the biased beliefs of the chroniclers. Cultural conflict has often been fought twice: once by imperial powers on the battlefield, and later by chroniclers attempting to establish historical records for posterity.

A major advantage of the comparative method of analysis and of the phylogenetic approach is that these methods can construct hypothetical patterns of relative relationship even if huge numbers of missing taxa or cultural records are missing. In biology, for example, we do not need knowledge of all species that ever existed to hypothesize that the dinosaur species *Tyrannosaurus rex* is more closely related to a species of bird than it is to a species of plant. We can determine this pattern of relationship and represent it with a simple cladogram or phylogenetic tree because the dinosaur and bird share many more characteristics with each other than either does with the plant (bony internal skeleton, red blood cells, heart, legs). Similarly, as mentioned earlier, we do not need knowledge of all religions to hypothesize that Roman Catholicism is historically more closely related to Protestantism than it is to Orthodox Judaism. Roman Catholicism shares more basic characteristics with Protestantism than it does with Orthodox Judaism (e.g., the believed significance of the man named Jesus).

Pattern analysis along with historical records allow us to hypothesize general evolutionary relationships, even without having many intermediate forms. This is critically important because we will never have every evolutionary variant of what

GENERALIZED SCALE OF RELIABILITY AND CONSISTENCY FOR HISTORICAL RECORDS OF RELIGION

More reliable

Personal observation (most reliable)
(personally witnessed events within one's lifetime)

↑ ↑

Electronic media (recorded audio and video)
(beginning in 20th century CE with radio, TV, and internet)

↑ ↑

Printing press (mass production with archival consistency)
(beginning 15th century CE with mechanically printed documents)

↑ ↑

Handwritten records (single copy documentation)
(beginning 3200 BCE in Mesopotamia with beginning of literacy)

↑ ↑

Professional oral transmission (early effort for consistency)
(generations of specially trained oral transmissionists, prior to literacy and handwritten records)

↑ ↑

Random stories, legends, and folklore (random lore)
(beginning before Professional Oral Transmissions)

↑ ↑

Archaeological artifacts only (objects w/o explanation)
(possibly beginning as early as 91,000 BCE)

↑ ↑

Totally unsupported speculation (least reliable)

Less reliable

we are studying—whether the subjects of analysis are cultural traditions or biological species. The compiling of historical information indicative of relationship is a cumulative ongoing effort. As time passes, continued research fills gaps in our knowledge (e.g., missing groups and character data, and in the case of cultural evolution, missing written records). As our accumulation of historical data increases, our evolutionary hypotheses become more detailed and robust as maps of historical diversification. The method of deciphering history through testable and continually revised hypotheses is part of what makes historical analysis a science.

This thirty-foot-high stone and bronze monument in Washington, DC, was built in 1890 to commemorate a general in the American Revolutionary War (Marquis de Lafayette). If such a structure was found in the preliterate archaeological record, we might be tempted to interpret the figure at the top as a divinity. The reliability of historical records generally declines as you go back in time (see chart on page 37).

If we go back far enough in the history of the most general characteristics of supernaturalism (i.e., animism, anthropomorphic supernaturalism, ancestorism, matristicism, or polytheism), the assumption of a single common ancestry for each of those particular characters is questionable. Each of those extremely general characteristics likely developed well before organized interaction formed among major social groups of humans. Such interaction would have been necessary for the intercultural transmission of ideas (i.e., the "memes" mentioned earlier) and the eventual formation of cross-cultural ideological lineages. Multiple "independent inventions" of animism, ancestorism, matristicism, and general polytheism would have been likely because of the complete cultural isolation of groups in which they developed, and because of the fundamental generality of those beliefs. For that reason, the phylogenetic focus of this book is on the known branches of what I call Organized Religion (see the trees on pages 44 and 486). The diversification of Organized Religions was much more specialized and transpired mostly after the beginning of large-scale cross-cultural influence among humans. Intercultural transmission of ideas became common because of trade, warfare, mass migration, and other social interactions. Religious elements within even the most rudimentarily connected cultures would have had the opportunity to be transmitted back and forth. After all, it would only take a single immigrant or visitor to transmit an idea.

In spite of the many challenges in reconstructing the early history and evolution of religions, we should not be discouraged. Just as most evolutionary biologists propose unambiguous patterns of relationship as hypotheses meant for testing and improvement, so too should cultural evolutionists. Just as we use paleontology and comparative biology to analyze millions of years of biological evolution, we can use archaeology and comparative sociology to analyze thousands of years of human cultural evolution, even where no contemporaneous records exist. This will, no doubt, be controversial. Just as philosophical tensions often have existed (and still do) between paleontology and biology over methods of interpreting the deep history of life, philosophical tensions and sensitivities also have existed (and still do) between archaeology and social anthropology over methods of interpreting the deep history of culture.[39,40] And just as the relationship between generalists and specialists in the biological sciences is sometimes strained, the same is true for generalists and particularists in the social sciences. Such controversies, intellectual challenges, and compromises are part of what makes science work as effectively as it does.

Some of my preliminary classifications and trees will be falsified over time. Most of these general hypotheses will need further phylogenetic study by specialists. As I reiterate numerous times in this book, the unambiguity (clarity) and falsifiability (testability) of these hypotheses are part of what make them scientific. They are meant to be a starting point for continued study. The fear of controversy, or of possibly being wrong, is not something that stops scientific investigation. If early comparative systematists from Aristotle to Linnaeus had given up in their efforts to identify and classify natural groups of plants and animals for fear of being wrong (and they made many major mistakes in detail), we would never have developed the comprehensive understanding of biological diversity that we have today. Like all complex scientific fields of study, the more we examine a subject, the more robust our evolving methods of analysis, hypotheses, and theories become (differences between hypothesis and theory discussed in **Note 2G**). We have barely begun to understand the complex evolution, diversity, and interrelationships of known Organized Religions, just as biologists have a long way to go in understanding the complex evolution, diversity, and interrelationships of the earth's biota. Also, in general, the further back in time we go, the less we know. The job of evolutionary analysis is an ongoing one that gradually improves our understanding of the deep past but is never finished.

EXPLORING THE DIVERSITY OF RELIGIONS WITHIN AN EVOLUTIONARY CONTEXT

The history of religions, like other aspects of human culture, is long and complex. Like the evolution of biological groups, the evolution of religions is an ever-branching pattern of diversification, not a universal ladder of progressive improvement. The process of evolution is sometimes gradualistic, with barely perceptible changes occurring over long periods of time (a process termed "phyletic gradualism" by evolutionists). In other cases, it appears to move with abrupt change over relatively short periods of time in between periods of relative stasis (a process termed "punctuated equilibrium" by evolutionists).[41] Some social scientists, such as History of Religion specialist Robert Karl Gnuse, have tried to apply such evolutionary process theories to the evolution of religion.[42] In cultural evolutionary studies, just as in biological evolutionary studies, the first challenge is to distill broad patterns of relationship from an effectively infinite, yet incomplete, amount of diversity. Then we can propose explanatory process theories for the origin of those patterns. As biologist Peter Richerson and anthropologist Robert Boyde wrote, paraphrasing the evolutionary biologist, Theodosius Dobzhansky, "Nothing about culture makes sense except in light of evolution."[43]

In chapters 3–11, I look at historical interrelationships among the known major Organized Religions of the world. I provide a series of phylogenetic tree hypotheses to frame the discussions of each chapter and provide summary comments in chapter 12. What this study lacks in detail, I hope it makes up for in breadth. I hope that it encourages others to test, revise, and expand the hypotheses presented here. I have tried to cover as many major branches of Organized Religion as possible, but branches with the most available information generally get the most coverage. That unequal coverage is not meant to be a relative value judgment of any particular religion, culture, or race. It is only an artifact of information availability, time limitations, and possibly unconscious bias from my Western perspective as well. Our perspectives are inevitably shaped to some degree by our own cultural development.

It is impossible to know where the belief in, or the practice of, religion in its most elemental form began. Many historians believe that the belief in the supernatural began with human's expanding cognitive ability for abstract thought in the Stone Age.[44] Therefore, it is within the broadest context of supernaturalism that I begin to look at the earliest development of religion in the next chapter.

MAIN SECTION

THE HISTORICAL DIVERSIFICATION (EVOLUTION) OF RELIGIONS

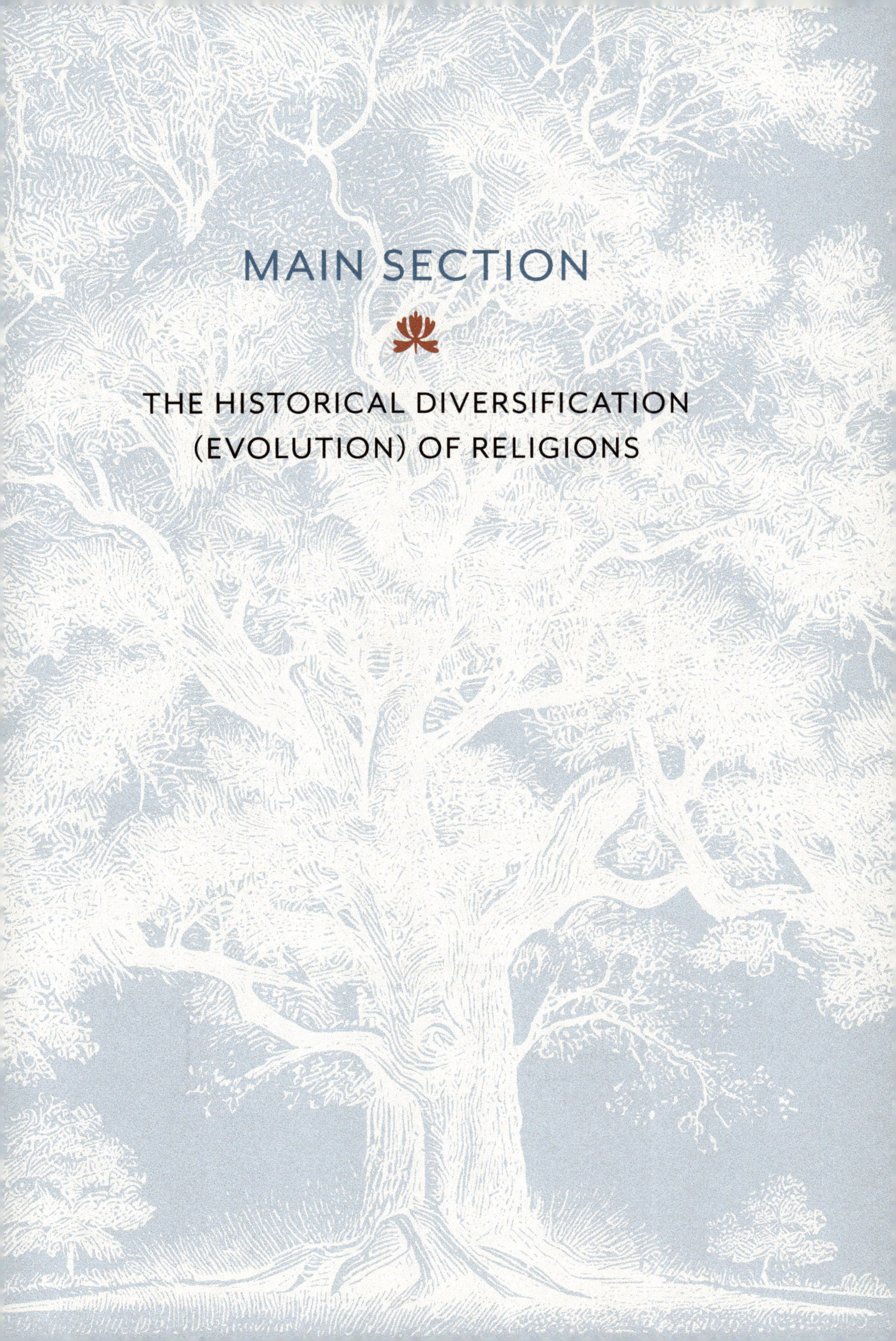

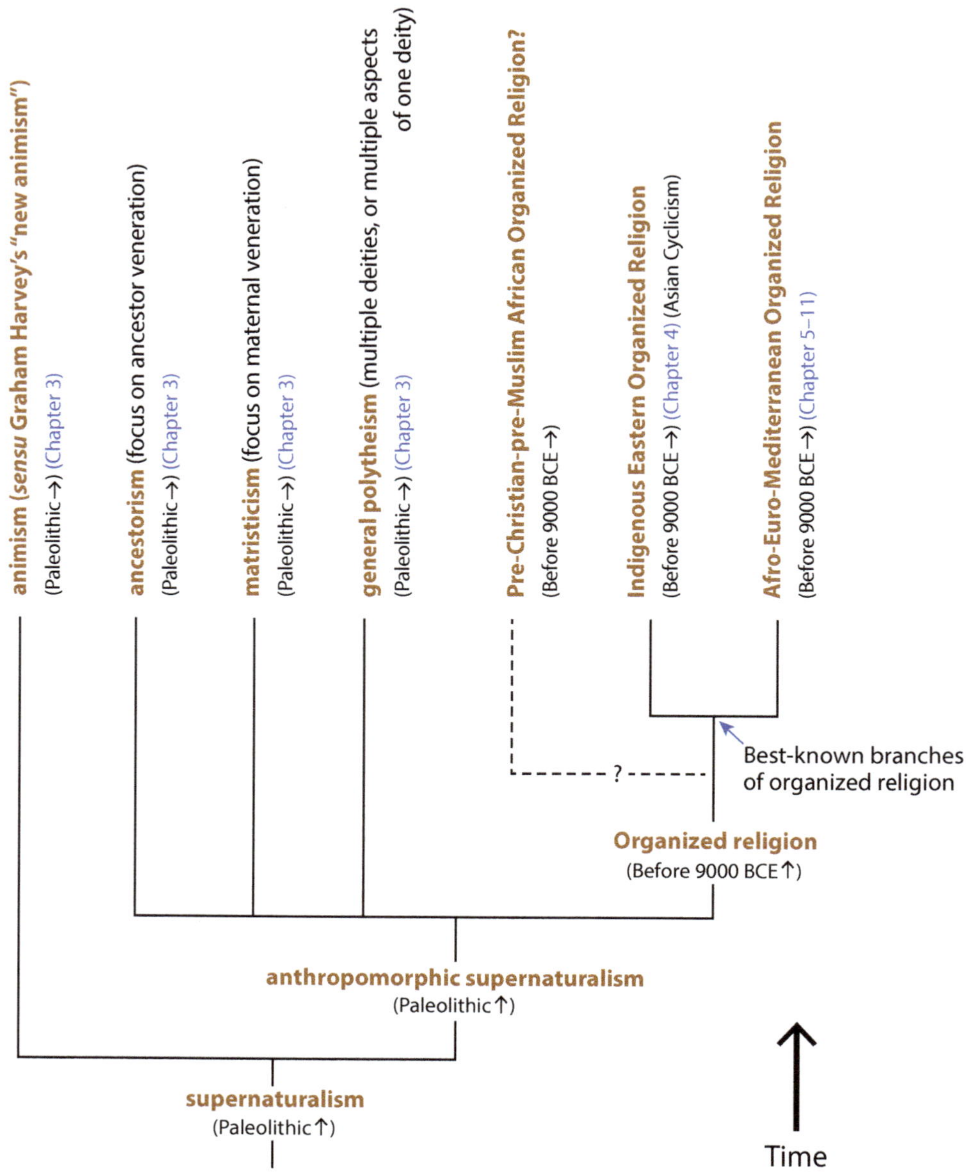
animism (sensu Graham Harvey's "new animism")
(Paleolithic→) (Chapter 3)
ancestorism (focus on ancestor veneration)
(Paleolithic→) (Chapter 3)
matristicism (focus on maternal veneration)
(Paleolithic→) (Chapter 3)
general polytheism (multiple deities, or multiple aspects of one deity)
(Paleolithic→) (Chapter 3)
Pre-Christian-pre-Muslim African Organized Religion?
(Before 9000 BCE→)
Indigenous Eastern Organized Religion
(Before 9000 BCE→) (Chapter 4) (Asian Cyclicism)
Afro-Euro-Mediterranean Organized Religion
(Before 9000 BCE→) (Chapter 5–11)
?
Best-known branches of organized religion
Organized religion
(Before 9000 BCE↑)
anthropomorphic supernaturalism
(Paleolithic↑)
supernaturalism
(Paleolithic↑)
Time

3

EARLY SUPERNATURALISM AND THE DEVELOPMENT OF ORGANIZED RELIGION

Religion is one tree with many branches.

—MOHANDAS GANDHI

Supernaturalism is the belief in spirits, souls, or other supernatural elements. It represents the most general characteristics of religion, and its most elemental natures are depicted by the phylogenetic tree on the opposing page. Eventually, those early elements coalesced and formalized into Organized Religion. Organized Religion eventually diversified into countless codified subgroups. The numerous branches of Organized Religion are expanded as phylogenetic trees of their own throughout chapters 4 through 11 of this book. The roots of all of those trees reconnect again at the beginning of chapter 12. As noted in chapter 1, each phylogenetic tree in this book is meant to be a testable, preliminary hypothesis of a relationship open to future revisionary work. My work is a synthesis of published information from many experts, historical texts, and theorists in the social sciences. I leave it to History of Religion specialists who are also comparativists to test and revise the various trees illustrated here, and to further expand many of the terminal branches of those trees.

In this chapter, I begin with a general discussion of early *supernaturalism* and its early foundational elements: *animism*, *anthropomorphic supernaturalism*, *ancestorism*, *matristicism*, and *general polytheism* (see phylogenetic tree of supernaturalism at the beginning of this chapter). These elements first arose not as organized religions, but as basic characteristics of human nature in the early evolution of supernatural beliefs. Because they originated in many different places before much intercultural transmission occurred among groups of humans, these elements may have evolved multiple times independently. They are not capitalized as proper nouns on my phylogenetic tree because the detailed diversity of their individual subbranches

Timeline for certain events and artifacts relevant to the early development of religion

Development of writing and literacy in the New World by the Olmec civilization	~900 BCE
Early staff-god artifact made in South America (Norte Chico)	~2250 BCE
Beginnings of writing and literacy in the Old World by the Sumerian civilization of Mesopotamia	~3400 BCE
Baghor 1 Shrine, constructed in what is today central India, suggestive of early *Eastern Organized Religion*	~9000 BCE
Temple of Gobekli Tepe, constructed in what is today Turkey, suggestive of early *Western Organized Religion*	~9130 BCE
Highly ceremonial human burials, suggestive of either *ancestor veneration* or *leader veneration* in Eurasia at the Sungir archaeological site in Russia (see text)	~26,000 BCE
Earliest known introduction of humans into the "New World" (i.e., the Americas), at first by way of the Siberian land bridge (Beringia) until its disappearance around 9000 BCE	~26,000 BCE
Earliest known "Venus sculpture," skillfully carved out of mammoth ivory, suggestive of matristicism (found in Hohle Fels archaeological site in Germany)	~35,000 BCE
"Lionman" figure skillfully carved out of mammoth ivory with the head of a lion and the body of a human, suggestive of *anthropomorphic supernaturalism* (found in Germany)	~35,500 BCE
Early records of ritualistic human burials in the Middle East with the use of red ochre and associated grave goods (archaeological sites in Israel)	~91,000 BCE
Early signs of mortuary ritual in eastern Africa (from the "Middle Awash" archaeological site of Ethiopa)	~155,000 BCE
Humans begin to spread from Africa to the Middle East and Eurasia	~200,000 BCE
Earliest known humans (*Homo sapiens*) evolve in Africa (earliest known skeletons from Morocco)	~300,000 BCE

(e.g., the Ojibwe, Maori, Aboriginal) are beyond the scope of this book (although a few of them are explored by specialists elsewhere[1]. These elements are meant only as generalized characteristics or vernaculars thought to have been highly influential to the earliest development of more formally structured major traditions, termed here as Organized Religions (which are the focus of this book).

THE RISE OF EARLY *RELIGION* AS A FORM OF SUPERNATURALISM

Supernaturalism may be as old as the human species, perhaps even older.[2] I use the term supernaturalism as a most general working definition of religion, although in the words of British anthropologist, Robert Dunbar, "the definition of religion is probably the single most fiercely debated topic in the study of religion."[3] Some anthropologist even suggest that the word religion "should be dropped- at least in all but the personalist sense."[4] The English founder of cultural anthropology, Edward Tylor, referred to religion, in its most basic sense, as a "doctrine of spiritual beings."[5] Psychologist Matt J. Rossano describes religion as "beliefs or actions predicated on the existence of supernatural entities or forces with powers of agency that can intervene in, or otherwise affect, human affairs."[6]. I take a pragmatic approach to the issue. As discussed in the Preface, the word "religion" is an anthropological one, not a theological one.[7] As history of religion scholar Jonathan Z. Smith stated, the term "is created for the scholar's analytical purposes by his imaginative acts of comparison and generalization."[8] I take this generalist approach, but exclude group traditions that lack clear supernatural elements. I also omit some groups that clearly began within a nonsupernatural context (such as certain groups mentioned in **Note 3A**). My pragmatic definition more or less follows that of anthropologists Ryan Williams and Donna Nash, who define religion as "a set of beliefs that structure the interaction between a community and the supernatural forces it perceives in the world,"[9] or anthropologist Charles Stanish, who defines religion as "a systematic body of metaphysical beliefs shared by a group of people."[10] Group support is important, because it is part of human nature to want to have one's beliefs validated by others, and a community of people with similar religious beliefs is one way to provide such validation.[11,12] Group support also helps create longevity of tradition for belief systems.

We will never know exactly when the first supernatural ideas were born, nor will we probably ever know much detail about the origin and earliest development of Organized Religion. Most of religion's early development took place well before the beginning of human literacy. Before the invention of writing around 5,400 years ago (discussed further in **Note 3B**), we have only fragmentary indications of religion based on rare archaeological sites as interpreted by anthropologists, sociologists, psychologists, and philosophers. Because of the numerous gaps in the archaeological record, the dates of origin for many early religious groups may be significantly older than the minimum date estimates given on my phylogenetic trees. Even where archaeological remains exist, interpreting religious belief systems from them can be extremely challenging. As I mentioned, the mere act of defining religion is notoriously controversial and the basic classification of even modern religious branches is onerous.[2,13,14] However, the challenges for the evolutionary study of religions are similar to the challenges in the evolutionary study of living organisms. The difficulty of defining "religion" or "supernaturalism" in the social sciences is not unlike the difficulty of defining "species" or "life" in the biological sciences.[15,16,17]

ABSTRACT THOUGHT AS A FOUNDATIONAL ELEMENT LEADING TO SUPERNATURALISM

The ability for abstract thought must have gone back further in human history than supernaturalism. Such thoughts are not necessarily of a religious nature, but they were necessary cognitive abilities for the concept of supernaturalism and religion to have developed. The general ability for abstract thought in its most basic form was likely already developed in the earliest humans, because it appears to be present in the closest living relatives to humans, the chimpanzees.[2,18,19] Chimps clearly exhibit the ability to make and use tools and to show concepts of self and grief. They also form cohesive community groups of more than fifty members and show altruistic behavior. Some chimps can even produce abstract art showing a rudimentary recognition of symmetry and composition (see the next figure). Similarly, *Homo erectus*, who made a variety of stone tools, appears to have etched geometric patterns on shells over half a million years ago, demonstrating early cognitive and neuromotor control.[20]

Homo sapiens (the modern human species) took this a step further with what is sometimes called the *Cognitive Revolution* (when humans are thought by some historians to have gained the ability to speak and create myths).[21]. Early records of cognitive development in humans include nonfigurative line drawings and cave paintings depicting figurative (recognizable) objects. The creation of figurative imagery, in particular, represents an expanded ability for abstract thought. It is impossible to know how early this ability developed in humans. The length of time that cave paintings can be preserved is limited, particularly in tropical areas. Archaeologists generally assume that environmental conditions limit the time such artwork can normally last to be about ten thousand years, although exceptions are found in Europe, Russia, Indonesia, and Africa. Early examples include a painting

Painting produced by a chimp named Congo (1954–1964) who made around four hundred drawings and paintings during his lifetime under the supervision of the English zoologist and ethnologist, Desmond Morris. Although Congo produced many abstract works, he never produced a discernable figurative image of anything.

of a wild pig from Indonesia that is at least 45,500 years old,[22] and a painting of a wild bull from nearby Borneo that is dated to be around forty thousand years old[23] (both are illustrated in the next figure). Possible older paintings simply did not last because they were worn away by the elements or have yet to be discovered. Non-figurative cave drawings, consisting only of cross-hatched lines, have been reported from much earlier times. These include the seventy-three-thousand-year-old drawings from Blombos Cave in south Africa.[24] The Cognitive Revolution in *Homo sapiens* was eventually associated with other intellectual developments including trade, agriculture (farming and animal husbandry), increasingly complex language skills, and increasingly organized forms of religious belief.[21] Collectively, this was tied to the evolution of larger cooperative coalitions and the rise of urban civilization (see "civilization" *sensu* **Note 3C**).

Early psychologists proposed various scenarios for the earliest development of religion. Some claimed it helped satisfy the need many people have for existential

Cave drawing and paintings by Late Pleistocene Age *Homo sapiens*. (*Top*) Abstract drawing engraved on ochre from Blombos Cave, South Africa; dated at seventy-three thousand years before present. (*Left*) Original artifact and (*right*) rendering of etched inscription. (*Bottom left*) Cave painting of a wild pig, about four feet in length. From Leang Tedongnge Cave, Indonesia, and dated at 45,500 years before present. (*Bottom right*) Cave painting of a wild bull, about five feet in length. From Lubang Jeriji Saleh Cave, Indonesia, and dated at forty thousand years before present (Aubert et al., 2018). The bottom two figures are two of the oldest-known recognizably figurative paintings by humans. Older ones likely existed but were worn away by the elements over time.

direction. Others claimed it had to do with coming to grips with one's own mortality, and providing simple and comforting answers to seemingly unanswerable questions about where we come from and why we are here. The famous neurologist and founder of psychoanalysis, Sigmund Freud (1856–1939), was an atheist. He proposed that the belief in gods arose from parental figures in childhood and a yearning for a powerful protective father figure.[25] As early human children became adults, they projected the sense of their parents on deities, which provided a similar sense of security and protective comfort. One of Freud's early disciples was Carl Jung (1875–1961), a fellow psychoanalyst. Jung saw religion as providing a healthy way to contemplate one's place in the universe and to prepare for death. He saw religion as a noble human response to the complexity and depth of life. He studied a number of religious groups, including various kinds of Old World Hard Polytheism, Hinduism, Buddhism, Taoism, Biblical Demiurgism, Traditional Christianity, and even the occult, looking for foundational characteristics common to all of them. He subsequently proposed that the primordial purpose to all religions was to meet the spiritual divine.[26,27]

Early sociologists and anthropologists also had a lot to say about the origin of religion, and I mention only a few here. French anthropologist and philosopher David Emile Durkheim (1858–1917) was one of the architects of sociology as an academic discipline. He defined religion as "a unified system of beliefs and practices relative to the sacred."[28] He saw religion as an emotional force that originated in early hunter-gatherer societies as those groups grew in size. Over time, this force became ritualized and religion came to distinguish between the sacred and the profane.[2] This seems not far from History of Religion scholar F. Max Muller (1823–1900), mentioned earlier, who was an early proponent of the "nature-myth" school.[29]. This theory suggests that religion came about as humans were awed by the great forces of nature. Similarly, German theologian and comparative religionist Rudolf Otto (1869–1937) proposed that the earliest religions emerged when people experienced frighteningly mysterious phenomena, such as great storms, volcanic eruptions, or earthquakes. He termed these phenomena "mysteries that cause trembling and fascination."[27]. The nature-myth school proposed that people personalized the unseen forces of nature and eventually regarded them as deities (much like what is referred to here as animistic polytheism). Paleolithic cave paintings, carvings, and stories of early people were attempts to express their wonder and link these great mysteries to their own lives and to the lives of their clan members.[25] This early reverence for natural forces is thought to have preceded any desire to more formally explain creation as the act of anthropomorphic deities.

None of the theories about the origin of religion (at least in the broad way religion is defined here) can be proven or contemporaneously documented because they would have happened tens of thousands of years before literate time. Modern humans (*Homo sapiens*) appear to have evolved from other human-like species about three hundred thousand years ago in Africa[30]. Around two hundred thousand years ago, humans had begun to spread from East Africa to the Arabian Peninsula and the Eurasian land mass. Around 155,000 years ago, they had left signs of mortuary ritual in the Middle Awash archaeological site in East Africa.[31,32] Humans were burying their dead by at least ninety-three thousand years ago in what today is Israel (discussed later in this chapter). Highly ceremonial burials with elaborate adornments began by at least 26,000 BCE, based on discoveries from the archaeological site of Sungir (also spelled Sunghir) in present-day Russia (illustrated next).[33]

Numerous anthropomorphic carvings as old as forty thousand years before present are known from archaeological sites around the world (discussed later in this chapter). Such objects are sometimes used as evidence of religion. However, they do not

One of the earliest clear records of ceremonial burial of a human, from the Sungir archaeological site in Russia (26,000 BCE). Note the heavily adorned skeletal remains with thousands of ivory beads. The skeleton was also covered with red ochre, an important material associated with ritualistic human burials in ancient time.

tell us what people were thinking. They could just as likely have been toys, art objects, or images of war heroes. As I mentioned, the earliest written records of human religious beliefs (or of anything else) are less than six thousand years old (**Note 3B**).

Although my working definition of religion in its broadest form is *the shared belief in the supernatural* ("supernaturalism"), some authors use a much more limited definition, equivalent to what I refer to as Organized Religion[31] or even Traditional Christianity.[34] One reason that the term "religion" is so hard to define is that some authors seek to emphasize only parts of it.[14] Most studies are less interested in cross-cultural comparativism, and many historians restrict their definitions to describe the specific major branch of religion that they are most familiar or most comfortable with. Entire books have been written on the difficulty of defining religion.[35] I do not have a definitive way to define this term other than by clarifying how I use it (as discussed near the beginning of this chapter). For the sake of consistency, I chose the most general anthropological definition of religion because my goal was to make broad, cross-cultural, comparisons.

Early characteristics of supernaturalism included the basic beliefs of *animism* and *anthropomorphic supernaturalism* (discussed in following sections of this chapter). Both of those ideological concepts incorporate the basic idea of spirits or souls, and they are among the broadest characteristics or philosophies of religion. We will probably never know for sure whether animism or anthropomorphic supernaturalism developed first. However, early cultural anthropologists, such as Edward Tylor (1832–1917), believed it to be what he termed *animism*.

ANIMISM AND *NEW ANIMISM*: A DOCTRINE OF UNIVERSAL SPIRITUAL ESSENCE

The term *animism* was coined by Tylor, in 1871.[5] As originally used, the term had two uses. The first was objectively descriptive. It referred simply to the belief in a spiritual essence or a soul, and the belief that both living and nonliving things could have such an attribute. It referred to the "deep lying doctrine of spiritual beings" and was offered by Tylor as "the groundwork of the philosophy of religion."[5] So far so good. Unfortunately, Tylor also used the term as a subjective, qualitative assessment of indigenous cultures. He considered animism as a sign of ignorance and the religion of "lower races."[5] This judgmental use of the term was tied to his colonialist notions of Western superiority, and it was highly insulting to the indigenous peoples who practiced animistic traditions.[1] Tylor, who was a Quaker (Quakerism discussed in chapter 9), claimed animism was a "backward" and "mistaken" form of religion, and that its practitioners were merely confused about the difference between people and things. He, like later Spencerian cultural evolutionists (discussed in chapter 2), viewed animists as savages at the bottom rung of a cultural ladder of universal progress leading up to an idealized Christian civilization in the Western world.[5] Later, this view was seen not only as racist but also as a fundamental misunderstanding of evolutionary theory that confused evolution with some sort of universal ladder of "progressive improvement."

Unfortunately, the derogatory colonialist interpretation of basic animism by Tylor, Spencer, and others became exaggerated early in the history of social anthropology. One of the earliest anthropologists who fought back against its racist component was the founder of North American anthropology, Franz Boas (1858–1942). Boas and his students, including Margaret Mead (1901–1978), worked to show that the so-called primitive races whose supernatural belief systems were animistic were as culturally sophisticated within their respective social environments as Western civilization was within its environment. Every aspect of culture must adapt to the particular natural and sociological environment in which it exists. Those environments vary in both time (they are always changing) and space (they differ geographically). Boas and many of his students effectively pushed back against the racist connotations of early cultural evolutionism. Eventually, however, their backlash went a bit too far by pushing back against the use of comparativism in the social sciences at all. The strict "particularist" exclusion of historical or group-level comparativism began a controversy that still exists in anthropology to this day. It created a serious challenge for many broadly scoped studies of culture, such as the global history of religions through time. Boas, and the rise of particularism in the social sciences, is discussed further in **Note 3D**.

It is my contention that comparativism cannot be ignored in historical research. Without some level of comparativism, the scientific study of deep history and diversity is not even possible (**Note 3E**). In the words of History of Religion scholar Oliver Freiberger, comparativism "is indispensable for a deeper analytical understanding of what we call religion."[36] At its extreme, the anticomparativist approach to cultural history is analogous to the special creationist (antievolution) approach to biological history. In both cases, the willingness to recognize evidence of relationship among different cultural groups faces strong, emotionally enhanced resistance. As discussed in chapter 1, comparativism should not be qualitative or biased. Just because one group of belief systems or a characteristic (e.g., animism) precedes another in time, or just because it appears as a lower branch on a particular phylogenetic tree, it does not mean that the earlier or lower branch is in any way inferior. It simply means that the branch has a deeper, more broadly fundamental, history within the context that it is given (**Note 3F**).

To redefine animism in more of an anticolonialist context, religious studies scholar Graham Harvey used the name "new animism" to distinguish it from the colonialist way animism was first used (what Harvey refers to as "old animism").[1] Harvey describes new animism as similar to the Ojibwe belief that "the world is full of persons, only some of whom are human."[1] In other words, new animism is the belief that most everything that exists has a spiritual essence, more or less going back to the objectively descriptive first part of Tylor's definition without the judgmental colonialist embellishments—that is, there are rock persons, tree persons, and so on. None is qualitatively superior or inferior to any others, and all are deserving of respect. I recognize "animism" in the sense of Harvey's "new animism" because of its absence of prejudicial connotations. It is simply a philosophy that attributes a spiritual essence or "soul" to inanimate objects, plants, animals, and natural phenomena. In some ways, animism's pragmatic respect for nature and lack of exclusionism fits well with the needs of today's world. Some social anthropologists now distinguish two different types of new animistic spiritualism: *basic animism* (in which the uniqueness of each spiritual force is emphasized) and *pantheism* (in which all spiritual forces and components of the universe are collectively synonymous with God).

Animistic principles are often fused with *totemism*, a system of belief in which humans are said to have kinship or a mystical relationship with a spirit being, such as that of an animal or plant. The word "totem" is derived from an Ojibwa word meaning, "he is a relative of mine."[37] The entity, or totem (representing animals, plants, or other natural things), is thought to interact with a given kin group or an individual and can serve as their emblem or symbol.[38] The most familiar example of a totem to people in North America is the totem pole of various Pacific Northwestern cultures. Totem poles were not worshipped, but they inspired great respect and held much spiritual significance. Many indigenous peoples of the upper northwest coast of North America (e.g., the early Inuit and the Haida) have believed that the animal and watchmen figures on the totems are spirit guides that protect them and their dwellings (discussed again in chapter 5). Today, certain peoples such as the indigenous Siberians (e.g., the *Chukchi*) also practice a combination of totemism and animistic polytheism.[27,39]

Shamanism is sometimes incorrectly considered to be a distinct branch of animistic religion. Although its history goes back several millennia, shamanism is an ideological characteristic that has developed independently in a number of different religions. It involves a specific practitioner (the shaman) who reaches an altered state of consciousness to perceive and interact with a spiritual world, and channel transcendental energies into this world.[40] The shaman's role is to communicate with agents in the spiritual world to ask for favors on behalf of the community that they serve. Shamanism, in its most general sense, has developed in a variety of different religious belief systems, including new animism, early polytheism, and certain branches of Asian Cyclicism, such as early Shintoism.

Through post-Columbian history, most branches of animistic polytheism were subverted through Western or Muslim proselytization and assimilation. Most early indigenous animistic religions of Africa were displaced by Christianity or Islam as a result of centuries of conquest and proselytism (**Note 3G**). Even better documented examples of cultural displacement took place in the Americas. There, indigenous animistic religions were overwhelmed by colonial conquest and aggressive Christian proselytism. The indigenous cultures of the Americas were decimated from the fifteenth through the nineteenth centuries by immigrating Europeans and their descendants who claimed the land of the New World as their own (see **Note 3H** and chapter 5). It was an overwhelmingly powerful mix of Euro-hypernationalism and invasive expansionism, enhanced by religious fervor.

Some animistic belief systems (many of which considered here as branches of Animistic Polytheism in chapter 5), persisted into the modern world, although their early diversity and historical interrelationships are poorly understood. Any thorough discussion of animism and new animism's modern diversity and evolution is beyond the scope of this book. Most previous studies of strongly animistic belief systems have focused on only a few select branches, such as certain Inuit peoples of the Arctic region,[41] the Maori people of the New Zealand,[1] the Ojibwe communities of Canada,[1,38] or the Aboriginal peoples of Australia.[1,27,38,42]

NEW ANIMISTS AS ENVIRONMENTALISTS

Animists do not see themselves as separate from their environment or from other aspects of their culture. They have an innate respect for nature and see all organisms as spiritually and functionally connected within the ecosphere.[1,27] As a result, eco-animistic groups began to develop in the late twentieth century in response to global environmental concerns. One such group is *New-Age Gaianism*. This belief system was originally based on a scientific theory called the *Gaia hypothesis*, named after the Greek Earth goddess, Gaia. This hypothesis was first clearly proposed in the early 1970s by a chemist in the United Kingdom named James Lovelock (1919–2022) and American micro-biologist Lynn Margolis (1938-2011). The Gaia theory or Gaia principle (as it has been called by subsequent authors) holds that all living organisms interact with their inorganic surroundings on earth to form a synergistic, self-regulating system of life. This idea was eventually extrapolated to a New Age religious ideology, in which the biosphere (*Gaia*) is seen as a pantheistic superorganism made of smaller synergistic organisms, much like a plant or animal species is as an organism made of synergistic cells. New age Gaianists see a "dark green" spirituality in nature. This belief system is characterized by the conviction that "nature is sacred, has intrinsic value, and is therefore due reverent care."[43] Eco-animism contains the elements of respect for all life on earth and their global interactions.

Recently, animistic principles have been embraced by many environmentalists around the world. Today, even multinational organizations show elements of eco-animism. The United Nations General Assembly proclaimed April 22, 2009, as Mother Earth Day. Elements of eco-animism are also popularized in modern North American culture. Two of the most popular American movies of all time have an eco-animistic foundation: *The Lion King* and *Avatar* (**Note 3I**).

In addition to animism, other early major characteristics contained within much of supernaturalism is *anthropomorphic supernaturalism* (the group that contains the rest of supernaturalism, as shown on the phylogenetic tree on page 44). Anthropomorphic supernaturalism involved anthropomorphization of the spirit world.

ANTHROPOMORPHIC SUPERNATURALISM: SEEING THE SPIRITS IN OUR OWN IMAGE

Anthropomorphic supernaturalism (applying human-like characteristics to supernatural entities) likely evolved very early in human history. Reverence for both ancestors (*ancestorism*) and mothers (*matristicism*) may explain the appearance of early deities with anthropomorphic features. Early archaeological evidence suggesting anthropomorphic supernaturalism includes the "Lion Man" figure from an archaeological site at Hohlenstein–Stadel cave in Germany, dated at between thirty-five and forty thousand years before present. Although the gender of the figure is today most often referred to as male, some anthropologists have proposed it may instead represent a female.[44] This therianthrope figure was sculpted with great technical skill, carved from mammoth ivory with a flint knife. It is thirty-one centimeters tall and would have taken many hours to make. It has the head of a cave lion and a body that is part human and part lion. It stands upright with its legs apart and arms to both sides. It shows wear on its body indicating it was rubbed repeatedly, possibly as part of a ritual. The figure's gaze, like its stance, is powerful and directed at the viewer, as though it is watching and listening. It is the oldest-known human-made sculpture of a human-beast combination, and possibly an early element of anthropomorphic supernaturalism.[44]

The use of anthropomorphic figures as revered objects of iconic significance appears to have been an influential stage in the early development of Organized Religion (e.g., statues of gods and various other divine figures). Some Abrahamic religions later rejected such a practice. Islam, much of Rabbinic Judaism, early Traditional Christianity, and some Protestant branches consider these figures to be forms of idolatry (a sin of idol worship).

The "Lion Man" artifact from Stadel Cave in Germany, sculpted forty thousand years ago from mammoth ivory. It is an early example of a *therianthrope* (a mythical part-human, part-animal creature) and measures one foot in height. This is the oldest-known, humanly carved, anthropomorphic sculpture. In the collection of the Ulm Museum in Germany.

Early forms of anthropomorphic supernaturalism probably included *ancestorism* (the belief in ancestral spirits empowered with supernatural powers in the natural world) and *matristicism* (veneration of maternalism, sexuality, or fertility). Those characteristics clearly influenced the early development of Organized Religion.

ANCESTORISM: THE VENERATION OF ELDERLY WISDOM AND RESPECT FOR THE DECEASED

Ancestorism is the general belief that the spirits of the dead continue to dwell in the natural world and have the power to influence the fates and fortunes of the living. Victorian anthropologist Herbert Spencer believed ancestorism to be "the root of every religion."[45]

Hunter-gatherer societies have often been characterized as egalitarian. Their religious beliefs are thought to have included ancestor veneration, but not in moralizing high gods.[31] An early pragmatic origin of ancestorism seems intuitive. In preliterate nomadic societies, experience and wisdom handed down from elders would have been necessary for survival, and a large-scale social governing system did not exist. Skills for hunting, gathering, child-rearing, and safety of the group would have all been critical. Those clans that appreciated the value of knowledge handed down from experienced group elders probably would have been the most successful, and thus would be an example of Darwinian selection at the tribal level. This success would have logically led to growing esteem for clan elders and ancestors.

Archaeological evidence of possible ancestorism dates back at least ninety-three thousand years ago (although dating the origin of any ancient social innovation remains highly speculative). The archaeological sites in the Skhul and Qafzeh caves of Israel contain several intentionally buried human skeletons with the ritualistic use of red ochre (a natural iron-oxide earth pigment) on the remains. The burials also contain grave goods (items buried with the body suggesting some form of supernatural belief associated with the dead).[46,47] The use of red ochre to stain the bones may have been meant to mimic blood and supported the belief that the color red was a reanimating force. Blood rites and blood sacrifices are common in many later religions, associating supernatural power of high potency with blood. Red ochre is still used today by hunter-gatherer populations around the world for ritualistic purposes. Possible evidence of grave goods with the Skhul burials included one human skeleton with the mandible of a wild boar in its arms.[48] These intentional, possibly ceremonial, burials suggest some form of veneration of the dead. However, without knowing the mind of the people of the time and place, this remains conjectural. As discussed in chapter 2, it is highly speculative to describe an emerging spiritual belief system based on archaeological artifacts alone.

Ancestor veneration has continued to be part of many later religions. It is still widely practiced in East Asia, where family ties continue long after death. The Yellow Emperor *Huang di* (also spelled *Huang-ti*) is said to have lived from 2698 BCE to 2598 BCE. He is today revered by many in China as the deity ancestral to all native Chinese (Han) people.[49] During rituals, people pray to their divine ancestors and ask for blessings. Some branches of ancestorism in China believe that each person has three souls. After a person dies, one soul goes to the grave, another goes to the ancestors, and the third remains in the descendants' home to be worshipped.[49] Ancestorism is not always a matter of "ancestor worship." It is most often a matter of ancestral spirits helping sustain the living family while being sustained by it in return. It is also believed that if ancestral spirits are neglected, they can become uncaring or even vengeful. Rituals are therefore developed to ensure proper respect to and consultation with spirits of the ancestors.[37]

Many examples illustrate ancestorism's influence persisting as important elements of Organized Religions, including many of the indigenous eastern branches of Asian Cyclicism (particularly its branches of Taoic Religion) as discussed in chapter 4. In many African religions, ancestor spirits are thought to bring health, wealth, and children if they are pleased, or disease and childlessness if they are not.[27] As mentioned, some authors contend that elements of ancestorism are universal to all religions.[50]

Another early type of anthropomorphic supernaturalism was *matristicism*, a term used here for veneration of female fertility and motherhood.

MATRISTICISM: THE VENERATION OF MOTHERHOOD

Matristicism, also known as the *Cult of the Mother Goddess* by some authors, focuses on maternalism and human fertility as being sacred.[25] Archaeologically, this was first suggested by numerous thirty-seven- to twenty-thousand-year-old statuettes called "Venus figurines" These iconic carvings have been found in archaeological deposits from southern Europe to Siberia and represent some of the earliest works of prehistoric art.[51] Many anthropologists have suggested that these carvings represent an early spiritual acknowledgment of the importance of fertile, reproductive females to nomadic groups of Stone Age people. The small figurines are hand carved out of stone, ivory, bone, or clay. They are characteristically rotund, often emphasizing parts of the female body associated with sexuality and fertility. Of more than two hundred known specimens, 60 percent of them are clearly identifiable as female and most of the rest are ambiguous with regard to sex.[52] Figurines clearly identified as male from this period are extremely rare. During that period, the male contribution toward reproduction may not yet have been fully realized.[37]

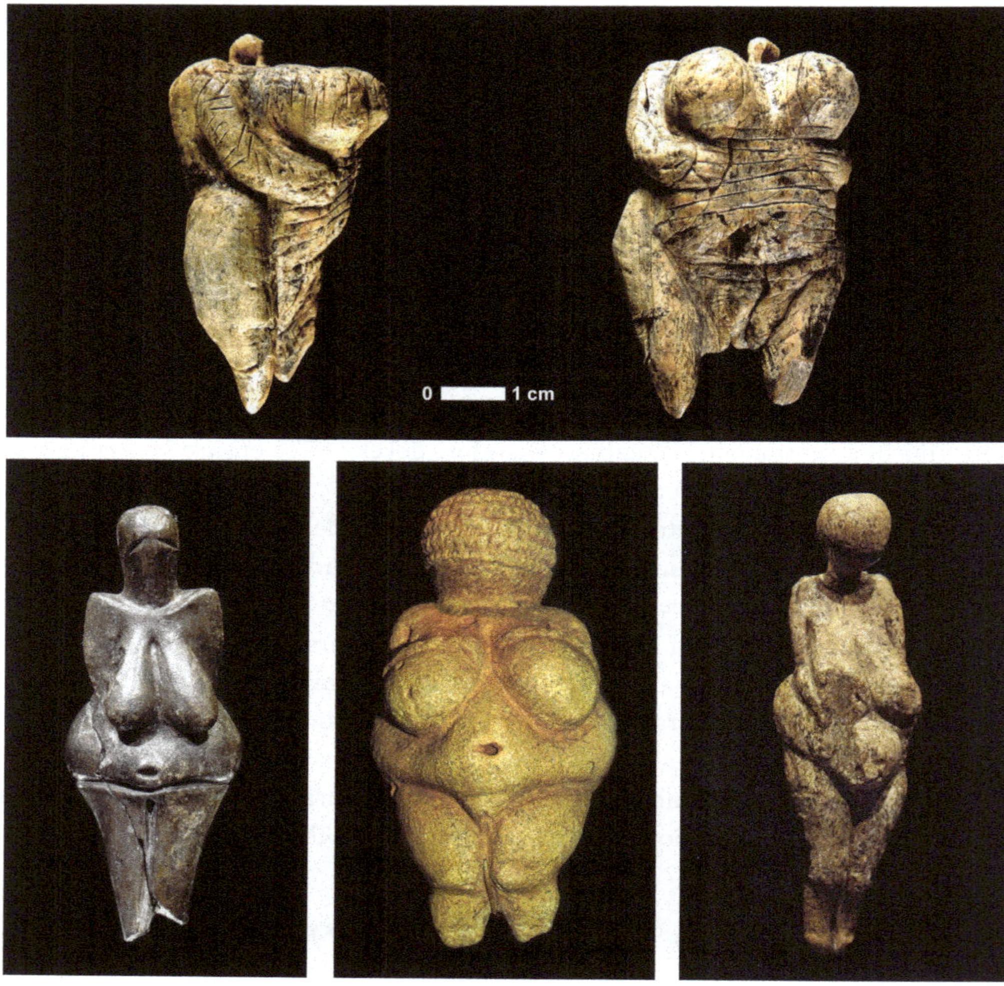

A few examples of "Venus figurines" from Stone Age Eurasia. (*Top*) *Venus of Hohle Fels*, the oldest-known Venus figurine, dated at around 35,000 BCE. Carved from mammoth tusk; 2.4 inches in height. Discovered in a German cave in 2008. In the collection of the Urgeschichtliches Museum Blaubeuren, Baden-Wurttemberg, Germany. (*Bottom left*) *Venus of Dolni Vestonice*, made of fired clay, 4.4 inches in height, dated around 27,000 BCE. In the collection of the Moravske Zemske Museum, Brno, Czech Republic. (*Bottom center*) *Venus of Willendorf*, found in Austria. Carved from limestone; 4.3 inches in height. Dated at around 27,000 BCE. In the collection of the Naturhistorisches Museum in Vienna, Austria. (*Bottom right*) *Venus of Kostenki*. Carved from mammoth tusk around 20,000 BCE. Found near Kostenki, Russia. In the collection of the Archaologisches Museum, Hamburg, Germany.

The earliest-known undoubted example of a Venus figurine is a carving of a woman made from mammoth tusk dated at about thirty-seven thousand years before present found in Germany (see above figure). Archaeologists refer to the sculpture as "Venus of Hohle Fels." In place of a head, this small sculpture has a loop suggesting that it was worn as a pendant. Some of the other small Venus figurines also appear to have been worn as jewelry. Some were marked with red ochre and others were decorated with markings associated with lunar cycles. Some social scientists believe the Venus figurines may have served an iconic religious function, representing a "Mother Goddess"[25] or multiple mother goddesses[53,54] and that they were "female figurines with likely cultic significance"[55]. They have inspired quasireligious movements in some modern feminist circles.[56] In contrast, other archaeologists have dismissed them as erotic art or "paleo-porn."[57]

It is impossible to know the purpose of early Venus figurines with absolute certainty, because written records did not begin until tens of thousands of years later. However, the intuitive logic for interpreting these objects as religious icons is enticing. In hunter-gatherer societies, social groups were small clans based on kinship. Some entire groups could have been from a single ancestral mother, and the clan's long-term success was ultimately dependent on female fertility and successful child-rearing. Even after the child was born, its early survival was dependent on nursing from the mother. Infant mortality would probably have been extremely high because of disease and challenging living conditions, so reproductive success and a reverent appreciation of it would have been a selective advantage for a clan.

With the emergence of early civilization, polytheism, and formalized pantheons of deities or great spirits, the scope of matristicism became more complex than simply an overriding cult of fertility. Mother goddess figures were absorbed into the pantheons of many different cultures as major deities, including the pantheons of Sumer (as *Inanna*), Babylon (as *Ishtar*), Canaan (as *Anat*), ancient Israel (as *Asherah*), Egypt (as *Isis*), Greece (as *Aphrodite*), Rome (as *Venus*), India (as *Shakti*), Japan (*Amaterasu*), Tibet (as *Tara*), and many others.[58] Eventually, male deities became the most numerous and powerful gods in most pantheons. The rise of increasingly more patriarchal pantheons is thought by some historians to reflect a more patriarchal social structure for post–Stone Age cultures in general, with a growing importance of male warriors, male food producers, and male governing bodies in civilizations focused on expansion (a theory that remains controversial[58,59,60]).

Vestiges of matristicism have remained within some branches of Abrahamic Monotheism. For example, the Roman Catholic Church still practices high veneration of Mary, mother of Jesus, as somewhere between saint and goddess, although usually in a subordinate role to that of the male Trinity. Some early records of the fourth century mention an extinct branch of early Christianity called *Collyridianism*,[61] in which Mary allegedly reached the level of a godly aspect, as part of a Trinity that included the Father, the Son, and the Mother (the Virgin Mary)

(**Note 3J**). In general, however, the formal authority of women in Abrahamic Monotheism, and religion in general, appears to have declined over time (discussed further in chapter 12).

Matristicism as a distinct belief system exists today in the form of Wiccanism (although some modern Wiccans also worship a horned god in addition to the goddess). Modern Wiccans originated in England in the 1950s. It is a duotheistic belief system that generally holds a goddess-god pair of deities as the focus of worship (the Lady and Lord of the Isles), and some branches have become monotheistic and focus only on the goddess. Today, Wiccanism has several branches and more than one million members, including both men and women.[62] The Wiccan terms "witch" and "witchcraft" should not be confused with the way the terms are used in devil-worshipping belief systems, such as Satanism and Luciferism. Historically, confusion with such "black magic" cults has resulted in persecution of Wiccans, especially in predominantly Christian countries.[62] Wiccans do not worship the devil. For many Wiccans, feminism is an all-important concept, and they profess to be healing themselves from the past wounds of patriarchy. One reason witchcraft is believed to have been historically opposed by men is because it empowered women as spiritual leaders.[37] Many Wiccans identify with the idea of religious persecution during the sixteenth and seventeenth century when thousands were executed for witchcraft, mostly by hanging. This dark period ranged from the early witch trials in Europe through the Salem witch trials in colonial Massachusetts, although the vast majority of executions took place in Europe.

Another type of anthropomorphic supernaturalism to have developed with probable animistic roots is *general polytheism*.

EARLY *GENERAL POLYTHEISM*: THE MANY DEITIES AND SPIRITUAL ASPECTS

The word "polytheism" comes from the Greek meaning "many deities." I have capitalized this word only when it is used to refer to a formal branch as used on my phylogenetic trees (e.g., Old World Hard Polytheism). Otherwise, it is used to simply refer to the supernatural characteristic of recognizing multiple spiritual beings (*hard polytheism*) or multiple aspects or avatars of a single spiritual being (*soft polytheism*). Records of early general polytheism indicate that it often included deities representing the sun, moon, rivers, animals, and plants, and often had a strong animistic influence.[5,63] It is tempting to speculate about the historical connections between animism and polytheism. For example, the notion of each tree having a spirit could intuitively have evolved into trees collectively governed by a "forest god." General polytheism (perhaps originating in Africa with the earliest development of human societies) influenced some of the earliest-known forms of Organized Religion (as termed here), such as Indigenous Eastern Organized Religions (such as Asian Cyclicism) and Indigenous Afro-Euro-Mediterranean Religions (such as Old World Hard Polytheism). However, not enough is known about the earliest origin of general polytheism to consider it to have had a monophyletic origin, or to have developed more than once independently. Therefore, like *ancestorism* and *matristicism*, *general polytheism* is treated as more of a characteristic than as a distinct branch of Organized Religion (see tree of supernaturalism at the beginning of this chapter).

The possible signs of general polytheism extend further back in time than any written records. Some of the earliest archaeological ruins that are thought by some authors to possibly indicate an early polytheistic form of Organized Religion are the ruins of Gobekli Tepe from around 9130 BCE. This site was discovered in 1994 in southeastern Turkey. It is thought by some historians to be the earliest-known major temple built by humans. It is a megalithic structure that includes an enormous central structure where excavation continues to this day (see page 64). It appears to have been used for more than a thousand years before it was abandoned. The structure includes around two hundred massive, T-shaped stone pillars, arranged in circles. Many of the pillars have artistically carved images of animals on them.[64,65] Each pillar weighs up to twenty tons and fits into sockets hewn out of solid bedrock. A totem pole structure carved from stone was also discovered there. Carved human skulls also suggest the possibility of ancestorism or a skull cult.[66] At the time the structure was built, most humans lived in small nomadic bands or tribes hunting wild animals and foraging for edible plants. A large concentration of hunter-gatherer clans and chiefdoms must have lived in the region with some element of collective organization, but no record exists of an organized class or governmental system. This social system has been referred to as a "complex stateless society" by some anthropologists.[10]

Moving the enormously heavy pillars of Gobekli Tepe, thousands of years before the invention of the wheel, would have taken the massively coordinated efforts of many people. No signs indicate that agriculture existed in the area at the time. The Gobekli Tepe would have been one of the world's earliest major population centers and may have been a rare exception in which an early major society predated the broad development of agriculture. The scale and the animistic/totemistic nature of the structure at Gobekli Tepe also suggests a possible early form of Organized Religion. However, not enough is known about it to include it as a distinct branch on any of my phylogenetic trees. As discussed in chapter 2, it is speculative to interpret details about a religious ideology based only on archaeological artifacts.

Gobekli Tepe archaeological site in southern Turkey, dated at 9130 BCE. Each vertical pillar weighs up to twenty tons and fits into sockets hewn out of solid bedrock, making up one of the earliest known megalithic building structures. (*Top*) Excavation site with black silhouette of an adult man for scale. (*Bottom*) Individual pillars (*right and left*) and a totem (*middle*). Some of the images carved on the pillars and totems are suggestive of animistic polytheism.

Some of the temple's pillars are stylized anthropomorphic figures, with human faces and arms carved into them. They are thought by some anthropologists to represent either distinct spiritual beings (hard polytheism) or aspects of a single spiritual being (soft polytheism). Animal images also may reflect an element of animism. Without any written records, of course, we have no way to be certain. Archaeologist Klaus Schmidt (1953–2014), who saw this as a temple, wrote, "The pillar statue in the center of these enclosures represented very powerful beings. If gods existed in the minds of Early Neolithic people, it is highly probable that the T-shape is the first known monumental depiction of gods."[64] This region of the world is near where we later have clear evidence of Old World Hard Polytheism from ancient Mesopotamians in the ancient civilization of Sumer.

THE EARLY ORIGIN AND DEVELOPMENT OF *ORGANIZED RELIGIONS*

Through the Neolithic and afterward, nomadic kin-based groups of humans began to form larger, more complex class-based societies with multilevel bureaucracies. This development eventually made possible a broad range of social innovations. *General social and economic innovations* included different hierarchical forms of governance, coordination of large work forces for massive construction projects, organized armies for defense and conquest, and organized systems of market trade and commerce. *Technological innovations* included the inventions of metal tools, wheels, group-tested medicines, and complex architecture. *Innovations for sustenance productivity and health* included farming, animal husbandry, and socialized medicine. Finally, *innovations in communication* included the development of writing, public education, and early literacy. The development of bureaucratic, class-based organization also had a downside, which included an increase in inequalities between sexes, social classes, ethnic groups, and entire societies.[31] It also included the development of large-scale warfare, and highly competitive, often violent, struggles for dominance. The early competition among empires was often reflected in the competition among state religions.

With the development of more complex social organization, Organized Religion (which includes all formally institutionalized religions) developed more formalized approaches to supernatural belief systems. Its rituals became systematically established with official doctrine, specific sites of worship, spiritual guidance specialists, and a codified set of rules and practices. Some of its gods became increasingly moralizing, and many branches took an increasing concern with the afterlife. Most branches of Organized Religion, in general, came to increasingly recognize founders or other highly inspirational historical figures with much reverence. Historical records of today's major branches of Organized Religion are relatively abundant (in contrast to those of early forms of pure animism, ancestorism, matristicism, or early general polytheism), because much of their diversification took place after the invention of literacy and writing. With the development of writing, sacred texts started to be produced as stable archives of whatever the scribes (often claiming divine inspiration) had to say. These texts were generally faith-based documents of personal beliefs, mythologies, and legendary stories, rather than objective records of history (**Note 3K**). As written documents, however, they at least provided some level of transgenerational consistency beyond oral transmission.

Early Organized Religion may have been a significant force in motivating social order and cooperation in early human collectives. Societies that had a communally cooperative, well-managed, hierarchical structure, likely would have been more successful than those that did not. That structure was provided, in part, by elements of Organized Religion, including (1) associated rituals and laws defining group affiliation; (2) certain codes of ethical behavior; (3) expanded social cohesion; and (4) deeper, more widespread acceptance of governing powers.[67] Organized Religion may have helped reinforce the civic altruism and prosocial behavior necessary to convince a large population to share food, join the military, pay taxes, recognize governing authority, and maintain social order.[68,69,70] Such willingness to subordinate immediate self-interests to the interests of the broader group would have been a selective advantage for those early societies that possessed such traditions. Some authors argue that the idea of moralizing gods was necessary to build large-scale human societies (a view sometimes referred to as the "big gods" hypothesis), although this hypothesis has been disputed by others.[71,72] That particular controversy is not the focus of this book. My interests are primarily on the historical interrelationships among different religious ideologies over time, which I propose with a series of interconnected tree hypotheses in the following chapters.

Many early urban centers (e.g., Sumer, Ancient Egypt, various empires of Greece and Rome) were theocracies (governments led by a person who was believed to be under direct divine guidance). As early governments became increasingly hierarchical, so went the pantheons of gods and goddesses taking on the personalities and social class hierarchies of human kingdoms and empires. The pantheon of a king god with subordinate gods came to mirror the structure of a royal court and government (discussed further in chapter 5). The Greek Philosopher Aristotle recognized this correlation in the fourth century BCE. In his philosophical tome *Politics*, he wrote, "all people say that the gods had a king because they themselves had kings, for men create gods after their own image." Many kings became respected as representatives of the gods. The threat of supernatural punishment for disobedience to leaders further enhanced civil allegiance and social cooperation.[68,69]

THE EARLY DIVERSIFICATION OF *ORGANIZED RELIGION*

The best-known Organized Religions of today have their deepest roots in one of two major centers of ideological origin, dating back more than eleven thousand years ago and expanding rapidly by the Bronze Age (3300 BC–1200 BCE). As termed here, one branch was *Indigenous Eastern Organized Religion*, which appears to have begun on the Indian subcontinent and various other parts of Asia with what I call early *Asian Cyclicism*. The other branch (also as termed here) was *Afro-Euro-Mediterranean Organized Religion*, which began mostly in West Asia, Northern Africa, and the greater Mediterranean region (based on most existing records to date). As yet, convincing evidence does not exist as to which branch evolved first, or if they both evolved from an earlier, yet unknown, form of Organized Religion from somewhere else in the world, such as southern or central Africa. They may both have roots in the hypothetical Proto-Indo-European (PIE) culture discussed in chapter 5. (many branches are missing on this basal part of my tree.) However, because humans first evolved in Africa before expanding into Eurasia, it is likely that both Indigenous Eastern Organized Religion and Afro-Euro-Mediterranean Organized Religion share at least some core elements of African supernaturalism. Therefore, Indigenous Eastern Organized Religion and Afro-Euro-Mediterranean Organized Religion are shown as sister-groups on my provisional tree. Their order of presentation in this book is not meant to imply that Asian Cyclicism necessarily evolved before Old World Hard Polytheism, or vice versa. Collectively, these two major centers of ideological origin gave rise to tens of thousands of derivative schools, sects, and denominations. Before getting to Afro-Euro-Mediterranean Organized Religions in chapters 5–11, I briefly review branches of the more poorly understood group that I refer to as Asian Cyclicism (which includes the bulk of early Indigenous Eastern Organized Religions as defined here). I also provide preliminary trees of relationship for some of its various subbranches.

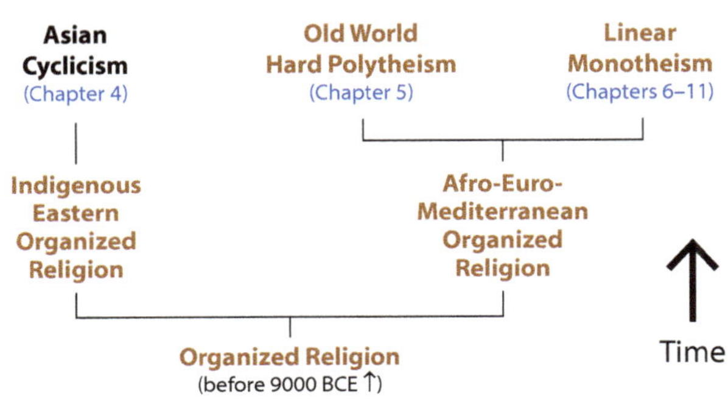

The earliest roots of Asian Cyclicism, like those of Afro-Euro-Mediterranean Organized Religion, appear to have begun as a form of animistic, matristic hard polytheism.[25,27,37,73] Much of it eventually became a form of soft polytheism in practice, or deemphasized the importance of gods altogether. Although it is difficult to say whether Eastern or Afro-Euro-Mediterranean Organized Religion arose first, as both branches evolved, each was influential on the traditions of the other. In the next chapter I begin my review of Organized Religion with Asian Cyclicism.

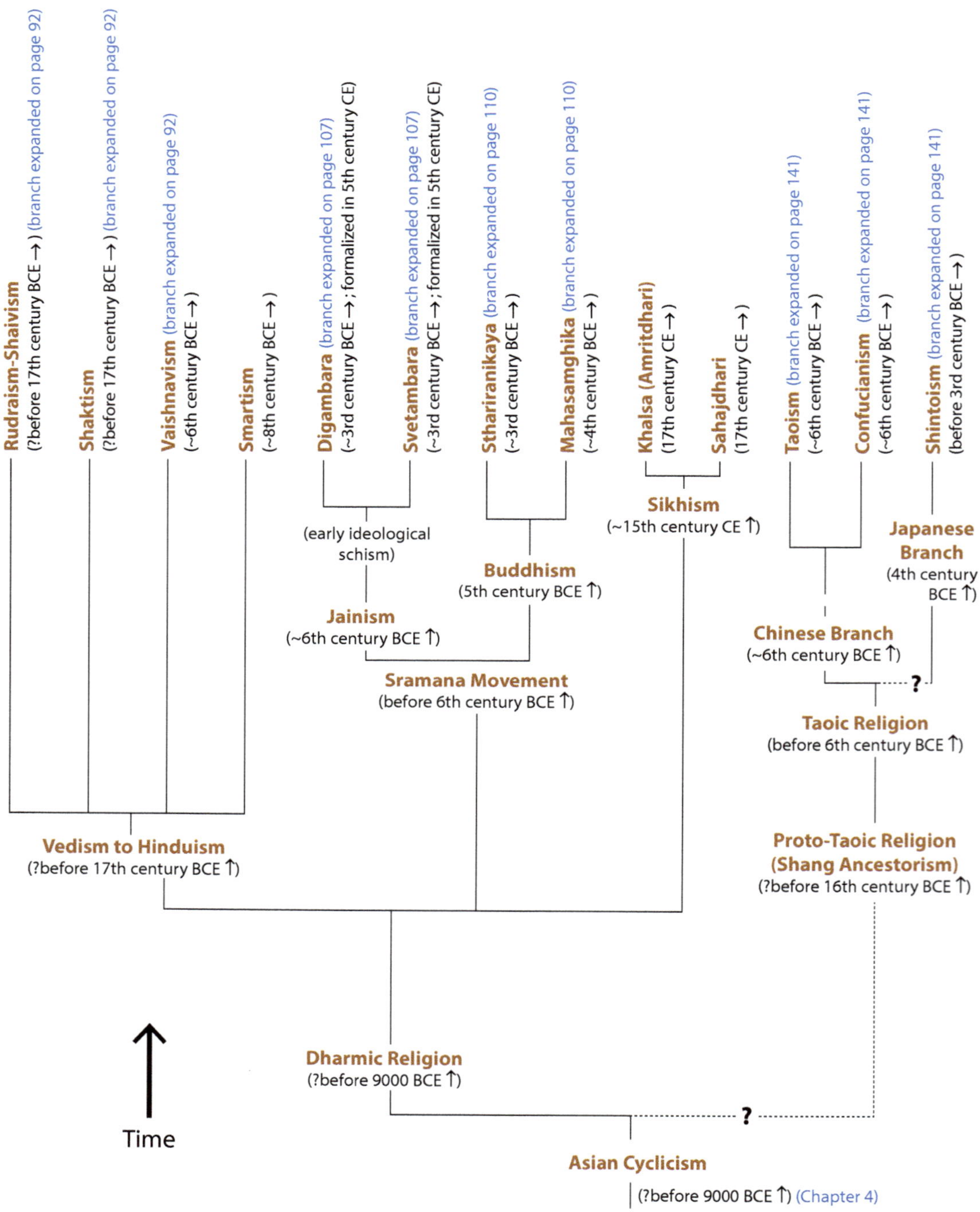

Rudraism-Shaivism (?before 17th century BCE →) (branch expanded on page 92)
Shaktism (?before 17th century BCE →) (branch expanded on page 92)
Vaishnavism (branch expanded on page 92) (~6th century BCE →)
Smartism (~8th century BCE →)
Digambara (branch expanded on page 107) (~3rd century BCE →; formalized in 5th century CE)
Svetambara (branch expanded on page 107) (~3rd century BCE →; formalized in 5th century CE)
Sthariranikaya (branch expanded on page 110) (~3rd century BCE →)
Mahasamghika (branch expanded on page 110) (~4th century BCE →)
Khalsa (Amritdhari) (17th century CE →)
Sahajdhari (17th century CE →)
Taoism (branch expanded on page 141) (~6th century BCE →)
Confucianism (branch expanded on page 141) (~6th century BCE →)
Shintoism (branch expanded on page 141) (before 3rd century BCE →)
(early ideological schism)
Sikhism (~15th century CE ↑)
Japanese Branch (4th century BCE ↑)
Buddhism (5th century BCE ↑)
Jainism (~6th century BCE ↑)
Chinese Branch (~6th century BCE ↑)
?
Sramana Movement (before 6th century BCE ↑)
Taoic Religion (before 6th century BCE ↑)
Vedism to Hinduism (?before 17th century BCE ↑)
Proto-Taoic Religion (Shang Ancestorism) (?before 16th century BCE ↑)
Time
Dharmic Religion (?before 9000 BCE ↑)
?
Asian Cyclicism
(?before 9000 BCE ↑) (Chapter 4)

4

INDIGENOUS EASTERN ORGANIZED RELIGION AND ASIAN CYCLICISM

Well, now everything dies, baby, that's a fact
But maybe everything that dies, one day comes back
—BRUCE SPRINGSTEEN

Asian Cyclicism (as I use this term) evolved within early Asian culture as an indigenous form of Eastern Organized Religion. Among the groups studied here, it is hypothesized to be the sister group of Afro-Euro-Mediterranean Organized Religion (chapters 5–11). Many of its followers believe that time is cyclic rather than linear. Some believe that their nonphysical essence regenerates through near-endless cycles (e.g., samsara, reincarnation, transmigration, or some other form of rebirth). Others focus on dualistic cycles of existence (breathing in and out, day and night, seasons of the year) whose interactions are believed to have given rise to the universe and its diverse forms. Some branches of Asian Cyclicism believe that even the gods (or whatever divine beings they recognize) and the universe go through cycles of death and reincarnation. In contrast, most of Afro-Euro-Mediterranean Organized Religion (*sensu* tree on page 67) sees time moving in a straight line. You are born once, die once, and then either cease to exist completely or move on to an eternal afterlife. A few small branches of Afro-Euro-Mediterranean Organized Religion secondarily developed a belief in reincarnation (e.g., the *Basilidian* and *Cathar* branches of *Biblical Demiurgism* in chapter 10, and the *Druze* and *Alawite* branches of *Islam* in chapter 11). However, those branches developed this belief independently, as an ideological convergence that was perhaps influenced by Asian Cyclicism.

In general, Asian Cyclicism professes to follow the natural order of the universe as a philosophy of life. Most branches emphasize a practice of *orthopraxy* (correct ethical and liturgical conduct) rather than a universal form of *orthodoxy* (rigidly "correct" belief, relentless faith, and grace from God, of the sort favored by Afro-Euro-Mediterranean Organized Religion). Much of it employs mind-body meditative and physical practices to enhance personal introspection and build connection between the spiritual and material worlds. In the Dharmic branches (as discussed later in this chapter), such practices include *yoga* and *zazen*. In the Taoic branches of Asian Cyclicism, those practices include *Tai chi*. Asian Cyclicism appears to have early roots in polytheism,[1]often as soft polytheism with various aspects of a primary deity. However, many of its modern branches put less importance on a God or gods.[2]

As the name suggests, Asian Cyclicism rose primarily in Southern Asia (what is today India and Pakistan) and perhaps also West Asia. However, the name is not meant to represent a geographic group. It is instead meant only as a term for religious ideologies that appear to have originated in that general region. Not all religions that evolved in Asia are branches of Asian Cyclicism (e.g., *Siberian Animistic Polytheism* discussed in chapter 5). Also, many branches of Asian Cyclicism today flourish outside of Asia.[3] A few later branches of Asian Cyclicism (e.g., Buddhism and Jainism) deny being a religion at all because they have no creator god. Others are monotheistic because they revere only one supreme divinity (e.g., Sikhism). I have included those branches in Asian Cyclicism because they have supernatural properties that appear to have evolved from earlier polytheistic belief systems with some element of ideological cyclicity. Their monotheistic characteristics constitute an example of evolutionary convergence. The early evolution of Asian Cyclicism appears to have predated religions such as Christianity and Islam by thousands of years. Its roots extend into preliterate civilizations or into civilizations for which we have yet to decipher their language (e.g., the Harappan script discussed later in this chapter). The religions of those very early Eastern civilizations are poorly known and therefore are not covered in detail.

CLASSIFYING THE SUBGROUPS OF *ASIAN CYCLICISM*

Asian Cyclicism is a difficult belief system to classify systematically or to understand historically. There was much ideological cross-influence among its different subbranches. Many of its branches consider themselves philosophies of life rather than religions (in the typical Western sense of the word). Although Western culture usually distinguishes between religion (a supernatural belief system) and philosophy (the fundamental nature of knowledge), many practitioners of Asian Cyclicism do not.[4] Most branches of Asian Cyclicism also tend to be less rigidly regulated or concerned with carefully worded creeds than Western religions.

Timeline for certain events, people, and artifacts of Asian Cyclicism

India (primarily as Hindu) and Pakistan (primarily as Muslim) gain independence from Britain	–1947
Life of *Mohatma Gandhi*, Hindu icon of religious tolerance	–1869–1948
British Empire takes control of Indian subcontinent	~1757–1947
Emperor Xuanzong declares *Taoism* state religion of China	~8th century CE
The rise of Hinduism and various Muslin conquests on the Indian subcontinent result in the decline of Buddhism there	~7th–16th centuries
Emperor Wu makes *Confucianism* state religion of China	~2nd century CE
Mahayana Buddhism has spread through much of East Asia through missionaries and the patronage from a number of ruling monarchs over time	~1st century BCE
Buddhism undergoes major expansion and diversification under the patronage of Ashoka the Great, Indian emperor of the Maurya Dynasty	~3th century BCE
Buddhism develops in India as a later part of the *Sramana Movement*, after founder Gautama Buddha	~5th century BCE
Sramana Movement begins in India, leading to *Jainism*	~6th century BCE
Formalized *Taoism* develops after founder named Laozi	~6th century BCE
Proto-Taoic principles develop in *China*, following the rise and fall of various emperors of the Shang Dynasty	~1600 BCE
Beginning of *Vedic period* as nomads in *India* begin to compose the Rig Veda and codify *Hinduism*	~?1700 BCE–1500 BCE
Harapan civilization of Indus River Valley with a seal possibly showing the Hindu god Pashupati in a yoga pose	~2350 BCE
Possible shrine to a goddess in central India at the Baghor 1 archaerological site, suggestive of early *Indigenous Eastern Organized Religion*	~9000 BCE

South Asians did not develop a strong sense of themselves as members of a distinct religion until there were other religions against which they needed to define themselves.[5] Major branches, such as what is commonly classified as *Hinduism*, have no single founder or core iconic figure. Many branches ignore historical relationships to earlier faiths, or deny them completely. Sikhs, for example, claim Sikhism does not have roots in any other religion, even though its founder was from a Hindu family and Sikhism was not formally established until around the fifteenth century CE.[6] Jains claim independent origin of their religion by insisting it is the world's oldest. Some Jains go as far as professing their religion to be millions of years old to support that claim![7] (The earliest well-documented occurrence of Jainism is actually less than three thousand years old.[8]) Some Hindus claim their religion arose independently in India without outside influence. They also claim that it was Hinduism that ultimately gave rise to the other early religions of the Middle East and Europe.[5] Such fundamentalist views are not distinct to branches of Asian Cyclicism, of course. Many religions prefer to claim exclusive origin, philosophy, or superiority, rather than recognizing shared inherited characteristics and historical influence between their faith and others.

There is an enormous range of specific beliefs among the hundreds, perhaps thousands, of different subbranches of Asian Cyclicism. These include hundreds of different deities or aspects of particular deities. There is also an extreme range of practices and rituals. For example, some sects of Jainism are so nonviolent they cannot even plow the soil for fear of killing worms or boil water for fear of killing organisms too small to see. In contrast, some branches of Shaktism have historically practiced large-scale animal sacrifice, and allegedly even human sacrifice, to appease Kali, the ferocious aspect of the mother goddess Shakti (discussed later in this chapter). We are far from understanding the detailed diversity and early evolution of Asian Cyclicism, or whether or not it was ultimately a derivative of Old World Hard Polytheism (e.g., chapter 5). Future comparative work by specialists in early Eurasian religions may better resolve this. The largest branch of Asian Cyclicism recognized here is *Dharmic Religion*, which has a deep history on the Indian subcontinent.

DHARMIC RELIGION AND ITS SOUTH ASIAN CORE OF MYSTERIOUS ORIGIN

Dharmic Religion (also sometimes called *Indian Religion*) today includes the branches commonly classified as Hinduism, Jainism, Buddhism, and Sikhism (as shown on the phylogenetic tree on page 70). It is thought by historians to have originally coalesced on the Indian subcontinent through a blending of migrant central

Asian beliefs with indigenous Indian pre-Vedic ideologies. It later adopted more formalized traditions with a large body of Indian religious texts called the Vedas.[5,9] Eventually, various Dharmic lineages spread from India into Eastern Asia, particularly China, Korea, Vietnam, and Japan. Some branches of Dharmic Religion are not deity-centric. Because they still maintain general elements of supernaturalism, I define them as religions. Many Dharmic Religions that revere deities generally do so in a muted manner, seeking ways to transcend beyond deities by turning inward with personal introspection and meditative practices, such as yoga.

Some of the most graphic and convincing early signs of Dharmic Religion on the Indian subcontinent are from the third millennium BCE. These consist mostly of archaeological artifacts from the Harappan civilization of the Indus River Valley dated at around 2350 BCE.[5,9,10] This valley formed in what is now Pakistan and northwestern India. The Indus River Valley region was one of three of the earliest cradles of civilization in the Old World, along with the Fertile Crescent region of Egypt and Mesopotamia, and the ancient river valley regions of China (the Yellow, the Yangtze, and the Liao). Archaeological artifacts from present-day Pakistan and India reveal an urbanized Harappan civilization thought to have existed from around 2500 BCE to around 1500 BCE.[5]

An earlier precursor to the Harappan civilization was the Mehrgarh civilization, which dates back at least to 7000 BCE.[11,12] Given its location, it is possible that elements of Mehrgarh religious belief systems were incorporated into the Harappan culture. However, this is speculative. The Mehrgarh appear to have been the earliest center of agriculture in South Asia. Although they appear not to have had a system of writing, they left many clay figures, both female and male, thought by some archaeologists to have ritualistic significance.[13] We know little else about the Mehrgarh people nor details of their religious traditions.

The roots of Asian Cyclicism on the Indian subcontinent may have gone back even further than the Mehrgarh. One possibility is the remarkable Bhimbetka cave paintings preserved near present-day Bhopal, India. Some of them are as old as around 8000 BCE, and they depict a variety of animals and include a unicorn-like creature that later shows up in Harappan artifacts.[5] Another possibility that has been suggested is an approximately eleven-thousand-year-old artifact from the upper Paleolithic archaeological site of Baghor 1 in central India. This artifact is thought by some archaeologists to have been part of a shrine to an Indian goddess, possibly an early version of the goddess Shakti.[14,15] The specific identification of the goddess is highly speculative based only on archaeological remains of a lightly painted symbolic stone bearing triangular laminae, which tribal people in the region today revere as a form of goddess.[15] The artifact was found centered on a stone platform that is thought by some anthropologists to have been a shrine.[15,16,17] If this was a

Artifact from the Baghor 1 site in Northern India found on a platform interpreted by some anthropologists to have been a shrine. The artifact is thought to be between ten and eleven thousand years old. Suggested by some archaeologists as representing an ancient goddess, possibly Shakti (Kenoyer et al., 1983; Chattopadhyaya, 2016). Original in the collection of the Departmental Museum of Ancient History, Culture, and Archaeology, University of Allahabad, India. (*Left*) Entire artifact reassembled in anterior view. (*Right*) Reverse side of angular stone in the base of the artifact prior to reassembly.

matristic shrine to an Indian goddess, it may possibly have been an early element of Dharmic Religion that was practiced by the paleolithic hunter-gatherers of the time. The phylogenetic connections of this and other pre-Harappan artifacts to Dharmic Religion are tenuous and in need of more definitive evidence.

The archaeological record of the Harappan culture is much better preserved than that of earlier cultures, and it is much more convincing as a Dharmic precursor. The Harappan society was sophisticated and technologically advanced. By 2600 BCE, some Harappan communities had grown into major urban centers of the region, in huge mudbrick cities along the Indus River.[12] They had trade contacts with Mesopotamia, Crete, and possibly Egypt.[5] Sometime around 1500 BCE, the Harappan civilization suddenly disappeared. Reasons for its disappearance remain unknown, but one mythical explanation attributes its collapse to a Great Flood.[5] This was but one of the many civilization-destroying Great Flood myths among different religions and cultures throughout history. The reason such myths are so common from this time period is likely because most early civilizations were concentrated within major river valleys and flood plains. A once-in-a-millennium flood could effectively end an entire civilization.

The Harappan people were literate, and there are many examples of their written text from archaeological sites. Unfortunately, the Harappan language died out thousands of years ago, and there is no "Rosetta Stone" to translate it. Consequently, linguists and archaeologists have so far been unable to decipher the

Comparison of the Pashupati seal with various Hindu gods. (*Top left*) Postage-stamp-size Pashupati seal dated as 2350 BCE from the Harappa archaeological site, Indus River Valley. Thought by some historians to represent the Vedic god *Pashupati* (avatar of the god *Shiva*). In the collection of the National Museum of India in New Delhi. (*Top right*) Enormous carved stone statue of *Shiva* in Bijapur, India, unveiled in 2006. Statue is eighty-five feet high, and it weighs 1,500 tons. (*Bottom left*) Three-headed stone carving of the Hindu Trimurti (Divine Triad of the Supreme Hindu Gods; *Brahma* the creator, *Vishnu* the preserver, and *Shiva* the destroyer). From the sixth century CE of Karnataka, India, and in the collection of the CSMVS Museum in Mumbai, India. (*Bottom right*) Three headed bronze statue of Hindu Trimurti from the twelfth century AD, Angkor, Cambodia. In the collection of Walters Art Museum, Baltimore, Maryland.

old texts.[11] The Harappan civilization left a number of important archaeological sites, such as the one at Mohenjo-Daro in Pakistan. A postage-stamp-size soapstone artifact called the Pashupati Seal was discovered there in 1928 and was dated to be from around 2350 BCE. One interpretation of the image on the seal is that it represents one of the early polytheistic gods of Dharmic Religion, Pashupata (although this theory is debated).[18] The seal shows a male figure in a Yogic pose with horns on his head who is surrounded by animals (upper left image on page 77). The figure possibly may have had three faces.[19] Pashupati, meaning "lord of animals," is thought by some anthropologists to be an aspect of the Vedic god Rudra or the Hindu god Shiva who is often depicted as having three to five heads. Shiva is also associated with yoga.

The Pashupati seal is thought by some archaeologists and historians to reflect Dharmic Religion in the form of early Rudraism or Shaivism).[5,10,11,19] I date the earliest elements of Dharmic Religion and Asian Cyclicism as "before 9000 BCE" based on the questionable archaeological evidence from the Baghor 1 site in central India (mentioned earlier). As I have stated, most early dates of origin on my phylogenetic trees are conjectural estimates. Precise times of origin for preliterate branches of religion will always be estimates, at best.

In addition to the Pashupati seal, the layout of Harappan cities provides another possible sign of Dharmic Religion. The Harappan archaeological sites lack major temples of worship of the kind found in contemporary Egyptian and Mesopotamian archaeological sites. The Indus River Valley civilization possessed the requisite engineering skill to build such structures; they evidently chose not to build them. Religious ceremonies of the Harappan culture may have been personalized and largely confined to individual homes or open air.[11] This may be an early reflection of the more personally oriented and introspective nature of Dharmic Religion as compared with the Mesopotamian and Egyptian religions of the time.

PHILOSOPHICAL FRAMEWORK THAT DEVELOPED WITHIN DHARMIC RELIGION: COSMIC LAW AND ORDER

Eventually, synapomorphic elements of general importance to most Dharmic Religions came to include the following: (1) *dharma* (duty and maintaining order and balance in nature), (2) *karma* (the consequences of one's actions and service to society), (3) *samsara* (reincarnation and the cycle of rebirth), and (4) *yoga* (as an aid to meditation). The origin and precise meaning of these elements, however, varies through time and among the different Indian subcultures.[5]

Dharma, a Sanskrit word (spelled *dhamma* in Pali and Dharam in Punjab) is hard to define in Western culture.[20] It generally means religious law, justice, and righteousness.[5] In early Vedic texts of the first millennium BCE (early *Dharmasastras* and *Dharmasutras*), dharma encompassed the laws of nature, the duties to be performed in accordance with those laws, and the predicted results from fulfilling these duties. In later texts, the meaning became more complex and personalized.

In Hinduism, dharma came to signify righteous behaviors that are in accord with the order of life in the universe.[5,9] These include duties, rights, laws, virtues, and "the right way of living."[21] Dharmic duties and expectations vary according to one's caste or profession. It is not morally judgmental in the Western sense. A person's dharma could entail being a perfect priest, a perfect warrior, a perfect merchant, or a perfect laborer. In Buddhism, dharma means "cosmic law and order" and applies to the teachings of The Buddha about the nature of reality.[22] In Jainism, dharma is a doctrine pertaining to the purification and spiritual transformation of human beings. In Sikhism, the word dharma means righteous concern for others and proper religious practice.[23]

Karma is a spiritual principle of cause and effect, where the intent of an individual's actions can affect his or her future in this life or in future lives. Good deeds with good intent contribute to good karma and future happiness. Bad deeds contribute to bad karma and future suffering. The concept somewhat resembles the Hellenistic school of philosophy referred to as *Stoicism*, founded by the Greek philosopher Zeno of Citium (334–262 BCE), which asserts that the practice of virtue is both necessary and sufficient to achieve happiness and good spirit.

Samsara (reincarnation) is a concept that is only vaguely indicated in the earliest Vedic literature (second millennium BCE), but more clearly in the Upanishads (500 BCE, discussed later in this chapter).[5,10] Although samsara is a fundamental belief of Dharmic Religions (Hinduism, Jainism, Buddhism, and Sikhism), it is not seen in the romantic context sometimes expressed in the West.[24] In most branches of Dharmic Religion, it means a person could come back as a human, a dog, or even an insect or worm. The ultimate goal is to escape an aimlessly mundane cycle of birth, death, and rebirth to reach *moksha* (also known as *nirvana*, *mukti*, or *kaivalya*) and a profound state of well-being.[5] The samsara doctrine is tied to the karma theory in modern Dharmic Religion. A few small branches of Hinduism later rejected the concept of samsara (e.g., the *Lokayatas* of the sixth century BCE to the twelfth century CE).[5]

Yoga is one of the oldest Dharmic practices of Dharmic Religion, indicated already in 2350 BCE by the Pashupati Seal mentioned earlier and illustrated on page 77. Yoga is a critical part of most Dharmic Religions.[25,26] However, some branches of Sikhism later rejected yoga, claiming that Yogic practices are in contradiction to the teachings of the Sikh Gurus. Yoga is much more than the physical exercise program that it represents today for much of the Western world. In Dharmic Religion, it also has a fundamental spiritual core. Its meditative properties are thought to facilitate enlightenment and spiritual release. The word yoga means "joining," referring to union with the divine (reaching God). According to a tolerant view in Hinduism, different sorts of people need different spiritual paths to reach God, and any one or more of them can work. Possible paths include several main types of yoga and a number of more minor ones.[10] The first is *Jnana Yoga* (Knowledge Yoga). This type is thought to be mainly for philosophers and intellectuals, and involves introspection, gnosis, and self-realization, facilitated by the study of sacred Hindu texts. The second path is *Karma Yoga* (Action Yoga). This type of yoga emphasizes good and useful work, particularly good deeds performed without a desire for reward. The third path is *Bhakti Yoga* (Devotion Yoga). This type of yoga emphasizes various expressions of love and devotion. These can be directed to one's spouse, parents, gods, or spiritual teacher (guru). The fourth path is *Raja Yoga* (Royal Yoga). This type of yoga promotes meditation (sitting quietly, turning one's thoughts inward, and calming one's mind). A word or brief phrase called a mantra (such as the sacred syllable *Om* or *Aum*) is sometimes used to clear the mind. A significant part of Raja Yoga is *Hatha Yoga* (Force Yoga). This is the physical exercise form of yoga that people in the West normally associate with yoga. It involves special stretching and breathing exercises that are often used in conjunction with meditation.

Yoga gurus from India introduced yoga to the West beginning in the late nineteenth and early twentieth century. By the 1980s, it had become popular as a meditative program of exercise across much of the world. This led to a backlash by several fundamentalist religious groups. In 2008, Malaysia's top Islamic body passed a fatwa prohibiting Muslims from practicing yoga, saying its practice was blasphemous. Its Hindu origin makes it controversial in some other Islamic countries as well. Fundamentalist Christian groups also shun yoga for religious reasons. Some consider yoga to be incompatible with their religious background, considering it a part of the New Age movement inconsistent with Western Christianity. An article online by the *Christian Ministries* (accessed January 3, 2018) called yoga "a pagan religious activity created to tap into demonic spiritual forces." In 2003, the Vatican issued a ninety-page document called *A Christian Reflection on the New Age* that was mostly critical of Dharmic approaches to spirituality. It warned that concentration on the physical aspects of meditation through practices such as yoga "can degenerate into a cult of the body" and "moral deviations." Even yoga, it seems, stirs up tribalistic fears among different religions competing for the hearts and minds of followers.

Most Dharmic Religions are generally noted for their kindness to animals, and many branches endorse vegetarianism as a way of life. Some Hindus eat meat. However, of those that do, most will not eat beef, because the cow is considered a particularly sacred animal. In India, cows often wander the streets, and cars, trucks, and taxis take care to drive around them.[10] Monkeys and mice alike are well fed in temples, and the divine entities, and avatars (earthly manifestations of deities), are often depicted with animal features (e.g., part fish, part cow, part elephant). The Dharmic belief in reincarnation plays a major role in this. It sees cows and other animals as beings that are in one of many reincarnations of cyclic spiritual development. It also believes that each being's existence will include, or has already included, a human stage. This brings a feeling of closeness to nonhuman forms of life. Some Hindu branches, such as the Chipko movement, extend this principle to the protection of trees. They reason that before people were animals, they were plants (highlighting trees). They advocate for protection of forests and actively protest the construction of dams that might drown trees.

Ironically, for such a largely nonviolent group of traditions, the swastika is a widely used sacred religious symbol for most of Dharmic Religion, including Hinduism, Jainism, and Buddhism (卍). However, for thousands of years, its original symbolic meaning was markedly different from its later misappropriation by Nazi Germany. In most Dharmic Religions it is an auspicious symbol of divinity and spirituality. Its four arms symbolize the four realms of existence in which rebirth occurs according to Jainism: humans, heavenly beings, hellish beings, and nonhuman organisms of the earth (plants and various animals).[27] In Buddhism (e.g., figured on page 126), the swastika is a symbol of Buddha's footprints, or the *wheel of dharma*.[9] At first, even the Western world used the swastika as a symbol of auspiciousness and good luck. It was not until the 1930s that it was subverted to become the German Nazi symbol of Aryan (white, non-Jewish) racial purity and superiority. As a result, the swastika has since become stigmatized in Western culture with ideas of racism, hatred, and even mass genocide. The misappropriation of a religious icon by extremists is not unique to the case of the swastika. Groups within the Ku Klux Klan have long used the Christian crucifix image in "cross-lightings" during gatherings, and recent websites on the internet have used the Jewish Star of David as a satanic cult symbol.

The earliest Dharmic Religions to evolve appear to have been primarily animistic and matristic forms involving the worship of spirits dwelling in stones, animals, trees, rivers, mountains, and stars, or goddesses. Some of these spirits were good, while others were bad, and great spiritual skill was the only way to control them. Among the earliest elements of more formalized Dharmic Religion is what is sometimes termed *Proto-Vedism* or *Brahmanism*.

PRE-VEDIC DHARMIC RELIGION: *PROTO-VEDISM* AND THE HISTORICAL TRANSITION TO *HINDUISM*

The word *Hindu* is derived from the Sanskrit name *Sindhu*, referring to the Indus River. Although Hinduism undoubtedly contains older, pre-Vedic roots (e.g., from the Harappan civilization of the Indus River Valley, discussed earlier, and possibly from earlier hunter-gatherer cultures), it didn't begin taking formal shape until sometime early in the *Vedic period* (1700–500 BCE).[5,12] Many historians contend that during this time in India there was a fusion of pre-Vedic indigenous belief systems with early Vedic elements introduced by invaders from the north and west called Aryans. This theory is called the Aryan migration theory.[9,10,28] The syncretization or merger of pre-Vedic and Vedic religious characteristics, along with the guidance of a controlling Brahmin priesthood, eventually developed into early forms of what became known as Hinduism (as the group is understood here). The details of this early evolutionary development are poorly known, and even the idea of Hinduism being a single monophyletic group is today questioned by some authors.[9] Most sects of the early Vedism have yet to be distinctly identified, classified, and understood in a phylogenetic context. Such detail is beyond the scope of this book. However, my classification recognizes Hinduism as a key putative group for discussion. It also summarizes a few of its major branches, based on simplified sources referred to in the text.

Much of Proto-Vedic culture in India was tribal and nomadic until around 600 BCE when settlements began replacing caravans and eventually evolving into villages. Historical reconstructions so far have been interpreted largely from a complex set of sacred texts, including the Vedas (meaning "knowledge") and later documents such as the Puranas (meaning "ancient"). The Vedas contain the oldest-known deciphered scriptures of Dharmic Religion. They encompass the most sacred hymns of praise to the divinities of the ancient Indo-Iranian people from the Vedic period. These sacred hymns channeled ideologies that eventually became foundational to early Hindu belief systems. Although the earliest surviving written versions of the Vedas date back to only around the eleventh century CE, oral transmission of early Vedic sections is believed to date back to at least between the seventeenth and fifteenth century BCE.[5,10,29]

There are four canonical Vedas.[5,10] The *Rig Veda* (Knowledge of Verses) is the oldest of the four. It contains hymns about the spiritual basis of Vedism, Hindu cosmology, and praises for deities.[5,9,10] The *Yajur Veda* (Knowledge of Sacrifice) builds on the foundational verses in Rig Veda with instructions for ritualistic chants and sacrificial prayers. The *Sama Veda* (Knowledge of Songs) is the Veda of hymns for chanting. Last, the *Atharva Veda* (Knowledge of the Fire Priest) is the Veda of magic spells, medicine, and folk healing. It is one of the earliest texts dealing with medicine. A later supplement to Atharva Veda that focused on healing and medicine is called the *Ayur Veda* (Knowledge of Life).[5,30]

The Rig Veda has a verse predicting the eventual diversification of branches within Dharmic Religion. It says "Ekam Sad Vipra Bahudha Vadanti," which translates into English as "There is but One Reality, but the wise (i.e., the awakened ones) speak of it in a variety of tongues" (*Rig Veda* 1.164.46). That line may also describe a form of soft polytheism; one spiritual universal represented by numerous aspects.[31] The Rig Veda planted seeds for a plurality of spiritual belief systems that later diversified within Dharmic Religion into many branches.

The Vedas are written in an ancient language called Sanskrit, one of the oldest Indo-European languages for which substantial documentation exists.[12] This language originated at least as early as the beginning of the second millennium BCE. It continues to this day to be used as a ceremonial language in Hinduism, Jainism, and Buddhism. In the late eighteenth century, Western scholars recognized that Sanskrit was linguistically related to Greek and Latin.[10,12] This and other shared characteristics suggest an early historical relationship to Greco-Roman culture. Many of the early gods mentioned in the Vedas or later Hindu texts correspond to the gods of ancient Greece and Rome.[10] These include many gods related to natural forces, such as storms, fire, and energy, much like the pantheons of Old World Hard Polytheism (some of which are discussed in **Note 4A**). The Vedic god Indra was first known as "King of all gods."[5,32] Indra closely resembles the Greek god Zeus and the Roman god Jupiter who, like Indra, led their respective pantheons and used a thunderbolt as their weapon of choice. The numerous shared linguistic and ideological characteristics among Vedic and early European religious belief systems may reflect early intercultural cross-influence, or even shared ancestral roots we have yet to identify from the anthropological record.

We do not know much about the authors of the Vedas. The earliest parts of the Rig Veda (books 2–7) are thought to have been composed in the Punjab region of India beginning between 1700 BCE and 1500 BCE.[5] They are thought to have been influenced by central Asians who invaded the Indus Valley and imposed their beliefs on the indigenous people of the subcontinent (the Aryans mentioned earlier in this section).[2] The rest of the Rig Veda and the other three Vedas are thought to have been composed later.[5,20] In Vedic belief system, the focus of sacred scripture has been on ideas rather than on the authors. This focus allows one to look at the message without being influenced by the messenger. The early Vedas were passed down through oral transmission from generation to generation. Even after writing was possible, the Veda was long thought of as being too powerful of a document to be put into written form. Preserving it exclusively in oral form by select transmissionists prevented its use by enemies or the unworthy.[5] Chosen students would rigorously memorize Vedas, word for word, syllable by syllable, by continuous recitations and chants. Remarkably all 1,028 hymns and 10,600 verses of the Rig Veda were passed down this way for more than one thousand years before being written down as a formal document.[34] Consequently, it is difficult to know for sure how oral transmissions of the Vedas may have changed over dozens of generations before their first written transcription. Although written forms of the Vedas exist today, oral transmission is still practiced and given principal authority in many cases. Each of the four Vedas has three or four major subsections, the last of which (the Upanishads) was written for the common people.

The Upanishads were of particular importance in the development of introspective meditative traditions for the common people. They also contained codified ethical standards for laypeople that were adopted by Hinduism as the ten *Yamas* (briefly listed in **Note 4B**). Earlier Vedic texts focused on the priestly class and reclusive ascetics who lived in the forest. In contrast, the Upanishads were broadly directed, indicating that people of any class or occupation could achieve enlightenment and spiritual release. The Upanishads relayed that any person who took the time to get the necessary experience could be a spiritual master.[10] They clarified the concept of samsara (reincarnation) as well as the concept of moksha (eventual release from the countless cycles of samsara).[10] A number of other popular post-Vedic texts also influenced the development and expansion of Hinduism to a broader base of people, such as the *Epics* (the Mahabharata Epic and the Ramayana Epic) and the *Puranas.* (The Vedas, Epics, and Puranas are discussed further in **Note 4C.**)

Although some authors equate Vedism and Brahmanism with what they term as Hinduism,[35] others recognize these as stages of a gradual transition.[9] Brahmanism combined pre-Vedic South Asian belief systems with Vedic texts (e.g., the Upanishads and the Puranas). It was a phase that placed great importance on the Brahmin priest class.[9] It explained the Vedic verses in a context of a growing urban society expanding southward, from the Indus River Valley to the southern tip of India.[5] However, use of the term Brahmanism in the literature is confusingly inconsistent. "Brahmanism," like "Vedism," is used by some authors as a term functionally synonymous with Hinduism. It is used by others as a phase or characteristic leading to what formally became Hinduism.

HINDUISM: THE OLDEST OF TODAY'S MAJOR ORGANIZED RELIGIONS

Hinduism (as I use the name) is the third largest major religion in the world today, after Christianity and Islam. It is by far the oldest of those three belief systems, and adherents refer to it as *Santana Dharma*. Today, Hinduism includes a diverse complex of subschools (known as *sampradayas*), beliefs, and practices. It has no known single prophet, founder, or core inspirational figure (e.g., no Buddha, Jesus, or Muhammad equivalent), although our knowledge of its origin is limited by its great age. It has no supreme leader (e.g., no pope equivalent). Some followers practice more than one Hindu denomination at a time, a practice some authors call

Hindu polycentrism.[36] For these and other reasons, it is a religion that is difficult to define within a Western context. Many Hindus resist classification (as do members of many other religions). Many Hindus consider Hinduism to be not just a religion, but instead a union of reason and intuition that cannot be defined and can only be experienced.[9,35] Some historians and followers consider the term "Hinduism" to be a colonial construct invented by European scholars.[37] I use the name here only to identify a hypothetical group of indigenous Indian supernatural belief systems that share certain ideological characteristics and/or history.

Hindus generally have not engaged in colonial conquest or empire building, other than a few early states within South Asia (e.g., the Guptas dynasty of 300–500 CE). Hinduism has not generally been a proselytizing religion.[5] Hindus have more often the victims of conquest or at least of controlling occupation by non-Hindu cultures (e.g., by the Portuguese Catholics who conquered Goa in 1510, by the Islamic Mogul Empire in 1526 to 1707 and by the Christian British Empire, beginning in the mid-eighteenth century). By 1813, Christian missionaries from the British Empire were allowed to operate on the Indian subcontinent where Evangelical English officers taught young Indian students to be ashamed of their ancestral religion.[5] This was met with limited success, and by 2011, only 2.3 percent of India self-identified as Christian, and about 80 percent still self-identified as Hindu (**Note 4D**).

British control of the Indian subcontinent began with the British East India Company (1757–1857) followed by the British Raj (1858–1947).[5] Hindus in India began to develop a strong sense of their indigenous religions once there were other religions against which they needed to define themselves.[38] The word "Hinduism" was originally not a term of self-identification for Hindus. It was not coined until around 1830, when it was adopted by Indians who opposed British colonialism and wanted to distinguish Indian religion from outside religious groups. It was also used by English writers in the nineteenth century to categorize the religions that had evolved from Vedism and Brahmanism. It was sometimes even used as a nonmonophyletic vernacular name for all non-Muslim Indians.

The Indian subcontinent gained its independence from the British in August 1947, with two separate countries: India (with Hinduism as the primary religion) and Pakistan (with Islam as the primary religion). The various hypotheses concerning the early history and interrelationships of Hindu subbranches (particularly the subbranches of Shaivism and Vaishnavism) are confusing and inconsistent. There have been many, often conflicting, attempts by historians to sort out this confusion.[5,20,28,29,39] Many practicing Hindus view academic scholars who study religion with distrust and distain, partly as a backlash against a long history of the European colonialism of India through the mid-twentieth century. Academics who have written controversial things about Hinduism have reported receiving hate mail and even death threats.[40] As with most major religions with a fervent following, writing about Hinduism from a broad, secular comparative perspective can be a risky endeavor.

My phylogenetic tree of Hinduism (page 92) is a simplified composite of a few basic sources.[9,29,31] However, these sources are only a few among many conflicting classifications of Hindu subgroups. Systematic, comparative analysis of the many Hindu subschools is still a work in progress, and any clear consensus on their detailed interrelationships has yet to materialize. It also comes down to how and at what stage you define the "beginnings" of Hinduism, because historic roots connect most of the religious belief systems at one level of generality or another.

Some Hindu branches worship many different gods, whereas others believe in one supreme divinity with different aspects (different forms of a Supreme God) or with different avatars (different earthly manifestations of a deity). Even the individual aspects of a Supreme God often have their own multiple aspects. A divine triad or trinity (as three revered aspects of a deity), is identified in some branches of Hinduism as the *Trimurti* (Sanskrit for "triple form").[9,10] Initially, the Hindu Trimurti consisted of *Brahma* the creator (later replaced by *Devi*, the Great Goddess), *Vishnu* the sustainer, and *Shiva* the destroyer/preserver (see the bottom figures on page 77).[31,32,41,42] Over time, each of these deities had many different godly aspects of their own.[5] Many other branches of Asian Cyclicism (e.g., examples from Buddhism and Taoism, discussed later) also have divine triads. In its broadest sense, the concept of a divine triad (three manifestations or "modes of being" of a deity), and the symbolic importance of the number three, is widespread among Organized Religions.[43]

The four basic goals of a person in Hinduism (called the *Purusartha* or *Purusha-arthas*) include the following: (1) *dharma* (righteous moral values and duty), (2) *artha* (prosperity and economic success), (3) *kama* (sensual pleasure, best known in the West by the ancient sex manual Kama Sutra), and ultimately (4) *moksha* (transcendental state attained because of being released from samsara and the cycle of rebirth).[5] Hinduism holds that these four goals are necessary and sufficient for a fulfilling and happy life. All must be pursued, although dharma is more important than the second and the third, and moksha is the ultimate goal (which might not be attained until after hundreds or even thousands of life cycles).

A later aspect of many Hindu sects was strict vegetarianism. This may not have always been the case. Some historians of Hinduism contend that the oldest Dharmic belief systems did not object to the eating of meat for certain occasions.[5] However, such a practice created difficult philosophical issues. The Hindu concept of reincarnation included most animals, such as cows, in the same sphere as humans. The great recycling of souls meant that the animal you killed and ate might have been your great-grandmother in a previous life. Some Hindu sacred scriptures clearly warn against the eating of meat, particularly against eating the meat of cows (e.g., *Atharva Veda* 10.1.29 and *Yajur Veda* 13.42). The Hindu prohibition against eating meat was eventually reinforced by influence from Jainism and Buddhism's strict prohibition against the killing of animals.

Cows, in Hinduism, gained particular significance. They became treated with special respect as motherly giving animals who were like members of the family. Bovine respect in some parts of Hindu India grew to the extreme where the slaughter of cows for meat became a capital offense.[5] It also became a lightning rod for violence between Muslims who still slaughtered cattle for food and commerce, and Hindus who forbade killing cows for any reason. Between 2010 and 2014 alone, twenty-four Muslims in India were lynched or otherwise killed by angry Hindu mobs for violence toward cows. Some were killed just for *eating* beef.[38] Today, this particular violence against those who promote the slaughter of cattle has its own name: "cow vigilantism." The cow sensitivity in Hinduism ultimately gave rise to the globalized metaphorical term, "sacred cow," for something or someone that must not be criticized.

THE CASTE SYSTEM OF HINDUISM: FROM THE BRAHMIN TO THE UNTOUCHABLES

A darker side to orthodox Hinduism is the development of a religiously sanctioned caste system. This formalized class system involves a strict social hierarchy that is dictated at birth. General class systems, of course, are not unique to Hinduism. Western societies have had upper, middle, lower, and even slave classes throughout much of their history. As late as the nineteenth century in the southern United States, the Christian *Southern Baptist Convention* preached for the enslaved (primarily African American) to accept their place and obey their masters (discussed later in chapter 9). Major features of the Hindu caste system that distinguish it from other forms of class discrimination include the following: (1) a strong hereditary prescription that has been in effect for thousands of years, (2) its complex formal systemization, and (3) its connection to reincarnation and karma. Many Hindus justify the caste system by reasoning that people are born into lower castes because of bad deeds in a previous life, and people are born in higher castes because of good deeds they did in a former life.

The Hindu caste system is claimed by some proponents (primarily upper-caste members) to have originated with the Vedic god of creation, Brahma. Although the system is briefly mentioned in the Rig Veda,[5,10] it was not completely codified until post-Vedic Hindu texts such as the Mahabharata Epic (**Note 4C**). It is also outlined prominently in the authoritative book on Hindu law called the *Manusmriti*, which was published in Sanskrit sometime between the first and third century CE.[5,9] The Manusmriti was verbally translated from Sanskrit to English by Sir William Jones, during the British rule of India (the British Raj) (Jones, 1796). The text acknowledged the caste system as the basis of order and regularity of society. Many lower-caste Hindus reject it as upper-caste (Brahmin) propaganda.[9] British colonialism of the nineteenth and early twentieth centuries used the Manusmriti and the caste system to strengthen the administrative apparatus of its colonial state in India.[20,44]

At the top of the caste hierarchy are the elite *Brahmin*.[5] The Brahmin consist mainly of priests, teachers, and intellectuals. They are venerated as having more spiritual purity than any of the other castes. They alone are capable of performing critical religious tasks, such as oral transmission of the Vedas. This was the caste from which Vedic priests came. For centuries, Indian cultural control was exclusively in their hands. They were supposedly created from Brahma's head. Next in the caste hierarch are the *Kshatriyas*, or the warriors and rulers. The traditional function of the Kshatriyas is to protect society by fighting in wartime and governing in peacetime. They were created from Brahma's arms. The third rung of the caste ladder is the *Vaishyas*. Hindu religious texts originally assigned Vaishyas to roles in farming and cattle rearing. Over time, they came to be landowners, traders, and money lenders.[45] It is their responsibility to provide sustenance for those of a higher class, because they were of lower class. They were created from Brahma's thighs. The lowest caste of the four recognized castes is the *Shudras*, who are servants responsible for menial jobs. They were created from Brahma's feet. Over time, the major castes were further divided into thousands of subcastes, each based on the person's specific occupation.

There is effectively a fifth major caste category below the Shudras called the *Dalits*, a name derived from the Hindu word meaning "oppressed."[5] Dalits (formerly known as the "untouchables") are literally outcasts; a group deemed so unworthy that it does not fall within the official caste system. This system assumes that you were born as a Dalit because of grave offenses in a past life, so whatever harm befalls you in this life is deserved. Dalits have historically been denied access to basic health care and education. They are shunned from society, left to do jobs considered "unclean," such as waste disposal, toilet cleaning, and cremation.

The caste system was outlawed in the Indian constitution of 1950, although the system continues to be practiced in parts of India. In the early 1980s, an Indian affirmative action group called the *Mandal Commission* pushed to offer nearly half of all government and educational places to members of underprivileged castes. This was such a contentious issue for upper-caste Hindus that Brahmin priests and teachers burned themselves to death in protest against extending this consideration to the Dalits.[5] More recently, to prevent harassment, assault, and discrimination against the Dalits, the government of India enacted the *Prevention of Atrocities Act* on March 31, 1995. However, clerically sanctioned human rights abuses against the Dalits have continued into the twenty-first century. According to a report by *National Geographic*, a random sampling of 2003 headlines in mainstream Indian newspapers included many violent instances of Dalits being murdered, tortured, and raped because of their caste position.[46] In many parts of India, Dalits are still not allowed to drink from the same wells as the upper castes, attend the same temples, or even wear shoes in the presence of an upper caste. India's Dalits live in constant fear of being publicly humiliated, paraded naked, beaten, and raped with impunity by upper-caste Hindus seeking to keep them in their place. Merely walking through an upper-caste neighborhood can be a life-threatening offense.[47]

The Dalit class, previously known as the "untouchables, is by no means a small population. In 2003, there were 160 million members of the caste in India. Mohandas Gandhi called them *Harijans* ("Children of God") and long worked for their emancipation. Over time, the name "untouchable" was regarded as condescending and offensive and was replaced with Dalit or the "Scheduled Caste." Things improved when Kocheril Ramen Narayan was elected as president of India from 1997 to 2002. He was the first Dalit to occupy a high office in the country. Nevertheless, the caste system persists today.

HINDUISM AND CHALLENGES OF PEACEFUL INTERRELIGIOUS COEXISTENCE

In 1867, the pioneer of Comparative Religion studies, Max Muller, wrote, "Hinduism is a decrepit religion, and has not many years to live."[48] That was not one of his better predictions. Today, Hinduism is the largest branch of Asian Cyclicism and the third largest religion in the world, with more than one billion followers.[3]

In general, Hinduism is one of the least proselytizing religions. It is not in conflict with most other religious belief systems, with the possible exception of Islam. Some Hindus with peaceful intent have tried to promote the peaceful coexistence of Hinduism with all other religions of the world. Mohandas Gandhi (1869–1948) was a liberal Hindu (also known by the reverent title, "Mahatma") who was greatly influenced by the Jain doctrine of nonviolence. He became a dedicated, charismatic advocate of peace, writing, "It cannot be said that Sikhism, Hinduism, Buddhism, and Jainism are separate religions. All these four faiths and their offshoots are one. Hinduism is an ocean into which all the rivers run. It can absorb Islam and Christianity and all other religions, and only then can it become the ocean."[49] Gandhi occasionally even read selected verse from the Quran (sacred book of Islam) during prayers in the Hindu temple. He considered himself an equally good Hindu, Jain, Christian, and Muslim.[50] During his prayer meetings, he always included verses from the Quran. He reminded people that it was folly to look upon one religion as better than another. His statements such as "all religions are one" and "God has no religion" were noble pluralistic efforts intended to stop the never-ending violence in the name of religious chauvinism or nationalism and to champion Indian causes against social and economic discrimination.

Gandhi became a folk hero to thousands and to some he became even more. Some people began to deify him.[5] They reported that spiders were writing his name in their webs. They also saw Gandhi's face in bottles of kerosene and in the rising moon and sun. The concept of "Gandhism" began to materialize. Gandhi himself finally put a stop to the mythologizing of his image. In 1936, he publicly proclaimed: "There is no such thing as 'Gandhism,' and I do not want to leave any sect after me. I do not claim to have originated any new principle or doctrine. I have simply tried in my own way to apply the eternal truths to our daily life and problems . . . The opinions I have formed and the conclusions I have arrived at are not final. I may change them tomorrow. I have nothing new to teach the world. Truth and nonviolence are as old as the hills."[51]

Gandhi had no intent of being revered as a deity. However, religion eventually takes on a life of its own in the eyes of its believers. Being a symbolic force of a transformative religious belief system can generate great influential power. It can also be dangerous, or even fatal, for the person at the heart of such a transformation. On January 30, 1948, on his way to address a prayer meeting, Gandhi was assassinated by a Hindu nationalist extremist named Nathuram Godse, who murdered him for his complacence toward Muslims and his advocacy of religious pluralism.[52]

The conflict between Hindus and Muslims continues to the day. As I wrote this chapter in December 2019, violent protests had broken out across predominately Hindu India. This conflict was over a new citizenship law that excludes Muslim immigrants from fast-track paths to citizenship, while facilitating it for people of other non-Hindu religions. Soon after the bill's passage, five people were killed and hundreds were injured across the country in violent clashes between Muslims and India's Hindu nationalist forces. The current Indian government claims to be a secular entity by defining Hinduism as a "way of life" rather than a religion. Its clash directed at Muslims in the country appears to challenge that assertion. Hinduism is only one of many ideologies that sometimes claim to be the proper way of life rather than merely a religion. However, it involves a clear religious nature, as I define the term "religion."

THE EVOLUTION (DIVERSIFICATION) OF HINDU SUBBRANCHES

Hinduism is divided into four major branches. These include *Rudraism-Shaivism*, *Shaktism*, *Vaishnavism*, and *Smartism* (see the phylogenetic trees on pages 70 and 90). These descriptions refer to core ideological principles. Note that each of these major branches, particularly Vaishnavism and Shaivism, has many subbranches (some of which are on my phylogenetic tree on page 92), each with its own belief system, ranging from extreme ascetism to less extreme syncretized traditions. As I have mentioned, there are many different interpretations of the nature and interrelationships of these subbranches. Much work is needed to decipher a clear phylogenetic pattern among them.

RUDRAISM-SHAIVISM AND THE ROOTS OF HINDUISM (AND YOGA)

Rudraism-Shaivism (or just *Shaivism*) appears to have begun with the godly aspects of *Rudra* and eventually *Shiva*. It is a henotheistic belief system in that it recognizes either Rudra, Shiva, or some aspect of them as the Supreme God of the pantheon.

Rudra ("howler") was the wild Vedic storm god of destruction and fertility, and a Supreme aspect from the Rig Veda (1700–1500 BCE).[5] Shiva began as a lesser, more nurturing aspect of Rudra. In some early Hindu sutras, Shiva is known as part of the joint divinity Rudra-Shiva.[9] By the second century BCE, Rudra's supremacy began to give way to his lesser aspect Shiva. At that time, we also find the earliest textural records of Shaivism, with the appearance of its earliest branch, the Pashupatas.[29] With the rise of *Shaivism* (also spelled *Saivism* by some authors), Rudra became the lesser aspect of Shiva.[9] In post-Vedic literature, Shiva has the combined destructive and constructive powers of ultimate recycler and rejuvenator of all existence, and He is the Supreme God.[53]

As mentioned earlier in this chapter, some scholars trace elements of Rudraism-Shaivism to the pre-Vedic Harappan civilization of the Indus River Valley or earlier. The Harappan seal artifact shown on page 77 is dated as being from 2350 BCE, and some archaeologists claim that it shows the Vedic god *Pashupati* (lord of animals and an aspect of the god Shiva) in a yoga pose.[10,54] This interpretation of the seal artifact is controversial and rejected by some scholars.[5] It is always challenging to base cultural hypotheses on archaeological objects alone and to interpret ancient objects without decipherable written text (as discussed in chapter 2). Other scholars have noted the similarities between Shiva and Greek and European deities, suggesting an Indo-European link.[55] The contrasting aspects of Pashupati,

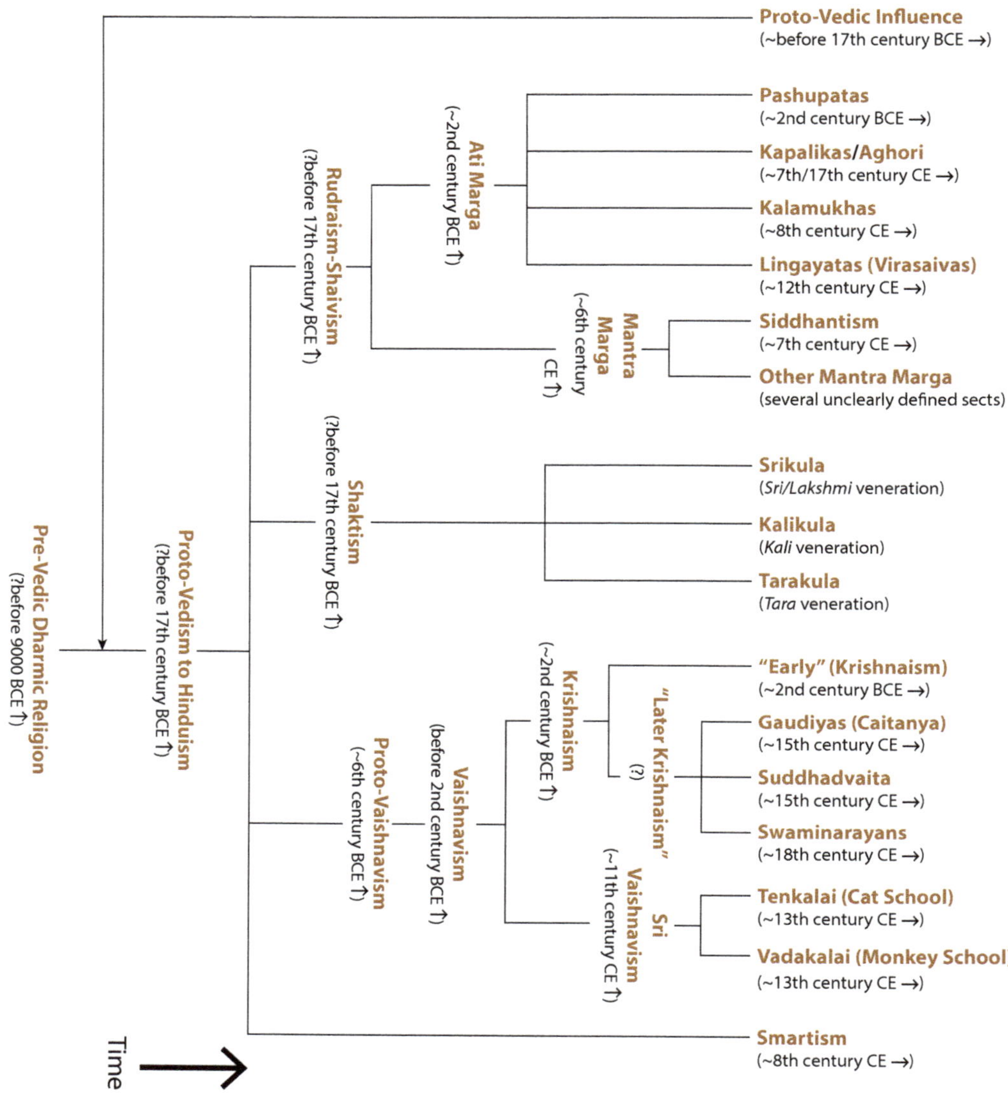

such as being fierce or blissful, depending on the situation are similar to those of the Greek god Dionysus. Both gods have iconic associations with a bull, snakes, anger, bravery, and carefree life. Ancient Greek texts in the time of Alexander the Great (356–323 BCE) call Shiva the "Indian Dionysus."[56]

Modern Shaivism took more formal shape during the first millenium CE.[53,57] Over time, it evolved into many different schools that still exist today, as well as several that are extinct.[31] According to a 2010 estimate, Shaivism is the second largest group of Hindus, with 252 million followers, or 26.6 percent of all Hindus.[3] Many emperors and kings of parts of the Indian subcontinent were strong supporters of Shaivism, helping to preserve its strength as a major Hindu branch.[58]

According to some authors,[31] Shaivism has two main subbranches, each with many subbranches of its own.[29,31] One subbranch is *Ati Marga* (also spelled *Atimarga*) thought to have evolved in early Vedic time. The other subbranch is *Mantra Marga* (also spelled *Mantramarga*) thought to have evolved later. Both have certain ascetic subbranches that revere celibacy, living away from dense civilization, wearing animal skins for clothing, and covering themselves with ashes.[31] Adherents to the practices of Shaivism are almost always men.[31]

ATI MARGA: THE EARLY VEDIC APPROACH TO SHAIVISM ON THE "EXTREME PATH"

Ati Marga (known as the "extreme path") originated in early Vedic time. Its subbranches include the *Pashupatas* (second century BCE), the *Kapalikas* (sixth century CE), the *Kalamukhas* (eighth century CE), the *Lingayatas* (twelfth century CE), and some not shown on the phylogenetic tree on page 92. Although I place the early roots of Ati Marga and its earliest branch, the Pashupatas, in early Vedic time, some authors recognize its more formalized origin in the first or second century CE with a founder named Lakulisha. He is credited by some as the author of the principal Pashupata text, the *Pashupata Sutra*. He was later considered by adherents to be an incarnation of the god Shiva.[5,57]

The *Pashupatas* (also spelled *Pasupata*) began as a reclusive, highly ascetic group. They are celibate, and they practice highly antisocial behavior.[9,31] This behavior includes snoring in public, acting if as mad, making inappropriate advances toward women, and taunting others. Their rituals were performed to oppose the purity commonplace in more conservative branches of Hinduism. Their thinking was that because the bonds between an individual and his community are strong, extra efforts are needed to cut them off. Once completely freed from social bonds, Pashupatas could resort to yoga and a number of austerities in hopes of acquiring magical powers. Such austerities included staring at the sun, standing on one leg for months at a time, and lying on a bed of nails. Their basic doctrine is sometimes referred to as the "servitude of souls to God." Their hope is to achieve immunity from extreme pain on their path to union with Rudra/Shiva. The Pashupatas were particularly influential between the seventh and fourteenth centuries CE on the Indian subcontinent, but they have been a relatively minor element since the eighteenth century.

The *Kapalikas* were another ascetic subsect of Saivism that flourished between the seventh and fourteenth centuries CE, but it has few adherents today. They have always lived on the edge of society, naked and covered with ashes of cremated corpses and carrying a human skull which they use as a bowl.[9,31] They seek possession by a powerful deity to achieve perfection (*siddhi*). Their practices have allegedly included the consumption of human corpse flesh (necro-cannibalism) and scatological substances.[9] The Kapalikas were scorned and feared by the Pashupatas. This subsect has all but vanished today, although most of their traditions are carried on by a later ascetic sect derived from then: the *Aghori*.[9,29] This later group originated in the seventeenth century CE.

The *Kalamukhas* were another celibate ascetic sect whose supernatural belief system was closely related to that of the Pashupatas. Like most of the Ati Marga, they were ascetics. Like the Kapalikas and Aghori, they covered themselves in cremation ashes.[9] Many of their core elements may have later been adopted by another Ati Marga subsect called the Lingayatas.

The *Lingayatas* (also known as the *Virashaivas* or *Veera Shaivas*) make up a sect that originated in southern India during the twelfth century CE, who worship Shiva in the form of a stone pillar called a lingam.[9,31] This movement was initiated by a Brahman named Basava (1106–1167 CE) who carried out systematic attacks on Jains, Buddhists, and Vaishnava Hinduism, while sponsoring Shaivism in the form of the Lingayat subsect. They adopted many social reforms, including the rejection of caste distinctions and acceptance of women as equals. They also reject the authority of the Veda and its Brahmin guardians. Most followers are not celibate, including many of its priests. They are strictly vegetarian and currently number about eight million, mostly in Karnataka, India.[29]

In addition to Ati Marga, the other major branch of Rudraism-Shaivism on my phylogenetic tree is *Mantra Marga*.

MANTRA MARGA: THE LESS ASCETIC APPROACH TO SHAIVISM

Mantra Marga subbranches of Shaivism include the *Siddhantism* (seventh century CE) and several others not shown on my phylogenetic tree.[9] Most followers of modern Shaivism are not ascetics, so the less ascetic branch, Mantra Marga, has become the more popular. However, the phylogenetic tree on page 92 is only one of several different interpretations of the historical diversification of Shaivism. Many other classifications, names, and diagnoses have been proposed.[9,54,57,58] There is no strong consensus as of yet for the systematic organization of the numerous sampradayas (Dharmic denominations). Listing all of the various existing classifications of Shaivism subgroups is a complex problem that I do not approach here.

The feminine side of the god Shiva is often embodied by the goddess Shakti, core of the next major branch of Hinduism: *Shaktism*.

SHAKTISM: FEMALE-CENTRIC HINDUISM AND WORSHIP OF THE MOTHER GODDESS

Shaktism (also spelled *Saktism*) is an English name for the female-centric religious tradition of worshipping *Shakti* (*Sakti*) as a personification of supreme power.[9,29] Although Shakti is worshipped as a Supreme Goddess (*Devi* or *Mahadevi*) in Shaktism, she is more than an individual goddess. She represents a dynamic feminine quality that all Hindu goddesses possess. She is venerated according to many different aspects, ranging from the gentle and loving *Sri* (also known as *Lakshmi*) to the terrible and threatening *Kali* (also known as *Dakshina Kalika*). As the Great Divine mother goddess, Shakti is the source, essence, and substance of everything in the great cycle of existence, from creation to destruction.[59] As discussed earlier in this chapter, the goddess Shakti may have predated the Vedic formalization of Hinduism by thousands of years, based on what some archaeologists believe may be a shrine from central India dating back to around 9000 BCE.[14]

Sri/Lakshmi is the goddess of love, beauty, joy, and prosperity. She is thought in much of Shaktism to be the consort (wife) of *Vishnu*, the Supreme Being within Vaishnavism. Along with the goddess *Parvati* (mother goddess, consort of *Shiva* the Supreme Being in Shaivism) and the goddess *Saraswati* (goddess of learning and the arts, consort of the god *Brahma* the creator god of the Trimurti), Lakshmi completes the Tridevi (divine triad of goddesses that is the feminine version of the Trimurti discussed earlier in this chapter).

Kali is the ferocious goddess of power, violence, and destruction. She is also known as the devourer of time.[5,9] Human sacrifice to Kali was purportedly practiced during the early development of Shaktism all the way into the nineteenth century.[36] The goddess is often shown wearing a necklace of human skulls and fanged teeth dripping with blood.[10] Although human sacrifice was not part of the original principles of Vedism or the Vedas and was outlawed in the mid-nineteenth century, human sacrifice to Kali is still occasionally reported by the media even in twenty-first-century India.[60] Such practice is extremely rare today and not condoned by most followers of Shaktism. In Nepal, animal sacrifice of goats, chickens, and water buffalo, occurs en masse during the three-day-long festival to the goddess

Two of many aspects of the Hindu goddess *Shakti*, who is often portrayed as the consort of Shiva. (*Left*) Nineteenth-century painting of *Sri/Lakshmi*, embodiment of love, beauty, joy, and prosperity. In the collection of the Maharaja Fateh Singh Museum, Gujarat, India. (*Right*) Nineteenth-century painting of the terrible and threatening *Kali*, embodiment of power and violence. Both paintings by the famous Indian artist, Raja Ravi Varma (1848–1906).

Gadhimai. Gadhimai is the Hindu goddess of power and an aspect of the Kali (in Hinduism, there are even aspects of aspects). In 2009, it was estimated that more than 250,000 animals were sacrificed during this controversial event.[61] In contrast, most branches of Hinduism find the event abhorrent, and many forbid the killing of animals even for food. It is a good example of ideological diversification, a process that inevitably occurs within all major religions over time. The image of Kali's extended tongue is well known to Western popular culture. It was appropriated in the late twentieth century by the popular music group, the Rolling Stones, to create their iconic logo (the lips, open mouth, and protruding tongue). The logo became the one of most famous icons in the history of popular music.

Most authors believe that Shaktism, as we know it today, began in the age of the Vedas (1700–1500 BCE)[5] and reached its present form around 700 CE with the early compiling of its central scriptures (the Sakta Tantras).[32] It has continued to expand and diversify ever since then, and as of 2010, there were an estimated thirty million followers of Shaktism, or 3.2 percent of all Hindus.[3] Shaktism can be divided into many schools involving aspects of deities far too numerous to list here. The two largest of these are the *Srikula* branch (family of Sri) and the *Kalikula* branch (family of Kali).[9] The *Srikula* branch (strongest in South India) worships the gentle and beautiful aspects of Shakti. Most notable is the gentle, loving Sri/Lakshmi (mentioned earlier) in the form of *Tripurasundari* (also known as *Lalita*).[9] In contrast, the *Kalikula* branch (strongest in northern and eastern India) worships the more threatening aspects of Shakti, such as the fierce *Kali* (the doomsday goddess also mentioned earlier). A smaller third group (the *Tarakula* branch) advocates the worship of the goddess Tara, who is a goddess that was possibly introduced into Hinduism from Tibetan Buddhism.[9,29] Tibetan Buddhists also revere her as the Mother of all Buddhas.[9]

Another major branch of Hinduism on my phylogenetic tree (page 92) is *Vaishnavism.*

VAISHNAVISM: THE WORSHIP OF VISHNU AND THE DIVINE TRIAD

Vaishnavism (also spelled *Vaisnavism* by some authors) is one of the major post-Vedic branches of Hinduism. It appears to have emerged by the sixth century BCE[29] as a fusion between non-Vedic and Vedic sects, beginning with the worship of a proto-Vaisnava deity called *Vasudeva.*[29] The followers of Vaishnavism (called Vaishnavas) eventually began to worship the male deity *Vishnu* (also spelled *Visnu*) or one of his ten incarnations (especially *Krishna*) as the Supreme God. In the Vedas, Vishnu was a god of little importance, but he subsequently became a major deity and member of the Hindu Sacred Triad called the Trimurti (mentioned earlier in this chapter).[9] For Vaishnavas, he is the supreme deity. One of Vushnu's most revered avatars is *Matsya,* the "fish savior" and source of the Hindu Great Flood story. Matsya rescues Manu, the ancestor of the future humanity, from a great flood by warning him to build a great boat. It is one of many stories thought to possibly be related to the great flood stories of Mesopotamian/Sumerian Polytheism and Abrahamic Monotheism.

Vaishnavas are deeply devotional, and their religion is rich in aspects of Vishnu, temples, and scriptures. They prophesize a messiah named *Kalki* who will be Vishnu incarnate.[5] Kalki will vanquish the wicked in an apocalypse-like scenario and restore order to humanity. Vaishnavism evolved from the merger of several ancient theistic belief systems, in which the various deities were integrated as different aspects or avatars of the same god, Vishnu.[28] The emperors of the Gupta dynasty (300–500 CE) favored Vaishnavism over other religions, helping its early success. However, they also patronized other Dharmic Religions, such as Buddhism and Jainism.[5,58] According to 2010 estimates, Vaishnavism is the largest of the four major branches of Hinduism shown on the phylogenetic tree on at the beginning of this chapter.[3] It has 642 million followers, or 67.6 percent of all Hindus, and is somewhat polycentric during festivals. During such events, some of its members recognize the deities of Shaivism, Shaktism, and Smartism, while those branches also then revere Vishnu in return.[20] There are dozens of subbranches of Vaishnavism today, differing in which sacred texts or aspects of Vishnu they prioritize. Different authors have classified these subbranches in a variety of different ways (discussed further in **Note 4E**).[31,57,58] The implied classification of Srinivasan (2011) modified somewhat by Bowker (2003) is shown on page 92 to illustrate some basic patterns of diversification. Systematic classification of the numerous subgroups of Vaishnavism is extremely complex, and there is no consistent, broadly inclusive, classification. As with Shaivism, listing all previous classifications of Vaishnavism is an extremely complex problem, and I list only a few basic branches. Two major groups that developed within Vaishnavism include *Krishnaism* and *Sri Vaishnavism*.

KRISHNAISM: THE WORSHIP OF THE SENSUAL DEITY, *KRISHNA*

The term *Krishnaism* has been used to describe the Vaishnavistic branches that focus on *Krishna* (also spelled *Krsna* in Sanskrit), the god of love, compassion, tenderness, and protection. He is thought by most followers to be the eighth avatar of Vishnu. He is also one of the most popular and revered of all Indian divinities. His name dates back to the Rig Veda,[9] and he may have originally been an object of fertility worship.[10] Early stories report that he had many women cow herders (*gopis*) as mistresses and wives (sixteen thousand by some counts of the Mahabharata[5]). Krishna's flute playing was said to attract young women who he enchanted with his erotic songs. In later branches of Krishnaism (after the eleventh century), volumes were written by poets describing Krishna's main consort as the goddess *Radha*. The most famous of these poets was Jayadeva (1170–1245), who wrote the Gita Govinda.[62] That poem evokes the joy of lovers being together, as well as the pain of their separation. It also emphasizes Krishna's love for Radha.

The early worship of Krishna (provisionally termed *Early Krishnaism*) is ancient, dating before the first century CE, but almost nothing is known about the earliest sects. They may possibly have involved Vasudeva, father of Krishna.[9] What is provisionally called *Later Krishnaism*, eventually evolved into several better-known subbranches,[9,31] including the *Gaudiyas* (also called *Caitanya* or *Chaitanya*), the *Vallabhas*, and the *Swaminarayans* (also known as the *Satsangis*) (see the tree on page 92).

The *Gaudiyas* movement was founded by an ascetic named Sri Caitanya Mahaprabhu (1485–1533) who grew up in the region known today as Bangladesh. He became renowned for his ecstatic devotion expressed in dance and song, and followers believed he was the combined avatar of Krishna and Radha. He was a major influence on the development of devotion to Krishna and revered by many Indian followers as a saint. One of the modern subbranches of the Gaudiyas to evolve and become popular in Western culture during the 1960s and 1970s was the *Hare Krishna* (the International Society for Krishna Consciousness, ISKCON). It was founded in New York by swami Prabhupada (1896–1977) and named after the sixteen-word Vaishnava mantra from the Kali-Santarana Upanishad. Hare Krishna was adopted by a number of Western celebrities, including George Harrison of the iconic pop rock group, the Beatles. His song "My Sweet Lord," released in 1970, features this mantra, which was also included in the famous Broadway musical *Hair* that opened in the late 1960s.

The *Suddhadvaita* branch (also spelled *Shuddhavaita*) was founded by the intellectual, Vallabha (1473–1531).[9,31] He called himself the reincarnation of the Hindu god Agni. He rejected ascetism as well as practices of self-denial, such as fasting. Some scholars and later Vaishnavas described this approach as lax and a way of enjoyment leading to irreligious and corrupt tendencies.[31] This less-ascetic tradition helped initiate a later branch of Krishnaism in response, the *Swaminarayans* (also known as the *Svaminarayans* or *Satsangis*). The Swaminarayans were founded by a high-caste Brahmin known as Lord Swaminarayan (1781–1830). He sought to "correct" the ways of the Suddhavaita, returning to an emphasis on ascetism, puritanical worship, and fully absorbed meditation. Swaminarayan eventually claimed to be the true incarnation of the Supreme God, Krishna, claiming to have taken human form to bring salvation to his followers.[5,9,31] By the 1950s, the Swaminarayan movement had spread from India to Britain, becoming the largest Indian religion in that country.[9]

Another major branch of Vaishnavism is *Sri Vaishnavism*.

SRI VAISHNAVISM: CATS AND MONKEYS

The *Sri Vaishnavism* branch reveres the god Vishnu, and the name derives from the prominence it also ascribes to his consort Sri (Lakshmi).[9,29] Although it traces some of its roots back to the Vedas, its central philosopher was the Indian social reformer, Ramanuja (1017–1137 CE). He drew various strands of proto–Sri Vaishnavism together into a coherent doctrinal whole.[29] He is considered a Hindu saint by many followers. Sri Vaishnavism is open to women and members of lower castes, but the priests and other spiritual leaders have been almost exclusively from the Brahmin caste.

During the thirteenth century, a schism occurred within Sri Vaishnavism resulting in two new subbranches. One is the *Tenkalai* (also known as the *Cat School*). The other is *Vadakalai* (also known as the *Monkey School*). The schism was over what was meant by a verse from a fundamental Hindu text. The Bhagavad Gita (Chapter 18, Verse 66). It reads: "Abandon all duties and surrender to me. I shall absolve you of all sins. Do not fear."

The Cat School interpreted this to mean that surrender to the Lord is total and requires no effort on the devotee's part (just as kittens allow the mother cat to pick them up by the scruffs of their neck and move them without any effort on the kitten's part). Grace and liberation are freely available, if only the devotee has faith. The Monkey School contends that grace and liberation must be earned by a person actively clinging to god (just as a baby monkey must cling to its mother as she moves through the trees).[5] The Vadakalai School has also been more insistent on retaining distinctions of caste and purity than the Tenkalai School.[29]

In addition to Shaivism, Shaktism, and Vaishnavism, the fourth major branch of Hinduism on my phylogenetic tree of Hinduism (page 92) is *Smartism.*

SMARTISM: A SYNCRETIC BRANCH OF HINDUISM WORSHIPPING MANY GODLY ASPECTS

According to many authors, and as illustrated on my phylogenetic trees, Smartism is a fourth major branch of Hinduism.[20,31,42] In contrast to many other Indian religious belief systems, it generally involves a nondualistic interpretation of the Vedic Upanishads, emphasizing oneness between the Supreme God and the universe, rather than the interpretation of God and the universe as two different substances.[5] Followers of Smartism make no distinction between the individual soul and the soul of God. This characteristic varies in the other three major branches of Hinduism and warrants further comparative study.

The Smarta movement was a Hindu belief system connected with the great South Indian philosopher and Brahmin, Adi Sankara (788–820 CE). It began as a syncretic movement, bridging elements of Vaishnavism with elements of Shaivism and integrating ideas from the Vedas with those of post-Vedic literature (**Note 4C**). Smartism also involves a nonsectarian synthesis of deities from the other major branches of Hinduism. These generally include *Vishnu* and *Shiva* (described earlier), *Ganesha* (an elephant-headed deity worshiped by Hinduism as well as Jainism and some branches of Buddhism), *Surya* (a solar deity of Hinduism), and *Parvati* (an aspect of the goddess Devi).[9,31] In Smartism as a whole, all five deities (or divine aspects) are worshipped as equal aspects of God. Unlike the other three main Hindu branches of my phylogenetic tree, Smartism generally takes a nonhenotheistic approach, recognizing no single god as supreme. Its followers can choose any god as their *istadevata* (deity of choice). They follow a philosophical, meditative path, emphasizing humanity's oneness with God through understanding. Like many Hindu branches, they accept the Vedas as a reliable source of religious guidance, but they developed and expanded their practice and supernatural belief system with post-Vedic texts as well.[1,20]

SUMMARIZING THE EVOLUTION (HISTORICAL DIVERSIFICATION) OF HINDUISM

Hinduism has become one of the world's most successful branches of Organized Religion, with nearly a billion followers by 2010.[3] Much remains to be done in identifying and classifying its many detailed subgroups (*sampradayas*). Many practicing Hindus do not claim to belong to any particular denomination or sampradaya. Nevertheless, distinctly different core ideologies and group traditions exist. As I have mentioned, this book is focused on the historical diversification and interrelationships of different *religious ideologies*, not of culture as a homogenous entity. It is also not focused on the collective beliefs of every individual person. Individuals may sometimes have unique combinations of different beliefs taken from different, distantly related sampradayas. This is not unlike many adherents of Western religions who practice a personal, modified version of the branch that they self-identify with. At the level of the individual person, ideological micro variation becomes extremely complex, much like the complexity of genetic variation among individual organisms in biology (discussed in chapter 1).

More comparative research is needed to test, modify, or strengthen this phylogenetic hypothesis for Hinduism. Given that the earliest putative members of this group were non-Vedic, even the general monophyly of this group needs to be verified. Future research requires a more comprehensive synthesis of historic records as well as a determination of which shared group characteristics are synapomorphies (characters uniquely derived for a specific group) and which ones are plesiomorphies (examples of convergence, parallelism, or secondary imitation that are independently derived in different groups and not unique to a single group). This will, no doubt, be difficult because of the sensitivities involved in any comparative study of religions. As with most major religions, unbiased reconstruction of historical diversification is often at odds with the emotionally charged conviction of fundamentalists who believe that they know the ultimate truth, and that their beliefs are beyond comparison or question. If, however, an unvarnished knowledge of history is the goal, historians need to look at both differences and similarities among group traditions in an unbiased manner.

Sometime during or before the sixth century BCE, a major schism developed within Dharmic Religion as a move away from certain Vedic concepts, such as Brahmanism, and away from the Brahmin caste system. It began with the *Sramana Movement.*

SRAMANA MOVEMENT: ESCAPE FROM BRAHMIN DOMINANCE

The *Sramana Movement* (or Sramana) rejected the divine origin of the Vedas. It also rejected the high-caste privilege of the high priests of Hinduism, the Brahmins. Although Sramana rejected the high-caste privilege, it did not reject the caste system entirely.[63] The Sramana movements began with a number of ascetic monks who renounced married and domestic life, adopting a path of extreme self-discipline and abstention from all indulgence to achieve spiritual liberation (moksha or nirvana). The rise of this movement was considered "heretical" by the Brahmin priestly class, who saw their role being diminished.

Jainism, *Buddhism*, and a few smaller branches (e.g., *Ajivikas*, *Ajnanas*, and *Carvakas*) arose and diversified as a result of Sramana, encouraged by the gradual decline of the Brahmin priest influence. This led to a shift of focus within Dharmic religious life from the external rituals and massive animal sacrifices of early Vedism to internal spiritual quests and the mental sacrifice of Sramana.[33] The Sramana movement developed a belief that it was possible to attain liberation from the cycle of rebirths and deaths (samsara) through spiritual and ethical disciplines.[64] It retained and reinforced the principles of yoga. It diversified and collectively grew into a major branch of Dharmic Religion with more than a half billion people by 2010.[3] Originally, the movement did not specify supernatural beings. However, it later came to incorporate them.[65]

Jainism and *Buddhism*, as practiced today, were thought to have arisen during the sixth and fifth century BCE from the Sramana movement (see the phylogenetic tree on page 70). Some subbranches continued to venerate various Hindu deities (e.g., Vishnu, Shiva, Krishna, Rama, Ganesha, and others).[142] The most iconic inspirational figures for Jainism (Mahavira) and Buddhism (Siddhartha Gautama) are thought by some authors to have been contemporaries.[66] As with all religious branches with BCE origins, the dates of origin are speculative and controversial.

JAINISM: ULTIMATE PACIFISM AND A BELIEF IN AN UNSURPASSED ANTIQUITY

The date for the origin of what we call *Jainism* (also called *Jain dharma* by adherents) is unknown. Although the name "Jain" was not officially used until the late nineteenth century,[67] its belief system as distinct from Hinduism began in or before the sixth century BCE with the early development and formalization of the Sramana movement.[3,8] Vardhamana Jnatiputra, also known as Lord Mahavira (sixth century BCE), is the most revered of all Jain teachers.[9,68] However, he is generally not considered to be the founder by many Jains and is instead thought to have been a highly influential transmitter of much earlier doctrinal elements.

Some Jain historians and archaeologists contend that the roots of Jainism go back at least as far as the ancient Indus River Valley civilization (third millennium BCE),[69] while many Jain adherents contend that the teachings of Jainism are eternal.[66] Some Jains believe that a savior named Rishabhanatha (Rishabha) was the true founder of Jainism millions of years ago. He is sometimes described as 1,200 feet tall and millions of years old.[69,70,71] He is also credited with inventing fire, cooking, science, charity, the arts, and the establishment of marriage. Fantastic beliefs about founders are not exclusive to Jainism, of course (**Note 4F**).[8] Rishabhanatha is believed by many followers to have been the first in a succession of twenty-four omniscient spiritual teachers (also called *jinas* or *tirthankaras*) who "displayed the way for salvation across the ocean of suffering and existence."[9] Lord Mahavira is thought by Jains to have been the twenty-fourth in that succession. The teachings of the tirthankaras were passed down by oral transmission for more than a thousand years before they were ever written down in texts that survived to the present day. The oldest surviving sacred texts of Jainism, the *Jain Agamas*, were not transcribed until centuries after Mahavira's death in the fifth century BCE. Earlier texts predating the Agamas, called the Purvas, are believed by Jains to have been lost.

Like the other Dharmic Religions, Jains share the concepts of karma and the cycle of rebirth (reincarnation).[72] Jains do not believe in eternal heaven, hell, or judgment day. Their ultimate goal, as in many other Dharmic Religions, is to finally achieve spiritual liberation (moksha or nirvana) from the endless cycles of rebirth. If someone has led a particularly bad life, he or she could be reincarnated in the next life as a worm or a dung beetle. Like Buddhism, but unlike much of Hinduism, Jains reject the belief in a creator god, the authority of the Vedas, a rigid caste system, and the authority of the Brahmin priests. They can choose which deities or divine aspects (if any) to worship. Like Hinduism, but unlike Buddhism, Jains believe there is a soul (*Jiva*) at the core of every person as well as in all things. They also have hyperanimistic concerns, believing that every animal has a soul, as well as every tree, rock, or drop of water. Even something as small as a bacterium, a particle of dust, or a molecule of air has a spiritual essence or soul. As mentioned in the beginning of this chapter, Jains are not permitted to cut down trees or till the soil for fear of killing worms. Jains are strict vegetarians. Their strict adherence to vegetarianism eventually influenced older branches of Dharmic Religion, such as Hinduism (discussed earlier). Jains are also permitted to have sex only for the purpose of reproduction within marriage.

Jain monks and nuns lead a particularly ascetic life. They are not permitted to use furnaces for fear of accidently burning flies or other insects. They are not even permitted to boil water for fear of killing microorganisms. They are also not permitted to eat root vegetables such as potatoes, onions, or garlic, because tiny organisms might be injured when the plant is pulled up, and also because the tuber's ability to sprout is seen as a characteristic of a higher life-form.[73] Jain monks and nuns are celibate and forbidden to form attachments to worldly possessions. They are expected to wander homeless, without families or emotional longings. Many Jain monks also sleep on the hard ground rather than bedding, avoid bathing, and avoid tooth brushing (particularly during fasting). The also do not use a razor, so the hairs on their face and head must be plucked out individually by hand. Many monks also carry a broom made of fallen peacock feathers for removing small insects out of the way of any harm, and a gourd for carrying water.[74] In the most extreme cases, some Jain ascetics take the vow of *Sallekhana*, in an effort to discontinue harming the souls of the food they eat, the water they drink, or even the air they breathe. This vow involves voluntarily fasting to death by gradually reducing the intake of food, liquids, and finally air (the latter resulting in death).[8] Of all the Dharmic Religions, the monks and nuns of Jainism have some of the strictest, austerity-driven, ascetic traditions of any religion. However, the severest ascetic restrictions and rules of Jainism do not apply to the laity. Most Jains are laypeople who opt to enjoy some worldly happiness and postpone the ascetic life of a monk for a future lifetime. This is just as well, for human civilization would not survive long without such basics as farming, commerce, heating of homes, clothing, and, last but not least, sex for reproduction.

The evolution of Jain belief systems occurred mainly within mendicant orders (sects or distinct groups of ascetic monks and nuns whose livelihood was dependent on alms from the laity). These groups were competitive with each other in terms of their standards of behavioral purity, education, and levels of modernity. Evolving differences leading to schisms and ideological diversification included views on ascetic practices, canonical texts, and gender. The laity (nonmendicant followers) generally supported one or more of the mendicant groups. Early group diversification within Jainism, as well as precise classification of most early sects, is unclear and lacking in consensus. It is clearest within the mendicant orders. Little or no demographic data are available for the Jain laity.[67] I rely primarily on the interpretations of History of Religion scholar, Peter Flügel for my classification and phylogenetic tree of Jain belief systems (page 107).[67]That tree includes only a few of the major branches. Many of the distal tips of that tree could be subdivided much further, given that Jainism has well over one hundred distinguishable branches (sects, orders, sanghas, or gacchas).[67]

As early as the third century BCE, the early major schism began to grow within Jainism between the *Digambara* (including about four-fifths of all Jains today) and the Svetambara.[3,9] Both branches, of course, claim to be the original (i.e., orthodox) Jains (followers of Mahavira).[8] According to the Digambara, they are the original followers of Mahavira, and the Svetambara branched off later in the second century BCE. However, according to the Svetambaras, they are the original followers, and the Digambara branched off in the first century CE. Neither version of the schism is found in any early text. The earliest documentation of the Digambara story is from the tenth century CE, and the earliest documentation of the Svetambara story is from the fifth century CE.[8] Other small branches do not clearly fit into either group, as far as currently known (e.g., *Yapaniya* (*Gopya*) *Sangha* on my phylogenetic tree on page 107), but they became extinct by the late Medieval period.

Although the date when the schism between Digambara and Svetambara began is uncertain, it was formalized many centuries later at the Assembly of Valabhi (fifth century CE), which was attended only by Svetambaras.[9] The fundamental differences between the two branches have to do with rules for their ascetic monks and disagreements over what represents sacred scripture. The most obvious difference is that monks of the Digambara (meaning "clothed in air") are forbidden to wear any clothing, whereas the Svetambara (meaning "clothed in white") are permitted to wear white clothing.[9]

The *Digambara* monks believe that clothing ultimately leads to possession and attachment, which subsequently prevent moksha (nirvana, or liberation from suffering, craving, and ignorance). Because women cannot practice nudity (a cultural given), they cannot follow the Digambara requirement of rejecting clothing to reach moksha. The Digambara monks therefore believe that women must gain enough karmic merit to be reborn as a man before they can achieve spiritual liberation. They also abandon all forms of possession, sexual desire, and attachment to achieve moksha. They may not even possess a bowl for eating. They receive food-alms in their cupped hands that they eat while standing, and they are allowed to eat only once per day. The Digambaras also dispute the authenticity of surviving versions of the Agamas, as opposed to the Svetambaras who accept them as their foundational texts. Although Digambara Jains currently have no official monastic orders or schools, it has many distinct organizational subgroups,[67], some of which are shown on my phylogenetic tree on page 107. For example, in the fifth century CE, a schism developed within the Digambara, which resulted in two major subgroups. Jain writers refer to these subgroups as the *Mula Sangha* and the *Heterodox Digambara*, each of which evolved into numerous subgroups (e.g., sanghas and gacchas; see page 107).[67] Reasons for the schisms included disagreements over the eating of certain plants, water temperature for bathing, and certain practices in agriculture and trade.[67] Much further work is needed to more clearly distinguish these branches and determine their interrelationships based on their ideological differences and special shared similarities.

In contrast to the Digambara, the *Svetambara* monks are allowed to wear simple white garments. They are allowed to have a few possessions, such as a small bowl, in which they can accept food-alms. The Svetambara monks can also eat more than once a day. They believe that women, even though clothed, can achieve moksha as women. A number of their beliefs about the history of Mahavira are in conflict with Digambara beliefs.[9] For example, the Svetambaras hold that Mahavira was born with a miraculous change of wombs, with his embryo being conceived in a Brahmin woman, but transferred to a non-Brahmin who gave birth to him. They depict Mahavira as a pleasure-loving prince who experienced a sudden conversion at the age of thirty. They also believe he was married. In contrast, the Digambara believe that Mahavira had a normal gestation period within a single woman, was always full of pious insight, and was never married.[9]

Today, four-fifths of all Jains in India are Svetambara. They can be categorized into three main subgroups: the *Murtipujaka* (meaning "image-worshippers"), the *Sthanakvasi* (a revival of an earlier group called the *Lonka Gaccha*, of which we know virtually nothing about), and the *Svetambara Terapanthis*.[67] The Murtipujakas differ from Sthanakvasis and Svetambaran Terapanthis in that they worship images of tirthankaras (the twenty-four legendary, ancient spiritual teachers). They eventually divided into many subsects (gacchas) between the eleventh and sixteenth centuries (see tree on page 107). In contrast to the Murtipujakas, the Lonkas, Sthanakvasis, and Svetambara Terapanthis (sometimes collectively referred to as *Amurtipujaka*) found the worship of images to be spiritually dangerous and omitted the practice. They were also more ascetic than the Murtipujakas with regard to certain practices, such as the use of breathing filters. Ascetic Sthanakvasis called maharaj sahib, keep their mouths covered with a square of white cloth to minimize inhaling (and thus harming) small insects and other airborne life-forms. They wander the countryside and depend on receiving food from the laity to survive.

The early evolution of Jainism, and the historical interrelationships among its many subgroups, remain obscure. Reasons for this include Jainism's ancient age, its complex history of diversification, a lack of contemporaneous written records, and the scarcity of comprehensive comparative studies. Most systematic studies to date have been only exploratory in nature.[67] Published lists of Jain sects and subschools are often organized chronologically or somewhat randomly, rather than phylogenetically. This does little to resolve the confusion about their historical diversification.[75]

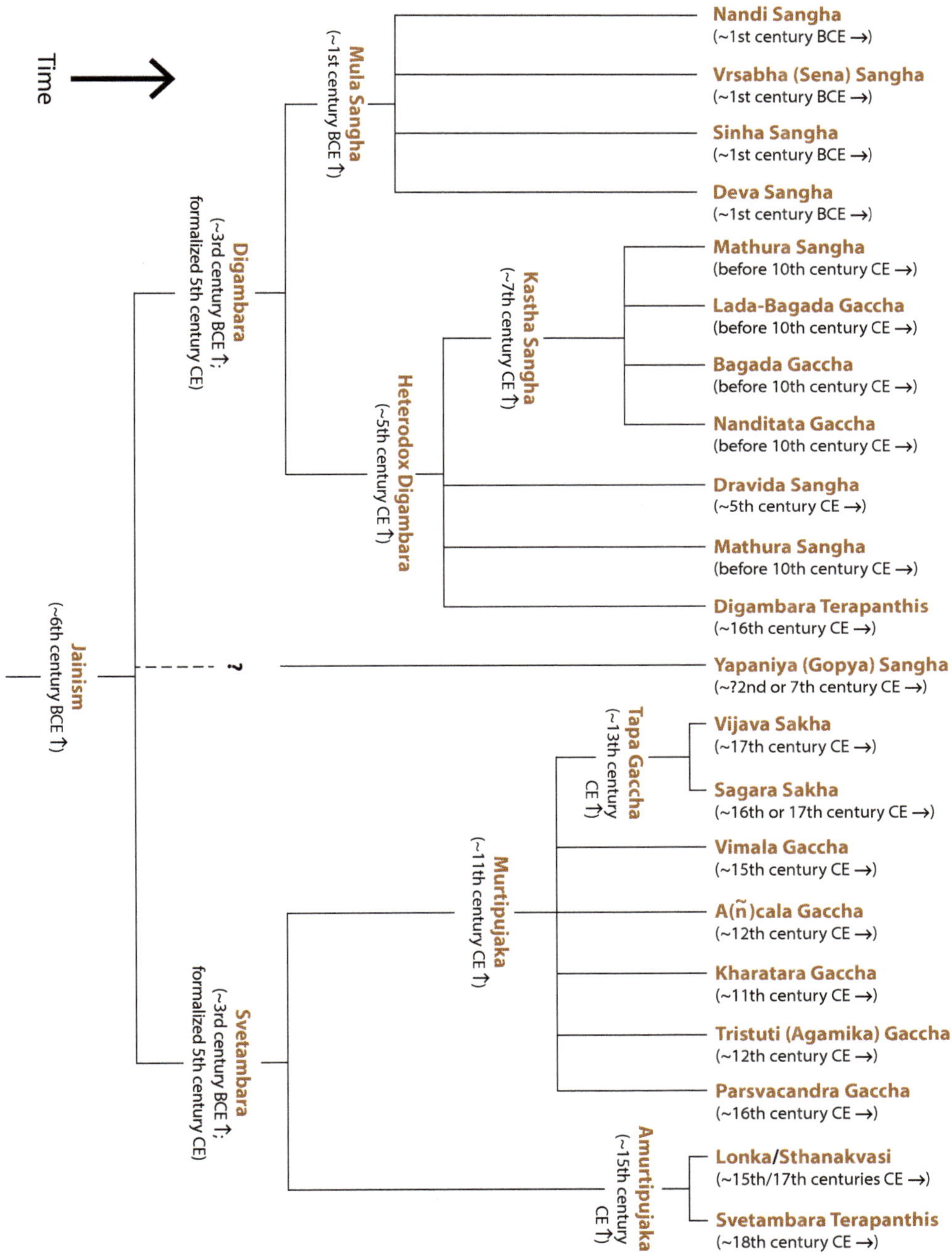

Time
Jainism
(~6th century BCE ↑)
Digambara
(~3rd century BCE ↑;
formalized 5th century CE)
Mula Sangha
(~1st century BCE ↑)
Nandi Sangha
(~1st century BCE →)
Vrsabha (Sena) Sangha
(~1st century BCE →)
Sinha Sangha
(~1st century BCE →)
Deva Sangha
(~1st century BCE →)
Heterodox Digambara
(~5th century CE ↑)
Kastha Sangha
(~7th century CE ↑)
Mathura Sangha
(before 10th century CE →)
Lada-Bagada Gaccha
(before 10th century CE →)
Bagada Gaccha
(before 10th century CE →)
Nanditata Gaccha
(before 10th century CE →)
Dravida Sangha
(~5th century CE →)
Mathura Sangha
(before 10th century CE →)
Digambara Terapanthis
(~16th century CE →)
?
Yapaniya (Gopya) Sangha
(~?2nd or 7th century CE →)
Svetambara
(~3rd century BCE ↑;
formalized 5th century CE)
Murtipujaka
(~11th century CE ↑)
Tapa Gaccha
(~13th century
CE ↑)
Vijava Sakha
(~17th century CE →)
Sagara Sakha
(~16th or 17th century CE →)
Vimala Gaccha
(~15th century CE →)
A(ñ)cala Gaccha
(~12th century CE →)
Kharatara Gaccha
(~11th century CE →)
Tristuti (Agamika) Gaccha
(~12th century CE →)
Parsvacandra Gaccha
(~16th century CE →)
Amurtipujaka
(~15th century
CE ↑)
Lonka/Sthanakvasi
(~15th/17th centuries CE →)
Svetambara Terapanthis
(~18th century CE →)

Orthodox Jains have been uninterested in documenting the detailed history of their belief systems, and historically, they have been resistant to publication of their sacred texts.[76] Much work is needed to resolve the interrelationships among the numerous Jain subgroups.

Between four and five million people are practicing Jains today. Most (more than 95 percent) live in India along the Ganges River region, where Jains make up less than 1 percent of that country's population.[3] Small Jain communities also exist in North America, Africa, and Europe. The younger, but much larger, sister group to Jainism, and the other major branch of the Sramana Movement, is *Buddhism*.

BUDDHISM: THE MIDDLE WAY AND MOST SUCCESSFUL BRANCH OF THE SRAMANA MOVEMENT

Buddhism is the largest branch of religion to evolve from the Sramana movement. Globally, it was far more successful than any other school from that movement. It is less ascetic, more of a missionary religion than Jainism, and it had the patronage of many kings, emperors, and other powerful rulers over much of its early history. It originated in India in the fifth century BCE, where it flourished for centuries. During the first millennium BCE, it had already begun to make contributions toward the field of medicine and to help people understand and deal with illness. Medicine, health, and healing have continued to be a central part of Buddhism since its origins.[30,77,78] Because of its strong emphasis on meditation and mindfulness, secularized versions of Buddhist tradition have even been adopted by many people, including a number of psychologists.[77]

Buddhism eventually began to dwindle on the Indian subcontinent between the seventh and thirteenth centuries CE because of a number of factors, including the popular resurgence of Hinduism, shifts of imperial patronage away from Buddhism, and Muslim conquest into northeastern India and Pakistan. Hindu belief systems assimilated some Buddhist ideologies, and one even transformed the figure of The Buddha into an avatar of the Hindu god Vishnu (see page 112).[5] Some historians report that by the twelfth century, Christian influence on Buddhism also was increasing and one version of The Buddha was Christianized as Saint Josaphat.[79] By the thirteenth century, Buddhism had nearly disappeared from its country of origin. However, by then it had greatly expanded throughout much of East Asia.

The early evolution and diversification of Buddhism is poorly understood. The characterizations and classifications of many of its schools are often vague and almost always controversial. No phylogenetic studies of Buddhism include all schools, both extinct and extant. This is partly due to the large number of those schools (numbering in the hundreds) and partly due to the ancient age of early Buddhism.

It is also due to the inconsistent way many branches have been diagnosed. I take a provisional look at the major putative groups and a few representative subgroups. I have arranged them in three trees that broadly expresses hypotheses of historical relationship based on implications from various authors. Page 70 shows where I put Buddhism within my phylogenetic tree of Asian Cyclicism, and pages 110 and 127 show a rough hypothesis of how major subschools of Buddhism are interrelated (based on a few general sources in the literature given at the base of both trees[9,10,80]). I reiterate that these trees are provisional and add that the dates on the early branches are rough estimates based on a range of dates given in the literature (discussed further in **Note 4G**).

Buddhism arose in eastern India following the teachings of Siddhartha Gautama, later known as Gautama Buddha or The Buddha. We don't know what he looked like. The earliest images of him were not created until centuries after his death. Gautama was the family name of his clan. Siddhartha was his personal name, which in Sanskrit means "he who achieves his goal." Proposed dates for his birth and death vary, depending on the author, but most estimates place his birth in the sixth century BCE and his death in the fifth century BCE.[5,9,10,80] The early history of Buddhism in India was filled with schisms leading to much diversification even before the first century CE. Most of the earliest branches are extinct today or merged with other religious belief systems.

Details of Siddhartha Gautama's life and teachings are controversial and difficult to interpret because they were never written down until several centuries after his death. He never declared himself a prophet or divine emissary, although he did eventually claim to be spiritually awakened.[81] Based on archaeological evidence, the impact of Buddha during his lifetime appears to have been limited. Nevertheless, I somewhat arbitrarily peg the start of Buddhism shortly after Gautama's final years, in the fifth century BCE.[9,10,66,80] It was not until the third century BCE that Buddhism began its major expansion and formal diversification under patronage from Ashoka the Great, ruler of the Maurya Empire (discussed later in this section). Long after The Buddha's death, followers elevated Buddha to the status of "divine being." The earliest general writings of Buddhist philosophy appear in the *Edicts of Ashoka*, inscribed in stone during the third century BCE.[9,10] The earliest written canonical texts of Buddhism are the *Tipitakas* (also called the *Pali Canon*) transcribed during the late first century BCE. Tipitaka is a word, in the Pali language of ancient India, meaning "three baskets." The first basket is the *Vinaya Pitka* (monastic rules), the second basket is *Sutta Pitaka* (The Buddha's teachings about the nature of reality or dharma), and the third basket is *Abhidhamma Pitaka* (further analysis of The Buddha's teachings). Many later schisms within Buddhism were over different interpretations of what should be included in the Tipitakas.[80]

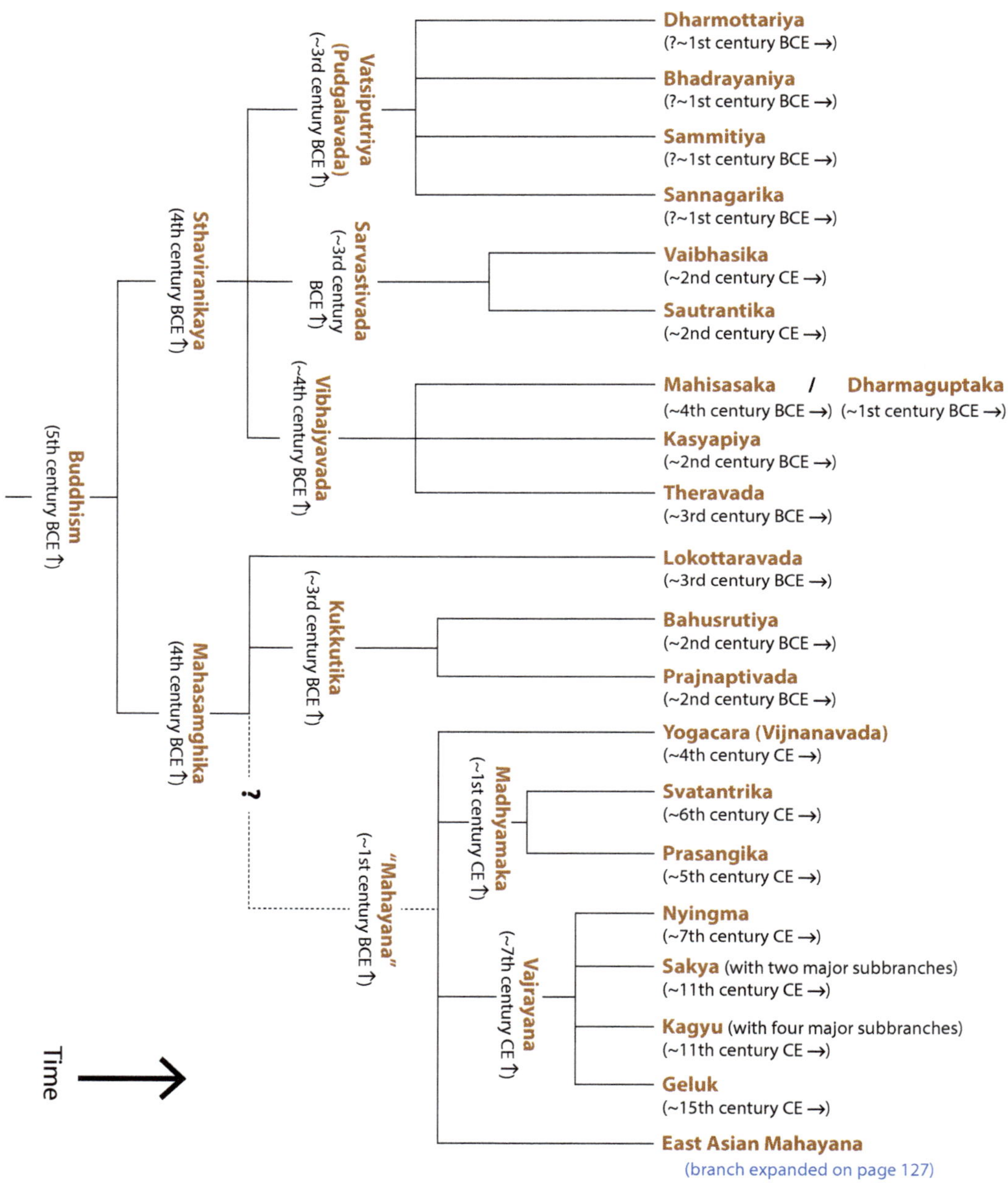
Buddhism
(5th century BCE ↑)
Sthaviranikaya
(4th century BCE ↑)
Vatsiputriya
(Pudgalavada)
(~3rd century BCE ↑)
Dharmottariya
(?~1st century BCE →)
Bhadrayaniya
(?~1st century BCE →)
Sammitiya
(?~1st century BCE →)
Sannagarika
(?~1st century BCE →)
Sarvastivada
(~3rd century BCE ↑)
Vaibhasika
(~2nd century CE →)
Sautrantika
(~2nd century CE →)
Vibhajyavada
(~4th century BCE ↑)
Mahisasaka / Dharmaguptaka
(~4th century BCE →) (~1st century BCE →)
Kasyapiya
(~2nd century BCE →)
Theravada
(~3rd century BCE →)
Mahasamghika
(4th century BCE ↑)
Lokottaravada
(~3rd century BCE →)
Kukkutika
(~3rd century BCE ↑)
Bahusrutiya
(~2nd century BCE →)
Prajnaptivada
(~2nd century BCE →)
?
"Mahayana"
(~1st century BCE ↑)
Yogacara (Vijnanavada)
(~4th century CE →)
Madhyamaka
(~1st century CE ↑)
Svatantrika
(~6th century CE →)
Prasangika
(~5th century CE →)
Vajrayana
(~7th century CE ↑)
Nyingma
(~7th century CE →)
Sakya (with two major subbranches)
(~11th century CE →)
Kagyu (with four major subbranches)
(~11th century CE →)
Geluk
(~15th century CE →)
East Asian Mahayana
(branch expanded on page 127)
Time

According to legend, Siddhartha was a prince of the Sakya tribe of Nepal. He grew up in a royal palace shielded from all the ills of the world. Gradually, he began to see things in the outside world surrounding the palace, such as sickness, old age, death, and wandering ascetics. This made him begin to ask questions like, "How did I get here?" So began one of the most impactful midlife crises in history! When he was twenty-nine years old, he left his comfortable home life to seek the meaning of the suffering he saw around him. After seven years of wandering northern India living the life of an ascetic, and with an increasing use of mindful meditation, the legend states that he had a "Great Awakening" and reached the special inner peace of moksha (nirvana). From that moment on, Buddhist scripture says Siddhartha Gautama became "The Buddha," which like the term "Christ" is a title rather than a proper name. In Buddha's case, it meant "one who has awakened to the truth" rather than "messiah."[9,41] He became an incarnation of many mythological Buddhas that had preceded him. The number and names of previous buddhas vary from branch to branch within Buddhism, but Gautama is thought of by most as "The Buddha" and founder of the faith, and the earliest historically chronicled buddha. Gautama Buddha is said to have wandered the plains of northeastern India for more than forth-five years, teaching dharma ("the path"). The Romanized Sanskrit term *dharma* (*dhamma* in the Pali language and *dharam* in the Punjabi language) has many meanings throughout the phylogenetic tree of Asian Cyclicism. In Buddhism, it refers largely to the teachings of The Buddha.

Gautama Buddha taught throughout regions of eastern India, attracting a modest following, including monks and nuns drawn from every caste. He taught that people could solve the problem of human suffering on their own, without recourse to high priests, God, or divine revelation.[41] He refused to designate a successor so there would never be a Buddhist pope. He believed that the teachings he left would help people find their own way and felt that we are all in charge of our own Great Awakening. Buddha's followers memorized accounts of his life, teachings, and monastic rules soon after his death. This information was passed down through oral transmission for nearly four hundred years before they were ever transcribed in the Tipitakas. Reliance on oral transmission over twenty succeeding generations raises obvious questions about the accuracy of such information. As with so many other religions dependent on oral transmission over long periods of time, such questions are usually countered with mystical answers about divine guidance or perfect memorization techniques to ensure the accuracy of transmitted detail.

The tradition that Buddha taught was a compromise between sensual indulgence and the severe asceticism found in Jainism.[82] His teachings are generally centered on what is called the "Middle Way." He is said to have come to his enlightenment after living two extremes: the first of which was his early life as a wealthy prince, and the second of which was his seven years of wandering northern India as a severe ascetic. He ultimately concluded that there must be a middle way. Buddha's beliefs can be summarized with three basic assumptions: First, no matter how many material comforts one has in life, they cannot protect one from the pain of suffering. Second, denial of all material comforts and living a life of asceticism (severe abstinence from any indulgence) also does not protect one from suffering. Third, each person needs to find a balance and moderately disciplined lifestyle that takes account of his or her individual circumstances. Buddha's teachings are further expanded in a series of *Four Noble Truths*, and the *Noble Eightfold Path* (listed in **Note 4H**). The organization of teachings in groups of four or eight is thought to have been conceived as an aid in memorization.[10] It also led to the eight-spoked wheel as the symbol of Buddhism.

The earliest forms of Buddhism had no particular god. Although The Buddha did not claim to be divine or to be personally associated with gods, after his death, some branches of Buddhism evolved that were full of deities, multiple hells, and multiple heavens. The Buddhist branches that incorporated deities considered none of them to be a supreme creator god. Buddhism shows much influence from Hinduism as well as from other religions. It has generally accepted syncretism as a

The Buddha, revered in both Buddhism and some later Hindu branches. (*Left*) Buddha statue from the fifth century CE in Pradash India. In the collection of the Sarnath Museum, Uttar Pradesh, India. (*Right*) Buddha as avatar of the Hindu god Vishnu. Stone carving from a thirteenth-century Hindu Temple, in Karnataka, India.

legitimate means of spreading elements of its practice and influence.[83] The Buddha came from a Hindu background influenced by both Hinduism and Jainism. As mentioned earlier, a branch of Vaishnava Hinduism later even considered Buddha to be one of the ten avatars of the Hindu god Vishnu.[5] All Buddhists share a spiritual goal of reaching personal enlightenment and nirvana.

At its canonical core (e.g., the original Pali canon) much of Buddhism acknowledged the existence of deities and their worship without strongly promoting them. Some branches of Buddhism incorporated certain devas, gods, and yaksa (forest spirits) from Hinduism. Later branches of Buddhism in China were influenced by Taoic Religions in the form of ancestor veneration (discussed later in this chapter). Hindus, Jains, Taoists, and early Buddhists all had their own versions of similar stories, to the point at which it is not possible to say in which of the four branches some of these stories originated.[5]

Early Buddhism developed the idea of a prophesized future, and a future Buddha, named Maitreya, would finally appear in about thirty thousand years.[9,10,60] He would come after the decline and eclipse of Buddhism to reinstate cosmic law and order (Buddhist Dharma) and inaugurate a golden age. Later branches (e.g., Mahayana Buddhism in the second century CE) came to revere a Sacred Triad (the *Trikaya*), which was modeled around three modes of being for Buddha. These were the *Dharmakaya* (the cosmic, divine essence of Buddha), the *Nirmanakaya* (Buddha's earthly physical manifestation), and the *Sambhogakaya* (supernatural incarnations of Buddha in the heavens, or "perfect bliss bodies").[10]

The focus of most Buddhist traditions is not worship. Instead, it is concerned primarily with the liberation of the individual from suffering in this world. It teaches you to want what you have and to not want what you do not have. A doctrine of rebirth is also central. In those parts of Buddhism that involve deities, even those deities eventually die and are reborn.[84] A fundamental question is, "What is it that is reborn in the cycle of reincarnation?" That question has been a subject of scholarly studies within Buddhism since ancient times, particularly in reconciling the Buddhist rebirth doctrine with its doctrine of no soul. Explanations range from *vijnana* (a person's consciousness or "life force" is what is reborn) to *avacya* (an inexpressible process is what facilitates rebirth).[80] Most branches of Buddhism other than the *Vatsiputriya* (discussed later) reject the notion of a personal unchanging self or soul (atman) that is reborn. They instead subscribe to a more complex explanation of human reincarnation involving the "anatman" (a belief that all things are constantly changing and going through cycles of rebirth, but the process involves no permanent soul in a person) (discussed further in **Note 4I**).

Buddhism is unlike many religions in which "original sin" is seen as a primal offense against a Supreme Being and the origin of human suffering. It does not believe in such a concept. In Buddhism, the cause of human suffering is often seen as ignorance and bad deeds in a previous life. Buddhism also does not believe in an eternal afterlife. Most Buddhists, like their Confucian and Taoist counterparts, are focused on the here and now rather than on an afterlife of a heaven or hell. The ultimate goal in orthodox Buddhism is to move beyond the repetitive cycles of samsara (reincarnation and rebirths). The path to ending these cycles is to move beyond craving and clinging to things. Such desires result in *karma* (actions in this life affecting your place in the next life). It is the accumulation of karma that continues the cycle of reincarnation. The cessation of craving and clinging eventually results in the cessation of karma, the cessation of samsara, and finally the attainment of *nirvana* (permanent peace of mind and a cessation of rebirth). Buddhism, like all religions, continues to evolve and ideologically diversify. Although many followers of Buddhism (particularly monks and nuns) continue to consider personal liberation from samsara and reaching nirvana to be the ultimate spiritual goal, today many lay Buddhists focus on accumulating merit and good karma through good deeds, ethical moral conduct, and certain rituals, simply to gain better rebirths.[85]

Buddhism's early proliferation was entwined with the expansion of political empires, like so many other major religions through history. By the third century BCE, the Maurya Empire had become the largest political entity ever to have existed on the Indian subcontinent by way of a number of aggressive military campaigns. One of its emperors, known as Ashoka the Great, ruled from 268 BCE to 232 BCE.[9,66] Although he eventually renounced war, at that point, he already had taken control of (thereby uniting) most of the Indian subcontinent. He adopted Buddhism during his reign, and his patronage transformed Buddhism from a small, localized tradition to a major religion that spread throughout much of India, and eventually far beyond its borders.[5] His conversion to Buddhism and his imperial support helped Buddhism become one of the world's largest religions. It was much like Roman emperor Constantine's conversion to Christianity gave imperial support to the rise of Christianity some six centuries later (chapter 8). We can find many such examples of coevolutionary collaboration between major religions and large empires through history (discussed further in chapter 12). Ashoka's engraved inscriptions on cave walls, boulders, and stone pillars (the *Edicts of Ashoka* mentioned earlier) are the earliest-known written records of Buddhism. Although Ashoka supported other branches of the Sramana movement, such as Jainism and Ajivikas,[5] and warned against disparaging people with non-Buddhist religious beliefs,[86] he was Buddhism's greatest early patron. He supported a program of proselytism, sending Buddhist missionaries (Dharma Bhanaks) throughout India and later to Egypt, Greece, Central Asia, and Eastern Asia.[87] In China, Buddhism eventually came under the patronage of other emperors. Some of them went as far as to declare Buddhism as the state religion. Similar monarchical support of Buddhism also occurred at times in Japan. Some historians claim that at its peak, Buddhism was adopted by close to one-third of the world's population.[41]

In spite of its global expansion, the rise of Buddhism in India had already begun to slow by the second century BCE when the Maury Empire was overthrown by a general named Pushyamitra Shunga (185–149 BCE). Pushyamitra later proclaimed himself as the new emperor, and proclaimed the region as the Shunga Empire. He was a Hindu Brahmin, and the region's monarchical patronage shifted from Buddhism back toward Hinduism with renewed attention to Vedic order. Pushyamitra Shunga, however, still tolerated Buddhism and Jainism.[5] Centuries later, during the reign of the Vardhana Empire, North Indian Emperor Harsha (560–647 CE) converted to Buddhism late in his life and sent Buddhist missions to China. By then, however, Hinduism had become entrenched in India and Buddhism continued to fade on the Indian subcontinent.[5]

By the sixth century CE, the Huns invaded northern India and suppressed Buddhism further.[87] In the seventh and eighth centuries, a resurgence of Hinduism and Jainism spread throughout the Indian subcontinent, with Hinduism assimilating certain parts of Buddhism. However, by the seventh through thirteenth centuries CE, Muslim conquest of India resulted in the vast destruction of Buddhist monasteries, temples, and sacred images. Although Hindu and Jain branches largely survived that period of Muslim rule in India, Buddhism did not. By the late fourteenth century, only a few isolated Buddhist communities were left on the subcontinent.[88] By 2011, only 0.7 percent of the Indian population self-identified as Buddhists.[17] However, by those times Buddhism had already migrated to East Asia where it had flourished, expanded, and continued to diversify. Today, Buddhism has more than a half-billion followers, more than 99 percent of which live in East Asia. Despite its near disappearance from India as a distinct religion, Buddhism exerted ideological influence on the continued development of Hinduism on the subcontinent, reducing the emphasis on authoritative ritual and moving it toward a more monastic philosophy.

The first major schism in Buddhism (sometimes called the Basic Schism) occurred over interpretation of disciplinary rules. This schism occurred during the fourth or third century BCE[9] during the second or third Buddhist Canonical Council in ancient India and resulted in two major branches. One was a conservative minority group that considered itself to be the keeper of The Buddha's original teachings, called the *Sthaviranikaya* (Sanskrit for "School of the Elders"). The other was a more liberal majority group that held a broader perspective regarding The Buddha's word, called the *Mahasamghika* (Sanskrit for "Great Congregation").

STHAVIRANIKAYA: INCORRUPTIBLE ARHAT PERFECTION

The first major schism within Buddhism came in the fourth or third century BCE[9] It was the result of doctrinal differences clarified at either the second or the third Buddhist Canonical Council (councils listed in **Note 4J**). Although there is a paucity of known doctrinal detail concerning these early schools, some reasons for the schism had to do with the way to achieve nirvana. Those who reached this "perfect being" were called *arhat* (or the Pali *arahant*). The more conservative *Sthaviranikaya* Buddhists believed that the arhat is perfect, incorruptible. Their concept of the arhat was an idealized entity, esteemed for detached wisdom and unworldly living. They believed that the arhat had overcome all ignorance and temptation, and in that sense, he or she was no longer human and would not be reborn.[9] This belief was in opposition to the less orthodox *Mahasamghika* Buddhists who believed that the arhat was still subject to temptation, not yet free of all ignorance, and still making progress.[9]

The Sthaviranikaya also formalized the rules of the *Vinaya* to govern the Buddhist monastic community and attempted to make them obligatory.[9,80,89] It clarified the following activities as violations of monastic practice: (1) The carrying of salt in an animal horn. (2) Eating when the shadow of the sundial is two fingerbreadths past noon. (3) After eating, traveling to another village on the same day to eat another meal. (4) Holding several monastic assemblies within the same temple (Sima) boundary during the same two-week period. (5) Making an ecclesiastical decision with an incomplete assembly. (6) Citing precedent as a justification for violating monastic procedures. (7) Drinking milk whey after mealtime. (8) Drinking unfermented wine. (9) Using mats with fringe. (10) Accepting gold or silver (money). The Mahasamghika chose not to recognize the Sthaviranikaya list of violations, which fueled a major schism.

The Sthaviranikaya split from the Mahasamghika branch in the fourth century BCE, and by the third century BCE, it had given rise to three major branches, including the *Vatsiputriya* (also called *Pudgalavada*), the *Sarvastivada*, and the *Vibhajyavada* (see the tree on page 110). Each of these gave rise to numerous other branches. Very little is known about the detailed doctrines of these numerous early belief systems. Most of them, other than Theravada, are extinct. A detailed comparative accounting of the many extinct subbranches is beyond the scope of this book, but such a study would be useful in better understanding the early evolution of Buddhism. I rely on the implied branching pattern of various scholars of Buddhism, such as Robert Buswell and Donald Lopez, in *The Princeton Dictionary of Buddhism*,[80] as well as a few others.[9,10,90,91] My phylogenetic tree of Buddhism summarizes this broad hypothetical pattern of diversification.

THE *VATSIPUTRIYA* (*PUDGALAVADA*) BRANCHES OF STHAVIRANIKAYA: LARGELY EXTINCT

The *Vatsiputriya* branch of Sthaviranikaya Buddhism appears to have evolved in the third century BCE.[9] It is named after its founder, Vatsiputra, from the third century BCE, who lived in Kosala (today East India). This branch flourished for centuries in early India as one of the most popular mainstream Buddhist branches in that region, until the eleventh century CE. It was deemed heretical by other Buddhist branches for its belief in the existence of a continuous soul, called a *pudgala*, similar to the Hindu atman.[80] The Vatsiputriyas believe that the pudgala transmigrates from lifetime to lifetime in one reincarnation after another. They argue that the pudgala is necessary for personal continuity over the course of a single lifetime and for karmic continuity over the course of multiple lifetimes.[80] The pudgala is believed to disappear once nirvana is attained. No other major Buddhist school accepts the doctrine of the pudgala.[9]

The Vatsiputriya evolved into at least four subbranches by the first century BCE (see the tree on page 110). These included the *Dharmottariya*, the *Bhadrayaniya*, the *Sammitiya*, and the *Sannagarika*.[80] Original records of these subbranches were mostly lost, other than a few later Chinese translations. Although some were influential on the development of later Buddhist branches, most of these subschools, as originally conceived, died out in medieval times. Little is known about these subschool in their original form.

THE *SARVASTIVADA* BRANCHES OF STHAVIRANIKAYA: THE THREE PLANES OF TIME

The Sarvastivada appears to have originated as early as the third century BCE, but no later than the first century CE.[80] It became one of the most influential branches of Indian Buddhism for centuries, flourishing throughout Northern India and Central Asia until the seventh century CE. Its geographic location along major trade routes led it to becoming a major influence on early East Asian Mahayana Buddhism.[80] The name Sarvastivada is a Sanskrit term that means, "The theory of all that exists." It is a doctrine that claims all dharmas (truly existing things) exist simultaneously in three time periods: the past, present, and the future. This view was in opposition to the idea that dharmas were impermanent and moved through successive periods of origination, continuance, decay, and extinction. Sarvastivada explains how dharmas exist in three time periods at the same time and still appear to undergo change.[80]

The Sarvastivada are best known for the massive eight-thousand-page volume of monastic code called the *Mulasarvastivada Vinaya*, which is still used by Tibetan

Buddhists today. This volume was first composed in Sanskrit in the first century CE. It contains 253 rules and regulations for ordained monks and 364 rules for nuns, as well as five precepts (basic ethical obligations) for lay practioners.[80] The early history of Sarvastivada Buddhism, authorship of its important texts, and details about its derivative branches are poorly understood and controversial. However, some authors claim that a schism occurred within Sarvastivada around the first or second century CE resulting in a split between the mainstream school called *Vaibhasika* and dissenters called Sautrantika.[80,84] The last of the three major schools of the Sthaviranikaya is the *Vibhajyavada*.

THE *VIBHAJYAVADA* BRANCHES OF STHAVIRANIKAYA: SURVIVED MAINLY BY THE *THERAVADA*

In Sanskrit, *Vibhajyavada* means "distinctionist," which refers to their "doctrine of distinctionism" used to determine which dharmas exist and which do not. Like the Vatsiputriya and the Sarvastivada, the Vibhajyavada is thought to have emerged as a distinct branch of Sthaviranikaya on the Indian subcontinent. It originated in the fourth or third century BCE.[80] In contrast to Sarvastivada, as well as to other mainstream Buddhist schools that believe that all dharmas exist in all three time periods (the past, present, and future), Vibhajyavada made a distinction among the past, present, and future dharmas. Vibhajyavada contend that while present dharmas obviously exist, no future dharmas exist, and the only past dharmas that exist are past actions that have yet to produce their karmic effects.[80] The Vibhajyavada belief system is also diagnosed by a number of other ideological details concerning dharmas and The Buddha. For discussion of such detail, I refer readers to *The Princeton Dictionary of Buddhism* by Buswell and Lopez.[80]

There were at least four major subbranches of Vibhajyavada (see the tree on page 110). These included the *Mahisasaka*, the *Dharmaguptaka*, the *Kasyapiya*, and the *Theravada*. Some of these subschools are largely extinct today. The *Mahisasaka* school was active in parts of what are now central and northern India. It appears to have emerged more than a century after the death of The Buddha. The *Dharmaguptaka* school evolved from the Mahisasaka school by the late second or early first century BCE. It was active in Sri Lanka and on the northwest Indian subcontinent until the seventh century CE.[80] This school did much to spread Buddhism beyond India to regions of Iran, Central Asia, and China. It is still a belief system of some East Asian monks today. The *Kasyapiya* school was founded by an Indian sage by the name of Kasyapa. It came into existence in northern India in the second century BCE and became extinct centuries later. These lineages differed on various issues, including which works should be included in the Tipitakas (sacred text of

Buddhism), the exact nature of the dharmas, and certain details about The Buddha. The precise relationships and detailed characteristics of the ancient subschools of Vibhajyavada (and the rest of Sthaviranikaya) are controversial. However, the best-known branch of Vibhajyavada is the *Theravada*, which flourishes today.

Theravada (Pali for "Way of the Elders") is the dominant form of all Sthaviranikaya Buddhism. It is one of its oldest branches, dating back to at least the third century BCE. By the eleventh century CE, it had become the dominant form of Buddhism in Sri Lanka and Burma, where it was patronized by kings as the state religion. By the fourteenth century, it had become the dominant form of Buddhism in Cambodia, Thailand, and Laos.[80] It eventually came to be known as "Southern Buddhism." Theravada Buddhism in countries such as Thailand, where it is the state religion, shows a strong influence from Hinduism and animistic folk beliefs.[9,10] The use of the name "Theravada" as a term of self-designation for members of this group goes back in time at least as far as the third century CE. However, the name does not appear to have come into common use until the twentieth century.[80]

The Theravadins claim to adhere to the original, unchanged teachings of The Buddha. Like the rest of the Sthaviranikaya, Theravadins see their mission as protecting the literal interpretation of Buddha's teachings from being altered or modified.[10] They are vested in the defense of the Pali Canon (the most complete Tipitaka) in its original form.[9] They were among the first to have recorded Buddhist canon in written form rather than relying solely on oral transmission.[92] Theravadins accepted that laypeople could eventually attain nirvana but believed that the ascetic life of a monk was a more certain path. This idea was also related to the Sthaviranikayin belief that the arhat, who has reached nirvana, is a perfect being. Unlike Mahasamghika Buddhism (discussed later) whose more altruistic ideal is becoming the bodhisattva focused on enlightenment to save all other beings from suffering, the Theravada ideal is the arhat, focused on his or her own detached enlightenment.[10,80]

The conservatism of Theravada was politically appealing to rulers for its moral rigor and discipline.[10] The relative success of Theravada rose, fell, and ultimately rose again with changing monarchies over time.[9] There are, no doubt, many distinct subsects of Theravada today across eastern Asia. However, they have yet to be clearly diagnosed, compared, and analyzed in a phylogenetic context. Theravada grew to become one of the major branches of modern Buddhism and is the primary survivor of Sthaviranikaya today.[9] For this reason, some authors even use the name as synonymous with the Sthaviranikaya (discussed later in this chapter), although such use ignores much of the early history of that belief system.[80,89]

The sister group to the Sthaviranikaya, and the other branch that resulted from the first major schism in Buddhism, was the *Mahasamghika*.

MAHASAMGHIKA: ALTRUISTIC BUDDHISM AND THE IMPERFECT ARHAT

The other early branch that resulted from Buddhism's first Basic Schism in the fourth or third century BCE was the more liberal branch, the *Mahasamghika* (see the phylogenetic tree on page 110). Like the Sthaviranikaya, the Mahasamghika arose in India. They distinguished themselves from the Sthaviranikaya by their conviction that the arhat was not spiritually infallible. They believed that he or she was still subject to sexual temptation and doubt and was not yet free of all ignorance.[9,89] The Mahasamghika also characterize themselves as distinct from Sthaviranikaya by seeing Buddha as supermundane (divine) and by accepting popular religious beliefs and practices that give a greater role to the laity.[9] They are not as strict in their laws and rules, and they ignore the ten formalized violations of Sthaviranikaya monastic practice. Although they hold a more liberal interpretation of The Buddha's teachings than the Sthaviranikaya, the Mahasamghika believe that their views are closer to The Buddha's original intentions.

The Mahasamghika also have a somewhat more altruistic vision of the path to nirvana. The Sthaviranikaya are generally focused on becoming an *arhat* and achieving nirvana (spiritual liberation) for themselves, whereas the Mahasamghika instead promote becoming a *bodhisattva* whose aim is to attain full enlightenment for the benefit of all living beings as a road to Buddhahood. They see the limited goal of the arhat as somewhat selfish.[9] Many Mahasamghika believe that the course of the bodhisattva results in a superior nirvana.[80] The Mahasamghika evolved into a number of subschools even before the beginning of the first millennium CE. These subschools included the *Lokottaravada* and the *Kukkutika* (both of which eventually became extinct, or near so, leaving little in the form of written records) and the more successful *Mahayana*.

THE *LOKOTTARAVADA* AND *KUKKUTIKA*: BUDDHA AS NO MORTAL BEING

The ancient schools of Lokottaravada and the Kukkutika are very poorly known and effectively extinct today.[80] Their founders are believed by practitioners to go back to the time of The Buddha, but documentation of this is sketchy at best. The *Lokottaravadins* (also known as *Transcendentalists* or *Ekavyavaharika*) taught that The Buddha was a transcendental (spiritual) being of limitless wisdom and celestial power.[9] They believed that even his seemingly mundane acts, as well as every one of his utterances during every one of his lifetimes, was concerned with transcendental matters. The Lokottaravada are best known today as the branch that composed the sacred *Mahavastu*, interpreted as a biography of all previous existences of The Buddha. That text claims that even though The Buddha often appeared to function as an ordinary person, he remained in a constant state of meditation and was always transcendent.[9,80] Some authors see Mahayana Buddhism as rooted in the Mahavastu,[80] reflecting a close relationship with the Lokottaravada.

The *Kukkutika* (also spelled *Kaukkutika*) derives its name from the Buddhist monastery founded by King Ashoka in the third century BCE. In contrast to the Lokottaravada, the Kukkutika interpretation of Buddhist scripture asserts that Buddha taught not only transcendent truths but also employed provisional concepts to frame his teachings for his audience.[80] Kukkutika later gave rise to at least two subbranches sometime around the second century BCE (although there is no precise date for this).[9] One was the *Bahusrutiya*, and the other was the *Prajnaptivada*. No original text of these schools is known to have survived, but later teachers of Sthaviranikaya reference lost texts concerning these branches as distinct sects.[9]

MAHAYANA: THE "GREAT VEHICLE" AND REFORM BUDDHISM

The largest branch thought by some authors to be rooted in the Mahasamghika is the "Mahayana", which contains more than half of all Buddhists today. The inclusion of Mahayana within Mahasamghika is controversial, and its provisional placement is uncertain and indicated by a question mark on my phylogenetic tree (page 110). I am following early conventional views on its placement,[89,90,93] while also acknowledging conflicting views on the origin and monophyly of Mahayana.[94,95,96] Part of the problem, as I see it, involves a question of monophyly for the group (thus the quotation marks around the name on my tree). Is it a single monophyletic group derived from a common ideological ancestor, or did part of it evolve from Mahasamghika, and another part independently evolve from Sthaviranikaya? If the latter is true, then perhaps "Mahayana" is historically two independent groups that are not each other's closest ideological relatives and must be classified separately. The similarities between these groups may be only secondarily derived analogies or homoplasies. The adequate answer to this question is well beyond the scope of my work. I leave clear resolution of this issue to historians much more knowledgeable than I about Buddhist schools, who might be willing to use a phylogenetic approach to investigate this issue further.

Mahayana Buddhism first arose in India before spreading into other parts of Asia. It was a reform movement attempting to counter the factionalism and strife that had developed within more traditional branches of Buddhism. Some historians refer to it as the Buddhist equivalent of Christianity's Reform Movement in sixteenth-century Europe.[59] Among the earliest-known Mahayana scriptures are the *Prajnaparamita Sutras*, which were composed in India from around 100 BCE to 600 CE.[10,97] Some historians suggest even earlier dates for the origin of Early Indian Mahayana in the southern part of the Indian subcontinent.[98]

As with the Mahasamghika, Mahayana Buddhism maintains that pursuing only the release from one's personal suffering (such as the arhat in Theravada) is too narrow and selfish of an aspiration, because it does not also promote helping others in their efforts to find nirvana.[99] Mahayana proposes that one should do more than introspective meditation for the sake of one's self. It emphasizes compassion and altruism, and advocates postponing one's entrance into nirvana to save other suffering beings. In Mahayana, a person who does this is known as a *bodhisattva*. This is unlike the use of the same term in Theravada, in which the title is used to identify a person who is about to reach nirvana ("Buddhahood").[9]

A key theme of the Mahayana Sutras, following that of Mahasamghika, is to work toward Buddhahood as a bodhisattva, rather than focusing only on your own rise to Buddhahood like the arhat. The name *Mahayana* is Sanskrit for "Great Vehicle," which is how its adherents self-identify. Historically, non-Mahayana Buddhists have been referred to as the *Hinayana* (the "lesser vehicle"), a pejorative term coined by the Mahayanists that is sometimes mistakenly interpreted as a particular branch of Buddhism.[9,80] Because the name Hinayana is used for a group that is defined only by what it is not (i.e., "not Mahayana"), and because parts of that group are more closely related to Mahayana than to other non-Mahayana sects, Hinayana is considered to be a non-monophyletic group. Therefore, it is not included on my phylogenetic trees. The name *Nikaya* Buddhism is sometimes used as a less derogatory substitute for Hinayana. That term, however, is non-monophyletic for the same reasons.

The origins of Mahayana remain the subject of scholarly debate, but it is generally agreed that it began as a minority movement on the Indian subcontinent.[80] Some authors contend that movement began with disconnected cults moving away from more conservative monastics and asceticism. Ideologically, it is thought by some historians to represent a move away from the archetypal monastic male organization detached from family life and common people. It is instead a move toward the firmer inclusion of nonmonks, women, and the married laity.[10] The view of Mahayana as prolaity is debated, however. Some historians contend that Mahayana in India never intended to be a separate Buddhist branch but instead was simply an expression of Mahasamghika compassion. It continued to emphasize the role of the bodhisattva and the importance of compassion for all living beings as a preeminent expression of wisdom. As I mentioned, it promoted the idea that the Mahayana practitioner should delay his or her own nirvana until all other beings were liberated.[9] Early Mahayana appears to have coalesced as a distinctive subgroup of the

Mahasamghika by at least the first century BCE in India. During the first and second centuries CE, the teaching of Early Mahayana began to spread into China by Buddhist monks who traveled the overland route across central Asia.[9] By the fifth through ninth centuries CE, Early Mahayana had evolved into many distinctive East Asian subbranches, particularly in China, Vietnam, Korea, Japan, and Tibet. The Mahayana school is the largest branch of Buddhism existing today. According to a 2010 survey, 53 percent of all Buddhists belong to Mahayana, 36 percent belong to Theravada, and the remaining 11 percent include all the rest of the world's Buddhist schools combined.[3]

EARLY MAHAYANA SCHOOLS: *YOGACARA* AND *MADHYAMAKA*

The earliest development and evolution of Mahayana belief systems in India is poorly understood. The dates given on my phylogenetic trees on pages 110 and 127 are provisional estimates based on the work of others. The Mahayana literature includes a number of canonical works purported to be the words of The Buddha. However, the Mahayana Sutras were passed down only as oral transmissions by specialized monks known as "Dharmareciters" (dharmabhanakas) long before they were ever written down.[100] At first, interpreting the authority of the Mahayana Sutras as the word of The Buddha was controversial. Later Indian schools that accepted their authority included the *Yogacara* and the *Madhyamaka*.

The *Yogacara* (also known as the *Vijnanavada*) was one of the early major branches of Mahayana that evolved in India. It was founded by Asanga (320–390 CE), an Indian scholar who converted from Sthaviranikaya Buddhism late in life.[9,80] Asanga is said to have spent twelve years in a meditation retreat, after which he had a vision of a future buddha, Maitreya, who instructed him in Mahayana and Yogacara doctrine.[80] The main postulate of Yogacara Mahayana is that consciousness itself is the only reality. It also postulates that the apparent diversity in the physical world is only a product of instability and obscuration in an individual's field of consciousness.[9] The broader exposition of Yogacara doctrine was largely the work of Indian scholastics in the fourth and fifth centuries CE. It was extremely influential in the development and evolution of later East Asian Mahayana branches.[80] It is probably best known as emphasizing the study of perception and consciousness through the meditative practice of yoga.

The *Madhyamaka* (also known as the *Sunyavada*) was another main branch of Mahayana that evolved in India. This school was founded by the Indian philosopher Nagarjuna (150–250 CE) in the second century CE. As with the Yogacara, the distinctive combination of ideological characteristics that define Madhyamaka are complex.[80] The Madhyamaka are best known for their doctrine of emptiness, which professes that all things are empty of intrinsic existence in nature. It is a complex tenet recognizing up to twenty kinds of emptiness, including the emptiness of emptiness. The detailed comparative philosophies involved with this and other branches of Mahayana are extremely complex and better addressed in a comparative study focused on the religious belief systems of Buddhism. Eventually, Madhyamaka diversified into two main subbranches in India. The first was the *Prasangika*, founded by the Buddhist priest Buddhapalita (470–550 CE). It regards itself as the more orthodox, original form of Madhyamaka. The second branch is the *Svatantrika*, founded by the Indian philosopher, Bhavaviveka (500–570 CE). These two subbranches differed in their philosophical interpretations of reality, in details about how to understand the nature of emptiness, and in the requirements of the arhat to achieve liberation.[80]

VAJRAYANA (*TANTRIC MAHAYANA BUDDHISM*) AND THE DALAI LAMA

Vajrayana is also known as *Tantric Buddhism* because its ideas and practices are taken from the Indian Buddhist *Tantras*. The Tantras are sacred texts pertaining to esoteric traditions (e.g., Yogic practices, chants, and imagery) rooted in both Hinduism and Buddhism. The Buddhist Tantras were partly derived from earlier Hindu Tantric sources.[20] Vajrayana is sometimes called Tibetan Buddhism because of its tremendous success in that country. However, it originally evolved on the Indian subcontinent and only later spread to Tibet, East Asia, and other Himalayan regions. Historians report that a king named Songtsen Gampo (605–650 CE) made Vajrayana the state religion of Tibet in the seventh century CE. He was instrumental in the promulgation of Buddhism throughout his kingdom.[10,80] During his reign, many Buddhist texts were translated from Sanskrit to Tibetan.

The *Vajra* (the Vehicle of the Diamond Lightning Bolt) symbolizes Vajrayana. The Vajra is a club-like, ritualistic weapon with a ribbed spherical head symbolizing the romanticized properties of a diamond (clarity and wisdom) and a lightning bolt (strength and enlightenment). Vajrayana is thought by most authors to have evolved from Early Mahayana,[9,10,41] or possibly from a syncretization of Mahayana elements with Tibetan shamanism and animism (sometimes called *Bon*).[9,10] Therefore, I included it as a branch of the Mahayana phylogenetic tree. Today Vajrayana has ten to twenty million followers, principally in Tibet and Mongolia, but also in parts of Nepal and Himalayan India.

According to various history of religion scholars, the Vajrayana branch of Mahayana Buddhism is now represented by four major branches:[9] *Nyingma* (the oldest branch), *Sakya* (with two major subbranches), *Kagyu* (with four major subbranches), and *Geluk*. The Geluk (also spelled Gelug) is the best known in the West because it is the branch to which the Tibetan Dalai Lama belongs. He has become one of the best-known people on the planet. In today's world in which celebrity is so highly valued, he is a spiritual superstar. The current incarnation of the Dalai Lama won the Nobel Peace Prize in 1989 for his nonviolent struggle on behalf of the Tibetan people. All Dalai Lamas trace their repeatedly reincarnated lineage to the first Dalai Lama, Gendun Drub (1391–1475).[9] Gendun Drub was the nephew of Tsong Khapa (1357–1419), the monk who founded Geluk Buddhism. When a Dalai Lama dies, his reincarnation is sought worldwide. A special delegation of monks consults with an oracle to find out in what part of the world the Dalai Lama will be reincarnated next. The delegation then travels to that site, taking with them objects used by the previous Dalai Lama. They mix the Dalai Lama's objects with similar objects to test the possible reincarnation for legitimacy. If the child chooses the right objects, he is identified as the new Dalai Lama, and "the one." The current (and fourteenth) Dalai Lama, Tenzin Gyatso, was born in 1935 and identified as the next Dalai Lama when he was only four years old. He was forced to flee Tibet in 1959 after the Chinese communists began a brutal repression against the Buddhists. As I write this, he still lives in exile in northern India.[10]

According to an animistic element of Tibetan Buddhism and Zen Buddhism, all beings (including plant life, inanimate objects, and metaphysical entities) may be considered sentient beings. As with other branches of Mahayana, Vajrayana teaches that the motivation to achieve Buddhahood and nirvana should be for the benefit of all sentient beings and not just for one's self. It even advocates postponing one's own entry into nirvana to save other suffering beings. Vajrayana is still poorly understood in the Western world. Many of its scriptures and rituals have never been formally ordered or systematized.[97] Teachings are often transmitted from teacher to student in a secretive manner. Many practitioners refuse to divulge information about the practice or sources of information.[97]

EAST ASIAN MAHAYANA BUDDHISM, BEGINNING IN CHINA

Although ultimately rooted in India, Mahayana had its greatest success in East Asia, where most of its major subschools were founded and flourished. I group the major schools of Mahayana that evolved in East Asia with the provisional name "East Asian Mahayana Buddhism" (see the phylogenetic tree on page 127). Further work on these schools within a phylogenetic context may better resolve the historical interrelationships among these diverse belief systems.

As mentioned earlier in this chapter, Buddhism entered China as early as the first century CE with missionaries from India who had traveled the overland route across

The *Trikaya* (Divine Triad of Mahayana Buddhism) in the Shanyuan Temple of Liaoning, China. It portrays the three aspects of Buddha: the *Dharmakaya* (Buddha's divine essence), the *Nirmanakaya* (Buddha's earthly physical manifestation), and the *Sambhogakaya* (supernatural incarnations of the Buddha in the heavens). In Buddhism, the swastika (as seen here on the middle statue) represented the wheel of Dharma long before it was misappropriated by the German Nazi Party as a symbol of racial purity and superiority (discussed on page 81).

central Asia.[9] Other religions already there included Taoism and Confucianism, and these three schools often influenced each other. Mahayana Buddhism appealed to a wide spectrum of people in China because of its acceptance of local cults, rituals, and practices that were already part of the preexisting culture. It also offered both philosophical and spiritual insight not already present in Chinese culture. It also appealed to many regional monarchs as a way to maintain the interests of their state and bolster their administrative power. In one way, East Asian Mahayana also took iconism to a new level. In honor of The Buddha, followers are known to have built some of the world's largest statues (e.g., see the figure on page 135).

I divide *East Asian Mahayana* into the following four major groups: (1) *Early Chan* (which later diversified into *Mature Chan, Thien, Son,* and *Zen*), (2) *Pure Land/Jodo Shinshu* (which eventually diversified into *Nishi Honganjiha* and *Higashi Honganjiha*, (3) the *Lotus School* (which eventually diversified into *Tiantai, Tendai,* and the eighteen or more schools of *Nichiren*), and (4) *Huayan* (which diversified into *Hwaeom* and *Kegon*). These hypothesized interrelationships are illustrated on my phylogenetic tree on page 127.

EARLY CHAN AND ITS MANY DERIVATIVES

Early Chan began with Mahayana Buddhism's introduction into China from India during the first century CE. It was introduced through trade routes between India and China, and it eventually spread through other parts of eastern Asia. Early Chan tradition was strongly influenced by elements of Taoism.[9,10,80] Some of Chan's earliest recruits were converted Taoists (Taoism is discussed later in this chapter).[101] Early Chan Mahayana adopted Taoic elements of ancestorism by observing reverent memorial services for departed ancestors.[9] One of Early Chan's most charismatic teachers was a legendary Indian monk named Bodhidharma who supposedly arrived in China in the fifth century CE. He eventually came to be known as the patriarch of formalized Chan Mahayana.[10,80] According to one legend, he once engaged in motionless meditation (zazen) for nine years.[9] Eventually, other legendary patriarchs materialized within the Chan communities.

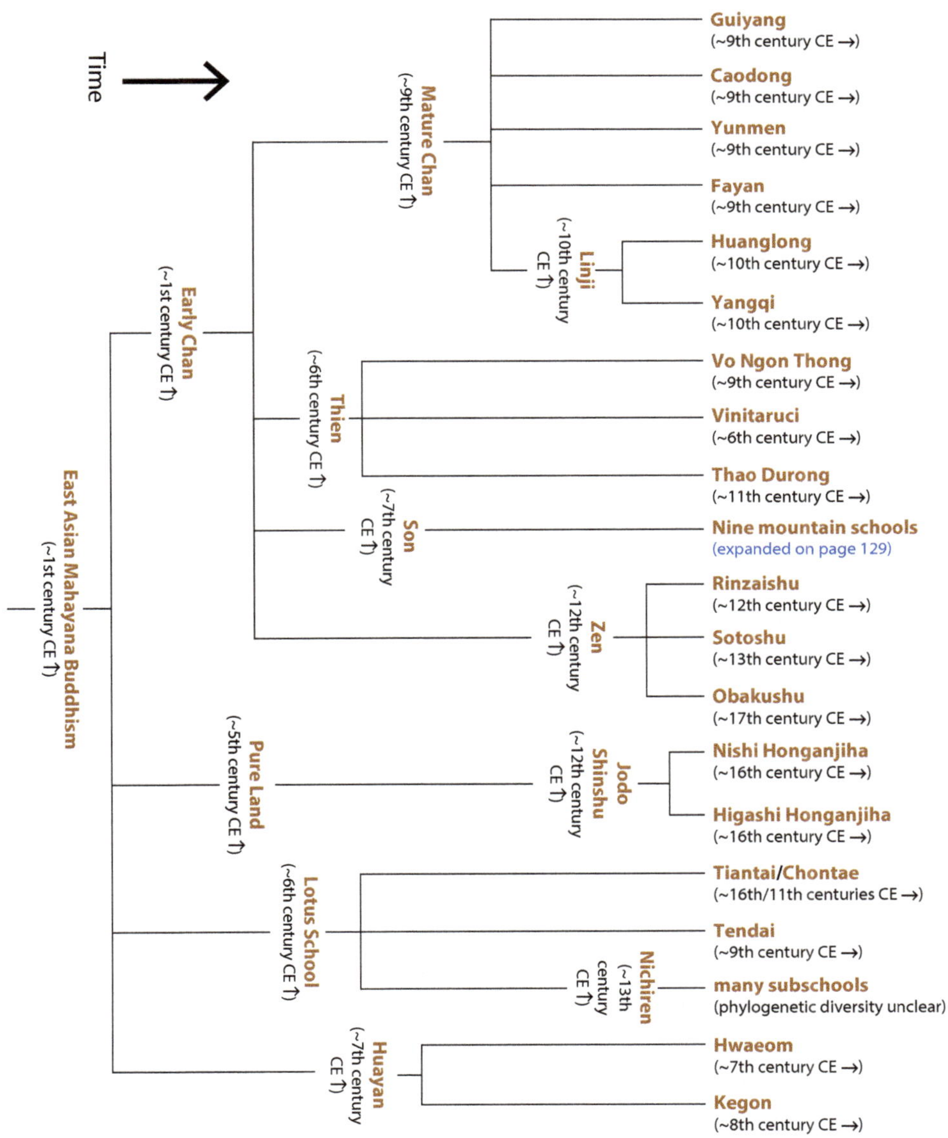

Although Early Chan ideology spread and diversified throughout much of East Asia, it also continued to evolve within China into what I have termed as *Mature Chan*. By the ninth century, polemic ideological differences had grown between disparate Buddhist communities in China, each claiming an orthodox high ground. Disputes over doctrinal details within Chan Mahayana led to a number of schisms and the diversification of new subbranches, each with their own patriarch (see the tree on page 127). Some of these new schools evolved in a phase referred to as the "five houses and seven schools" of Mature Chan Mahayana.[80] This first involved the development of five subschools, including the *Guiyang* (now extinct), *Caodong* (which influenced the rise of Sotoshu Zen in Japan), *Yunmen* (whose core elements were partly adopted by the Linji branch in the eleventh century), *Fayan* (extinct), and *Linji*. Later, the Linji split into two new subbranches called *Huanglong*, and *Yangqi*, although the Huanglong did not survive as a separate practice past the thirteenth century. For a brief description and historical detail of those Chan subschools, I refer readers to *The Princeton Dictionary of Buddhism* by Buswell and Lopez.[80]

As Mahayana was developing in China as *Early Chan*, it was also developing in Vietnam as *Thien* (also called *Thien Uyen Tap Anh*). Some authors attribute its formal arrival in Vietnam to the late sixth-century patriarch, Vinitaruci,[80] although others claim that it had already begun to take form well before Vinitaruci arrived. Vinitaruci is said to have studied in China before arriving at the Phap Van monastery in Vietnam around 580 CE. Thien Mahayana is less ritualistic than other Mahayana schools and is one of the more difficult schools to differentiate and characterize. It is thought to have eventually evolved into three branches, all named after their founders.[80] These branches were the *Vinitaruci*, the *Vo Ngon Thong*, and the *Thao Durong*.

At least by the seventh century CE, the teachings of the Early Chan School were introduced in Korea where it became known as *Son Mahayana* (also spelled *Seon*). It became the state religion during the Koryo dynasty (918–1392 CE), and by the twelfth century, it came to dominate Korean Buddhism. During this time, eighty thousand wooden blocks were made for the printing of all known texts of the Korean Tripitaka (the Buddhist canon with a few later additions).[10] After the first set of blocks was burned during a Mongol invasion, a second set was finished in 1251 CE. This set was called the *Tripitaka Koreana*, and it is used to this day.[80] The South Korean temple named Haeinsa, built in 802 CE, still houses the blocks. It was designated as a UNESCO World Heritage Site in 1995.

Son Mahayana eventually diversified into what became known as the Nine Mountain Schools.,[80] Each of the nine schools was named after the mountain on which the monastery of its founder was built. The differences that distinguished these schools were often clearly known only to monastic adherents. Most of these early schools eventually disappeared, particularly during the years where Confucianism replaced Son Mahayana as the state religion beginning in the late fourteenth

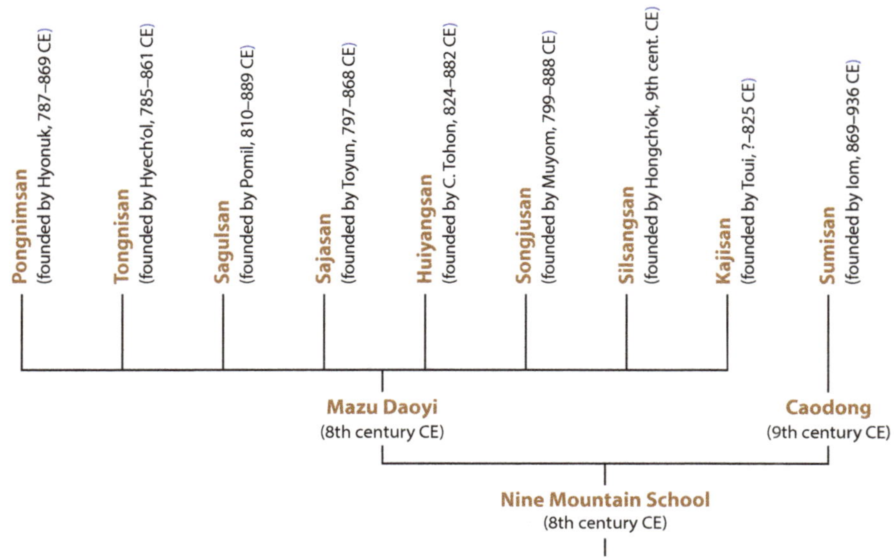

century. However, even as the changing imperial and aristocratic power began to identify with Confucianism, most common people remained adherents of Son Mahayana Buddhism.[10] The main surviving lineage of Son Mahayana today is the *Chogye Chong*, which is thought to have evolved from Kajisan, one of the original Nine Mountain Schools and the largest Buddhist denomination in Korea today.[80]

Chan teachings came to Japan at least by the twelfth century CE where it become known as *Zen*. The name "Zen" is the Japanese pronunciation of the Chinese word "Chan." Although the core of Zen was Chan, it was modified with strong influences from Taoism, Confucianism, and Son Mahayana. It has its own meditative exercise, analogous to yoga, called *zazen*. This exercise is regarded as a means of insight into the nature of existence and achieving enlightenment.[9] It involves sitting in an upright position on a pillow with legs crossed, mouth closed, and eyes open or closed softly. It must be done in silence in a quiet room. Breathing is slow and regular while you mentally free yourself from all attachments to the material world (e.g., desires, judgments). This state is maintained for a several hours in the morning and evening.[10] Zazen is thought to be the gate to great rest, joy, and awareness.[9,10]

The Mahayanistic element of Zen concerned with setting others on the "path to enlightenment" has occasionally been misused to counter the more pacifist traditions of Theravada. In Zen, more aggressive sentiments exist, such as "It is a good thing to strike down evil people and exhibit the righteous" and "It is good to kill the evil that is the enemy of Buddha's doctrine, shaking a sharp sword to destroy demons in order to instruct our truth . . . it is good to kill a small being and a few beings in [order] to bring round a big being and numerous beings."[102] These sentiments were used by Japan's militant Zen schools in the second Sino-Japanese war of 1937–1945 to engage Buddhists in the war against China. Melding Zen Buddhism with a strict utilitarian logic allowed Japan to declare the violent conflict to be a "Holy War," accusing Communism to be "the enemy of doctrine for Japanese Buddhism."[102] The Sino-Japanese war with the invasion of China resulted in as many as twenty million dead and fifteen million wounded Chinese and Japanese people, most of whom were civilians. However, China and Japan were not the only countries to have politicized Buddhism in the twentieth century. At least since the 1950s, the United States, as well as other Western countries, have tried to use Buddhism in their psychological cold war against communism in Southeast Asia, particularly in China. Even Buddhist Monks have been brought into the seemingly never-ending conflict. One monk who became famous in the 1970s is famously quoted as comparing the killing of communists to slaughtering fish, saying, "There is certainly demerit in killing the fish, but we place it in the alms bowl of a monk and gain much greater merit."[102] Even Buddhism, it seems, can be politicized by governing powers to influence people toward violence.

Currently, the three major subsects of Zen practiced in Japan are the *Rinzaishu*, the *Sotoshu*, and the *Obakushu*.[80] The *Rinzaishu*, which distinguished itself in the twelfth century, was the first branch of Zen to develop. It began as a form of *Linji* Chan Mahayana brought to Japan by Eisai (1141–1215 CE). Eisai began in Japan as a student of *Tendai* (a Japanese branch of Mahayana discussed later in this chapter). He eventually became dissatisfied with the state of Buddhism in Japan, and in 1168, he went to China to study Buddhist traditions. He returned to Japan in 1191 and built the first Rinzaishu Zen temple.[9] This branch is noted for its audacious forms of Zen training that include shouting, striking, and dynamic exchanges between master and disciple to invoke an emotional crisis in the student, leading to a spiritual breakthrough.[9] Rinzaishu has evolved into more than a dozen subsects throughout Japan.[80]

The *Sotoshu* branch of Zen formed in the thirteenth century and developed into one of the largest Buddhist branches in Japan. It focuses on zazen (a meditative technique discussed earlier) to achieve a state of concentrated clarity and alertness, free from any specific content. Sotoshu was strongly influenced by the Caodong branch of Mature Chan in China (a connection that warrants further study for possible revision of my tree). Much of the tremendous success of Sotoshu was due to the implementation of the mandatory parish system (*Danka Seido*) during the early seventeenth century by the feudal military government of Japan. This system required every household to register as a member of a local Buddhist temple in an effort to stamp out the spread of Christianity that had begun infiltrating the country. It was an effective measure, and Christianity was eventually all but wiped out of the country (at least officially). By the mid-eighteenth century, there were more than 17,500 Sotoshu temples in Japan.[80]

The *Obakushu* branch of Zen evolved in Japan during the seventeenth century. This branch was founded by the Chan Mahayana master, Yinyuan Longxi (1592–1673 CE), who immigrated to Japan from China to escape the succession of wars and political turmoil following the collapse of the Ming dynasty.[80] Other branches of Mahayana Buddhism materialized in Japan before Zen, one of which is called the *Pure Land School.*

PURE LAND BUDDHISM/JODO SHINSHU AND THE CULT OF AMIDA

Like Chan, some of the earliest ideological roots of Pure Land Buddhism can be traced to the Indian subcontinent. It has been referred to as "the cult of Amida" because followers venerate the Buddha Amida (also called *Amitabha*), principal Buddha of the Pure Land lineages.[89] However, details of its earliest development are poorly known. *Pure Land* became the English term for this belief system after it arrived in East Asia, and it is grouped here as a form of East Asian Mahayana.

In China, the patriarch of Pure Land Buddhism was a monk named Tan Luan (476–542 CE).[9,10] His legend describes him receiving a vision. That vision is said to have shown him that complete devotion to the Amida would result in rebirth in Amida's *Pure Land*, the equivalent of the Western world's paradise. Part of the ritual of this school eventually became to chant a short phrase each day from Mahayana scripture that meant, "Praise to Amida Buddha." The ritual began as a monastic practice, but eventually spread to the laity.[10] In Japan, the Pure Land movement was established by a monk named Honen (1133–1212 CE). Honen was first trained at Enryakuji and became a Tendai monk at the age of thirteen.[9,19] At the age of forty, he converted to Pure Land teachings based on his readings of old Chinese texts. He started a new school of Pure Land Buddhism in Kyoto that became known as *Jodoshu*. The older established Buddhist schools expressed extreme opposition to Honen's teachings, eventually leading to his exile from Kyoto in 1207. He was finally pardoned in 1211 and died a year later. One of his disciples, Shinran (1173–1262 CE), eventually diverged from mainstream Jodoshu, founding a new subschool called *Jodo Shinshu*.[9,10] Shinran had been ordained as a Tendai monk at the age of nine. He eventually became a disciple of Honen and converted to the Pure Land movement and later founded his own branch of that school. Shinran and his school rejected the necessity of monastic rules and residence. He married, had children, and emphasized that simple trust in the saving power of the Buddha was the way to salvation. Honen's Jodoshu school relied on one's personal efforts (*jiriki*) to attain salvation, whereas Shinran's Jodo Shinshu school relied on the power and grace of the Buddha Amida (*tariki*) for that purpose.[80]

In the centuries to follow, contending factions evolved within Jodo Shinshu, but by the fifteenth century, a powerful abbot named Rennyo (1415–1499) reconsolidated Jodo Shinshu as a single institutionalized branch of Mahayana. Eventually, the pressure toward diversification increased once again, and there was another schism within the Jodo Shinshu branch over obscure political issues. This resulted in two descendant schools that evolved in the sixteenth century: the *Nishi Honganjiha* and the *Higashi Honganjiha* (also known as *Otaniha*). Both remain powerful and influential in Japan. Today, Pure Land Buddhism is collectively one of the most successful branches of East Asian Mahayana, second only to the Chan schools.

THE LOTUS SCHOOL: FROM CHINA TO KOREA AND JAPAN

I introduce the term "Lotus School" to include the branches of *Tiantai* and *Tendai* Mahayana. It began in China with *Tiantai*[9,10,89] and evolved as a distinctively Chinese interpretation of Indian Buddhist sutras that had become available as Chinese translations during the sixth century CE. Its founder was Zhiyi (538–597 CE), and the school was named after Tiantai mountain in China, where he lived. The *Lotus Sutra* (a shortened English name for the *Saddharmapundarika Sutra*) was central to Zhiyi's teachings as it would be for the later emerging *Tendai school* in Japan. This sutra, composed by unknown authors in first- and second-century India, was perhaps the most influential of all Mahayana sutras.

Lotus Sutra doctrine proposes many new interpretations on the nature of The Buddha and other traditional beliefs. In it, The Buddha is no longer regarded as a mere mortal, but rather as a sublime being with supernatural powers.[9] He preaches in a mythological paradise surrounded by thousands of followers. The doctrine also teaches that the true Buddhist labors extensively along the bodhisattva path for the liberation of all beings, not just her- or himself.[9] By the early seventh century, Tiantai was one of the leading schools of Chinese Buddhism, although its influence would wax and wane over the years. Its practices and doctrines also influenced most other schools of East Asian Buddhism. It was later introduced to Korea, where by the eleventh century, it had developed into an independent school known as the *Chontae* (*Ch'ont'ae*).[9,80] This school was formally established by a royal Korean prince who became a Buddhist monk, named Uichon (1055–1101). He founded the sect in 1097 to quell certain interbranch conflicts that had developed within Buddhist schools of Korea.

By the ninth century, the Lotus School spread to Japan through the efforts of a Buddhist monk named Saicho (766–822 CE).[80] The name "Saicho" in Japanese means "most pure." Saicho, as a court priest, had an association with the Japanese emperor Kanmu (735–806 CE). The emperor had high hopes for a new form of Buddhism in Japan to bridge the ideological rivalry between Yogacara and Madhyamaka

Buddhists. He granted a petition by Saicho to travel to China to study Tiantai doctrine and to bring back texts to Japan. Saicho left for China in 804 CE and stayed for nine and a half months. After he returned, he founded a slightly modified Japanese version called *Tendai*. He introduced it at Enryakuji, a temple he had established in Japan before going to China. With the Japanese emperor's approval, he ordained one hundred disciples in 807 CE.[9] Although Tendai was similar to Tiantai, in practice, Saicho modified it by introducing Japanese esoteric practices known as *Shingon*, or the "true word." Tendai monks were among those who incorporated Japanese Shinto spirits (*kami*) into Tendai practice by acknowledging them as manifestations of the Buddha.[80] The Tendai school was a major historical influence to other Buddhist schools in Japan. The founders of Rinzaishu Zen and Jodo Shinshu were both trained at the Tendai temple Enryakuji, as was the founder of another branch of East Asian Mahayana called *Nichiren*.

Unlike most of the major schools of Mahayana, which began in India or China, Nichiren began in Japan. Its early elements came from a Tendai monk named Nichiren (1222–1282 CE). Like the founders of Jodo Shinshu and Rinzaishu Zen, Nichiren had his early training in a Tendai monastery. He studied Pure Land and Zen as well.[9,10] Eventually, he combined elements of the different schools he had studied to help form new ideas, summarized in his 1260 essay *Rissho ankoku-ron*. Some of his more radical ideas critical of the government and other Buddhist schools got him exiled twice and even sentenced to death once, although the execution was later stayed. According to Nichiren Buddhist lore, the executioner's sword was struck by lightning just before it struck Nichiren's neck.[9] Like Tiantai and Tendai Buddhism, Nichiren Buddhism considers the Lotus Sutra to be its central text.

Although *Nichiren Buddhism* was built around core concepts from Nichiren the man, he is not considered the founder of the formalized school. After he died in 1282, one of his senior disciples, Nichiko (1246–1332 CE), founded the first temple to defend the teachings of Nichiren. The Nichiren school came to revere Nichiren as the religious founder (*shuso*), and Nichiko as the school founder (*haso*).[9] Over time, the school evolved into at least eighteen Nichiren subschools, although the clear classification and phylogenetic relationships of them remains unclear.[9] Two of the best known of these are the *Nichiren Shoshu*, and the *Soka Gakkai*.[10] Nichiren Buddhism was confined mostly to Japan until after World War II.

HUAYAN: BUDDHIST CONFLICT WITH TAOISM AND CONFUCIANISM IN CHINA

The *Huayan* (or *Flower Garland*) School is an early, influential school of East Asian Mahayana. It first flourished in China during the Tang dynasty, and its chief scripture is the Avatamsaka Sutra, another highly influential Mahayana sutra of East Asia. The school doctrine teaches that Buddhahood is coextensive with everything there is, and that reality is thoroughly interpenetrating. There is no "self." The details are complex and explained in depth by philosophers Bryan Van Norden and Edward Zalka.[91]

The founding and early development of Huayan is traditionally attributed to a series of five Chinese patriarchs, the last two of which are not so much founders of the school but rather are revisers or modifiers of traditions established by the first three.[9,80,91] The first patriarch was a monk named Du Shun (557–640 CE). He was responsible for the early establishment of Huayan studies in China. The second patriarch was Zhiyan (602–668 CE), who was a disciple of Dashun. He is thought to have established some of the early doctrines of the Huayan. The third patriarch was Fazang (643–712 CE), a disciple of Zhiyan. He is credited with further rationalizing Huayan doctrine for wider acceptance in Chinese society. This made him the patriarch most recognized as the formal organizer of the Huayan school.[9] The fourth patriarch was Chengguan (738–839 CE), who studied under a student of a disciple of Fazang. Finally, the fifth patriarch was Guifeng Zongmi (780–847). He had spent many years studying Confucianism, Taoism, and Chan before becoming a student of the fourth Huayan patriarch, Chengguan. He was also known as a patriarch for the short-lived *Heze school* of Chan Buddhism.[91]

The success of the Huayan school in China was tied to imperial support and was successful for a number of years. However, conflict with the Taoists and Confucians ultimately did not end well for the Buddhist schools in China. Taoists and Confucians encouraged the Tang emperor Wuzong to launch the Great Anti-Buddhist Persecution of 842–845 CE.[91] Buddhist monks and nuns were forcibly defrocked, temples were destroyed, and Buddhist property was seized. It was a great setback for all of Buddhism in China, and Huayan all but disappeared. Buddhism recovered in China during the Song dynasty in the tenth through twelfth centuries, but it was primarily Chan and its derivatives that revived there. During this period, Chan influenced the development of many branches in Korea and Japan as well.

Before the Great Anti-Buddhist Persecution in China, the Huayan belief system had already spread to other parts of Asia. It had been transmitted to Korea in the seventh century by Uisang (625–702 CE), a disciple of the patriarch Zhiyan in China.[91] The Korea subbranch became known as *Hwaeom* (also spelled Hwaom).[80] This branch became a well-established, independent doctrinal school during the Goryeo dynasty. The Huayan belief system spread to Japan in the eighth century, where it became known as *Kegon*. There, it began when an ex-Yogacara monk named Ryoben (689–773 CE) invited the Korean monk Simsang, a disciple of the Huayan patriarch Fazang, to bring Huayan teachings to Japan. Kegon later adopted elements from Vajrayana and other branches of Mahayana, which further distinguished it as an independent branch of Mahayana whose practice continues to this day.

BUDDHISM IN SUMMARY

Buddhism encompasses an extremely diverse group of beliefs with a poorly known early history that originated in India. Most of its early branches, as originally conceived on the Indian subcontinent, are today extinct. By the eleventh century,

Enormous statue of *Vairocana*, a cosmic Buddha of Vajrayana and East Asian Mahayana Buddhism, standing on top a Buddhist Monastery in Henan Province, China. This statue, called the *Spring Temple Buddha*, is the largest Buddha statue in the world. Built from 1997 to 2008, it stands more than five hundred feet tall (more than thirty-six stories). The staircase to the entrance has 365 steps.

Buddhism had virtually disappeared from its country of origin (less than 1 percent of the population, or just under nine million members in 2010). However, thanks to its early spread and influence outside of India, by the twentieth century, Buddhism had more than a half-billion followers in eastern Asia (more than 99 percent of all the world's Buddhists).[3] As discussed earlier in this chapter, major reasons for its rapid spread and diversification through eastern Asia included early Buddhist missionary efforts from India, and the periodic patronage by various ruling monarchs in China, Japan, and other parts of eastern Asia.[9]

The main supernatural belief systems of Buddhism in the world today belong to three surviving branches. The largest branch is the *Mahayana*, which is most prominent in East Asia and is known as "Eastern Buddhism." By 2010, it included 53.2 percent of all practicing Buddhists.[3] The second-largest branch is the *Theravada*, which is most prominent in Southeast Asia and was previously known as "Southern Buddhism." In 2010, it included 36 percent of all practicing Buddhists.[3] The third-largest branch is the *Vajrayana*, which is most prominent in the Indo-Tibetan region and also known as "Northern Buddhism"). In 2010, it included 5.7 percent of all practicing Buddhists.[3] Together, Theravada, Mahayana, and Vajrayana are sometimes referred to by adherents as the "Three Vehicles" of Buddhism. Other Buddhists see them as a single vehicle, representing only the variety of strategies that the Buddha used to reach different audiences.[80] To more fully understand the evolutionary history of Buddhism, it is necessary to acknowledge both individual and shared characteristics among these many different subgroups.

Long after the Sramana branch of Jainism and Buddhism split from the Vedism/Hinduism branch of Dharmic Religion, another Dharmic branch called *Sikhism* developed.

SIKHISM: RECONTEXTUALIZING THE VEDAS BUT RETAINING THE CASTE SYSTEM

This Sikhism movement is thought by many scholars to have been a recontextualization of Vedism in India[103] It was partly in response to the arrival of Islam[20] and partly a rejection of certain Vedic concepts. Guru Nanak (1469–1539) is recognized as the holy founder of Sikhism.[5] However, Sikhism has older ideological roots (see the phylogenetic tree on page 70). It has basic ideological connections with Hinduism and Sramana, but it lacks close connection to any specific Hindu or Sramana subgroup. It was a branch of Asian Cyclicism that evolved monotheistic characteristics. Some authors suggest an ideological relationship with Vaishnavism.[5] Details of Sikhism's earliest roots and closest relationships are yet unresolved, and so it is ambiguously rooted on my tree as a general Dharmic branch (page 70).

Nanak was the first Sikh Guru who contributed to the great collection of Sikh spiritual teachings. He is said to have been a prophet bearing revelations from God (*Ik Onkar*) after a mystical three-day experience.[9] Following his death, there was a succession of nine gurus. Two of them, Arjan (1563–1605) and Tegh Bahdur (1621–1675), were tortured and executed by Muslim rulers for proselytizing Sikhism and blasphemy against Islam.[5,104] The last of the gurus, Guru Gobind Singh (1666–1708), christened the collective teachings of the previous gurus as the *Adi Granth* (also known as the *Guru Granth Sahib*). This became the primary sacred scripture for Sikhism. The 1,430-page text consists of six thousand line compositions, which are poetically rendered and set to a rhythmic ancient north Indian form of music. The preeminent spiritual site for Sikhism, and the tradition's most important pilgrimage site, is the Golden Temple, In Amritsar, India. It houses the Adi Granth, and over 150,000 people visit this Holy Shrine each day.

In theory, the Vedic concepts rejected by Sikh scripture include idol worship, bathing rituals, gender inequality, and the caste system. In practice, however, the caste system still exists in Sikhism. In some respects, the Sikh caste system is as problematic as the Hindu caste system.[105,106] It is currently a point of controversy within the religion, although some progress has been made. In 1953, a law was passed to reserve 20 of the 140 seats on the SGPC (the Sikh Parliament of India) for the members of the Sikh Dalait caste.[105] Sikhism also rejects the traditional Vedas. Nevertheless, Sikh scripture contains hundreds of references to the Vedas and retains numerous Hindu spiritual concepts (e.g., karma, dharma, samsara, mukiti, and maya). Sikhism recognizes a number of Hindu gods and goddesses but only as alternate names or aspects of "The Almighty One."[107] The Supreme God in Sikhism is known as *Ik Onkar* (also spelled *Ek Onkar*), the Absolute One.[108] Sikhism represents a move toward monotheistic thinking within Asian Cyclicism that evolved independent of Abrahamic Monotheism (i.e., an independent evolutionary parallelism or convergence). Because of its ideological origin, and also because of its belief in the cyclic nature of existence (samsara), I still included it as a branch of Dharmic Religion.

After being baptized into what is referred to as the *Khalsa* order (also known as *Amritdhari*), Sikhs are expected to wear the *five articles of faith* (commonly known as the "five Ks"), which distinctively identifies the wearer as Sikh. These include *kesh* (the practice of leaving the hair and beard uncut, sometimes under a turban), because hair is considered by Sikhs to be a gift from God; *kangha* (a comb for grooming and symbol of cleanliness); *kirpan* (a symbolic dagger or sword signifying courage in the defense of right); *kara* (a circular steel bracelet to symbolize God as never ending); and *kachera* (cotton shorts customarily worn by men enabling swift action in war).[9] Sikhs who are not baptized and do not observe the five Ks are often called *Sahajdhari* Sikhs. Some authors treat the Khalsa and Sahajdhari as distinct Sikh branches or sects.[108]

Although the recognized founder of Sikhism, Guru Nanak Sahib, was born in a Hindu family, Sikhs adamantly disclaim any relationship to Hinduism. They have a history of violent conflict with both Muslims and Hindus. To establish their independence, they often make the claim that Sikhism does not have roots in any other

The Darbar Sahib (later known as the Golden Temple), first built in the sixteenth century in the center of a large man-made pool deemed sacred by Sikhs. It is the holiest shrine in Sikhism and is located in Amritsar, India. (*Top*) Close up photo of the current temple sanctum, rebuilt after several periods of destruction and rebuilding, first by the Sunni Mugal Empire and later by various other entities. The copper exterior of the present sanctum was overlaid with gold leaf in 1830, leading to the name "Golden Temple." (*Bottom*) Aerial view of the temple sanctum centered in the sacred pool.

religion or ideology.[6] While modern Sikhism is neither Hindu nor a product of the Sramana movement, its historic roots place it with those groups in the great Dharmic family of belief systems. Just as Jesus was once a practicing Jew preaching to other Jews (see chapter 7), Nanak Sahib was of Dharmic origin. Jesus and Nanak Sahib each eventually became the seed of a divergent movement. In an evolutionary context, you need only to follow the historical branches of Sikhism, Hinduism, and the Sramana movement back in time to find that they eventually converge on a common source (tree on page 70).

Sikhs consider an individual's religion to be the result of the culture in which they were brought up, whether Hindu, Muslim, Christian, or Sikh. They believe that anyone who worships a god follows the same path as Sikhism. Like most other branches of Dharmic Religion, they generally do not aggressively attempt to convert people of other faiths. Sikhism is one of the youngest of the major world religions. Even so, by 2010, it had grown to about twenty-four million members,[3] the great majority of whom live in the Indian state of Punjab. Since its origin, Sikhism has evolved into more than a dozen different branches.[109] Historical diversification (not yet well understood in a phylogenetic sense) resulted from those who began to believe in alternate lineages of gurus or to have different interpretations of Sikh scriptures.[109] Sikhism is the final subgroup of Dharmic Religion that I cover.

SUMMARIZING THE DHARMIC BRANCH OF ASIAN CYCLICISM

The branch of Asian Cyclicism that is defined as Dharmic Religion began in central and southern Asia, possibly as far back in time as before 9000 BCE. It eventually evolved into a number of major branches, including Hinduism, Jainism, Buddhism, and Sikhism. As it spread throughout the region, it was adopted by more than one-third of the Asian population and more than half of the south-central region of the continent.[3] Another major branch of religion that I provisionally include in Asian Cyclicism is *Taoic Religion.*

TAOIC RELIGION: THE EARLY RISE OF ASIAN CYCLICISM IN THE FAR EAST

As poorly known as the earliest origin of Dharmic Religion is, the origin of *Taoic Religion* is no better. It is provisionally placed at the base of my phylogenetic tree of Asian Cyclicism (page 70). I indicated its placement with a question mark to emphasize the need for further phylogenetic testing of that hypothesis.

Formalized Taoic Religion appears to have begun at least as early as 600 BCE in China.[9,10,42] It was possibly built around earlier (*Proto-Taoic*) principles of Chinese ancestorism from the Shang dynasty (1600–1046 BCE) as well as even earlier animistic elements.

The early Taoic schools of China have few surviving records, and even the records that exist can be confusing (this lack of records is discussed further in **Note 4K**). Historians of Eastern culture often do not sufficiently distinguish religion from other aspects of culture. That can obscure our understanding of the independent evolutionary patterns of religious belief systems. It has been said, "China is a country without religion and the Chinese are a people who are not bound by religious superstition."[110] Such opinions assume a different definition of religion than the one I use in this book. As discussed in chapters 1 and 3, notions of what qualifies as religion vary. One person's personal philosophy is another person's myth and yet another person's religion. However, the innate spiritualism and ritual present in the Taoic belief system certainly qualify it as a religion as understood here.

I have assembled a provisional tree of Taoic belief systems based on a few simple published sources to illustrate at least some of the rudimentary diversification within this group (see the tree on page 141). However, this tree is highly provisional and simplified. Much detail is missing, and many of the dates are ballpark estimates. Some of the group names used on the tree (e.g., *Koshinto* and *Sinoshinto*) are merely efforts to retroactively organize poorly known early religious branches. Even with such shortcomings, this phylogenetic tree provides a rough hypothetical picture of ideological relationships and complexity. Substantial comparative work is needed by specialists of East Asian religions to produce a comprehensive phylogenetic tree of denominations that more clearly encompasses the evolution and detailed interrelationships within this ancient group.

The poor understanding of Far Eastern religion in Western society is sometimes blamed on the Western inability to grasp the complexities of East Asian culture and the Chinese language.[111,112] This limited understanding, however, is also due to a number of other major factors. For one thing, the predicted ancient age of the earliest Proto-Taoic roots suggests that the earliest contemporaneous records of their origin are yet unknown. Another factor is the common use of non-monophyletic vernacular group names. For example, little has been done in a rigorous comparative way to unravel the diversity and historical relationships of the numerous religious belief systems now collectively referred to as "Chinese folk religions." Many of those traditions have little or no unbiased documentation concerning their ideological origins or specific relationships. Chinese folk religion is a vernacular name representing a non-monophyletic category of religious traditions united only on the basis of geography, historical period, and other nonreligious aspects of culture (further discussed in **Note 4L**). It is an extremely diverse assortment of belief systems collectively held by more than fifty minority groups speaking well over two hundred different minority languages.[113] The specific diversity of ideologies involved are in need of a much more rigorous systematic study by specialists, within a comparative context, before they

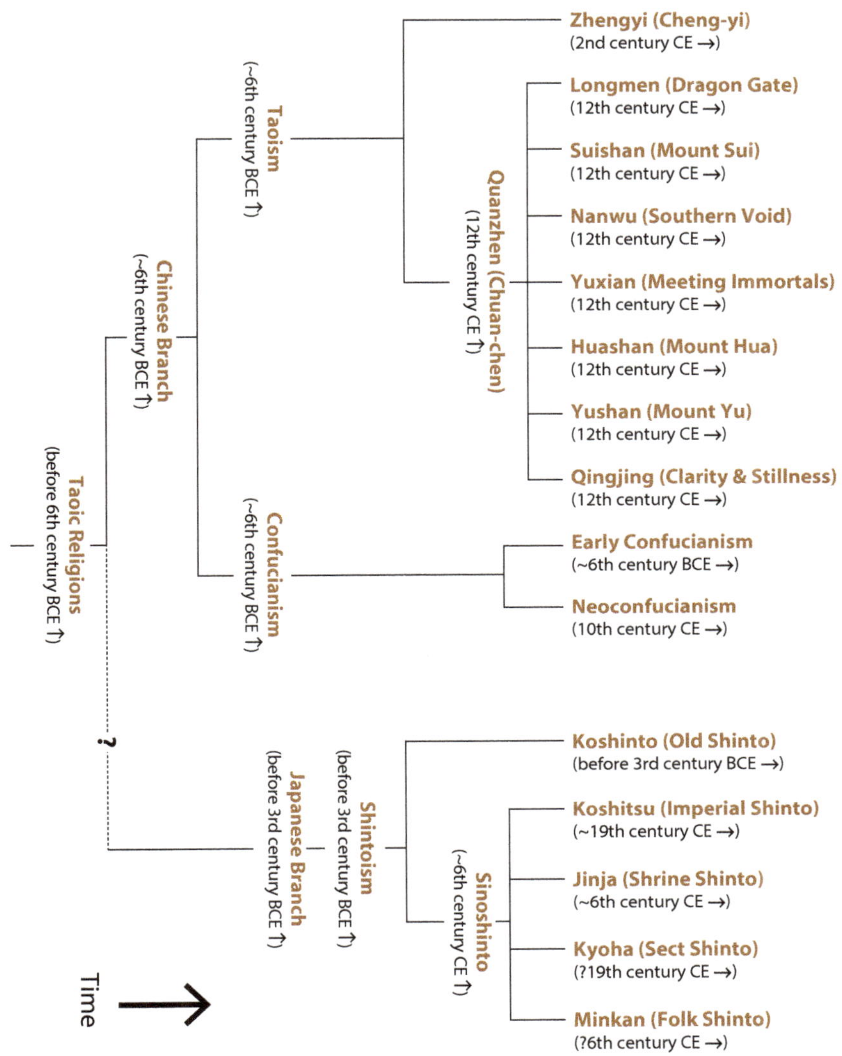

can be confidently included among phylogenetic trees, such as the ones in this book. Some of its putative subgroups may not belong within the Taoic branch at all and may be better classified within a Dharmic Religion or elsewhere. The composition and phylogenetic diversity of Chinese folk religions is in need of rigorous phylogenetic study, and such an analysis is well beyond the scope of this book.

Much like Dharmic Religions, Taoic Religions have historically focused on the present life more than an afterlife. Making "this-worldliness" a priority is key to their philosophy.[114] Like Dharmic Religions, many Taoic branches also include a cyclic philosophy.[115] The ancient Taoic school of *yinyang* (discussed later in this chapter) expresses itself as balanced, often cyclic, forces of nature. Several of these forces consist of opposite energies in a constant cycle of interaction from which the universe and all its diverse forms emerge.[9] Examples include the transition between day and night, the act of breathing in and out, the waves and undertow of the ocean, and the cycle of the seasons.[10] In addition, Taoic and Dharmic Religions both share a belief in five basic forces of nature, although with differences in two of them. Most Dharmic Religions (best summarized in the Upanishads) believe the five forces to be *earth, water, fire, air,* and *space.*[116] Most Taoic Religions believe the five forces to be *earth, water, fire, wood,* and *metal.*[116]

Another thing that differentiates Taoic Religions from early Dharmic Religions is its stronger emphasis of honoring ancestors. Filial piety and the veneration of ancestors goes back at least as far as the Shang dynasty.[10,117] Maintaining the reverent relationship between dead and living kin is a priority in most Taoic Religions. Taoic Religion also lacks the formal perception of samsara (reincarnation) and moksha (nirvana) present in Dharmic Religion (at least before the influence from Buddhism). Nevertheless, the number of shared ideological characteristics between Dharmic philosophy and Taoic philosophy convinced me to provisionally place those belief systems together as sister groups on my phylogenetic tree of Asian Cyclicism (page 70).[116]

Early elements of Proto-Taoic *yinyang* are indicated on "oracle bones" from the Shang dynasty (seventeenth to twelfth centuries BCE). These inscriptions on turtle and other animal bones do not specifically mention the words yin or yang, but they do mention related words that became fundamental to the concept (e.g., *qi,* for the numinous and flowing substance that constitutes and animates all phenomena and whose imbalance causes disease).[118] Proto-Taoic yinyang elements may have had roots in even earlier civilizations, such as those of the Yellow River, Yangtze, or Liao Valleys, although we know very little about the supernatural belief systems of those early cultures. Very few records of East Asian belief systems exist before the fourth century BCE.

As with other major branches of Organized Religion, the major religions of China followed the rise and fall of ruling dynasties, kingships, and empires. The early Chinese people historically believed that a "Mandate from Heaven" (the *T'ien-ming*) gave their rulers the right to be king or emperor.[9] If a ruler was overthrown, this was interpreted as an indication that the ruler was unworthy and had lost the mandate. Consequently, each newly installed dynasty could bring significant cultural change, including new directions for religious beliefs. The symbiotic relationship between Taoic Religion and imperial power waxed and waned over the centuries. As one branch became the official state religion, others would be banned or suppressed.

Although the Chinese branches of Taoic Religion and Buddhism were often political rivals, they deeply influenced each other.[119] In the early fifth century CE, a syncretized amalgam of the two was the official religion of the Wei dynasty.[72] The seventh emperor of the Tang dynasty, Xuanzong (emperor 712–756 CE) declared Taoism as the state religion of China. Other religions were prohibited; their monks, nuns, and priests were persecuted; and their temples were destroyed. This era of suppression ended with the decline of the Tang dynasty. Taoism once more came to compete with the growing popularity of Buddhism and Confucianism. Most emperors after that time were lenient of multiple religions. Neither Taoism, Buddhism, nor Confucianism would ever again have empire-wide state control.[88]

The many branches of *Taoic Religion* share core concepts of spirituality, world order, and the veneration of ancestors. They focus more on this world than on an afterlife. Retention of one's good name after death is also highly valued.[88] These branches profess to follow the path of Tao (meaning "way"). Their goal is to become one with the primal force of Tao, thereby achieving spiritual immortality. Tao is seen as the ultimate principle of the universe and beyond the capacity of ordinary knowledge and human intellect.[120] Briefly, what *Tao* (also called *Dao*) represents to adherents is that it is the fundamental principle and source

Doumu, mother goddess of the Big Dipper whose stars are believed to be her children. She became queen of heaven in Taoism but was originally of Indian origin as the Buddhist goddess *Marici* (Pregadio, 2008). Doumu is often shown with anywhere from eight to eighteen arms. This is an image of a porcelain statue from the eighteenth-century Qing Dynasty in the Asian Art Museum of San Francisco, California.

that underlies all things in the universe.[9,42,116] It is similar to the Indian Vedic concept of the *Brahman* representing "the divine substance that connects all things in the universe."[5,116] Another way to think of Tao is as the cyclic rhythm of nature. The *yin* and *yang* doctrine (also referred to as the *yinyang*) stipulates that every action (*yin*) creates a counteraction (*yang*). The agents of *yin* (dark, cold, female, negative) complement the agents of *yang* (light, hot, male, positive). It is a dualistic balance of opposites. It is also a cyclic progression of day and night, the four seasons, and even the act of inhaling and exhaling.[10] Taoic beliefs hold that the universe and all of its diverse forms emerged from the interacting energies of yin and yang.[9]

Taoic Religion includes many branches from China as well as some from Japan, Korea, and Vietnam. I focus on the *Chinese Branch* and the *Japanese Branch* of Taoic Religion, as shown on my phylogenetic tree on page 141.

THE *CHINESE BRANCH* OF TAOIC RELIGION

The *Chinese Branch* of Taoic Religion consists largely of *Taoism* (*Daoism*) and *Confucianism*. It appears to have predated the *Japanese Branch* of Taoic Religion called *Shintoism* by centuries. Although some authors contend that Taoism and Confucianism are social structural systems or ethical systems rather than organized religions, that view is an oversimplification of culture within those branches. I included Taoism and Confucianism in this study of religious belief systems because they include sacred texts, rituals, and distinguishable supernatural beliefs that are represented in a variety of subschools. The Chinese branch of Taoic Religion values the concept of *te* (or *de*), a virtue meaning integrity and wholeness in Taoism, or morality and sound character in Confucianism.[121]

Each of the many branches within these two major schools has its own deities, traditions, myths, and forms of ancestor veneration. The lack of clear detail and systematic information may be partly an artifact of early Western bias in the systemization of religions. Roman Catholic missionaries in the sixteenth century contended that Confucianism was not a religion to make it culturally compatible (i.e., noncompetitive) with Roman Catholicism.[122] Many early efforts to classify religions of the world during the late nineteenth and early twentieth centuries were little more than thinly veiled efforts to demonstrate the superiority of Western Christianity.[123] The lack of systematic clarity may also be due to the rarity of broad comparative studies that include Taoic Religions with other world religions in an unbiased, comparative context. The Chinese language did not even have a word for "religion" before the Japanese introduced it in the late nineteenth century through translation of European works.

TAOISM: SEARCHING FOR HARMONY IN THE UNIVERSE

Formalized versions of *Taoism* (or *Daoism*) were established in mainland China by at least the sixth century BCE.[116] Taoists trace themselves back to the legendary figure and Chinese philosopher named Laozi (571 BCE–?), whose name means "Old Master," and who is author of the influential *Tao Te Ching*.[10] Taoism began as a religion that emphasized living in harmony with the *Tao* (*Dao*) or the *way*. The two interchangeable English spellings are due to a problem of transliteration from Chinese to English. Although the most common transliteration of the tradition is "Taoism," the name is actually pronounced "Daoism." Taoism is one of the five religious doctrines officially sanctioned in the People's Republic of China in 1982. (The other four are Buddhism, Roman Catholicism, Protestantism, and Islam.)

The interaction of *yin* and *yang* mentioned in the previous section is particularly important to Taoic belief systems. That interaction is thought by Taoists to direct the destinies of living things and events. Popular Western culture sometimes simplifies the yin as bad and the yang as good, but this is an inaccurate usage of the terms.[10] These two opposites are the constituents of all things. Although seemingly destructive to each other, these forces are instead cyclically interconnected. Each defines the other through their alternating interactions in the physical world. To breathe out, one must first breathe in. Day follows night, which follows day, again and again. Taoism emphasizes natural order, and it is thought that a healthy life is one in which the forces of yin and yang are balanced.

Tao can be interpreted as a sort of doctrine. It is believed by adherents to be the source, pattern, and substance of everything that exists in the universe, even though it cannot be fully understood. Unlike Confucianism, Taoism focuses more on effortless natural order and the individual rather than on social order and a harmonious society.[10] It emphasizes the "Three Jewels of the Tao": Right belief, Right knowledge, Right conduct.[22] Most branches of Taoism are polytheistic. The deities of Taoism are not fixed. They are subject to change like the rest of the cosmos. They can be promoted or demoted over time. Deities from local cults and other religions can be absorbed into the Taoist pantheon. Taoism developed its own divine triad consisting of the *Three Pure Ones*, which are thought by some historians to have ideologically evolved from the divine triad of Mahayana Buddhism (the Trikaya).[10] The Three Pure Ones are regarded as the highest Taoic divinities. They include *Yuanshi Tianzun*, the highest lord of heaven representing the primordial Tao; *Lingbao Tianzun*, lord of treasures who is associated with transmission of Taoist insight; and *Daode Tianzun*, whose earthly manifestation was the legendary character *Laozi*.

(*Left*) Stone carving of a Taoist Divine Triad, *The Three Pure Ones*. Probably made between 528 and 532 CE in northern China. In the collection of the Musee Guimet, Paris, France. (*Right*) Silk manuscript of Tao Te Ching from the second century BCE. In the collection of the Hunan Province Museum.

Laozi (Lao-Tzu) was an ancient Chinese philosopher who is thought to be one of the earliest teachers of organized Taoism. Whether or not he actually existed is disputed. However, the work attributed to him—the *Tao Te Ching*—is the most influential of all Taoist texts.[124] The oldest-known written parts of the text date to the fourth century BCE, although Laozi is thought by many historians to have lived in the sixth and fifth centuries BCE and to have composed it then. After the Hebrew/Christian Bible, the Tao Te Ching is the world's most frequently translated book.[3] It has undergone many different reinterpretations continuing to the present day.[120] The origin of Taoism may predate the Tao Te Ching, because in chapter 15, the book already refers to the "ancient masters" of the Tao.

The quests for long life and achieving spiritual immortality are some of the strongest goals of Taoism.[120] Followers have a number of practices involving physical exercise. Much like yoga and zazen of Dharmic Religion, Taoism has *Tai chi*, a series of ballet-like slow arm and leg motion exercises thought to aid in balance, blood circulation, and meditation. Taoists believe these exercises to contribute to longer life by helping a person feel the flow of nature and attain spiritual purity. Taoists of the past, similar to the Buddhists, went to great lengths in their search for longevity. In ancient China, some experimented with alchemy, hoping to create an immortality potion. Because gold does not rust, some attempted to make it into a drinkable form, thinking it would prevent aging and provide certain curative powers. Some even added elements such as mercury to the potion (another practice similar to that of the Buddhists).[10,77] Although drinking heavy metals is fortunately no longer practiced, vestiges of the search for life-prolonging elixirs still exist in the Chinese culture, which also has great interest in pills and medicines to prolong life. Some traditional Chinese medicines, such as ginseng, ginger, and garlic, appears to have genuine medical benefits.

Historically, Taoism is not known to have become highly organized until the development of the *Zhengyi* (*Cheng-yi*) lineage (historically called by a number of other names, including *The Way of the Five Pecks of Rice, The Way of the Heavenly Masters*, or *The Orthodox Way*). This lineage is said to have evolved at the end of the second century CE. It was initially led by Zhang Daoling (34–156 BCE), who claimed to have had a vision from Laozi.[9,10] Little is known about the early history and diversification of Zhengyi Taoism.

By the twelfth century CE, the monastic *Quanzhen* branch (also known as *Chuan-chen* or *The Way of Complete Perfection*) evolved in Shandong. This branch grew during the thirteenth and the fourteenth centuries to become the largest, most important Taoist denomination in Northern China. Throughout Chinese history, Taoism was proposed numerous times as the state religion, although after the seventeenth century, it fell from favor. It was often intertwined with Zen Buddhism, Confucianism, Feng Shui, Chinese astrology, Chinese alchemy, and the martial arts. By 2010, Taoism had at least eight and a half-million followers[3] and many subbranches (see the tree on page 141). In modern times, with concerns about global warming and environmental pollution, the Taoic practices of valuing the earth and of acting in harmony with nature have been gaining popularity in the Western world.

Wang Chongyang (1130–1170 CE) is one of the founders of the Quanzhen branch of Taoism. According to some sources,[125] Wang had seven disciples (six men and one woman), each of whom created an independent subbranch of Quanzhen Taoism (tree on page 141). The most successful of these branches was the *Longmen* (or *Dragon Gate*) sect founded by Qiu Chuji (1148–1227 CE). A famous story about Qiu is about when he met with Genghis Kahn in 1222. Kahn is best known as a famous warrior-ruler and one of history's most famous conquerors of all time, but he was also a shamanistic animist who was tolerant of other religions. Qui successfully persuaded Kahn to exercise restraint against Taoic followers during Kahn's brutal conquest of Eurasia.[126] Kahn even honored Qiu with the title "Spirit Immortal." With the support of such a powerful ruler as Genghis Kahn, the Longman school grew and continued to evolve.[125] It eventually became strongly influenced by both Buddhism and Confucianism.

The other six main branches of Quanzhen Taoism included the *Suishan* (or *Mount Sui*) sect founded by Liu Chuxuan; the *Nanwu* (or *Southern Void*) sect founded by Tan Chuduan; the *Yuxian* (or *Meeting the Immortals*) sect founded by Ma Danyang; the *Huashan* (or *Mount Hua*) sect founded by Hao Datong; the *Yushan* (or *Mount Yu*) sect founded by Wang Yuyang; and the *Qingjing* (or *Clarity and Stillness*) sect founded by Sun Bu'er. Some of these branches eventually drifted to extinction, but the *Longmen* branch saw a resurgence in the seventeenth and eighteenth centuries, and it ultimately evolved into many subbranches.[127]

Today, the surviving branches of Taoism are routinely classified mainly as either Zhengyi or Quanzhen (each with many subbranches).[126,128] As mentioned earlier, the detailed phylogenetic relationships among most Taoic subsects remains unclear. The lack of historical clarity for the early diversification of Taoism is due in part to the somewhat cavalier way that the names "Zhengyi" and "Quanzhen" have been applied by governing bodies in China, obscuring a much more complex history. Early records of Taoism were also clouded by Buddhist intellectuals and Confucian-trained scholars who devalued the tradition. Sometimes the lack of historical clarity about that period of Taoic diversification is dismissed with the claim that early Daoism was not an organized religion.[10] That excuse is most likely a reflection of how little we know or understand about the early evolution of Taoic belief systems. The next branch of early Taoic Religion I discuss is *Confucianism*.

CONFUCIANISM: BRINGING SPIRITUAL REINFORCEMENT TO SOCIAL ORDER

Confucianism (also sometimes known as *Ruism*) is a branch of Taoic Religion that was highly influential in Chinese government and society for much of the past two thousand years.[9,121] Unlike Taoism, which emphasizes *natural* order, Confucianism emphasizes *social* order. It promotes a harmonious system linking humankind to society and heaven.[129] Social harmony results from every individual of society knowing his or her place in the order, playing their part well, and recognizing the high importance of family and reputation. In this system, one must promote ethical behavior, loyalty, and reverence for one's ancestors and rulers. Confucianism professes that men and women are social beings by nature and that they are fundamentally good, improvable, and even perfectible. They can become one with heaven through contemplation of harmonious social order. Unlike Taoism, which focuses simply on fitting in with the forces of nature (yin and yang), Confucianism emphasizes scholarly study, training in virtue, and improvement of human interactions. Where Taoism asks, "What kind of person should I be?" to fit in with a natural order that can never be fully understood or significantly altered, Confucianism asks, "What can I do?" to better understand natural order and create a better society.[9,10]

Confucianism is pantheistic in that it is not based on the belief in a personal god existing independent of the physical universe. Instead, it sees the universe as a physical manifestation of God. Confucianism is unconcerned with "salvation" in the afterlife, and it sees etiquette and ritual as solutions to the world's ills.[41] The goal of Confucianism is to have long life, health, and wealth in this world, rather than being rewarded in the next one.[72] After death the main goal is retention of one's good name.[88] Where Taoism sees the Tao as a primal force that cannot be questioned or understood and must simply be followed, Confucianism emphasizes a strict scholarly and scientific study of such forces as important. Confucianism also maintains that training in virtue is necessary to enable the Tao to manifest itself in a human being.[10,42] It has been described as a "sociopolitical doctrine having religious qualities."[65] Other historians see Confucianism as something that began as a philosophy and only later also developed as a religious system, partly in response to the success of Buddhism in China.[9]

I consider Confucianism to be a religion, even if it is not primarily theistic. It includes rituals with supernatural undertones, clearly performs the basic psychosocial functions of religion, and evolved as a form of Taoic Religion. In the sixteenth and seventeenth centuries, the Roman Catholic Jesuits classified Confucianism as a nonreligious ethical system to make it compatible (nonheretical) with their faith. However, by the early eighteenth century, the Roman Catholic Dominicans and Franciscans argued vigorously that Confucianism was a form of idolatry, creating a dispute within the Roman Catholic Church known as the "Rites Controversy." Pope Benedict XIV eventually settled the matter by declaring a ban on Confucian rituals in 1742.

Confucianism developed from the earlier referenced *Hundred Schools of Thought* from the Chinese philosopher K'ung Ch'iu, later known as Confucius (551–479 BCE).[130] He considered himself to be a messenger of recodified theological values from the Shang (1600–1046 BCE) and Zhou (1046–256 BCE) dynasties. Like so many historical figures retroactively designated as founders of a particular religion, Confucius historically claimed he was "a transmitter, not a creator" of the Confucianist belief system (*Analects* 7.1). He had two main ideals.[10] First, he wanted to produce "excellent" individuals who could be social leaders. Second, he wanted to help create a harmonious society that would nurture excellent individuals. He never claimed to be a divinity, but centuries after his death, Confucius was often depicted as a demigod.[131]

Ancestor worship, duty to family, sacrifice, and ritual were considered by Confucius to be more critical for social harmony than emphasizing otherworldly sources of spiritual value. The family unit is the primary social unit, with devotion to parents as the most important of all. As in most other Organized Religions, women often occupied a lower status than men. The "Three Feminine Obediences" doctrine of Confucianism tells women to always remain obedient to the men in their lives (to her father as an unmarried child, to her husband as a wife, and to her sons as a widow).

Confucius did not leave behind any written works.[131] The *Analects* are the primary texts of Confucianism. Followers of Confucius compiled these texts a generation or two after his death. *Analects* 10.11 describes how Confucius always took a small part of his food and placed it in small sacrificial bowls as an offering to his ancestors. Although Confucius appears to have believed in a supreme deity, he preferred to teach a doctrine of humanism open to the transcendent.[9] Confucianism considers the ordinary activities of human life, particularly those having to do with human relationships, as manifestations of the sacred.[10] Later branches of Confucianism (e.g., *Mohism*) deviated from the exclusive focus on ancestors and advocated extending such reverence to all beings.[9] Its founder, Mo Tzu (470–380 BCE), held that most social problems arose because people's love and respect was graded and partial.[10]

The growth of Confucianism was significantly enhanced during the Han dynasty (206–220 CE) when it was made the state religion of China by Emperor Wu (157–87 BCE).[3] After the collapse of the Han dynasty in the third century CE, Confucianism saw a relative decline because many people were turning to Taoism and Buddhism.[131] By the tenth century CE, however, Neo-Confucianism (a somewhat vague Western term) arrived as a growing movement to bring Confucianism back to its traditional roots. By the eleventh century, it became the dominant religion of China.[9,130] Its influence spread throughout government and eventually the Analects were integrated into civil service tests. Prospective government officials had to master Confucian classical texts. Officials were selected based on perceived moral qualities because they were charged with setting a moral example for the people. Confucianism embedded itself deeply within other aspects of Chinese culture and social order. Everyone was expected to know their place (father, mother, scholar, ruler, subject) and to do their best to maintain it. A hypothesis among political scientists and economists called the *Confucian hypothesis* proposes that Confucianism played a latent role in the development of the rigorous work ethic of modern-day East Asia, which led to the rapid modern industrialization of China, Japan, South Korea, Taiwan, Hong Kong, and Malaysia.[132] Over time, Confucianism and Neo-Confucianism evolved into a number of subbranches, although they have not yet been clearly and consistently delineated.

At the beginning of the twentieth century, official support for Confucianism began to fall with the collapse of the Qing dynasty.[10] Confucianism and Neo-Confucianism went into serious decline with the loss of imperial support and attacks by new cultural movements culminating in outright suppression. The peak of anti-Confucianism came with the Chinese Cultural Revolution of 1966–1976. The pressure eventually let up in the late twentieth century, but by that time, the percentage of Confucianists in mainland China had been severely reduced. By 2010, Confucianism and Neo-Confucianism in mainland China made up far less than 0.1 percent of the population (behind Buddhism, Taoism, Christianity, and Islam). In the same year, the country with the strongest Confucian influence was South Korea, with 82 percent of all surviving Confucianists.[3]

The People's Republic of China has become somewhat more tolerant of Confucianism in the twenty-first century, although it is still not one of the five religions officially recognized by the government. Confucianism as an important part of China's cultural heritage has received some increasing recognition.

JAPANESE SHINTOISM AND THE HOLY EMPEROR

More than two thousand years ago, a polytheistic branch of religion began to develop in Japan that became known as *Shintoism* (a denominational name used mainly by academics, but rarely by adherents). The earliest history of Shintoism is unclear, and its placement as a possible ideological branch of Taoic Religion is provisional (indicated by a question mark on my phylogenetic tree on page 141). As currently understood by historians, Shintoism had no identifiable founder, no early sacred scriptures, and no early formal doctrine.[9] Although the term "Shinto" wasn't coined until the sixth century CE,[9] Shintoism's systematized worship of *kami* (Japanese animistic spirits or deities discussed later in this section) extends at least as far back as the fourth century BCE.[133] Later institutionalized forms of Shintoism were strongly influenced by Buddhism, Taoism, and Confucianism from China. The detailed records of early Japanese religious belief systems remain obscure, and even the use of the name "Shinto" remains controversial.[134]

Shintoism can be broken down into five basic groups.[9,10] These include the following: (1) early, pre-Buddhist, pre-Taoist, phase of Shinto called *Koshinto* (or *Old Shinto*) with many subsects; (2) *Koshitsu Shinto* (*Imperial Shinto* or *State Shinto*); (3) *Jinja Shinto* (*Shrine Shinto*) with many subsects; (4) *Kyoha Shinto* (*Sect Shinto* or *Shuha Shinto*) with at least thirteen subsects; and (5) *Minkan Shinto* (*Folk Shinto* or *Minzoku Shinto*), including many subsects.

I include Shintoism provisionally as a Taoic branch of Asian Cyclicism on my phylogenetic tree (page 141) because of the general characteristics that it shares with other Taoic schools (e.g., strong emphasis on ancestor veneration, historical veneration of emperors, prolific use of shrines, less concern with the afterlife than the present life). The name *Shinto* is a combination of the Chinese characters *Shen* (divine being) and *Tao* (way). However, if I have correctly classified Shintoism as a Taoic derivative, then it appears to have lost the strong cyclic nature that is basic to most other branches of Asian Cyclicism, although a yinyang influence was brought to Japan from Korea at some point before the seventh century.[135] Reincarnation became a significant part of some Shinto subbranches only after the sixth century CE, because of the strong influence of Buddhism.[72] After further study, early Shintoism, or at least certain early subgroups of what we currently include in Shintoism, might turn out to show a closer affinity to one or more non-Taoic branches of Organized Religion.

The indigenous core of Old Shinto (*Koshinto*) on the Japanese archipelago was tied to veneration of *kami* and ancestor worship. Although it probably originated before the third century BCE,[133] the details of its origin and earliest history are clouded in the mists of Japanese prehistory.[9] Although humans lived on the island of Japan as of thirty thousand years ago, we know virtually nothing of the earliest cultures there. The period between 30,000 BCE and 400 BCE is vast, and scholars debate what happened during that time based on precious little evidence. The earliest archaeological evidence of Japanese prehistory that could reasonably be linked to the veneration of kami is from the *Yayoi* period, around 400 BCE.[136] This was shortly after agriculture and iron technology had been introduced to the Japanese archipelago, and well before there was any centralized Japanese authority among its many tribes and kingdoms. The Yayoi people are thought to have originally come from China and Korea, which may explain some of the core similarities between Shinto and certain Chinese Taoic traditions.

The Japanese reverence for the *kami* is thought to have originated from early animistic influence.[72] Based on archaeological evidence, early traditions saw kami as spiritual aspects of nature that generally were not anthropomorphized (e.g., mountains, the weather, and crops). The kami were initially imagined as invisible spirits without permanent dwelling places. It is thought that the worship of kami in shrines eventually led to their increased anthropomorphization with human traits.[133]

Most kami are not all-powerful or all-knowing, but most are wise and regarded with the utmost respect in Shintoism. Some are the spirits of ancestors who were able to embody the virtues of kami in life. Some were past emperors. Many kami have extraordinary powers that they use to help humans (e.g., the kami of wisdom, *Omoikane*), whereas others are destructive and inflict hardship (e.g., the kami of smallpox, boils, and contagious diseases, *Hoso-no-kami*). Some live in heaven and some live on earth. All kami require worship. The main qualifications for something to be considered a kami are relatively broad, including most anything that inspires awe in the Japanese people.[137]

Early kami rituals consisted of asking the kami for help with the growth and harvest of crops, particularly of rice. Such rituals were a symbol of power and strength for the early emperors.[138] Ancestral spirits of humans were continuously added to the mix along with other deities or aspects of deities. The total mix resulted in a vast pantheon of spiritual entities. Each village had its own special group of divine forces protecting it. A text from 712 CE on Japanese mythology called the Kojiki, estimated there to be as many as eight million different kami.[139] Two kami are even creator gods (something lacking in most of the rest of Asian Cyclicism). Some kami are more anthropomorphized than others. Among the most important were the following personified deities.

- *Izanagi* and *Izanami* were the Shinto creator gods who created the world and the rest of the gods. They were brother and sister, husband and wife, and first man and woman. They were the eighth pair of primordial beings before creation. One day when Izanagi washed his eyes, the sun goddess *Amaterasu* came from his left eye, the moon god *Tsukiyomi* came from his right eye, and the storm god *Susanoo* came from his nose.[127] The three new kami became known as the "three precious children" (i.e., the divine triad of kami offspring from Izanagi).
- *Amaterasu* was the sun goddess, ruler of heaven, and most important divinity in the Shinto pantheon. The first emperor of Japan, Jimmu, was believed to be Amaterasu's descendant, as were later emperors, encouraging veneration of them as *arahitogami* (kami who are also human).
- *Tsukiyomi* was Amaterasu's brother, and god of the moon who rules the night. He killed *Ukemochi*, goddess of food, because he was disgusted by the way she produced food (by spitting it out or defecating it). Amaterasu was so angered by Tsukiyomi's violent behavior that she vowed never to see him again. This is why the day and the night are never seen together.
- *Susanoo* was Amaterasu's other brother and storm god. In some stories, he was impetuous and eventually banned from heaven by the other gods for bad behavior. In other stories, he is considered a hero rather than a testy villain.
- *Ninigi-no-Mikoto* was grandson of the goddess Amaterasu and the great-grandfather of Japan's first emperor, Jimmu. He was sent from heaven to bring rice and order to earth. His divine ancestry established the supreme position of the Japanese Imperial Family.

Through the early centuries CE, indigenous religions of the Japanese archipelago were greatly influenced by mainland Asian traditions. The level of ideological syncretization between proto-Shinto and Buddhism during the fourth through sixth centuries CE eventually led to the development of institutionalized Shintoism (*Sinoshinto* on my phylogenetic tree on page 141). Organized Shintoism never went very far beyond the Japanese archipelago. Its strong connection to imperial power, and its veneration of ancestors, are two of its most prominent characteristics. It focuses on tangible benefits that promote life in the human world. The earliest major texts of Shinto beliefs are from the early eighth century CE. They include the *Kojiki* (the *Record of Ancient Matters* mentioned earlier)[140] and the *Nihongi* (Chronicle of Japan, written in 720 CE).[141] These ancient texts begin with the Japanese story of creation. They also describe various kami and provide an institutionalized mythology for Shinto religious tradition. They thoroughly glorify imperial power and the Japanese emperors (beginning with Jimmu). Jimmu (supposedly from the 6th century BCE) was the legendary (possibly mythological) first emperor who was proclaimed to be a descendant of the sun goddess Amaterasu. He and all Japanese emperors that came after him were to be venerated as living gods descended from the goddess.[9] It is one of many examples of how religion and politics have often been symbiotically entwined

During the mid-seventh century CE, a legal code called *Ritsuryo* was adopted to formalize a central bureaucratic government in Japan with the emperor at its head and a caste system below. Some scholars consider the implementation of the Ritsuryo system as the real origin of institutionalized Shintoism.[133,136] The caste system of Ritsuryo placed government officials and samurai at the top, and slaves and undertakers at the bottom. It wasn't until 1871 that the Shinto caste system was officially abolished. However, the Ritsuryo veneration of imperial power continued with an imperial office in charge of worship and ritual called the *Jingikan* (the Ministry of Kami Ritual). The Ritsuryo system was a strong theocracy, intimately aligning imperial rule with indigenous religious tradition. It exemplified one of history's strongest symbiotic relationships between religion and imperial power.

Shinto was made the state religion of Japan in 1868 during the Meiji Restoration (the formal establishment of a single Japanese empire). This brought order and consolidated the political system under a single emperor of Japan. The country was previously divided by a number of Shogun military dictators. State Shintoism encouraged extreme reverence toward the emperor. Even into the early twentieth century, Shintoists continued to believe that the Japanese imperial lineage was descended from the sun goddess *Amaterasu*.[42] This belief continued to fuel an immense sense of patriotism and national loyalty. In one of the most extreme examples, the State Shinto veneration of the emperor was used during World War II to

Jimmu, legendary first emperor of Japan, descendant of the Shinto sun goddess, *Amaterasu*. Also believed by orthodox Shinto in the past to be the ancestor to all subsequent Japanese emperors. Image from late-nineteenth-century Japanese artist, Tsukioka Yoshitoshi (1839–1892).

create an elite group of suicide bombers who flew bomb-filled planes into American warships. They were named after the Shinto god of wind, *Kamikaze*.[9] A Shinto legend says that the god Kamikaze blew away Mongol invasions when they tried to attack Japan in 1274 and 1281 CE. The emperor used the story of that legend to inspire young Japanese pilots to become "living bombs" in the twentieth century. The suicide pilots (known as Kamikaze pilots) saw themselves in the service of a living descendant of God. As sometimes happens, the close connection between imperial governance and religion resulted in a lethal manifestation of extremism.

Shintoism remained a state religion until the end of World War II, when the occupying military government, led by the United States, severed the intimate ties between the government of Japan and that religion. At the request of the Allied occupation force, Emperor Hirohito was obliged to deny that he was a divinity, which opened the way for the new Japanese constitution prohibiting the government from becoming embedded in religious matters. After the formal connection between the government and Shintoism was officially severed, the number of Shinto followers decreased significantly. In 1910, Shintoism had included just over 7.61 million followers which was about 15 percent of the population at that time.[3] By 2010, Shintoism had decreased to around 2.76 million followers, which was about 2.2 percent of the population. Japan, however, still has more than one hundred thousand Shinto shrines.[3] Post–World War II, Shinto is sometimes referred to as *Shrine Shinto* or *Sect Shinto*. Many Japanese people today practice a mix of Shinto and Buddhist belief systems (e.g., Shinto traditions for weddings, and Buddhist traditions for funerals).[9]

FINAL THOUGHTS ON *ASIAN CYCLICISM* AS A SISTER GROUP TO *AFRO-EURO-MEDITERRANEAN ORGANIZED RELIGION*

This chapter has taken a broad, somewhat breviate look at what I term *Asian Cyclicism*, summarized by the trees on page 70 and elsewhere throughout this chapter. This group includes *Dharmic Religion*, which first arose in South Asia (the Indian subcontinent), and *Taoic Religion*, which first arose in East Asia (China). The provisional placement of Taoic Religion and Dharmic Religion as sister groups needs further study and testing, because little is known about the earliest roots of these two major belief systems (as indicated by the question mark on my phylogenetic tree).

Many branches of Asian Cyclicism have a historical core of animistic polytheism (worshipping many gods or aspects of a single god or worshipping aspects of natural forces). This may have originally developed out of earlier animistic/polytheistic roots in Southern or Western Asia.[5] It may even have had roots in the hypothetical Proto-Indo-European culture that is the focus of the next chapter. Some

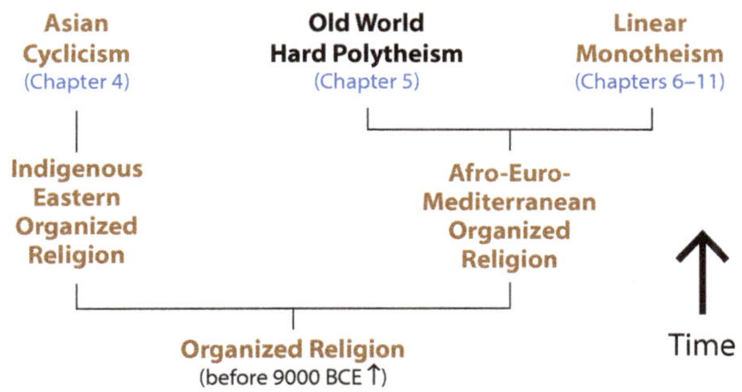

anthropologists report possible Dharmic beliefs to be as old as 9000 BCE,[14] and intricate cave paintings of a variety of animal-like forms, including mythical beasts such as the unicorn, go back more than thirty thousand years on the Indian subcontinent where Dharmic Religion evolved.[5] However, it is difficult to interpret belief systems from archaeological artifacts alone. Virtually nothing is known about the supernatural belief systems of the earliest Asian people, and there is still much to learn.

Asian Cyclicism is one of the earliest-known forms of *Indigenous Eastern Organized Religion*. I term its sister group as *Afro-Euro-Mediterranean Organized Religion* (see the accompanying phylogenetic tree), which appears to have begun as what I term *Old World Hard Polytheism*, which is the subject of the next chapter.

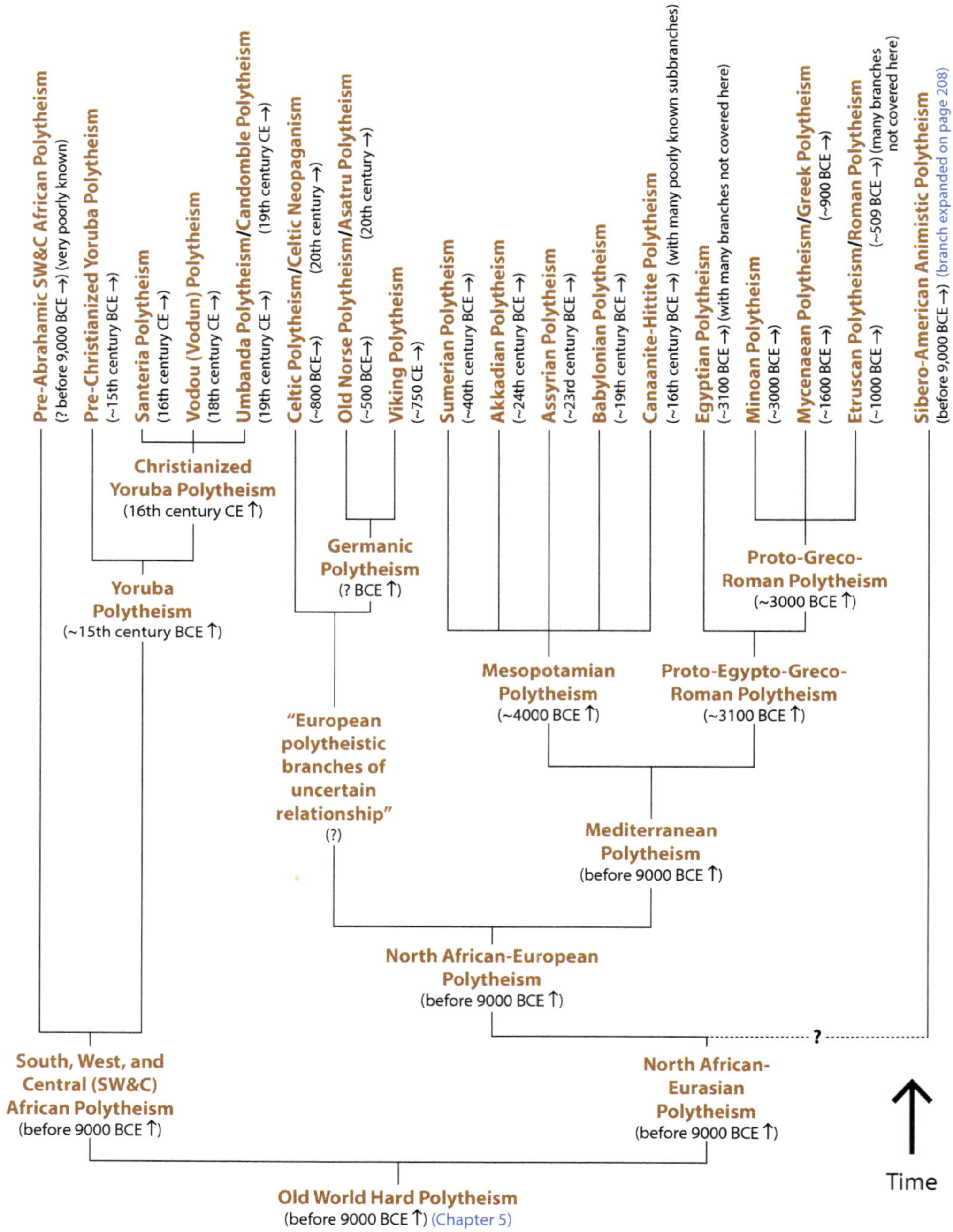

Pre-Abrahamic SW&C African Polytheism
(? before 9,000 BCE →) (very poorly known)
Pre-Christianized Yoruba Polytheism
(~15th century BCE →)
Santeria Polytheism
(16th century CE →)
Vodou (Vodun) Polytheism
(18th century CE →)
Umbanda Polytheism/Candomble Polytheism
(19th century CE →) (19th century CE →)
Celtic Polytheism/Celtic Neopaganism
(~800 BCE→) (20th century →)
Old Norse Polytheism/Asatru Polytheism
(~500 BCE→) (20th century →)
Viking Polytheism
(~750 CE →)
Sumerian Polytheism
(~40th century BCE →)
Akkadian Polytheism
(~24th century BCE →)
Assyrian Polytheism
(~23rd century BCE →)
Babylonian Polytheism
(~19th century BCE →)
Canaanite-Hittite Polytheism
(~16th century BCE →) (with many poorly known subbranches)
Egyptian Polytheism
(~3100 BCE →) (with many branches not covered here)
Minoan Polytheism
(~3000 BCE →)
Mycenaean Polytheism/Greek Polytheism
(~1600 BCE →) (~900 BCE →)
Etruscan Polytheism/Roman Polytheism
(~1000 BCE →) (~509 BCE →) (many branches not covered here)
Sibero-American Animistic Polytheism
(before 9,000 BCE →) (branch expanded on page 208)
Christianized Yoruba Polytheism
(16th century CE ↑)
Germanic Polytheism
(? BCE ↑)
Proto-Greco-Roman Polytheism
(~3000 BCE ↑)
Yoruba Polytheism
(~15th century BCE ↑)
Mesopotamian Polytheism
(~4000 BCE ↑)
Proto-Egypto-Greco-Roman Polytheism
(~3100 BCE ↑)
"European polytheistic branches of uncertain relationship"
(?)
Mediterranean Polytheism
(before 9000 BCE ↑)
North African-European Polytheism
(before 9000 BCE ↑)
?
South, West, and Central (SW&C) African Polytheism
(before 9000 BCE ↑)
North African-Eurasian Polytheism
(before 9000 BCE ↑)
Time
Old World Hard Polytheism
(before 9000 BCE ↑) (Chapter 5)

5

AFRO-EURO-MEDITERRANEAN ORGANIZED RELIGION, BEGINNING WITH OLD WORLD HARD POLYTHEISM

Gods always behave like the people who make them.

—ZORA NEALE HURSTON

Afro-Euro-Mediterranean Organized Religion is the subject of chapters 5–11. This diverse group of organized traditions appears to have begun with what I term *Old World Hard Polytheism* (the subject of this chapter). As discussed in chapter 3, hard polytheism is the belief in the existence of multiple gods or goddesses as independent entities that are more than mere aspects or avatars of a single god. Institutionalized hard polytheism is first clearly documented in Northern Africa and Eurasia (hence the formal name Old World Hard Polytheism at the base of the tree on the opposing page).

Old World Hard Polytheism generally lacks the cyclic philosophy of samsara (rebirth) or yinyang present in Asian Cyclicism (chapter 4). In most of Old World Hard Polytheism, you are born once and you die once. After death, you either cease to exist completely or you move on to an afterlife with varying degrees of desirability. Afterlife can take the form of eternal joy, eternal suffering, or eternal boredom, depending on the ideological branch. If we accept that Gobekli Tepe (page 63) represents an early occurrence of *Mediterranean Hard Polytheism*, then this ideology dates back at least to the tenth millennium BCE. Based on its widespread pre-Columbian presence in both the Old World and New World (a logic further explained in **Note 5A**), it could well be even older.

Theoretically, elements of Old World Hard Polytheism arrived in the New World (i.e., the Americas) along with the earliest humans, making New World Hard Polytheism a descendant of Old World Hard Polytheism. Although *Homo sapiens* evolved in the Old World by at least three hundred thousand years ago,[1] our species appears not to have begun arriving in the New World until around thirty thousand

Timeline for certain events, theories, and artifacts of Old World Hard Polytheism

Animistic Polytheism in the Americas begins to be largely displaced by Christianity, as a result of European conquest and colonization between the 15th and 18th centuries CE	~1492 CE↑
Roman Polytheism begins to be displaced after Roman Emperor Theodosius I and co-emperors Velentinian and Gratin make *Nicaean Christianity* the State Religion of the empire and declare polytheism as heretical and illegal	~380 CE↑
Roman Polytheism grows and expands by symbiotically intertwining, first with the powers of the Roman Republic, and then with the Roman Empire. As it grows, it "Romanizes" deities from other branches of *Ancient Mediterranean Polytheism* (e.g., *Egyptian, Mycenaean, Greek,* and *Eutruscan*)	~509 BCE↑
Grecco-Roman Polytheism develops, influenced by *Minoan, Egyptian,* and other earlier polytheistic traditions	~1600 BCE↑
Date of the earliest known archaeological artifact of the Andean Staff God. From the ancient Norte Chico civilization of present-day coastal Peru	~2250 BCE
The oldest known texts of Egyptian polytheistic traditions (the Egyptian Pyramid Texts) are carved into the stone walls of pyramids in Lower Egypt	~2350 BCE
Early Sumerian texts written on clay tablets mention several gods of the Mesopotamian pantheon	~2500 BCE
Oldest known Mesopotamian temple; built by ancient Sumerians to the god Enki	~4000 BCE
Land connection between Asia and North America disappears (*Beringia*, the land connection that brought the first humans to the Americas at least twenty eight thousand years ago)	~9000 BCE
Gobekli Tepe, oldest known human-built temple, was built in what is today Anatolia Turkey, suggestive of early *Mediterranean Polytheism*	~9130 BCE

years ago.[2] Significant numbers of humans did not arrive there until around fifteen thousand years ago. The earliest migrants are thought to have come from Siberia, probably bringing their indigenous religions with them (discussed later in this chapter).

Each branch on the phylogenetic tree of Old World Hard Polytheism (tree on page 158) was influenced by other aspects of the regional culture in which it originated. We know very little about the earliest denominational diversity within Old World Hard Polytheism, and we don't know how those belief systems self-identified. Consequently, most of the branches on the phylogenetic tree at the beginning of this chapter are provisionally named after the cultures or regions from which these branches are best known, rather than after distinct religious doctrine or iconic inspirational figures. Those branches represent only highly simplified and condensed versions of the diversity that actually existed. However, they provide a way to begin a systematic discussion of the most basic early diversification of Old World Hard Polytheism. Much further descriptive, comparative, and systematic work is needed to clarify the vast ideological diversity within this group (discussed further in **Note 5B**).

Some anthropologists and historians believe that early hard polytheism may have originated from animistic beliefs, in which many things (e.g., the sun, moon, animals, trees, and forces of nature) had a spiritual essence or soul.[3,4] Some theoretical psychologists similarly propose that the families of gods "were natural intellectual developments from the more generalized animisms and pantheisms of the prehistoric past."[5] This theory would explain why many early forms of hard polytheism are difficult to distinguish from pure animism. With time and increasing anthropomorphism, the deified entities developed more human-like features, increased in number, and eventually became hierarchically organized. Special gods, or sometimes a single god, often claimed supremacy of the pantheon (the god assemblage). Higher-level gods were often associated with the sun and became depicted with sun-like halos or glowing crowns. For many early forms of anthropomorphized supernaturalism, the father persona generally became associated with the sky, and the mother persona generally became associated with the earth.

Eventually, much of Old World Hard Polytheism took a supermarket approach to deities. You could choose from many different gods and pantheons, including deities for music, arts, love, death, heroism, politics, science, farming, water, and hundreds of other endeavors or natural wonders. The gods could be cruel, kind, mischievous, or indifferent. When you had a particular problem, you could pray to a particular deity who was a specialist with your area of need. This trend toward deity specialization influenced later religions that considered themselves monotheistic. Even the nuns in the Roman Catholic grade school that I attended as a child suggested praying to Saint Anthony to find something lost, to Saint Christopher for safe travel, to Saint Michael for protection from danger, and so on. Then, of course, Saint Nicholas was patron saint of children. His legendary habit of secret gift-giving gave rise to the myth of Santa Claus, by way of the Dutch *Sinterklaas*. Perhaps a belief in spiritual specialization is part of human nature.

The term "*pagan*" was coined from Latin slang in the fourth century CE. Nicene Christians used it as a pejorative for followers of Old World Hard Polytheism (as well as for practitioners of several other non-Abrahamic religions). It was a "*them* versus *us*" term that early Christians used as part of Christian self-definition. It was also a somewhat-elitist term, given that the Latin meaning was "peasant." In a sense, the popularization of the word came to represent a singular, long-lasting victory of Abrahamic Monotheism over Old World Hard Polytheism.[6,7] Early Christians were not alone in the practice of recognizing their faith as truth and in lumping other religions together in a derogatory manner. Jews use the term *goy* for those who are not Jewish, and Muslims use the term *kafir* for those who are not Muslim. The ancient Greeks used an early form of the term atheist (*atheos*) as a derogatory term for those who were impious. Later Greek literature more explicitly used the term for "lacking in the belief in the gods." By the fifth century CE, Chalcedonian Christians appropriated that term and changed its meaning to refer to someone who did not believe in the Abrahamic God. This all points to an underlying exclusivist element of Organized Religion with which society struggles: a persistent *in-group* exclusiveness that is often tied to prejudice against *out-groups*.

This chapter focuses on major subgroups of Old World Hard Polytheism through time to illustrate its broadest diversity. However, my phylogenetic trees represent only a preliminary attempt to do this, with minimal supportive characters, limited contemporaneous records, and many missing pieces. The historical relationships indicated on these trees rely heavily on information from the History of Religion literature cited in the text, which is often short on unambiguous shared character information to supplement implied patterns of relationship. I leave the more detailed comparative study of the many rituals, doctrines, pantheons, state gods, and sacred scriptures to specialists in the comparative study of polytheistic religions.

Many of the earliest forms of Old World Hard Polytheism are unknown or cannot easily be placed into an evolutionary context with other religions. Records of these belief systems in their original, unadulterated form are severely lacking. Existing records are mostly those that have been blended (syncretized) with, or strongly influenced by, the religious traditions of migrants, invaders, and colonizers. This is particularly true for three provisionally placed large branches on my phylogenetic tree: *South, West & Central African Polytheism*; *European polytheistic branches of uncertain relationship*; and *Sibero-American Animistic Polytheism* (see page 158).

Many people in the Congo, for example, are baptized and attend services in a Christian Church, yet they still believe in witches and perform sacred rituals to their ancestors to ward off evil spirits.[8]

Old World Hard Polytheism was the major religion of the world for thousands of years until Abrahamic Monotheism eventually eclipsed it. The evolution of Old World Hard Polytheism is as at least complex as that of Asian Cyclicism (discussed in the last chapter) or of Linear Monotheism (examined in chapters 5-11). The constant processes of innovation, reorganization, syncretization, and assimilation among the different pantheons of gods and practices would have created countless subbranches over time. Most of these ancient belief systems were lost to history without leaving any clear, distinguishing records. I present a simplified generalization.

THE POORLY UNDERSTOOD *SOUTHERN, WESTERN, AND CENTRAL AFRICAN POLYTHEISM* (NON-EGYPTIAN, AFRICAN POLYTHEISM)

The hard polytheism that arose in Africa is subdivided into two broad groups on my phylogenetic tree at the beginning of this chapter. The better-known group is *Egyptian Polytheism*, a subcategory of *Mediterranean Polytheism* discussed later in this chapter. The other, lesser-known group is the subject of this section: *Southern, Western, and Central African Polytheism* (*SW&C African Polytheism* for short). SW&C African Polytheism includes the forms of African Polytheism other than those of Egypt. Its diversity, as presented, is woefully incomplete because of the paucity of broad comparative knowledge about the pre-Christian polytheistic belief systems from the SW&C African region. I name this group only provisionally, as a questionably monophyletic group of convenience, for the sake of representation.

There are no known ancient books or contemporaneous written records for the earliest forms of SW&C African Polytheism.[9] There is very little interpretable archaeological information about these forms of polytheism as well. Early cave paintings of therianthropes (mythical beings which are part human and part non-human animals) over thirty thousand years old have been reported from southern Africa.[10] The intended nature of this imagery cannot be established, because it was well before the existence of any decipherable written language. Because the earliest humans came from this region, and because forms of hard polytheism were so widespread in Africa (particularly in the north), it is reasonable to infer that hard polytheism had deep roots on the continent, and perhaps the therianthropes were part of them. The current information that we have about most early forms of SW&C African Polytheism (or early indigenous religions of SW&C Africa, in general) is speculative and often subjectively distorted. The known written records of precolonial SW&C African religions were reconstructed and transformed after centuries of Christianization and Islamization in that region. Today, the dominant religions of the northern half of the African continent are branches of Islam, and the dominant branches of the southern half of the continent are Christian. Many details about the original religious belief systems of Africa have either been lost or modified by the chroniclers of history. As anthropologist Robert Horton warns, many studies of African religions have been distorted by Western Christian theologians who are "strongly influenced, first and foremost, by their own Christian faith."[11] The relationships of early SW&C African Polytheism are in need of much further study and discovery by specialists in the field of African religions who are willing to look for unbiased, pre-Westernized records of the early indigenous belief systems.

Some authors even object to treating African belief systems *as* religions, based on supposed moral objections.[12] However, such objections are often due to the author equating classification with colonialist progressivism, or to the author mixing elements of supernatural belief with other aspects of culture. The word "religion" is largely a Western term; however, the way I use the term (as "supernaturalism" or "supernatural belief systems") provides a consistent analytical and nonjudgmental context for the broad comparative study of human belief systems. With this focus on shared similarities and historical connections rather than differences, data-based hypothetical patterns of historical relationship can be discovered. Hopefully, with deeper comparative studies, and with new discoveries of pre-Christianized, pre-Islamized forms of SW&C African Polytheism, our knowledge about its diversity and group interrelationships will improve.

Many indigenous religions and their unadulterated historical records were lost or transformed to monotheism through centuries of removing entire African communities for the slave trade. As early as the eighth century CE, Arabia was enslaving massive numbers of people from eastern Africa. Later, Western Europeans began enslaving even larger numbers of Africans for labor in the New World. Historians estimate that between 1450 and 1900, the Europeans forcibly removed more than twelve million people from central Africa, and sent them to the Americas and the Caribbean to be enslaved.[13] The result of aggressive colonialism, the slave trade, and cultural assimilation, resulted in the loss of many or most indigenous African religious traditions. Those who were enslaved often were forced to adopt the Judeo-Christian monotheistic beliefs of the slaveholders in an attempt to quash native African religions. By the nineteenth and twentieth centuries, there was a cultural backlash to this by Black nationalists. New syncretized groups, such as *Rastafarianism* (or *Rastafari*), arose in the 1930s, to assert the importance of African heritage within the Abrahamic context that they syncretized with. Although monotheistic, Rastafarianism identifies indigenous Africans as the true biblical Jews, superior to whites. It identifies Africa as the Jewish homeland.[14] They reject many aspects of European culture, and reject more traditional Christian churches as "Babylon." Rastafarianism is largely confined to Jamaica, although elements of its message is carried around the world by traditional reggae music.

A few pockets of SW&C African Polytheism resisted complete assimilation and survived to modern times, although they often were still influenced or modified by some degree of Christian or Islamic influence. One better-known example, whose history goes back thousands of years, is the highly animistic *Yoruba Polytheism*.[15] The homeland of the Yoruba belief system is Nigeria, although forms of it eventually spread to Brazil, the Caribbean, and a few cities in North America.[16] It is a branch of religion that believes in reincarnation.[17] Its pantheon included six hundred deities, with four hundred benevolent powers referred to as the *Orisa*, and another two hundred malevolent powers referred to as the *Ajogun*.[18] The apex of the Yoruba pantheon is known as *Olodumare* (also known as *Olorun* or *Ol-orun*). For most Yoruba, this Olodumare is androgynous, being at the same time both male and female. Olorun was not originally the creator god, but instead is a remote, distant entity who rules indirectly from on high and is not in direct communication with any humans. Later Christianization efforts on Yoruba beliefs turned Olodumare into the Abrahamic creator god, with a male persona.[18]

Two of the Yoruba deities responsible for the intercommunication between humans and the pantheon are *Eshu* and *Orunmila*. Together, these two enable communication between orun (heaven) and aiye (earth). *Eshu* is the god of chance, accident, chaos, and unpredictability. He is also a messenger to Olodumare and reports to him on the actions of humans and the other gods. Eshu is a highly popular god, and his likeness can be found in almost every Yoruba home.[17] People invoke him to harm their enemies, and he is known for his wickedness as well as for his ignorance. He often stirs up trouble, an aspect of his character as divine trickster. He was later deemed *Satan*, *Lucifer*, *Demiurge*, or *Iblis* by various Christian missionaries and Islamic dawat in an attempt to discredit him.[19] *Orunmila*, the orisha of wisdom, oversees communication in a different direction than Eshu. He is the deliverer of messages from the heavens to humans.[17] For some Yoruba, Orunmila is second only to Olodumare within the pantheon, whereas for other Yoruba, Eshu is second in the pantheon. These ideological subbranches have yet to be clearly defined.

Other gods include *Shango*, the divinity of thunder and lightning whose servants are the wind and the rainbow. He represents divine wrath and is much feared. He can also provide peace, prosperity, and protection against danger.

Osun (also known as *Oshun*) is his wife and the goddess of love, beauty, and the River Osun. She is the custodian of fertility and the goddess of destiny. *Obatala* is the creator god who fashioned the first humans out of clay so that Olodumare could give them the breath of life. The pantheon has many other divinities, numbering as many as 1,700, according to some authors.[19,20] Anthropomorphic iconography of deities, however, appears to have been extremely rare within Yoruba Polytheism.

Although the Yoruba religion was originally a form of animistic hard polytheism with a Supreme God and a large pantheon, it was strongly affected by the slave trade in Africa and the resulting Judeo-Christian and Muslim influences. One such example is the *Santeria* religion that evolved in Cuba during the sixteenth century as a syncretization between elements of Yoruba Polytheism from Africa and Roman Catholicism introduced by Spanish colonialists.[21] Later branches that evolved from Yoruba Polytheism under the influence of Roman Catholicism included *Umbanda* and *Candomble* (in Brazil in the nineteenth century) and *Vodun* or *Vodou* (in Haiti in the eighteenth century). To retain elements of their traditional beliefs, syncretized branches of Yoruba often treated their pantheon of lower deities as if they were Christian saints. They also designated a Supreme God who was also a creator god, such as *Bondye* in Vodou.[22] For the first half of the twentieth century, the Roman Catholic Church launched an ineffective campaign, aided in 1941 by government forces, to destroy Vodou temples.[14] Then in 1957, the Duvalier dynasty courted Vodou for political support in its rule of Haiti. Today, the numerous branches of Yoruba religion collectively have about twenty million members, most of which still show strong influence from Roman Catholicism.

We know very little about most of the polytheistic branches and pantheons in the SW&C African region, other than many of them were highly animistic. Indigenous polytheistic branches in Africa were deemed "heathen" by early Christian and Muslim missionaries, who did much to convert, and subsequently eliminate clear records of, the original indigenous belief systems. Even the modern diversity of indigenous religions in Africa is undoubtedly much higher than we know, particularly given the private nature of many social groups there, and the fact that the indigenous African people speak more than two thousand languages.[23] Details of that diversity, particularly in its original, unadulterated indigenous form, are well beyond the scope of this book. Broadly comparative work with SW&C belief systems warrants much further research. It may contain elements of not only the mother religion of Old World Polytheism, but also certain roots of Organized Religion in general, because modern humans first evolved in Africa.

The sister group of SW&C African religion, *North African-Eurasian Polytheism*, has a much denser contemporaneously transcribed record.

NORTH AFRICAN-EURASIAN POLYTHEISM: AN EARLY GLOBAL WAVE OF SUPERNATURAL IDEOLOGIES

North African-Eurasian Polytheism (the formal name introduced here) includes two major subbranches: *North African-European Polytheism* and *Sibero-American Animistic Polytheism* (as illustrated on the tree at the beginning of this chapter). Although I have theorized both to have originated from elements of Old World Hard Polytheism, they diverged substantially from each other because they became geographically isolated from each other by the Pacific Ocean beginning around eleven thousand years ago, and remained so for several thousand years. Human interchange between Eurasia and North America was curtailed after the massive land bridge known as *Beringia*, which connected Siberia to North America for tens of thousands of years, submerged far below the surface of the Pacific Ocean (discussed later in this chapter). Loss of this land connection resulted in geographic isolation between new- and old-world cultures for thousands of years, until the fifteenth century when large seafaring ships opened an easier conduit for intercontinental migration once again. In general, much of New World Animistic Polytheism appears to have remained or become more animistic than North African-European Polytheism. In contrast, the pantheons of most of North African-European Polytheism (e.g., Greco-Roman) became more anthropomorphized and hierarchically organized very early in their history and eventually were largely replaced by monotheistic belief systems during the mid- to late first millennium CE. I provisionally place *North African-European Polytheism* and *Sibero-American Animistic Polytheism* as sister groups because of their hypothesized common evolutionary origin from early *North African-Eurasian Polytheism* (see the phylogenetic tree on page 158).

The branches of North African-European Polytheism are much better known than the early branches of Sibero-American Animistic Polytheism and New World Animistic Polytheism. This is partly due to the remote sparseness of early Siberian civilizations and the scarcity of records regarding them. It is also partly an artifact of the post-Columbian conquest of the New World by old-world Christian forces. Indigenous religions in the Americas, as well as records of their early history, disappeared or were radically changed as they gave way to European missionaries forcefully proselytizing Western Christianity. To create a Roman Catholic Empire in the New World, sacred spaces, traditions, and histories of indigenous religions (e.g., the polytheistic branches of the Inca, Maya, and Aztec) were systematically transformed or eliminated. Historical records of early New World Animistic Polytheism were lost or altered to present them as savage or demonic.[24] It was similar to the loss of historic records of early African religions as a result of conquests of

Western Christianity and Islam. It was also similar to the loss of historic records of ancient Siberian religions as a result of Islamic and Mongolian conquests in the sixteenth century[25] and forced conversion to Eastern Orthodox Christianity by the Russian conquest in the seventeenth and eighteenth centuries.[26] Historically, forced conversion and assimilation have been common strategic elements to control the hearts and minds of conquered people. Most surviving historical records are the products of the conquerors. Historical records of pre-Columbian indigenous New World religious diversity have been largely lost or have been selectively preserved to emphasize its darker elements (e.g., human sacrifice). In the words of History of Religion scholar Wendy Doniger, "Often the future is shaped not by what we remember, but what we forget."[27] Before exploring *Sibero-American Animistic Polytheism* much later in this chapter, I will begin with the better-known *North African-European* Polytheism and its main derivatives.

NORTH AFRICAN-EUROPEAN POLYTHEISM: A HYPOTHETICAL IDEOLOGICAL ANCESTOR

The study of *North African-European Polytheism* includes the study of both preliterate and postliterate societies. Deciphering the aspects of preliterate cultures can be extremely challenging, because without written records, we are dependent on oral transmission and archaeological artifacts. Both of these sources are often insufficient to accurately document the details of cultural history, religious or otherwise. In a few cases, hypothesized ancestral societies and their cultures have been reconstructed by archaeologists, based on shared characteristics among more recent cultures. Scholars that do this are sometimes referred to as archaeo-reconstructionists.

In anthropology, many early archaeo-reconstructionists were comparative language specialists known as *archaeo-linguists*. They reconstructed theoretically ancestral languages (and other major aspects of culture) by identifying words and roots that are common to many different languages (i.e., linguistic synapomorphies).[28,29] A classic example of archaeo-reconstruction relevant to the radiation of early Organized Religions is the hypothetical Proto-Indo-European (PIE) civilization. PIE was proposed as an ancestral Eurasian culture whose theoretical common language dates back as far as 9,500 years (or perhaps even farther).[30,31] This language diversified over time and evolved into more than 440 different Indo-European languages still used today. The homeland of the PIE culture is hypothesized by some historians to have been in the region of today's southern Ukraine and Russia, although this is still hotly debated.[28] The PIE culture left no archaeological record. Its hypothesized existence is a reconstruction based on common linguistic elements of the Hindustanis, Mesopotamians, Greeks, Romans, Etruscans, Indians, Celts, Norse, and a few other cultures. These linguistic elements are all hypothesized to have evolved from an ancestral PIE mother language. The original PIE hypothesis based on comparative linguistics was later supported by comparative archaeogenetic studies and DNA analysis of ancient bones.[32] Perhaps one day, the discovery of a significant archaeological site left by the PIE will further validate this hypothesis.

Just as the hypothetical *PIE language* was reconstructed by looking for common shared linguistic elements of later Indo-European cultures, a hypothetical *PIE religion* was reconstructed based on common shared religious characteristics of later Indo-European cultures.[29] The PIE religion is hypothesized to have influenced the early development of both early Indigenous Eastern Organized Religion and Afro-Euro-Mediterranean Organized Religion (*sensu* tree on page 157). This deduction is based on the later widespread occurrence of hard polytheism around the Indo-European region. PIE deities were reconstructed as composites of widely shared gods and goddesses among the later Indo-European branches of religion. Although the names for these hypothetical PIE deities were the creation of scholars, their substance was deciphered by noting recurring themes of divinity among the many different pantheons of later Indo-European religions. For example, the name *Dyeus* was proposed by PIE reconstructionists for a hypothetical sky god that later became the Mesopotamian god *Anu*, the Egyptian god *Horus*, the Greek god *Zeus*, the Roman god *Jupiter*, the Etruscan god *Tinia*, the Vedic god *Dyaus*, and the Celtic god *Latobius*. Similarly, the deity *Hausos* was proposed as the deity of dawn that later became the Mesopotamian goddess *Aya*, Egyptian god *Khepri*, Greek god *Eos*, Roman god *Aurora*, Etruscan goddess *Thesan*, Vedic god *Usas*, Celtic goddess *Brigid*, and Norse god *Dellingr*. The PIE religion is hypothesized to have had an underworld and an afterlife and to have instructed that to reach the underworld, one had to cross a body of water (a belief shared by Mesopotamian, Egyptian, Greek, Roman, Vedic, Celtic, and Germanic religions). The early evolutionary history of the PIE religion is unknown. Although archaeologists believe that the early development of the common PIE language took place around 8,500 years ago, the roots in Organized Religion could possibly be significantly older than that (e.g., the archaeological site of Gobekli Tepe discussed in chapter 3, and mentioned at the beginning of this chapter).

From this point, I subdivide the North African-European branch of polytheism into subbranches: *European Polytheistic Branches of Uncertain Relationship* (including *Celtic* and the *Germanic* branches of polytheism) and *Mediterranean Polytheism* (including *Mesopotamian, Egyptian*, and *Proto-Greco-Roman Polytheism*) as shown by the phylogenetic tree on at the beginning of this chapter). The detailed ideological diversity and interrelationships of many of these (e.g., the branches within Mesopotamian Polytheism) are poorly known and have yet to be clearly delineated. However, their pantheons and practices diversified and changed over time. The scarce records that do exist indicate a diverse assemblage of evolving supernatural belief systems.[33,34]

EUROPEAN POLYTHEISTIC BRANCHES OF UNCERTAIN RELATIONSHIP: CELTIC AND *GERMANIC POLYTHEISM*

Two ancient forms of Old World Hard Polytheism with rich later histories but poorly known origins are *Celtic* and *Germanic* (the latter often referred to as "Germanic Paganism"). They left no written documentation of their beginnings or early relationships to other contemporaneous religions, because writing was ritually prohibited. Interpretation of their belief systems is based on archaeology.[35] Both Celtic and the Germanic Polytheism probably had early roots in the PIE group. They were heavily Greco-Romanized beginning around the first century BCE, and heavily Christianized in the fifth through twelfth centuries CE. I provisionally include them in a non-monophyletic group of convenience that I call *European Polytheistic Branches of Uncertain Relationship*, so named because of uncertainties of where they fit within the phylogenetic tree of Old World Hard Polytheism (page 158).

CELTIC POLYTHEISM: THROUGH THE LOOKING GLASS OF SYNCRETIZATION

Even after the development of writing, the early Celts preferred oral transmission as the primary means of recording their religious belief systems and deities from generation to generation. The written information we have on Celtic Polytheism dates only from well after its Romanization and Christianization, supplemented with scant archaeological evidence.[36,37] The Romans referred to the process of absorbing and modifying other cultures in their own image as "the civilization of barbarians." The result was Celtic Polytheism recast as a Celtic-Roman mix that perceived the non-Roman aspects of Celtic culture as barbaric. Later, the process of Christianization was even more thorough in its obliteration of both native Celtic and syncretized Celtic-Roman elements. It relied heavily on *Interpretatio Christiana*—the practice of transforming native polytheistic rituals, imagery, and places of worship into Christianized forms.[38] The conversion of a Celtic ruler often was followed by the compulsory baptism of his subjects and a rewriting of his kingdom's cultural history. Some historians suggest that some deities of polytheism were transformed into Christian saints to accelerate the process of proselytization (e.g., Celtic goddess *Brigid* into Saint Brigid), although this suggestion is controversial.[39,40,41] Other Celtic gods were demoted to humans with supernaturally gifted characters, such as Merlin and Cuchulain.[42,43]

In a letter to the first bishop of London dated at 601 CE, Pope Gregory I wrote:

> The temples of the idols [deities] in that nation ought not to be destroyed; but let the idols that are in them be destroyed; let water be consecrated and sprinkled in the said temples, let altars be erected, and relics placed there. For if those temples are well built, it is requisite that they be converted from the worship of devils to the service of the true God; that the nation, seeing that their temples are not destroyed, may remove error from their hearts, and knowing and adoring the true God, may the more freely resort to the places to which they have been accustomed.[44]

Statues of some of the old gods had their names changed to the names of Christian saints. Original building infrastructure was retained as though it was created for the Christianity in the first place. Our records of history for overtaken religions, like those for Celtic Polytheism, are often obscured with the aftereffects of assimilation by other cultures. With those qualifications in mind, some of what we have recorded about Celtic Polytheism is as follows.

The Celts were a people of Western Europe whose culture was first documented with artifacts from around the eighth century BCE through the seventh century CE. Some authors consider the origin of Celtic culture to have been much earlier.[45] Dating the origin of Celtic Polytheism is highly speculative at best. The Celts dominated much of western and central Europe in the first century BCE. The earliest forms of Celtic religion were highly animistic and generally did not envision deities in anthropomorphic forms but rather as spirits of nature.[46] Trees had special spiritual significance. Early Celtic Polytheism was like many other branches of polytheism in that it became more anthropomorphized over time, eventually containing more than two hundred deities.[47] Celtic Polytheism had a variety of burial practices indicating a belief in the afterlife, such as burying their dead with food, weapons, and ornaments. Their warriors were headhunters. Archaeologist Barry Cunliffe surmised that the Celts held special "reverence for the head" and "to own and display a distinguished head was to retain and control the power of the dead person."[47]

The images of ancient Celtic deities that exist are largely from Roman conquerors who often modeled the imagery after Greco-Roman gods such as Zeus/Jupiter or Apollo. By 54 BCE, Caesar's army had invaded Britain, and the indigenous Celtic belief systems were beginning to transform due to outside influence. By 61 CE, Roman conquest and assimilation had resulted in much of Celtic Polytheism being refashioned in the image of the Roman pantheons of gods and goddesses. Eventually, with the Christianization of Rome by 450 CE, Nicene Christianity morphed much of the Romanized Celtic religion into a monotheistic state, further clouding its historical roots.

According to Greek and Roman accounts, a powerful priestly caste of magico-religious specialists known as *druids* in the Celtic religion,[41] were the intermediaries between the people and the divinities and were considered wise. They were also transmissionists—that is, they were charged with remembering religious lore and tradition and for passing it down verbally from generation to generation. Some druids were men, and others were women. They predicted the future and presided over blood sacrifice. The name druid means "knowing the oak tree" and refers to the fact that early Celtic rituals were performed in the forest. Druids had great power in early Celtic society. They helped choose kings and controlled the laws and social rules. As far as is known, the Celts had no temples before Romanization of their religious belief system in the first century BCE. Once Rome absorbed the Celtic religion, the practice of building religious temples began. Archaeologists have discovered many of these in Britain as well as in Gaul. While best remembered as religious leaders, the druids were also legal authorities, adjudicators, medical professionals, and political advisers.

With regard to the druid blood sacrifice, Greek and Roman writers (including Julius Caesar) referred to the druids as practitioners of human sacrifice.[48] According to Caesar, those who were found guilty of criminal offences by the Celts were prioritized for use as sacrificial victims. When criminals were in short supply, however, regular citizens were acceptable as victims. One form of sacrifice reported by Caesar was burning victims alive in a large wooden effigy, today known as a "wicker man." It was a large, hollow, wooden framework in the shape of a standing man. It was reported to have been filled with the bodies of living animals and humans, and then torched. Other sacrifices were made by drowning, hanging, and burning, respectively (known as the threefold death). Caesar also reported that people enslaved by and dependents of high-ranking Celts would be burned along with the body of their master as part of the funerary rite. Some of these reports were possibly embellishments resulting from Romanization or Christianization. Historians note that ancient Greeks and Romans often embellished barbaric acts of foreign cultures to confirm their belief in the superiority of Rome and Greece.[49] Because the Celts wrote nothing down about their religious activities, the Greco-Romanized accounts are unfortunately the only contemporaneous written records we have. The Celtic druids are reported to have been literate, but they were forbidden by their own doctrine from recording their history in written form. Archaeologists have unearthed many beautiful and sophisticated objects of art, ornamented weaponry, chariots, and even indications of written language.[41] In spite of what the Greek and Roman writers would have us believe, the Celts were more than barbaric headhunters.

Of the hundreds of Celtic gods and goddesses, many were specific to local regions.[47] To speak of a single "Celtic pantheon" is to greatly oversimplify the diversity of Celtic religious belief systems. Celtic pantheons were decentralized.[50] Their pantheons were not strictly hierarchical the way most other polytheistic pantheons were. As noted earlier, the Celts typically did not depict their gods in human form until they were overcome by the Roman conquest in 58–51 BCE. After that, the different polytheistic branches in Celtic Gaul were significantly transformed and nearly all indigenous records of the pre-Roman Gallic religions were lost. This process resulted in the development of syncretic Celtic-Roman forms of polytheism. Celtic supernatural belief systems continued to transform through Christianization during the fifth and sixth centuries CE, during which time polytheistic elements were phased out. The Celtic people who lived the farthest away from Rome and who maintained political and linguistic identities (such as the Irish-Celts and the Welsh-Celts) fared better in maintaining vestigial records of their early supernatural belief systems. Some of their surviving records fueled a new polytheistic movement in the twentieth century called *Celtic Neopaganism*. Much research is needed to reconstruct the detailed evolutionary history of early Celtic religions.

The stories of more than three hundred Celtic deities have been lost to history through Romanization, Christianization, Syncretization, and other forms of cultural subjugation; all we know of most of them is their name. I mention only a few of the better-known deities. The best-known Celtic deities appear to be from the Irish pantheon, and a few of these may have been pan-Celtic.[41]

- *The Dagda* was perhaps the most important of all Celtic gods. To some he was known as the "father of all" and comparable to the Norse god Odin. He was a chieftain druid who was associated with strength, wisdom, fertility, time, weather, and control over life and death. He held a great club, which suggests parallels with the hammer of Thor. Like many of the Irish gods, he was a generalist, rather than being narrowly focused on only one aspect of nature. He was eventually killed in battle. Even the greatest Celtic gods were mortal.
- *Morrigan* was the frightful goddess of war, death, and fertility. She was also a shape-shifter who could appear as a black crow, an old hag, or a beautiful maiden. She was the most feared deity in the Irish Celtic pantheon. She was often presented by a divine triad of Celtic war aspects (the three Morrigna), including *Badb* (war goddess who appears as a crow), *Macha* (another goddess of war), and Nemain (goddess of sovereignty who confers sovereignty on a king by mating with him).[41]
- *Cernunnos* was one of the better-known Celtic gods, thanks to several recovered pieces of artwork. He was a horned god with the ears and antlers of a stag (page 174). He was also known as *Lord of Nature* and is linked by some historians to the Hindu god *Pashupati*.[41]

Image on the Gundestrop cauldron, a silver vessel excavated in Denmark dated between 150 BCE and 1 BCE. Thought by archaeologists to depict the Celtic god of the beasts, *Cernunnos*, who is often represented as a therianthrope with the body of a man and the antlers of a stag. In the collection of the National Museum of Denmark, Copenhagen, Denmark.

If a central Celtic story about the creation of the universe ever existed, it was lost.[50] Another major branch of North African-European Polytheism whose detailed origin and relationships are slightly better understood, but still unclear, is *Germanic Polytheism*. This branch included a number of subbranches with roots in the PIE culture, such as Old Norse Polytheism and Viking Polytheism.

GERMANIC POLYTHEISM: *OLD NORSE POLYTHEISM* AND *VIKING POLYTHEISM*

Germanic Polytheism encompasses a number of different supernatural belief systems, including two particularly well-known branches: *Old Norse Polytheism* and the *Viking Polytheism* (tree on page 158). Germanic Polytheism, like Celtic Polytheism and Greek Polytheism, may have had roots in PIE Polytheism, although details of its origin are unclear. It was a religion of the ancient Scandinavian people. Although it is probably best known as the fierce, seafaring *Vikings* of the eighth through eleventh centuries CE, its *Old Norse* branch of religious culture goes back much further. The early roots of Old Norse Polytheism took hold at least as early as the Scandinavian Iron Age (500 BCE)[51] or perhaps even earlier. As with Celtic Polytheism, contemporaneous historical records of Old Norse Polytheism are mostly absent or highly speculative. Although Nordic culture extends back well into the Bronze Age, we know virtually nothing about their early pantheon and religious practices. The Norse had an alphabet called the runic alphabet used around 150–1500 CE on stones called "runestones." Those stones were mostly memorials to dead people, occasionally with vague hints about Old Norse mythology. They contain little detail about the religious belief systems. This was possibly due to the Old Norse context of their supernatural beliefs as pragmatic ritual, rather than the abstract language of theology and philosophy. Its practitioners did not recognize a term for "religion" as we understand it today.

Details of Old Norse Polytheism were passed down only through oral transmission for centuries until they were finally written down in the thirteenth century by Christian authors who were clearly influenced by their supernatural beliefs. One of the earliest-known detailed written records is from native historian Snorri Sturluson (1179–1241). He was the Icelandic author of the *Prose Edda*, a three-part manuscript which includes the *Gylfaginning* (a narrative of Norse mythology), the *Skaldskaparmal* (a book of poetic language), and the *Hattatal* (a list of verse form). His is the broadest-known contemporaneous text of Old Norse religion. Other authors with conflicting Western Christian beliefs wrote about Old Norse Polytheism from a much more hostile perspective.[52]

Our sparse picture of Old Norse Polytheism is further complicated by the delay of serious research into its history. There appears to have been little popular interest in Old Norse Polytheism until the romantic movement of the late nineteenth and early twentieth centuries of Europe. This period inspired a wide range of artistic expression influenced by Old Norse Religion ranging from German opera by Wagner (*Der Ring des Nibelungen*) to famous novels by J. R. R. Tolkien (*Lord of the Rings* and *The Hobbit*). It was eventually also used by certain right-wing nationalist groups in the early twentieth century, including Nazi Germany, who usurped some of its symbols and references.[53] One famous example is the letter from the Old Norse Rune alphabet that symbolizes the sun. A modified version of it was used as a symbol for the Nazi SS. Another example is the swastika, which was a symbol used in Old Norse to denote power, authority, and an element of spirituality long before it was commandeered by the Nazis.

The evolutionary history of Old Norse Polytheism is complex and in great need of further systematic study. It included many different branches over the centuries and was a conglomerate of related customs and beliefs, most of which we know very little about. I give only a few highlights and generalities, emphasizing Old Norse deities and supernatural belief systems mentioned in popular culture today. Most of this information is based on works by the anthropologist Anders Andren, and on the early writings of Snorri Sturluson.[53,54,55]

Before Christianization in the eleventh and twelfth centuries, Old Norse Polytheism included a rich pantheon of gods, giants, dwarves, elves, and land spirits. The three major groups of supernatural entities in Old Norse Polytheism included the *Jotnar* (singular form Jotun), the *Aesir*, and the *Vanir*. The *Jotnar* was the oldest group, both powerful and uncivilized. They are usually defined as "giants" or the "Frost Giants." They have been equated with the evil Fomorians of Irish Celtic Polytheism. The other two groups were pantheons of gods: the *Aesir* and the *Vanir*. Several of the *Aesir* gods are portrayed as fictional characters in popular Western culture today, such as Thor, Odin, and Loki. They were the sky, warrior, and ruler deities representing qualities of leadership, combat skill, and craft. The *Vanir* were the original pantheon of gods, more ancient than the Aesir. They were associated with earth, fertility, wisdom, and an ability to see into the future. They represented the natural forces that helped humankind and were as powerful as the Aesir. The Aesir and the Vanir were said to have engaged in a prolonged war until realizing that they were equally powerful and that the war was unwinnable. This led to a pact of peace enabling them to rule together.

There are several different Old Norse stories of human creation. One begins with *Ymir*, ancestor of all the giants of Jotunheim (land of the Jotnar). The Jotnar ruled the cosmos, which began in a state of chaos. Over time, Ymir turned evil. The Aesir god Odin (part descendant of the Jotun clan himself) and his two brothers Vili and Ve, formed a warrior trinity to fight and overthrow Ymir and to gain control of the cosmos. They killed Ymir, and so much blood flowed from his wounds that all the rest of the Jotnar drowned except for one family. This family survived only by building an ark to keep them safe until the flood levels receded. Once the survivors landed safely back on land, they eventually went on to create a new generation of giants. Odin and his brothers did not seek to destroy all of the Jotnar. Their primary aim was to keep the Jotnar in check so that the cosmos could remain in an orderly balance. The Old Norse seemed to have an underlying philosophy of balance, somewhat like the yinyang in the Taoic branches of Asian Cyclicism (chapter 4).

From the giant Ymir's immense dead dismembered body, Odin and his brothers created the earth (which they call Midgard or Middle Earth) and humans to inhabit it. Ymir's blood became the sea, his bones became the rocks and hills, and his hair became the trees. With Ymir's skull they made the sky, and with his brains they made the clouds. The earth was created in the shape of a disk. Odin and his brothers set up Ymir's eyelashes around the edges of the disc to protect the humans from future generations of giants.

Although the Old Norse gods were amazingly powerful, they were not invincible or immortal. They had a clear notion of an "end of times" that would one day come for most everyone. It was called Ragnarok, a time when even most of the gods would die. During this apocalyptic event, the worlds of both god and man would suffer great cataclysms. However, the Norse viewed this great ending as one that would pave the way for a new beginning. The Norse view of time (or at least the way it is today presented) often seems to blend the past, present, and future together. Stories of Ragnarok, for example, are sometimes told as prophesies, and other times they are told as though they have already happened.

Although there were many gods and goddesses within the Norse pantheons, the gods of Aesir are the best known. They lived in Asgard, one of the nine worlds of Old Norse Polytheism. According to Sturluson's account, the gods of Asgard included a special group of twelve good gods overseen by *Odin*, their chief, plus a trickster named Loki. Odin is best known as the "Allfather," god of wisdom, healing, royalty, knowledge, sorcery, and the runic alphabet. He is described as having a long beard and one eye, having sacrificed the other one to drink from the well of cosmic knowledge. It is an old theme that we hear in many religions through history: wisdom comes at a price. Although Odin is known as the father of Thor and the ruler of Asgard, Thor eventually eclipses him to become the most powerful god.

Thor was the first-born son of Odin. He was the strongest of all gods and associated with thunder, lightning, and strength. He was the protector of humankind. Most of his myths have to do with giant slaying.[51] His greatest possession was his hammer, *Mjolnir*. It had a massive head and short handle and it was shaped somewhat like a cross. When he hurled it, it always hit his target and came back to him like a boomerang. Mjolnir became the symbol of fertility at weddings, as well as the symbol for resurrection at funerals. After Christianity came to Scandinavia in the ninth century, worship of Thor continued for centuries more. Thor's hammer became the most widespread symbol of Norse Polytheism. Some historians believe its growing use was a response to the Christian symbol of the cross.[56] During the ninth through eleventh centuries in Scandinavia, pendants of Mjolnir resembling the Christian crucifix were commonly worn in resistance to Christianization efforts. Because of the similarity of appearance, followers of Thor-revering Norse religion were less noticeable and therefore were less likely to be persecuted by authorities.[57]

Loki is usually known as the handsome but malicious trickster god and sometimes as the god of darkness. He is the son of one of the giants, but is counted among the Aesir pantheon of deities.[51] He often got the Aesir pantheon into trouble, and then got them out of it with his trickery. Loki's children ranged from malicious to demonic. They included the god-eating monster wolf named Fenrir who eventually kills Odin at Ragnarok, the sea-serpent Jormungandr who eventually kills Thor, and his daughter Hel who was goddess of the underworld domain by the same name. Hel (with one l) was the place in Old Norse mythology where those who did not die a heroic or notable death went after they left the world of the living.

(*Left*) The Stenkvista runestone in Sodermanland, Sweden; probably carved around the ninth or tenth century CE. It depicts Thor's hammer (*Mjolnir*) along with text in the Old-Norse runic alphabet. (*Right*) A painting by Marten Eskil Winge from 1872 depicting Thor's battle with the giants (the *Jotnor*) wielding his hammer.

By some accounts, it was a destination for evil men, much like hell in Abrahamic Monotheism.

There are several ancient stories of Thor and Loki doing battle with each other at one time, and then teaming up together to fight others at other times. Such stories have been widely adopted in popular movies today, although the stories are embellished with a significant degree of artistic license (e.g., *Thor, Thor: The Dark World*, and *Thor: Ragnarok*). For example, in the movies Loki is generally portrayed as the adopted son of Odin, Thor dates an American astrophysist, and both Thor and Loki survive Ragnarok.

The final battle between the gods was foretold to be Ragnarok. It is an apocalyptic face-off between two groups. One included most of the gods of Aesir together with select, fallen human warriors, all led by Odin on the side of good. The opposing group included the evil Jotnar from the land of the giants, led by Loki. In the end, the battle would claim the lives of Odin, Thor, and Loki, along with most of the participants. The sun, moon, and the nine worlds of the gods would all be destroyed, and the earth burned to ash and sunk into the sea. A human couple,

Lif and Lifthrasir would survive to repopulate the world after it arose once again from the sea. Ragnarok is described in detail in the tenth-century Old Norse poem *Voluspa*, and in Sturluson's thirteenth-century *Prose Edda.*[58]

Ritualistic blood sacrifice was supposedly a cornerstone of Old Norse Polytheism. Animal sacrifice was the most common, although there were also examples of human sacrifice (or perhaps sacral death penalty is a better term). People convicted of certain crimes were sentenced to death and ritually sacrificed to the gods. Their execution was done by hanging or by ritualistic drowning in special temple wells.[59] Old Norse simply combined their form of capital punishment with religious ritual. The concept of sacrifice with the Old Norse was a pragmatic one. It was barter for something specific in return. It was the main way people interacted with the gods. The deities were highly personified with human-like personalities, and they were seen as regularly intervening in human affairs. People treated their godly interactions as they would treat interactions with other people. They assumed that to get something from the gods, they needed to give something useful to the gods in return.[60] The Old Norse religion had several ideas about the afterlife with four different realms to welcome the dead. However, at least as reported by later Western Christian chroniclers, the religion did not adhere to the belief that moral behavior affected an individual's afterlife destination.[52]

Some historians believe a relationship existed between the Norse religion and their sociopolitical structure. Before the Christianization stage of Old Norse religion that occurred during the Viking Age (approximately 750 CE to 1066 CE), the relationship between the chieftain and his warriors reflected a clear degree of reciprocity. A warrior who fought bravely and was loyal to his chieftain was rewarded with a share of the spoils taken in battle. Both parties participated voluntarily and held dignified positions despite their huge differences in rank. This changed with the coming of medieval Christianization. Christians were supposed to serve God unconditionally as his servants without any expectations of reward in this world. As Old Norse religion became more hierarchical, so too did the ruling authority. The chieftain became king and was now to be served by his subjects unconditionally. What had previously been a contractual understanding of mutual benefit became a decree or commandment to a subordinate.[60] Religion and politics were highly influential on each other, as they have been in so many other times and places throughout history.

During the Viking Age, the Old Norse people left Scandinavia and settled in other regions throughout Northwestern Europe, taking their religious beliefs with them. They even colonized Iceland, where *Thor* became the most popular god of the pantheon and Odin fell somewhat out of favor.[52] However, the expansionism of Norse/Viking Polytheism was short lived because it eventually came into conflict with the proselytic efforts of Western Christianity.

When Western Christian missionaries first arrived in Scandinavia, Christianity had already become the dominant religion throughout most of Europe. However, convincing the people of the Scandinavian mainland to convert totally to Christianity turned out to be a challenge. Missionaries found it difficult to convince the Norse people that polytheism and Christianity were mutually exclusive.[52] Old Norse, as all good polytheists, were glad to add Jesus to their pantheon, along with Odin, Thor, and the rest. This, of course, was not acceptable to the Christian missionaries. The Norse wanted to keep much of their own indigenous religion, but the Christians were persistent in their mission to defeat all forms of hard polytheism, even if it meant resorting to violence. The two sides fought and burned each other's churches down. Finally, in 995 CE when Olaf Tryggvason became the Christian king of Norway, he began a mass compulsory conversion program. He forced high-status Norwegians to convert, and he destroyed polytheistic temples. He executed polytheists, whom he referred to as "sorcerers."[52] Although Old Norse Polytheism gradually faded, much later it saw the formation of a revivalist branch in Iceland and other parts of Scandinavia called *Asatru*, officially founded in the 1970s.

Within the European groups of polytheism, the historical relationships and diversity of the Celtic and Germanic branches remain poorly understood. The polytheistic branches of the Mediterranean region are much better known.

MEDITERRANEAN POLYTHEISM AND THE RISE OF THE RULING CLASS AS DIVINE

Ancient *Mediterranean Polytheism* eventually gave rise to *Mesopotamian Polytheism* and *Proto-Egypto-Greco-Roman Polytheism* (see the tree on page 158). Given possibilities suggested by the archaeological site of Gobekli Tepe discussed in chapter 3, Mediterranean Polytheism could have roots going back more than eleven thousand years.

By the fourth millennium BCE, *Mesopotamian Polytheism* and *Egyptian Polytheism* began to develop hierarchical ruling classes, large-scale exploitation of human energy, and administrative divisions of labor. As their civilizations evolved, so did their pantheons of gods. The religion of each city-state had a Supreme God with a variety of specialized subordinate positions that clearly reflected a royal court. A reciprocal understanding of gods and governing leaders eventually led to a belief that ruling monarchs had divine connections. A type of government developed in some regions that was known as a *theocracy*, or rule by "divine guidance." The "will of the gods" was carried out by rulers, some of whom were eventually revered as divinities themselves. Within the ancient polytheistic branches of the Mediterranean region, close association developed between the gods and kings, gods and pharaohs, and gods and emperors.

The earliest-known major branch of *Mediterranean Polytheism* to develop was *Mesopotamian Polytheism* (tree on page 158).

MESOPOTAMIAN POLYTHEISM AND THE EARLIEST-KNOWN WRITTEN RELIGIOUS TEXTS

The term Mesopotamia means "land between the rivers." It refers to the Tigris-Euphrates River region, corresponding to modern-day Iraq, Kuwait, Northeastern Syria, Southeastern Turkey, and Southeastern Iran. Ancient Mesopotamia flourished for several thousand years, incorporating at least four major civilizations (the Sumerians, Akkadians, Assyrians, and Babylonians) and influencing the biblical Canaanite-Hittite lineage mentioned in ancient Ugarit texts. It was a uniquely diverse crossroads and melting pot for vastly different groups of people over thousands of years, unlike the later more uniform civilizations. such as those of Egypt, Greece, and Rome.[61] The diverse elements of Mesopotamian civilization were connected through a common cuneiform script (one of the earliest-known writing systems in the world), highly conservative works of religious art, and many shared deities.[61] Many of the Mesopotamian civilizations, as well as countless subdivisions within each of those civilizations, had their own state god, pantheon, and polytheistic belief system (only a few of which are delineated). Constant processes of differentiation, assimilation, and syncretization resulted in an ever-increasing diversification of belief systems.

The people of Mesopotamia created some of the earliest-known dynastic empires by the third millennium BCE. These empires thrived until 539 BCE when the last of them, the Babylonian Empire, was conquered by Persia. The existence of Mesopotamian Polytheism (as well as of other aspects of Mesopotamian culture) goes back to at least the fourth millennium BCE, based on limited archaeological evidence, including various inscriptions on clay tablets.

The earliest well-known Mesopotamian civilization was called Sumer, which was settled at least by the fourth millennium BCE in the southernmost region of ancient Mesopotamia.[61,62,63] At the time, Sumer was not the hostile desert that exists in the region today. Instead, it was a fertile valley of grasslands and marshes along the Tigris and Euphrates Rivers, where Sumerian farmers could grow crops to support the earliest-known urban center of humanity. This was a true cradle of Eurasian civilization, together with later population centers forming in Egypt, the Indus Valley, and the great river valleys of China. In terms of religious tradition, it was an ideological ancestor to most of the rest of Mediterranean Polytheism.[61] Some historians believe that Mesopotamian religious tradition was highly influential in the early development of Abrahamic Monotheism.[64]

Sumerians are credited with a number of major inventions critical to the development of human society. These include the wheel, the plow, modern units of time (hours, minutes, and seconds), large-scale agriculture, a codified judicial system, government bureaucracy, and large-scale urbanization. However, their most important invention for documentation of human history was their invention of the earliest writing system, more than five thousand years ago. The Sumerians are the earliest-known literate people. They produced the oldest written records of polytheism, or of anything else for that matter, on clay tablets. One of the oldest religious texts known from Sumer is the Kesh Temple Hymn. It consists of a set of inscribed clay tablets, which scholars typically date around 2500 BCE or older.[65,66] The tablets contain an ode to the temple and to the deities represented within it. Sumerian temples are even older than the oldest writing. Among the oldest-known Sumerian temples is one for the god Enki dating back at least as far as 4000 BCE.[64]

As indicated at the beginning of this chapter, we are far from understanding the true diversity (different traditions or "denominations") of Old World Hard Polytheism. This is also true for the diversity within Mesopotamian Polytheism, because Mesopotamia includes three thousand years of history and at least five different major civilizations.[61] To my knowledge, no comprehensive, detailed attempt has been made to classify its hundreds of different subbranches of religion. I discuss only the most general characteristics of Mesopotamian religious belief systems.

In the beginning, the Mesopotamian gods were few in number and highly animistic; they were associated with forces of nature, emotion, and the value of knowledge. Over time, the pantheon grew and became more anthropomorphized and organized into a complex hierarchy. It eventually included a king, groups of special importance just under the king, and thousands of lesser deities for which, in many cases, we know nothing but their names. Among the many Mesopotamian gods, the primary deities were the Seven Gods who Decree the Fates.[64] These were *An*, *Ninhursag*, *Enlil*, *Enki*, *Nanna-Sin*, *Utu*, and *Inanna*.

There are many different interpretations of each Mesopotamian god, and these stories changed over time. Some aspects of these seven primary deities mentioned were as follows:

- *An* (also known as *Anu*) was initially the Supreme God, sky god, and prime facilitator of creation.[61] According to most Mesopotamian belief systems, An and his consort, the goddess *Ninhursag* (also known as *Ki*), were the parents to many major gods of the Mesopotamian pantheon (the *Anunnaki* or *Anunna*).[64] An was also part of a divine tetrad, including the four most important gods: *An* the father, his consort *Ninhursag*, and his two sons *Enlil* and *Enki*.[64] An gradually lost his prominence over time and lost much of his power to his son Enlil.

- *Ninhursag* (also known as *Ninmah* or *Ki*) was the Sumerian mother goddess of fertility and creation. She created a female deity named *Ninti* to heal one of Enki's ribs after he fell sick from eating forbidden flowers from the garden. Some historians believe that this story influenced the later biblical story of creation in *Genesis*, in which Eve is created from one of Adams ribs, and eating from the forbidden tree in the garden brings on the curse of death to humankind.[67] Some historians suggest that together with An as father, Ninhursag may have been mother to the rest of the Mesopotamian pantheon.[64]
- *Enlil* (also known as *Ellil* or *Elil*) was the god of air, storms, and foreign lands. Initially, he was not represented anthropomorphically, but instead depicted only as a horned hat. Later, he was revered as one of the most important gods of the Mesopotamian pantheon where the other gods might not even look upon his splendor.[61] He rose to supreme prominence by the twenty-fourth century BCE, taking An's place. He became known as the Granter of Kingships.[64]
- *Enki* (later known as *Ea* in the Babylonian and Akkadian pantheons) was the god of water, knowledge, and magic. He is often depicted as a bearded man with a horned cap. In some Mesopotamian stories, Enki and the earth goddess *Ninhursag* (or *Ninmah*) created the first humans from clay. Enki is also associated with the Sumerian version of a Great Flood story, similar to the story of Noah in Abrahamic religions. Enki instructed Ziusura (also spelled Ziusudra) to build a large boat for an impending flood from a storm that lasted seven days. Unfortunately, most of the rest of the story was lost. The tale of Ziusura is known only from a fragmentary, Sumerian tablet written in the seventeenth century BCE.
- *Nanna-Sin* (also known as *Nanna-Suen*, *Sin*, *Nanna*, *Namrasit*, and *Inbu*) was the god of the moon and wisdom, and son of Enlil. His name was written simply as the number 30, which is the number of lunar days.[61]
- *Utu* was the god of the sun, justice, truth, and morality. In his role as the sun god, he was believed to emerge from the eastern doors of heaven each dawn and ride across the sky in his chariot before returning to heaven each evening through another set of doors in the west. He was a link between the gods and the human kings, an early monarchistic element, and special protector for some of the later, particularly heroic kings.[61]
- *Inanna* (also known as *Inana*) was the ancient Mesopotamian goddess of love, beauty, sex, desire, fertility, war, combat, justice, and political power. According to one Mesopotamian story, she was daughter of the moon god Nanna-Sin and twin sister to the sun god Utu. She is often depicted with bird features, such as wings or bird-like feet. Babylonians and Assyrians later worshiped her as the goddess *Ishtar*.

Wall relief from the ninth century BCE depicting the god *Ashur* from the Mesopotamian pantheon (specimen in the collection of the British Museum in London). In Mesopotamian Polytheism, Ashur's symbolic image was often represented as a *Faravahar* (the image of a deity standing in a winged sun, as shown here). The Faravahar image was later also adopted by Zoroastrians as a primary symbol of their religion (chapter 6).

Another god named *Ashur* (also spelled *Assur*), whose origins are poorly known, was identified as an equivalent of Enlil by some Mesopotamian branches.[61] He began as the state god of Assyria, but advanced in status over time, assuming many different levels of importance among the different Mesopotamian branches of polytheism.[61] He eventually became head of the Assyrian pantheon. His image was commonly depicted as a Faravahar (see the figure above), which was later used as a symbol for Zoroastrianism (see page 268).

Generalized protector deities in Mesopotamian Polytheism were called *lamassus*. These creatures were depicted as having the head of a human and the winged body of a bull or a lion. Many images of these creatures are preserved in archaeological sites as massive figures carved out of solid rock. The lamassus was a symbol of power that was often placed at the entrance to a palace or temple.

By the third millennium BCE, Mesopotamian deities were becoming increasingly anthropomorphized. Although they were usually seen as immortal, they were also often seen to have human-like vulnerabilities. They could become sick to the point of death, be wounded, and occasionally even be killed.[64] Over time, pantheons grew larger. Successive waves of settlers and conquerors that came into the region introduced their own gods and goddesses, which were readily integrated into existing pantheons. There seemed to be little or no prejudice against incorporating new

Three *lamassus* (Mesopotamian protective deities) at an archaeological site in Nimrud, Iraq. A lamassus is a therianthrope, with a human head on a winged body of a bull or lion. These three were carved in the ninth century BCE in the ancient city of Nimrud, Iraq. They were destroyed in 2015 by Islamic fundamentalists (ISIL) who deemed them to be un-Islamic and blasphemous.

gods from the outside (which is one of the more inclusive characteristics of Old World Hard Polytheism in general). The nomads who invaded Mesopotamia from the north or the east brought their water gods and sand gods with them. People who came from high mountain regions brought their gods of thunder and lightning. It was a diverse mosaic of the divine. The number of gods eventually numbered more than three thousand.[61] Some were great gods in control of the sun, earth, and sky. Others were minor gods in charge of more trivial things, such as the pickax or brick molds.[64]

The Mesopotamian pantheons often reflected changing political environments. The rise and fall of various gods in the pantheon generally followed the rise and fall of individual city states, each of which had its own patron god. After the Babylonian king Hammurabi conquered Sumer in the eighteenth century BCE, Enlil's importance as Supreme God began to wane. He was eventually supplanted by Marduk, the national god of Babylon. *Marduk* then also became head of the Sumerian pantheon. His power grew until many or most other gods of the pantheon were swallowed up to become mere aspects of Marduk. The growing hierarchical arrangement of the pantheon paralleled a growing hierarchical arrangement of government structure and a quest for intellectual order.[3] The growing dominance of Marduk may have provided influence for the later Abrahamic story of creation by El or Yahweh.

Marduk is said to have created the earth, the stars, the sea, plants, animals, and humanity to serve him. After that, he rested. Mesopotamian Polytheism eventually found minor deities unnecessary for human well-being, setting a path toward Linear Monotheism.

There are hundreds of stories about the rise, fall, and shifting responsibilities of Mesopotamian gods as well as their hierarchical organization over the centuries. The increasingly hierarchical organization of the pantheon was accompanied by the increasing importance of their temple priests and functionaries. Chief priests began to be treated like royalty. Religious conviction was growing that the kings and rulers of city-states (known as the *ensi*) could ensure long lives, well-being, and prosperity for themselves as well as others.[64] By the fourth millennium BCE, pantheons had become structured into monarchical hierarchies, with a national god as head of the pantheon. Mesopotamian government took the form of individual city-states, some of which became theocracies (governments in which priests, priestesses, and royalty ruled in the name of the national God or gods).[64] The Mesopotamian gods, their mythological framework, and their incorporation into government, eventually led to the diversification of Mesopotamian Polytheism into several subbranches.[68] These included Sumerian Polytheism, Akkadian Polytheism, Assyrian Polytheism and Babylonian Polytheism. Sumerian Polytheism is thought to have been theologically formative to the Biblical patriarch Abraham, before he migrated to Canaan (Palastine) in the second millennium BCE. Many historians believe Semetic Canaanite Polytheism was highly influential to the early rise of Abrahamic Monotheism.[69,70] Abraham is believed to have initially worshipped the Canaan god El who was the High God of the Canaanite pantheon (discussed further in chapters 6 and 7).

In the early first millennium BCE, an enormous ninety-one-meter (nearly three-hundred-foot) temple tower was built in the name of the Supreme God Marduk. It was called the *Etemenaki ziggurat* (Temple that is the foundation of the heavens and the earth). This tall, pyramid-like structure was destroyed with the city of Babylon in 689 BCE. It was rebuilt by Nebuchadnezzar II but was demolished again by Alexander the Great in 331 BCE. This tower is thought to have inspired the biblical story of the Tower of Babel (*Genesis* 11:1–9).

Mesopotamia was a region prone to violent flooding from the surrounding Tigris and Euphrates Rivers. Likely as a result of this flooding, Mesopotamian Polytheism includes one of the earliest-known accounts of a Great Flood myth, which may have been influential to the development of flood stories in later religions, such as that of Noah in Abrahamic Monotheism (chapter 6).[71] The Mesopotamian flood epic is told in tablet 11 of the *Epic of Gilgamesh*, which is thought to have been composed around 1800 BCE. It involves a patriarch named *Utnapishtim* (I'll call him *Ut* for short) and a number of gods from the Sumerian pantheon. The great Mesopotamian gods An,

Enlil, Ennugi, and Enki grew unhappy with the state of humanity and decided to return the earth to its precreation state. They planned to do this by causing a great flood to annihilate all humans as well as most other living things. Enki decided to save a representative sample of life-forms to repopulate the earth after the deluge. The humans that Enki chose to survive the deluge included Ut, his family, and a few select friends. Enki commanded Ut to build an ark large enough to hold him, his family, friends, provisions, and "all the beasts and animals of the field." Enki's additional instructions to Ut specified that the ark be built from wood and measure 120 cubits in length by 120 cubits in width (14,400 square cubits compared with 15,000 square cubits of the Noah's ark story in *Genesis*). When the boat was finished, Ut, his family, and the animals entered the ark, and the door was sealed. After that, the long rains began, unleashed by the storm gods and the god of thunder. Then the deep floods came and covered even the highest mountains. Life outside the ark perished. Eventually, the ark finally came to rest on the high peaks of Mount Nimush (also known as Mount Nasir), located in what is today Iraq Kurdistan. After the ship rested for seven more days, Ut sent out a dove to see if it would find suitable land, but it found none and returned to the ark. He then released a swallow, and it too came back to him. Finally, he released a raven, which was able to find land and did not come back. Now feeling safe, Ut, his family, and all of the animals on the ark disembarked to the land, and over time, they repopulated the earth with their descendants.

Mesopotamian Polytheism tended to be a relatively pessimistic religion. It saw the gods as unpredictable beings of great power, and it saw humans as the servants of the gods whose job was to carry out the will of the gods and appease them.[64] If the gods were unhappy, they took it out on the people by sending disasters and hardship. The afterlife in Mesopotamian religion was no picnic. Immortality in respective comfort was reserved only for certain gods and kings. The afterlife for common people involved descent into a gloomy netherworld as a ghost, where improving ones' general level of comfort (or at least lessening the level of discomfort) was largely dependent on offerings from descendants. Otherwise, the dead were doomed to an eternity of consuming dust in a "shadowy version of life on earth."[72] The Mesopotamian netherworld was neither a place of punishment nor a place of reward. It was simply the only destination for departed human spirits. Although the idea of the Mesopotamian netherworld varied over time, its description generally ranged from dark and dismal, to simply a dulled version of life on earth. The general level of Mesopotamian pessimism was probably a result of the harsh living conditions of Sumerian civilization. This version of eternity was in stark contrast to the later Egyptian beliefs of a glorious life to come, leading to their practice of embalming and mummification. The Egyptian view of the afterlife, like the later Greco-Roman view, was more optimistic, at least for those thought to be of purest heart and members of the monarchy.

PROTO-EGYPTO-GRECO-ROMAN POLYTHEISM AND THE HOPE FOR A BETTER AFTERLIFE (FOR SOME)

The polytheistic religions of Egypt, Greece, and Rome had a number of similarities. They each increasingly strengthened the ties between the monarchy and the gods. The chief deity and most other deities in the pantheons were men, but the pantheons also had a number of important goddesses. These religions all borrowed heavily from each other's traditions and regularly assimilated some of each other's gods. In addition, the anticipated afterlife among believers (at least among the elite ruling faction) was more optimistic than it was for the Mesopotamians. It ranged from not quite as bad to downright glorious, as long as certain rituals were adhered to and the deceased was a member of the right social class. I term the two major subbranches of this group *Egyptian* Polytheism and *Proto-Greco-Roman* (PGR) Polytheism.

EGYPTIAN POLYTHEISM AND THE DIVINITY OF THE PHARAOH

Ancient Egypt has ties to many different religions, from polytheistic branches worshipping many gods to the biblical beginnings of Linear Monotheism. In this section, the focus is on Egypt's polytheistic branches. The different subgroups were historically numerous, but today they are not well distinguished from each other. The pantheons of Egypt evolved in many ways: sometimes with the development of a complex new deity, sometimes with the expansion or reassignment of a particular deity's powers, and sometimes by adding the gods and goddesses from the pantheons of other cultures.

Geographically, Egypt is part of Africa. However, ancient Egypt was culturally more closely related to other Mediterranean cultures, such as Mesopotamia, Greece, Rome, than it was to most of the rest of Africa. The Nile River Valley of Egypt, together with the Tigress and Euphrates River Valleys of Mesopotamia, made up the so-called Fertile Crescent. This region is often referred to as the "cradle of civilization." Mesopotamia and Egypt were already engaged in trade with each other by the fourth millennium BCE, and their two religious cultures developed a number of syncretic similarities.[73] The Egyptian branches of Old World Hard Polytheism followed Mesopotamian beliefs and practices in building elaborate temples to their gods. Like the Mesopotamians, their ever-evolving pantheons involved more than 1,500 gods.[74] Also like the Mesopotamians, the Egyptians constructed immense architectural structures of religious significance. The Mesopotamians built ziggurats, which were bases for massive temples, and the Egyptians built pyramids, which were elaborate tombs for the pharaohs. Both cultures had the monarchistic religious beliefs that their leaders had special connection to the gods that was reflected in their leader's accomplishments and control of great wealth and worldly power. The development of writing, monumental architecture, monarchistic rule, and pantheons of hierarchically organized gods all had ancient records in Mesopotamia. Later, those cultural aspects seem to have continued to evolve in Egypt. Egypt absorbed much of its early governing and general religious tradition from Mesopotamian influence.[75]

In Egypt, as in Mesopotamia, religion and the ruling class were intimately connected and different states emphasized different gods.[76,77] The complex evolution and diversity of Egyptian Polytheism is well beyond the scope of this book, so my account is abbreviated. Like Mesopotamian Polytheism, Egyptian religions emphasized people's interaction with gods and goddesses who were in control of nature. Some of the earliest-known religious texts in the world are the Egyptian Pyramid Texts from 2353–2323 BCE, which were carved on the walls of the pyramid for the pharaoh and king named Unis.[78] They contain hundreds of utterances and spells to help the deceased king resurrect in the next world. They also reveal a great deal about what the ancient Egyptians believed about the afterlife, and eventually were expanded into the Egyptian Book of the Dead for the broader Egyptian public.[79]

Egyptian Polytheism was generally more optimistic than early Mesopotamian Polytheism. The gods generally were seen as well disposed toward humanity. Egyptians believed that their gods protected them from a world filled with chaos and disorder. Egyptians were also more invested in funerary practices and the afterlife than the Mesopotamians. In Egyptian Polytheism, death was only a temporary interruption to life, and the dead could have a comfortable afterlife. Various funerary practices, procedures, and rituals were carried out to ensure this comfort, including mummification and burial objects for the deceased. Egyptians preserved the body of the dead so that it could be used by the deceased person in the afterlife. In the earliest days of Egyptian religious development, bodies were simply buried in the desert and naturally mummified by the arid conditions. Eventually, artificial mummification came into use to preserve the departed's former physical shell. Only the rich and ruling elite could afford the best mummification services, and the common people continued to bury their dead in desert graves. The Egyptians also buried furniture, food, clothing, and other everyday items along with the deceased for use in the afterlife. Protective amulets were included to provide the soul of the deceased with added protection against evil.

Egyptian Polytheism underwent a complex evolution beginning in the third millennium BCE with its ever-changing succession and diversification of gods and goddesses. Like the pantheons of Mesopotamia, most Egyptian gods and goddesses were anthropomorphic and exhibited human-like behavior. They could be benevolent or cruel, generous or greedy, wise or petty. They had family battles among themselves.

Although generally immortal, they could be injured, grow old, or occasionally even die (often to be resurrected). Most of the earliest gods were anthropomorphized versions of natural phenomena, such as the sun and the Nile River. The dazzling sun rose and set each day in a predictable manner; the river rose and shrank each season giving life to the fertile Nile River Valley. The river provided the topsoil each spring when it flooded and gave water for the growth of wheat and other abundant crops. Ancient Egyptians believed that its unfailing magnificence was the work of the gods.

I mention only a few of the most notable Egyptian deities, but much has been written about the hundreds of Egyptian deities. A note on names: The Egyptian alphabet of the time had only consonants. Historians added vowels later, so both spelling and pronunciation of godly names vary from author to author.

- *Re* (also known as *Ra*) was an ancient creator sun god, and a primary deity in Egyptian Polytheism.[80] He was one of Egypt's most ancient gods and is often depicted as having a falcon's head wearing the fiery disk of the sun. The Egyptians believed that during the day, Re traveled across the underside of the sky, and at night, he passed through the underworld, where he had to fight and defeat the serpentine god of chaos, *Apep*, and meet with *Osir* for regeneration to be reborn each day at dawn. At various times and in various branches of Egyptian Polytheism, Re fused with other gods to form greater ones. These included the syncretized gods *Khnum-Re* (*Re* syncretized with *Khnum*, creator god of all living things on a potter's wheel), *Min-Re* (*Re* syncretized with *Min*, the personified generative force of nature), *Sobek-Re* (*Re* syncretized with *Sobek*, representing the power of the Nile to fertilize the land), and, most prominently, *Amun-Re* (discussed next).[76,79]
- *Amun* was a supreme deity and the state god of Egypt during the New Kingdom (1550–1069 BCE).[79] He was part of a divine triad called the "Theban Triad" as the father, together with *Mut* the mistress of all gods as the mother, and *Khonsu* the moon god as their son.[81] He eventually syncretized with the sun god *Re* to become *Amun-Re* "King of the Gods" during the height of Egyptian civilization.[77,80] The uniting of these two powerful gods was seen as a symbolic unification of Upper and Lower Egypt and consolidation of the priesthood. Amun-Re's temple, Karnak, was the largest religious complex ever built by humans, covering about two hundred acres. Amun-Re's influence eventually spread beyond Egypt, and he even later became identified with the Greek god Zeus.[80]
- *Mut* (also known as *Maut* or *Mout*) means mother in the ancient Egyptian language. She was the preeminent mother goddess of Thebes who had many aspects throughout Egyptian history.[79,80] She is a primal deity associated with the waters from which everything is born. She is generally depicted as a slender woman wearing two crowns, but alternatively as a cobra, cat, cow, lioness, or vulture. Her many titles include "Mother of the Gods," "World Mother," and

The *Theban Triad*, as depicted on the wall of the mortuary temple of Ramses III, Medinet Habu, Egypt. Built during the late second millennium BCE. From left to right: *Amun*, *Mut*, and *Khonshu*.

"She who gives birth but was not herself born." Mut was the consort of Amun, and part of the Egyptian divine triad that included her, Amun, and their son Khonsu.[80]

- *Horus* also became one of the major gods of the Egyptian pantheon as lord of the sky, ancient creator god, and the symbol of divine kingship.[79,80] He is most often depicted as a falcon or a man with a falcon's head. He has had many different aspects through history, sometimes recognized as separate gods. He also has syncretized or fused with many different gods at one time or another, including *Sopdu* (god of the sky and eastern border), *Khonsu* (lunar god), and *Matu* (the falcon god). He was a pantheon unto himself.

- *Osiris* (also known as *Osir*, *Asar*, or *Ausar*) was the god of death and the afterlife, and his domain was Duat, the Egyptian underworld.[79,80] He was also known as the god of transition, resurrection, and regeneration. According to the many different stories about Osiris,[77] he began as the first-born son of *Geb* (god of the world) and *Nut* (goddess of the sky). He married his sister Isis (a beautiful mother goddess) and took over rule of Egypt. *Osiris*'s brother *Set* envied the power of his elder brother. In one of the long and complicated interpretations of his death, Set killed Osiris by suffocating him in a large chest, threw the chest into the Nile, and took over the throne. Early ideas about Osiris are thought to have been influenced by earlier stories of the Mesopotamian god Marduk.[80]
- *Isis* (sister-wife of Osiris) was one of the major goddesses of Egyptian Polytheism. She was daughter of *Geb* and sister of *Osiris*. She was considered the symbolic mother of every pharaoh.[77] By the *first* millennium BCE, *Isis* and *Osiris* had become two of the most widely worshiped deities.[74] Isis is depicted sometimes with a human head, and other times with the head of a cow. In the Hellenistic period (323–31 BCE), she also came to be worshiped by the Greeks and Romans, along with the Greco-Egyptian god, *Serapis*. Cleopatra (69–31 BCE), one of the few known female Egyptian pharaohs, claimed to be the living embodiment of Isis. The cult of Isis was eventually widely absorbed by Rome in the late first century BCE, ending only with the rise of Nicene Christianity in the fifth century CE.
- *Anubis* was known by the name "The divine embalmer" and was the main Egyptian god associated with mummification and the afterlife. He was usually depicted as a canine or a jackal-headed man. He was often colored black to represent death. Anubis was once Lord of the entire underworld, but Osiris eventually replaced him. Anubis became the attendant for the scale used in the "weighing of hearts" ceremony at the Egyptian gates of judgment (described in chapter 125 of the Egyptian Book of the Dead). In the mummification process for the deceased, the heart was kept within the body so it could travel with the deceased into the underworld. When the departed soul reached the entrance to the Netherworld, Anubis would weigh the heart of the person (representing the person's thoughts and emotions) on a scale against the feather of truth and justice (*Maat*). If the heart was judged pure and weighing no more than the feather, then the person could enter the afterlife and go before *Osiris*, where a lush, pleasant part of the Netherworld awaited them. In the case of royalty, their passage would be ascent to the sky to dwell among the stars.[82] If, however, the heart was judged impure by weighing more than the feather, it was devoured by a beast-like goddess named *Ammit* (also spelled *Ammut*). Once the heart was eaten, the person was annihilated from existence. Some historians believe ancient depictions of this eternal punishment influenced medieval perceptions

Therianthropic gods of Egypt, as illustrated in the Egyptian Book of the Dead. Papyrus copy from the eleventh century BCE. (*Left*) Three characters from left to right: *Thoth*, the ibis-headed god of writing and scribe to the gods; *Anubis*, the jackal-headed god of embalming and attendant of the scale for the weighing of hearts; and deceased young woman being prepared for final judgment. (*Right*) Black-skinned character on the left is *Osiris*, god of death and the underworld. Standing behind him is his sister-wife, the mother goddess *Isis*. Both pieces are in the collection of the Field Museum in Chicago, Illinois.

The mythological Egyptian weighing of hearts ceremony, reconstructed in a Field Museum diorama. The jackal-headed *Anubis* weighs the heart of a deceased individual in a small pot against the weight of a feather, as the scribe *Thoth* records the proceedings. Awaiting is the crocodile-headed creature, *Ammit*, with the front legs of a lion and the back legs of a hippo. If the heart fails to be lighter than the feather, it will be devoured by *Ammit*, and the deceased individual will be forever annihilated.

of hell in early Christian texts.[83] It is a concept of Judgment Day that coalesced long before the origins of Abrahamic Monotheism.

Like the Mesopotamians, the Egyptians had thousands of gods (far too many to mention here). In Egyptian religion, the pharaohs were also considered gods. They were integral links between humanity and the rest of the pantheon, working with the gods to create divine order and justice. The worship of the major gods in Egypt led to the construction of more than one hundred pyramids for the pharaohs and their queens, beginning in the twenty-seventh century BCE. Some of these pyramids are immense. Their function is thought to have been helping the deified pharaohs reach their exalted position in the afterlife. The shape of these massive structures is thought to represent the descending rays of the sun. Egyptian Polytheism is another major piece of the evolutionary history of religion. It is much more complex than can be adequately addressed here, and for further general reference, I suggest *Cultural Atlas of Ancient Egypt* by John Baines and Jaromir Malek[84] and *The Routledge Dictionary of Egyptian Gods and Goddesses* by George Hart.[85]

Egyptian Polytheism greatly influenced the evolution of other branches of polytheism in the Mediterranean region, most notably the *PGR* (Proto-Greco-Roman) branch including the polytheistic branches of Greece and Rome.

PROTO-GRECO-ROMAN POLYTHEISM AND THE RISE OF WESTERN CIVILIZATION

The development of PGR religions began in Greece and its nearby islands, although it also was influenced by earlier Egyptian Polytheism. Based on archaeological evidence, the influence from Egypt (and vice versa) dates back at least as far as the second millennium BCE. Other early influence came from the Minoan civilization of the Greek Aegean Islands (3000–1100 BCE) and the Mycenaean civilization of the Greek mainland (1600–1100 BCE).[85] Greek ceramics and written records are well known from ancient Egyptian archaeological sites, and Egyptian art objects and written records are well known from ancient Greek archaeological sites.[86,87,88,89] Egyptian hieroglyphs were also used as a model for Minoan and Mycenaean pictographic writing systems.[85] The Greek historian Herodotus (486–465 BCE) noted Greek-Egyptian god equivalents of the Greek *Zeus* with the Egyptian *Amon*, the Greek *Dionysus* with the Egyptian *Osiris*, and the Greek *Hephaestus* with the Egyptian *Ptah*, among other cross-culturally connected deities. Greece's best-known ancient king, Alexander the Great (356–323 BCE), claimed that he was the son of the syncretized Egyptian-Greek god *Zeus-Amon*. Perhaps the most vivid example of a broadly interconnected Egyptian-Greco-Roman deity is the goddess *Isis*, who had worshippers across the Mediterranean by the first century BCE.[78] It was often a struggle for early Christianity to eclipse her worship.[80]

Although the early development of PGR Polytheism was strongly influenced by Egyptian beliefs,[90] the Greek and Roman pantheons featured fewer gods than those of either Mesopotamia or Egypt. Nevertheless, the PGR gods still numbered in the hundreds. I mention only a few of the major deities. For a much richer, more detailed presentation of Greco-Roman gods, see the huge tome written by Mark Morford, Robert Lenardon, and Michael Sham called *Classical Mythology*.[85]

For much of its history, PGR Polytheism included twelve supreme deities who ruled from one mountaintop or another. The PGR pantheons, in general, included more anthropomorphized and human-like deities than the Egyptian or Mesopotamian pantheons. Many of their deities appeared to have deeper-set human traits, emotions, strengths, and weaknesses. Their pantheons consisted mostly of family groups of gods and goddesses with incestuous interconnections. Like the earlier polytheistic religions of Mesopotamia and Egypt, the PGR religions continued becoming more monarchistic with time. By the Hellenistic period, monarchies were already closely associated with the religious life of the kingdoms they ruled.[91] Kings often claimed decent from gods. The Imperial Cult of ancient Rome identified its emperors and certain family members as having divinely sanctioned authority. Church and state were not separated; they were virtually the same. Public servants were often elected to serve as both priest and government official. Deceased emperors could be elected as a state divinity by the senate. In one of the best-known examples, in 42 BCE, the Roman emperor Julius Caesar was officially deified by decree of the Roman Senate.

PGR Polytheism was the dominant branch of religious belief systems for the Western world from the sixth century BCE until the end of the fourth century CE when the emperor of Rome, Theodosius the Great, instituted Christianity as the state religion and banned the practice of hard polytheism in Rome (discussed further in chapter 8).

MINOAN/MYCENAEAN/GREEK POLYTHEISM: DEPICTING GODS WITH IDEALIZED HUMAN IMAGERY

In Ancient Greece, beginning with the Mycenaean culture and influence from Minoan culture, Western philosophy, science, democracy, and the arts all began to develop a robust and broadly influential profile. Religion was a fundamental aspect of early Greek culture.[90] Although the Greeks borrowed certain religious elements from Egypt, classic Greece was independently creative and developed its own highly iconic characteristics. Of all the polytheistic deities through history, the gods of Greek Polytheism are some of the most familiar to popular Western culture.

The Greek gods were individual personified deities with human emotions and rich personality traits. They were more than just different aspects of a single god. The Greeks were some of the first to regularly depict most of their gods and goddesses with realistic, if not idealized, human imagery. The Egyptian, Mesopotamian, and earlier polytheistic religions depicted even their anthropomorphic deities as often having animal features or stylized nonhuman attributes. In contrast, the depictions of most deities by the Greeks were presented as highly realistic, beautiful men and women with humor and dignity. The Romans would later adopt this style. The Mycenaean, Minoan, and Egyptian religions contributed many essentials to the development of Greek religion. As the famous historian, Will Durant, observed, "by a hundred channels the old civilization emptied itself into the new."[92]

Over time, Greek Polytheism evolved into hundreds of different cults or denominations that have yet to be clearly distinguished. In the temples, priests attended male deities and priestesses attended female deities.[90] Many pantheons were tied to particular city-states and varied primarily in who the chief god was. Others varied according to what their particular practices of worship were, their myths about creation and the afterlife, or which gods were members of the pantheon. I present but a few examples.

The Greek poet Hesiod (eighth and seventh century BCE) was one of the earliest written sources of information about ancient Greek Polytheism. His poem "Theogony," which translates as "the genealogy or birth of the gods," describes three generations of Supreme Gods and Goddesses: the *Primordial Deities*, the *Titans*, and the *Olympians*.[93] Each succeeding pantheon overthrew the previous one through war and battle for domination. Within each pantheon, all of the gods were related to each other through bloodlines. Sometimes births resulted from mating between two gods, and other times birth was the result of parthenogenesis (asexual reproduction from one parent only). The first gods of Greece were largely tied to natural processes and phenomena. They represented the most basic concepts such as "nothingness," "earth," or "sky." The early Greek gods focused on the creation of other gods rather than things in the mortal world or on anything to do with humans. Even after the first humans were created, the god's concern with them ranged from fondness, to apathy, to general dislike.

THE *PRIMORDIAL DEITIES*: BUILDERS OF THE UNIVERSE FOR GREEK POLYTHEISM

The *Primordial Deities* of Greek Polytheism were collectively the first generation of Greek gods and goddesses. They represented fundamental forces of nature and generally were not worshipped. Their many different interpretations vary over time and place. I present only a few particular examples. In general, the Primordial Deities began with *Chaos* (sometimes simply primordial nothingness and not a god at all, and sometimes the god of nothingness and disorder).[85] Chaos had no gender and ruled alone until *Gaia* (the primal mother goddess of the earth and ancestral mother to all life, also known as *Ge*) sprang from the nothingness.[94] Shortly thereafter, Chaos gave birth to many other entities, including *Tartarus* (god and place of the deepest underworld), *Eros* (god of love and fertility, later called *Cupid* by the Romans), *Erebus* (god of darkness and the gloomy underground cavern through which the dead had to walk on their way to the underworld), and *Nyx* (goddess of night). Nyx was the consort of Erbus and mother to more than a dozen deities, including the gods and goddesses of day, air, destiny, friendship, sarcasm, and deceit. The Primordial Deities were the ones who built the universe, as the Greeks knew it. Gaia gave birth to *Uranus* (god of the sky), the *Ourea* (nine goddesses of the mountains), and *Pontus* (god of the sea). Gaia later mated with her son *Uranus* to produce many gods and goddesses along with an assortment of monsters including the *Cyclops* (the one-eye giants) and the *Hecatoncheires* (human-like giants with one hundred hands and fifty heads). Inbreeding was apparently not always a great idea, even with the gods. The first twelve deities she produced with Uranus eventually became the second generation of deities, and they were known as the *Twelve Titans*.[85] They would eventually overthrow the Primordial Deities to assume godly rule over the Greek world.

THE *TWELVE TITANS* AND THEIR OVERTHROW OF THE PRIMORDIAL DEITIES

The original *Twelve Titans* are sometimes considered the first formally organized pantheon of Greek gods. They became the supreme deities of the Greek pantheon after overthrowing the Primordial Deities. The Twelve Titans included six male and six female deities: *Cronus* was the god of time and the leader of the Titans. *Rhea* was goddess of fertility and sister-wife of Cronos. *Mnemosyne* was the goddess of memory. *Theia* was the goddess of sight. *Phoebe* was the goddess of prophecy. *Themis* was the goddess of justice and wisdom. *Tethys* was the goddess of the primal font of freshwater, and the mother of all the rivers and seas of the world. *Oceanus* was brother and husband of the goddess Tethys. He was the father of all the rivers and seas of the world. *Hyperion* was the god of heavenly light. *Coeus* was the god of intellect. *Crius* was the god of heavenly constellations. Finally, *Iapetus* was the god of mortality. Cronos overthrew his father Uranus by castrating him, which enabled him and his eleven siblings (fellow Titans) to seize control of the cosmos from the Primordial Deities. The Twelve Titans settled on top of Mount Othrys in central

Greece and produced many additional gods and goddesses over time. Cronos and his wife Rhea later gave birth to a particularly important group of deities: Poseidon, Hades, Hestia, Demeter, and Zeus. These gods, eventually known as the *Olympians*, would ultimately lead to the Titans' downfall.

THE *TWELVE OLYMPIANS* AND THEIR OVERTHROW OF THE TWELVE TITANS

Just as Cronus overthrew his father Uranus and the Twelve Primordial Deities to seize control of the cosmos; Zeus overthrew his father Cronus and the Twelve Titans to do the same.[85] Zeus and his siblings fought and defeated the Titans in a ten-year-long war called the Titanomachy. After Zeus and his kin claimed ownership of the cosmos, he put Hades in charge of the underworld, Poseidon in charge of the sea, and he took the sky for his own domain. He added a few more deities to his group to make an even dozen and moved to Mount Olympus, about 110 kilometers north of Mount Othrys. These deities became known as the Twelve Olympians. They replaced the Twelve Titans as the Supreme Gods and Goddesses of the Greek Pantheon. Mount Olympus is the highest mountain in Greece, echoing the ascendance of the Olympians above the Titans of Mount Othrys (only the sixty-first highest mountain in Greece). The Titans were imprisoned in a dungeon, deep in the underworld. The god Atlas, who was the nephew of Cronus, had sided with the Twelve Titans in the war. For that, he was condemned by Zeus to hold up the sky on his shoulders for all time. Zeus's brother Hades stayed in the underworld rather than joining the Olympians. There, he guarded the incarcerated Titans as well as the souls of dead humans who crossed the River Styx. The name *Hades* eventually became synonymous with the underworld.[94] Later, many Christians would also adopt the name as a euphemism for hell.

One ancient writer who wrote extensively about the Olympian pantheon was the legendary Homer. He lived in the seventh century BCE and wrote the famous epic poems the *Iliad*, which was 15,693 lines in length, and its sequel the *Odyssey* that was 12,110 lines long.[95] Homer summarized the names and relationships among the Olympian gods and goddesses, as well as their heroic deeds, specific powers, and responsibilities. He portrayed them in all their combative glory as well as their petty fears and vengeful natures. They were shown to be promiscuous, incestuous, and generally not interested in the fate of humans. Homer was the first to portray the Twelve Olympian gods and goddesses as above all others. He specified Mount Olympus as their living place where they would rule as a council. The names of the Twelve Olympians are familiar even in Western culture. The following characteristics are based mainly on ancient works by Homer and on the classics scholars

Walter Burkert[96] and Jenny March[94]). I refer the reader to those sources for greater depth on these and other Greek deities.

- *Zeus* was the king of all Greek gods, including the other Olympians.[85,90,94,96] He was supposedly born in Crete and was the god of the sky, lightning, thunder, law, and justice. He was brother or father to all the other Olympians. To say Zeus was a womanizer would be an understatement. He had a wife who was the goddess *Hera*, plus at least seven more divine consorts, including close relatives with whom he had children. He also had many additional lovers who were mortals. Many of his children became gods in the Greek pantheon. *Dionysus* was his son born of a mortal named Semele, which did not go over well with his wife *Hera*. Hera is often portrayed as jealous of Zeus's philandering, and she facilitated the death of Semele and others who were in her bad graces. For a time, Zeus gave a nymph named *Echo* the job of distracting Hera from his numerous affairs by talking incessantly. When Hera discovered the deception, she cursed Echo to forever repeat the words of others (her name is the source of the word "echo"). Even the most powerful gods of Greece were not generally portrayed as models of good moral behavior.
- *Hera*, sister wife of Zeus, was the Greek goddess mainly of women, marriage, family, and childbirth.[85] She is portrayed as often trying to get revenge on Zeus's many lovers and children. Her temples from the eighth and sixth centuries BCE are some of the earliest-known monumental Greek temples. She is thought by many historians to have been worshipped on the Greek mainland before most of the better-known gods, even Zeus.[90] The many conflicts between Hera and Zeus may reflect philosophical shifts between the old religions, in which a woman was the powerful Earth Mother and Queen, and the later, strongly male-dominated pantheons of Zeus and the Olympians.
- *Poseidon* was Greek god of the sea, earthquakes, and horses.[85,94] He was the brother of Zeus, and the legendary island of Atlantis was his domain. His symbol was the trident, which he used to spear fish and stir up storms. He was the chief deity for the Greek cities of Thebes and Pylos.
- *Demeter* was the Greek goddess of agriculture, and a great sustainer of life for men and beasts alike.[85,94] She was another sister and wife of Zeus. Their daughter *Persephone* became the wife of Hades and queen of the underworld.
- *Athena* was Greek virgin goddess of wisdom, war, crafts, and skill.[85,94] She was another child of Zeus and was also his favorite. She had no mother. Zeus was experiencing an enormous headache one day, and he asked his son Hephaestus for help. Hephaestus obliged and whacked his father on the side of the head with a large hammer. Out popped Athena, fully grown wearing a full suit of armor

with weapons in hand. She became the guardian of Athens, where the Parthenon served as her temple.

- *Apollo* was the Greek god of the sun, reason, intelligence, music, and the arts.[85,94] He is the son of Zeus and the goddess Leto, and the twin brother of the goddess Artemis. He is the only god to appear in both the Greek and Roman supreme pantheon under the same name (page 205).
- *Artemis* was the Greek goddess of the hunt, wild animals, chastity, childbirth, and the moon.[85,94] She is often depicted as a huntress carrying a bow and arrows. She was daughter of Zeus and the goddess Leto, and twin sister to Apollo. Her name became *Diana* in Roman Polytheism. Her origins are very old, and she is thought to have ideologically evolved from earlier earth-mother-goddess religions.
- *Ares* was the Greek god of war, and the son of Zeus and Hera.[85] He represented the raw violent aspect of war, in contrast to his sister Athena, who represented the strategic thoughtful side of war. His sons are *Phobos* (god of fear) and *Deimos* (god of terror). His sister/lover *Enyo* was the goddess of discord.
- *Aphrodite* was the Greek goddess of erotic love, beauty, and procreation.[94] Her name is the source of the word "aphrodisiac." She was also the patron goddess of prostitution. She was the daughter of Zeus and the Titan goddess Dione, or possibly the daughter of Uranus.[85] In Greek mythology, Aphrodite once had a feud with the goddesses Athena and Hera that resulted in the Trojan War. Her cult in Greece was strongly influenced by the Phoenician-Egyptian goddess *Astarte* and the Mesopotamian goddess *Ishtar*.[96,97]
- *Hephaestus* was the Greek god of fire, volcanoes, stone working, and metal working.[94] He was son of Zeus and Hera. He was the blacksmith for the Olympian gods and made all their weapons. He was worshiped primarily in Athens, where a temple was devoted to him. He fashioned the arrows of *Cupid* and was generally a peace-loving god. He was married to *Aphrodite*, although she was often unfaithful to him with *Ares* (among others).
- *Hermes* was the Greek god of travelers and thieves.[85,94] He was known as the divine trickster who outwits other gods for his own satisfaction and occasionally for the sake of humankind. He was also the emissary of the gods. He is the son of *Zeus* and the mountain nymph Maia. He moved freely between the worlds of the mortal and the divine, and he conducted souls into the afterlife.
- *Hestia* and *Dionysus* are each said to have occupied the twelfth position at one time or another on Olympia. There is some uncertainty about the nature of this transition. *Hestia* was the Greek goddess of the hearth and its sacred fire. She was goddess of family, home, and state, and another virgin goddess of chastity.[85] She was the sister of Zeus and was the first to occupy the position of Twelfth Olympian. *Dionysus* was the god of the grape harvest, winemaking and wine, fertility,

theater, and religious ecstasy, who later replaced Hestia. The Mycenaean Greeks worshiped him as early as 1500 BCE. Although his origins are controversial, he is thought to be the only Olympian of mortal birth. One possible explanation for the transition between the two deities is that Hestia was said to be passive and self-effacing.[98] By later surrendering her seat to the more forceful Dionysus, she prevented heavenly conflict.

The earliest creation stories in Greek Polytheism have nothing to do with humanity. For a long time, these stories focused on gods creating other gods who themselves then created other gods. Humans were said to have gradually entered the picture much later, almost as an afterthought. The many different detailed versions of how this happened represent the different ideological branches within Greek Polytheism. One general version goes like this:[85] During the rule of the Titans on Mount Othrys, the nephew of Cronus, *Prometheus*, created the first human out of clay. Hesiod describes five successive ages of the beginnings and development of humanity in his poem "Works and Days" written sometime between 750 and 650 BCE.[85,93] These stages were the *Golden Age*, the *Silver Age*, the *Bronze Age*, the *Heroic Age*, and the *Iron Age*. (These names should not to be confused with the way they have sometimes been used by archaeologists to describe the technical ages of developing civilizations.)

The *Golden* ("good and noble") *Age* of men occurred during the rule of Cronus and the Titans. During this time, men lived among the gods, mingling freely among them. There were no women. Men did not have to labor to feed themselves because food was in abundance and was free for the taking. It was a time of peace and harmony. Men lived to a very old age and retained their youthful appearance. Death was peaceful, like going to sleep, and after that, their spirits lived on peacefully as guardians of the mortals. The Golden Age ended when the Titan god Prometheus, creator of humans, presented humankind with the gifts of fire and the arts. Zeus, who had no love for Prometheus's creations, became outraged at this and punished Prometheus by chaining him to a rock in the Caucasus, where an eagle would rip open his abdomen each day and feast on his liver. As a god, Prometheus would heal each night, but the next day the eagle would return to rip him open and eat his liver again, creating a chain of eternal suffering. Without women to procreate, the humans of the Golden Age were doomed to eventual extinction.

Adding to the mix, Zeus instructed his son Hephaestus to create the first woman out of clay, the beautiful maiden Pandora. The gods entrusted Pandora with a jar (sometimes portrayed as "Pandora's box" in popular literature) that she was forbidden to open. However, her uncontrollable curiosity got the better of her and she opened the box, thereby unleashing all manner of evil into the world. Hesiod's version of the Golden Age of men and Pandora's story eventually influenced the early development of Abrahamic theology. The parallels between Pandora and Eve are obvious. Each myth indicates that the first woman in the world was created after the first man. Each shows the woman being first to succumb to the temptation of evil and leading the man astray (helping fuel a general decline for the authority of women in much of Organized Religion that will be summarized in chapter 12). Each myth facilitates a fall from a state of plenty and ease, to a state of suffering and death as a punishment for transgression of divine law, often instigated by a woman. Although the human inhabitants of the Golden Age eventually died out in physical form, their benevolent spirits continued to roam the earth.

The *Silver Age* was the second attempt at jump-starting humanity. The Olympian gods created a new generation of humankind, this time with both men and women. These humans were said to be inferior to those of the Golden Age, and they needed to live one hundred years with their mothers to grow up. As adults, they lived only a short time and spent most of that time in conflict with each other. They also refused to honor the gods properly. Zeus eventually lost patience with this stage of humankind and destroyed it. After death, they became the "blessed spirits" of the underworld.

The *Bronze Age* was the third attempt to fashion a suitable race of humanity. Zeus had become bored, and so created this race out of Ash trees. They were a hardened, tough people, with war as their purpose and passion. They had armor, weapons, and homes forged in bronze. They worshipped the destructive nature of *Ares*, god of war. Rather than eating bread, they ate the hearts of their enemies. The men of this age were consumed with rage and killed each other in large numbers. They blatantly sinned against the gods. Their dead were damned to dwell in the "dank house of Hades," never to see light again. Zeus became so angered by the hubris and cannibalistic savagery of this stage of humankind that he sent a great flood to destroy them, except for Deucalion and his wife Pyrrha.[85] Before the flood, the gods had instructed Deucalion and his wife to build a boat to survive the flood. During the flood, they landed on Mount Parnassus, the only place spared by the flood. This was yet another example of an early Great Flood myth predating that of Noah in the Old Testament.

After the floodwaters receded, Deucalion and Pyrrah, with the help of Zeus, repopulated the earth and began the *Heroic Age* of humanity. This resulted in an honorable race with many heroes who respected the gods. Among them were several well-known demigods (half-god offspring from unions between humans and gods). These included Achilles, Hercules, Theseus, Perseus, and Helen of Troy. Most of the beings from this age died in wars, such as the *Trojan War* and the *Seven Against Thebes* war. The souls of those who died during this age went to a special place in the underworld called Elysium, where the souls would remain forever blessed and happy. Zeus eventually pardoned his father Cronus (defeated leader of the Twelve Titans) from his prison in the darker side of the underworld to become ruler of the righteous and significant souls in Elysium. However, it seems that all good things (or in this case, all ages of humanity) come to an end. The Heroic Age of humanity eventually died out in the mortal world after six generations.

Finally, Zeus created the fifth, and current, age of man: the *Iron Age* (in its mythological sense, not to be confused with the archaeological time period for when iron weapons and tools begin to appear). Hesiod paints a bleak picture of this fifth age of humanity. During this time, humans live a life of stress and hard labor. Morality has declined, children dishonor their parents, and brother fights against brother. It is an age of lies and guiltless brutality. As a result, the myth predicts that the gods will one day forsake humankind once again. Zeus will eventually return to destroy this race, just as he has done to the four previous ages of humankind. This prediction has some similarity to the apocalyptic "end of times" prophesies of many religions today.

In Greek culture, the evolution of the gods was seen as separate from the evolution of human thought. Natural philosophy and science developed to provide rational (natural) explanations for the workings of the world without the reliance on supernatural beliefs or gods. Many hypotheses proposed by early Greek philosophers about the nature of the universe were later falsified. However, the use of logical analysis and hypothesis testing by these early philosophers led to the rise of the scientific method. Early challenges to religion by philosophers were often risky (as is sometimes the case today). In 399 BCE, Socrates was brought to trial for using his Socratic method—that is, finding the truth through questioning, considering opposing beliefs, and then modifying one's previous beliefs. This method included an early somewhat agnostic tone toward supernaturalism. After his trial, which lasted a day, Socrates was found guilty of disbelieving in the gods and corrupting the young people. For this he was sentenced to death. He later died in accordance with his sentence by drinking poison.

In the Western world, we owe much of our cultural core to ancient Greece. Today's best-known international sporting event, the Olympics, was originally dedicated to the Olympian gods and the head of the pantheon, Zeus. It began in 776 BCE as a challenging series of athletic contests among the different Greek city-states. The event was held in Olympia, home of Mount Olympus and the main temple to Zeus. Even after Rome sacked Greece in 146 BCE, the games continued. Then in 393 CE the Roman emperor Theodosius suppressed the Olympic games as part of a campaign to impose Christianity as the state religion, because the games were seen as part of Rome's polytheistic past. It took 1,500 years until the Olympics resumed in 1896. That event was held in Athens, Greece. From then on, the location of the games switched from country to country around the world every four years, becoming the international event that it is today.

ETRUSCAN/ROMAN POLYTHEISM: TRANSFORMATION AND EXPANSION OF THE GREEK PANTHEON

In its early history, Roman religion was a form of hard polytheism, but not a particularly original form. Although there were eventually hundreds of gods in its pantheon, very few of them were strictly Roman. As the Romans extended their dominance throughout the Mediterranean world through conquest, assimilation, and *Romanization*, their general policy was to absorb the deities and incorporate various religious traditions of other civilizations rather than try to eradicate them. This was an important factor of their successful expansionism. It was a strategic way to promote social stability among the varied people within the growing empire.[99] The result was the development of countless Roman subbranches of Old World Hard Polytheism over time, many of which we know little or nothing about. The following is a highly abbreviated summary.

Some of the earliest religious belief systems and gods thought to have been assimilated by Rome were from an ancient people known as the Etruscans. The Eutruscans were the first great historical civilization of the Italian peninsula, and their early development appears to have been influenced by Greek culture.[100] They lived in a region of west central Italy formerly known as Etruria where they persisted from at least the ninth through third centuries BCE. They were united by the Etruscan language, a lost dialect. By the third century BCE, the Etruscans, elements of their culture, and the territory of Etruria had all been absorbed by the Roman Empire. Their language, as well as the religious culture, remains mysterious and poorly understood today, and nobody knows precisely where the Etruscans originated. Unlike the state-organized civilizations of Greece and Rome, they left no major contemporaneous literary records.[100] Some of our understanding of Etruscan Polytheism is interpreted from sparse archaeological evidence. Other information is based on biased literature written by Roman authors after Etruria was incorporated into the Roman Empire.[100,101]

Many Roman gods were absorbed or influenced by the Etruscan pantheon. The Etruscan god *Tinia* was a version of the earlier Greek god *Zeus* and the later Roman god *Jupiter*. The Etruscan *Uni* was a version of the earlier Greek god *Hera* and the later Roman goddess *Juno*. These and many more Etruscan deities were transitional links between the Greek and Roman pantheon.[100] Other deities usurped by the Romans included *Orcus*, the Etruscan god of the underworld (and inspiration for Tolkien's orcs in his *Lord of the Rings* trilogy). In addition to the Etruscan deities that they incorporated into their pantheon, the Romans appropriated deities from Egyptian Polytheism, Greek Polytheism, Celtic Polytheism, and Persian Polytheism, as they conquered or battled with those civilizations.

Romanization of twelve Greek deities of Greece (the Olympians) into twelve Roman deities (the Dii Consentes)

Greek deities		Roman deities
Zeus (god of the sky)	→	Jupiter (god of the sky)
Hera (goddess of women and wife of Zeus)	→	Juno (goddess of women and wife of Jupiter)
Poseidon (god of the sea)	→	Neptune (god of the sea)
Demeter (goddess of agriculture)	→	Ceres (goddess of agriculture)
Athena (goddess of wisdom, warfare, and the arts)	→	Minerva (goddess of wisdom, warfare, and the arts)
Apollo (god of the sun, light, music, poetry, medicine, & prophesy)	→	Apollo (god of the sun, light, music, poetry, medicine, and prophesy)
Artemis (goddess of the hunt and childbirth; twin sister of Apollo)	→	Diana (goddess of the hunt and childbirth; twin sister of Apollo)
Ares (god of war; son of Zeus and Hera)	→	Mars (god of war; son of Jupiter and Juno)
Aphrodite (goddess of love, beauty, pleasure, prostitution, and procreation)	→	Venus (goddess of love, beauty, pleasure, prostitution, and procreation)
Hephaestus (god of fire, volcanoes, and smithery)	→	Vulcan (god of fire, volcanoes, and smithery)
Hermes (god of travellers and thieves; also a devine trickster)	→	Mercury (god of travelers and thieves; also a devine trickster)
Hestia/Dionysus (goddess/god transition of 12th position going from goddess of the hearth, architecture and domesticity to god of wine and ecstacy)	→	Vesta/Bacchus (goddess/god transition of twelfth position going from goddess of the hearth, architecture, and domesticity to god of wine and ecstacy)

The Romanization of the Greek god *Zeus* (god of the sky, thunder, and justice) into the Roman god *Jupiter* (god of the sky, thunder, and giver of victory and peace). (*Left*) Nineteenth-century engraving of the Statue of Zeus in the Temple of Zeus at Olympia (Garland, 2013). The statue was forty-three feet tall, constructed in the fifth century BCE by Greek sculpture Phidias. It, along with the rest of the temple, was destroyed between the fifth and sixth century CE. (*Right*) Statue of *Jupiter*, twelve feet tall, sculpted in marble in the first century CE near Rome by an unknown artist. Statue was moved to the Hermitage Museum of Saint Petersburg in 1862 As you can see, Jupiter appears as a barely transformed version of Zeus.

The greatest influence on Roman Polytheism, however, was from Greece. Roman Polytheism adopted many of the major Greek deities.[85] The Greek gods were much more anthropomorphic than the native Italian gods, and the Greek gods and legends were more artistically expressed with more power. The polytheistic religions of Greece and Rome eventually merged so thoroughly that they are often spoken of as the seamless Greco-Roman lineage. This is shown graphically by the twelve major deities of each civilization (page 205). What the Greeks called the Twelve Olympians effectively evolved into what the Romans later called the *Twelve Dii Consentes*.

The twelve supreme deities of the Greek Olympians and the Roman *Dii Consentes* exemplify the frequent spiritual significance of the number twelve among religions of the Old World throughout history. The significance of twelve for the special deities of Olympia and the Dii Consentes also followed the Twelve Titan deities of earlier Greek Polytheism. Old Norse Polytheism had twelve special gods overseen by Odin. There were the twelve tribes of Israel, twelve apostles in Traditional Christianity, and twelve imams of Shia Islam. According to the *Book of Revelations* in the New Testament of the Christian Bible, twelve gates of heaven are guarded by twelve angels. There were twelve aspects of the god Shiva in Hindu Shaivism and twelve Divine Generals in some of the Buddhist denominations. Going back to the first millennium BCE, Mesopotamian astronomers of Babylon created the twelve signs of the Zodiac, which often were associated with Sumerian deities. The number twelve has had spiritual significance throughout the history of Organized Religion. At least some of these shared characteristics among different religions must be due to interdenominational influence or cultural inheritance, both of which are mechanisms of cultural evolution.

Roman Polytheism was one of many important and highly influential branches of North African-European Polytheism. It later became known for its imperially driven demise and replacement by Christianity (the history of that transition is covered in chapters 7 and 8). This brings to a close my discussion of North African-European Polytheism, a major part of my phylogenetic tree of Old World Hard Polytheism shown on page 158. The proposed sister group to North African-European Polytheism on my phylogenetic tree is *Sibero-American Animistic Polytheism,* although that theory is in need of further testing (indicated with the question mark on my tree).

SIBERO-AMERICAN ANIMISTIC POLYTHEISM AND ITS SPREAD TO THE NEW WORLD

My phylogenetic tree of *Sibero-American Animistic Polytheism* on page 208 includes two sister group branches: *Siberian Animistic Polytheism,* (which evolved in Siberia and continued to evolve and diversify there after the Pleistocene) and *New World Animistic Polytheism* (which was brought to the Americas with the earliest human immigrants from Siberia and eventually diversified into countless New World subbranches). As with other major polytheistic groups presented here, this tree includes only representative samples of the diversity that existed and is in much need of further study.

Although modern humans (*Homo sapiens*) evolved in the Old World by at least three hundred thousand years ago in Africa, they are not known to have reached Eurasia until around 170,000 years ago, or northeastern Siberia until around thirty-two thousand years ago.[102] Between twenty-eight and eleven thousand years ago, early human immigrants began to enter the New World from northeastern Siberia, by way of a vast trans-Pacific land bridge called *Beringia* (also known as the

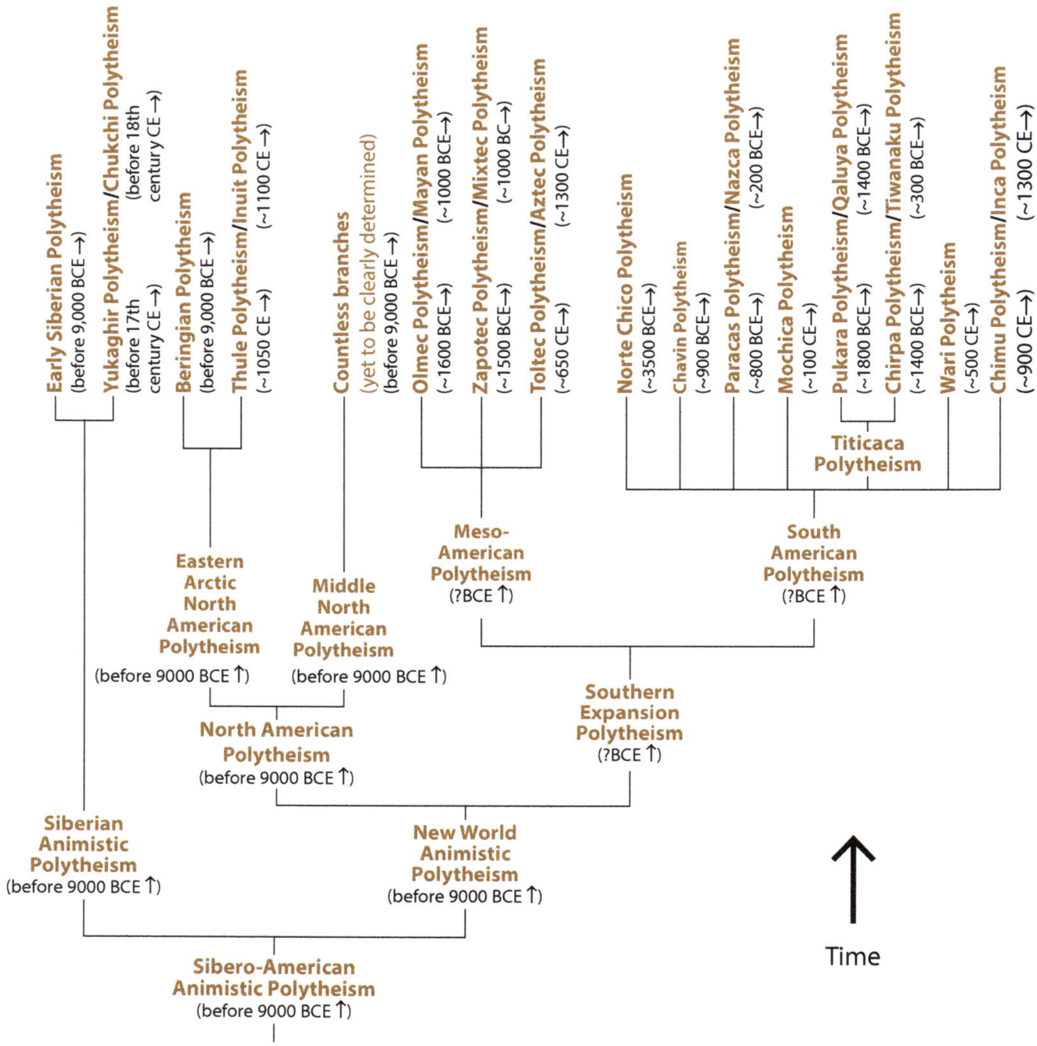

Bering Land Bridge).[2] This land mass existed during the Late Pleistocene Ice Age because sea level was several hundred feet lower than it is today.[103] The lower sea level was due to vast amounts of water being tied up in the glacial ice sheets on the continents. Beringia was massive and was more than six hundred miles wide (north and south) during the glacial maximum (see the figure on next page). Its topography was a vast tundra plain, presumably conducive to hunting and foraging, as well as fishing along the coast, which would have facilitated human migration into the Americas.

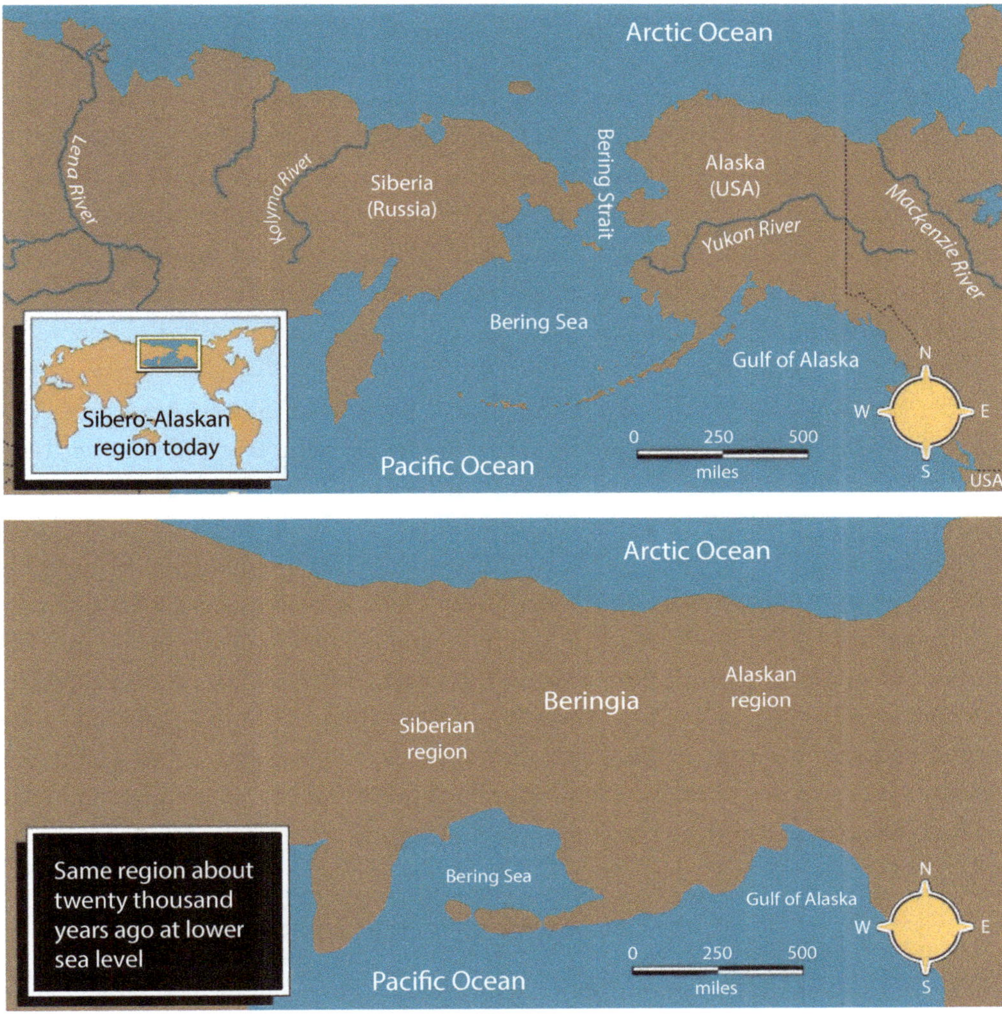

The Upper Paleolithic land bridge, Beringia. (*Top*) The Siberian-Alaskan region today. The Siberian region of Asia is separated from Alaskan region of North America by more than fifty-three miles of ocean between one hundred and one hundred and sixty five feet deep (the Bering Strait). (*Bottom*) The same region approximately twenty thousand years ago when sea level was several hundred feet below what it is today, exposing a massive land connection called *Beringia* between the Siberian region of Eurasia and the Alaskan region of North America. Parts of this land connection persisted until around 9000 BCE (Morritt, 2011).

I theorize that when the earliest human immigrants arrived in the New World, they brought elements of *Siberian Animistic Polytheism* with them. This seeded what would eventually become an independent evolutionary branch of polytheism in the New World (North America, Mesoamerica, and South America), which I term

New World Animistic Polytheism. The extreme divergence over time between the Animistic Hard Polytheism that originally came to the New World and the Animistic Hard Polytheism that remained in the Old World was ensured by the disappearance of Beringia over eleven thousand years ago, which resulted in the geographic isolation between the Old World and the New World for more than ten thousand years (discussed later in this chapter). This is a classic example of what evolutionary biologists refer to as *allopatric speciation*. Allopatry is a mode of independent diversification that occurs when two populations become physically isolated from each other for long periods of time, preventing interchange and creating independent evolutionary trajectories.

We know relatively little about the ancestral forms of Sibero-American Animistic Polytheism. The question mark on my phylogenetic tree reflects the theoretical possibility that the earliest ideological roots of *Sibero-American Animistic Polytheism* may have been an animistic combination of both early Old World Hard Polytheism and early Asian Cyclicism. It also reflects the possibility that a significant number of Paleo-Arctic North Americans may have come from Europe (the much-disputed *Solutrean theory*).[104] Much work is needed to clarify the origin and earliest diversification of *Sibero-American Animistic Polytheism*. We will probably never completely understand the early complexities involved, because they are hypothesized to have begun more than eleven thousand years ago. As with all evolutionary studies of deep history, the further back in time we go, in general, the more missing links and empirically unanswerable questions we encounter.

SIBERIAN ANIMISTIC POLYTHEISM

Little is known about the origins of the early polytheistic religions in Siberia. People inhabited the Siberian region as early as 45,000 BCE.[105] Sculptures of animals and anthropomorphic figurines, as well as symbolic ornamented pendants, are known from archaeological sites in Siberia dated at 24,000–15,000 BCE. Although these ancient people are sometimes collectively called Paleo Siberians, they included many different subcultures, each with their own languages and religious beliefs. Some evidence indicates that they had shamans in charge of communicating with the spiritual world that practiced animistic totemism and that the supernatural beings that they recognized were largely animistic spirits. The world's oldest-known wooden sculpture is an artifact from the Ural region of Siberia referred to as the Shigir Idol, which was most recently dated to be 12,500 years old.[106] Some anthropologists believe it is the top half of what was originally a seventeen-foot totem pole or spirit pole. It was amazingly preserved thanks to being buried in the highly unusual acidic, antimicrobial environment of a peat bog. Sometime after its discovery in the late nineteenth century, it appears to have further deteriorated, and several pieces of its lower portion were lost. However, its original condition after excavation was recorded with drawings by archaeologist Vladimir Tolmachev in the early twentieth century (one of which is shown in the figure on page 212).[107] Later well-preserved ancient Siberian totem poles include one carved by the Nanai people of the Amur River Basin, illustrated (and referred to as an "idol pole") by the Geographer E. G. Ravenstein in 1861.[108] The widespread occurrence of wooden totem poles among early Pacific Northwest American cultures may originally have been due to influence from early Siberian immigrants. Although totem poles have yet to be recorded prior to the 18th century in North America, this may be due to the fact that outdoor wooden sculptures such as totem poles normally begin deteriorating after only a century due to the climate of the northern Pacific coastal region.[109] The earliest totem poles in Siberia, North America, and other parts of the Pacific Rim are likely to have long since disappeared.

Many surviving indigenous Siberian religions (e.g., the Chukchi mentioned in chapter 3, and the Yukaghir) are both highly animistic and totemistic (in the general sense of the word).[110,111] They rarely give consistent names to their spiritual beings or animistic deities, and generally do not provide a concise classification. That may explain why most early North American branches of polytheism were also highly animistic. Shamans in Siberia, as well as early North America, functioned as go-betweens for humans and spirits of plants, animals, living people, and the dead. Early Sibero-American Animistic Polytheism is thought to have generally recognized good spirits, bad spirits, and master spirits in its highly animistic pantheon. They also valued the spirits of departed ancestors. However, these religions generally appear to have lacked a clear recognition of the spirits as gods with clearly defined, broadly accepted characteristics. This may simply be an artifact of poor historical records.

Original records of early indigenous Siberian religions by the people who self-identified with those traditions were either lost, hidden, or destroyed. It is also possible that they never existed. Most remaining records of ancient Siberian religions are documents from the eighteenth through twenty-first centuries CE and were written by people trying to extinguish the indigenous religious culture. The Mongolian ruler Kuchum Khan forced conversion to Islam in Siberia in the sixteenth century.[112] Nearly two centuries later, in 1742, Christian Russian Empress Elizabeth ordered the slaughter of all indigenous men, enslavement of their women and children, and erasure of the indigenous "pagan" culture. Major Dimitry Pavlutsky led the war from 1744–1747 "with the help of Almighty God" to kill the men and enslave the women and children as booty.[113] As with the early indigenous populations of the Americas and Central Africa at the hands of European colonialists, the early

Shigir totem pole (Siberia) **Nani totem pole (Siberia)** **Haida totem poles (British Columbia)**

Totemism and wooden totem poles. (*Left*) The 12,500-year-old Shigir pole from the Ural Mountain region of Siberia, amazingly preserved due to highly unusual environmental conditions preventing decomposition for thousands of years. It is the world's oldest-known wooden sculpture (although there were, no doubt, earlier ones that were never preserved). The figure shows the current state of the specimen on exhibit in the Sverdlovst Regional Museum in Sverdlovst Russia, which has deteriorated somewhat since its original excavation in the late nineteenth century. Next to it against a light gray background is a drawing of the sculpture made shortly after its excavation showing its original condition (mentioned in text). (*Center*) The 200-year-old "idol-pole" from the Nani people of the Amur region of Siberia, as drawn by German-English geographer Ernst Ravenstein (1861). (*Right*) Wooden totem poles on exhibit in the main hall of the Field Museum in Chicago. They were made in the late nineteenth century by the Haida culture living on an island off the Pacific Coast of British Columbia. Carved from red cedar, these sculptures include a stack of animistic supernatural beings resembling bear, birds, frog, and other animals with a protective watchman at the top. The totems, which stand outside of Haida dwellings, are believed to watch over the land as spiritual guardians.

indigenous populations of Siberia were victims of selective ethnocide and erasure of indigenous historical records by Russian Christian invaders. This erasure contributed to our poor state of knowledge about early Siberian Animistic Polytheism.

Some early Siberians immigrated across Beringia into the New World, bringing their animistic supernatural beliefs with them. Elements of these beliefs evolved into what I call *New World Animistic Polytheism*, which eventually expanded throughout North America, Mesoamerica, and South America.

NEW WORLD ANIMISTIC POLYTHEISM AND THE FIRST HUMAN SETTLEMENTS IN THE AMERICAS

As previously discussed, the earliest elements of *New World Animistic Polytheism* (*sensu* the tree on page 208) likely began as a derivative of *Siberian Animistic Polytheism* from the Old World. These elements likely came to North America with the first humans to arrive there, who were from Siberia and Northeastern Asia. These immigrants first began to arrive at least twenty-eight thousand years ago[2] over the massive land bridge of Beringia. They began to be present in the Americas in significant numbers by around fifteen thousand years ago.[103,105,114] The records of early human migration to the New World have been reconstructed by archaeologists and historians based on a combination of archaeological materials.[2,115] genetic evidence (comparative DNA studies of indigenous people of the world),[114] and linguistics (comparative language studies of indigenous people of the world).[105]

By eleven thousand years ago, the warmer climate melted much of the glacial ice sheets, greatly raising sea level and submerging Beringia 120 feet below the surface of the Pacific Ocean.[105] The permanent disappearance of the land bridge put a damper on intercontinental interchange by humans for more than ten thousand years, until the late fifteenth century CE, when large seafaring ships once again opened a major conduit to the Americas for conquesting invaders from the Old World. During the period of geographic isolation between Siberia and North America, human interchange would have been extremely limited, at best, via dugout canoes and small animal-skin-covered boats across the fifty-one-mile-wide Bering Strait. The result was that New World Animistic Polytheism evolved independently from polytheism in the Old World for thousands of years. Over time, this resulted in major, allopatric differentiation between New World and the Old World religions.

The supernatural belief systems that the first immigrants brought with them to the Americas from Siberia included animistic pantheons in which animals and even inanimate objects had special power. Later, waves of humans from the Alaskan region moved to more favorable climates farther south and west, giving rise to most other Native American cultures across North America, Mesoamerica, and South America. By 14,600 years ago, human occupation of the Americas had already spread southward to Monte Verde, Chile.[114] By nine thousand years ago, it had reached what is now Tierra del Fuego at the southern tip of South America.

As human cultures subdivided, evolved, and diversified in the Americas, so too did their supernatural belief systems. I divide *New World Animistic Polytheism* into two major branches (see the phylogenetic tree on page 208): *North American Polytheism* (including the *Middle* and the *Eastern Arctic* branches) and the *Southern Expansion Polytheism* (including the *Mesoamerican* and the *South American* branches).

EARLY *NORTH AMERICAN POLYTHEISM* AND THE LOSS OF CULTURAL RECORDS THROUGH CULTURAL ASSIMILATION

The indigenous religions of North America include some of the least-well-known supernatural belief systems, particularly regarding the Middle North American region. These belief systems were once extremely numerous. Before the arrival of European explorers and eventual colonization, Middle North America (the region that became the continental United States, southern Canada, and Northern Mexico) had more than two million natives divided into more than a thousand different tribes, and several hundred different languages. Most tribes probably had their own particular religious traditions. Even today, there are small remnants of 562 federally recognized tribes within the United States. Unfortunately, most precolonial information about Native American religions was lost, destroyed, or misrepresented. Very few comprehensive, un-Christianized records of pre-Colonial Native American religions exist. The lack of such information is due to a number of factors. First, written records by indigenous people of North America are lacking before the eighteenth century. Second, the population of Native Americans severely declined from the sixteenth century to the nineteenth century, resulting in a loss of verbally transmitted records. Third, cultural tradition was lost due to forced assimilation of many remaining indigenous people of North America (as discussed in chapter 3) and the intentional destruction or misrepresentation of indigenous records. Although many pre-Columbian Native American religious belief systems appear to have been forms of animism without distinct deities, some of them also appear to have been forms of animistic polytheism with pantheons of animistic spirits.

I provisionally divide *North American Polytheism* into two provisional branches (phylogenetic tree on page 208). One is ancient *Eastern Arctic North American Polytheism* (polytheistic beliefs of the early human immigrants into the North American Arctic). The other is the provisionally named *Middle North American Polytheism* (polytheistic beliefs of indigenous people from what are now the lower forty-eight states of the United States, southern Canada, and northern Mexico). Much comparative research is needed on the basic composition, diversity, and history of these groups before their historical relationships can be adequately understood.

EASTERN ARCTIC NORTH AMERICAN POLYTHEISM, AND THE EARLY PALEO-AMERICANS

The earliest humans to cross the Bering Land Bridge into North America by land or by small boats along its shore arrived tens of thousands of years ago. Human marks on mammoth bones, dated at twenty-eight thousand years before present, have been reported from the Bluefish Cave archaeological site in the Yukon Territory of northwestern Canada.[2] Human footprints dated at between twenty-one and twenty-three thousand years of age also have been reported from as far south as New Mexico.[116] The earliest immigrants to the New World (sometimes referred to as Beringians) were hunter-gatherers, and they are known from a number of archaeological sites.[115] The Arctic Beringians left no hard clues to their religious beliefs. It is believed, however, that their traditions included versions of highly animistic polytheism, because those elements were prevalent in Siberia and became extremely widespread among later pre-Columbian cultures across the New World. Although the Arctic Beringian culture diversified and rapidly moved throughout the Americas to give rise to countless other groups of Native American people, the original populations of Beringians that remained in the Arctic eventually disappeared between five and six thousand years ago, presumably having moved south to better conditions.[117] Thousands of years later, other waves of seafaring Siberian immigrants arrived in Arctic North America in small boats. These immigrants gave rise to later arctic cultures, such as the *Thule* and their ideological descendants the *Inuit*.[110,118,194] Some early Inuit pantheons contained elements of highly animistic polytheism, such as the sea goddess *Sedna*[118] (also known as *Takanakapsaluk*[3]). Her severed fingers became the whales and seals that the Inuit depended on for food. Inuit religious belief systems were strongly influenced by Western Christian missionaries during the eighteenth and nineteenth centuries.

MIDDLE NORTH AMERICAN POLYTHEISM: THE GREAT SPIRIT AND THE ESSENCE OF NATURE

Most of the early supernatural belief systems of the indigenous people of Middle North America (the region that became the continental United States, southern Canada, and Northern Mexico) appear to have been highly animistic forms of polytheism, especially if we consider the *sacred spirits* of Middle North American tribes and cultures to be analogous to the *deities* of belief systems elsewhere. Most North American deities are based on forces of nature.[16] Their sacred spirits are rich with the symbolism of seasons, weather, plants, animals, earth, water, sky, and fire, and they have many different combinations of spirits and practices. In general, most Native American religious belief systems maintain that the living are intimately connected with the souls, or spirits, of nature and the dead. Commonly deified spirits, among various tribes, include a *Great Spirit*. This supreme deity or spirit goes by many different names, including *Wakan Tanka* of the Lakota, *Gitche Manitou* of the Algonquin, and *Unetlanvhi* of the Cherokee. Lesser spirits include spirits of the *hunt*, the *sun*, and the *moon*. Even the four directions of the wind each have a dedicated spirit.[118] Some religious traditions had shamanistic elders, called medicine men, who carried out sacred rituals. Most early indigenous people of Middle North America showed reverence for ancestors, although some, such as the Apache and Navajo, feared them.[16]

Artifacts of the *Clovis* culture, dated mostly around 11,000 BCE to 9000 BCE, are thought to represent one of the earliest-known widespread cultures in the Americas. This culture is characterized primarily by a multitude of the distinctive "Clovis" stone spear points. More than ten thousand Clovis points have been found, scattered throughout more than 1,500 sites covering much of North America.[119] DNA analysis of 12,700-year-old human bones from a Clovis site in Montana revealed a close genetic connection of the Clovis people to Asia.[114] The Clovis people were likely descendants from the earlier Paleo-Americans who moved south and may have been ancestral to most of the Native American cultures that evolved later throughout the Americas. Virtually nothing is known of Clovis religious beliefs other than they sometimes buried their dead with offerings of Clovis points (and presumably certain perishables) on behalf of their deceased. The Clovis points are unknown from Siberia and therefore a probable invention by early North American immigrants. The Clovis people appear to have been nomads, leaving no sign of major urbanized settlements.

It was not until much later, during the Formative Period (beginning around 1500 BCE), that Middle North Americans changed from a nomadic hunting and gathering culture to a sedentary agriculture-based culture with the development of large villages, pottery, and weaving and an enhanced set of religious ceremonies and rituals. Many eastern and southeastern tribes believed in a sky god or Great Spirit as the creator, a great mother goddess, and various other spirits for the sun, moon, thunder, various animals, and other facets of nature. Many wise bird-like spirits inhabited the heavens. Like so many other forms of hard polytheism around the world, most Native American branches had a trickster deity known for trickery, mischief, and deceit.[114,118] Some ancient cultures in the Ohio and Mississippi River Valleys collectively referred to as "mound builders" (1800 BCE–1600 CE) created huge effigy mounds in the form of animals. Some of these mounds surpassed even the Egyptian Pyramids in size. One example is the Great Serpent Mound in Ohio that was built around 300 BCE by the Adena culture (1000–200 BCE). It is more the 1,300 feet long and is the largest serpent effigy in the world. The effigy mounds, pottery, and carved figures all hint at an animistically polytheistic practice (as hypothesized on my phylogenetic tree), but more detailed historical information about the culture and its religious beliefs has yet to be discovered.

Our knowledge of most ancient indigenous (pre-Christianized) *Middle North American* religions is so poorly known that any phylogenetic presentation of its subbranches is well beyond the scope of this book. One of the oldest-known possible signs is a 10,600-year-old human mummy from Spirit Cave in Nevada. It was carefully wrapped in a rabbit skin blanket and a woven mat of long reeds.[120,121] Nothing is known, so far, about the spiritual traditions, if any, originally associated with that mummy, which was embalmed purely by natural forces. It was repatriated to a Native American tribe who reburied it in 2018, after a legal battle over custody of the remains.

Relatively little is known about the hundreds, or perhaps thousands, of branches of Middle North American Polytheism that must have existed over time. The main way pre-Columbian records of early indigenous belief systems were passed down by the indigenous people was through oral transmission. This tradition relied on face-to-face teaching over hundreds of generations. As Native American populations were decimated over the centuries by Western conquest (**Note 3H**), many of these cultural traditions were lost as a result of the physical attrition of the people who transmitted their cultural history. By the sixteenth century, the indigenous populations of Middle North America, along with those of Mesoamerica, were some of the most seriously affected by the European conquest. Historians estimate that in 1491, before the arrival of Europeans, as many as twenty-five million indigenous people were living in Mexico. By 1600, fewer than one million were left.[122]

Loss of unbiased information about early branches of Middle North American Polytheism also resulted from biased transcription by Christian chroniclers as well as from forceful proselytization efforts on Native American children removed from their homes and taken to "Residential Schools" during the mid-seventeenth through early twentieth centuries (**Note 5C**). Perhaps archaeological-reconstructionists, working with Native Americans and new archaeological discoveries, will one day be able to piece together more about lost pre-Columbian supernatural belief systems and their historical interrelationships. Currently, we have little unbiased information about them.

The more southern branches of New World Animistic Polytheism that developed (*Mesoamerican Polytheism* and *South American Polytheism*) have a much better-known history and archaeological record, possibly because their pre-Columbian urban centers were larger and more consolidated. Also, more of their archaeological and written records survived to the present day.

SOUTHERN EXPANSION POLYTHEISM IN MESOAMERICA AND SOUTH AMERICA

Once human populations were established in North America, they eventually began to spread southward into Mesoamerica and South America. The earliest well-represented cultures in each region were usually located near the coast. This location suggests that before the establishment of agriculture, marine foraging helped facilitate the southward expansion. In a few regions, humans eventually established enormous urban centers. Where the so-called cradles of civilization in the Old World were Ancient India, China, and the Middle Eastern Fertile Crescent, in the New World, they appear to have been regions within Mesoamerica and Peru.

The *Southern Expansion* of *New World Animistic Polytheism* evolved into many major subbranches (as per tree on page 208). As one civilization conquered another, it would incorporate gods from the cultures they defeated into their pantheons, which expanded and grew more complex over time. They included highly anthropomorphized deities as well as animistic spirits with great power. Shamans helped make proper connections with the spiritual world. Temples were situated in the center of the cities, and they were often built on top of huge stepped pyramids to honor the gods. Many people lived to serve the gods because they believed that the gods had created them to do so.

HUMAN SACRIFICE

A darker side to the Southern Expansion of New World Animistic Polytheism was human sacrifice, which reached elevated levels not seen before or since (discussed in later sections of this chapter). Admittedly, many historians believe that the number and violent nature of sacrificial rituals was embellished by the Spanish and other Christian chroniclers in their efforts to stamp out indigenous polytheistic beliefs and justify the conquest. Also, many religions (most major religions) have had human sacrifice somewhere in their history (although to a lesser degree). Even Abrahamic Monotheism recalls root influences of human sacrifice. In the Torah/Old Testament of the Judeo-Christian Bible, Abraham was ready to sacrifice his son Isaac to his God until he was stopped by an angel (*Genesis* 22:1–13; discussed further in chapter 6). Other biblical verses suggest that child sacrifice may have been part of the earliest Abrahamic religions (discussed further in **Note 5D**). The general concept of Christianity is based largely on God sacrificing his only son Christ through the Crucifixion. Although human sacrifice largely disappeared in the Old World by the late first millennium BCE, the practice in some cultures of Mesoamerica and South America continued, apparently reaching a whole new level (as transcribed by Christian chroniclers).

Blood sacrifice of living creatures was seen as a powerful form of nourishment for the gods, and human sacrifice was the ultimate offering. It was offered to the gods in gratitude for their benevolence. It appears to have been an essential part of early Mesoamerican and South American Polytheism. Self-sacrifice was part of bloodletting tradition. People would offer thorns to the gods that were tainted with their own blood from piercing their tongue, ear lobe, or genitals. More extreme rituals culminated in the sacrifice of human life. Such sacrifice was thought to be necessary to sustain the universe and to repay the gods. Most victims other than prisoners went willingly to their deaths. They were often worshiped as though they were especially holy. In certain cases, victims saw themselves as providing an altruistic service to their community as atonement for the sins of their people.[123] Human sacrifice may also have been an expression of political power. The social control hypothesis proposed that ritualized human sacrifice was used by social elites as a religiously sanctioned means of terrifying underclasses into obedience.[124] As with so many branches of Organized Religion, the rise and power of Southern Expansion Polytheism was often intertwined with the rise and power of the governing empires where it flourished.

By any modern standard, human sacrifice seems horrific. Often it doubled as a form of capital punishment, and the victims were often war captives or prisoners. However, in some cases, victims were effectively volunteers. It seems inconceivable that anyone would volunteer to have their own beating heart ripped out of their body. To understand why anyone would willingly do this, however, we must consider the ancient cultural context. People believed that the gods would abandon everyone in the community unless some sort of human sacrifice was made. Some willing victims may have looked at self-sacrifice as an extreme form of community altruism. In return, you not only entered the upper levels of heaven but also your family and friends would survive with continued benevolence from the gods. Perhaps it was the equivalent of a soldier falling on a grenade to protect his or her comrades, or even martyrdom. In the modern world, human sacrifice for the sake of religious belief became largely an extinct practice, although it has occasionally been replaced with modern variations, such as suicide bombings, holy wars, and capital punishment for certain crimes. Militant action that results in many deaths is often believed to be following "divine guidance." Religious belief systems are powerful forces in shaping subjective aspects of morality. Religion is an aspect of culture with the potential for good or bad, for life or death. The respect for nature, community altruism, and relative tolerance of diversity were positive elements of New World Animistic Polytheism, just as aspects of charity, honesty, and forgiveness became positive elements of much of Abrahamic Monotheism in the Old World. We may never know about other positive elements of Southern Expansion Polytheism. Much of its history was chronicled by Western powers of conquest, intent on subverting the indigenous religions and portraying them as savage, and heretical.

I divide the *Southern Expansion Polytheism* into two major subbranches, although there was certainly cross-influence between them. One was *South American Polytheism* (focusing on the Andean region of Peru and Bolivia) and the other was *Mesoamerican Polytheism.*

MESOAMERICAN POLYTHEISM: CORN, SACRIFICE, AND THE FEATHERED SERPENT DEITY

Mesoamerica includes an area of what is today the southern half of Mexico and most of Central America. The pre-Columbian civilizations of that region included the most complex urban societies in the Western Hemisphere, representing the New World's earliest development of agriculture, large cities, and hierarchically organized government. Mesoamerica is also one of three known places in the world where writing appears to have originated independently (in addition to Mesopotamia and China).

Mesoamerican Polytheism is documented with archaeological records going back to the second millennium BCE as well as sparse written records beginning in the first millennium BCE. Particularly significant are archaeological traces from the Olmec civilization beginning around 1600 BCE. After that, we have the increasingly informative records from the Zapotecs, the Maya, the Mixtecs, the Toltecs, and the Aztecs. The dates on my phylogenetic tree (page 208) are speculative generalizations averaged from various published sources and intermittent archaeological records. These dates are meant to provide only a rough degree of temporal context. The roots of these traditions extended into earlier forms about which we know little or nothing. The phylogenetic trees in this book are primarily about illustrating broad, hypothetical patterns of historical relationships rather than identifying precise times for the origin of preliterate ideologies.

An important multicultural population center of pre-Columbian Mesoamerica was the ancient city of Teotihuacan, located in central Mexico about thirty miles west of Mexico City. It was the largest pre-Columbian city in the Americas. At its height, it was populated with more than two hundred thousand people.[125] It was known as a great religious center that exerted strong cultural influence throughout northern Mesoamerica. Archaeological evidence indicates that it was a multiethnic city with distinct neighborhoods of Zapotec, Mixtec, and Maya. It was first settled around one hundred BCE by people of unknown origin,[125] and it became well known as a multicultural civilization that built many of the most significant Mesoamerican pyramids. Starting around 550 CE, the city was systematically sacked and burned. By the eighth century, it was effectively gone. Later Mesoamerican civilizations, such as the Aztecs of the thirteenth century, considered Teotihuacan to be a sacred place where the gods met to discuss matters of creation. They called it the *Abode of the Gods*, and they absorbed many of the Teotihuacan deities into their pantheons.[125]

Most Mesoamerican religions followed a sacred calendar year of 260 days. It was used for the purpose of divination to foretell the future. These religions also recognized an independent solar calendar year of 365 days. Every fifty-two years (a cycle called the "calendar round"), the solar calendar and the sacred calendar ended on the same day. According to many Mesoamerican religions, it was foretold that at the end of one of the fifty-two-year cycles, the gods would destroy the world in an apocalypse. On the day after each of these cycles when the Mesoamericans woke to find their world still there, it was a day of thanksgiving.

Other shared characteristics among most of the Mesoamerican branches of polytheism were reflected in their religious structures and monuments. Temples of worship and sacrifice were built with a similar architectural style: truncated pyramids with sanctuaries on top. Their polytheistic beliefs and pantheons also shared an animistic emphasis on agricultural cycles. Food was survival, and the gods controlled its production. Maize (corn) was a major element of concern, often representing the heart of a successful harvest, and most Mesoamerican pantheons had a corn god. According to some authors, corn made up about 90 percent of all calories consumed by early Mesoamerican civilizations.[126] Beans were equally important, and squash and peppers were grown as well. Corn and beans were symbiotically crucial sources of carbohydrates and protein for the survival and growth of early Mesoamerican civilizations.

In Mesoamerican religions, the jaguar was deified as a symbol of war, warriors, and the underworld. A number of deities commonly displayed jaguar characteristics. Mesoamerican pantheons shared many other deities, if not by name, then in spirit. One of the most widespread and important of these was the feathered serpent god, *Quetzalcoatl*, depicted in the next figured. His parents, the creator gods *Ometecuhtli* and *Omecihuatl*, were also present in most Mesoamerican branches, although sometimes by different names.

The ubiquitous feathered serpent god of Mesoamerican pantheons. (*Top left*) Stone carving dated between 800 BCE and 400 BCE from the Olmec civilization. Monument 19, La Venta archaeological site, Mexico. This is one of the oldest-known representations of the feathered serpent god, and it is on the collection of the Museo Nacional de Anthropologia, Mexico City. (*Top right*) The Aztec *Quetzalcoatl* eating a man; drawing from a sixteenth-century CE Aztec codice. In the collection of the Bibliotheque Nationale de France, Paris. (*Bottom*) Steps on the side of the *Temple of the Feathered Serpent* pyramid in the multiethnic (Zapotec, Mixtec, and Maya) city of Teotihuacan, central Mexico. Constructed during the second century CE. Note the numerous feathered serpent heads adorning the pyramid.

Another common element of Mesoamerican Polytheism is the *World Tree,* which spiritually connects both the horizontal and vertical aspects of the universe (although this is a broader cosmological concept that is also present in certain Siberian and Indo-European cultures). The horizontal plane representing the middle world (surface of the earth) encompasses the four cardinal directions (north, south, east, and west) plus a center. Each of the four directions was often associated with a specific deity. The vertical connection is made in the form of a World Tree, which passes through the center of the middle world with its roots in the underworld and its upper branches holding up the heavens. In Mesoamerica, this tree is often envisioned as a Ceiba tree, which is a tall native tree there.[125]

In Mesoamerica, the pre-Columbian empires of the Maya and Aztec people were among the largest and most successful civilizations of their time. They left huge archaeological sites attesting to their rich cultural histories. Most of our knowledge of Mesoamerican Polytheism comes from the writings of its conquerors and later from writings of clerics whose mission was to convert the masses to Roman Catholicism. Consequently, most records have been Christianized to varying degrees. Examples include a number of lengthy historical documents called *codices* that were coproduced by native Mesoamerican priests and Spanish missionaries. They include the *Popul Vhu*, which were passed down for several generations only by oral transmission. They were eventually written down around 1550 and were translated into Spanish by a Roman Catholic Friar named Francisco Ximenez in the early eighteenth century. There is also the *Florentine Codex*, a 2,400-page text on the Aztec culture written between 1540 and 1585 by a Roman Catholic Friar named Bernardino de Sahagun. He spent fifty years studying the Aztecs, and although his observations are filtered through his strong Roman Catholic beliefs, it is still thought to be one of the most reliable documents known on Aztec religion. He has been called the father of American ethnography.

Spanish conquistadors and clergy often worked to purge "pagan" ideas of indigenous Mesoamerican and South American religions in the New World.[127] They worked diligently to destroy indigenous polytheistic belief systems by tearing down temples and monuments, destroying cultural records, and occasionally torturing or killing indigenous people. They set out to "cleanse" the native souls and convert them to Roman Catholicism, often by holding mass baptisms. They saw their mission as a divine one to keep millions of souls from eternal damnation. In the end, their efforts often resulted in a Christianized hybrid religion. Today, many Mesoamerican and Andean belief systems still combine aspects of polytheism, ancestorism, and Western Christianity into a single tradition.[128]

Previously, many anthropologists thought that widely shared cultural characteristics among Mesoamerican cultures (including those of religions) were inherited from the Olmec or Proto-Olmec "mother culture."[129] More recent work, however, indicates that instead of a single mother culture, several cross-influential groups may have developed together in Mesoamerica.[130]

OLMEC AND *MAYAN POLYTHEISM*: THE EARLY RISE OF CIVILIZATION IN MESOAMERICA

Roots of the Olmec civilization are likely to have extended into earlier small Mesoamerican farming and fishing cultures of southern Mexico, at least as far back as 5100 BCE.[131] We know virtually nothing about the religious belief systems of proto-Olmec cultures there. Olmec civilization, which began to flourish around

1600 BCE, was the first major Mesoamerican civilization. Although this civilization left a number of stone sculptures, buildings and temples, its religious belief systems remain a mystery because of the scarcity of any clear written records. Although the Olmec civilization appears to have developed the earliest writing system in the Western Hemisphere (900 BCE), these writings remain largely untranslated.[132] The civilization disappeared around 300 BCE. We do not even know what they called themselves. The name Olmec was given to them by the Aztecs. Their heartland was in what are now the Mexican states of Veracruz and Tabasco. At their peak, their influence reached as far as present-day Nicaragua. They were the earliest-known Mesoamericans to make great buildings and huge sculptures of stone and to start widespread Mesoamerican traditions including animistic gods, ceremonial (possibly ritualistic) ball games, and the ritual drinking of chocolate. They and the Maya were among the first in Mesoamerica to build great pyramids for their gods and rulers. The Olmec civilization was influential to later cultures in the region. Most of our theories about their religious beliefs has been reconstructed based on archaeological objects. The Olmec people left many fine artistic renderings, mostly in the form of stone carvings, pottery, and cave paintings. Although some of these renderings are thought to represent their gods, their most famous stone carvings, aptly named the "Colossal Heads," are not thought to represent gods. According to current consensus, they instead are thought to be images of rulers.[131]

Olmec religious rituals are thought to have been performed by full-time shamans and priests. They built pyramids to honor their dead or gods. Based on archaeological reconstructions, the Olmec believed that the universe had three layers or "worlds" (like most religions). For the Olmec, these worlds included an upper world (the heavens), a middle world (surface of the earth), and a watery underworld (the World Tree cosmology referred to earlier in this chapter). Evidence suggests that Olmec belief systems practiced human blood sacrifice in the form of ritual bloodletting with the use of thorns and other sharp objects. It is unknown how important the sacrifice of human life was in their rituals. Details about the Olmec pantheon are still preliminary. Several gods of later Mesoamerican cultures also appear on Olmec pottery, indicating they likely had a polytheistic belief system. The sexes and names that they gave individual deities have been lost over time. It is unknown whether some of their deities are completely independent entities or merely different aspects of a single deity. Several of these deities, for example, are given jaguar features, a common Olmec theme. The number of different subgroups of Olmec Polytheism is unknown, and the study of Olmec religion is still developing. Based on preliminary studies of archaeological objects, and the comparison of depicted Olmec deities with the deities of later Mesoamerican cultures, following is part of the Olmec pantheon:

- *Feathered serpent deity* (the avian monster) was eventually depicted throughout Mesoamerica, but was first known with the Olmec.[133] This deity was later known as a powerful god with a variety of possible aspects, depending on the time and culture. These included being the god of the sun, sky, war, earth, wind, wisdom, vision, thieves, gambling, or various forces of nature. He was often associated with divine rule. The feathered serpent deity would be later known as *Kukulcan* by the Yucatec Maya, as *Gucumatz* or *Tohil* by the Quiche Maya, and as *Quetzalcoatl* by the Toltecs and Aztecs.
- *Jaguar deity* (the were-jaguar) was thought of as the god of rain, storms, and lightning. It featured an anthropomorphized body and a head with jaguar features. Deities with jaguar and serpentine characteristics were common among most Mesoamerican pantheons. The Olmec jaguar deity is seen by some anthropologists as antecedent to the Toltec and Aztec god *Tlaloc*, the Maya god *Chaac*, and the Zapotec and Mixtec god *Cocijo*.
- *Dragon deity*, also known as the *Earth Monster*, had flame eyebrows, fangs, and a bifurcated tongue. It was often shown with a combination of human, crocodilian, bird, or jaguar characteristics, and its mouth was sometimes open as a cave. It was associated with earth, agriculture, fertility, rain, water, fire, and kinship. It was possibly ancestral to the Maya god *Itzam Na* (high god and creator deity) and the Aztec crocodile god *Cipactli*. It was one of the most commonly depicted Olmec deities.
- *Maize deity* was associated with maize (corn), which was a main crop of Mesoamerican agriculture and critical for the survival of many early civilizations.[129] This deity was often depicted as a human-like figure with a stalk of corn growing out of its head. The element of maize was common among all later Mesoamerican pantheons, including the Maya god *Hun H'unahpu* and the Aztec god *Centoetl*.

There are a few other even lesser-known deities, but currently most knowledge of the Olmec pantheons has been lost to history. The number of deities in the Olmec pantheon appear to have been less than that of many later branches of Mesoamerican Polytheism. Although we know little of its detailed history, it was clearly an influence on later Mesoamerican branches of polytheism, including those of the ancient Maya, Mixtec, Zapotec, and Aztec people. The Olmec religious belief systems appear to have been a critical element in shaping the belief systems of another better-known civilization, the Maya. Some anthropologists believe that the earliest Maya may have originated from Olmec people who were forced out of their homeland to the west by the collapse of the great Olmec city, San Lorenzo.[126] Others contend that before 700 BCE, the Olmec and Mayan civilizations (including their polytheistic beliefs) were transitional, interacting, and difficult to distinguish from each other.

The Maya civilization is the longest lived of all known Mesoamerican civilizations, beginning at least as early as 1000 BCE,[134] reaching a peak between 250 CE and 900 CE, and with a few surviving pockets in existence today. It had one of the better-known religions of pre-Columbian Mesoamerica (its records discussed briefly in **Note 5E**). Although their great stone cities were largely abandoned after 900 CE, the Maya polytheistic belief system was never completely wiped out like that of the Olmecs, or as completely transformed by Western Christianity as that of the Aztecs.[135] Today, around two million Maya still are living in Mesoamerica, many who practice a syncretized combination of animistic polytheism and Western Christianity. Members of the Roman Catholic Trinity, the Virgin Mary, angels, saints, and the devil have merged in various ways with indigenous gods and goddesses of the indigenous pantheon.

Maya religion is rich with stories and personalities. According to some Maya creation beliefs, the god Huracan created the earth and the sky. At first, both earth and sky were connected with no space in between. To make a space for life, a Ceiba tree was planted. The tree grew roots into the underworld, and its branches grew into the upper world pulling the sky up from the earth. The tree's growth created a space on earth for the creation of life. First, the plants and nonhuman animals were created. The gods were not satisfied with creatures that could not speak to honor them, however, and so they created humans.

Many different variations of Mayan Polytheism have different Supreme Gods and different stories about the creation of humans. Some believed that several gods were involved in the creation of humans. One major version of this story of creation says that there have been three different creations of humans and that we are part of the third.[176] In the first creation, people were made of mud. They could speak, but they could not think or move. The gods eventually destroyed the mud people with water. They then tried a second creation of humans with men made from wood and women made from reeds. This group could move and think, but they had no souls and did not honor the gods appropriately. Once again, the gods were dissatisfied, and they destroyed the second phase of human creation. The few survivors of that apocalypse became monkeys. Finally, the third creation effort resulted in modern-day humans. This time they were created from maize dough and blood from the gods. At first, they were deemed too wise by the gods, which almost brought on yet another apocalypse. However, the god Huracan clouded their minds and eyes so they would become less wise. And so here we are. Not quite as smart, but we are still surviving. The Maya believe that even this third stage of creation will eventually be destroyed and replaced by a fourth phase, for which we have yet to know the details of. What an interesting slant on the evolution of humans! As I stated, however, there are several different versions of the creation story depending on time and place within Mayan history.[136]

The Maya World Tree understanding of cosmic levels was more complex than simply earth, heavens, and the underworld. The Maya expanded those levels into multiple sub-levels. The middle world was the earth's surface (much like the previously described Olmec belief). However, the upper world consisted of thirteen steps of the heavens where most of the gods lived, and the underworld had nine levels called Xibalba, which roughly translates as "place of fear.".[136] The indigenous (pre-Columbian) Maya view of the afterlife is poorly known. In post-Columbian translations of the Popol Vuh (possibly a somewhat Christianized context), Xibalba was a horrific place ruled by twelve Maya death gods who presided over all forms of human suffering. There appeared to be not much chance of escaping the place after death. If you were not already a god, reaching the levels of the heavens usually involved enduring a violent death, such as voluntary sacrifice, suicide, or complications in childbirth. According to earlier chroniclers, the idea of enjoying a heavenly afterlife by not doing bad things in life was not originally part of the Maya supernatural belief system. Most Maya hopes and prayers were about advocating for a better life in this world rather than in the next. Today, the possibility of a happier afterlife is somewhat more optimistic in the syncretized version of Maya religion (Mayan Polytheism combined with Western Christianity).[136]

Blood was seen as the vital cosmic energy that came from the gods and must be returned to them through sacrifice.[137] This act was performed through autosacrifice (nonfatal, self-inflicted bloodletting) and human sacrifice (involving the fatal execution of people). Autosacrifice was practiced mainly by Maya royalty and was thought to add to the ability to communicate with ancestors and perhaps even with the gods. The gods demanded blood because of their contribution to the initial creation of humans in which the gods had given their own blood. The tools used to draw blood were typically thorns or stingray spines adorned with symbols to show their religious significance. Drawing blood was often a painful experience and was a ritual probably inherited from the Olmecs. One method was to draw a thorn-studded rope through a hole in the tongue to scatter blood over Maya icons. Men would sometimes pierce the penis rather than the tongue.[138]

Human sacrifice was also practiced by the Maya, although not at the scale that the Aztecs are reported to have practiced it. Maya human sacrifice was reserved for special occasions and generally involved the use of political prisoners for the offering. When a king or queen would ascend to the throne, they would typically sacrifice royals or elites of an enemy state (a Mayan version of a firing squad). These sacrifices were performed in many ways, but the three main methods were decapitation, beating-heart extraction, and drowning the victim in a *cenote* (a natural well). Human sacrifice was also practiced to ensure military victories, rain, or a fruitful harvest.

The Mayan pantheon featured many deities. Different Mayan sects also had different pantheons and deity prioritizations. Some authors contend that during some phases of Mayan history, the different forms all became different aspects of a single sacred quality.[139] However, for the majority of early Mayan history, Mayans appear to have practiced some form of hard polytheism. Most authors believe the idea of a one-and-only Mayan god in Mesoamerica was due to later influence from Christian missionaries.[140] Many deities were depicted with combinations of human and animal characteristics. A few examples of Mayan gods are as follows, although considerable variation existed among the different sects in the way that each of them was revered or referred to over time:[136,137,138]

- *Kukulcan* was the great-feathered serpent god.[137] He was the Mayanization of the Aztec god *Quetzalcoatl* and considered to be one of the most important gods in the Mayan pantheon. In some sects, he was one of the thirteen deities who shaped the world and created human beings. He had different significance in different regions and time periods of Mayan civilization.[137]
- *Hunab Ku* for many Mayans was the first creator god, who after creating the world remained in the highest stratum of the sky without concerning himself with the lives of humans. He had no body, but was nevertheless venerated without form. His son, *Itzamna*, did have a body and eventually became the supreme celestial god of the Mayan pantheon.[137]
- *Huracan* (*Hurakan*) was referred to as the Heart of Heaven and the god of wind, storms, lightning, and fire. The Quiche Maya held him to be the creator of the earth, and the god who created humans from maize dough and blood of the gods during the third human creation event. He also participated in the first two less-successful creation attempts mentioned earlier.
- *Chaac* (*Chak, God B*) was the god of water (rain, rivers, lakes, and the sea). He demanded blood sacrifices for the rain he provided. Young men and women were lowered into wells (cenotes) to drown for this purpose.[136] Chaac was depicted as a therianthrope that was part human and part reptile and he was both worshiped and feared. The Maya believed that Chaac was the god who discovered corn (also known as maize), when one day he broke open a rock and there it was. Corn was one of the central animistic aspects of most every Mesoamerican pantheon.
- *Cizin* (*Kizin*) was a god of death, earthquakes, and the lord of the underworld. His name meant "stench," and he lived in Xibalba (the underworld) with the other death gods. He was associated with the smell of purification and even flatulence (as the gods were involved in *every* aspect of Mayan life). He was often depicted as a dancing human skeleton smoking a cigarette. He burned the souls of the dead, and in postconquest times he was closely identified with the

Christian devil.[136] He was responsible for keeping the souls of evil people in the underworld where they would be subjected to his eternal torments.

Late in history around the time of the Spanish conquest in the sixteenth century, the Maya added a "Unified God" also referred to as "The Only God" named *Hunab-ku*. The other gods were demoted to mere aspects of him. Most historians today believe that he was introduced after the Spanish conquest or that this was a Mayan name for the Spaniards' Christian god.[125,127] The Spanish invaders did much to obscure historical records of early Mayan religious traditions. In 1562, Spanish missionary priest Diego de Landa sought to hasten the conversion of the Mayas to Christianity by destroying a library of their hieroglyphic books on a blazing pyre—all regarded by him as works of the devil. In the words of anthropologist Robert Hall, "The world had possibly not seen such a loss of written knowledge since the burning of the library of Alexandria in Julius Caesar's day."[24] This helps explain why we know so little about the early Mayan belief systems today.

Unlike the branches of Old World Hard Polytheism that remained in the Old World, the branches that first spread to the New World primarily evolved as non-theocratic organizations, with an independent, vocational priesthood.[126] In the area of Mexico known today as Oaxaca, other branches of Mesoamerican Polytheism were also developing at least as early as the first millennium BCE: the *Zapotec* and the *Mixtec*.

ZAPOTEC AND *MIXTEC* POLYTHEISM

The Zapotec and Mixtec were composed of two of the earliest major Mesoamerican cultures in the region today known as Oaxaca. They were culturally similar and coexisted as neighbors for centuries. They largely merged by the thirteenth century.

Zapotec culture predated the Mixtec and had cultural links with the earlier Olmec and Maya civilizations. Although Zapotec civilization began as a series of small autonomous sedentary communities sometime before 1500 BCE, their first major population center, Monte Alban, was founded around 500 BCE. The Zapotec became the dominant culture in the region by at least 200 BCE.[141] Their general religious traditions must have begun coalescing early in their history, so I somewhat arbitrarily use 1500 BCE as their origin on my phylogenetic tree (page 208).

Originally, the indigenous Zapotec religion was a form of animistic polytheism, like most indigenous Mesoamerican religions. The Zapotec also incorporated elements of ancestorism, occasionally invoking recently departed spirits to intervene on their behalf with the gods. They had a hierarchical priesthood. Their pantheon included a number of gods and goddesses, many of which were associated with agriculture, rain, light, earthquakes, and the harvest. To the Zapotec, their deities were the direct cause of everything that happened.[142]

Most Zapotec religious ceremonies followed the agricultural seasons. Like the later Maya and Aztec, human sacrifice was part of their religion, including heart extraction. Sacrificial events had large audiences involving feasting, drinking, music, and dancing. Autosacrificial bloodletting rituals also took place, with participants offering blood taken from their earlobes, tongues, noses, and genitals. Child sacrifices are said to have been done on occasion to ensure the harvest.[142] Many of the Zapotec gods and stories later became part of Aztec Polytheism. We know of somewhere between fifteen and twenty Zapotec deities, including the following:[142]

- *Quetzalcoatl* was the ubiquitous feathered serpent god common to most Mesoamerican branches of polytheism.
- *Cozaana* (*Piyetao*) was the creator god and the god of hunting and animals. His consort was the goddess Pitao *Hulchaana.* He was often depicted as a male with a T-shaped ornament in his nose.
- *Cociyo* (*Cocijo*) was the god of rain, lightning, chili peppers, and the planted field. He was depicted as a man, sometimes with jaguar or serpent features, similar in appearance to the Mixtec and Aztec rain gods. During periods of drought, human sacrifices of children and adults were made to him.
- *Pitao Cozobi* was the god of maize (corn) and his name literally meant "bountiful harvest." In pre-Hispanic times, human sacrifices were offered to him to ensure a successful harvest.
- *Xipe Totec* (Tlatlauhquiis) was the god of spring, planting, flowers, and rebirth for later Zapotec, Mixtec, and Aztec. He was the patron of jewelers and combat. Xipe Totec sacrificed himself by tearing out his eyes and skinning himself alive so corn could be grown and humans would have food. The skinning was symbolic of an ear of corn when it is shucked before eating. The gods saw him as the ultimate sacrifice for the benefit of humankind. In the month of March, a festival honoring Xipe Totec involved human sacrifice. In it, sacrificial victims (presumed prisoners) were skinned alive and either priests or the warriors who captured the prisoners wore the victim's skin in a ritualistic ceremony.
- *Pezeelao* was the god of death, the underworld, and the hereafter. His consort in the underworld was *Xonaxi Quecuya.* The Aztec and Mixtec counterpart of Pezeelao was *Mictlantecuhtli.* Pezeelao was depicted as having a human body and a skull or animal-like head. A sacrificial knife often formed his nose and his hair had two instruments for autosacrificial bloodletting protruding from it. Around his neck was a necklace with a human heart as a pendant.

Zapotec cities often each had their own particular patron deity. Their pantheons later incorporated deities from the Mixtec and Aztec religions. Over time, the Zapotec religion is reported to have become more similar to those branches.[142]

Much of what we know about the Zapotec pantheon comes from a generalized sixteenth-century work by the Spanish Dominican linguist Juan de Cordovain, reinterpreted by modern-day anthropologists.[142] During the Spanish conquest, the conquistadors destroyed the Zapotec temples and used their stones to build Roman Catholic churches. This resulted in a certain amount of syncretization between the Zapotec and Western Christian religions, and a few odd traditions in which Roman Catholic saints were simply substituted for the Zapotec gods. Remnants of the Zapotec civilization have survived to the present with close to one million people located mostly in Oaxaca. Most Zapotec people are Roman Catholic (or some syncretized version of Roman Catholicism and Zapotec Polytheism). Early in its pre-Columbian history, the Zapotec became close to their neighboring civilization, the Mixtec.

The Mixtec civilization began forming by at least 1000 BCE, although its roots may have extended into farming communities as old as 1500 BCE. By 300 BCE, the Mixtec urban states were in place.[130] The early Mixtec religion began as an animistic form of polytheism that included a reverence for deceased ancestors. Its followers worshiped the forces of nature, spirits, life, and death. The personifications of these spirits and forces were the essence of their deities. Although they were graphically symbolized as "beings," they were not worshipped in the sense that other religions worshipped their gods. The Mixtec deities were instead forces to be sensed, appeased, and possibly controlled through proper religious ritual and sacrifice. If the rains were late or the crops were failing, the proper sacrifices and rituals were made to maintain the balance among humanity and nature. To disrespect or neglect these principles would lead to disaster. It was a pragmatic arrangement focused on trying to improve things in the present life. The supernatural realm was merely an extension of the natural world. Little thought was given to an afterlife or spiritual direction for moral or ethical behavior.[130]

Each community had its own patron deity. Like other branches of Mesoamerican Polytheism, they practiced blood sacrifice rituals ranging from autosacrifice (pricking their own ears and tongues) to animal sacrifices and human sacrifices. The ever-present Mesoamerican practice of heart extraction was also part of their ritual, as was flaying both adult and child sacrificial victims alive (although again, based largely on reports from Christian chroniclers). The priests who performed such sacrifices began training around the age of seven. Following are a few examples of the indigenous Mixtec pantheon (although each was known by a variety of different names depending on the region and dialect).[143,144,145] For a more complex and comprehensive look at Mixtec deities, see Jansen and Jiminez.[145]

- *Ehecatl* was the god of the wind and aspect of *Quetzalcoatl*, ubiquitous feather serpent god of Mesoamerica. In the Mixtec pantheon, Ehecatl had four aspects of his own: the north, south, east, and west winds.
- *Xipe Totec* was the god of spring, planting, flowers, and rebirth. He was also a god in the Zapotec and Aztec pantheons. As described in the Zapotec pantheon, he was known as the "flayed god" and supposedly associated with a tradition of human sacrifice involving skinning victims alive. Some authors interpret the flaying as a metaphor for cleaning the fields and preparing for the new agricultural season.
- *Dzaui* (*Dzahui*, *Dzavui*) was god of rain and often was shown as a man with goggle eyes and jaguar-like teeth. Child sacrifices were made to him during times of drought and sickness and at planting time.

As their dominance waned, the Zapotec and Mixtec civilizations became involved in constant battle with other indigenous civilizations, including the *Toltec* and the *Aztecs* (discussed later in the next section). The Mixtec eventually merged with the Zapotec by around 1250 CE, partly through a history of ruling Zapotec elites intermarrying with Mixtec lords.[141] The Zapotec and Mixtec struggled to keep the Aztecs from gaining control of their trade routes to Guatemala and other lands to the south. The Aztecs finally triumphed over the Zapotec and Mixtec people in 1458 and ruled over them for decades until the Spaniards began to overthrow the Aztecs starting around 1521. The Zapotec and Mixtec tried to form an alliance with the Spaniards, which worked for a while. Starting in 1528, however, the Dominican friars established a permanent base in Antequera, Oaxaca. Armed with the fiery zeal to convert the masses and eradicate the indigenous religions, Roman Catholic missionaries arrived in ever-increasing numbers. As in other areas of Mesoamerica, books and records were destroyed to erase the older pagan belief systems. The Spanish invaders exposed the Zapotec and Mixtec to a number of old-world diseases. Smallpox, chickenpox, diphtheria, influenza, scarlet fever, measles, typhoid, mumps, and influenza all took an immense toll. By 1650, the native population had declined from around 1,500,000 people to around 150,000. Much knowledge about Zapotec and Mixtec culture was all but erased from history through the combined effects of cultural indoctrination and imported disease. The same would later happen to the Toltec and the Aztec.

TOLTEC AND *AZTEC POLYTHEISM*: THE LAST MAJOR STAND OF INDIGENOUS MESOAMERICAN RELIGION

The Toltecs were said to have existed in central Mexico from before 650 CE to around 1200 CE. Toltec and Maya civilizations overlapped for centuries, and there was a significant cultural interchange in central Mexico. At first, the Toltecs consisted of a number of loosely connected small communities. Over time, they developed an effective government, incorporating smaller tribes and communities from the north. By the ninth century, they formed an effectively unified military confederacy that included a practice of collecting tribute from people living in the region. Scholars debate the details of the Toltec expansion, but some claim that it was a military enterprise that was driven, in part, by the need to capture prisoners to serve as sacrificial victims.[146] Human sacrifice became an important part of the Toltec tradition; it was greater than that in many earlier Mesoamerican religions, although not quite elevated to the level that would be practiced by the Aztecs. One of the most common subjects of Toltec art is the *Chacmool,* a stone sculpture associated with human sacrifice. It depicts a god in the form of a human male (possibly Chac) reclining with his head and knees raised, and a bowl resting on his stomach. The bowl is thought to have been used to hold the hearts of sacrificial victims captured in war.[146] Much of what we know of the Toltecs comes from the *Florentine Codex: General History of the Things of New Spain* written by the Spanish Franciscan Catholic friar, Bernard Sahagun, between 1540 and 1585.

At their peak between the ninth and twelfth centuries, the Toltec came to occupy territory from what is now the southwestern border region with the United States down through the Yucatan Peninsula.[146] The capital of the Toltec Empire was Tollan. Most anthropologists today believe the remnants of Tollan are preserved at the archaeological site known as Tula. Tollan was dominated by a sacred ceremonial center located in the center of the city, consisting of pyramids, temples, ball courts, and other structures surrounding an airy plaza. The Toltec civilization developed a form of monarchistic polytheism that was ruled by priest kings. The monarchistic element appears to have evolved independently in the New World because its new-world predecessors are believed to have been branches of animistic, less monarchistic forms of polytheism. Some of the reported similarities between Toltec Polytheism and ancient Mediterranean Polytheism may be the result of translational bias. As mentioned earlier, most early written records of New World Animistic Polytheism were recorded by Old World Christian historians and missionaries (e.g., the Spanish Roman Catholic missionaries in Mesoamerica, and the Anglican Puritan missionaries in New England). Such transcriptions were clearly biased against polytheistic belief systems in general. In fact, many Christian missionaries refused to even acknowledge the existence of indigenous Native American religion.[147]

The later Aztecs saw the Toltecs as their cultural predecessors. They appear to have put the Toltecs on an idealistic pedestal and incorporated many of their deities and religious practices. The Aztecs claimed to be descended from the mighty Toltec kings. Most of what we know of the Toltecs has been filtered through generations of Aztec oral transmissions and pictographs that were later translated by Spanish missionaries, primarily Friar Sahagun. You could say most records of Toltec history were first Aztecized and later Christianized, making it a challenge to separate unbiased historical accounts from biased mythical ones. Nevertheless, I am simply comparing the existing consensus accounts of these various supernatural belief systems. Some anthropologists have questioned whether the Toltecs actually existed as a distinct culture or if they were only a mythical invention of the Aztecs.[148] The human mind has a tendency to glorify legendary figures and periods of the past. Today, most anthropologists agree that an actual Toltec civilization was real, although existing written records (as translated by the Spaniards) may reflect a variety of Aztec embellishments and Christian bias.

The Toltec pantheon appears to have contained some deities that were also recognized by the Aztec and Mixtec. However, most existing comparative literature on this is confusing or vague. Based on limited information, following are a few general examples of Toltec deities:

- *Ometeotle* was the first deity of the Toltec and the Aztec. He created himself and thus was outside of time and space. He lived in Omeyocan, the highest of the thirteen levels of heaven. He was a dual deity with both male and female aspects (*Ometecuhtli* and *Omecihuatl*), the parents of all creation.[125] Images of Ometeotle were extremely rare, as were temples or records of specific veneration.
- *Ometecuhtli* and *Omecihuatl* were the godly aspects of the god Ometeotle and the dual creators of all humans as well as some gods. They were in charge of sending a soul to earth each time a new child was born. Recognized by both the Toltec and the Aztec, they resided in the highest (thirteenth) level of the heavens. One of their sons was Quetzalcoatl, the feathered serpent god.[125]
- *Quetzalcoatl*, the ubiquitous feathered serpent god, was a supreme deity for most Toltec subgroups. He was the god of many things, which varied among different Toltec subbranches. He was the god of air, wind, clouds, fertility, agriculture, medicine, thieves, and gambling. The Toltecs built a great pyramid temple to him in their capital city, Tollan.[125]

- *Tezcatlipoca* was another major god of the Toltecs and Aztecs.[125] He was another son of *Ometecuhtli* and *Omecihuatl* and was often pictured as the divine antagonist of Quetzalcoatl.[125] He was associated with darkness, jaguars, and shamanic magic.

Archaeological evidence indicates that the capital of the Toltec civilization met a violent end sometime after 1100 CE. Excavations reveal evidence of widespread fire and destruction throughout the site.[146] Centuries after the fall of the Toltec kingdom, the Aztecs rose to prominence. This would be the last of the great pre-Columbian Mesoamerican civilizations. It was the last indigenous barrier to the colonialization from Spain, whose driving goal was to transform Mesoamerica into "New Spain."

The Aztec Empire dominated what is now central and southern Mexico from around 1300 CE until the Spaniards conquered them in 1521. Aztec society had a hierarchical class structure that was strictly enforced. Rulers, priests, and warriors were at the top. Those who were enslaved were at the bottom. Religion pervaded their lives. Much of their version of Mesoamerican Polytheism was influenced by the Maya, Mixtec, Zapotec, and especially Toltec religions. To the Aztec, the Toltec were the originators of all culture. Like them, the Aztec model of worship had nothing to do with considering gods to be role models or pushing humanity in a humanitarian direction. It was strictly a system of payment to the gods for rain, crops, healthy children, and success in battle. The major form of payment was blood sacrifice. The Aztecs not only practiced animal sacrifice on a large-scale, but also continued the long Mesoamerican tradition of human sacrifice and appear to have taken it to a greater level than any previous Mesoamerican culture. They considered human sacrifice to be highest level through which they could repay their debt to the gods and ask for continued benevolence.

The Aztecs, like the Maya, believed that the cosmos consisted of three general levels, some of which contained multiple sublevels.[125] The highest general level was the upper world or heavens, which contained up to thirteen sublevels. The second level was the middle world (the earth's surface). Finally, the third level was the underworld (with nine sublevels). Each of the thirteen levels of the heavens was inhabited by diverse gods, often depicted as conjugal pairs. The highest level (the thirteenth) was Omeyocan, and it was ruled by the supreme creator god and his aspects. After death, human spirits usually did not make it into the heavens unless they endured a violent death, such as through heroic acts of battle, suicide, death in childbirth, or as a victim of human sacrifice. The last of the four options helped fuel the rise of human sacrifice in the later stages of Mesoamerican Polytheism. Most people who died a normal death were thought to be sent to the underworld, where they would have to endure four years of torturous challenges in its eight upper levels before reaching the ninth and lowermost level for their eternal rest. The Aztecs called this final level of the underworld Mictlan, after the death god *Mictlantecuhtli* who resided there. Mictlan became known as a place where the dead eventually experienced extreme boredom. Originally, the Mesoamerican heavens and underworld were not the same as the Christian heaven and hell, although the Spaniards in their efforts to convert the indigenous people sometimes tried to convince them otherwise.[135]

The most common form of human sacrifice in the Aztec tradition was heart extraction. The heart was thought to be the center of a person's vital energy and something the gods would eat or otherwise absorb. The victim would be tied face up on a sacrificial stone and the priest would quickly cut open his chest or abdomen with a sharp obsidian or flint knife. The heart would then be ripped out of the body while it was still beating and held toward the sky in honor of the sun god.[123] If a courageous warrior was sacrificed, the corpse was sometimes eaten by attending warriors and bystanders. Other forms of human sacrifice included being decapitated, shot full of arrows, drowned, burned alive, or buried alive.[139] Most sacrificial victims other than prisoners offered themselves up willingly. They believed this would ensure their place in the polytheistic equivalent of heaven while at the same time helping to ensure good weather, good crops, and good fortune for their community.[123] Before, during, and after the killing, audience members could also participate by piercing and bleeding themselves in autosacrifice. Blood was the key. It was shed as a mix of payback and atonement to the gods to keep the universe running smoothly and as gratitude for the god's generosity. One of the creation stories of the Aztecs involved the god Quetzalcoatl who offered blood extracted from a wound in his genitals to give life to humanity.[123]

Different gods demanded different classes of human sacrifices. Sometimes the sacrifices were prisoners or those who were enslaved. Other times they were willing volunteers from the community. Usually the victims were adults, but sometimes they were children. Some sacrifices were made at the end of elaborate rituals. One such ritual was for the god *Tezcatlipoca*. In preparation for this ritual, a youthful male victim would spend an entire year impersonating the deity as his living incarnation on earth. During that time, he was treated like a god, given four beautiful women, and offered access to most anything he desired (presumably other than his escape). At the end of the year, it was time to pay the piper. He would then surrender to the priests who would tie him to the sacrificial alter and cut out his beating heart to offer to the gods. The day of terminal sacrifice was accompanied with great public feasting and celebration.[123]

The reported scale of human sacrifice by the Aztecs is astounding. Mesoamerican anthropology specialist Ross Hassig estimated that at the reconstruction of the Great Pyramid of Tenochtlan in 1487, the Aztecs sacrificed well over ten thousand people over the course of four days.[149] Estimates of the scope of human sacrifice by the Aztecs range between 20,000 and 250,000 people annually.[150] In 2018, a single Aztec ceremonial structure was discovered in Mexico City called the *tzompantli* that was constructed almost entirely out of the skulls of thousands of sacrificed men, women, and children.[151] The archaeologists who excavated the site found the victims were evidently in good health and had been residents of the region for months before they were sacrificed. When considering the huge scale of this human sacrifice, it may have been more than religion that drove up the numbers.[150]

The Aztec civilization was under the combined pressure of extremely dense population and nutritionally limited food resources. The development of plant domestication and farming beginning around 6000 BCE had swollen the human population of Mesoamerica far beyond what it was when hunting and gathering was the only method of acquiring food. Aztec population estimates range as high as twenty-five million people. Farming provided abundant carbohydrates, such as corn, squash, avocados, tomatoes, and guava. Corn was particularly important because it could be ground into flour. For protein, however, they were primarily dependent on their bean crop. If that crop failed, the people risked starvation as a result of malnutrition. They lacked animal husbandry and domesticated species for meat protein. Modern sources, such as cows, pigs, sheep, goats, and chickens, were old-world animals that had not yet arrived in the New World with the Europeans. In addition, most of the local game animals had been hunted to extinction by the time of the Aztecs. When the bean crop failed, cannibalism may have helped fill the need for protein, using the tens of thousands of human bodies resulting from sacrifices to the gods.[150] Although cannibalism possibly explains the scale of human sacrifice in the Aztec tradition, most likely, it had a religious element to it as well.

The Aztec religion evolved from earlier forms of Mesoamerican Polytheism. It revered many of the Toltec gods, as well as some from the pantheons of the Maya, Mixtec, and Zapotec. The Aztec population adopted most of their beliefs and gods from other cultures they conquered or met through trade. They also added some of their own gods, resulting in a pantheon of hundreds of gods and goddesses. By the fifteenth century, the chief Aztec god was Huitzilpochtli, one of the rare original Aztec gods. A few predominant members of the Aztec Pantheon are as follows, although the way that different Aztec subgroups interpreted them and others varied over time:[125]

- *Ometeotle* (from the Toltec pantheon) was the supreme dual creator of the universe for the Aztrecs.[125] The duality was expressed by two distinct aspects. One was the male aspect *Ometecuhtli* (or *Tonacatecuhtli*) who was associated with the sun. The other was the female aspect *Omecihautli* (or *Tonacacuhuatli*) who was associated with the earth and fertility. Together the two were thought to have peopled the earth. Both aspects merged to produce a series of lesser deities, including corn deities, rain deities, and a variety of deities associated with generative and destructive properties. This led to a diverse pantheon of gods and goddesses.

- *Huitzilpochtli* was the sun god and warrior god. The Aztec religion was heavily enmeshed with military themes. War was the place where "the jaguars roared" and death on the battlefield was called the "flowery death."[125] Huitzilopochtli became the most important god to those who lived in the Aztec capital of Tenochtitlan. He was a new god that originated with the Aztecs, not a god assimilated from the Toltecs or any other culture. His name meant "hummingbird of the left," reflecting their belief that hummingbirds were the souls of dead warriors. He usually was shown wearing a hummingbird helmet. The Aztecs built a great temple to Huitzilpochtli in Tenochtitlan. It was destroyed by the Spanish conquest in 1521. Spanish soldiers carried away a colossal stone image of Huitzilopochtli reported to have been at the temple, which remains lost to this day.[125]
- *Coatlicue* (*Teteoh innan*) was the earth-mother goddess, ruler of life and death, and mother of the god *Huitzilopochtli*.[125] She was a ferocious goddess depicted as having claws for hands and feet wearing a necklace of human hearts and skulls. To the Aztec, Coatlicue and her family represented the cosmos. She was the earth, her son *Huitzilopochtli* represented the sun, her daughter *Coyolxauhqui* represented the moon, and her 399 other children represented the stars. Huitzilopochtli eventually killed Coyolxauhqui and his 399 other sibling deities in defense of his mother, who the other siblings had planned to kill. This event was periodically ritualized by the Aztecs through human sacrifice of masses of women and captured warriors. They were sacrificed at the top of the Templo Mayor near Huitzilopochtli's shrine after which the bodies were rolled down the temple steps toward a colossal stone image Coyolxauhqui.[125]
- *Mictlantecuhtli* was the god of death and the king of the lowermost level of the underworld. He usually was depicted as a blood-spattered skeleton or a person wearing a toothy skull. The Aztecs saw skeletal imagery as symbolizing fertility, health, and abundance as well as reflecting the symbolic links between life and death. The worship of Mictlantecuhtli sometimes involved ritual cannibalism, with human flesh being consumed in and around the temple.[152]
- *Quetzalcoatl* was the ubiquitous feathered serpent god also revered by the Toltec and Zapotec. His parents *Ometecuhtli* and *Omecihuatl* were important members of the Aztec pantheon, as they were in pantheons of most branches of Mesoamerican Polytheism. In some traditions of Aztec religion, Quetzalcoatl was brother to *Huitzilpochtli*, and they arranged the universe and created fire, the human race, the calendar, and numerous other things of importance. They also helped create and destroy the cosmos four times.[125]

Many other primary gods of the Aztec were inherited from the Toltec, Mixtec, and Zapotec religions.

When Hernando Cortes (1485–1547) and the Spanish conquistadores first arrived in 1519, the Aztec openly welcomed him and his soldiers because they believed Cortes to be the returning god Quetzalcoatl and his soldiers to be his attendants.[153] This belief changed as it became clear to the Aztecs that the intent of Cortes and the Spanish invaders was to conquer their civilization, both physically and culturally. A long and bloody war ensued. Cortes had a few hundred Spanish soldiers and more than a thousand local allies from indigenous populations with no love for the Aztecs. He was also aided by a plethora of diseases that the Spaniards brought with them to the New World. Amazingly, he ultimately managed to defeat the entire Aztec Empire with a decisive battle in the Aztec capital of Tenochtitlan, a city of more than two hundred thousand people. After his victory in Tenochtitlan, the city was renamed "Mexico City" because the Spaniards found the older name to be too difficult to pronounce.

In 1524, three years after the Spanish invaders defeated the Aztecs, twelve Franciscan missionaries from Spain arrived in Mexico. There they began a "spiritual conquest" to "defeat" polytheism throughout Mesoamerica and convert the indigenous populations to Roman Catholicism.[154] The historical aftereffect of this was the extinction of many indigenous pre-Columbian supernatural belief systems, like those of the Aztecs. The Christianization of Mesoamerica also involved the loss of most pre-Spanish records of Mesoamerican religions. The historical records that we have today are largely the product of centuries of historical reconstruction begun by Christian chroniclers, followed by anthropologists, archaeologists, ethnologists, and linguists. Some vestiges of Aztec Polytheism were syncretized into Mesoamerican Christianity. An example is the Roman Catholic Church of Our Lady of Guadalupe in Mexico.[16] This church was dedicated to the mother of Jesus and was built on the site where the Aztec mother goddess *Tonantzin* had previously been worshiped. Following the Spanish conquest of 1519–1521, the Spaniards destroyed the temple to Tonantzin and built a Christian church on the same spot. Many indigenous people believe Our Lady of Guadalupe to be a continuation of Tonantzin rather than the Roman Catholic Virgin Mary. Her image also served as a unifying force when Mexicans fought Spain for their independence in 1810.

The history of polytheism in the New World also extended into South America where it is best known from the Andean region.

SOUTH AMERICAN POLYTHEISM AND THE MYSTERY OF ANCIENT MUMMIES

When the earliest people migrated to South America from the north is unknown, but by at least twelve thousand years ago, they had already reached the southern tip of the continent.[155,156] Although we do not know exactly when the earliest religious traditions first appeared in South America, at least some of the earliest settlers probably had a supernatural belief system. Based on the early widespread occurrence of animistic polytheism throughout pre-Columbian South and Central America, the religious traditions of the early settlers probably involved multiple animistic deities with close connections to the natural world. Most of what we know about ancient South American religion comes from Andean and Pacific coastal civilizations of Peru, and the Lake Titicaca Basin extending into Bolivia. There are seven-thousand-year-old mummies of the Chinchorro culture in what is now northern Chile, two thousand years older than the earliest mummies from Egypt. Unfortunately, we have no details about their religious significance. There were also a number of early communities in the Amazonian, Uruguay, and Ecuadorian regions, but we know virtually nothing about the earliest supernatural belief systems of people from those areas with few exceptions (e.g., *Baniwa* Polytheism from the Amazon Basin[157]). Consequently, the only indigenous (pre-Christianized) South American religious traditions covered here are the better-known Andean branches, mostly from Peru and northern Bolivia.

Substantial pre-Hispanic written records of South American religion do not exist. The existing information consists of objects from a few archaeological sites and second- or third-hand written accounts by missionaries. Although the information is meager, anthropologists have reconstructed a rough picture of early Andean belief systems. Most Andean religious traditions were highly animistic, with people, animals, and objects, all infused with a life force or spiritual essence. Some appear to have had a strong element of ancestorism (veneration of ancestors).[158] Like the Mesoamerican gods, Andean deities were often centered on agriculture or aspects of nature. Also like the Mesoamericans, several Andean cultures built pyramids in honor of their gods. Human sacrifice was an important part of South American Polytheism, going back more than four thousand years to the archaeological site of Cerro Sechin in northern Peru.[159] I present only a highly generalized introduction to what was a rich diversity of animistic polytheism throughout South America. Some of these groups probably had dozens or even hundreds of subbranches, most of which remain to yet be clearly distinguished. I begin with the earliest evidence of revered deities in South America from the cradle of civilization on the South American continent.

EARLY URBAN CIVILIZATION AND THE ORIGIN OF THE SOUTH AMERICAN STAFF GOD

The ancient civilization of Norte Chico, sometimes also called Caral-Supe, was located near present-day midcoastal Peru in four river valleys. It is known as the cradle of Andean civilization. It is thought by anthropologists to contain the oldest urban center in the Americas, predating even the Olmec civilization of Mesoamerica by more than one thousand years.[160] Although people were already living in the region by 9200 BCE, the earliest known urban center of Norte Chico was the city of Huaricanga in the Fortaleza River Valley, dated at around 3500 BCE.[155] By 2600 BCE, the city of Caral in the Supe River Valley had become a major urban center within the Norte Chico region, with monumental architecture and irrigated farming.[161] It contained more than three thousand inhabitants at its peak and was surrounded by more than a dozen smaller communities within the Supe River Valley. Details about the age, history, and even what to call the Norte Chico culture have been a source of passionate controversy within academic circles during the twenty-first century (discussed in **Note 5F**). Nevertheless, the consensus is that the people of Norte Chico were a significant force in the Andean landscape for more than 1,200 years and that they helped provide a cultural foundation for later cultures in the Andean region over the next several thousand years.[162]

Evidence regarding Norte Chico religious tradition is extremely limited. The people of Norte Chico (also known as Caral) had no writing, and there is no recorded text of verbally transmitted records. Nevertheless, Norte Chico left a number of pyramids or terraced hills that some anthropologists believe were places of communal spiritual exaltation.[163] Peruvian archaeologist Ruth Shady interpreted the design of these structures and other buildings of Caral to have a special relation to the sun, moon, stars, and mountains, reflecting a Three World Andean concept, much like what was discussed earlier for Mesoamerica. The three worlds were *Kay Pacha* (the world where we are), *Hanan Pacha* (the higher world where the stars are), and *Ukhu Pacha* (the underworld where the ancestors now reside).[160] This concept has also been reported in other Andean cultures, such as the Inca.[164] In later Andean religions, these worlds were each associated with a specific god. Shady identified the Caral society as a polytheistic theocracy.[160] Therefore, I provisionally included the Norte Chico belief system on my tree of South American Polytheism (page 208). Their culture appears to have differed from Mesoamerican culture and other South American cultures in it lacks evidence of warrior glorification, human sacrifice was rare, and the basis for the ruler's power was primarily a theocratic use of tradition and ceremonialism.[163] It appears to have been largely a peaceful theocracy.

Norte Chico also left what may be the earliest-known image of a god from the Americas: a carving of a deity on a gourd from the archaeological site in the Pativlca River Valley site dated to 2250 BCE.[165] It is a forward-facing anthropomorphic figure with fanged teeth in standing position holding a staff in each hand. This figure appears to be the ubiquitous Andean "staff god", which appears in many later Andean

Norte Chico Staff God

Chavin Staff God

Wari Staff God

The ubiquitous staff gods from various early Andean cultures. (*Left*) Staff god from the *Norte Chico* culture (Based on a gourd etching more than 4,200 years old; redrawn here.) (*Middle*) Staff god from the *Chavin culture*. (Based on a carving from between 800 BCE and 200 BCE on the Raimondi Steele monolith.) (*Right*) Staff god from the *Wari culture*. (Based on a pot-shard image from the ninth century CE excavated from Conchopata; redrawn here.) In the collection of the Museo Historico Regional Hipolito, Peru. Also see page 247 for the staff god from the *Tiwanaku culture*.

pantheons, including those of the Chavin, the Mochica, the Wari, the Tiwanaku, and the Chimu. To date, the gourd specimen is the oldest image of a god from the Norte Chico region, although the exact age of the artifact is controversial.[155,163]

Clearer understanding of Norte Chico religion awaits further discoveries. Later South American religious belief systems are better known, such as those of the Chavin.

CHAVIN POLYTHEISM: THE CONTINUING GROWTH AND DIVERSIFICATION OF SOUTH AMERICAN RELIGIONS

The Chavin civilization was another early culture in the northern Andean highlands of present-day Northern Peru. They were most prominent from around 800 BCE to 200 CE.[163] Like the Norte Chico, they did not use the wheel, had no written language, and maintained no army. Unlike the Norte Chico people, however, the Chavin civilization left a robust record of artwork depicting probable deities, including sculpture, pottery, textiles, and paintings. Their art was the first to have a widespread recognizable Andean style sometimes referred to as "Mythical Realism."[166] This style often reflected the importance they placed on human-animal connections as well as how they viewed their connection with the supernatural realm. (Note that if you turn the Chavin staff god image upside down in the above figure, you can see the anthropomorphic face becomes the face of a crocodile.) The Chavin appear to have been an agricultural society not particularly concerned with an afterlife. Eventually, however, their religious belief system is thought to have had a major influence throughout the Andean region.

Chavin imagery indicated a small pantheon of possible deities. The most important and central of them is represented by the *Lanzon*, a fourteen-foot-high granite temple sculpture depicting what is interpreted as a supreme deity.[167] It appears to be a man with feline features and claws for toes and fingers.[168] Numerous other anthropomorphic supernaturals are depicted in Chavin stone sculptures and other artwork. As previously mentioned, the Chavin also had a staff god. The staff god was a common element of many Andean pantheons (see figures on pages 242 and 245). Their carved stonework depicted a vivid iconography of serpents, raptorial birds, and felines. Chavin art also included many images of cacti that shamans are thought to have used as a psychedelic drug in religious ceremonies.[168] The presence of small mortars and snuff spoons support this hypothesis.[167] The most important Chavin religious site was its namesake, Chavin de Huantar, in the Mosna Valley, where the Lanzon is located. This site was used for more than five centuries as a pilgrimage by people of the Andean region. Little else is known about their religious activities. However, it is thought that Chavin art, primarily in the form of textiles and sculpture, helped spread elements of their particular brand of animistic polytheism to later Andean civilizations. The Chavin were among the earliest people in South America to have an organized priesthood and temples. It has even been suggested by some historians that they spread their influence with proselytizing missionaries.[168]

PARACAS AND *NAZCA POLYTHEISM*: TECHNOLOGICAL ADVANCEMENT AND HEADHUNTING

The Paracas culture appears to have developed in Southern Peru along the northwestern shore of Lake Titicaca. It was most prominent in the region from around 800 BCE to around 200 BCE.[169,170] At first, the Paracas appear to have been strongly influenced by earlier Chavin religious belief system, based on the iconic imagery they produced on textiles and pottery. Later in their cultural development (300 BCE), they had replaced much of the Chavin pantheon with new deities, including the *Oculate Being* (a large-headed creature with owl-like eyes).[170] Some of the Paracas imagery of supernatural entities on textile and pottery included anthropomorphic figures with feline or other animistic embellishments. Some of these iconic figures were later adopted by the Nazca into their pantheon.[170] Ritualistic taking of human trophy heads was part of the Paracas belief system, as was cloth-wrapped mummified human burial. Little else is known about their religious belief system.

The Nazca people (also spelled Nasca) lived in the valley along the southern coast of Peru. They flourished between 200 BCE and 600 CE.[169] Although they are recognized as a distinct civilization based on certain technological advances, such as the development of farming with a complex irrigation system, their religious practices are thought to have been a continuation of those from the earlier Paracas culture.[169] Like all other ancient Andean civilizations, the Nazca lacked writing. Interpretation of their religious belief system is based on artifacts of their material culture (e.g., images on pottery and textiles). The Nazca are well known for immense images they carved into the flat landscape known as "geoglyphs," although some of the earliest ones may have been carved by the Paracas.[170] Some images are up to 1,200 feet long and 500 feet wide; they are so large that they are recognizable only from the air. They used mathematical extrapolations to design and construct these enormous figures. It is thought that they were used for ritual walkways.[168] The geoglyphs have attracted much attention because of their sheer size. The fact that they can best be seen from the air led some authors in the late twentieth century, with little scientific training but a lot of imagination, to speculate that they were constructed by aliens from another planet or by Nazca in hot air baloons.[170]

Nazca religion was based on the belief that humans and nature are related in an active, sacred relationship.[171] The spiritual lives of the Nazca revolved around attempts to understand and control forces of nature that affected droughts, rain, earthquakes, flash flooding, and agricultural success. The pantheon of Nazca Polytheism was animistic and largely centered on agriculture. The Nazca had no temples or official priests.[170] Shamans were the intermediaries between humans and the spirit world, often with the help of hallucinogenic drugs. Nazca pottery images indicate powerful gods of nature and mythical beings, combining anthropomorphic features with killer whales, cats, serpentine creatures, or other less recognizable forms. Warrior images are also frequent, often depicted with human trophy heads.[169] The Nazca, like the Paracas, practiced headhunting. Their trophy heads had holes made through the forehead so they could be carried on a rope. It is not certain whether these heads were taken as trophies of war or as objects of religious significance. No evidence suggests that human sacrifice played an important role in Nazca religious ritual, unlike the nearby contemporaneous *Mochica civilization.*

MOCHICA POLYTHEISM: WARRIOR PRIESTS AND BLOOD RITUALS

The Mochica civilization (also called the *Moche* after their capital city) began in the Moche Valley of Northern Peru. They reached their height between around 100 CE to 750 CE and were contemporaries of the Nazca civilization that lived farther down the Pacific coast. The Mochica built extensive trade networks that expanded their borders, and they became one of the major early civilizations of South America. Their religious tradition, like many Andean cultures, was influenced by the earlier Chavin and Paracas traditions.[172] The Mochica were known to worship several gods, and their rulers were warrior-priests.

Human sacrifice was part of Mochica Polytheism, although its smaller scale and the rituals surrounding it were unlike its practice in Mesoamerica. There are no signs of heart extraction, and the main means of sacrifice was throat slashing or decapitation. Sacrificial rituals involved costumed elites, monumental settings, and ritual consumption of human blood. Victims were either losers of ritual battles among local elites or warriors captured in territorial battles. Evidence of human sacrifice of up to, but no more than, seventy victims at a time have been found.[173] Ritual combat and shedding one's life for the good of the gods and society appears to have been thought of as an honorable demise.[168]

Knowledge of the Mochica pantheon is sparse and somewhat sketchy, but the following deities have been interpreted from Mochica artwork:[172]

- *Aiapaec* (the *Decapitator*) was a fearsome chief deity often shown with one arm or hand holding a knife and the other hand holding a decapitated human head. This deity figured prominently in rituals surrounding human sacrifice. Some authors believed Aiapaec to have been revered as a creator god. (The name Aiapaec means "maker" in the Mochica language.)
- *Wrinkle Face* was another main Moche god. He, like Aiapaec, was an anthropomorphic, fang-faced deity. His main characteristic was his wrinkled face, depicted by lines on his face, and a belt with snake heads at the ends. He was often depicted as being in combat.[172]

Other Mochica deities (or at least aspects of certain deities) included the Rayed Deity (a deity with a conical hat surrounded by rays), and a warrior god resembling an owl. Some authors have mentioned a feline staff god, with large canines, clawed feet and hands, and cat-like eyes.[174] The members of the pantheon changed over the course of Moche history. According to some authors, as in most of ancient North African-European Polytheism, the Moche shared a common pantheon but different regions favored different gods.[172] The names and detailed descriptions of most Mochica deities are confusing and unclear because most interpretations of the Moche pantheon are based on a few archaeological structures and looted artifacts.

The Mochica civilization went into decline during the sixth century CE, possibly because of extreme climate change. Based on studies of ice cores from the Andes mountains dated between 536 CE and 594 CE, the Moche region went through several years of intense El Niño flooding followed by many years of serious drought.[168,175] This is thought to have led to diminished faith in the imperial pantheon and the old order. During the reign of the Mochica in and around the Moche valley, there were many other civilizations in the Andean region. One was the *Tiwanaku*, which started just south of Lake Titicaca in what is now Bolivia.

TIWANAKU POLYTHEISM: THE RISE AND FALL OF ANOTHER MAJOR SOUTH AMERICAN CIVILIZATION

Although the city of Tiwanaku was first settled around 400 BCE, the Tiwanaku Empire was not well established until around the fourth century CE, when it occupied parts of what are now Bolivia, Peru, and Chile.[168,176] This was one of the most important pre-Inca civilizations, and by the seventh century CE, it had become the first fully developed imperial state in the Titicaca Basin.[177] Its administrative center was in the city of Tiwanaku near the southern shore of Lake Titicaca, which was thought by the Tiwanaku to be the center of the world. The Tiwanaku population grew to immense size for its time, more by religious expansion and trade than by military conquest.

The early Tiwanaku absorbed elements of religious tradition from earlier cultures of the Titicaca Basin, such as the Qaluyu (1400–500 BCE), the Pucara (or Pukara) (1400–400 CE) and the Chiripa (1400–100 CE). In 1988, Central Michigan University's archaeologist Karen Mohr Chavez identified what she hypothesized to be a Pan-Titicaca Basin religious tradition that she named *Yaya-Mama*.[174] Lacking written records of any such religion, she based this exclusively on archaeological evidence , including the artistic style of the sculpture, certain ritual paraphernalia, and a style of architecture. Yaya-Mama represented the emergence of a new elite ideology associated with the sociological and political structure of the Titicaca region.[177] The proposed characteristics that Chavez lists have not been considered in a broadly comparative way with other Andean characteristics, such as recognition of a staff god, ritualistic details, and imperial involvement. Further work is yet needed to verify the significance of Yaya-Mama as a religious tradition (discussed further in **Note 5G**).

Reconstructing the details of Tiwanaku ancient history is challenging for a couple major reasons. First, most of the available information is based on oral transmission through the Incas, which was later transcribed by Roman Catholic missionaries of the Spanish Conquest. Second, Tiwanaku archaeological sites have been largely destroyed, further complicating things. These sites were subjected to extensive looting and damage for nearly a thousand years, starting with the original revolt of people at the end of the Tiwanaku Empire. This looting eventually included amateur "pot hunter" excavations, vandalism, quarrying stone from monuments for modern construction projects, and even target practice on archaeological sites for military personnel.[178]

In spite of these challenges, excavations have revealed stepped pyramids and temples of worship containing cut stone blocks and carvings of immense size. Some of these are estimated to weigh more than sixty-five tons. They were amazingly transported from quarries many miles away. Also included are numerous carvings, such as the *Gateway of the Sun* (figured below). This structure was carved out of a single block of granite and weighs more than ten tons. Many of the carvings feature the front-facing Andean staff god without fangs, as well as other rayed-head deities.[179] The later Tiwanaku people thought that the staff god created a race of giants that moved the massive stones.

The Tiwanaku conducted human sacrifice as part of their religious rituals. At least some of the victims were from other societies, possibly prisoners of war.[176] The method of sacrifice involved disembowelment, cutting the victims into pieces, and leaving the butchered pieces to the elements and wild carnivores. Tiwanaku religion appears to have been a form of animistic polytheism with deities centered on agriculture. Some indication of ancestorism is given by preservation, use, and reconfiguration of mummy and skeletal remains, similar to that practiced by the later Inca.[180] Few details about the Tiwanaku pantheon have been deciphered, although the main

Tiwanaku Staff God

Tiwanaku Staff god and the *Gateway of the Sun* monolith. (*Left*) This monolith, nearly ten feet tall and thirteen feet wide, was carved from a single block of igneous volcanic rock by craftsmen of the Tiwanaku civilization. Note the carving of the staff god *Viracocha* over the entryway. (*Right*) Drawing of the *Viracocha* relief from the monolith by Charles Wiener, 1880.

deity appears to have been the staff god, *Viracocha*, which is shared by several other Andean civilizations. He was the Supreme God and creator of the Andean people. For some Tiwanaku belief systems, Viracocha created people out of rock. He was also later believed to have created the giants that moved the massive stones of their monuments, but he grew unhappy with them and created a flood to destroy them.

By 800 CE, Tiwanaku had ten to twenty thousand people.[181] By 1000 CE, however, the region began to suffer the effects of severe climate change. Years of drought eventually resulted in the water level of Lake Titicaca, as well as levels of surrounding water tables, dropping by more than fifty feet. Hydraulic systems that had previously watered critical cropland were left high and dry, resulting in massive crop failure. The environmental crisis catalyzed social tensions that had been developing within the society over sociopolitical and demographic changes.[181] These forces eroded commonly held values and the sense of state. People eventually revolted against elite subgroups and even the religious belief systems. Monuments, buildings, and places of ritual and worship were destroyed and abandoned. By 1150 CE, Tiwanaku was gone.[176]Another Andean civilization that was largely contemporaneous with the Tiwanaku was the Wari.

WARI POLYTHEISM: IT'S CHICHA TIME

The Wari (Huari) civilization flourished in what is now Peru between 500 CE to 1000 CE.[140] They were contemporaneous rivals of the Tiwanaku Empire to the south. The Wari Empire absorbed many Tiwanaku religious beliefs into their cosmology as a pragmatic means of cultural expansion. Consequently, many shared traditions of the two civilizations laid a broad ideological groundwork for the later Inca.[179,182] The Wari built temples in the familiar style of the Tiwanaku to syncretize their religious establishment, although unlike the Tiwanaku, they did not build megalithic pyramids.[179] By 800 CE, the Wari had incorporated many Tiwanaku religious rituals into the Wari state institutions and had adopted some of the same religious iconography. Some of this shared iconography, however, had different meanings in each of the two cultures.[164] The Wari also incorporated the Paracas and Nazca practice of trophy-head taking, although whether or not these were some sort of religious rituals, or simply battle trophies, is debated.[183]

Early in their history, the Wari ritualistically cannibalized deceased tribal or family members. This was not an act of violence, but rather was a sign of respect and part of their tradition. By consuming the flesh of the deceased, they believed that they also consumed their spirit to literally reincorporate the person into the tribe or family. Such cannibalism was considered by the Wari to be far more respectful and compassionate than permitting the remains to decay into the earth.[184]

Like the Tiwanaku, the Wari religion was a form of animistic polytheism with the staff deity as a supreme icon of power (page 247).[185] The Wari occasionally transformed the staff into a stalk of maize, perhaps to remind others of how they benefited from farming.[155] The Wari recognized both a male and a female version of the staff god[179]. Secondary deities were anthropomorphic beings with various animal features, such as wings, beaks, and claws. Like the Tiwanaku and Nazca, the Wari shamans and elites used hallucinogenic drugs in their rituals.[185] Details of the Wari pantheon have yet to be clearly deciphered, although the creator deity is thought to have been *Pachacamac* (*Pacha Kamaq*), and his consort was *Mama Pacha* (also called *Pacha Mama* or *Pachamama* by various Andean cultures). As with so many major religious belief systems throughout history, Wari religious traditions had a strong interconnection with imperial power. Archaeological evidence suggests that the Wari Empire manipulated local supernatural belief systems and used religion to express its power and legitimize its domination. As they conquered new territories, they absorbed religious elements from those territories to strengthen their imperial power and influence.[185]

The Wari fermented a beer-like beverage called *chicha* that was consumed during rituals. They practiced human sacrifice by decapitation[182] and smashed pottery and ceramic vessels as another form of sacrifice. Victims of human sacrifice were often foreigners rather than residents of the Wari heartland. Like the Tiwanaku civilization, the Wari appear to have become another victim of climate change. By 1000 CE, their capital city was abandoned. Although both the Tiwanaku and the Wari had virtually disappeared by the twelfth century, they and the Mochica collectively left their influential mark with the later Chimu and Inca cultures.

CHIMU POLYTHEISM AND CHILD SACRIFICE

The Chimu civilization succeeded the Moche civilization in the Moche Valley of present-day Trujillo, Peru. It dominated much of the north coast of Peru between 900 CE and 1470 CE. It was eventually conquered by the Inca Empire. The Chimu had a structured society with a ruling aristocracy that they believed descended from the stars. Blood sacrifice was important to Chimu religious traditions. They sacrificed animals, such as llamas and they practiced human sacrifice. They are perhaps most famous for their mass sacrifice of children. Chimu religious belief systems drew some of their icons and traditions from the earlier Mochica culture.[186]

The Chimu may have turned to children when they believed that the sacrifice of adults was not sufficient to stop certain environmental disasters. Some gods and specific situations, such as drought, were believed to demand children as sacrifices. The general reasoning behind child sacrifice was the idea that the more precious the sacrifice was, the more devout the person making it was. The Chimu believed that their sacrificed children would become deified. Some of these child sacrifices were massive. In 2011, a 550-year-old Chimu archaeological site in northern Peru was found containing the largest-known child sacrifice in history. It contained the remains of two hundred baby llamas and one hundred forty children along with two adults.[187] The children were between five and fourteen years of age and all had deep slashes along their sternum and broken rib cages indicating heart extraction. Their faces were smeared with red ochre.[187]

Chimu Polytheism, like most Sibero-American Polytheism, was animistic, with the following four principal deities in their pantheon:[186]

- A *Staff God*, like other Andean staff gods, was front facing with a staff in each hand. Unlike the staff gods of most earlier cultures, the Chimu staff god was more human-like, lacking the large canines and clawed hands and feet.[186]
- A *Plumed Headdress Deity*, like the decapitator god of the Mochica (*Aiapaec*), was most often depicted with a knife in one hand and a severed human head in the other. He appeared to be the primary god concerned with sacrifice. As the name suggests, this deity was distinguished by its plumed headdress, with two or more plumes falling on either side of his cap.
- A *Chimu Goddess* had two aspects: one as the goddess of the moon (*Si*) and the other as the goddess of the sea (*Ni*). The Chimu believed the moon to be even more powerful than the sun because it was visible both day and night.
- A *Moon Animal* was an animistic deity reinforcing the sacred Chimu concept of unification between the sea and the moon.

The Chimu relied on shamans to connect them to the gods and spirits. They also believed that the social classes were preordained by the gods. One of their creation stories describes separate creations for each class. The royal and noble males were spawned from a golden egg, the noble women from a silver egg, and all commoners from a copper egg.[186] It was yet another example of religion being used to strengthen the ruling monarchy and perhaps to suggest the superiority of the male gender as well.

By the late fifteenth century, the Chimu were conquered by the Inca. The Inca Empire subsequently absorbed parts of the Chimu religion and many of their deities into the Inca Polytheism.

INCA POLYTHEISM AND LATER CHRISTIANIZATION BY THE SPANISH CONQUEST

The Inca people are one of the best-known pre-Columbian civilizations of the New World. They were the last major native empire to develop in South America before European invasions introduced foreign culture, foreign religion, and devastating

Machu Picchu, the iconic archaeological site of Incan civilization, located high in the Peruvian Andes. It was built in 1450 CE, abandoned a century later, and rediscovered in 1911 by an American archaeologist.

foreign disease. They arose in the Peruvian highlands sometime in the thirteenth century.[164] At the height of their expansion, they had conquered much of western South America in a swath 2,500 miles long and 200 miles wide, from Ecuador to central Chile. Their capital city and administrative center was located in Cusco (Inca name Qusqu), in south-central Peru. The Inca Empire was one of the world's largest, until its final decimation by Spanish colonialism in the late sixteenth century.[155,168]

One of the best-known rulers of the Inca Empire was *Manco Capac*, who lived sometime in the thirteenth to fourteenth century. He was the mythical founder of the Inca civilization in Cusco and was its first king. Another well-known ruler was *Pachakuti Inka* (1418–1472), the ninth in a succession of Cusco kings. He is credited with elevating the Inca civilization to "empire" status.[155,164] One of his estates was the famous archaeological site of *Machu Picchu*, the most familiar and iconic place of the Inca civilization. Through Pachakuti Inka's conquests, the Inca were able to expand the kingdom of Cusco to a much farther-reaching empire. By the mid to late fifteenth century, they had become the largest empire in the pre-Columbian Americas, and by the sixteenth century, the Inca were possibly the largest empire in the world.[155,168] Finally, *Atahualpa* (1502–1533), the last of the great, independent emperors of the Inca, was the one to witness the empire's massive downfall (discussed later in this chapter).

As with other pre-Columbian new-world cultures, the Inca left no written records other than a system of knotted cords called *Quipus* (Khipus), also used by the earlier Wari culture. These may represent a largely untranslated binary system of recordkeeping reminiscent of today's computer languages.[155,164] The recorded history of the Inca was first transcribed from oral transmissions handed down over many generations by official Inca "memorizers." It is possible that Quipus were used as memory devices, much like rosaries guiding prayer-like incantations of their history. Roman Catholic chroniclers, during the conquest, were the first to transcribe oral histories of the Inca. Many of these writers were under pressure to portray the Inca religion as harshly cruel to legitimize the Spanish conquest in the face of continued resistance by the indigenous population.[188] For its first 350 years, the study of Inca culture relied almost exclusively on documents produced by Spanish Catholics and Christianized Andeans during the first century after the conquest.[189]. In addition to intentional distortions by Spaniard chroniclers, many historians believe that the gulf between the Spanish and the Inca cultures was so great that the Spaniard chroniclers simply did not understand what they were being told by the Inca storytellers.[164]

In addition to incorporating many Chimu elements into their own religious tradition, the Inca also held the Tiwanaku culture as sacred. Some anthropologists suggest that the Inca reconstructed many of the destroyed monuments of the Tiwanaku to build a park that reinforced the Inca sense of imperial identity.[168] Inca religious belief systems also had a number of basic differences from other Andean belief systems.[189] First, their textiles and pottery had far fewer depictions of supernatural beings, and they produced few public images of such beings (e.g., statues or stelae). Hallucinogens were evidently not as predominant in religious ceremonies as they were in Nazca, Wari, and Tiwanaku traditions. Trophy-head ritual activity was not practiced as it was in other Andean societies. Much like the Chimu, many human sacrificial victims of the Inca were children or adolescents.

The Inca, like the Andean cultures that came before it, had a formal state religion that emphasized ritual and hierarchical organization rather than mysticism and spirituality.[164] According to existing records, the Inca claimed their rulers were descended from the sun god *Inti*. The Inca elite thought of themselves as divinely selected to dominate the empire.[189] Their polytheistic belief system had roots in many earlier religions, including those of the Chavin, Nazca, Mochica, Tiwanaku, and Chimu.[180,188] As with most branches of hard polytheism, the Inca happily incorporated deities from other religions into their pantheons. They were tolerant of diverse beliefs, as long as the Inca gods were also respected. As they expanded their empire, the Inca would often take idols of newly conquered provinces and place them in honored positions in Cusco, thereby facilitating the cultural assimilation of the conquered people.[188] This was a strategy also used by polytheistic religions in the Old World, such as the Greeks and Romans discussed earlier in this chapter.

The Inca view of cosmology was based on the dualistic principles of reciprocity.[164] Similar to the yin versus yang or the deity versus universe concepts of Asian Cyclicism discussed in chapter 4, the Inca saw the world as filled with contrasting forces, such as dark versus light, hot versus cold, and male versus female. These forces complemented each other, providing an equilibrium in which life could exist. Like the Tiwanaku and other Andean religions before, the Inca believed that Lake Titicaca was the origin from which the world began.[190] The Incas only briefly mention the creation of humankind, but when they do, it is also associated with the Lake Titicaca region. The scarcity of information about the Inca creation beliefs may be an artifact of what was, and what was not, accurately recorded. It was difficult for the Spaniard Roman Catholic chroniclers to focus on creation stories in the New World, because the Andean version of Adam and Eve competed with that of the Old World and Genesis. Reluctant to contemplate such heresies, and not able to speculate on how descendants of people created by God in the Old World could have possibly reached Peru, it was easier for the Catholic chroniclers to ignore the issue.[188]

Inca Polytheism, like other forms of hard polytheism, had strong elements of animism and ancestorism.[164] Each of the many different branches of Inca Polytheism had its own pantheon. Details of the diversity within the branches of Inca Polytheism have yet to be clearly presented because of incomplete and confusing historical records. The Incas worshipped *huaca* alongside their formal pantheon of gods. Huaca could take the form of animistic spirits embodied in stones, mountains, water, or anything great or unusual in nature. They also could take the form of sacred places or mummies that embodied the spirits of ancestors. The dead exerted enormous influence over the living. The Inca guarded and cared for the mummies of important dead ancestors almost as if they were still alive, somewhat like in Egyptian Polytheism of the Old World. Mummies of important rulers were fed each day, provided with drink, and even provided with entertainment. These mummies continued to own property and estates that they had owned in life. In return, the ancestral spirits of these mummies were believed to intervene from the spirit world to provide for the living, particularly with regard to providing the fertile land and the water necessary to sustain life.[164] Water was sacred to the Inca.

The Inca pantheons became the largest of the Andean region, in part because of the number of deities they absorbed from other cultures incorporated into the Inca Empire over time. Following are some of the more common major deities:[164,191]

- *Viracocha* (also spelled *Wiracocha*) was shared by many Andean civilizations, but the Inca elevated him to the Supreme God. He was the creator god who caused the sun to emerge from Lake Titicaca each day. He created all other gods, as well as humans and all animals.[164,168] Viracocha also introduced a trinitarian-like element into Inca religion with some Inca branches recognizing a divine triad of Viracocha aspects (Viracocha and his two sons) as the heart of Inca creation.[191] It is difficult to discern whether this idea originated within Inca Polytheism, or if it was a later artifact of Christianization. The historical records of the Incan divine triad were written by Catholic chroniclers who themselves were trinitarians. After creation, Viracocha was no longer seen as taking an active role in human affairs.

- *Inti* was the sun god, and most important servant of Viracocha. He was eventually elevated to patron god if the Inca. The Inca emperors (*Sapa Inca*) were thought to be descendants of Inti by some and descendants of Viracocha by others. Inti was often portrayed as a golden disc with a human face surrounded by rays. Some branches of Inca Polytheism considered Inti to be the son of Viracocha and to have had a wife named *Mama Quilla*, a moon goddess.
- *Pacha Mama* (*Earth Mother*) was an Inca earth goddess inherited from earlier Andean religions. She was also a fertility goddess who presided over agriculture, mountains, and earthquakes. After the Spanish conquest, the Roman Catholic figure of the Virgin Mary became associated with that of the Pacha Mama for many indigenous people throughout the Andean region of South America. Today, the Earth Mother goddess is still revered in the Andean region in the guise of the cult of the virgin.[164]
- *Supay* was the god of death, evil spirit, and ruler of the Inca underworld (*Ukhu Pacha*). Children were sacrificed in his name to keep him happy. During the Spanish conquest, Roman Catholic missionaries used the name Supay to refer to the Christian devil.

Other gods were included in the various Inca pantheons, including *Pachacamac* who was added from pantheons of the Wari and earlier Andean pantheons. Pachacamac was a creator god, in some ways redundant with Viracocha. The Inca coopted him because of his importance to other Andean religions. They believed that he would add to the prestige of the Incan pantheon.[164] Over time, the many different Incan branches each had their own pantheon. We will probably never know the true diversity. In general, the stars, weather, planets, moon, and sun all signified important deities.

The Inca made animal and human sacrifices to some of their deities. Most common sacrifices were llamas, guinea pigs, other foods, and chicha (the fermented beer-like drink mentioned earlier in this chapter). The most important and solemn sacrifices were those of human beings, often made during times of national distress from natural disasters. The Inca practice of human sacrifice, called *Quapaq hucha*, focused on the sacrifice of children who were considered physically perfect, because they were thought to be the purest of beings.[164,189] War captives were also sacrificed. Defeated rulers and generals, along with the most perfect men and women from the newly conquered provinces, were brought to the Inca capital, Cuzco, where they would be sacrificed to the sun god as thanksgiving for victory. This was more than an offering to the gods; it also cemented Inca conquests and discourage resistance.[183] Inca ritual also valued the practice of mummification. The bodies of deceased rulers and important people were often mummified, allowing their bodies to be worshipped within the palaces. Inca Polytheism was in many ways a conglomeration of the Andean religions that came before it. However, the Inca, their indigenous culture, and their polytheistic belief system was relatively short lived.

In spite of the immense size of the Inca Empire, a small number of Spaniards with the help of a number of South American locals who had no love for the Inca, eventually defeated the empire in the name of the king of Spain and Roman Catholicism. The majority of people who survived in the region were those who converted to Roman Catholicism or at least to some syncretized version of Inca Polytheism with imbedded elements of Western Christianity. In 1532, Spanish conquistador Francisco Pizarro (1476–1541), representing the Holy Roman Emperor of Spain, along with only 168 Spanish soldiers and some South American enemies of the Inca that the Spaniards had managed to enlist to their cause, managed to capture the Inca emperor, Atahualpa. This was an amazing feat because Atahualpa at the time was in the middle of his empire of millions of people and in command of tens of thousands of warriors (although many of his best troops were engaged in a civil war with his brother Huascar of Cuzco).[192]

As colorfully recounted by Jared Diamond in his book *Guns, Germs and Steel*, based on Spanish records, Pizzaro entered the Inca city of Cajamarca at the friendly invitation of Atahualpa. Accompanying Pizzaro was a Catholic friar named Vincente de Valverde. He was first to speak to Atahualpa, and required him "in the name of God . . . to subject himself to the law of our Lord Jesus Christ and to service of His Majesty the King of Spain."[193] The friar gave Atahualpa a copy of the Christian Bible to examine. When the Inca emperor opened it and looked at it uncomprehendingly, he finally threw it on the ground, perhaps in frustration. This enraged the friar, and he begin to shout to the hidden armed soldiers that had accompanied him: "Come out Christians! Come at these enemy dogs who reject the things of God. The tyrant has thrown my holy book of laws to the ground. . . . March out against him, for I absolve you!"[193] Then the slaughter began.

The Spaniards had advance weaponry of guns, steel, metal armor, and horses. The Inca had no horses, steel, or armor, and their few weapons consisted of wooden clubs, bronze weaponry, stones, and slingshots. Most of them were unarmed and not expecting conflict. The armed warriors had also been given bad military intelligence about the Spaniards.[192] They believed that the Spaniards were ineffective fighters unless on horseback and that the horses were powerless at night. They also believed that the Spaniard's rifles and canon were capable of only two shots each and that the Spaniard's swords "were no more dangerous than women's weaving battens."[192] This left them both surprised and unprepared for Pizarro and his men. The Spaniards reportedly killed thousands Inca warriors in the raid and kidnapped Atahualpa. Pizzaro explained to the captive emperor: "We come to conquer this land by [our king's] command, [so] that all may come to a knowledge of God and of His Holy Catholic Faith . . . in order that you may know Him and come out from the bestial and diabolical life that you lead. . . . Our Lord permitted that your pride should be brought low and that no Indian should be able to offend a Christian."[156] Pizzaro later demanded a large room full of gold and silver from the Inca in exchange for a promise to return their ruler. After receiving history's largest-known ransom, Pizzaro reneged on his promise to the Inca and executed Atahualpa.

There is no indication that Atahualpa ever disavowed his faith in the Incan patron god Inti, or in the Incan pantheon. However, he accepted a Christian baptism on the day of his execution to avoid being burned at the stake and that his body might be given customary funeral honors.[192] In the years that followed, the Spaniards brought devastating disease, more advance weaponry for which the Inca had no defense, and many more soldiers. Lacking a plan for orderly succession of power, Atahualpa was the last independent ruler of the entire empire. A series of more regional rulers, some who were puppets of the Spaniards, attempted to bring the empire back, or at least nurse the regional remnants along, but this did not last for long. The last Inca stronghold was conquered by the Spaniards in 1572, and the Inca became another cultural victim of the Spanish conquest.

As in Mesoamerica, the population of indigenous South American people plummeted from disease, Spanish aggression, and overwork from slavery. The aggressive colonization of the New World by Europeans was one of history's most dramatic examples of ethnocide. It decimated the population of indigenous people, and in the process it erased most knowledge of their indigenous cultural history. Historians estimate that in 1491 CE there were as many as between 75 and 145 million native inhabitants in the New World. By 1550, after fifty years of Spanish conquest and the spread of diseases from Europe that indigenous Americans had no immunity to, it is estimated that only about ten million natives remained.[122] Before the Christianization of South America, polytheism was extremely diverse, probably containing hundreds of different subgroups over time and in different regions of the continent. Today, the vast majority of people in the region are either Roman Catholic or they practice a syncretized mix of Roman Catholicism and indigenous Andean Polytheism.

NON-ANDEAN SOUTH AMERICAN POLYTHEISM

Farther east in South America (e.g., the Amazon Basin) the history and diversification of ancient indigenous supernatural belief systems are much more poorly known. One of the few groups studied in much detail is *Baniwa Polytheism* of the Northwest Amazon, briefly mentioned earlier in this chapter, and described by religion scholar Robert Wright.[157] A detailed comparative analysis of central and eastern South American groups is beyond the scope of this book, because so little has been published on most of it, and I could not locate any broad comparative studies. Hopefully, as more information about these groups materializes in the future, and it is included in some sort of comparative analysis, we will better understand the early historical interrelationships of religions from this region.

HISTORICAL SUMMARY OF OLD WORLD HARD POLYTHEISM AND ITS GLOBAL EXPANSION

The roots of Old World Hard Polytheism are likely to have originated in Africa and Eurasia at some unknown time well before 9000 BCE. One of its eastern-most branches, *Siberian Animistic Polytheism*, eventually spread to the New World with the early arrival of humans from Siberia sometime between twenty-eight and eleven thousand years ago seeding a branch of Old World Polytheism in the New World that would eventually evolve into what I have termed *New World Animistic Polytheism*. Around eleven thousand years ago, the New World and the Old World became geographically isolated from each other for thousands of years because of the loss of the trans-Pacific land bridge (Beringia). During that time of isolation, the polytheistic beliefs on both sides of the Pacific diverged significantly. *North African-European Polytheism* and *Siberian Animistic Polytheism* continued to

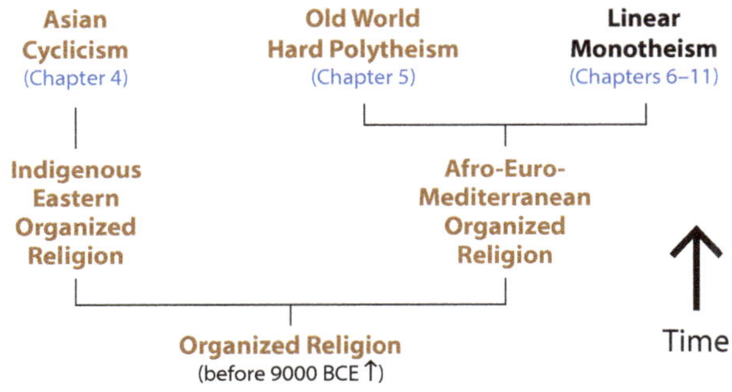

develop in the Old World (trees on pages 158 and 208), and *New World Animistic Polytheism* continued to develop independently in the New World (tree on page 208).

Hard polytheism (as North African-European Polytheism and New World Animistic Polytheism) eventually evolved into countless branches, making up the vast majority of all religions of the world for thousands of years. Later, hard polytheism would become largely displaced by *Linear Monotheism,* first in the Old World and much later in the Americas. Most records for the earliest evolution of Old World Hard Polytheism were lost to history. I place Linear Monotheism as the sister group of Old World Hard Polytheism (see tree on previous page).

Centuries of proselytism, conquest, and imperial support helped Linear Monotheism form the dominant group of religions today. This evolution of Linear Monotheism is the subject of the next six chapters.

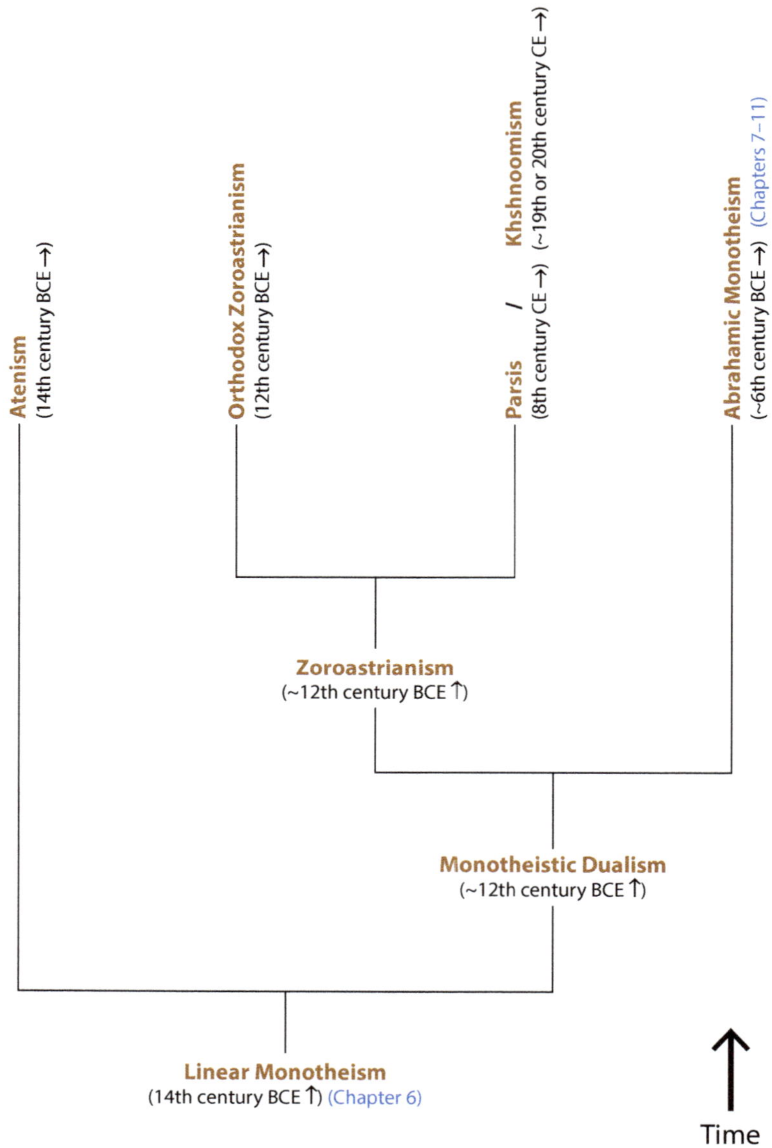

Atenism
(14th century BCE →)
Orthodox Zoroastrianism
(12th century BCE →)
Parsis / Khshnoomism
(8th century CE →) (~19th or 20th century CE →)
Abrahamic Monotheism
(~6th century BCE →) (Chapters 7–11)
Zoroastrianism
(~12th century BCE ↑)
Monotheistic Dualism
(~12th century BCE ↑)
Linear Monotheism
(14th century BCE ↑) (Chapter 6)
Time

6

LINEAR MONOTHEISM

Oh sole God, like whom there is no other . . .

—GREAT HYMN TO ATEN, FOURTEENTH CENTURY BCE

Linear Monotheism appears to have evolved in the Middle East from Old World Hard Polytheism (**Note 6A**). It evolved from a gradual reduction of gods in the pantheon and the increasing importance of a Supreme God. It became fully monotheistic when it transformed into a belief in *only* one God. I use the word "linear" to distinguish it from schools of Asian Cyclicism that independently evolved monotheistic elements (e.g., Sikhism in chapter 4). According to most of Linear Monotheism, an individual's time as linear (one birth, one life, one death, and finally an eternal existence in the afterlife). This is in contrast to most of Asian Cyclicism (discussed in chapter 4), which emphasizes cycles, either as a series of repetitive reincarnations (e.g., samsara) or a cyclic balance of nature (e.g., yinyang).

In Linear Monotheism, as with other major forms of Organized Religion (Asian Cyclicism and Old World Hard Polytheism), a clear symbiotic linkage often has been made with an imperial monarchy, dynasty, kingship, or other governing entity. Early attempts of state-sponsored Linear Monotheism began in Egypt at least as early as the fourteenth century BCE, as *Atenism*. Later, in the twelfth century BCE, it arose in Persia (modern-day Iran) as *Zoroastrianism*. Finally, in the seventh or sixth century BCE, in the region of Israel, Abrahamic religion became monotheistic, putting Linear Monotheism on the road toward becoming the world's largest branch of religion (chapters 7–11).

Today, Linear Monotheism has billions of followers and includes over half of the world's population. As shown by the phylogenetic tree on page 260, I first divide early Linear Monotheism into two branches: *Atenism* (a short-lived early attempt)

Timeline for certain events, people, and beliefs relevant to early Linear Monotheism

Zoroastrianism steadily declines after the mid-seventh century, due largely to the Muslim conquest of Persia (modern day Iran) →	~7th century CE
True *Abrahamic Monotheism* (named after Abraham, patriarch of the Israelites), develops from the *Yahweh alone movement*, elevating Yahweh from chief god of the Israelite pantheon to the *one and only God* →	~6th century BCE
Zoroastrianism takes hold in the Middle East. Over the next thousand years it expands into surrounding regions to become one of the world's leading religions thanks to collaboration with three successive Iranian empires →	~6th century BCE–7th century CE
Monotheistic Dualism develops in the form of *Zoroastrianism* with clearly defined realms of good versus bad, heaven versus hell, one Supreme God *Ahura Mazda* (with many aspects), and a devil known as *Angra Mainyu* →	~12th century BCE
The prophet *Zoroaster* is born in pre-Islamic Persia (modern day Iran), later to be recognized as patriarch of *Zoroastrianism* →	~1200 BCE
Akhenaten dies, along with his monarchial influence, and Atenism soon follows to extinction about a decade later, with Aten demoted to just another god in the Egyptian pantheon →	~1335 BCE
Atenism is formalized as a monotheistic religion when Akhenaten becomes sole ruler of Egypt and declares Aten to be the *one and only God* →	~1352 BCE
Pharaoh Amenhotep IV promotes the sun *god Aten to chief god* of the Egyptian pantheon. He also changes his name to Akenaten (meaning "beneficial to Aten"), formalizing ties between his monarchy and the new chief god →	~14th century BCE

and *Monotheistic Dualism* (a much more successful group containing *Zoroastrianism* and *Abrahamic Monotheism*).

ATENISM: AN EARLY ATTEMPT TO TRANSFORM EGYPTIAN POLYTHEISM INTO MONOTHEISM

Atenism began in Egypt in the mid-fourteenth century BCE. For more than fifteen centuries before that, the Egyptians had worshiped a diverse pantheon of gods and goddesses, including a sun god known as *Aten* (first mentioned in the nineteenth or twentieth century BCE.[1] The religion of Egypt was historically tied to the monarchical management of the country, with the pharaoh being central to both. Egyptian Pharaoh Amenhotep IV was only a few years into his reign when he raised the status of Aten to chief god of the Egyptian pantheon. This replaced the former chief god of the pantheon, Amun.[2] Around the same time, the pharaoh officially changed his own name from Amenhotep (meaning "Amun is content") to Akhenaten (meaning "beneficial to Aten"). This aligned his status with the chief god and greatly elevated the importance of Aten, although the deity was still part of a large pantheon of lesser gods.

Then, in the ninth year of his reign, the pharaoh proclaimed another radical transformation of the Egyptian religion. Aten would no longer be merely the chief god of the pantheon. Instead, he would be the *only* true god of Egypt and the sole deity for all people.[3] All idols and images of other gods were forbidden as the monotheistic prohibition of idolatry was introduced. Images of the old chief god Amun were ritualistically defaced in temples throughout Egypt. Aten as supreme deity was neither masculine nor feminine.[4] All creation was said to have emanated from "it," who was not depicted in an anthropomorphic (human) form but instead as rays of light extending from a solar disk. Pharaoh Akhenaten commanded that Aten be referred to as "the Aten." *Atenism* became the earliest-known major form of Linear Monotheism. This move was not popular with the old priests of Amun and the rest of the old pantheon, who now found themselves out of power.

Akhenaten continued to manipulate his own stature along with that of Aten. He declared himself the son of Aten, and that only he and his wife Nefertiti could worship Aten directly.[3] All other people would worship the pharaoh and his wife as the only intermediaries to Aten. This was a significant promotion for the royal family. Akhenaten manipulated religion for political ends to concentrate power in his hands and the hands of his monarchy.[3] The intertwined coevolution between imperial power and major branches of Linear Monotheism would be repeated many times throughout history, with more long-standing success in religions such as early Christianity and Islam. Although the concept of "one God" put more power in the hands of the pharaoh who claimed to represent him, the short time and limited nature of imperial sponsorship for Atenism was apparently not sufficient to enable its survival past the fourteenth century BCE.

Akhenaten's manipulation of Egyptian religion upset the social balance and harmony between the Egyptian people, the priests, and their old pantheon of gods. Conservative Egypt became dissatisfied with many of the pharaoh's ideas, and the financial, political, and moral corruption became overbearing. It was Akhenaten's power and drive that sustained Atenism. After he died in 1335 BCE, his successors began to systematically expunge records of Atenism, Akhenaten, and those who had supported the movement.[1,3]. Atenism was branded as heresy (the "Amarna Heresy"),

Akhenaten, Nefertiti, and their three daughters, together with the god *Aten* who is depicted as a solar disc with many rays ending in hands. Limestone carving measuring about twelve inches high, dated at around 1340 BCE. Discovered in 1912 in Amarna Egypt, and currently in the Neues Museum (the Egyptian Museum of Berlin, Germany).

and Akhenaten was posthumously reviled as a heretic. The return to polytheism after Akhenaten's death came so swiftly and completely that a little over a decade later it seemed as though the cult of Atenism had never existed. Worship of Amun and the rest of the pantheon returned. The priests of Amun even persuaded Akhenaten's young son and future pharaoh, Tutankhaten, whose name meant "the living image of Aten" to change his name to Tutankhamun, "the living image of Amun." In the modern world, Tutankhamun is one of the best known of all Egyptian pharaohs, popularly known as "King Tut." He began to restore the old temples and presided over the early return of Egypt to the old pantheon, the reinstatement of the old priests, and a return to a better sense of social order in Egypt. It is likely to have been a heavy burden for this boy who became pharaoh by the age of eight and died before his twentieth birthday.

Few details of Atenism theology remain (in part due to the efforts of Akhenaten's successors who were thorough in their destruction of historical records), but Atenism appears to have been a form of nondualistic Linear Monotheism that predated clear concepts of heaven, hell, angel, devils, and Messianism. It also predated the idea of divine judgment after death based on one's moral character in life resulting in either the reward of eternal paradise or the punishment of hell. These elements later developed most successfully within *Monotheistic Dualism*.

MONOTHEISTIC DUALISM: THE CONFLICT BETWEEN GOOD AND EVIL

I define *Monotheistic Dualism* as a branch of Linear Monotheism with clearly opposing elements of good and evil that are in competition with each other for the souls of humanity. This should not to be confused with the non-conflicting form of dualism in Asian Cyclicism (mentioned in chapter 4). Monotheistic Dualism generally holds that *good* will ultimately overcome *evil* at some point, at least by the apocalyptic end of times. It effectively, however, remains in a dualistic state of conflict until that time. The theology of Monotheistic Dualism went beyond earlier religions that had an apathetic overworld and underworld, where gods demanded worship and sacrifice in return for providing basic natural necessities for survival, such as rain, food, and sunlight. Monotheistic Dualism established clear divisions between the opposing sides of order and chaos, good and evil, angels and devils, and heaven and hell. The souls of humans who were on the side of good in the material world would receive clear rewards in an afterlife, and those who were evil would face eternal punishment. Even if you suffered severely in this world, your good deeds and faith would provide eternal bliss for you in the afterworld. This paved the way for the appearance of martyrdom, and religious zealots who did not fear death.

The pain of torture in this world was believed to be small compared with the possibility of eternal bliss or eternal hell. No one was martyred for Zeus or Aten, but the martyrs for religions of Zoroastrianism, and later in Abrahamic Monotheism, were numerous. Although evil and chaos continually challenge the order and moral fiber of the universe, it is prophesized in most branches of Monotheistic Dualism that good will eventually triumph over evil.

Such a clear moral division between good and evil, heaven and hell, and a good god and his evil opponent (the Ahriman or the devil) was lacking in Atenism and in most earlier religions as well. Versions of hell vary among the diverse branches of Monotheistic Dualism, ranging from temporary stops to cleanse the soul before entering heaven (e.g., Zoroastrianism and early Abrahamic Monotheism), to places of eternal damnation and suffering (branches of Traditional Christianity and Islam). Unlike Atenism and many non-monotheistic religions, the major branches of Monotheistic Dualism are associated with sacred texts (e.g., the Avesta, Tanakh, New Testament, and Quran). Monotheistic Dualism is also associated with one or more male *prophets*, *divine messengers* (*messiahs*), or *patriarchs* at their core. A prophet or a divine messenger is an individual in human form who was contacted or sent by a divine being to speak to humanity on the deity's behalf. He serves as an intermediary with humanity to deliver divine messages and teachings. He is nearly always a male of virgin birth (or more precisely, virgin conception), and is often eventually martyred. The further back in time the life of a prophet or founders go, the more mysterious and embellished the stories of their history becomes. Monotheistic Dualism proposes that the world is not how God wants it to be and that emissaries from the divine (prophets, patriarchs, or saviors) have come to guide humanity to a more virtuous place.

The charismatic religious patriarchs whose life history we know the least about are often the ones who become most venerated, deified, and mythologized over time. What may have originally been intended as metaphorical prose becomes proselytized as literal history.[5] It seems to be part of human nature to look for spiritual saviors and embellish their legends. I suspect that if Mohandas Gandhi, for example, had been born two thousand years earlier, or when literacy was nonexistent or just starting to take hold, by today his image may have evolved into that of a miraculous prophet or iconic inspirational figure of a major religion. As mentioned in chapter 4, even during Gandhi's lifetime, people tried to deify him against his wishes.

Monotheistic Dualism may go back as far as the twelfth century BCE. It can be divided into two early patriarchal branches: *Zoroastrianism* (with the one God of Creation known as *Ahura Mazda*) and *Abrahamic Monotheism* (with the one God of Creation known as *Yahweh*, *Jehovah*, *El*, *Yaldabaoth*, *Allah*, *El Shaddai*, *God*, *G_d*, or even "he who should not be named").

ZOROASTRIANISM: LINEAR MONOTHEISM FINALLY STICKS

Zoroastrianism is the earliest-known branch of Monotheistic Dualism (as I define it). It was a much more successful attempt at institutionalizing Linear Monotheism than the earlier, short-lived Atenism. Most authors believe it dates back to a prophet in the twelfth century BCE named Zoroaster.[6,7,8,9] In addition to being one of the world's oldest surviving religions, Zoroastrianism is the oldest-known monotheistic religion with a broadly recognized prophet. The primary priests of Zoroastrianism are men and are referred to as *mobad*. Women who serve as assistants are referred to as *mobedyar*.

Zoroaster (also known as Zarathustra to his followers) was an inspired poet who lived in pre-Islamic Persia (the Iranian plateau).[8,9] The exact dates of his life are uncertain. Most historians today believe that he lived sometime around the twelfth century BCE.[6,7,8] However, some authors (e.g., psychologist William Uttal) believe that he lived much later, perhaps as late as the seventh or sixth century BCE.[10] At the other extreme, some Parsis (modern-day Zoroastrians in India) believe that he lived much earlier, dating him as early as sixtieth century BCE, following Greek texts that misinterpret ancient Iranian sources.[6] The intent of the Parsis who believe he lived earlier is evidently to claim that Zoroaster was the first of the world's great religious prophets.

Among some present-day, orthodox Zoroastrians, Zoroaster is seen as a manifestation of the divine who was born of virgin conception.[6,8] At age thirty, he is said to have received a vision of God (Ahura Mazda) leading him to preach the Zoroastrian message.[11] Some historians contend that Zoroaster simply codified a religious ideology that had existed centuries before him.[10] He was married several times and had several children. He lived into his late seventies, after which he was supposedly killed by invaders while he was praying.[6]

Zoroaster is said by followers to have prophesized the virgin conception and birth of a divine being whose place of birth would be indicated by a blazing star. He told his followers: "When you behold the star, follow it wherever it leads you. Adore the mysterious child, offering him gifts and profound humility. He is indeed the Almighty Word which created the heavens. He is indeed your Lord and everlasting King."[12] The Persian Zoroastrian myth may have influenced the later Christian legend about the virgin conception and birth of Jesus Christ. In the New Testament, *Matthew* (2:1–21) describes three "wise men" or Magi from the East (interpreted by many scholars as being from Persia) who followed a star to Bethlehem (modern-day Palestine) to see the Christ child. When they arrive, they offered precious gifts to the infant and knelt down to pay homage to him.

Unlike Atenism, Zoroastrianism had clearly defined opposing realms of good versus evil, angels versus devils, and heaven versus hell. Although some authors interpret Zoroastrianism's mention of multiple sacred beings or "yazata" and "amesha spentas" as suggestive of polytheism,[13] many authors, as well as traditional Zoroastrians, interpret these as mere spiritual entities akin to angels, archangels, or mere aspects of a single God[9,11,14] (as I have interpreted it). The one supreme Zoroastrian God is the all-knowing *Ahura Mazda* (also known as *Ohrmazd*, or *Wise Lord*).[14,15] He is thought to be the creator of both the spiritual and the material worlds and the one God proscribed in Zoroastrian scripture to be worthy of worship. Those who worship one of his sacred aspects are said to be worshipping him only indirectly.[14] (Similarly, in trinitarian Christianity, those who worship Christ or the Holy Spirit are often said to be worshipping God indirectly). An evil counterpart and adversary to Ahura Mazda is known as *Angra Mainyu.* This counterpart eventually became known as an equivalent of the Christian and Islamic devil, and he is the personification of evil.[2] In the oldest Zoroastrianism scripture, hell is also inhabited with evil spirits called daevas, which are thought to be false gods.

The best-known symbol of Zoroastrianism is the *Faravahar*, a winged sun image with a male entity of uncertain identity standing in a disc. Early Zoroastrians adopted the winged disc part of the symbol from earlier polytheistic belief systems of Mesopotamia (compare the Zoroastrian faravahar in the figure below with the Mesopotamian god Ashur on page 184). Although many modern theories hypothesize about what the symbol originally meant and its origin (e.g., personal spirit, divine grace, or emulation of the Mesopotamian god *Ashur*), no one actually knows what the Faravahar meant to the earliest ancient Persians. Another common symbol of Zoroastrianism is the *Adar*, a cauldron containing the sacred fire representing the divine goodness of God (illustrated below). Fire is the central iconic symbol

Common symbols of Zoroastrianism are (*left*) the *Adar*, also known as the sacred fire, and (*right*) the *Faravahar*, which was the best-known symbol of Zoroastrianism and was inscribed on many ancient Zoroastrian temples. Note the resemblance to the symbol for the much earlier Mesopotamian god Ashur, on page 184.

within the Zoroastrian fire temples.[11] Priests who enter these sacred places of worship wear veils to prevent contamination of the holy fire by their breath or saliva. Fire also became important to early Abrahamic Monotheism. According to the Old Testament, God spoke to Moses from a fire within a bush that did not burn the bush (*Exodus* 3:1–5). *Hebrews* (12:29) also states that "Our God is a consuming fire."

The sacred scripture of Zoroastrianism is the *Avesta*, which is said by many followers to have been received by Zoroaster from God. Part of it, the *Gathas* or the *Old Avesta*, is thought by historians to have been authored by Zoroaster. The Avesta consists of a series of hymns and psalms that were passed down by way of oral transmission for many generations until they were finally written down. Additional parts were incorporated into the Avesta during the pre-Islamic Sasanian epoch (224–651 CE) resulting in what is known as the *New Avesta*. In Sasanian society (pre-Muslim Iranian/Iraqi society), the roles of women were restricted and limited. This belief system influenced early Islam when it arose later. Between the Old Avesta and the New Avesta, some of the religion's beliefs transformed. For example, in the New Avesta, an increased level of autonomous evil was associated with the Angra Mainyu. In Zoroaster's time Angra Mainyu was simply the "angry spirit" and the opposing disorder to Ahura Mazda's order. Post-Zoroastrian scripture eventually introduced the concept of Ahriman (the devil) as an enhanced evil personification of the Angra Mainyu.

The oldest surviving written text of the Avesta is a manuscript transcribed in the Avestan language dated 1323 CE.[16] Although the Avestan language has yet to be fully translated, it shows some similarities to the Vedic Sanskrit used by the earlier Indo-Iranian people for the Rig Veda.[13,16] Given the potentially immense time between the life of Zoroaster and the first written version of the Zoroastrian canon, it is difficult to know how much of this represents the Old Avesta, and how much is the product of revision (including influence from other religions).

According to the Avesta, the souls of those who help good triumph over evil will be rewarded, and the souls of those who do not will be punished.[11] Among the most important values and elements of "goodness" are honesty, truthfulness, generosity, care of the poor, fairness, and acting with moderation. The category of evil includes the traits of greed, hate, slander, and envy. Zoroastrianism may have been influential in the later development of Abrahamic Monotheism and its many subbranches. Some of the Zoroastrian ideas that translated into early Abrahamic Monotheism include the concepts of free will, messianism, Judgment Day, the resurrection, the human soul, heaven, hell, angels, and the end of times (apocalypticism). Some of the major differences included, for example, the very strict concept of Zoroastrian monotheism. As Christianity expanded from the fourth century CE onward, Zoroastrians began to look at Christianity with contempt for its trinitarian belief systems,[17] even though it previously had its own different aspects of God early in its history (the yazata discussed earlier in this section).

Earth and fire are both sacred in Zoroastrianism, but dead bodies are not. They are seen as a spiritual pollutant, and the place where the Zoroastrian devil (Angra Mainyu) and his forces are most powerful.[6] All dead matter is polluting, but the corpse of a righteous person is especially so, because death is seen as a significant (although temporary) victory of evil over good. The religion developed a ritual for disposal of the dead involving open exposure to the elements to avoid polluting the earth with the burial of the corpse, or profane fire by the cremation of the corpse. This is *dakhma*, or "the rite of exposure." Evidence of this practice goes back at least as early as the ninth century BCE.[13] The dead are laid out in the open for carrion eating wild dogs and vultures to consume.[18,19] Care in moving a corpse is taken so not to touch it with bare skin, because anyone who touches a corpse is said to be polluted even "to the end of his nails, and unclean thenceforth, forever and ever."[20] Some corpses are eaten by scavengers down to the bones in a few hours by this method, although most take much longer. Eventually, the practice led to the construction of huge stone structures called "towers of silence." These towers were built atop hills in the desert, where they were managed by Zoroastrian priests far from population centers. Some of these structures were massive, with three concentric rings for the bodies. The corpses of men are laid out in the outer ring, women in the middle ring, and children in the inner ring.[19] Once a body was picked clean and the bones bleached by the sun (a process that could take a year or more), the bones were then transferred to a pit at the center of the tower containing lime to disintegrate them further. The remaining material was then filtered through sand and coal before being washed out to sea. In Iran, the practice was gradually phased out in the early to mid-twentieth century and was finally outlawed in the 1970s. The continued growth of cities eventually resulted in the large towers becoming too close to major population centers. Since the prohibition of these towers in Iran, dead bodies have instead been buried in graves lined with cement to avoid polluting the earth. However, the towers of silence are still used today in India by the *Parsis* branch of Zoroastrianism (e.g., illustration on page 271). In twenty-first-century India, the population of vultures drastically declined by more than 99 percent because of the inadvertent poisoning from a drug developed for livestock. The drug was banned in 2006, but the few surviving birds were too few to consume all the dead. The Parsis have resorted to solar concentrators to accelerate decomposition of bodies and a captive breeding program for vultures to try to remedy the problem.[21]

The doctrinal institutionalization of Apocalypticism (end-of-times and final judgment scripture) first appeared in Zoroastrianism, before it became more fully developed within Abrahamic Monotheism. In Zoroastrian eschatology, it was

Interior of a Parsi Zoroastrian *Tower of Silence*. From an "old print" reproduced by M. Shoemaker (1912). Although much of his writing reflects the typical colonialist attitude about cultures other than his own, his drawing illustrates what has been reported by many historians and Indian scholars describing the Tower of Silence inner sanctum. Although this scene may be disturbing to many Abrahamic Monotheists, many Zoroastrians are equally disturbed by the idea of polluting the sacred earth with a buried human corpse, or polluting sacred fire with the cremation of a human corpse.

prophesized that one day a three-thousand-year struggle between good and evil will be fought, punctuated by evil's final assault. As a prelude to the Zoroastrian end of times, the sun and moon will darken, and humankind will lose its reverence for religion, family, and elders. The world will fall into a long winter, and the evil son of *Angra Mainyu* (the "antisavior") will be born to terrorize the world. In response, a savior known as *Saoshyant* will be born to a virgin for the final showdown between good and evil.[6] Saoshyant will raise the dead—including those in both heaven and hell—for final judgment. Then, he will return the wicked to hell to be purged of bodily sin. Next, all will wade through a river of molten metal in which the righteous will not burn. Heavenly forces will ultimately triumph over evil, rendering it forever impotent. All humans will then live in eternal paradise.

Although Zoroastrianism has fewer than two hundred thousand followers today (less than 0.003 percent of the world's population),[22] it was once one of the world's major religions[6,7]. As with so many religious belief systems, its early success was tied to its collaboration with powerful empires and emperors beginning in the sixth century BCE. Around 550 BCE, Cyrus the Great founded the first Persian Empire, which became the largest empire the world had yet seen. Under his rule, Zoroastrianism became the state religion. Zoroastrianism thrived through three successive Iranian empires: the Achaemenid Empire (550–330 BCE), the Parthian Empire (250–224 CE), and the Sasanian Empire mentioned earlier (224–651 CE). At its peak, Zoroastrianism came to dominate an enormous region, extending from Egypt and Greece to northern India, with Iran at its center. Zoroastrianism remained the state religion of the empire for much of the next thousand years. At one point, it even briefly attempted to convert or suppress followers of competing religions (e.g., Jews and Christians).[6]

Zoroastrianism's widespread success came to an end in the mid-seventh century CE, when the Zoroastrian Sassanian Empire was defeated by the Islamic Rashidun Caliphate (discussed in chapter 11). Once the Persian region came under Muslim control, state-sponsored religion shifted to Islam. The subsequent transition of imperial power to Islam saw the steady decline of Zoroastrianism in Persia. The religion became increasingly suppressed by the Islamic state, and Zoroastrians in Persia had to pay extra taxes to retain their religious faith. Over time, most Persian Zoroastrians converted to Islam.

During the eighth to tenth centuries CE, many Zoroastrians escaped from Persia to the west coast of India to preserve their religion and escape continued Muslim pressure and oppression.[6,15] These were the people who became known as the *Parsis* who established some of their own traditions (discussed earlier in this chapter with regard to the towers of silence). Over a thousand years later, a subgroup developed within the Parsis called *Khshnoomism* (also Ilm-Khshnoon, meaning "Path of Knowledge")[23] This subgroup of Parsis Zoroastrianism was influenced by Hinduism and by India's Independence movement from British rule. It was founded by Behramshah Shroff (1858–1927) who claimed to have received a special message from a secret race of Zoroastrian masters in Iran. It emphasized vegetarianism and reincarnation and was inclined more toward the occult than other sections of Zoroastrianism. Although adherents prefer to describe themselves as a "movement" than a sect, the movement represents an ideological move from the earlier, more orthodox Parsi tradition.

By the beginning of the twenty-first century, the Parsis numbered only around sixty thousand.[6] Although the religions of India are mostly Hindu and Muslim, the Parsis have maintained Zoroastrian core beliefs.[24] They also spread from India to North America, England, and Australia. Even so, by 2010, the entire world population of all Zoroastrians numbered only about 197,000.[22] Unlike later forms of Linear Monotheism, such as Christianity, Zoroastrians are thought by many historians to only rarely have been aggressive proselytizers of non-Zoroastrians into their faith, particularly after the seventh century and the loss of imperial support.[9] In spite of Zoroastrianism's decline and small following today, it was influential in the early ideological development of *Abrahamic Monotheism.*

ABRAHAMIC MONOTHEISM: THE RISE OF THE MIGHTY YAHWEH

Abrahamic Monotheism includes a number of religious belief systems that began to believe in one God only (e.g., *Yahweh* or *Jehova* for most of Judeo-Christianity, and *Allah* for Islam). Although Abraham was thought not to have been a strict monotheist (and therefore not strictly an Abrahamic Monotheist), many centuries later, the god that he came to regard as a Supreme God above all others became the one and only God of early Abrahamic Monotheism.

Abrahamic Monotheism, as a product of a growing "Yahweh alone movement," is thought to have formalized around the sixth century BCE from earlier forms of Abrahamic polytheism (discussed in the next chapter).[23,25]. Some branches of Abrahamic Monotheism later reincorporated soft polytheistic or animistic elements (also known as soft monotheism[26]). The most widespread modern example of this reversion is the divine triad of in Nicene Christianity, known as the Holy Trinity (the Father, Son, and Holy Spirit as three aspects of God, with the Holy Spirit animistically represented as a bird; discussed further in chapters 8 and 9).Another example of soft polytheism to emerge from Abrahamic Monotheism is the Divine Dyad of *binitarianism.* Examples of this include the *Macedonians* of Traditional Christianity (the Father and the Son as two aspects of God; discussed in chapter 8) and the *Biblical Demiurgists* (the Monad and the Demiurge as two aspects of God; discussed in chapter 10). These soft polytheistic groups are historically rooted in early Abrahamic Monotheism and claim to be monotheistic in principle.

The major branches of Abrahamic Monotheism share a number of important patriarchs, including three particularly prominent early figures discussed later in this chapter: Noah, Abraham, and Moses. In a sense, they are synapomorphies of Abrahamic Monotheism, although they vary in importance from branch to branch. Each major subgroup of Abrahamic Monotheism has its own canonical texts thought by followers to contain important information and guidelines concerning their religion's origin and development (the Written *Torah* for Rabbinic Judaism, the *Old Testament* and *New Testament* for Christianity, the *New Testament* and various apostolic gospels for Biblical Demiurgism, and the *Quran* for Islam). However, these texts were made by ancient peoples in cultural settings and contexts radically different from those of today. Consequently, their meanings are often difficult to interpret in modern times. They were originally interpreted in a variety of different ways, often reflecting the perspective of the theologian or historian doing the interpreting. They were also handed down over many generations as oral transmissions before they were written down for the first time.

Dating the origin of Abrahamic Monotheism is both controversial and historically complicated. One of the challenges of looking at world religions in an evolutionary context is teasing out the religious aspects (e.g., Second Temple Judaism) from broader cultural aspects (e.g., Jews and the secular history of the Israelites). Abrahamic religions begin with the ancient Israelites. These people originated as a group of Semitic-speaking tribes of the ancient Near East who settled in part of Canaan during the second millennium BCE. Early Israelite supernatural beliefs included a small pantheon of deities that may have been influenced by Canaanite/Babylonian Polytheism. Canaanite/Babylonian Polytheism was influenced, in turn, by earlier Sumerian beliefs from Mesopotamia.[27]The early cultural divisions of the Israelites by the twelfth century BCE are often equated with the Hebrew biblical narrative of the twelve tribes of Israel, whose common patriarch was Abraham's grandson, Jacob. Each of these twelve tribes was supposedly founded by one of Jacob's twelve sons.

Early in Israelite history, a god by the name of *El* became particularly important as the Supreme God.[2,25] In fact, the name Isra*el* means "May El preserve." Psalm 82 in the Hebrew Bible and the Christian Old Testament also describes what appears to be a god named *Yahweh* in conflict with other deities in the "Divine Assembly of El" (El's pantheon).[28] It is unknown when and where the Israelites first discovered Yahweh and transformed him into the Supreme God. Some authors suggest that he was originally a warrior god or storm god worshipped by the ancient Midianites of a region called Midian (modern-day Jordan).[2,13,25,28]. Egyptian texts from the thirteenth century BCE mention worship of Yahweh by the seminomadic Shasu peoples of the Transjordan region in the Middle East, some of which may have given rise to the early Israelite people.[29] Most nonbiblical (historical) details of Yahweh's ideological early history are today lost. Conflicting theories by various historians about the origin of the names El and Yahweh during the time of Abraham are often passionately debated, based on conflicting faith-based beliefs and biblical references.

The name El was sometimes used as a generic noun by early Israelites to mean god.[30] The names El and Yahweh became ideologically syncretized over time.[25]

The Israelites eventually came to revere Yahweh above all other gods and as the national god of Israel.[2,30] Historians believe that many early Israelites followed a form of *henotheism* or *monolatrism* (worshipping Yahweh exclusively, but accepting the existence of other gods). In the eighth century BCE *Book of Hosea*, one of the earliest Jewish prophets named Hosea instructed the Israelites to know (meaning be faithful to) no god but Yahweh. However, Hosea never denied the existence of other gods nor did he instruct anyone other than Israelites to worship Yahweh.[2,25,30] Similarly, several phrases of the Torah/Old Testament refer to other gods. *Deuteronomy* 10:17 says, "For the Lord your God is the God of gods, and the Lord of lords, a great god, a mighty and a terrible," and *Exodus* 15:11 says, "Who is like you, Yahweh, among the gods?" Even the early commandments given in *Exodus* hint at the early Israelite belief in multiple gods. *Exodus* 20:3 says, "Thou shalt have no other gods before me." Note that this does not deny the existence of other gods; it simply forbids placing them before Yahweh. Yahweh was already given special prominence well before the Israelites transitioned into strict monotheism.

Some scholars and historians of religion also suggest that the early Israelites of Monarchic Israel (1050–930 BCE) recognized a mother goddess named *Asherah* as the queen and consort of El or Yahweh.[30,31] Biblical archaeologist William Dever reported excavating tombs from the eighth to seventh century BCE with Hebrew inscriptions mentioning "Yahweh and his Asherah."[29] Asherah was also the queen and consort of the earlier Mesopotamian sky god An (Anu). The "Asherah pole" was a sacred tree or pole icon placed at Canaan religious sites to honor her.

Over time, Yahweh's status grew. He became more than an anthropomorphization of some natural force and more than the Supreme God of a pantheon. Rather than being a storm god, a warrior god, or even king god, he became the god of nothing in particular, yet the God of all things. He became a realm apart.[28] A Yahweh-alone movement developed and grew in strength (discussed in chapter 7). It was often at odds with followers of another member of the Canaan pantheon, the storm god Baal.[28] The final phase leading to strict monotheism came sometime around the sixth century BCE, during the Babylonian exile of the Israelites. Yahweh became the one and only God, giving rise to *Abrahamic Monotheism*.[2,23,28,32,33,34] At that point, the existence of other gods was not only formally denied, the worship of other gods was treated with hostility. The Old Testament's book of Isaiah speaks of Yahweh clearly in monotheistic tones, quoting him as saying, "I am the first, I am the last; and there is no god but Me" (*Isaiah* 44:6). His name appears more than 6,800 times in the Torah/Old Testament; however, because early biblical script lacked vowels and it did not distinguish between the letters Y and J or the letters W and V, the name usually appeared as YHWH (translated as Yahweh) but also sometimes as JHVH (translated as Jehovah). (The spelling of the Abrahamic god's name is discussed further in **Note 6B**.) By the third century, BCE, the name Yahweh changed simply to "God" sometimes written as "G_d" because the name Yahweh, or even the capitalized form of the word "God," had become too sacred to speak or write.

THREE IMPORTANT SHARED PATRIARCHS AMONG ABRAHAMIC RELIGIONS: NOAH, ABRAHAM, AND MOSES

Biblically, the namesake of Abrahamic religions is Abraham a poorly known character from ancient Mesopotamia (present-day Iraq). Abrahamic religions share a number of special iconic figures, the historical reality of which is uncertain. Some may have been reinterpretations of earlier legendary figures in earlier cultures. One of the early legendary Abrahamic icons of particular importance is the grand patriarch of modern humanity, *Noah* (whose family is the only surviving lineage after the flood narrative in *Genesis*). Another was *Abraham* (as already mentioned, the revered source of the term "Abrahamic"). A third major patriarch was the prophet *Moses*, who was the symbol of Yahweh's (and later Allah's) laws. He was also the leader of the Exodus (the founding myth of the Israelites and their journey out of slavery in Egypt traditionally dated around 1250 BCE[11]). Rather than being true historic characters, some historians believe that these three iconic characters (as well as many others) originally represented metaphorical lessons about deeper issues. However, the later development of Rabbinic Judaism, Christianity, and Islam each translated these spiritual allegories into their own versions of alleged history, sometimes losing sight of the original meaning in the process.[5]

In *Genesis* 6–9, Noah was said to have been a ninth-generation descendant of Adam and Eve. God had grown unhappy with the state of humanity and decided to return earth to its pre-creation state for a fresh start. He planned to do this by causing a great flood to annihilate all humans as well as most other living things, with the exception of a few representative life-forms to repopulate the earth. The humans that God spared included Noah, his three sons, and each of their wives. God commanded Noah to build an ark (a great ship) large enough to hold his family, provisions, and two of each kind of animal (two of "every living thing" according to his command). The issue of terrestrial plants is not addressed, nor is the issue of the insects necessary to pollinate them. God's instructions specified that the ark be built from wood and that it measure three hundred cubits in length and fifty cubits in width (roughly thirty-four-thousand square feet of deck space, or roughly one-fifth the size of a modern aircraft carrier). This was pretty much a seat-of-the-pants set of instructions, because a cubit is an imprecise ancient measure of length based on the distance from the elbow to the tip of the longest finger. Seven days before the deluge, Yahweh told Noah to enter the ark with his family and all the animals.

After the doors were sealed, it rained for forty days and nights. The floods came, covering even the highest mountains. Most life outside the ark perished. Eventually the water began to recede and the ark came to rest on Mount Ararat, the highest mountain in what is today Turkey (although this location is debated by some scholars). Noah sent a dove out to see if it would find suitable land, but it found none and returned to the ark. Seven days later, Noah sent the dove out again, and this time it came back with an olive leaf in its beak indicating it had found suitable land where at least a tree had survived. Shortly after that, Noah, his family, and all of the animals disembarked from the ark. Yahweh told Noah to "multiply on the earth and be fruitful and increase in numbers on it" (*Genesis* 8: 16–17). Over time, Noah, his family, and the animals repopulated the earth with their descendants. Evidently, the trees and other plants of the earth had survived the flood. According to *Genesis*, Noah died at the age of 950 years.

The *Genesis* Great Flood story of Noah appears to have inherited many of its details from polytheistic flood stories predating Abrahamic Monotheism, such as the Mesopotamian Epic of Gilgamesh described in chapter 5. Other non-Abrahamic (and pre-Abrahamic) Great Flood stories, complete with arks to save a few chosen by the gods, can also be found in branches of Hinduism (page 97), Greek Polytheism (page 202), and other religions with ancient histories. The prevalence of so many Great Flood myths may be an artifact of where early civilizations lived. Early civilizations, such as that of Mesopotamia, were located in flood plains and river valleys where the environment was most suitable for early farming practices (the seed of early civilization). In those early days, an unusually massive flood like what we today call a "one-hundred-year flood" would seem like the end of the world to such communities, particularly if they were relatively isolated from the rest of the world. Such floods could wipe out entire civilizations except for a few survivors who could make it to other civilizations to recount the disaster. Such tales could have provided the roots from which great mythologies were born. Flood myths were enhanced by the discovery of aquatic fossils in inland regions. The ancient Egyptians, Greeks, and Romans all believed that the earth had been covered with water, citing shell fossils and fish fossils on mountaintops as evidence.[35]

The next iconic person of Abrahamic Monotheism, in order of appearance in the Bible, was its namesake, Abraham (also spelled Abram). Abraham is believed to have been born in the Mesopotamian city of Ur in the eighteenth or seventeenth century BCE.[11,27] He is believed to be the earliest common patriarch of the Israelites (also known as the Hebrews). Little is known about him other than biblical references (e.g., chapter 12 of *Genesis*), and no archaeological evidence empirically shows him to be anything more than a mythical or metaphorical figure.[11] Some scholars speculate that he may have been one of the wandering chieftains who had led their people from Mesopotamia toward the Mediterranean at the end of the third millennium BCE.[28] According to biblical myth and various passages from *Genesis,* he was a tenth-generation descendent of Noah and the great-great-great-great grandfather of Moses. He is venerated and considered to be a patriarch in all Abrahamic religions, although somewhat more so in early Judaic branches and Islam than in Traditional Christianity. In early Judaic branches, Abraham is recognized as the grandfather of Jacob, the common patriarch of the Jews as a people.[36] As mentioned, however, many historians do not think Abraham was a strict monotheist, at least not early in his life. In Islam, Abraham is a link in a chain of prophets that began with Adam (from Genesis of the Hebrew Bible and Old testament) and culminated with Muhammad.[36] Islam considers Abraham to be the perfect Muslim. In Christianity, Abraham is considered to be "our father in faith," but he is given less stature than in most of Rabbinic Judaism or Islam, because Jesus is the central figure of Christianity.

The Book of *Genesis* describes Abraham as a worshiper of El, the Supreme God of the Canaanite pantheon.[28] Other authors contend that during Abraham's lifetime, El was used as a word simply meaning "god" rather than referring to a specific deity, and so Yahweh and El may have been one in the same to Abraham.[30] Biblical references are vague and controversial regarding Abraham's early history. As mentioned earlier in this chapter, no archaeological evidence even clearly supports his existence.

According to *Genesis* (22:1–19) child sacrifice was also part of Abraham's story (as mentioned in chapter 5). The biblical passage relates that Yahweh commanded Abraham to sacrifice his young son Isaac. While he was preparing to cut his son's throat and burn his body in a sacrificial fire to God, the biblical account says one of Yahweh's angels stopped Abraham at the last minute and opted for the sacrifice of a ram instead. One of the more gruesome early aspects of many religions was the glorification of human sacrifice, often children, done in the name of a God or gods. Abrahamic Monotheism was no exception. According to biblical scripture, the early Israelites, including the grand patriarch of Abrahamic Monotheism, were still willing to practice child sacrifice like the earlier Canaanites (discussed further in **Note 5D**).[30]

Abraham served as a sort of ideological bridge between Mesopotamian Polytheism and early Abrahamic Monotheism. Many believe that he was eventually the first to recognize a monotheistic Abrahamic God (thus making him the first patriarch of Abrahamic Monotheism). Abraham may have originally believed in other Canaanite deities, such as the storm god Baal and the goddess Asherah. The biblical story of Abraham is long and complicated. He and his descendants are said to have been the ones to which God promised the land of Canaan (known today as Israel) for their homeland. Today, the Jews and the Muslims still battle each other over who is the rightful beneficiary of that promise, resulting in what seems to be never-ending conflict. According to *Genesis* (25:7–10), Abraham died at the age

Abraham, patriarch of Abrahamic religion and the Israelite people. Here he is depicted in a 1634 painting by Rembrandt about to offer his son Isaac as a human sacrifice to the god El (or Yahweh), only to be stopped by one of El's angels. Story is based on *Genesis* 22:1–12. Painting in the collection of the Hermitage Museum in Saint Petersburg, Russia.

of 175 years. There are no contemporaneous references to Abraham, despite the existence of many written documents on other religious matters from the Mesopotamian, Syro-Palestinian, and Egyptian cultures, where *Genesis* reports him to have lived. The Abraham story, like the Exodus, is a late literary construct that cannot be accurately dated.[37] Scholars believe it was composed when most of the Near East was under the control of the Ancient Persians (the late sixth century BCE). Parts of Abraham's story were passed down by oral transmission for many generations before finally being transcribed.

Moses is yet another iconic character of Abrahamic Monotheism. He was an early prophet who became the symbol of Yahweh's laws for Jews and Christians, as well as Allah's laws for Muslims. According to *Exodus* (6:2–3) he is also the first to have learned that the god who identified himself as *El* to previous prophets should be known as *Yahweh*. The story of Moses's birth and early childhood first appears in the Book of *Exodus*, which was written as early as the tenth century BCE based on oral transmissions that went further back in time. He, like Noah and Abraham, may not have been a historical figure. Instead, he may originally have been a mythological or metaphorical figure who was embellished and dramatically exaggerated to serve an ideological agenda.[2,5,29] Based on Abrahamic scripture, leaders of Rabbinic Judaism calculate his time of birth to have been somewhere between the early sixteenth and late thirteenth century BCE. His time was supposedly several generations after Abraham, but Biblical references regarding the exact number of generations are confusing and unclear. According to *Deuteronomy* (34:7), Moses was 120 years old when he died. His burial site is unknown. The importance of Moses to early Biblical Demiurgism is less clear. That belief system demotes the Old Testament's creator god to be an ignorant deity, secondary to a good "unknowable" Supreme God (discussed later in chapter 10).

Moses, like Abraham, is said to have come from a background of polytheism. Some passages in the Old Testament indicate that he had polytheistic beliefs even after escaping Egyptian slavery. In *Exodus* (15:11), he asks Yahweh, "Who among the Gods is like you Lord?" He later says, "Now I know that the Lord is greater than all other gods, for he did this to those who had treated Israel arrogantly" (*Exodus* 18:11).

Many Jewish and Christian believers ascribe the original authorship of the *Pentateuch* (the first five books of the Hebrew Bible and of the Old Testament of the Christian Bible) to Moses in the thirteenth century BCE. However, modern biblical historians believe it to have been composed by multiple authors between the ninth and fifth century BCE.[6] The five books include *Genesis, Exodus, Leviticus, Numbers,* and *Deuteronomy*. Originally, transmission of the volumes was oral, dependent on the passed down memories of oral transmissionists over many generations, with no written version until centuries later.[38] Moses of the Old Testament and the Tanakh Book of *Exodus*, as well as in the Quran, is the prophet who reportedly led his people, the Israelites, out of bondage in ancient Egypt to the "promised land" of Canaan. The beginning of the Exodus is traditionally given as 1250 BCE.[11]

The story of Moses also has a number of alternate interpretations. The famous neurologist and founder of psychoanalysis, Sigmund Freud, studied religions. His theory was that the biblical Moses was a priest of Akhenaten who fled Egypt after the pharaoh's death to perpetuate monotheism through different channels.[39] It has even been proposed that Moses and Akhenaten were the same person.[10] Certain elements of his story may also have come from a remarkably similar and earlier story from Mesopotamian Polytheism concerning Sargon of Akkad (described in **Note 6C**). Others suggest that Moses and the story of the Exodus was merely a metaphor for Jewish liberation.[29] In ancient scriptural stories of religion, it is often impossible to know where to draw the line between metaphor and literal history.

The story of Moses's birth and adoption is a familiar one in Western culture. The impact of his story is so great among the Abrahamic religions that he is broadly regard as one of their most sacred figures, and his story is one of the early defining characteristics of this branch. The story of Moses presents him as an Israelite (Hebrew) by birth, born in a time when the Israelites were an enslaved minority in Egypt. The pharaoh had become wary of the increasing population of Israelites, fearing their numbers might one day overwhelm the Egyptians. At one point, he ordered that newborn Hebrew babies be killed as a precaution to maintain Egyptian superiority. To save Moses from this fate, his birth mother left him in a reed basket beside the river. A princess of the Egyptian royal court discovered the basket and took Moses into her custody. She raised him under the protection of royalty (again, similar to the story of Sargon of Akkad, given in **Note 6C**).

Little is said about his early life in the Egyptian palace. The biblical story picks up later in his life when one day he kills an Egyptian for beating an Israelite slave. As a result, Moses had to flee to Midian (in the present-day area of the northwest Arabian Peninsula) where he encountered a burning bush. The fire was an angel of God telling Moses to return to Egypt and free his people, the Israelites, who had been enslaved. Like Zoroastrianism before it, early Abrahamic Monotheism saw fire as sacred, representing the light of God. The next part of the story, the Exodus, is well known in Western culture. Moses brings a series of plagues to the pharaoh until he is finally allowed to leave Egypt with the rest of the Israelites. When the pharaoh changes his mind and sends his army to bring them back, Moses parts the sea to let his people escape and then brings the sea crashing back down on the pursuing army of the pharaoh.

Much of the cultural success of Organized Religion depends on the power of the stories it tells, and the Exodus was an action-packed inspirational story that made it the focus of one of the most successful Hollywood movies of the 1950a: *The Ten Commandments* (1956). I remember watching it when I was a young kid, seeing its novel special effects for the time, and the giant waves crushing the army of the pharaoh. Whenever I hear the name Moses, I still see the face of Charlton Heston (the actor who played Moses). The saga of Moses is a great story that portrays him as charismatic but uncompromising. According to biblical scripture, however, he also was quite prone to anger and violence.

During the Exodus, Moses continued to lead the Israelites day and night until they came to Mount Sinai. There, he left them and climbed to the top of the mountain where he received the two stone tablets of God's laws (the Ten Commandments). God, who had previously been known by the name of El, then also revealed his true name to Moses as YHWH (pronounced as Yahweh). From then on, Moses is portrayed as having a strict monotheistic perspective with a contempt for idolatry. While Moses was on the mountaintop, trouble was brewing down below. He had been gone for quite a while, and eventually some of the people feared he might be dead, so they built an idol (a golden calf or bull, which was the traditional effigy of the older god El) to which they prayed. When Moses finally returned and saw the idol, he became so furious that he broke the stone tablets. He accused those who built the calf as having betrayed Yahweh. He then showed the darker side to his nature. He commanded that those who did not participate in the veneration of the idol to kill all those who did. According to *Exodus* 32:27–28, about three thousand people were executed.

Moses was not much into forgiveness or "turning the other cheek" in general. If the events in the Old Testament happened today, Moses would be prosecuted for war crimes (discussed further in **Note 6D**). There lies a conundrum of religious faith: When do you follow someone who justifies murder or genocide by saying that "God told me to do it"? And how is that ethically possible in a culturally diverse world? Such horrific stories of the Torah/Old Testament are part of what led Biblical Demiurgism (chapter 10) to reject the creator god of those scriptures as evil and in opposition to an unknowable Supreme God of good. However, the metaphorical mission of Moses was never one of mercy and tolerance. It was instead one of establishing clear and enforceable biblical rules for a specific ideology. The history or legend of Moses has numerous different interpretations, but all Abrahamic branches portray Moses as a symbol of God's law. He is particularly famous as having received the Ten Commandments (also known as the *Decalogue*) according to the Old Testament's Book of *Exodus*.

The ten laws or "commandments" that Moses reportedly transmitted from Mount Sinai are very much a reflection of the God of the Old Testament. They were also the seed of later monotheistic conquest. The first four commandments of *Exodus* are concerned with the veneration of Yahweh, and only the last six focus on human ethics and morality. The ethical laws in the book of *Exodus* are similar to the ethical and moral standards established by polytheistic religions a thousand years earlier. Historically, this point has rarely been acknowledged by leaders of Abrahamic religions, because they were generally loath to find any redeeming ethical value in the religions they were trying to replace.

One example from early polytheistic religion that may have been a template for the more ethical elements of *Exodus* (written around 700 BCE) is from the earlier

Egyptian *Book of the Dead* (written in second millennium BCE).[40] The one stylistic difference between these two versions is that in *Exodus*, the laws are presented as straight-up commandments ("Thou shalt nots"). In the Book of the Dead, the laws are instead in the form negative confessions one must provide to the gods while being judged in the afterlife ("I have nots"). The commandment from *Exodus*, "Thou shalt not kill," reflects the Book of the Dead's "I have not slain people." The *Exodus* commandment "Thou shalt not commit adultery" reflects the Book of the Dead's "I have not had intercourse with a married woman." The Exodus commandment "Thou shalt not steal" reflects the Book of the Dead's "I have not stolen." Finally, "Thou shalt not take the name of the Lord thy God in vain" reflects the Book of the Dead's "I have not cursed or blasphemed a god." The Book of the Dead also covers a number of other ethical points, such as "I have not uttered lies," "I am not deceitful," "I have not slandered," "I have not terrorized others," "I am not a man of violence," "I have not acted in arrogance," and so on. The Book of the Dead has forty-two negative confessions in all (listed in **Note 6E**).

One of the biggest general differences between the set of commandments in Abrahamic Monotheism and their earlier analogs in Egyptian Polytheism had to do with deity worship. In the commandments of *Exodus*, the Abrahamic God Yahweh demanded exclusive veneration in the first two commandments ("Thou shalt have no other gods before me" and "Thou shalt not make any graven images"). In contrast, earlier Old World Hard Polytheism had welcomed new gods into their pantheons with a sort of "the-more-the-merrier" approach, and they had no prohibitions against depicting them with imagery. In hard polytheism, as well as in the Dharmic branches of Asian Cyclicism, there are many paths to a spiritual end.

THE DEVELOPMENT OF MONOTHEISM IN ABRAHAMIC RELIGIONS

Compared to most earlier religions, early Abrahamic Monotheism was much more exclusivist and intolerant of alternate views, gods, or competing religious belief systems. As historian Robert Wright puts it, Abrahamic religions have had "a tendency to exaggerate their past specialness."[40] In Abrahamic Monotheism, there was only one acceptable path, and any deviation from that path had severe consequences. Yahweh (of the Old Testament and Torah) was portrayed as a powerful, jealous, sometimes violent God, with severe penalties for certain offenses. Leviticus 20 of the Torah/Old Testament considers cursing your parents, committing adultery, or being homosexual to be capital offences. Numbers 15 of the Torah/Old Testament prescribes stoning to death those who make the mistake of working on the Sabbath. Yahweh also had no mercy for those who might also worship any other god. The following is one of many passages from the fifth book of the Written Torah and the Christian Old Testament:

> If anyone secretly entices you—even if it is your brother, your father's son or your mother's son, or your own son or daughter, or the wife you embrace, or your most intimate friend—saying, "Let us go worship other gods," whom neither you nor your ancestors have known, any of the gods of the peoples that are around you, whether near you or far away from you, from one end of the earth to the other, you must not yield to or heed any such persons. Show them no pity or compassion and do not shield them. But you shall surely kill them; your own hand shall be first against them to execute them, and afterwards the hand of all the people. Stone them to death for trying to turn you away from the Lord your God, who brought you out of the land of Egypt, out of the house of slavery. Then all Israel shall hear and be afraid, and never again do any such wickedness. (*Deuteronomy* 13:6–11)

The Old Testament/Torah set the seeds of future conquests and obliteration of competing religious belief systems with passages like this:

> When the Lord your God brings you into the land you are entering to possess and drives out before you many nations—the Hittites, Girgashites, Amorites, Canaanites, Perizzites, Hivites and Jebusites, seven nations larger and stronger than you—and when the Lord your God has delivered them over to you and you have defeated them, then you must destroy them totally. Make no treaty with them, and show them no mercy (*Deuteronomy* 7:1–2).

> This is what you are to do to them: Break down their altars, smash their sacred stones, cut down their Asherah poles and burn their idols in the fire. For you are a people holy to the Lord your God. The Lord your God has chosen you out of all the peoples on the face of the earth to be his people, his treasured possession (*Deuteronomy* 7:5–6).

> Of the cities . . . which the Lord thy God doth give thee for inheritance, do not leave anything alive that breaths . . . completely destroy them (*Deuteronomy* 20:16–17).

These biblical passages, among many others, set the tone for centuries of religious conquest—from the early battles of Judaic groups in the Middle East to the Islamic conquests of India and North Africa, and from the Roman Catholic inquisitions in Europe to the Spanish Catholic conquests of the Americas. In many branches of Abrahamic Monotheism, violence has often resulted from conflicting interpretations of sacred scripture, and the conflation of national governance with elements of religious proselytism. In extreme cases, over the course of history, religious fundamentalists in league with imperial powers have used their chauvinistic views to promote ethnocide of opposing cultures.

LINEAR MONOTHEISM IN THE MODERN WORLD

Many scholars of religion today have written about the challenges of bridging human survival in the modern world with the violent aspects of religions established in ancient times.[41] Sacred texts were written long ago by people whose motives were the conquest and conversion of others to their way of thinking. All too often, religions developed an overly tribalistic, "us versus them" motivation that overshadowed whatever moral or ethical values were present. One casualty among early Christ-revering branches of Linear Monotheism was Biblical Demiurgism (discussed in chapter 10). Its followers believed that Yahweh as described in the Old Testament was overly cruel and jealous. They demoted him to the lesser status of "Demiurge," which was below a higher unknowable good god of wisdom called "The One" or the "Godhead." Competing factions of early Abrahamic Monotheism eventually crushed the Biblical Demiurgists out of existence.

Many other branches of Linear Monotheism have often been at violent odds with one another. Even though they are historically connected, they often choose not to recognize their historical relationships. They instead focus on their differences (e.g., Islam versus Judaism, Judaism versus Traditional Christianity, Islam versus Zoroastrianism, or Traditional Christianity versus Islam). They seem to ignore that no matter how different one is from one's cousin, the two will always be historically connected through a shared ancestry. Interdenominational rivalries among the branches of Linear Monotheism throughout history have often been catastrophic. As the size of human population has increased, the scale of those catastrophes also has increased. Finding better, more peaceful ways to interact in the modern world with more than eight billion people remains a challenge for religions and for our species in general.

The earliest development of Linear Monotheism was a process of trying to establish the believe in only one God, and who that God was. This development eventually led to *Abrahamic Monotheism*, which today contains nearly all followers of Linear Monotheism and well over half of all the world's religious population.[22] By the last few centuries BCE and the first few centuries CE, Abrahamic Monotheism began to try to establish proper ways to acknowledge and worship God as well as who his authentic prophets, teachers, or messiahs were. This effort led to many schisms and much branch diversification. The early diversification of *Abrahamic Monotheism* is the subject of the next chapter.

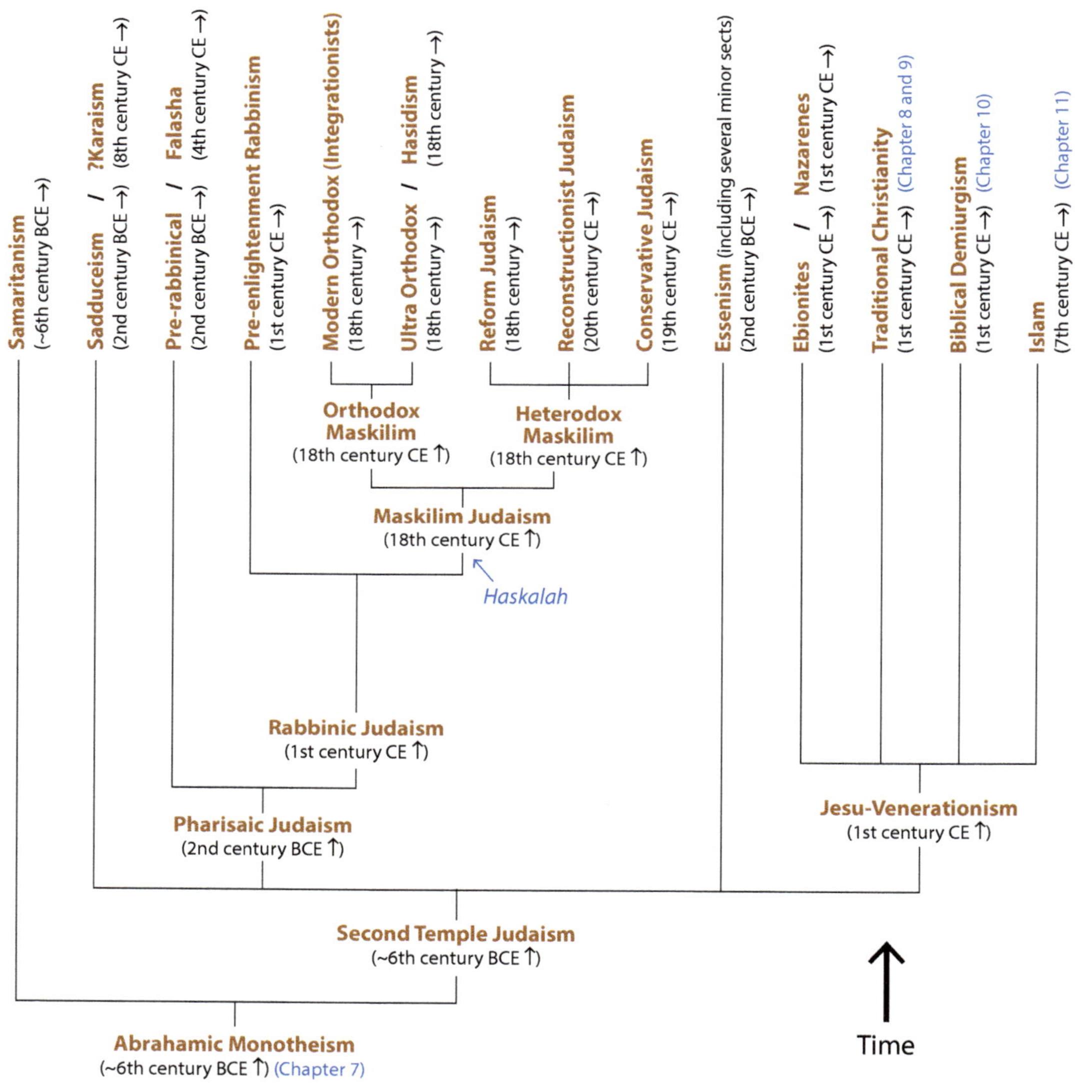
Samaritanism (~6th century BCE →)
Sadduceism / ?Karaism (2nd century BCE →) (8th century CE →)
Pre-rabbinical / Falasha (2nd century BCE →) (4th century CE →)
Pre-enlightenment Rabbinism (1st century CE →)
Modern Orthodox (Integrationists) (18th century →)
Ultra Orthodox / Hasidism (18th century →) (18th century →)
Reform Judaism (18th century →)
Reconstructionist Judaism (20th century CE →)
Conservative Judaism (19th century CE →)
Essenism (including several minor sects) (2nd century BCE →)
Ebionites / Nazarenes (1st century CE →) (1st century CE →)
Traditional Christianity (1st century CE →) (Chapter 8 and 9)
Biblical Demiurgism (1st century CE →) (Chapter 10)
Islam (7th century CE →) (Chapter 11)
Orthodox Maskilim (18th century CE ↑)
Heterodox Maskilim (18th century CE ↑)
Maskilim Judaism (18th century CE ↑)
Haskalah
Rabbinic Judaism (1st century CE ↑)
Pharisaic Judaism (2nd century BCE ↑)
Jesu-Venerationism (1st century CE ↑)
Second Temple Judaism (~6th century BCE ↑)
Abrahamic Monotheism (~6th century BCE ↑) (Chapter 7)
Time

7

THE EARLY DIVERSIFICATION OF ABRAHAMIC MONOTHEISM

I am Jehovah, and there is no one else.

—*ISAIAH* 45:5

A*brahamic Monotheism* began in the Middle East at least as early as the sixth century BCE. It expanded over time, evolving into thousands of subbranches and becoming religion's most successful attempt at the concept of one God and one God only. By 2010, it collectively included more than 55 percent of the world's population, mostly within its two largest branches: *Traditional Christianity* and *Islam*.[1] By the same time, its earliest forms (e.g., *Samaritanism*, *Sadduceeism*, and *Pharisaic Judaism*) had collectively dwindled to about 0.2 percent of the world's population.[1] The major early branching events of Abrahamic Monotheism are illustrated on page 286. Its diversity occupies the next five chapters in this book (where that emphasis is largely an artifact of the knowledge available to me).

Abrahamic Monotheism evolved within an ancient Israelite culture that was initially polytheistic with traces of Mesopotamian (Sumerian) influence.[2]

THE BEGINNINGS OF ABRAHAMIC MONOTHEISM AS "JUDAISM"

The earliest history of Abrahamic Monotheism is poorly known, and I touch on only a few highlights here (e.g., timeline on the next page). It began mostly as a series of what today generally are referred to as Judaic branches. Part of the challenge of reconstructing the earliest history of Abrahamic Monotheism is that before the first century BCE, its written records are not contemporaneous with the events they record. Instead, they consist of short pieces of handwritten text based on centuries of oral transmissions. Also, historical (archaeological) evidence supporting

Timeline for certain events, people, and beliefs relevant to Abrahamic Monotheism

Quran written down in completed form	~650 CE
Muhamad, founder of *Islam,* born in Mecca	~570 CE
Christianity becomes state religion of the Roman Empire	~380 CE
Traditional Christianity and *Biblical Demiurgism* develop	~1st and 2nd century CE
Final book of New *Testament* written (*Revelation*)	~100 CE
Destruction of *Second Temple* in Jerusalem by the Romans	~70 CE
Jesus, central figure of *Christianity,* is born in Nazareth	~4 BCE
Hebrew Bible (Tanakh)/Old Testament completed	~200 BCE
Dedication of *Second Temple* in Jerusalem	~515 BCE
Destruction of *First Temple* in Jerusalem by Babylonians followed by the rise of the Yahweh-alone movement and emergence of true *Abrahamic Monotheism*	~586 BCE
Assyrians take over the Northern Israelite Kingdom and the Israelites are exhiled from there	~722 BCE
Soloman dies and the Israelite Kingdom is split into a northern kingdom (the *Kingdom of Israel*) and a southern kingdom (the *Kingdom of Judah*)	~931 BCE
Completion of the Israelite's *First Temple* ("Solomon's Temple"), where Yahweh is chief God of the pantheon and the state god of the Israelites	~950 BCE
Yahweh first indicated as a warrior god in a pantheon where *El* is the supreme deity. Yahweh eventually replaces El as supreme deity after Moses, leader of the Israelites, receives the *Ten Declarations* (*Ten Commandments*) on Mount Sinai	~14th and 13th century BCE
Abraham, patriarch of the Israelites, is said to have been born in Ur, Mesopotamia. He is also said to have worshipped *El* as chief (but not monotheistic) god of a larger pantheon	~18th or 17th century BCE

biblical (belief-based) interpretation is often lacking. Another challenge for modern historians is the fact that many early branches of Abrahamic Monotheism were in violent competition with each other. They had little desire to find common ground, and little or no motivation to preserve accurate records of each other's traditions. Ancient historians were often religious fundamentalists who were prejudiced in their descriptions of competing supernatural belief systems.

"Judaism" has two different cultural contexts. One is *religious* and focuses on Judaism's monotheistic history (accepting the religious elements of the Torah and Abrahamic Monotheism). The other is more broadly *ethnic* and not as constrained by religious ideology. A 2010 public survey in Israel indicated that 42 percent of resident Jews consider themselves to be secular.[3] According to Jewish religion historian Daniel Boyarin, "Jewishness [in the ethnic sense] disrupts the very categories of identity, because it is not national, not genealogical, not religious, but all of these in dialectical tension."[4] The focus of this book, however, is only on the religious aspect of culture. Therefore, nonreligious belief systems, such as *Hiloni* (non-observant Jews) are not included on my phylogenetic trees.

As a religious ideology, the term "Judaism" is traditionally used as a non-monophyletic vernacular name for various followers of Abrahamic Monotheism who do not recognize Jesus a messiah or prophet. To be monophyletic (i.e., phylogenetically complete) Judaism would need to include the group that I refer to in this book as *Jesu-venerationism* (page 286). Some branches of what is traditionally referred to as Judaism (e.g., *Pharisaic Judaism*, *Sadduceism*, *Essenism*, and *Jewish Christianity*) appear to be phylogenetically more closely related to *Traditional Christianity* than they are to other "traditional Judaic" branches, such as *Samaritanism* (tree on page 286). Another way of saying this is that *Traditional Christianity* and other religions that revere Jesus of Nazareth have early roots in Judaic tradition. They are historically derived from "traditional Judaism" and therefore ideologically connected to it. Even Jesus was reported to have been Jewish. Therefore, my use of the term "Jew" or "Jewish" in this book is meant only as a traditional vernacular (and non-monophyletic) name for Abrahamic Monotheism excluding Jesu-venerationism (sensu the tree on page 286).

Ethnically, the early ancestors of the Jewish people were the Israelites.[5,6]. The Hebrew Bible (the Tanakh) describes the early Israelites as twelve tribes whose immediate patriarch was Abraham's grandson *Jacob* (who was later named *Israel*).[7] As mentioned in chapter 6, each of the twelve tribes is said to have been founded by one of Jacob's twelve sons. The Tanakh describes the Israelites as a variety of ethnic groups eventually united by their loyalty to the God of Moses (*El* or *Yahweh*). Most History of Religion scholars believe that the earliest Israelite religions were polytheistic, like their Canaanite precursors.[7,8,9] Archaeological evidence indicates a shift from Canaanite to a more formalized (yet still polytheistic) Israelite religion around 1250 BCE–1150 BCE.[5] The details and dates of this early history are controversial. According to biblical scripture, the monarchical organization of the Israelites began with loose confederation of the original twelve tribes called the *United Monarchy*, whose capital was Jerusalem. According to the Jewish bible, the monarchy was ruled by a succession of three great Israelite kings beginning with King Saul around 1025 BCE, followed by King David in 1013 BCE, and ending with King Solomon who reigned from 973 BCE until his death in 931 BCE.[10]

King Solomon supposedly oversaw the building of a national temple in Jerusalem known as Judaism's "First Temple" or "Solomon's Temple." It was completed around 950 BCE. The temple united all the Israelite tribes for some time, but high taxes by Jerusalem soon caused the people of the north to revolt. After Solomon's death, the United Kingdom split, with the northern Israelite tribes forming their own kingdom, the *Kingdom of Israel*, whose capital was *Samaria*. The remaining Southern kingdom became known as the *Kingdom of Judah* (a name later Romanized to *Judea*), whose capital was Jerusalem. The two kingdoms remained as separate states for more than two centuries, and this division weakened them both. In 722 BCE, the Assyrian Empire invaded and took over the Northern Kingdom and Samaria (in part of what today is known as the West Bank), and most Israelites of that region were exiled from their homeland. This effectively ended the Northern Kingdom. The *Samaritans* claim that they are descended from the Northern Kingdom Israelites.

In 586 BCE, the Babylonian Empire conquered the Southern Kingdom (Judah or Judea), destroying Jerusalem and the First Temple along with it. This added to the number of Israelites exiled from their homelands (the *diaspora*). Over the following centuries, the Israelites in Jerusalem went through periods of exile and return, and contraction and expansion. Over the centuries, Jerusalem eventually became a place that is holy to today's three major Abrahamic branches: Rabbinic Judaism, Traditional Christianity, and Islam.

The original patron god of the Israelites was El, thought to have been inherited from the Canaan pantheon and sometimes animistically represented as a bull.[5,9] As mentioned in chapter 6, the goddess Asherah is thought by some historians to have been revered as his consort. Other gods in the early Israelite pantheon included the storm god and divine warrior, Baal.[9,11] The beginning of *Genesis* is usually translated as "In the beginning, God created heaven and earth," but the Hebrew text for this line uses the word *Elohim* for God. Technically, Elohim is the *plural* form of Eloah. Consequently, it is conceivable that the line originally said, "In the beginning, the *gods* created heaven and earth."[6] El eventually became known as *Yahweh* (also known by the Latinized form *Jehovah*), a name originally used for a storm or warrior

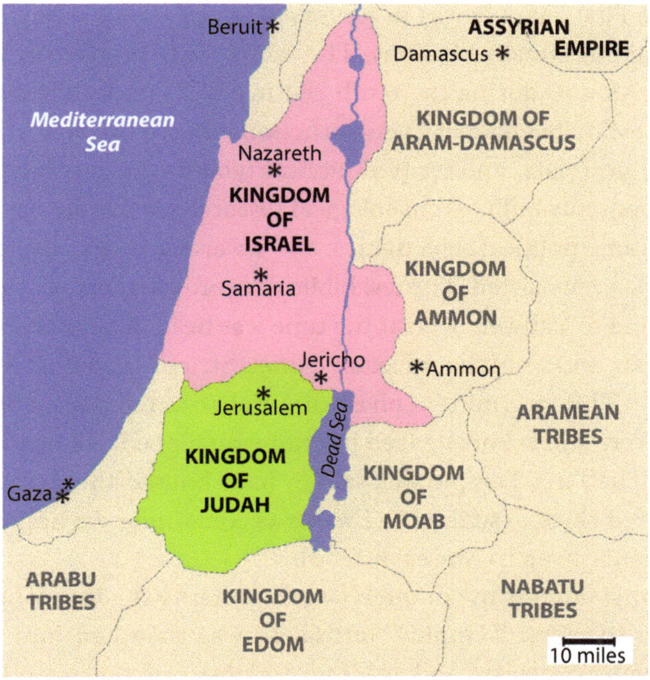

Soon after the death of King Solomon in 930 BCE, the *United Monarchy* of the Israelites split in two. The Northern Kingdom was known as the *Kingdom of Israel*, and its capital was *Samaria*. It existed as an independent state until 722 BCE when it was conquered by the Assyrian Empire. The Southern Kingdom was known as the *Kingdom of Judah* (or *Judea*), and its capital was *Jerusalem*. It remained an independent state until 586 BCE when it was conquered by the Neo-Babylonian Empire.

god possibly described in Egyptian texts dated to the thirteenth century BCE.[5,7,9] Yahweh eventually became chief God of the early Israelite pantheon and the monarchy made him the state god of the Israelites.[9,11] The idea of Yahweh as the one and only God was slow to take hold. There were hints of a "Yahweh-alone" movement in the Northern Kingdom before it fell in the eighth century BCE,[9] and a century or two later, Yahweh had absorbed the other deities in the Southern Kingdom.[9,12] As discussed in chapter 6, many historians believe that true Abrahamic Monotheism did not fully emerge until Judaism's exile to Babylon in the sixth century BCE.[6,7,8,13,14]

Early Abrahamic religion developed as a highly centralized form of temple-based ceremonial religion with a single temple managed by an elite group of temple priests. Literacy and writing were rare among the ancient Israelite common people, and sacred scripture had yet to be fully developed. Therefore, the temple priests were critical for conveying the religious belief system from generation to generation through oral transmission. Blood sacrifice of living animals was a key part of the ceremony. The First Temple was in Jerusalem, a city that is today claimed by both Israeli Jews and Palestinian Muslims. The temple was supposedly built on a place called Temple Mount, during the tenth century BCE, under the authority of King Solomon. The estimated date of its construction is based on biblical scripture rather than archaeological data, and written biblical records did not exist until centuries after it was supposedly built. Archaeological research on the site has been forbidden because of extreme political sensitivities, and no archaeological evidence finds that an actual temple even existed. The few biblical records that do exist say that the temple was dedicated to Yahweh, who at the time was head of a pantheon, including his consort, Asherah, and a number of lesser deities. Strict Abrahamic Monotheism had yet to emerge.[8,11] This intermittent phase of Abrahamic religion, in which gods other than Yahweh were still acknowledged but not worshipped, is sometimes referred to as *Yahwism*.[15] The First Temple was said to have housed the Ark of the Covenant (the gold-covered chest described in *Exodus* as containing the original stone tablets of commandments given to Moses by God).

After the Babylonian army conquered the Southern Kingdom of Judah in 586 BCE, they destroyed the First Temple.[16] Jerusalem was absorbed into the Babylonian Empire, and temple priests and leading members of the Israelite people were forced into Babylonian exile for the next fifty years. Their deportation added to an ongoing process of expelling Israelites from regions of the former United Monarchy.[16] The Babylonian exile incentivized the Israelites to assemble and organize their sacred texts.

The Persian Empire, led by the Zoroastrian King Cyrus the Great, conquered the Babylonians around 539 BCE.[16] The Israelites were then allowed to return to their homeland. For Abrahamic Monotheism, it was a period of restoration, although without a Judean Monarchy. Even though the state religion of the Persian Monarchy was Zoroastrianism, it was a tolerant system which allowed the Israelites to maintain their religious tradition. During this time, there was influence from Zoroastrianism on the development of early Abrahamic Monotheism, particularly with regard to ideas about good, evil, the end of time, and various mythological details.[10] The next two centuries were productive years for Hebrew literature and clarification of scripture. In the mid-fifth century BCE, a priest, scribe, and religious reformer named Ezra returned to Jerusalem leading a group of Babylonian exiles.[17,18] He is thought by historians to have been a descendant of a high priest of the First Temple, although details of his life are controversial.[16,18,19] Some historians also believe Ezra to be an author of the Book of Chronicles (the final book of the Ketuvim section of the Hebrew Bible that was also later used as an important part of the Christian Old Testament).

During exile, the Jews had developed the institution of the synagogue as a place of study and prayer. The *Second Temple* (also known as *Herod's Temple*) was eventually built in Jerusalem by King Herod I, to replace the one that had been destroyed. It was formally dedicated in 515 BCE.[10] The end of the Israelite exile coincided with their adoption of Yahweh as the only God.[20] All Canaan deities other than Yahweh were demoted to mere angels. This was the true beginning of Abrahamic Monotheism. During its early development it also assimilated certain Zoroastrian ideas about resurrection and eschatology (story of Judgment Day and various details about the afterlife).[16,21]

The Hebrew Bible, called the *Tanakh*, is basically equivalent to the Christian Old Testament. It began to take shape during the sixth century BCE and was thought by some authors to have been completed around 200 BCE.[10] The name TaNaKh includes an acronym of the first letter of each of the document's three major parts: the *Torah* (believed to be the Law of Moses), the *Nevi'im* (narratives of the Jewish prophets), and the *Ketuvim* (psalms, proverbs, songs, and comments by figures other than prophets). At first, the message of the Tanakh was delivered only through oral transmission by high priests. Eventually, small pieces of what would become the Hebrew Bible were written down in Hebrew (an alphabet that contains no vowels, much like the ancient Egyptian and Arabic alphabets). None of the original text is known to still exist today. The earliest-known reproduced copies are fragments of Hebrew script discovered among the Dead Sea Scrolls dated to the first century BCE. The earliest-surviving complete text of the Tanakh is the Leningrad Codex dated to 1008 CE, written in Hebrew. The *Torah*, also called the *Pentateuch*, includes the books of *Genesis*, *Exodus*, *Leviticus*, *Numbers*, and *Deuteronomy*.[22] The Written Torah also later served as the first five books of the Christian *Old Testament*. In the Islamic Quran, it is known as the Tawrat.[17]

The monotheistic God of the Torah (and of the Christian Old Testament) is known by several names, but most commonly as YHWH (most often pronounced as *Yahweh*). He is said to have presented ten statements inscribed on stone tablets to Moses, which serve as the core of Abrahamic law. Although Christians generally refer to these as the "Ten Commandments," other branches of Abrahamic Monotheism consider them to be "Ten Declarations," which function as categories for a much larger group of commandments.[23] Rabbinic authorities have deciphered 613 commandments from the Torah, although some of them are no longer applicable. For example, several have to do with the preparation of animals for blood sacrifice (e.g., animals to be slaughtered must be at least eight days old). Rabbinic Judaism no longer practices animal sacrifice. A few other commandments have to do with the buying, selling, and treatment of people to be enslaved. The most violent commandments promote retributional violence against nonbelievers and opposing cultures (Deuteronomy, chapters 7, 11, and 20). The monotheistic God of Abrahamic Monotheism rewards those who are good and punishes those who are wicked. In the Tanakh/

Old Testament, God is seen as a sometimes jealous and vengeful figure who actively interferes with people's lives, rewarding some (e.g., Noah) and punishing others (e.g., the Canaanites of Sodom and Gomorrah).[24] Times change, and with it so too does the understanding of ethics, morality, and interpretation of scriptural records.

Today, some of Rabbinic Judaism and much of Traditional Christianity view the figures of Noah, Moses, Abraham, and many others as symbolic or metaphorical figures, rather than historical ones.[10] Many biblical names are highly symbolic, such as Adam (meaning "man" or "humankind" in Hebrew) and Eve (meaning "life").[10] However, even at the beginning of the twenty-first century, approximately 50 percent of the United States electorate still believe the story of Noah.[25] In addition, the "sin of Eve" has been used by various branches of Abrahamic Monotheism to justify a lower social status for women in religion (discussed further in chapter 12). Separating biblical metaphor and parable from contemporaneously recorded historical events is often a challenge and is nearly always a source of controversy in the study of religions with ancient origins.[5]

Very early in the history of Abrahamic Monotheism, a major schism occurred, splitting the religion into two opposing factions that detested each other: *Samaritanism* and *Second Temple Judaism* (see the phylogenetic tree on page 286).

SAMARITANISM: UNDERDOGS OF EARLY CONFLICT WITHIN SECOND TEMPLE JUDAISM

The followers of Samaritanism claim to be descended from those Northern Kingdom (Samarian) Israelites who survived its destruction by the Assyrians in 722 BCE. The *Samaritan* branch shares many basic traditions with the *Judean* branch (*Second Temple Judaism*). Both branches worship Yahweh as the only God, and consider his nature as unknowable. Both follow the laws of Moses and Sabbath regulations, practice circumcision, and consider the Written Torah to be sacred scripture. Both Samaritanism and Second Temple Judaism recognize many of the same festivals, such as Passover. However, a number of differences evolved between these two branches early on. The Samaritans accept *only* the Written Torah, which they refer to as the original Torah. The Samaritan name translates to "Guardians of the Torah." They reject other sections of the Hebrew Bible and the Oral Torah. The Samaritans built their temple on Mount Gerizim near the ancient city of Samaria, whereas the Judeans built their temples on Mount Zion in Jerusalem (see the map on page 291).[17] The Samaritans reject Second Temple Judaism's prohibition against pronouncing the name of Yahweh and they follow stricter Sabbath regulations. A number of other differences emerged, as would be expected, as a result of this schism, which dates back to the sixth century BCE. In the eyes of Second

Temple Judaism followers, the early Samaritans also "diluted" the lineage of the Israelites by intermarrying with Mesopotamian and Syrian gentiles.

Because of ideological differences and the mixed ancestry of the Samaritans, the Judeans violently rejected them from mainstream *Second Temple Judaism*.[26] The Judeans eventually destroyed the Samaritan temple on Mount Gerizim (in the second century BCE[17]). Both Samaritanism and the Second Temple Judaism traditionally taught that it was wrong to have any contact whatsoever with the other. Each group has historically accused the other of being heretics, and violent confrontations ensued between the two branches throughout the early first century. As is often the case with religious hostilities, some of the most violent conflicts occur between historically related branches fighting for dominance. Eventually, Samaritanism all but disappeared, and by 2015, had only about one hundred surviving members.[65]

The story of the "good Samaritan" is embedded in popular Western culture, but its original source was the Christian New Testament, in the Gospel of Luke (*Luke* 24:25–37). It is an ancient tale acknowledging the refusal of Second Temple Jews and Samaritans to help each other. According to the gospel passage, Jesus is asked by a lawyer, "Who is my neighbor?" (Meaning "who should he love?"). Jesus responds with a parable about a traveler who is beaten, stripped of clothing, and left to be half-dead alongside the road. A couple of Judeans walk by, first a priest and then a common person. Both ignore the man. Finally, a Samaritan comes by and stops to help him. Jesus tells the lawyer that the man who helped the injured person is the one deserving of love. Whether he was Samaritan or Judean did not matter. Part of the message to the lawyer was that Judeans should take a more charitable attitude toward the Samaritans.

Historically, most Jewish authors have focused on the differences between Samaritans and Second Temple Judaism, rather than on their similarities. With but a few exceptions, many Jewish leaders today still do not even consider Samaritans to be Jews.[15] Israeli Rabbinic authorities, however, do accept Samaritanism as an early Judaic branch.[27] Even among those who refuse to include Samaritans within Judaism, many acknowledge that the two denominations have many distinctly shared traditions derived from shared ancestral ideologies. Based on the age and shared traditions of Samaritanism and Second Temple Judaism, I place them as sister groups at the base of my phylogenetic tree of Abrahamic Monotheism (page 286). Although Samaritanism all but disappeared, its sister group, *Second Temple Judaism*, continued to diversify. However, other than the group termed here as Jesu-venerationism, Second Temple Judaism never totally allied itself with the imperial power of Rome. As a result, Rabbinic Judaism ("traditional Judaism") was never very successful in conversion of the general population. Today it makes up only about 0.2% of all people who self-identify as religious (page 490).

SECOND TEMPLE JUDAISM: JUDAIC STRUGGLES WITH GREECE AND ROME

Second Temple Judaism arose in the Israelite's Southern Kingdom region formerly known as Judah (also discussed earlier in this chapter) with roots extending into the ancient Judean culture. It began around the sixth century BCE when the many Jews from Babylonian exile returned to build the Second Temple in Jerusalem (hence, the name I use for this branch). This branch evolved into a variety of denominations, many of which still exist. With the Israelite Monarchy and the Babylonians gone, the priests of the Second Temple led the Judean belief system and returned it to temple-based ceremonial religion with elite high priests and the blood sacrifice of live animals. The temple priests were responsible for dissemination of the Torah by way of oral transmission, because literacy was rare.[16]

In 332 BCE, the region was conquered once again, this time by Alexander the Great. It became part of the Greek Kingdom of Ptolemaic Egypt. Most Judaic supernatural belief systems gradually became Hellenized, with the intrusion of Greek traditions into many aspects of Judaic life. By 167 BCE, the Hellenistic King Antiochus IV attempted to push the Jews back to Greek Polytheism. He decreed the Supreme God of the Temple would change from Yahweh to Zeus and forbade circumcision and various other Jewish traditions. This new state of affairs did not last long, and in 165 BCE, the Jews rebelled, retook the temple, and returned to the way of Yahweh. The retaking of the temple by the Jews in 165 BCE is memorialized each year with the Jewish festival of Hanukkah. They retained their territorial autonomy for nearly a century until the Romans gained control of Judea and conquered Jerusalem in 63 BCE.[19]

Like the Greeks, the early Romans were polytheists. Judean opposition to Roman occupation grew over time. Some historians speak of the *Zealots* from the first century CE as though they were a branch of Second Temple Judaism, but they were not a religious denomination. They were instead a political movement that rebelled against being ruled by the polytheistic Romans and against the Hellenization of Abrahamic Monotheism. A splinter group of the Zealots, called the *Sicarii* (meaning "dagger-men"), formed in the early first century CE and consisted of assassins that focused on killing Jewish officials or high priests who were Roman sympathizers. They would conceal a dagger inside their long cloaks, get close to a targeted person in a crowd, fatally stab them, and then quietly slip back into the crowd before being detected.[16] The Sicarii were highly successful with assassinations and with kidnapping officials, such as the family members of high priests. Together, the Zealots and the Sicarii played a leading role in the Jewish war against Roman occupation that began in 66 CE. The Roman general Titus and his troops eventually destroyed the

(*Top*) Model reconstruction of the Second Temple (later known as Harod's Temple. The actual temple was located in Jerusalem, dedicated in 515 BCE, and marked the beginning of true Abrahamic Monotheism. It was destroyed during the Roman siege in 70 CE, along with much of Jerusalem. (*Bottom*) The last-surviving portion of the Jewish Second Temple, The Western Wall (also known as the *Wailing Wall*) is shown with people lined up praying in front of it. It is located in Jerusalem and is a most sacred place for Jews. It is also known by Muslims as the Buraq wall, where the Prophet Muhamad tied his steed al-Buraq before ascending to paradise (see the figure on page 443). Jerusalem is also believed by Christians to be near the location where Jesus was crucified, buried, and resurrected. Consequently, Jerusalem is a supremely holy place for all three major branches of Abrahamic Monotheism: Rabbinic Judaism, Christianity, and Islam.

Second Temple in 70 CE,[10,19] and 73 CE, the Romans had destroyed Jerusalem and won the war. Rather than submit to continued Roman Rule, the Sicarii retreated to a fortress at Masada on Dead Sea and committed mass suicide.[28]

Rome refused to allow another temple to be built, and the Jewish people sought other ways to retain their identity. With the end of the Second Temple, Abrahamic Monotheism began to move away from the centralized temple-based worship controlled by the high priest. The diaspora (the Second Temple Judaism outside of Israel) also expanded, with widespread development of synagogues and rabbis outside the temple for the study of and interpretation of scripture, and the need for the temple high priest faded.[10] An expectation of Jewish males to read and study the Torah led to increased literacy,[29] and the role of common teachers of the Torah continued to grow and disperse.

Certain authors use the term *Kabbalah* (Qabalah)as though it was a specific movement of Judaism.[30] Because it is not a monophyletic denomination, I do not include it on my phylogenetic trees. Instead, it is a collection of mystical traditions found in different Judaic subbranches.[18] These traditions encompass many esoteric teachings of Judaism dealing with the nature of God and Creation that evolved during the early Second Temple period through the Middle Ages. The Kabbalah has continued significance today particularly within Hasidic Judaism.[17,30] Much of its early development was influenced by Hellenistic (ancient Greek) philosophy.[18] The Kabbalah was most famously codified by the *Zohar* (Hebrew for "book of splendor"), which is said to have been written by Rabbi Moses Shem Tov de Leon of Castile (now northern Spain) in 1280 CE.[17] Parts of these writings are said by some scholars to be transcriptions of earlier works dating back as far as the second century CE.[17] One of the Zohar's mystical implications was that the Hebrew Bible was written in code that could be interpreted only by those who knew the code.[10]

The years of Hellenistic and Roman influence left its mark on early Second Temple Judaism, leading to the rise of several new branches in the Middle East. These included Sadduceism, Pharisaic Judaism, Essenism, and Jesu-venerationism. *Sadduceism* appears to have been the first of these.

SADDUCEISM: ARISTOCRATIC MONOTHEISM AND ARCHRIVAL OF THE PHARISEES

One of the earliest schisms within Second Temple Judaism was one that led to the splintering off of *Sadduceism*. The early history of the Sadducees is not well known, but most historians place its founding at least as early as the second century BCE.[10] What is known of the Sadducees is biased, because none of the writings of the Sadducees survived to the present day. What little we know about the Sadducees comes primarily from their archrivals, the Pharisees. According to those reports, the

Sadducees were high priests and wealthy aristocrats interested in maintaining the centralized temple hierarchy. They were based mainly in Jerusalem.[10] The religious responsibilities of the Sadducees included temple maintenance and performing temple-based ceremonial sacrifices. They were not popular with the masses because they tended to side with the ruling Roman monarchy and were too closely identified with the aristocracy.[18] The willingness of the Sadducees to collaborate with the occupation helped sustain them until the decline and the fall of the Second Temple. They were bound up with the temple cult, and the destruction of the Second Temple in 70 CE led to their demise.[17,18] No Jewish group today claims heritage from the Sadducees.[16]

Some authors have suggested that the poorly known Judaic branch of *Karaism* may be an ideological derivative of Sadduceism because of its similar rejection of the Oral Torah and rejection of Rabbinic Judaism. Other Judaic scholars, however, reject this notion as unfounded and based largely on a bias against both the Sadducees and the Karaites.[31] Karaism is a branch that arose in the eighth century CE in Babylon (modern-day Iraq) and had its golden age in Israel in the tenth and eleventh century CE. They have always been difficult to classify within the broader scope of Abrahamic Monotheism. During World War II, the German government pronounced that the Karaites were not even Jews, and they were spared from the Holocaust.[17] Their earliest history and ideological relationships remain unresolved.

Like the Samaritans, the Sadducees stressed the Written Torah as the authoritative scripture. They rejected the Oral Torah of Pharisaic Judaism and the observances it added concerning the afterlife and the spiritual realm. Their rejection of the Oral Torah meant they rejected the Pharisees' ideas of the afterlife, the immortality of the soul, angels, spirits, and rewards and penalties after death.[19] Sadducees followed a strict adherence to the law of the Written Torah and were known for their severity in dealing out severe punishment according to biblical law.[18] For example, their interpretation of the biblical principle of "an eye for an eye" was taken literally. If someone's action caused the loss of another's eye, tooth, or hand, the punishment was to remove the eye, tooth, or hand of the offender. In other, less-severe branches of early Abrahamic Monotheism, the offender could offer monetary compensation instead of bodily parts.

Sadduceism became extinct shortly after the destruction of the Second Temple.[17,18] This left the field open for their longtime opponents, the *Pharisees*.

PHARISAIC JUDAISM: EARLY MAINSTREAM JUDAISM FOR THE COMMON PEOPLE

Pharisaic Judaism (also known as *Pharisaism* or *Phariseeism*) arose in the second century BCE. Sadduceism had consisted of Second Temple priests and members of the priestly families living in Jerusalem who stressed the Written Torah.[10]

In contrast, Pharisaism saw itself as the inheritors of the reform traditions summarized by the priest and doctrinal scribe, Ezra (mentioned earlier in this chapter).[17] It was a formalization of tradition that laid the groundwork for *Rabbinic Judaism.* Rabbinic Judaism subsequently gave rise to most of what is thought of as "modern Judaism."[10,16] After the destruction of the Second Temple and the fall of Sadduceeism, Pharisaism became the foundational core for the growth and diversification of Rabbinic Judaism. Modern Jewish Rabbis are traditionally seen as descendants of the Pharisees. The Pharisees replaced the Temple-based focus of older Judaic branches with prayer and personal study of scripture.

Although they were contemporaries, the Pharisees were in constant conflict with the Sadducees. Unlike the Sadducees, the sacred scripture of the Pharisees contained laws, beliefs, and traditions in addition to the Written Torah. They believed that the Oral Torah was revealed to Moses at the same time as the Written Torah. They also believed in resurrection, destiny, angels, an afterlife with reward and punishment for acts in this life, and the sanctity of oral law, which the Sadducees vehemently denied.[17,19] Pharisaic Judaism led to most of what is thought of as Judaism today, beginning with early *Pre-Rabbinical Pharisaism* and *Rabbinic Judaism.*

PRE-RABBINICAL PHARISAISM: *FALASHA* AND THE GROWING DIASPORA

The *Pre-Rabbinical* phase of Pharisaic Judaism began as early as the second century BCE. The first century Roman Jewish historian Josephus (37–100 CE) estimated the population of Pharisees in Judah just before the destruction of the Temple to be around six thousand people.[32] The fall of the Second Temple saw the early anti-Semitic destruction of Israelite communities in Jerusalem and the death or bondage into slavery of most Israelites living there. However, the diaspora (i.e., the Israelite population outside the former Kingdom of Israel) was growing.

Some of Pre-Rabbinical Pharisaism spread into many small villages throughout Ethiopia by the early first millennium.[33,34] Ethiopian Jews, also known as *Falasha* or *Beta Israel* (meaning *house of Israel*)[17] eventually coalesced around 325 CE in response to Ethiopian Emperor Ezana, who declared Christianity as the state religion of the Ethiopian Empire. Jewish inhabitants refused to convert, and they eventually declared an independent state for the Jewish community. The Falasha claimed to be descended from Menelik, son of King Solomon and the Queen of Sheba.[17] Ethnographers hold that they descended from the ancient Agua (Agaw) tribes of the northern and central Ethiopian Plateau.[30] The origin of the Falasha has several different theories, and the exact date remains highly speculative. They are thought to be ancient in part because they only accept the five books of the Written

Torah as canonical, rejecting the Talmud and the Oral Torah.[10] Civil war between the Falasha and the Christian forces persisted for centuries until around 960 CE when the Falasha joined forces with polytheistic groups to conquer the Christian forces of Axum. They burned the Christian churches and imposed non-Christian rule. The conflict never really stopped, and the wars in Ethiopia continued through the centuries and even included Muslims on occasion.

Ethiopian Jews developed their own traditions over the centuries, remaining largely isolated from the later movements elsewhere among the Jews.[34] Not all authors agree with the ancient origin of formalized Falasha religion. Some propose a later origin between the fourteenth and sixteenth centuries resulting from Ethiopian Christians who adopted older Pharisaic traditions of the Torah and came to identify as Jews.[35] The history is still controversial. During the late twentieth century, the population of Falasha began to migrate to Israel in large numbers, where a large community now exists.[30]

Most historical details of Pre-Rabbinical Pharisaic Judaism are obscure and still in much need of further study within an evolutionary context. With the transition from a closed, central temple that was controlled by elite high priests to widespread numerous synagogues that were led by rabbis and the study of scripture by many people, the idea of *Rabbinic Judaism* proliferated.

RABBINIC JUDAISM AND THE GROWING IMPORTANCE OF THE RABBIS

Rabbinic Judaism (also called *Rabbinism*) began developing in the first century CE after the destruction of the Second Temple in Jerusalem. They accept the Oral Torah in addition to the Written Torah. They believe that it contains laws, statutes, and legal interpretations not included in the earlier Written Torah. The Oral Torah held rabbis as important Jewish leaders, even in the harsh conditions of exile. Formal Rabbinic Judaism was enhanced with the completion of the Babylonian Talmud around 600 CE.[10,19] After the Hebrew Bible, the Talmud is its second most important body of sacred literature. It holds the teachings and opinions of hundreds of rabbis over several centuries, and it became the founding text of Rabbinic Judaism.

Along with the growth of Rabbinic Judaism came a continuation of large-scale anti-Semitism. By the sixth century CE, the diaspora had brought Rabbinic Judaism to a variety of places outside of Israel. It was usually tolerated by early Islam, which at the time was the dominate form of monotheism in Spain, North Africa, and the Middle East.[10] Early Islam held that Pharisaic Jews, Christians, and even Zoroastrians were all the "people of the book" (the book meaning the Bible, New Testament, or the Avesta) and members of the same extended religious family as Islam. They were allowed to keep their religion in Islamic territory (although they sometimes had to pay an extra tax for doing so). Traditional Christianity, however, was often not as tolerant of Pharisaic Judaism or of competing Christ-revering groups (e.g., the Cathars and certain other Biblical Demiurgists).

At various times through history, the Christian Church blamed adherents of Pharisaic Judaism for the crucifixion of Christ. The term *Jewish deicide* refers to the historic belief by some Christians that the Jewish people as a whole were responsible for the murder of Jesus. Over the last millennium, the anti-Semitic slur "Christ-killer" has been used to incite violence against Jews. In the eleventh century CE, messages from the pulpit encouraged Christians to support the Crusades (so-called Holy Wars), which commenced from 1095 through 1291 CE against non-Christians. The Crusades also ran intermittently during the thirteenth through fifteenth centuries against various factions of Christianity that were competing with Roman Catholicism. In the end, the Crusades resulted in the massacre of tens of thousands of Muslims, Pharisaic Jews, and various Christian splinter groups such as the Cathars (discussed in chapter 10) who refused to convert to Traditional Christianity.[36] Pope Innocent III (1160–1216) presided over the Catholic Fourth Lateran Council of 1215, which stripped Jews of civil rights and forced them to wear special marks of identification so that no Christian would willingly deal with them.[37] A long period of intermittent anti-Semitic history followed that culminated with the Holocaust during World War II. Incredibly, it was not until the early 1960s when the Roman Catholic Church, under Pope Paul VI, finally issued an official statement (the *Nostra aetate*) to repudiate the belief in the collective guilt of the "Jewish people" for the crucifixion of Christ.[38]

In spite of a challenging history, Rabbinic Judaism survived. Today, it is an influential force in modern Western culture; even though its relative size remains relatively small (less than 0.2 percent of the world population).[1] Rabbinic Judaism contains two major branches: *Pre-Enlightenment Rabbinism* and *Maskilim* (proponents of the Jewish Enlightenment or Haskalah).

PRE-ENLIGHTENMENT RABBINISM: *ASHKENAZI*, *SEPHARDI*, AND THE CONTINUATION OF ANTI-SEMITISM

The detailed historical relationships of Rabbinic Judaism with other forms of Pharisaic Judaism between the first century and the eighteenth century CE are very poorly understood, especially for the period prior to the tenth century. This is partly because Jewish historians have often used ethnic or geographic divisions rather than religious characteristics to classify certain branches. For example, the classification of Ashkenazi and Sephardi Jews has little consistency to facilitate their comparison with other character-based groups of Rabbinic Judaism (e.g., the Orthodox and Heterodox branches of Maskilim Judaism). Some authors[10] refer to Ashkenazi and Sephardi as "*cultural-based divisions*" of Judaism (groups differentiated primarily on ethnicity and geographic origin) rather than true religious denominations. I therefore omit Ashkenazi and Sephardi from my phylogenetic trees as ideologically non-monophyletic. However, I briefly discuss them in the cultural spirit in which they were conceived because of their broader significance to Jewish history.

The *Ashkenazi Jews* (plural form *Ashkenazim*) were those Jews who established communities along the Rhine River in Western Germany and Northern France during the Middle Ages. The *Sephardi Jews* (plural form *Sephardim*) were those who established communities in Medieval Spain, Italy, Greece, and Portugal within kingdoms ruled by Muslims or Christians.[10,17] According to some historians, the term occasionally has been used as a name for all non-Ashkenazi Jews.[17] The roots of Ashkenazim and Sephardim go back at least into the sixth century CE, and each group developed its own separate language. The Ashkenazi developed the *Yiddish* language, a Germanic language with influences of Hebrew, Aramaic, and Slavic dialects. A rich culture of literature and songs in Yiddish appeared during its early history. The Sephardi developed the *Ladino* language, a Judeo-Spanish dialect. As is often the case, differences in language provided a degree of separation that promoted ideological divergence over time.

Although its roots extended back into the sixth century, the Ashkenazim more formally coalesced within Germany and Northern France around the tenth or eleventh century CE.[10,36] The term Ashkenazi comes from Ashkenaz, a biblical descendent of Noah. There is a long history of violent persecution against the Ashkenazi. Most scholars set the beginning of institutionalized anti-Semitism with the Rhineland Massacre in the late eleventh century and the beginning of the Christian Crusades.[7] On their way to reconquer the Holy Land from Muslims for Christianity, Crusaders from Germany and France slaughtered Ashkenazi communities along the Rhine Valley. The Crusaders deemed it an opportunity to avenge the death of Christ by slaughtering what they determined were descendants of those who killed him.[7] According to historian Leonard Levi, Crusaders considered themselves to be unworthy of redeeming the Holy Land from the Muslims until they first killed a Jew to avenge Christ.[37] Some historians deemed this period to be "the first Holocaust."[39] It was "a frenzy of hate and violence unprecedented before the advent of the technological age and the scourge of Hitler."[40] The majority of the Ashkenazi moved from Central Europe to Eastern Europe during the late Middle Ages to escape persecution. By the nineteenth century, those Jews who had stayed in Germany and those who returned to Germany had largely abandoned the use of Yiddish for their primary language and had adopted the German language for day-to-day existence as well as an attempt at cultural assimilation.[41] However, the growing persecution of Jews continued in Poland, Austria, Hungary, Russia, and Germany, culminating with the Holocaust in World War II.

The rise of German Nationalism was a devastating period for the Ashkenazim. When the Nazi Party came to power in 1933 there were 503,000 Jews living in Germany. By 1944, only 14,574 remained. Some fled or were deported, but as many as 180,000 were murdered or died because of persecution. The majority of German Jews murdered during the Holocaust were Ashkenazim. If all of Nazi and Fascist occupied Europe is included during this period, the Jewish victims of the Holocaust are estimated by some historians to have been as many as six million.[17]

Hitler, who was born and raised as Roman Catholic, attempted to use modified views of Christianity as a tool of aggression and cultural imperialism. He promoted what he called *Positive Christianity*. This sect mixed ideas of Christianity, racial purity, and Nazi ideology. It completely rejected the Old Testament of the Christian Bible and claimed "Aryanhood" (white non-Jewish) origin for Christ. As discussed in chapters 4 and 5, the Nazi Party also usurped hypothetically "Aryan" symbols (e.g., the swastika and the ss lightning bolt symbol) from much older Dharmic and Old Norse religions. In 1937, the Nazi Minister for Church Affairs called Hitler "the herald of a new revelation" and claimed that the Nazi Party was doing God's work.[42] Hitler became one of many twentieth-century political leaders who claimed divine guidance.

As had happened so many times before, through history, religion was used as a tool to strengthen a particular monarchical power (Hitler and the Nazi Party in this case) over the masses. In *Mein Kampf*, Hitler acknowledges, "Any violence which does not spring from a firm, spiritual base will be wavering and uncertain."[43] Hitler manipulated the power of religion to enhance his image above that of a mere politician. He also used religion to justify his persecution of the Jews. Hitler also proclaimed, "I am acting in accordance with the will of the Almighty Creator: by standing guard against the Jew I am fighting for the work of the Lord."[44] The resulting Holocaust resulted in the large-scale emigration of Ashkenazim to other parts of the world. Many people moved to Israel, the United States, Canada, Latin America, Australia, South Africa, and other parts of Western Europe.

The *Sephardi Jews* are those who trace their descent back to early Medieval time on the Iberian Peninsula (Spain and Portugal before 1492).[17]. After the Roman victories over the Jews in 70 and 135 CE, many Jews emigrated from Israel to Southern Spain. Their common language is Ladino, a type of Spanish mixed with Hebrew words and often written in Hebrew characters.[10] The name Sepharad means Spain in Hebrew. In general, the Sephardi were more orthodox than Ashkenazim and they have an older history. Although both the Ashkenazim and Sephardim base their religious practices on the tenets of the Talmud, they differ in interpretation of this text and certain practices.[17] For example, the Sephardi eat rice on Passover and the Ashkenazi do not.[30]

Like the Ashkenazim, the Sephardim were brutally persecuted at times, particularly by Catholic monarchies. Most of the Sephardim left Spain and Portugal by the late fifteenth century because of increasingly strict royal edicts of expulsion. This came to a climax with the Alhambra (Edict of Expulsion) from the Catholic monarchies of Spain, giving their Sephardi residents three choices: convert to Roman Catholicism, leave the country and all their lands behind, or be summarily executed.[45] Sephardim who left carried their language and religion to Morocco and countries across the Mediterranean region where they flourished for centuries. During World War II, during Nazi occupation of the former Yugoslavia, most Sephardim from Greece, the Netherlands, and Yugoslavia were deported to concentration camps where the majority perished. In recent times, most have immigrated to Israel where half of all Jews self-identify as Sephardim.[10]

For much of the period between the sixteenth and seventeenth centuries, followers of Rabbinic Judaism lived in isolation as second-class citizens in both the Christian and Muslim worlds.[18] In the late seventeenth century, a changing attitude began to develop among many Jews toward the secular world of their Gentile neighbors. Some adherents of Rabbinic Judaism became increasingly interested in participating in the social, economic, and cultural life of Gentile society.

MASKILIM JUDAISM AND THE GROWTH OF *HASKALAH*: THE JEWISH ENLIGHTENMENT

By the early eighteenth century, a major movement within Rabbinic Judaism was on the rise. It was called the *Haskalah*, or the *Jewish Enlightenment*. This movement attempted to incorporate secular education and culture into traditional Talmudic studies to elevate Jewish standing in European society. It also served to increase acceptance of Jews among non-Jewish society. Those who espoused the Haskalah eventually became known as the *Maskilim Jews*.[7,17] This group coalesced in the late eighteenth century, primarily among Ashkenazi Jews of Central and Eastern Europe. The movement also affected the Sephardic communities of Western Europe, although to a lesser degree.[45]

The Maskilim Jews were centered in Berlin but eventually spread throughout Europe.[47] Its proponents had two main aims. First, they sought to preserve certain elements of Jewish culture. This included a revival of the Hebrew language. Second, the Maskilim Jews strove to integrate the Jewish people into the broader range of Western society by adopting modern Western values, culture, and economic success. They sought compromise between Jewish conservatives who were traditionalists and progressives who sought major changes in Jewish tradition. When the movement began, many Jews lived in ghettos and poor settlements. In contrast, the Maskilim Jews were worldly Jews who were often wealthy merchants or intellectuals. They wanted to integrate overall Jewish culture with mainstream society and improve their quality of life. Maskilim Jews were particularly patriotic toward their host countries.[17]

One of Haskalah's early leaders was Moses Mendelssohn (1729–1786) who translated the Torah into German. This helped the Jewish people learn the local language and helped them assimilate into German society. Mendelssohn worked toward Jewish–Christian reconciliation and depicted Jesus as a Torah-observant rabbi who was loyal to Judaism. He declared himself a believer in religions of universal reason, of which Judaism was but one historical manifestation. He made it possible for Jews to join the Western world without sacrificing their broader Judaic culture. Maskilim Judaism is divided here into two major subbranches: *Orthodox Maskilim* and *Heterodox Maskilim.*

ORTHODOX MASKILIM: AN EARLY BACKLASH TO HASKALAH AND CHALLENGES WITH MODERN TECHNOLOGY

Orthodox Maskilim was the post-Haskalah branch of Judaism that was most resistant to change. It came into being as a resistance to certain reforms of Haskala and to preserve older traditions.[10] Orthodox Maskilim opposed the degree of modernity and secularization instituted by Heterodox Maskilim, and it arose as a backlash to this modernity in the late eighteenth century.[17] Orthodox Maskilim Jews define themselves as defenders of Torah and tradition. They accept only male rabbis and conduct all services in Hebrew.

Orthodox Maskilim retains many traditional practices and beliefs of earlier Judaic groups. Some examples include the following:[10] (1) Synagogues separate men from women in different seating sections. (2) A quorum of ten male Jews is necessary for a service to take place. (3) Services are conducted only in Hebrew and are led only by male rabbis. (4) Manual labor is forbidden on the Sabbath, including such activities as driving a car or even turning on an electric light. (5) Married women must cover their heads outside the home because a woman's hair is considered seductive. (6) A variety of dress, grooming, and dietary codes must be strictly kept.

The Sabbath, also known as the Shabbat, is the most important of Jewish holidays, coming each week from sundown Friday to sundown Saturday. It recognizes Saturday as God's rest on the seventh day after creating the world during the first six days. It is the holy day prescribed in the Jewish version of the Ten

Commandments ("Remember the Sabbath day, to keep it holy"), and it is the source of many seemingly anachronistic ideas that have evolved over time. For example, Orthodox Maskilim law strictly forbids a wide variety of activity on the Sabbath. This includes any personally performed work that produces a fire or sparks, which has been updated to even forbid completing an electrical circuit. Orthodox Maskilim Jews must unscrew the refrigerator light bulbs for the Sabbath to keep the light from coming on when the door is opened. They cannot use hot water on the Sabbath because it requires a hot water heater to turn on.[48] It even forbids pressing an elevator button on the Sabbath, which can potentially make life difficult for Orthodox Maskilim who live in or visit high-rise buildings on the Sabbath. The type of elevator they are permitted to use on the Sabbath must be one without buttons that simply stops on each floor sequentially without the need to press a button or throw a switch. This is called a Shabbat elevator. The Israeli Knesset passed a special law in 2001 requiring all residential high-rise buildings and all public buildings in Israel with more than one elevator, to install a Shabbat elevator.[49] Of course, one can use a few tricks in non-Shabbat elevators. An Orthodox Maskilim is allowed to wait for a non-Jew to enter the elevator and press a button in the hope that they might press the right button. They are not, however, even allowed to request that the non-Jew press a specific button. I do not mention the elevator example to make light of it. I only mention it as a prime example of the complex way religious belief systems often try to adapt their older practices to the ever-changing cultural (and technological) environment. Orthodox Maskilim Judaism arose in a time before elevators and electric power switches existed. As environments change, groups either find ways to adapt or fade to extinction. The continued parallel rise of the more modernized Heterodox Maskilim through the nineteenth century has only strengthened the Orthodox movement's resolve as a counter movement to Jewish modernization.

Theologians often divide Orthodox Maskilim Judaism into two subgroups, based on different social aspects. One is called *Modern Orthodox* (also called *Integrationists*) and the other is called *Ultra Orthodox* (also called *Haredim, separatists,* or *self-segregationists*).[10,18,30] The Modern Orthodox church seeks to play at least some role in civil society and accommodate their Orthodox faith into society. In contrast, the Ultra Orthodox prefer to live their traditional lifestyle apart from society, avoiding modern culture. They maintain their separate schools, dress codes, and cultural mores. One of the subbranches of Ultra Orthodoxy is *Hasidism* (the Hasidic Jews). Ultra Orthodox Judaism was nearly wiped out with the Holocaust. A 2010 survey by the Israel Central Bureau of Statistics indicated that even in Israel, less than a tenth of the population self-identifies as Ultra Orthodox, although that segment has considerable political power.[10,23]

In contrast to the Orthodox Movement of the Maskilim, the *Heterodox Maskilim* branch set out to fit more firmly into modern society and beyond Israel, assimilating more firmly into European culture.

THE *HETERODOX MASKILIM* BRANCH OF RABBINIC JUDAISM: MODERNITY OF THE JEWISH FAITH

The more culturally adaptive branches of Maskilim Judaism resulting from the Enlightenment are provisionally named as *Heterodox Maskilim*. This group traces its origins back to Germany.[10] Its development resulted in a schism within Maskilim Judaism during the eighteenth century during which time the *Orthodox Maskilim* established itself as distinct from *Heterodox Maskilim*. The two branches differ from each other in the way they interpret the *Halakha*, the collective body of Jewish religious laws derived from the Written and the Oral Torah. *Orthodox Maskilim* holds that the Halakha is divine law from the Torah and the word of God, so it is not subject to change. In contrast, *Heterodox Maskilim* Judaism does not accept that the Torah is the literal word of God and therefore allows revision of Jewish law to conform to modern ideas of ethics. For example, it allows certain activities on the Sabbath that are forbidden by the Orthodox Maskilim. It also allows for the ordination of women and rabbis can be homosexual.[23] The largest Heterodox Maskilim communities today are located in the United States. The branches of Heterodox Maskilim include the *Reform Judaism*, *Reconstructionist Judaism*, and *Conservative Judaism* (phylogenetic tree on page 286).

Reform Judaism (also called *Liberal Judaism*) is an early progressive branch of Heterodox Maskilim with early roots in the eighteenth century and influence from Moses Mendelssohn (1729–1786) discussed earlier. It began in Europe but flourished in the United States with an impulse toward integration and assimilation.[23] Reform Judaism seeks to reinterpret and modernize the Halakha by considering that most Jewish laws are outdated and counterproductive. Adherents believe that most Jewish rituals are no longer necessary. They hold that many of the traditional laws of Judaic religious culture are holdovers from before Abrahamic religions became monotheistic, and therefore they do not need to be followed. According to Reform Judaism, Jews should adopt the politics and general cultures of the countries in which they live. Many Orthodox Maskilim Jews consider these beliefs heretical. In the Reform synagogue, women and men do not sit separately and services are conducted in a variety of languages. Traditional ways of dressing and grooming have disappeared. In the Reform Heterodox movement, every traditional Jewish belief and practice can be questioned and revised based on ethical grounds.[10]

In 1963, a philosophic offshoot of Reform Heterodox Judaism called *Humanistic Judaism* was founded by the "atheist rabbi" Sherwin Wine (1928–2007).[23] It took secularism to the extreme, in a far more radical departure from traditional Abrahamic Monotheism than even taken by Reconstructionist or Reform Judaism. Rather than allow its followers to develop their own interpretations about the meanings of Jewish rituals and prayers that could still include theistic elements, Humanistic Judaism strove to create rituals that were purely nontheistic. Services for the Jewish holidays were reorganized according to secular humanistic philosophy.[50] Humanistic Judaism includes rabbis and various rituals, but it focuses only on the secular history and nonreligious culture of the Jewish people. It presupposes that people have the power and responsibility to shape their own lives independent of any supernatural authority. Although Humanistic Judaism is an offshoot of Reform Judaism, it is not strictly a religious belief system, so it is not included on my phylogenetic tree of Abrahamic Monotheism. Some Humanistic Jews consider their tradition to be a "religion" simply for its positive approach to life. For the purpose of this book, however, my definition of religion requires a supernatural element to be present.

Reconstructionist Judaism was the first "traditional Judaic" branch to evolve in the United States. It rejects many of Abrahamic Monotheism's theological claims. It considers all anthropomorphic descriptions of God to be metaphors. Angels, prophecy, revealed law, and the messiah are all considered to be symbolic entities. God is considered the amorphous power that makes personal salvation and purpose of life possible, rather than an anthropomorphized deity.[51] He is seen as a pragmatic influence that makes someone follow greater ideals[10]. Reconstructionist Judaism encourages Jews to become familiar with traditional Judaic traditions and embrace its rituals and prayers, but it also allows individuals to develop their own personal interpretations about the meanings of those rituals and prayers. It considers the Commandments within the Torah as tied to specific times and folkways, and thus they are subject to change as insights and values also change from generation to generation.[18] It rejects the concept of Jews (or any other group) as a "chosen people." Reconstructionist Judaism considers Abrahamic Monotheism as an ever-evolving cultural phenomenon.

Conservative Judaism (also sometimes called the *Historical School*) occupied a middle ground between the theological liberalism of *Reform Judaism* and the conservative traditionalism of *Orthodox Maskilim*.[30]. It was founded by rabbinic scholar Zacharias Frankle (1801–1875) in the late nineteenth century, who was unable to identify with either "old-fashion" Orthodox Maskilim or with "radical" Reform Judaism.[18] His approach to Judaism was as an evolving religious way of life whose laws could be observed and modified, when necessary, to fit the times.

This concludes the discussion of religions that evolved from the *Pharisaic Judaism* branch of *Second Temple Judaism* (see the phylogenetic tree on page 286). *Pharisaic Judaism* was a much more successful branch than *Sadduceism* or *Samaritanism*, and it includes most of what we think of today as "traditional Judaism". Another branch of Second Temple Judaism that arose in the second century BCE was *Essenism*.

ESSENISM: EXTREME ASCETICISM AND THE DEAD SEA SCROLLS

Essenism arose out of disgust with both the Sadducees and the Pharisees whom the Essenes believed to have corrupted Jerusalem and the Temple. They flourished into the first century CE but were never numerous. After the first century, they appear to have become largely extinct. They had an influence on early Christianity (as did several other religious branches of the time). Little is known about the Essenes. What we do know comes from three ancient historians. The first was a Jewish theologian named Philo (10 BCE–50 CE). The second was a Jewish general named Josephus (37–100 CE). The third was the Roman writer and philosopher, Pliny the Elder (29–79 CE).[10]

The Essenes lived an ascetic communal life of voluntary poverty. They avoided meat and wine, and animal sacrifice was forbidden. Their priests were celibate. According to Josephus, they were forbidden to swear oaths, strictly controlled their tempers, did not own slaves or weapons of war, and served as proponents of peace. They described themselves as "sons of light" who were fighting the forces of darkness, and they were said to have certain medical skills.[10] They believed the soul was imprisoned in its earthy body waiting to one day be released. Then the souls of the righteous would find an eternal reward in paradise to the far west, while the souls of the wicked would find unending punishment in the dark underworld. They were extremists with regard to cleanliness. They dressed only in white linen and were ritually immersed in stored rainwater each morning. Purity and cleanliness were so important to the Essenes that they refrained from defecation on the Sabbath.[48] (They must have really looked forward to Sunday mornings!)

The Essenes lived primarily in an isolated desert area near the Dead Sea, where they waited for the end times. They believed that among their contemporaries, they alone would be redeemed. The Dead Sea Scrolls, uncovered between 1947 and 1955, are thought by some scholars to have been remnants of an Essene library.[10] Dated at between the third century BCE and the first century CE, they contain some of the earliest-known copies or portions of the Written Torah/Old Testament. According to Jewishvirtuallibrary.org, several subbranches of Essenes included the

Hemerobaptists, the *Bana'im*, and the *Maghāriya*. Some historians report that the Essenes may have even influenced the teachings of Christianity's Jesus.[10] (John the Baptist is thought to have been an Essene, according to some authors.[7]) Another branch of Abrahamic Monotheism that began to coalesce in the first century CE was what I have termed *Jesu-venerationism*.

JESU-VENERATIONISM AND THE EARLY ROOTS OF CHRISTENDOM

Jesu-venerationism consists of the many branches historically rooted in Second Temple Judaism that venerate Jesus of Nazareth (*Yeshua* in Hebrew), or at least venerate a "spirit of Christ" within him. Most branches revere him as a messiah and/or prophet (e.g., Christianity, Biblical Demiurgism, and Islam discussed in chapters 7-11). Some branches revere him simply as a vessel for the spirit of Christ for much of his life, from just after his baptism to just before the crucifixion, and then again at the resurrection (certain branches of Biblical Demiurgism). Then other religions came to revere him as even more, ranging from the son of God to a trinitarian aspect of God himself (e.g., much of Traditional Christianity). In this book, the group Jesu-venerationism is defined simply as the traditions that venerated Jesus of Nazareth in one way or another.

Jesu-venerationism arose in the first century CE as a branch of Second Temple Judaism (**Note 7A**). At first, it scrupulously followed Jewish ritual, ceremony, and law. Early adherents venerated Jesus of Nazareth as a messiah who was prophesized in the Tanakh. Messianism is the belief in a prophesized savior whose mission is to spiritually liberate people. Messiahs are important elements of most major Organized Religions, including Hinduism, Buddhism, Zoroastrianism, Christianity, and Islam. Jesus is the most popular of all scriptural messiahs.[26]

Many historians belief that the historical Jesus was born around the year 4 BCE and died around 30 CE.[10,38,52] Apostolic references to his year of birth in the New Testament range from 10 BCE (*John*) to 6 CE (*Luke*).[53] The exact date of his birth is also unknown. The December 25 date is a rough consensus proclaimed by Nicene Christianity in 336 CE.[17] The date may have been chosen to overlay the December 25 festival of the Roman sun god, Sol Invictus. The sun god was an important deity to Roman Emperors (e.g., Constantine and Theodosius I, discussed chapter 8) who eventually became early supporters and ideological expansionists of early Christianity. Some Eastern branches of Traditional Christianity chose January 6 for the birth of Jesus, and the Armenian Church still observes this holiday on that day.[10,17] The place of birth for historical Jesus is also controversial. Christian mythology usually sets the birth of Jesus in Bethlehem, located in the modern-day West Bank of Palestine, about five and a half miles south of Jerusalem. Historians, however, believe that Jesus was born in the village of Nazareth, located in the Lower Galilee region about seventy miles north of Bethlehem. Today, Nazareth is the largest Arab city in Israel (map on page 291).[26,38,54]

Details surrounding the young life of Jesus (his first thirty years) are virtually unknown.[53] Existing portrayals were reconstructed well after the events are said to have happened. Such details were likely provided by disciples trying to fill in the blanks for a more appealing story. Jesus apparently did not begin his public life until he was in his late twenties. After only a few years of preaching, he was arrested in Jerusalem by the Roman authorities who considered him a threat to public order.[10] Biblical gospels describe that he was whipped, nailed to a cross, and crucified (all indicative of a degrading form of public execution). Death would have presumably resulted from shock, suffocation, and loss of blood. Most historians of Traditional Christianity believe that he was buried on a Friday. On the following Sunday, others who went to care for the body reportedly found his tomb empty. His disciples were convinced that he had returned to life, and the New Testament says that forty days later he ascended into the sky promising to return one day.

The sacred literature of Christianity (the New Testament) was written to be a presentation of "good news" (the literal translation of the word "gospel"). It was not recorded history. It was part of an evangelical campaign. Nearly all of what we know of Jesus as a historical person was reconstructed decades to centuries after his death. The information often has been modified over time to fit different times and branches of religion.[26] Even the image of Jesus was a guess that has changed many times through history (see the figure on next page). No one knows what Jesus looked like because no contemporaneous portraits exist (discussed further in **Note 7B**). In fact, detailed imagined portraits of him did not emerge until centuries after his death. Because the disciples in the New Testament did not highlight his physical characteristics, it is likely that Jesus did not have any unusual physical characteristics that stood out to his friends. As a Jew born in Nazareth in the early first century, we could surmise that Jesus would probably have been a dark-skinned Middle Easterner (Semite) with a short beard and cropped dark hair standing between five feet four inches and five feet seven inches tall. He probably weighed between 120 and 140 pounds.[54] Nevertheless, in Romanized Christianity and much of later Western civilization, the common imagery for Jesus evolved in a non-Semitic direction. In the West, Jesus is most often depicted looking as if he stepped out of a Medieval Roman portrait; complete with northern European features, long hair, and light skin tones. By the sixteenth century, artists such as Pietro Perugino (1446–1523) and Pierre-Jules Jollivet (1794–1871) were even depicting Jesus as a blond-haired Caucasian (see next page, bottom figure).

Changing Jesus iconography over time in the Western world. (Also see the figure on page 430 for another contrast.) (*Top far left*) One of the earliest-known depictions of Jesus, from around 232 CE. Discovered on the wall of an ancient Christian church in present day Syria. (*Top left center*) Fourth-century painting from the catacomb of Commodilla in Rome, showing one of the first bearded images of Jesus. (*Top right center*) Sixth-century painting of Jesus with somewhat lightened skin and deified to level of *Pantocrator* (ruler of the universe), a common image of Eastern Orthodox (Byzantine) Christianity. The slightly different facial expressions between the right and left side signify the two natures of Jesus: God and human. (*Top far right*) Nineteenth-century painting by artist Heinrich Hofmann showing Jesus as a fully Westernized image. (*Bottom*) Eighteenth-century painting by Pierre-Jules Jollivet (1794–1871) in Saint-Vincent de Paul Roman Catholic Church, Paris. Jesus is shown here as part of the Holy Trinity and depicted as a light-haired Caucasian.

Although Western imagery of Jesus predominates today, he is sometimes portrayed in non-Western cultures as black, Hispanic, or Asian (see page 430). Depictions of a crucified Jesus (the crucifix) did not begin to appear much until after the sixth or seventh century CE.[52] As with so many other personal details about Jesus, his historical image remains a mystery. However, his mythical image has evolved in countless directions. The challenge to historians interested in researching the image and history of Jesus has always been to see through the metaphorical and creatively embellished elements.[26]

At first, the differences between *Jesu-venerationism* and other branches of *Second Temple Judaism* were minimal. It was not until around the third or fourth century CE that each began to understand the other as "the other." Most traditional Jews rejected Jesus as a false messiah, while Traditional Christianity worked to differentiate itself from the rest of Second Temple Judaism. The gulf between Traditional Christianity and the other early branches of Abrahamic Monotheism progressively widened.[55] Today, the close relationship between the branches of Jesu-venerationism and the rest of Second Temple Judaism is generally deemphasized or even denied altogether. Nevertheless, you cannot change history. When we trace these branches far enough back in time, they converge (discussed further in **Note 7C**). Early Judaism and the roots of Traditional Christianity did not have a rigid demarcation.[56] It was all part of the early evolution of Abrahamic Monotheism.

Jesu-venerationism evolved rapidly. One of the earliest branches that did not last long was the one with the short-lived *Nazarenes* and *Ebionites.*

SHORT-LIVED INTERMEDIATE BRANCHES BETWEEN *PHARISAIC JUDAISM* AND *TRADITIONAL CHRISTIANITY*

The *Ebionites* and *Nazarenes* were two short-lived, closely related groups. Both branches maintained the ritual and ceremonial regulations of other Judaic branches and considered themselves to be Jews, not Christians. They rejected the Canonical gospels of the New Testament and the concept of Jesus being the son of God. Most of them rejected the idea of the virgin conception (also called the "Virgin Birth") of Jesus,[38] and they adhered to the Jewish Torah.[17] We know little about these two groups other than they were early Jews who accepted Jesus as The Messiah prophesized in the Tanakh. They were also in fierce competition with many other religions in the first century CE. Their competitors included other early branches of Abrahamic Monotheism, Greco-Roman Polytheism, Traditional Christianity, and Biblical Demiurgism.

The little historical information that we have about Ebionites and Nazarenes comes mainly from Christian authors of the second through fourth centuries, who were writing polemics against these movements as being heretical. The schism between the Ebionites and the Nazarenes developed over differences in how they perceived the person of Jesus.[57] The Ebionites venerated Jesus as The Messiah, and sometimes they are called Jewish Christians,[38] but at least some of them rejected his divinity, believing him to be merely human and not of virgin birth. Most thought God did not choose him until the time of his baptism when he received the Holy Spirit.[17] This belief is also found in branches of Biblical Demiurgism (see chapter 10).

Some accepted a limited divinity of Jesus beginning at birth but refused to ascribe to him any transcendental supernatural characteristics.[58]

The Ebionites did not think highly of Paul the Apostle (also known as Saul, Paul of Tarsus, or Saint Paul). Paul was the author of much of the New Testament, and he rejected much of Jewish law. Instead, the Ebionites claimed to follow the teachings of James, Jesus's brother. James became the head of the Christian Church in Jerusalem after Jesus's death and the departure of Peter.[17] He continued to follow Jewish law, as Jesus had also done. It is ironic that Ebionism, one of the earliest religious belief systems to recognize Jesus as The Messiah and the one associated with Jesus's own brother, was eventually deemed heretical by early Christian Church authorities. By the fifth century, this branch effectively became extinct, although some authors believe that at least part of it was absorbed into Islam, which shares some of their beliefs about Jesus as a mortal prophet and teacher.[6]

The Nazarenes were even closer to Christianity than the Ebionites in that they came to accept the divinity of Jesus, at least after he was baptized.[58] The name Nazarene was at first applied to all of Jesu-venerationism.[57] The earliest reference of the name Nazarene being applied to a person is found in the Christian New Testament (*Matthew* 2:23), where the name is used to refer to Jesus.[57] As a religious branch, we know even less about the Nazarenes than we do about the Ebionites, and the distinction between these two groups is not always made clear by the authors describing them. Like the Ebionites, the Nazarenes disappeared around the fourth century CE.

Although the Ebionites and the Nazarenes were both gone by the fifth century CE, attempts began once again in the late nineteenth century to reestablish forms of "Jewish Christianity," which claimed a deeper Jewish affiliation than Traditional Christianity. Many such groups evolved over the next century and are still in existence today, including *Beth Sar Shalom* (founded in 1894), *Beth Yeshua*, *Ariel Ministries*, *Emmanu-el*, and the best-known Hebrew Christian group, *Jews for Jesus* (founded in 1970). They teach that Jews who accept Jesus are not giving up their Jewish faith but rather they are witnessing its fulfillment.[58] Most Jewish authorities do not accept these groups as Jews, maintaining that a person that accepts Jesus as The Messiah is simply a Christian who has rejected Judaism. Israeli courts have also rejected claims by these groups to be regarded as Jews. Some of the conflict may also be influenced by nonreligious aspects of culture.

FURTHER DEVELOPMENT AND DIVERSIFICATION OF JESU-VENERATIONISM

By the late first century CE, Jesu-venerationism had diverged into two major subbranches: *Traditional Christianity* (see chapter 8) and *Biblical Demiurgism* (see chapter 10). Both of these branches developed elements of soft polytheism (believing in one Supreme God, but recognizing multiple aspects of him). Traditional Christianity developed its trinitarian entities (God the father, the Son, and the Holy Spirit as three godly aspects). Biblical Demiurgism developed its binitarian entities (the Supreme Unknowable God and the Demiurge). Both branches also had powerful archangels who were either creators or destroyers. Centuries later, the branch of *Islam* (see chapter 11) evolved within Jesu-venerationism, rejecting the soft polytheistic characteristics of Biblical Demiurgism and Traditional Christianity. Islam, as indicated by the Quran, still accepts Jesus as a prophet and The Messiah. By the early sixteenth century, *Reformation Christianity* (see chapter 9) developed as a branch of Christianity rejecting many tenets of Roman Catholicism.

A significant degree of overlap exists among the sacred scriptures of the traditional Judaic branches (*Sadduceism*, *Pharisaic Judaism*, and the *Essenism*), and the three major branches of Jesu-venerationism (*Traditional Christianity*, *Biblical Demiurgism*, and eventually *Islam*). For example, the biblical canon of Traditional Christianity includes the Old Testament, which is based on the Jewish Tanakh. The New Testament is what sets *Traditional Christianity* and *Biblical Demiurgism* apart from the rest of *Abrahamic Monotheism*. It contains the *Gospels* (Gospels of *Mark, Matthew, Luke*, and *John*), the *Acts* (accounts of the apostles' ministry after the death of Jesus), the *Epistles* (letters written by early Christian leaders), and *Revelation* (metaphors describing the state of the world and prophecies about its end). *Biblical Demiurgism* accepts most of Traditional Christianity's New Testament but rejects the Old Testament. *Islam* considers the Torah, Old Testament, and the New Testament to contain revelations from God, but considers parts of them all to have become corrupted and attempts to correct them with the Quran.

Unlike most other branches of Abrahamic Monotheism, most of Jesu-venerationism holds the notion of choice as part of membership. That is to say, you can choose to be a Christian, Muslim, or Biblical Demiurgist. In contrast, Orthodox and Conservative Judaism considers membership in their faith to be an elected calling of sorts. If you have a mother who was born as a member of a Judaic branch, then you are a Jew whether or not you practice the faith. If a mother was not born a member of a Judaic branch, Jewish law (Halakha) dictates that the child must undergo a complex process of conversion to become Jewish. Most branches of Pharisaic Judaism do not recognize the branches of Jesu-venerationism as Jewish.

As in the Christian New Testament, the Islamic Quran refers to Jesus as a messiah many times (e.g., *Quran* 4:171). The Greek word for messiah is "Christ" (used in Christianity) and the Arabic word for messiah is "al-Masih" (used in Islam).[17] In most branches of Biblical Demiurgism, Christ is seen as a separate spiritual entity that entered the body of Jesus at his baptism and then left him before the crucifixion (discussed further in chapter 10). As mentioned earlier, the historical life of Jesus is poorly known. The most direct information we have describing him or his teaching is based on apostolic gospels, composed decades after his death by believers in Traditional Christianity. Contemporaneous historical documents and archaeological artifacts are lacking. He left no personal writings. He never even left any personal indication that he intended to found a new religion. Tom Harpur, a former Anglican priest and author of *The Pagan Christ*, says he believes in a spiritual Christ but doubts that a historical Jesus ever existed.[52] Most historians, however, believe in a historical Jesus. The difficulty for the historian is separating the Jesus of history from the Jesus of faith and legend. Many biographers of the first few centuries were not objective chroniclers of history. Regarding Jesus, they wrote primarily about his putative character and his preached message. The mission of these early biographers was focused on ideological proselytism. They were mostly apocalyptic Jewish converts who believed that the second coming of Christ and the apocalypse were imminent. It was not recording history that preoccupied them; it was preparing for the end of times.[59]

The apostles Mark, Matthew, Luke, and John were the earliest sources of biographical texts about Jesus. (Paul's biblical writings were never of a biographical nature, **Note 7D**.) They were written several decades after his death, between around 65 CE and 100 CE. Collectively, these four texts form the four "Gospels" of the New Testament. Theologians believe that Mark was the earliest written of the four and that it was transcribed approximately four decades after the death of Jesus (the Crucifixion).[8] It is also thought to give the least embellished version of the historical Jesus. The later gospels of Matthew and Luke used Mark as a source when writing their verses.[26] Those verses are less constrained by historical memory and are somewhat more embellished.[8]Other than using Mark as a source, Matthew and Luke also appear to have used a "lost Gospel" called "Q" by historians.[26,60] Q is a hypothetical book whose existence is based on identical elements in Matthew and Luke that did not come from Mark. It also assumes that Luke did not use information from Matthew. Q is effectively a "missing link" whose hypothetical existence is based on shared characteristics among different biblical sources (shared passages between the gospels of Matthew and Luke that are thought to be based on a common source).

The Gospel of John came later than the rest (around 90–100 CE), with a less synoptic tone about Jesus and a more spectacular presentation of miracles performed by him. John also views human life as a struggle between the principles of light and darkness,[10] showing some similarities to Zoroastrianism (chapter 6) and Biblical Demiurgism (chapter 10). The biography of Jesus, as relayed in the gospels, is like the sacred scripture of many other religions. It likely has been embellished and romanticized over time, by the people who relay the stories, well after the iconic figure's death. The eventual result is engaging narrative that enhances evangelistic effort.

According to the Gospel of Mark (*Mark* 3:31–35, 6:1–4) Jesus had four brothers and at least two sisters, all of whom we hear very little about. Some Christian branches have excused these relatives as *step*brothers and *step*sisters to preserve the notion of his mother Mary's permanent virginity. According to the biblical narrative, Jesus's comment on his family was that it consisted not of his blood relatives but of "all those who hear the word of God and keep it" (*Mark* 3:31–33). The Gospels are silent on the boyhood of Jesus, other than to imply that in early life, he was a carpenter like his father (*Mark* 6:3). Most of his life was supposedly lived in Galilee, a region in northern Israel. Jesus apparently remained unmarried and had no children. He may have influenced, or been influenced by, early Essenes. The Essenes were a semimonastic Jewish movement of his era whose clergy remained celibate (as previously mentioned). The later adoption of celibacy by the clergy of certain branches of Traditional Christianity was supposedly justified by the belief that Jesus was celibate. One extremist Christian denomination that originated in the eighteenth century (the *Russian Skoptsy*), even suggested that Jesus was a castrate (discussed further in **Note 7E**).

Virtually nothing is reported about the personal history of Jesus until he began to preach publicly a few years before his death.[17] He was then viewed as a charismatic teacher, rather than a God. The Synoptic Gospels (supposedly written in 65–85 CE by Matthew, Mark, and Luke) do not clearly say that Jesus considered himself to be God. It is only in the later written Gospel of John (after 90 CE) in which it is first claimed that Jesus called himself God.[8] However, because earlier accounts refer to Jesus as a mortal man, the Gospel of John is thought by some historians to be more embellishment and metaphor than a historical record.[8,61] The references of Jesus as God in John may have been intended as a vernacular title, because at the time the Gospel of John was written, Roman emperors were also referred to as divinities.

When Jesus was thirty-two years old, he joined the Essene movement for which John the Baptist was a preacher. The Gospel of Mark relates that as soon as John baptized Jesus, he recognized him as The Messiah.[7] The experience was said to have inspired Jesus, who then cited a verse from the Torah, "The spirit of the Lord is upon me, because he has anointed me to bring me good news to the poor" (*Luke* 4:18; *Isaiah* 61: 1). He then proclaimed himself to be a prophet or messenger of God (but not a divinity) and set out on a quest to spread the word. The first important gatherings of early followers of Jesus were held in Jerusalem, which is a territory today claimed by both Jewish Israel and the predominantly Muslim Palestinian Authority.

Christianity did not develop as a religion until after the death of Jesus. As discussed, Organized Religions are the constructs of people. No matter how great the revered person was during his or her life, most of the laws, rituals, and behaviors inspired by these persons are later human constructs. Jesus was not a "Christian." He was a Jew.[60] No record exists of him wanting to create a "non-Judaic" religion. According to apostolic records about Jesus, he merely claimed to be fulfilling Jewish prophecy. According to consensus, as well as the gospels of Mark, Matthew, Luke, and John, Jesus effectively became a practicing rabbi who believed that Jewish scripture revealed the word of God. He preached that people should follow the commandments received by Moses in the Torah. He kept the basic Jewish food laws and laws about the Sabbath observance, and he attended synagogue meetings on Saturdays as part of the observance of the Jewish Sabbath (*Luke* 4:15). Some historians argue that Jesus was a Pharisaic Jew of the same school as Hillel the Elder (110 BCE–10 CE), much like Paul the Apostle.[12]

According to biblical passages, Jesus preached a pacifist message of peace, forgiveness, and turning the other cheek (*Matthew* 5:39), proclaiming, "Blessed are the Peacemakers" (*Matthew* 5:9) and "all who live by the sword will die by the sword" (*Luke* 22:49–51). He is credited with promoting what we think of today as the Golden Rule, saying, "Do for others just what you want them to do for you" (*Luke* 6:31). This rule reflects the earlier quote from Jewish sage, Hillel the Elder, who said, "That which is hateful to you, do not do to your fellow" (Babylonian Talmud, Shabbat 31a). No records suggest that Jesus promoted violence, and his overall message is said to have been one that emphasized peace, love, and forgiveness. Even the famous atheist, Richard Dawkins, wrote, "Jesus, if he ever existed (or whoever wrote his script if he didn't) was surely one of the great ethical innovators of history."[25]

It was not until centuries after the recorded death of Jesus that the early Christian Church, under the influence of imperial powers, grew in a militant direction subverting the original message of peace. Where early Christian leaders had once tried to stop followers from being soldiers, the Church during the latter half of the Middle Ages came to see conquest, warfare, and the Crusades as an effective means of expansionism (also discussed in chapters 8 and 9). Crusader mentality, at its worst, employed a belief that Western Christianity should be adopted by everyone. Those who would not convert could be enslaved or killed with the Church's blessing.[62] The early history of Crusaders was a religious conquest on a scale that dwarfed even modern-day jihadism. Although it failed to eliminate either Islam, earlier "traditional Judaic" branches, or competing branches of Christianity in the East (e.g., Eastern Orthodox, Oriental Orthodox), it left a permanent black mark on the reputation of Western Christianity. Early American writer Mark Twain (1835–1920), is quoted having said, "If Christ were here there is one thing he would not be–a Christian."[63] Given how various branches of Western Christianity evolved over the past thousand years with their so-called Holy Wars of conquest, persecution of dissenting branches, and sporadic pedophiliac scandals within the Church hierarchy this is probably true. "Christianity," is not the same as the person "Jesus of Nazareth." Instead, it is a diverse collection of transformed, often conflicting religious belief systems that all claim him to be their iconic inspirational figure.

Little is written about the parents of Jesus (Joseph and Mary) before the ninth decade CE.[54] Mary is said in the Bible to have been pregnant before she came to be with Joseph (*Matthew* 1:18). Rather than letting the story be tainted with any hint of illegitimacy, the condition is explained by many Christian branches, such as Roman Catholicism, as her being impregnated by the Holy Spirit (one of the trinitarian aspects of God). The birth of Jesus is subsequently explained as a virgin birth and an "Immaculate Conception" (*Quran* 3:47, which accepts Jesus as being of virgin birth as well, although without the existence of a Holy Spirit). A continuing tradition of uneasiness concerning human sexual activity within the Roman Catholic Church eventually came to include requiring a vow of celibacy for its clergy, and a prohibition against oral contraceptives in most cases for its laity (although most Roman Catholic women who have sex today have used oral contraceptives at one point or another during their lives[64]).

Traditional Christianity, Biblical Demiurgism, and Islam all generally believe that Jesus Christ came to this world to help humans find their way to the spiritual world (Heaven, Pleroma, or Jannah). Most of Jesu-venerationism also gives the "anti-Supreme God" (Satan, Demiurge, or Iblis) a relatively powerful, semi-deified, place of influence in the material world. This contrasts with many branches of Abrahamic Monotheism outside of Jesu-venerationism in which Satan is either a clear subordinate to God who is unable to act without his permission or is not real but merely a metaphor for sinful impulses.[17]

During the time of Jesus, the land of Israel was known as Palestine by its Roman occupiers.[10] The various early branches of Abrahamic Monotheism were in different states of mind about the occupation. The Sadducees accepted the Romans and maintained the Jewish Temple rituals. The Essenes lived an isolated, ascetic life in the desert near the Dead Sea believing that they alone would be saved at the soon-to-come end of the world. The followers of early Pharisaic Judaism spread out to synagogues beyond the temple and focused on keeping Jewish law, and the Zealots among them fermented rebellion to expel the Romans. It was a turbulent period, and many traditional Jews thought it represented the "end times" with a great battle coming, including a Royal Messiah to free humankind from its suffering in this world. They expected that this envoy from God to be a great king or military leader rather than a poor man who eventually would be executed by the Romans. Most of Jesu-venerationism accepted Jesus as the promised messiah, whereas traditional Jews maintained their anticipation of a Royal Messiah yet to come.

Some of the Jewish traditionalists alleged that Jesus was a false prophet guilty of blasphemy. They tended to dismiss Christianity, thinking that Christians knew God but denied him.[19] This slowed the rise of Christianity within the Abrahamic tree, until the fourth century CE, when the Roman Empire proclaimed Jesus Christ to be the true Messiah. Christianity was then declared as the state religion of the empire, and it began to expand much more rapidly.

Once the Roman Empire incorporated Christianity as its state religion and outlawed older polytheistic traditions in the fourth century CE, it was on the road to becoming the dominating religion of the Western world, and eventually the entire world. It first swept through the Roman Empire during the fourth and fifth centuries at a time when the empire encompassed the largest civilization the world had ever known. Even after the Western Roman Empire fell in 476 CE, Christianity in the Roman Church was so well established in the region that it continued to dominate and diversify through the Middle Ages. After the Americas fell to the Christian conquests from Europe during the fifteenth and sixteenth centuries, Traditional Christianity went on to become the world's largest religion. By 2010, the followers of Traditional Christianity and Islam included well over half of the world's population:[1] traditional Christianity accounted for 38.2 percent (2.2 billion followers) and Islam accounted for 22.5 percent (1.5 billion followers). The eventual dominance of these two religions was greatly enhanced by strong imperialistic support. It also included two of the most popular religious icons in the history of the world: Jesus and Muhammad. I could not satisfactorily resolve which of the four branches of Jesu-venerationism are ideologically most closely related to earlier branches of Judaism, and they appear on my phylogenetic tree as an unresolved tetrachotomy (tree on page 286).

In the next chapter, I look more closely at the largest branch of Jesu-venerationism: *Traditional Christianity*. I use the name *Traditional Christianity* to distinguish it from other branches of Jesu-venerationism that hold Christ in special reverence, such as Biblical Demiurgism, Nazarenes, Ebionites, and Islam. The shorter term "Christianity" is also used on occasion in this book as a less specific vernacular reference.

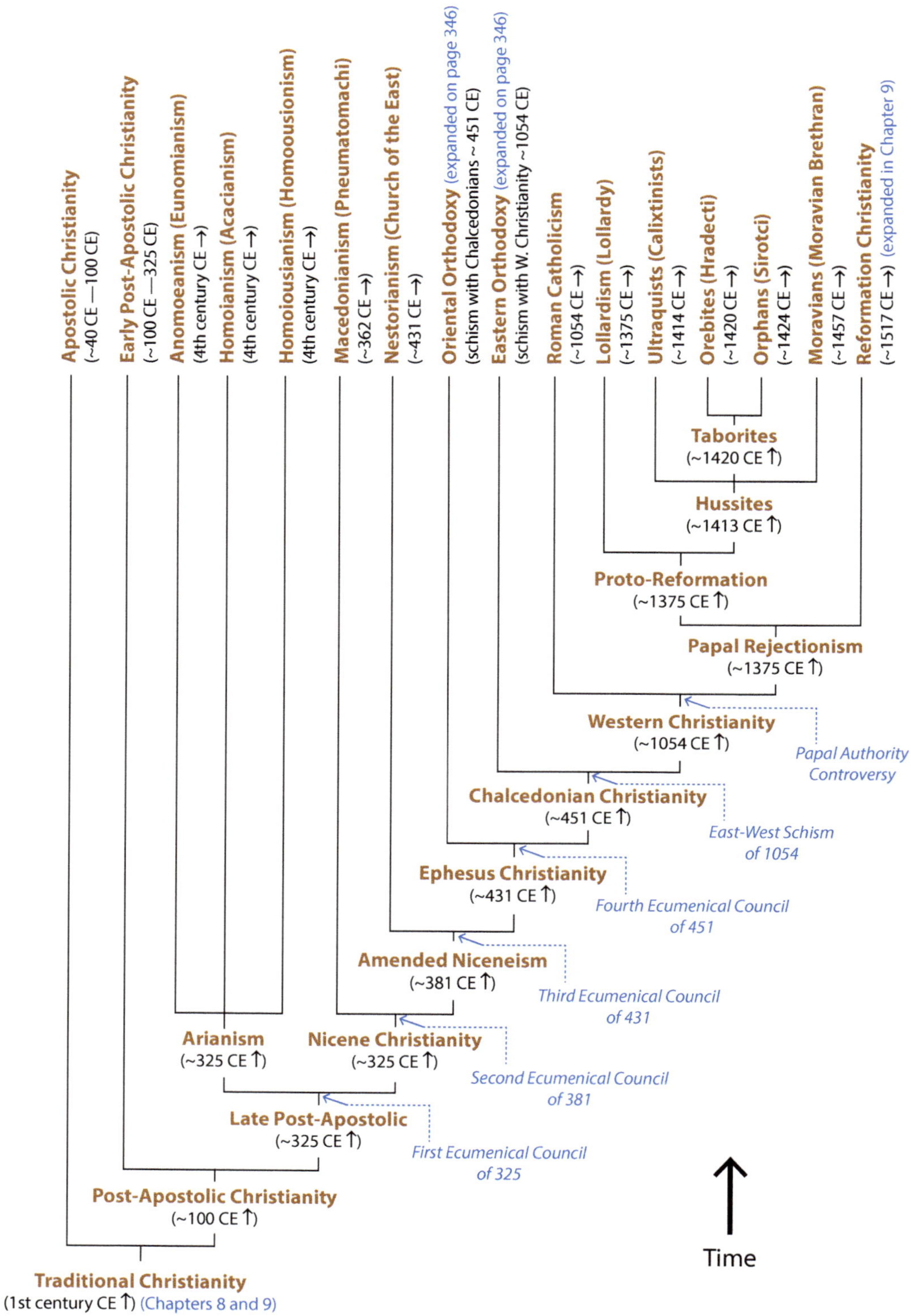
Apostolic Christianity
(~40 CE —100 CE)
Early Post-Apostolic Christianity
(~100 CE —325 CE)
Anomoeanism (Eunomianism)
(4th century CE →)
Homoianism (Acacianism)
(4th century CE →)
Homoiousianism (Homoousionism)
(4th century CE →)
Macedonianism (Pneumatomachi)
(~362 CE →)
Nestorianism (Church of the East)
(~431 CE →)
Oriental Orthodoxy (expanded on page 346)
(schism with Chalcedonians ~ 451 CE)
Eastern Orthodoxy (expanded on page 346)
(schism with W. Christianity ~1054 CE)
Roman Catholicism
(~1054 CE →)
Lollardism (Lollardy)
(~1375 CE →)
Ultraquists (Calixtinists)
(~1414 CE →)
Orebites (Hradecti)
(~1420 CE →)
Orphans (Sirotci)
(~1424 CE →)
Moravians (Moravian Brethran)
(~1457 CE →)
Reformation Christianity
(~1517 CE →) (expanded in Chapter 9)
Taborites
(~1420 CE ↑)
Hussites
(~1413 CE ↑)
Proto-Reformation
(~1375 CE ↑)
Papal Rejectionism
(~1375 CE ↑)
Western Christianity
(~1054 CE ↑)
Papal Authority Controversy
Chalcedonian Christianity
(~451 CE ↑)
East-West Schism of 1054
Ephesus Christianity
(~431 CE ↑)
Fourth Ecumenical Council of 451
Amended Niceneism
(~381 CE ↑)
Third Ecumenical Council of 431
Arianism
(~325 CE ↑)
Nicene Christianity
(~325 CE ↑)
Second Ecumenical Council of 381
Late Post-Apostolic
(~325 CE ↑)
First Ecumenical Council of 325
Post-Apostolic Christianity
(~100 CE ↑)
Time
Traditional Christianity
(1st century CE ↑) (Chapters 8 and 9)

8

TRADITIONAL CHRISTIANITY

We authorize the followers of [Saint Peter the Apostle] to assume the title of universal Christians; but as for the others, since in our judgement they are foolish madmen, we decree that they shall be branded with the ignominious name of heretics. . . . They will suffer in the first place the chastisement of the divine condemnation and in the second the punishment of our authority.

—EDICT OF THESSALONICA, 380 CE, BY THE COEMPERORS OF THE ROMAN EMPIRE

Abrahamic Monotheism had waited centuries for the prophesized Messiah. By the first few centuries CE, branches of *Jesu-venerationism* (see the tree on page 286) began to venerate a man named Jesus as a messiah, a prophet, a savior, a vessel for the divine, or even an aspect of God himself (as discussed in chapter 7). This development evolved in many directions from the start. By the late first century CE, Jesu-venerationism had given rise not only to the *Ebionites* and *Nazarenes* (as discussed in chapter 7) but also to *Traditional Christianity* (this chapter and chapter 9) and *Biblical Demiurgism* (chapter 10).

The story of Christianity's birth and earliest development is a retrospective and often biased reconstruction written long after the death of its central figure, Jesus. Jesus did not write any historical records. The vast majority of early Christian texts were lost or destroyed, particularly those that were not biased toward the predominant beliefs of later Traditional Christianity.[1,2]Surviving texts are mostly pieces of pious romance meant to engage followers, rather than to document historical facts.[1] They were repeatedly revised, translated, and reinterpreted over the centuries by "copyists" in charge of reproducing documents by hand one at a time. Along the way, these documents were sometimes revised, embellished, or censored by officials called "Church Fathers" (also known as heresiologists). Between the late first century and the mid-eighth century, the Church Fathers were in charge of defining and constraining the parameters of Traditional Christian doctrine (as they interpreted it) to weed out dissent and maximize its use for ideological proselytism.

Timeline for certain events, people, and beliefs of Traditional Christianity

Anglican Reformation in England	~1534 CE ↑
Protestant Reformation in Europe	~1517 CE ↑
European conquest/Christianization of the Americas begins	~1492 CE ↑
Collapse of Eastern Roman Empire (later referred to as the Byzantine Empire) as it falls to the Islamic Ottoman Empire	~1453 CE
The first of several Roman Catholic Crusades, or "Holy Wars to recover the Holy Land" begins	~1096 CE ↑
Roman Catholicism attempts to restrict the title of "Pope" for only the Supreme Bishop of Rome	~1073 CE
Split between *Eastern Christianity* (*Eastern Orthodox*) and *Western Christianity* (*Roman Catholicism*)	~1054 CE
Restrictions against veneration of icons relaxed by the Seventh Ecumenical Council of *Calcedonian Christianity*	~787 CE
Western Roman Empire collapses, but Christianity continues to develop there, particularly in Rome	~476 CE
Roman Empire grows too large to govern from one place, so it splits into an *Eastern Empire* and a *Western Empire*	~395 CE
Roman Emperor Theodosius I and co-emperors Valentinian and Gratin make *Nicaean Christianity* the State Religion of Rome. They then deem polytheists as illegal heretics	~380 CE
Trinitarianism formally adopted by *Nicaean Christianity*	~325 CE
Emperors Constantine and Licinius legalize Christianity within the Roman Empire with the *Edict of Milan*	~313 CE
Last book of the New Testament (*Revelations*) written	~100 CE
Last of original Twelve Apostles (John) dies	~100 CE
Earliest writings of New Testament begin	~50 CE
Death of Jesus (the Crucifixion)	~30 CE
Jesus of Nazareth is born	~4 BCE

What they chose to record (or not to record) shaped the bulk of historical records concerning the earliest development of Christianity. With those qualifications in mind, modern historians have mapped out a few basic historical scenarios by comparing a variety of sources. Distilling historical information from embellished myth and metaphor is a significant, and often controversial, challenge for History of Religion scholars.

Earliest Traditional Christianity emerged in the mid- to late first century CE in Roman Palestine. By the fourth century CE, it had become intertwined with the Roman Empire, which fueled its early growth. Traditional Christianity ultimately played a prominent role in the expansion of Western civilization throughout Europe and eventually to the Americas and other regions of the world. Traditional Christianity arose during a time when Jewish national destiny was declining, and it evolved within a Greco-Roman cultural mold.[3] I term its earliest phase as *Apostolic Christianity* (Christianity during the life of the apostles). The division between *Apostolic* and *Post-Apostolic* Christianity is somewhat arbitrary and is used mainly for the sake of discussion and lack of a character-based division. It is also an acknowledgment of how little we know about the earliest history and ideological diversity of Traditional Christianity.

APOSTOLIC CHRISTIANITY (40–100 CE): OBSCURE ROOTS AND CHALLENGING BEGINNINGS

Apostolic Christianity, as I use the name provisionally, covers the period from just after the crucifixion of Jesus until the death of the last of the twelve apostles, approximately from 30 CE to 100 CE (discussed further in **Note 8A**). Unfortunately, the detailed diversity and evolution of Christianity during this period remains virtually unknown. For those ancient times, even distinguishing Traditional Christianity from other early branches of Jesu-venerationism (e.g., Nazarenes, Ebionites, and Biblical Demiurgism) is difficult and somewhat subjective. Like paleontologists in the study of early biological evolution, the historian's study of early cultural evolution is challenged by missing data and hard-to-interpret sporadic historical records. We cannot expect every extinct branch of religion to have a robust historical record any more than we can expect every extinct lineage of organisms to have a robust fossil record. Nevertheless, scientific studies use whatever data they have to reconstruct hypotheses of deep history and evolutionary relationships. Hopefully, our knowledge about the evolution of Apostolic Christianity will improve with further discoveries and analysis. Until then, the name *Apostolic Christianity* is a chronological placeholder for a putative group in need of much further study.

Early Christians of the first few centuries did not read the New Testament because it did not exist. Few churches had been started, and the most common places of worship were in the home. It was a time of passionate ideological competition among different Christ-revering belief systems. Some of these systems were later deemed heretical by the Church Fathers of Traditional Christianity, and they faded to obscurity or extinction. Some early religious belief systems not included on my phylogenetic trees include *Docetism*, *Adoptionism*, and *Separationism*.[4] *Docetism* believed that Christ was a pure spirit and his physical human form on earth was only an illusion. *Adoptionism* believed he was neither divine nor born as the son of God, but that he was later adopted by God to be his son. *Separationism* believed that Christ and the human Jesus were distinct, different beings. Another major branch that has its own phylogenetic tree but that largely faded to extinction was *Biblical Demiurgism* (the subject of chapter 10). No doubt, many other mysterious belief systems included a Jesus or Christ figure, but their records also were lost to history. Church Fathers of Traditional Christianity deemed many of these records heresy and mandated their destruction. Other religious belief systems were lost because after their practitioners faded from existence, transmissionists and copyists discontinued reproducing accurate records of their history.

It is not easy to place where or when Apostolic Christianity became a recognized and established religion. It may have started with Jesus's brother James and other followers spreading the word in Jerusalem shortly after the crucifixion. We could also say it started a decade or so after the death of Jesus, by which time the apostles had been preaching Christian tenets beyond the Middle East. At that time, as many as one thousand people had converted to Christianity.[5] On the phylogenetic tree of Traditional Christianity on page 322, I place its early roots roughly in the mid-first century CE, shortly after the reported death of Jesus.

As discussed in chapter 7, early Christianity evolved from a Hebrew culture. At first, Traditional Christianity was starkly different from what it is today. For one, religious icons were not allowed, following the second commandment of the Judaic Bible, "Thou shalt not make unto thee any graven image." Images of the saints, God, Christ, angels, the Holy Ghost, or Mary the mother of Jesus were forbidden (or at least strongly discouraged), a practice referred to as *aniconism*. In addition, Jesus was also not an aspect of God in Hebrew culture. He was, at most, another link in a revered line of Jewish prophets. The earliest pre-Romanized Christians had no enthusiasm for a Holy Trinity or the deification of Jesus. Trinitarianism, and the deification of Jesus, would not be officially instituted until the Council of Nicaea, in 325 CE, convoked by the Roman Emperor Constantine. The use of incense in early Traditional Christianity was prohibited as "pagan." The day of worship was the Sabbath (Saturday) rather than Sunday. Infants were not baptized and adults could be baptized only between Easter and the Pentecost. There were no sacraments of Confession and Penance, Confirmation, Matrimony, Holy Orders, or Extreme Unction. Easter and Christmas were celebrated on different days of the year than they are today. There were no doctrines of papal infallibility or Immaculate Conception. The story of "virgin birth" did not appear until fifty or sixty years after the death of Jesus.[6] This may have been introduced to accommodate prophesies of earlier religions about a future savior of virgin birth (e.g., *Saoshyant* of Zoroastrianism, discussed in chapter 6).

The twelve apostles were the closest personal disciples of Jesus. The original twelve tribes of Israelites in the Jewish Bible may have inspired the number twelve in this entourage, or at least the Bible's rendition of it.[3] The number twelve had spiritual significance to many different religions throughout history (discussed on page 206). The twelve apostles (those personally acquainted with Jesus), along with James the brother of Jesus, Paul of Tarsus, and a few other disciples, became the earliest teachers, missionaries, and authors of the letters and assorted prose that eventually coalesced into the New Testament. By ten years after the death of Jesus, these early teachers had established a handful of Christian churches in the Middle East, North Africa (mostly Egypt), and other parts of the Roman Empire. Many modern-day branches of Traditional Christianity have roots in these traditions, although the relationships and diversity of Christian belief systems from Apostolic time have yet to be clearly deciphered. Early successful efforts to expand Christianity outside the Jewish world of the Middle East later helped preserve it from extinction when a polytheistic Rome crushed Jerusalem in 70 CE.[1]

Christianity was slowly taking root throughout much of the Roman Empire during the mid to late first century. Part of Christianity's success was being at the right place at the right time. The Jewish versus Roman clash of civilizations was occurring at the same time that the Roman population was losing faith in the ability of their numerous gods to protect them. By the late first century BCE, polytheistic Romans longed for a divine deliverer who would take them from constant war to an age of peace.[7] Furthermore, their old polytheistic gods did not promise much in the way of an afterlife for common people. Christianity, however, offered the possibility of eternal joy in a divine kingdom to all believers. It also preached certain moral standards that were more receptive to concerns of the common people, such as the "Golden Rule" and "love thy neighbor." This created a more effective and altruistic social network. For many polytheists of that period, Christianity was an attractive alternative to the competing spiritual options in Rome. Christianity also had pragmatic advantages over earlier Judaic branches in that it did not require giving up working on the Sabbath. It also did not require infant circumcision, a widespread tradition in the Near East going back to the third millennium BCE.[8] Circumcision is a surgical procedure where the foreskin is removed from a male's penis. Some scholars argue that the reason for the tradition was to mark any Jewish man apart from other men. According to the Hebrew Bible (*Genesis* 17:13), circumcision was enjoined by God upon Abraham, his descendants, and slaves as "an everlasting covenant." It is mandatory in Judaism and Islam, but was rejected by much of Traditional Christianity already in the mid-first century CE.

Jesus was a great spiritual teacher who had a profound effect on many people. In fact, he became what is probably the most influential person in history. However, what initially helped increase Christianity's sphere of influence beyond the Middle East was the work of its proselytizing emissaries. As a missionary religion, Christianity was much more passionately concerned with converting the masses than were other contemporaneous branches of religion. By the late first and early second centuries, Christian evangelizers began referring to Jesus as much more than a mortal man. They passionately promoted him with increasingly embellished narratives of divine power. Because monarchs of the time were often treated as divinities, Christian proselytizers reasoned that the divine figure of their faith should be respected as a monarch. By the late first century, Jesus began to be described as the "King of Kings" (*Revelation* 19:16). The deification of Jesus by some of Christianity would eventually become an unbridgeable ideological gulf between it and other branches of Jesuvenerationism, such as Biblical Demiurgism, Reformation Christianity, and Islam.

Already by the late first century, a more complex mythological narrative about Jesus began to take shape, including an increasing number of stories about miracles he performed. Miracles of Jesus were not reported during his presumed lifetime (4 BC–30 CE), nor were they reported in the earliest writings of the New Testament by Paul of Tarsus (the *Pauline Epistles* of 50 CE). Although Paul mentions the resurrection of Christ, he describes it as a spiritual event rather than one of a human body returning from the dead (1 *Corinthians* 15:42–44). Reports of specific miracles only began to appear several decades after the death of Jesus, in the Gospel of *Mark* (65–70 CE) and in later gospels (80–100 CE). This suggests that stories of miracles (e.g., controlling the weather, creating loaves and fishes out of nothing, turning water into wine, healing the sick, and raising the physical dead) were layered into the story of Jesus as expressions of an ultimate God experience.[6] It may show the dynamic power this man had on people, and the degree to which people were ready for such an individual. It also demonstrates how Organized Religions can take on lives of their own, well beyond their founding teachers, patriarchs, and figureheads.

Once the stories of miracles began to appear in early Christianity, they were retold repeatedly, until they became ingrained beliefs. More stories were added, such as miracles about singing angels, stars announcing earthly happenings, and even a fetus (that of John the Baptist in his mother Elizabeth's womb) leaping to acknowledge the anticipated power of another fetus (that of Jesus in his mother Mary's womb).[6] These details, many of which probably began as metaphorical lessons, gradually became accepted by many followers as literal historic truths. It is probable that some of these stories were never intended as documents of historical fact. As the Gospel of John states, "These things are written so that you may come to believe that Jesus is Messiah" (*John* 20:31).

The apostles were the primary missionaries of early Christianity in the first century CE, particularly Paul of Tarsus (5 BC–67 CE). Paul was not one of the original twelve apostles. He became a Christian a few years after the death of Jesus based on what he said was a vision, in which God revealed the spiritual presence of Christ. Before this vision, he had been vigorously persecuting Christians in Jerusalem. Among other things, he had approved the stoning death of Christianity's first known martyr after the crucifixion, Stephen of Jerusalem.[1] Once Paul converted, he seemed driven to make up for lost time and past sins with his proselytizing efforts. He was the earliest Christian writer. He did more to spread the reach of Christianity than any other apostle, and he is even credited by some historians as being a founder of Christianity as a formal religion.[9] This was all the more remarkable because unlike the original twelve apostles, Paul never met the historical Jesus, and he even spoke a different language (Greek rather than Aramaic).

Paul took advantage of his status as both a former Pharisee and a Roman citizen to minister to both Jewish and Roman audiences, particularly in Greece and other parts of the Mediterranean region. He saw himself primarily as a missionary to Gentiles (followers of Greco-Roman Polytheism) outside of Jerusalem, and he traveled more than ten thousand miles focusing on major cities.[2] Paul left the preaching of Christianity inside Jerusalem to other apostles. Paul argued fiercely against a popular notion of the time—that the Christian Messiah was sent only for the Jews.[1] Some apostolic missionaries, such as Peter the apostle, took converts only from other Judaic branches.[3] Paul, however, was perfectly happy to accept Christian converts directly from polytheistic belief systems. They tended to be easier to convert than the traditional Jews.[10] The Christian promise of a better afterlife, even for the common people, was a much more attractive alternative for most people than what Old World Hard Polytheism promised (see chapter 5). Paul wrote many letters to the churches that he founded and visited. These letters (the *Pauline Epistles*) make up the largest and oldest parts of the New Testament. Some of these writings date back to around 50 CE.[10]

Although Paul was extremely prolific, he never wrote anything about the historical Jesus. He does not mention the parables or other biographical details about Jesus Christ mentioned by later writers of the Gospels. When Paul writes about Christ, he speaks of him in a mystical sense, as a spiritual entity of human consciousness.[11]

His conception of Christ was also adopted by some groups outside of Traditional Christianity, such as *Biblical Demiurgism* (chapter 10).[12] Many followers of Biblical Demiurgism saw Christ as a spirit that possessed the man Jesus at his baptism and left him before his death at the crucifixion (a form of "Separationism" mentioned earlier in this chapter).

Although Paul was the most important evangelistic figure during the Apostolic phase of Traditional Christianity, he was never the Church leader. That role, according to Matthew in the New Testament, fell to Peter (1–68 CE). Peter was one of the original twelve apostles who personally knew the historical Jesus, and much of what we know about Peter comes from writings of the Apostle Matthew who looked to Peter for leadership. Of the twelve apostles, Matthew was the one most convinced of Peter's centrality to the early Christian Church. It is according to Matthews's gospel from the New Testament (*Matthew* 16:18–19, translated to English), that Jesus told Peter "Thou art Peter, and upon this Rock I will build my Church" and "I will give you the keys to the Kingdom of Heaven." These quotes were eventually used to establish the clerical hierarchy within the Christian Church of Rome. On the dome of St Peter's Basilica in Vatican City, under which Peter's remains are thought to be buried, the quote is written in Latin with letters six feet high. Theologians and historians debate the exact meaning of the quote. The words are presumably filtered through the eyes of Peter's devoted friend Matthew. The phrase is commonly interpreted as singling Peter out, among the apostles, as the founding spiritual leader of the Christian Church clergy. Roman Catholicism later claimed Peter to have been the first *pope* and *bishop of Rome.* (The title of Pope is further discussed in **Note 8B**.)

In the beginning, Christianity was loosely organized and did not have any priests, bishops, or popes; it had only elders and supervisors elected by the local communities. However, following the Roman Empire's hierarchical governing model, Traditional Christianity eventually developed a more hierarchical style of organization. It involved a number of population centers, each with a bishop in charge of lower-ranking clergy.[10] Apostolic and Early Post-Apostolic Christianity lacked a central authority for the whole of Christianity. Major centers existed in Rome, Greece, Egypt, the Middle East, and other Mediterranean regions, each with their own bishop, who was functionally a high priest.

After the last of the original twelve apostles died in 100 CE (John the Evangelist), the *Post-Apostolic* phase of Traditional Christianity began. I provisionally divide this phase into *Early Post-Apostolic* and a *Late Post-Apostolic Christianity*. The differences between these branches have yet to be clearly deciphered. Like *Apostolic Christianity*, these two names are chronological placeholders for further systematic study by historians who specialize in early Christianity.

POST-APOSTOLIC CHRISTIANITY (100–325 CE): EXPANSION IN ROME

In the first and second centuries, more than 90 percent of the Roman Empire practiced some form of hard polytheism.[13] Early *Post-Apostolic Christianity* was still only a small Hebrew tradition. As it became associated with Hellenistic (Greek) and Roman society during the second and third centuries, it began to grow exponentially for a number of reasons. First, all that was needed to be a Christian was to profess a faith in Christ. Once converted, your descendants were also raised as Christians. The effect was proportionately cumulative. Each year, more people converted from Old World Hard Polytheism to Christianity, but virtually no Christians converted back to hard polytheism. Conversion was effectively a one-way street. Additionally, the fertility rate of early Christians was higher than that of the Romans in part because of Christian doctrinal prohibitions against female infanticide, homosexuality, and birth control.[5,14] Furthermore, Roman society was severely lacking in social services, but early Christians helped one another through an exclusive mutual assistance network. The net effect was that Christianity grew rapidly throughout the world's largest empire of the time. Rodney Stark,[5] sociologist and controversial professor of comparative religion, calculated that in 40 CE there were only about one thousand Christians in the world (0.0017 percent of the Roman Empire's population). By 200 CE, the number reached 218,000 (0.36 percent of the empire's population). Those numbers would continue to grow exponentially for centuries, and by the late fourth century, Christians would make up more than half of the empire's population.[5,14]

The central scripture of Traditional Christianity, the New Testament, started to come together by the early second century CE. The last of its books (*Revelations*) was completed around 100 CE, and the last of the Epistles was completed around 125 CE.[10] At first, the writings consisted only of hymns, sermons, letters, and liturgies assembled over more than half a century. They were written in Greek by many different disciples, and they contained poetic memories and metaphorical parables about the life and teachings of Jesus. The writings encompassed the earliest evolution of Traditional Christian scripture. There were many different versions of it, because accurate reproduction of records was a challenge for the first 1,200 years. All known copies of the New Testament are reproductions, because none of the original complete text in Koine Greek survived to the present day. The New Testament is not primarily a history of actual events. Instead, the development of the New Testament over time was an evolving record of Christian belief systems and mythology.[11]

Before the invention of the Gutenberg printing press in the mid-fifteenth century, there was no effective way to reproduce identical copies of the Christian Bible.

Each copy had to be handwritten by a scribe who translated and copied an earlier version. Parts of early copies still exist. They range from credit card size fragments of text from the second century to relatively more complete sections from the fourth through early fifteenth century. Some parts of the New Testament were recopied thousands of times through history. Almost every successive version was filtered through the eyes of a different copyist, resulting in countless subtle differences. Sometimes the copies were translated from one language to another resulting in additional differences of interpretation. Over the generations, the successive chain of copyists became a driver of scriptural descent with modification (scriptural evolution). Of the surviving parts of the New Testament copied from Greek to Greek, none of the 5,300 versions is identical to the other.[15] Many of the textual contradictions between certain versions have never been resolved.[15] Nevertheless, the written pieces of New Testaments, along with the growth of literacy, represented an advance beyond oral transmissions alone. At least then, historical contradictions and ideological changes over time would be more clearly preserved.

From the second century onward, the substance of Christian doctrine continued to expand from one based on historical characters to more of an embellished narrative. The Church hierarchy added elements of romance, wonder, and nostalgia over time to enhance its image and emotional power. Iconography became more widely accepted, following subtly evolving Christian edits to the Ten Commandments of the Torah. They tucked the graven-image prohibition of the second commandment inside the first commandment rather than highlighting it as a freestanding element. They encouraged the use of images as long as they were of church-approved entities. To keep the number of commandments at ten, Western Christianity divided the "covetousness" commandment (the tenth commandment of the Torah) into two separate commandments (one for coveting the neighbor's wife and another for coveting the neighbor's goods).

Traditional Christianity's worship of Jesus as the long-awaited savior distanced it from earlier branches of Abrahamic Monotheism that were still waiting for a different kind of savior. Most "traditional Judaic" branches were expecting their messiah to be a Royal Monarch descended from King David (the second great king of the United Israelite Kingdom), rather than a poor carpenter's son who was executed by the Romans.[10] As time passed, the schism between Christianity and traditional Judaic branches increased, especially with the Sadducees, who rejected immortality of the soul and the possibility of physical resurrection. Christianity began to formalize doctrine that Jesus was God in human form (the doctrine of *Incarnation*), even though Jesus never clearly claimed to be God (discussed in the last chapter).[9]

The campaign of the Traditional Christian Church eventually became enormously successful. It was also tightly controlled. It was considered heresy to defy evolving church dictum, even as it changed over the centuries. Branches of the later church would excommunicate, persecute, and even burn people at the stake for the "sin" of heresy. Heresies could include clinging to older Apostolic practices that had been abandoned or developing interpretations of scripture that differed from the interpretations approved by the Traditional Christian powers in Rome. The Traditional Christian Church gradually developed traits of ideological chauvinism, nationalism, and imperialism.

Roman culture was at first slow to fully embrace Traditional Christianity. It had been polytheistic for millennia. Of the thousands of different branches, each had its own set of gods, often with its own Supreme God, and its own set of traditions. Hard polytheism in general was an *inclusive* set of religions that continually welcomed new gods and traditions. Followers were even happy to welcome the Traditional Christian God into their pantheons. There were no significant quarrels among the thousands of different polytheistic branches based on theological grounds. They did not have codified doctrines, complex ethical laws, or the need for accusations of heresy. Because of the additive, *inclusive* nature of Old World Hard Polytheism, the concepts of proselytism and mandatory rejection of old gods were meaningless. In contrast, Traditional Christianity was an *exclusivist* sort of religion, with extensively written sacred scriptures and laws requiring the rejection of all gods except the one Traditional Christian God. Its missionaries actively and passionately evangelized to convert non-Christians to Traditional Christianity and to have them reject their old beliefs and the gods of Rome. For its first few centuries, Christianity was seen by the imperial monarchy as a form of atheism whose disrespect to the Roman gods could bring great misfortune and anarchy.

In Greco-Roman Polytheism, prayers and sacrifices to the pantheon of Roman gods were necessary for survival in the material world (rain for the crops, freedom from disease, health of one's children, victory in civil wars). It was important to seek the gods' help for day-to-day existence in the present life and perhaps to put in a good word for the monarchy as well. Life in ancient Rome was hard. There were a multitude of famines, civil wars, and diseases. Even a tooth abscess could kill you. Only with the help of the gods, could someone survive the normal challenges of life on earth, and the help of the gods came in return for prayers and sacrifice (generally animal sacrifice). Most people were not concerned with an afterlife. In most polytheistic branches, dead meant dead. The gods did not reward or punish people in an afterlife for actions in the present life. A frequent inscription on the tombstones of Roman commoners, borrowed from the Greek Philosopher Epicurus, was "I was not; I was; I am not; I do not care."[16] The sentiment was humorously paraphrased in a quote attributed to Mark Twain by Richard Dawkins and others, saying, "I do not fear death. I have been dead for billions and billions of years before I was born, and had not suffered the slightest inconvenience from it."[17] Traditional Christianity's new offer of an eternal afterlife of bliss was novel and gradually became an attractive enticement to the Roman common people.

The early emergence of Traditional Christianity was at first a personal challenge to the monarchies' power and authority. Not much separation existed between church and state in polytheistic Rome. The emperors were treated like one of the deities, because it was thought that their rule was the will of the gods. Their success in battle proved their divine connection.[6] In contrast, the reprioritizations of Traditional Christianity threatened that the emperors would no longer be worshiped. The prospect of a comfortable afterlife for common people diluted their focus on the present life and perhaps their dedication to the emperor. Some of the Roman monarchy also objected to the spiritual elevation of a convicted criminal like Jesus (as judged by the Roman monarchy) over the Roman emperor himself. Traditional Christianity seemed to present a complex challenge for the empire. This led to an intermittent period of persecution against the early Christians that lasted more than two centuries.

Early persecution of Christians in Rome began in earnest after a great fire devastated much of the city of Rome in 64 CE. The emperor Nero blamed the Christians for the devastation, although some suspect that Nero may have started the fire himself. He already hated the Christians for what he saw as disrespect to the Roman gods and to the monarchy. The fire was an excuse for him to retaliate. Nero had many Christians in the city rounded up and arrested. They were tortured and executed. He crucified some (including the Apostle Peter), beheaded some (including the Apostle Paul), and even wrapped some Christians in animal skins and threw them to dogs. The latter form of execution was a precursor to what took the air of a sporting event later in Roman history. A few later emperors enthusiastically engaged in throwing Christians to the "wild beasts" in arenas as a public spectacle for the common people of Rome. Anger against the Christians by the monarchy continued in response to the Christian disrespect to the Roman gods and to the divine emperors. Some historians even suggest that the monarchy saw Christianity as a glorification of cannibalism, with the sacrament of the Eucharist seen as eating their God.[3]

The persecution of Christians in Rome escalated with Emperor Decius when he tried to fortify Roman Polytheism against the rise of Christianity in 250 CE. He ordered all citizens in the Roman Empire (except for the Jews, who were exempt) to perform sacrifice to the Roman gods in the presence of a Roman magistrate. The punishment for noncompliance was beheading, crucifixion, or the increasingly popular death by wild beasts. As a result, an unknown number of Christians were executed, including a bishop of Rome known by historians as Pope Fabian. The last and most severe persecution of Christians in the Roman Empire occurred in 303 CE by Emperor Diocletian and his three coemperors. Once again, the order was to offer sacrifice to the Roman gods, and the penalty for noncompliance was death.

However, even under severe conditions of persecution, Christianity survived and continued to grow. It had a few things going for it with the common people, not the least of which was the promise of a better afterlife accessible to those who accepted the faith. It lessened the power of the threat of execution, and even brought some people to welcome what they perceived to be their martyrdom. Such a death, if suffered in the right spirit, guaranteed entry into heaven.[1]

The imperial attitude toward Christianity made a radical change in the early fourth century CE. Constantine the Great (272–337 CE) became emperor of Rome in 306. He lived most of his life as a Roman Polytheist, and his father was one of the four coemperors responsible for the previously mentioned persecution by Emperor Diocletian. However, Constantine embraced the idea of Christianity once he started to believe that the Christian God helped him in battle. He eventually came to see himself as the emperor of the Christian people. Under his rule, Christianity began its transition from a suppressed, persecuted cult to the most favored and privileged of all Roman religions.[1] The shift from Roman persecution to Roman promotion of Christianity was profound. Where many of the first- through third-century emperors of Rome had seen Christianity as a threat and a form of political dissent, Constantine began to see it as the glue that might hold his fragmenting empire together.[10] He ordered the cessation of Christian persecution and formally legalized the religion. Together with coemperor Licinius, he issued the Edict of Milan in 313, which proclaimed equal tolerance for Christian and non-Christians within the empire. It officially decreed the end of the Christian persecutions.[1] In his efforts to unify Rome under a single religion, Constantine fused certain polytheistic symbols, dates, and rituals with Traditional Christian traditions to make Christianity all the more palatable to followers of Roman Polytheism. He also lavished the Christian Church with financial support and tried to unify Christianity and resolve issues that divided the Church. Constantine's efforts mark a pivotal point in the history of Traditional Christianity and paved the way for future Emperor Theodosius to make Nicene Christianity the state religion of the Roman Empire (discussed later in this chapter).

Many historians believe Constantine eventually converted to Christianity (it was his mother's religion), although others believe that he may have seen the Traditional Christian God or Christ as one in the same with the Roman sun god *Sol Invictus*.[13] He did not agree to be baptized until just before his death.[1] In 321, with the Edict of Constantine, the main day of Christian worship shifted from the Saturday to *Sunday*, named after the day of worship for the earlier Roman sun god. Vestiges of solar deification remained in Christianity as many early Christians continued to see

A gold coin showing Emperor Constantine with the Roman sun god, *Sol Invictus*, behind him. Struck in 313 CE by the Roman Empire. Some historians believe that Constantine may have thought the Christian god and Sol Invictus were one in the same.

the Christian God or Christ as a solar deity. The New Testament Book of *Revelation* 1:16 describes the face of Christ as "like the sun shining in full strength." By the fifth century, Pope Leo I had to tell church members to stop worshipping the sun.[11] To this day, many Christian images depict the sun behind the head of God or Christ or feature a bright halo resembling a solar disc.

Just as the Roman Empire became Christianized over time, Post-Apostolic Christianity became Romanized. Latin (the official language of Rome) became the primary language for prayer and ritual, displacing the Aramaic language used by Jesus and his closest disciples, as well as the Greek language used by Paul and the earliest authors of the New Testament. Soft polytheistic elements (trinitarianism, iconography, prayers to saints, archangels, and the Virgin Mary) gradually became institutionalized over time within the Church. Satan was recognized as ruler of an underworld now referred to as hell. Greco-Roman purification rites and sharing of sacred food and drink ceremonies may have influenced the development of Christian Baptism and Eucharistic rituals.[18] Western Christianity adopted a monarchical organizational system much like the Roman political monarchy. Where secular matters were ruled by the emperor, religious matters were ruled by the bishops of Rome who eventually became known as popes. The relationship between Rome and early Christianity became increasingly symbiotic over the centuries to follow, helping fuel the expansion of Christianity toward becoming the world's largest major religion. Emperor Constantine moved his empire even further along the course toward formal Christianization by convening the Nicaean Council (the first *Ecumenical Council*). This council produced the first uniform doctrine of Traditional Christianity. It linked the Roman Empire with the codification of Traditional Christian doctrine, and this is where I begin discussion of what I term *Late Post-Apostolic Christianity.*

LATE POST-APOSTOLIC CHRISTIANITY (325–FORWARD): COLLABORATION WITH THE ROMAN EMPIRE

Much of Late Post-Apostolic Christianity remained centered in the city of Rome. It was the ideal center in which to develop a monarchical episcopacy. After all, it could lay claim to the burial places of both Peter and Paul the Apostles, and it was a symbol of secular rule in the largest city of the Western world. Four bishops of Rome began to assume authority over other bishops of Post-Apostolic Christianity by the early fourth century,[10] although the idea of a single pope as head of the Church would not develop until much later. Constantine began supporting a greater spiritual role, as well as greater secular role, for the bishops of Rome. He also began construction of major palatial structures for the Church, including the original St. Peter's Basilica (commonly referred to as Old St. Peter's Basilica), in an area that would one day become the theocratic Roman Catholic state of Vatican City. The power of the Roman bishops did not significantly falter when Constantine moved the capital of the Roman Empire eastward to Byzantium in 325 CE for strategic reasons. Byzantium became known as "new Rome" and the name was later changed to Constantinople.[19] As a result, Constantine added the bishop from Constantinople to the group of supreme bishops.

By the fourth century CE, the major religious conflicts for Christianity were no longer between it and non-Christianity. They had instead become conflicts forming within Christianity among different evolving subbranches, each claiming to represent Apostolic Christianity. In response, Roman Emperor Constantine called for the first Ecumenical Council in 325 CE, the *Council of Nicaea,* in the hope of better unifying Christianity. This was an international convention of bishops and other representatives charged with developing a common set of tenets, rules, and doctrinal specifics for Traditional Christianity. The interest and support of the emperor also led to an increasing convergence of spiritual and temporal authority between the Christian Church and the Roman Empire. Constantine presided over the opening session of the council meeting and took part in the discussions. His hope was that in unifying the Christian Church, he might better unify the Roman Empire.[13]

The Council of Nicaea in 325 CE produced the *Nicene Creed*, which founded *Nicene Christianity* (see the tree on page 322) The creed, as later amended, was the first uniform doctrine of Traditional Christianity. The Nicene Council was the first of seven highly impactful Ecumenical Councils convened in the fourth through eighth centuries, which fueled the early evolution, expansion, and diversification of Traditional Christianity (listed in **Note 8C**). Many of the issues discussed at the council meetings were debated heatedly and passionately. What today might seem as minute conflicting details were seen by some church leaders to represent the difference between correct doctrinal interpretation and eternal damnation.[13] The resulting council decisions often cultivated intolerance and divisiveness within the Church, leading to schisms, excommunications, and banishments.

At the Nicene Council, one problem that Constantine dealt with was a bitter doctrinal dispute between *Arianism* (a small branch of Traditional Christianity unwilling to accept trinitarianism) and the bulk of mainstream Christianity backed by Rome. The meeting included more than three hundred bishops from around the known world. The council's rejection of Arianism demonstrated the fermenting divisiveness that existed within early Traditional Christianity. It also resulted in a schism between Rome-centered Nicene Christianity and *Arianism*.

ARIANISM: FORMALIZED SCHISM DOCUMENTING THE PRICE OF DISSENT

A Christian priest from Alexandria Egypt named Arius (256–336 CE) founded a branch that became known as *Arianism*. He broke from mainstream Christianity in the early fourth century CE by preaching the antitrinitarian view that Christ and God were not the same, because God was unknowable and Jesus was well known from his time on earth as a human. Because God was eternal and Christ came later, Arius concluded that Christ was clearly subordinate to God.[13] The intent was not to degrade Christ, but rather to present a savior who was more like the people, had historical substance, and participated in the human struggle toward virtue.[1] Although this might seem to be an innocent difference on its face, most of the council bishops at Nicaea found this not only unacceptable but also heretical. The Nicene Council proclaimed the trinitarian view that Christ was "of one substance with God." Christ and God were seen as equals, as was the Holy Spirit. They decided that Arianism's implication that Christ was somehow inferior to God was heresy. (The best of intentions sometimes cannot hold up against growing bureaucracy.) Of the 318 bishops that attended the council meeting, only two supported Arius in the end. Later, Arius and the two bishops who supported him were excommunicated and exiled.[13] Constantine, the convener of the Council, ordered all writings of Arius to be burned.

Arianism was ruled an illegal practice, and anyone caught hiding the writings of Arius could be condemned to death. Consequently, Arianism's remaining contemporaneous records are those written mostly by heresiologists trying to extinguish that particular belief system. Such records are often prejudiced distortions of truth and a challenge for historians trying to piece together an objective record of history.

By the fourth century, branch diversification had already begun within Arianism. This included *Anomoeanism* (also known as *Eunomianism*), *Homoionism* (also known as Acacianism), and Homoiousianism (also known as *Homoousionism*).[,20,21] Ideological differences mainly surrounded belief in a greater or lesser distinctiveness between the God and Christ.[21,22] Most of the sacred texts on these and other early Christian sects from this period were lost.[2] In the modern world, Unitarians and Jehovah's Witnesses still embrace certain core principles of Arianism. Once Arianism was rejected by mainstream Traditional Christianity in the fourth century, what remained in Traditional Christianity became known as *Nicene Christianity.*

NICENE CHRISTIANITY: THE CODIFICATION OF CHRISTIAN TRINITARIANISM

Some historians and theologians have referred to Nicene Christianity as the "Holy Catholic and Apostolic Church" because in Greek and early English, the word Catholicism means "universal doctrine." Others use the term Catholicism as synonymous with Roman Catholicism. This inconsistent use of the name Catholic is confusing for any broad comparative approach such as this one. Therefore, in this book, I use the term Roman Catholicism to refer to the Christians who continued to recognize papal authority after the beginnings of Proto-Reformation of the fourteenth century (discussed later in this chapter). I occasionally use the term "catholic" (small *c*) as a vernacular to refer to Nicene Christianity after it was formalized by the first Ecumenical Council in the fourth century (see **Note 8D**).

Nicene Christianity officially began after the convening of the first Ecumenical Council (the *Council of Nicaea*) in 325 CE and the formalized schism with Arianism over the Nicene issue of trinitarianism. Trinitarianism became a core principle of Nicene Christianity, and it was the focus of much controversy over the centuries resulting in a series of schisms. It is most graphically represented by a symbol known as the *Scutum Fidei*, also called the *Shield of the Trinity*. Although Christian trinitarianism was first codified in the fourth century CE, the Scutum Fidei symbol was first illustrated in the early thirteenth century. It was drawn by an unknown scribe to accompany a manuscript by French theologian, Peter of Poitiers.

The idea of trinitarianism further distanced Nicene Christianity from other branches of Abrahamic Monotheism. It was rejected by most "traditional Jews"

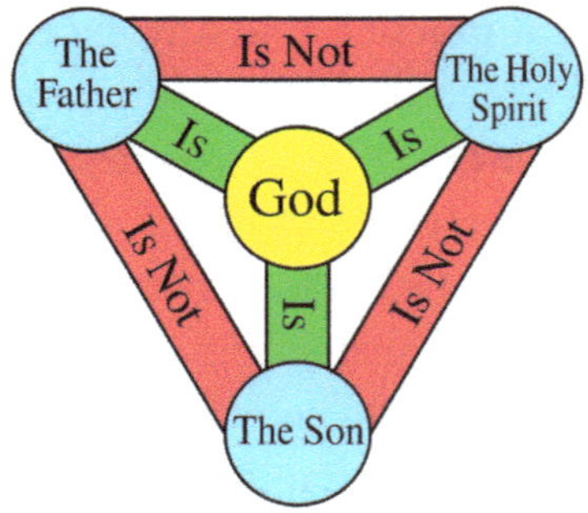

The *Scutum Fidei* or "Shield of the Trinity" symbol used by most of Nicene Christianity to explain Trinitarianism (the Divine Triad of Nicene Christianity). Latin translated into English here. The emblem represents the heraldic arms of God connecting his three aspects.

who saw it as a reversion to polytheism. Divine triads, in the most general sense, predated Christianity, and are found in many other religions (e.g., see "divine triads" in the index). When Islam arose in the seventh century, it too would reject trinitarianism as a form of polytheism. It was one thing to acknowledge Jesus as an active rabbi, prophet, or messiah; it was another to say he was an aspect of God.

Although the Council of Nicaea was the first Ecumenical Council of the Christian Church, it was not convened by a bishop or pope. It was instead convened by a Roman emperor, as were the next seven Ecumenical Councils during the fourth through ninth centuries. As mentioned previously, the council of Nicaea was convened by the Emperor Constantine I who had developed a strong interest in the success of Christianity. It was held in the eastern half of the Roman Empire, in Nicaea, about fifty-seven miles from the Greek city of Byzantium. The bishop of Rome, Pope Sylvester I, did not attend. The highest-ranking church authorities to attend were the Ecumenical Patriarchs from the East. However, the meeting began to organize the structure and continued growth of Christianity within the entire Roman Empire. Constantine, as well as numerous later Roman emperors who supported Christianity by convening Ecumenical Councils, were eventually glorified as saints within the Eastern Orthodox Church.

The Christian population continued to expand exponentially. As mentioned earlier in this chapter, the number of Christians in the Roman Empire during the early second century CE, was 218,000 (0.36 percent of the Empire's population). By the mid-fourth century CE, the Roman Empire's Traditional Christian population had expanded to more than thirty-three million people (56 percent of the population).[5,14] The Christian majority among Western religions would last to the present day, thanks in no small part to the collaboration with Roman Empire. In 380 CE, Emperor Theodosius I and coemperors Valentinian II and Gratian gave the final blow to Roman Polytheism. The three of them made Nicene Christianity the state religion of the Roman Empire with the Edict of Thessalonica (given in **Note 8E**). The edict also deemed polytheists as heretics, which was punishable by the state. Polytheistic ceremonies were banned, temples were closed, and idols were destroyed. By the end of the fourth century CE, the office of the bishop of Rome (the papacy) began to inherit some of the power and prestige of Roman emperors. The interconnections between popes and the monarchs became increasingly close and complex. It was a classic example of coevolution. By the time the empire began to fall apart in later centuries, Christianity had become strong enough to continue on its own path of imperial expansionism, interacting strategically with other monarchical powers.

GROWTH AND EVENTUAL DISAGGREGATION OF ROMAN POLITICAL POWER

Roman civilization had a long and powerful history and was key to the early history of Nicene Christianity. Historians often divide this time into three periods: the *Roman Kingdom* (750–509 BCE), the *Roman Republic* (509–31 BCE), and the *Roman Empire* (27 BCE–1453 CE). The Roman Kingdom began around 750 BCE. It was ruled by a series of powerful kings until their overthrow and the establishment of the Roman Republic in 509 BCE. The Republic was a form of government in which citizens voted for individuals to represent them in governmental processes, often guided by a constitution. During the period of the Republic, Roman control spread beyond the Italian peninsula to the rest of the Mediterranean world. The Republic eventually ended with the defeat of Mark Antony and Cleopatra's forces in 31 BCE. It was replaced in 27 BCE with the return to a more monarchical imperial system. Thus, the Roman Empire was officially born. Its first emperor was Julius Caesar's adopted son, Octavian, who was given the new title Augustus. He oversaw the early Romanization of Greek and other cultures, including the Romanization of Western religious beliefs and practices. The early united Roman Empire lasted more than three hundred years, and at its peak was the most extensive political structure in all of Western civilization.

The size of the Roman Empire eventually grew beyond its ability to govern all of its claimed territory from a single center. It nearly collapsed during the third century CE because of various problems, including invasion, civil war, plague, and economic depression. To reinforce stability, Emperor Diocletian divided the empire into two administrative halves. These are referred to today as the Western Roman Empire (containing the city of Rome) and the Eastern Roman Empire (later called the Byzantine Empire). Each half had a senior emperor (titled Augustus) and a junior emperor/successor (titled Caesar). This early system with four emperors was called a tetrarchy. Although the tetrarchy ended a little over two decades later, the ease-west division of the empire endured. Emperor Theodosius I became the sole emperor of Rome for a few years, but upon his death in 395 CE, the split between west and east became permanent.

The capital of the United Roman Empire had been the city of Rome from 27 BCE to 286 CE. Once the empire split, however, the effective capital (also called the *Imperial Court*) of the western provinces had to move several times for strategic reasons. The main problem was constant border conflicts with what were called the "Barbarian" tribes to the north (the Huns and various Germanic tribes, such as the Saxons, Franks, Vandals, and Goths). The capital city for the western half of the empire first moved from Rome to the well-fortified city of Mediolanum (modern-day Milan) from 286 to 402 CE to defend the empire from attacks from the north. After the northern frontier collapsed, the capital moved south to the better-defended city of Ravenna from 402 to 455 CE. It moved back to Rome from 455 to 473 CE and moved back to Ravenna from 473 to 476 CE. The Western Empire finally fell to attacks from the north and ceased to exist sometime between 476 and 480 CE.

The capitals of the Eastern Roman Empire (which was referred to as "New Rome" in its early days) generally occupied senior status over the capitals of the Western Empire after the split. The Eastern capital began in Nicomedia (286 to 330) and later moved to the Greek city of Byzantium (present-day Istanbul, Turkey). After the establishment of the Eastern capital in Byzantium in 330, the name of the city was formally changed to Constantinople after its founder, Emperor Constantine.[23] When Constantine inaugurated the new capital in 330, he decorated a column with figures from Greek Polytheism and Traditional Christianity to link classical religious belief systems with the state religion of Rome.[18,23] The Eastern Empire was the wealthier, better geographically positioned, and ultimately more powerful of the two sides from its inception. It became the principal half of the Roman Empire for most of its existence, and the earliest and strongest half to advocate for Christianization. The language of the region was Greek, the original language of the New Testament. However, the business of Roman government continued to use Latin, as in the Western Empire. The eastern half of the empire lasted a thousand years after the fall of the western half. It finally fell to the Islamic Ottoman Empire in 1453.[10,18] Citizens of the Eastern province always considered themselves to be Romans, even though that province did not contain the city of Rome. It was not until the sixteenth century, that historians began to refer to the Eastern Roman Empire as the Byzantine Empire.

As Nicene Christianity continued to evolve, a series of further schisms developed within it. The major ones will be listed in the rest of this and the following

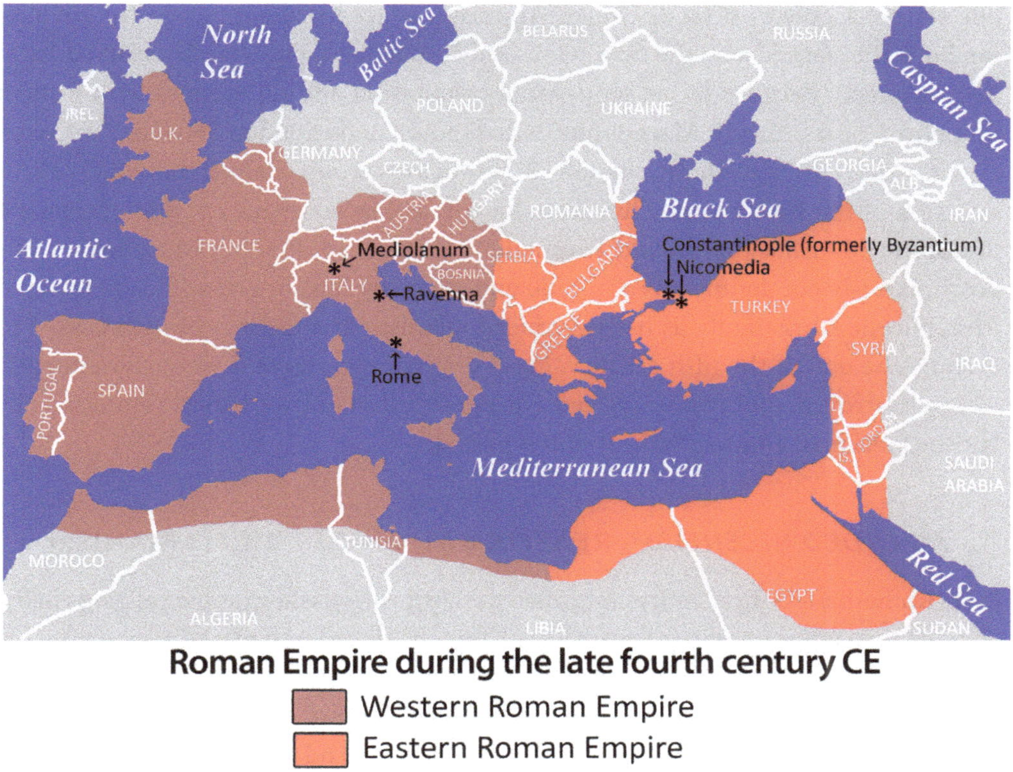

Map showing extent of the Western and Eastern divisions of the Roman Empire by 395 CE. Superimposed on modern national borders for geographic context. The empire first split in 286 CE, at which time the imperial capital of each half moved from Rome. The capital of the Western half began in Mediolanum and moved to Ravenna on 402 CE. The capital of the Eastern half (later called the Byzantine Empire) began in Nicomedia and was moved to the city of Byzantium in 324 CE. Byzantium was renamed *Constantinople* in 330 CE after the Emperor Constantine. Superimposed on modern national boarders for geographic context.

chapter. One of the first was between Macedonianism and the rest of Nicene Christianity (tree on page 322).

MACEDONIANISM (*PNEUMATOMACHI*): "FIGHTERS AGAINST THE SPIRIT"

The founder of *Macedonianism* (also known as *Pneumatomachi*), was Macedonius (?–362 CE), twice bishop of Constantinople. He was first bishop from 342 to 349 CE and again from 357 to 360 CE. Somewhat like Arianism, Macedonius did not accept trinitarianism. In particular, he rejected the doctrine of divine status for the Holy Spirit. He preached that the Holy Ghost was not a person or personality but simply a power of God. Unlike Arianism, Macedonius accepted the divinity for both the Father and the Son (God and Jesus). This doctrine is referred to as *binitarianism.* Because of his semi-Arian orientation and differences with Roman Emperor Constantius II, Macedonius was deposed and exiled in 360 CE. Two years after the death of Macedonius in 362 CE, a religious group evolved that followed his rejection of the Holy Spirit. They were known as Macedonians or "fighters against the Spirit."[1] During the second Ecumenical Council in 381 CE, Macedonius was posthumously condemned as a heretic and Macedonianism was officially rejected from Nicene Christianity as heresy. Macedonianism as a distinct branch ceased to exist by the end of the fifth century. Most original records of it were subsequently lost or destroyed, and remaining records consisted mainly of biased refutations by Nicene Christian Church Fathers of the period.

AMENDED NICENEISM: REAFFIRMATION OF TRINITARIANISM

Early Rome-based Christianity continued to reject various developing religious belief systems through succeeding Ecumenical Councils (listed in **Note 8C**), which fueled schism and early diversification of some branches and led to the extinction of others. The second Ecumenical Council (381 CE) was the *Council of Constantinople* held in Constantinople. It officially reaffirmed trinitarianism and its rejection of Arianism and Macedonianism from the imperial Church. After that, the majority of Nicene Christianity (*Amended Niceneism*) continued to embrace trinitarianism. Over time, various splinter groups continued to develop that again rejected the concept (e.g., *Unitarians* in the sixteenth century, *Jehovah's Witness* and *Mormons* in the nineteenth century, and *Oneness Pentecostals* and *La Luz del Mundo* in the early twentieth century). Some of these nontrinitarian branches are discussed in the next chapter.

By 431 CE, another schism was formalizing within Amended Niceneism. This time it resulted in a split between two branches that I label as *Nestorianism* and *Ephesus Christianity* (phylogenetic tree on page 322).

NESTORIANISM: DEBATING THE MATERNAL ROLE AND STATURE OF MARY, MOTHER OF JESUS

Nestorianism was a branch of Nicene Christianity that evolved through the efforts of a priest named Nestorius (386–450 CE) who was bishop of Constantinople from 428 to 431 CE.[1] He aggressively promoted the view that the Nicene doctrine of the Virgin Mary as the *mother of God* was nonsense.[1] He reasoned that God was the creator of time and not created within time. He considered Mary to be the mother of the human Jesus and that to overpraise her as mother of God was to revive the worship of a mother goddess.[1] He also contended that the divine nature of Christ did not enter Jesus until after his birth (a concept that would later be revived in some branches of Biblical Demiurgism (chapter 10). His radical distinction between two natures (human and divine) of Jesus Christ was in opposition to the Nicene doctrine that Christ's humanity and divinity were one in the same. Although these variations on a theme night not seem earth-shattering, they were serious and unresolvable to early mainstream Nicene Christianity. The Council of Ephesus (the third Ecumenical Council) condemned Nestorius as a heretic, removed him from his position as archbishop, and exiled him to Egypt. After he died, the emperor ordered all of Nestorius's writings burned, and all children bearing his name to be rebaptized and renamed.[1] This was mild punishment compared with what others would later receive for heresy. Thus, I set the official date for Nestorianism as a separate branch of Christianity on my phylogenetic tree at 431 CE, after the death of Nestorius.

What many historians believe to be a surviving early branch of Nestorianism is the *Church of the East*[1,20] (not to be confused with the *Eastern Orthodox* Church). The Church of the East adopted the ancient language of Jesus (Aramaic) as opposed to the Latin used by Roman-based Christianity or Greek used by Eastern Orthodox Christianity. It originally occupied the Middle Eastern region where Christianity began. However, once Islamic rule was established in the Middle East by the early Muslim conquests of the seventh century, the dominance of Christianity never returned there.

EPHESUS CHRISTIANITY: MARY AS THE MOTHER OF GOD

The third Ecumenical Council (431 CE) was the *Council of Ephesus*. It affirmed the Virgin Mary as the mother of God and that Jesus Christ had a single (human-divine) nature. This established what I call *Ephesus Christianity*, which officially rejected *Nestorianism* and its leader Nestorius, referring to him and his teachings as the Nestorian Heresy. Eventually, another Ecumenical Council was formed to clarify doctrine and settle disputes within Ephesus Christianity. The fourth Ecumenical Council in 451 CE, the *Council of Chalcedon*, resulted in a schism between *Oriental Orthodoxy* and *Chalcedonian Christianity* (see the phylogenetic tree on page 322).

ORIENTAL ORTHODOXY (*MIAPHYSITISM*): REBELLION AT THE FOURTH ECUMENICAL COUNCIL

The common name of *Oriental Orthodoxy* for this branch is somewhat misleading. In modern times, the term "Orient" (Latin for "east") is often used to refer to East Asia. However, in fourth- and fifth-century Rome, the Orient also referred to Middle

Eastern regions, such as Syria and Egypt. Over the centuries, differences continued to grow between what I term *Chalcedonian Christianity* (the branch tied to the Roman papacy) and what would eventually become known as *Oriental Orthodoxy* (also called *Miaphysitism* by some authors[1]). The two branches formally split from each other with the decisions of the fourth Ecumenical Council in 451 CE (the *Council of Chalcedon*). At that council meeting, the majority of bishops declared that although Jesus Christ was a single person, he existed "in two complete natures." In contrast, Oriental Orthodox contended that the human and divine natures of Christ are united without separation. They flatly rejected the Council's statement, that "the human and divine natures of Christ are separate." In fact, they considered the Council's stand to be a reversion to the Nestorian Heresy mentioned earlier in this chapter. Although these differences may seem esoteric, they were sufficient to result in two major branches of Ephesus Christianity to condemn each other and create the great *Chalcedonian Schism*. Each branch had its own pope: *Dioscorus* from Alexandria representing *Oriental Orthodoxy*, and *Leo I* representing Rome and *Chalcedonian Christianity*. They excommunicated each other over their differences, and the Oriental Orthodox Church would continue to reject all Ecumenical Councils after the first three.

Oriental Orthodoxy came to include several major subbranches (e.g., see above figure).[24] Although the formal schism between Oriental Orthodoxy and Chalcedonian Christianity was codified in 431 CE by the fourth Ecumenical Council, both branches claim earlier roots dating back to Apostolic Christianity. The details of ideological diversification for this group are in need of further detailed study. Oriental Orthodoxy continues to maintain its own pope equivalent (the Coptic Patriarch of Alexandria, also known as the archbishop of Alexandria), separate from the Roman pope of Western Christianity.

CHALCEDONIAN CHRISTIANITY, AND THE INCREASING PRESTIGE OF THE PAPACY

Chalcedonian Christianity continued to grow after its schism with Oriental Orthodoxy in 451, as did the power of its organizational leader. Pope Leo I was a Roman aristocrat who came to be known as Pope Leo the Great. In addition to issuing the *Tome of Leo*, which served as a foundation for the debates at the Council of Chalcedon, he is popularly known for having met Attila the Hun in 452 and turning him back from invading Italy.

Soon after the establishment of Chalcedonian Christianity in the mid-fifth century, new fractures had begun to develop. Theological differences between the eastern branch and the western branch included such matters as exactly how the Holy Spirit fit within the Trinity, whether or not to use unleavened bread for Holy Communion, whether priests could be married, and even which sort of iconic imagery of Christ was allowable (lamb versus crucified man).[11] Political differences also emerged concerning the relative power of the Church within the state, as well as the relative power of the state within the Church. In addition, the power of the Roman pope was becoming more controversial within the Traditional Christian Church as time went on.

The fall of the Western Roman Empire in 476 elevated the Roman pope's secular authority. The Church had begun to integrate itself into the social and civic structure of the Western Empire long before it fell. As centralized power of the Western Empire weakened, elements of governmental management often fell to the bishops. By the late sixth century, the Christian Church came to form five major centers of authority ("episcopal sees"). Four of these centers were located in the eastern part of the former empire (Alexandria, Antioch, Jerusalem, and Constantinople) and only one was located in the western region (Rome). The existence of several centers in the east kept any one of them from becoming a single ruling power equivalent to that of the West. In contrast, the Roman Church and the papacy had no competition for power in the west, allowing it to grow in authority and strength.[10] The Roman Church and its pope eventually asserted dominion over all Chalcedonian Christians and remained largely unopposed until the East-West Schism of 1054 CE.

The pope's influence steadily grew within the conglomeration of states that would eventually become known as Italy. (The historical consolidation of Italy is discussed in **Note 8F**.) The people of this region came to depend on the power of the papacy to protect them from barbarian invasions. The pope had the power to influence kings, emperors, and armies. The old Western Roman Empire had ended in 476 CE, but with collaborative support from the Roman papacy, a new Western monarchy started to develop by 800 CE, modeling itself after the old Western Roman system. The monarchy began with Pope Leo III crowning the Frankish King Charlemagne as Emperor, thus reviving the title for Western Europe and informally launching the *Holy Roman Empire* (discussed further in **Note 8G**). This new "empire" was a collection of tribalistic states in the modern-day regions of Germany, western France, Italy, and the Czech Republic. From its beginning all the way into the sixteenth century, the emperor of the Holy Roman Empire would be crowned by the pope in Rome. Without this, no king could call himself emperor. The power of the papacy grew to a comonarchical level. Sometimes the papacy was under the control of vying imperial factions, but at other times, the imperial factions were under the control of the pope. Although the papal office was charged with maintaining unity and order in the Church, it sometimes did the opposite. It became one of the largest drivers of schism and diversification within Chalcedonian Christianity, largely because of the controversy caused by certain popes or bishops of Rome.

PAPAL INDISCRETION: HELPING FERMENT THE EAST-WEST SCHISM OF 1054

As in most seats of absolute power that exist for many centuries, some monarchs eventually abuse that power. The Roman papacy was no exception. Papal abuses created an environment for major schisms and diversification within Traditional Christianity. Roman Catholics, who think of the papacy as the one led by the seemingly mild-mannered, charitably minded Pope Francis (2013–), should realize that the papacy was not always led by honorable people. Papal scandals were common between the ninth and sixteenth centuries. Several popes from this period became known as the "bad popes."[25] This was part of what led up to the East-West Schism of 1054, as well as Reformation Christianity in the sixteenth century (discussed later in this chapter and in chapter 9). Two classic examples of papal misconduct leading up to the East-West Schism include Pope Stephen VI and Pope Benedict IX.

Pope Stephen VI was pope from 896 to 897. He is best remembered for his vindictive conduct toward the remains of his predecessor, Pope Formosus (pope from 891 until his death in 896).[25,26] Pope Stephen harbored a special hatred for Formosus. In 897, he had the rotting corpse of Formosus, who had been dead for eight months, dug up, dressed in papal vestments, and put on trial for coveting the papal throne and violating the canon law prohibiting bishops to move from one diocese to another. A deacon was appointed to answer for the papal corpse. The mock trial (referred to as the *Cadaver Synod*) was presided over by Pope Stephen. To no one's surprise, the dead defendant was found guilty. The body was then stripped of its sacred garments. The three fingers of the corpse's right hand that had been used for blessing were cut off, and the body was thrown into the Tiber River. Pope Stephen annulled all ordinations of Pope Formosus. The body of Formosus was later retrieved by a sympathetic monk and reburied. Pope Stephen later paid for his conduct when supporters of Formosus rebelled. They deposed Stephen, stripped him of his papal regalia, and threw him in prison. He was later strangled there by an unknown assailant.[25]

Pope Benedict IX was an intermittent pope who reigned on three different occasions between 1032 and 1048.[26] He was the nephew of the two previous popes. His first election to the papacy was at the age of twenty, and his father is reported to have facilitated his election through bribery. He gained special notoriety for selling the papacy for 1,500 pounds of gold to finance his early retirement. Although he did not want to be pope, he did not want to give up the luxurious lifestyle that came with the position.[25]

Over the centuries, the Chalcedonian debate on matters of doctrine, governance, and papal authority could not be resolved between Eastern and Western parts of the Church, with both sides remaining inflexible in their dogmatic positions. The tensions eventually culminated in the bitter *East-West Schism of 1054* that split Chalcedonian Christianity into *Eastern Orthodoxy* and *Western Christianity* (see the phylogenetic tree on page 322). The Ecumenical Patriarch Michael Cerularius in Constantinople represented Eastern Orthodoxy and Pope Leo IX in Rome represented Western Christianity. Each leader eventually excommunicated the other. Relations between these two branches remained hostile for centuries after the split. It was not until 1965 that Pope Paul VI (representing Western Christianity in Rome) and Patriarch Athenagoras I (representing the Eastern Church in Constantinople) revoked the respective excommunication decrees. Despite the revocation of excommunication decrees, a significant sense of separation still exists between the two branches.[10]

EASTERN ORTHODOXY: GREEK-SPEAKING CHRISTIANITY

Eastern Orthodox Christianity (*Eastern Orthodoxy*) split from *Western Christianity* (the rest of Chalcedonian Christianity) in the eleventh century. Eastern Orthodoxy is sometimes known as the *Orthodox Catholic Church* (as opposed to the *Roman Catholic Church*) and it identifies with the original *Apostolic* church. Like many branches, it does not highlight denominational diversity within its ranks. Many adherents do not recognize themselves as a Christian "denomination." Instead, they claim to be the original core of Christianity, or the "Holy Orthodox Catholic Apostolic Church." They consider all those outside their group to be the departed "denominations."[27] As might be expected, Western Christianity believes the reverse:

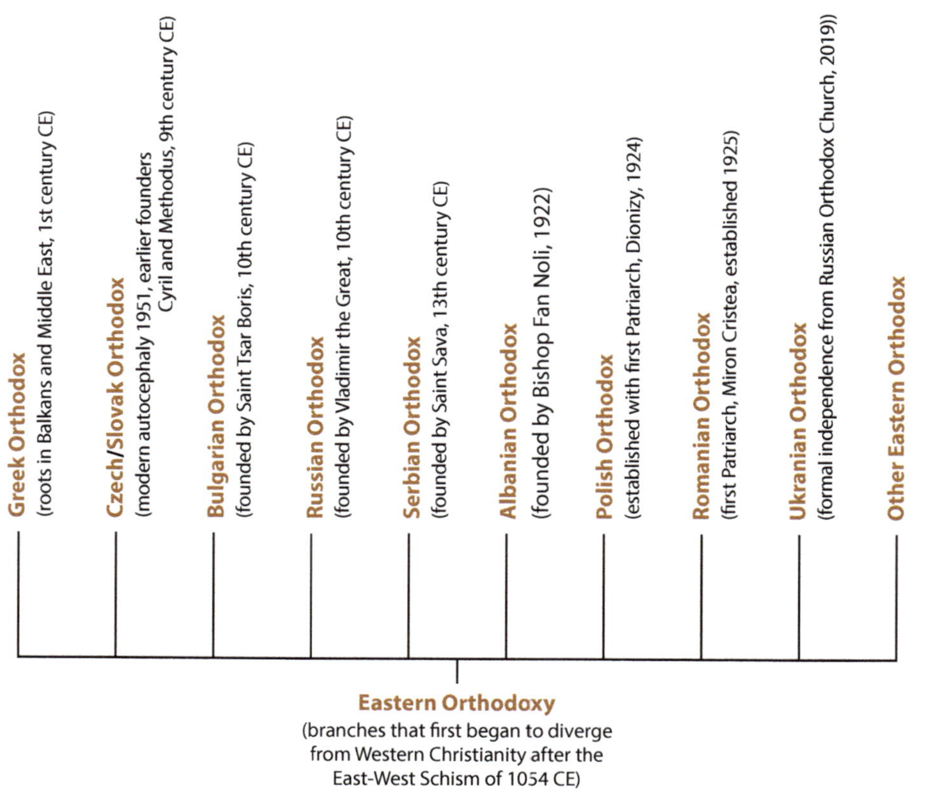

that it is they who constitute the true core of Christianity and that it is Eastern Orthodoxy that is the "denomination." Objectively, they are historically both denominations, or subgroups, of larger, older groups.

Although the formal recognition of the Eastern Orthodox Church as independent from Western Christianity is usually set at the *Great Schism of 1054*, the separation was a gradual process that developed over many years. Like Western Christianity, but unlike the Oriental Orthodox Church, Eastern Orthodoxy follows all seven of the first Ecumenical Councils, all of which were held in the Eastern Empire. Rather than a *Pope*, Eastern Orthodoxy has an *Ecumenical Patriarch* based in Constantinople (today known as Istanbul, Turkey) who is recognized as "first among equals" of bishops.[1] The language of Eastern Orthodoxy's early history was Greek, the original language of the New Testament, rather than the Latin used by Rome. *Greek Orthodox*, sometimes inappropriately used as synonymous with *Eastern Orthodoxy*, is only one of its many subbranches (e.g., see tree on previous page with ten of these subbranches).[27] Schism between some of these branches was originally initiated by competing nationalistic objectives rather than by ideological characteristics (e.g., the twenty-first-century schism between Russian Orthodoxy and Ukrainian Orthodoxy discussed further in chapter 12). The ideological diversity and historical interrelationships among these subbranches are yet in need of much further phylogenetic study.

The core of Eastern Orthodoxy was badly affected over the centuries by invaders who took over Constantinople, capital of the Byzantine Empire. In 1204, the city fell to the Western (Roman) Christian forces of the fourth crusade. With that, Latin forces sacked the city, and for the next several decades, they attempted to supplant Eastern Orthodoxy. Western Christianity spread throughout the region. Later, in the same century, Greek forces reclaimed the region. However, by the fifteenth century, Constantinople fell again, this time to the Muslim Ottoman Empire. The new Islamic ascendancy forbade the Christians to preach their faith. After that, various diasporic colonies preserved the Eastern Orthodox traditions.[27] Today, a peninsula in Northeastern Greece named Mount Athos remains an Eastern Orthodox theocratic state, under the ultimate jurisdiction of the Primate of Constantinople.[20]

Differences in language, geography, and other aspects of culture have helped maintain differences between the Eastern Orthodox Church and Western Christianity, as has the memory of past transgressions. Eastern Orthodoxy does not recognize the ultimate authority or infallibility of the pope in Rome, and it has less of a hierarchical clergy. Instead, it has a number of independent, regional branches (called autocephalous branches), each headed by its own patriarch. Eastern Orthodoxy also differs from Western Christianity in a number of doctrinal issues. Most important, it rejects Western Christianity's trinitarian concept of *Filioque* in which the Holy Ghost is equivalent in status to the Father (God) and the Son (Christ). Controversy boils down to two esoterically different visions of trinitarianism. The Eastern Orthodoxy follows the original Nicene Creed of 381 CE, stating that the Holy Ghost "proceeds from the Father" (the Father gave rise to the Son and the Holy Ghost). Western Christianity amended the phrase in 1054 to read that the Holy Ghost "proceeds from the Father *and the Son*" (giving the Son equivalent historical connection to the Father). The Eastern Orthodox Church found Western Christianity's revision of the Nicene Creed to be heresy. This example shows, once again, the level of fine subjective detail that has so often led to major schisms within Christianity. To this day, Eastern Orthodoxy and Western Christianity remain inflexibly separate and in conflict over the issue. Other differences between these two major Traditional Christian branches include that unlike Western Christianity, Eastern Orthodoxy does not enforce celibacy among its priests (they can marry) and they use leavened bread (bread with yeast) for Communion ceremonies.

The majority of Eastern Orthodox Christians are today found in Eastern Europe (e.g., the Russian Orthodox Church). A 2008 international survey by the Pew Research Center and the International Social Survey Program found that 72 percent of Russians self-identified as Eastern Orthodox (up from only 31 percent in 1991).[28] Eastern Orthodoxy is the second largest branch of Chalcedonian Christianity, and in 2010, it had more than 250 million followers.[29] Since the 1960s, the Eastern Orthodox Church has held conferences with the Oriental Orthodox Church and the Roman Catholic Church in the hope of theological reconciliation.[20] I term the sister group to Eastern Orthodoxy *Western Christianity* (phylogenetic tree on page 322).

WESTERN CHRISTIANITY: CELIBACY FOR PRIESTS, HOLY WARS, AND EARLY CHRISTIAN CONQUEST

After the formal separation of *Eastern Orthodoxy* from *Western Christianity* in 1054 CE, the Western branch continued to expand and diversify. The authority of the bishop of Rome also began to consolidate and strengthen. Before the eleventh century, the term pope had been used on occasion for many Western Roman bishops as well as the Eastern Orthodox Patriarchs. In 1073, however, Pope Gregory VII attempted to restrict its use to the supreme bishop of Rome. Today, the Roman Catholic Church recognizes this authority through its doctrine of *Papal Primacy*. This doctrine attributes "full supreme and universal power over the whole church" to the pope in Rome.[30]

Universal celibacy of the clergy was made compulsory in 1139 by the tenth Ecumenical Council in Rome. All clerical marriages were declared invalid. One of the major reasons for the rule of celibacy was that before it was implemented, married clergy could found dynasties and leave their property to family heirs rather than the Church.[1] Celibate priests were married only to the Church, ensuring the stability of Rome's treasury. Another reason given for the requirement of celibacy was an assumption that Jesus was celibate. That issue has been debated, however, given we know so little about the early life history of Jesus.[31] Although Reformation Christianity abolished mandatory celibacy for its clergy in the sixteenth century, it has been retained in Roman Catholicism to the present day.

The Western Christian belief that *purgatory* is an actual place evolved in the eleventh and twelfth centuries.[1] Historians have speculated that its prominence developed as a way to justify praying for the dead. After all, if God has already judged all departed souls to be in heaven or hell for eternity, what would be the point of praying for the dead? If, however, there was a holding zone such as purgatory, then there was the hope that prayers could accelerate a soul's ascendance into eternal paradise. The concept of purgatory also seeded an intricate liturgical industry of fundraising opportunities for the Church. *Indulgences* (a way to reduce one's amount of time in purgatory for past sins) could be had for a price. This arrangement sometimes took the form of a donor setting up a trust fund (a *chantry*) to employ one or more priests to perform services for the soul of a departed loved one. Other times, it took a form of a hard sell by the Roman papacy (e.g., Popes Alexander VI and Leo X, discussed later). The sale of indulgences was used to pay for expenses of the Roman papacy, such as rebuilding Saint Peter's Basilica. The sale of indulgences was one of many factors that led to the Protestant Reformation in the early sixteenth century (discussed later in this chapter and chapter 9). Reformation Christianity rejected the doctrine of Purgatory as unscriptural.[20]

The secular power of the papacy continued to grow to new levels by the eleventh century. Emperors, kings, and monarchs were sometimes subjugated as instruments of the pope. The Church of Rome initiated a period of increasing Christian conquest. It began in the Middle East, spread to several parts of Europe, and eventually reached into the New World. Some of the darker examples of this conquest were the *Crusades* of Western Christianity. The Crusades ran from the eleventh through the thirteenth centuries, primarily against Muslims and Jews. In the thirteenth through fifteenth centuries, it was waged against other competing factions of Abrahamic Monotheism as well. Other violent elements included the *Holy Inquisitions* (also referred to as the "Catholic Inquisitions"), which first started to take shape in the twelfth century and were more formalized later (discussed further in **Note 8H**). The early church often subverted its earlier message of peace. Where early Christian leaders had tried to stop followers from being soldiers, the later Church came to see warfare as an effective means of proselytism and expansionism.

It is difficult for Christians to acknowledge these periods of history. To learn from mistakes of the past, however, those mistakes must be clearly identified and honestly acknowledged.

The *Crusades* of Western Christianity have been described as "religious enthusiasm gone badly astray"[10] or "militant Roman Catholicism."[18] At first, these were simply called "Just Wars" (*bellum iustum*) by the Church in Rome. Then, in 1095, Pope Urban II elevated them to "Holy Wars" (*bellum sacrum*). The enemies of Western Christianity (particularly of Roman Catholicism) were branded as "enemies of God."[32] The sword became a new missionary instrument of the Church.[33] It was Roman Christianity declaring war on the unconverted and using the faithful as their soldiers. The Roman papacy bestowed rewards on the armies that did its bidding. Blasphemy was a capital offense. The pope promised his crusaders that to fight and die for Western Christianity would guarantee immediate entry into heaven. This promise of salvation was the beginning of using "papal indulgences" as currency, a practice that would lead to major schisms within the Church.[1]

The Roman papacy and imperial aristocracy were often entwined through Medieval times. Louis IX was the King of France from 1226 to 1270 and considered himself to be the "lieutenant of God on Earth." He led the seventh crusade to the Holy Land in 1248–1250, and died on another crusade to Tunisia. He dealt violently with blasphemy and heresy against the Roman Catholic Church, with penalties ranging from mutilation to execution of offenders. He is quoted saying that when a Christian "heareth the Christian law reviled" he should use his sword "to pierce the vitals of the reviler as far as it will go."[34] He was canonized as a saint in 1297 by Pope Boniface VIII. Such views about heresy were common in Medieval Christianity, supported even by the renown Catholic philosopher, Thomas Aquinas (1225–1274). Aquinas was one of Roman Catholicism's greatest thinkers and was influential on the development of Western thought in general. Unlike most of his contemporaries, he embraced ideas from Aristotle and was one of the world's great early philosophers. However, in modern time, he would have also been considered to be a violent fundamentalist. Aquinas held that heresy against Roman Catholicism (which he essentially equated with blasphemy) was even a greater offense than murder. He advocated for those convicted of heresy to "not only be excommunicated but even put to death."[34] Aquinas was canonized as a saint in 1323 by Pope John XXII and was declared "Doctor of the Church" in 1567 by Pope Pius V.[20]

After two centuries of effort and an enormous amount of bloodshed trying to take back the Holy Land from the Muslims, the effort eventually failed. Nevertheless, the Crusades continued throughout many areas of Europe, where Roman Catholicism mounted brutal campaigns against Muslims, various Judaic branches along the Rhine Valley,[9] Eastern Orthodox Christianity (discussed earlier in this chapter), *Hussites* and *Calvinists* (discussed in chapter 9), and Biblical Demiurgism (*Catharism* and *Bogomilism* discussed in chapter 10). Between 1492 and 1502, Christian crusaders of the *Reconquista* (the "Reconquest") reclaimed Spain and the rest of the Iberian Peninsula. They then exiled not only the Muslims but also the Sephardi Jews from the region.[9,10,18] As mentioned in chapter 7, the early history of the Crusades was a religious conquest on a scale that dwarfed modern-day jihadism. In the words of comparative religion specialist and former Catholic nun, Karen Armstrong, "*The first Crusaders . . . still had a very primitive conception of God and religion. Soldier saints like St. George, St. Mercury, and St. Demetrius figured more than God in their piety and, in practice, differed little from pagan deities. Jesus was seen as the feudal lord of the Crusaders rather than as the incarnate Logos: he had summoned his knights to recover his patrimony—the Holy Land—from the infidel. As they began their journey, some of the Crusaders resolved to avenge [the death of Jesus] by slaughtering the Jewish communities along the Rhine Valley.*"[35] Although the conquests failed to eliminate Islam, earlier Judaic branches, or most competing branches of Eastern Christianity on a global scale, it left a permanent black mark on the reputation of early Western Christianity. It was not until the twenty-first century that the Roman Catholic papacy began issuing clear apologies for the Crusades, calling them sinful.[36]

The idea of a Holy War persists even today. In modern Western civilization, the concept is most often associated with Islamic extremism (the violent side of jihadism) where it has been highly publicized and politicized in the West since the September 11 terrorist attacks and the jihadist destruction of the World Trade Center of New York. Certain Western actions that precipitated such acts of violence have often been downplayed or ignored in the West (discussed in chapters 11 and 12), as has the fact that this attack was one of politicized Islam rather than strictly religious Islam. The attack was in retribution for what their leader saw as past transgressions of the West against Muslims around the world.

The idea of a Holy War still exists in the West as well, such as in the fringe elements of Christian fundamentalism in the United States. It was reported by the Seattle Times in 2018 that five-term elected Washington state representative and Christian fundamentalist Matt Shea admitted to distributing a four-page manifesto titled "Biblical Basis for War."[37] The document attempts to justify Holy Wars for Christianity as "God's will." It justifies assassination of ungodly rulers as "just." It specifies "War is a fact of life" and that all people must agree to strict obedience to Christian Biblical law. It also says that those who do not submit to this law are an enemy who must be fought. Furthermore, it specifies that if the enemy yields, they must pay taxes and "booty" to the Christian warriors. If they do not yield, all males must be killed. (Other examples will be discussed again in chapter 12.)

By the fourteenth century, controversy was building in Western Christianity over papal abuse of authority and a number of other issues. The core of Western Christianity that has remained tied to the Roman papacy until the present day is *Roman Catholicism.*

ROMAN CATHOLICISM: ROME-CENTERED CHRISTIANITY CLAIMING TO BE THE MAINSTREAM

Roman Catholicism often sees itself as the only true catholic religion, but historically it is but one of several branches that consider themselves to be "catholic" in the most general sense of the word (discussed further in **Note 8D**). As previously mentioned, I use the "Roman" prefixed version of the term to refer to the part of Western Christianity that retained allegiance to the Roman papacy after the Proto-Reform movements of the fourteenth century began. Roman Catholicism traces its history to Jesus Christ and the apostles, as do all branches of Traditional Christianity. (Laying claim to the core inspirational icon is an important element of perceived legitimacy for most major branches of religion.)

Roman Catholicism is a branch of Western Christianity that could as easily have been named "Orthodox Western Christianity." It opposed efforts toward Christian Reform (*Proto-Reformation* and *Reformation Christianity*) that were building in the fourteenth through sixteenth centuries. Roman Catholicism was characterized by its tie to the papacy and the authority of the pope in Rome (although during the fourteenth century for a period of around seventy years, the papal office temporarily relocated outside of Rome). Reformation Christians came up with the name "Roman Catholicism" to identify it as an adversary. The papacy is the oldest continuing absolute monarchy in the history of the world. Even when Constantine transferred the capital of the empire from Rome to Constantinople in 330 CE, weakening Rome's civil authority, Rome's spiritual authority remained intact. Roman Catholicism has been highly successful in spreading its word, far surpassing the rest of Western Christianity in its proselytizing efforts. Its popes oversaw an expansion of Roman Catholicism during the fifteenth through seventeenth century that took it to every continent except Antarctica and Australia.[1] During the same period, other branches of Western Christianity were not looking much beyond European horizons. Those early efforts set the fortunes of the Roman Catholic Church that we see today. Recent studies indicate that Roman Catholicism had more than 1.16 billion adherents by 2010.[29]

The influence of Greco-Roman Polytheism on Roman Catholicism (as well as certain other Christian branches) is apparent, even though most Christian church officials would probably disagree. Having been raised as a Roman Catholic and having attended a Roman Catholic school for several years, I experienced the soft polytheistic elements of that religion firsthand, particularly with regard to trinitarianism (the three aspects of God) (**Note 8I**). Roman Catholicism maintains early Western Christianity's view of trinitarianism as "one God in three Divine Persons": the Father, the Son (Jesus Christ), and the Holy Spirit. The three equally divine persons are distinct, yet they are of one "substance, essence or nature."[38] Additionally, the religion condones the veneration of the Virgin Mary as mother of Jesus as well as veneration of the many different patron saints who can be prayed to for specific requests (e.g., Saint Anthony for the recovery of lost objects, Saint Christopher for safe travels; discussed in chapter 5) (discussed further in **Note 8J**). The *Intercession of the Saints* doctrine of the Vatican[38] condones the act of praying to various saints for help (a practice rejected by Rabbinic Judaism, Islam, and certain other Christian branches). Roman Catholicism features thousands of patron saints, ranging from Saint Ambrose (patron saint of beekeeping) to Saint Zeno (patron saint of fishermen). It even has patron saints for things such as gambling (Saint Cajetan) and beer brewing (Saint Arnold of Soissons). Although many Roman Catholics do not consider such beliefs as polytheistic, the veneration of saints resembles a form of polytheism. It effectively has a Supreme God plus a number of lesser supernaturals to pray to for lesser requests. The explicit veneration of saints to gain their intercession in spiritual matters is referred to by some historians as the *cult of saints*.[9,20] This was one of the practices that was rejected by Reformation Christianity in the sixteenth century.

Another example of Christian branches resembling earlier polytheistic ideas is their description of the early battle of the archangels for dominance of the heavens. These stories sound much like the battling deities in the pantheons of Greece or Rome. Finally, we come to the issue of the devil. Traditional Christianity as well as other Abrahamic subgroups recognize a distinct spirit of evil: Satan. Much like polytheism's gods of theft, trickery, lust and war, sickness, and death, Satan is an essential part of the doctrine and the yin to the yang of God. Satan is said to be lord of the underworld; invested with supernatural powers of evil, even though he is subordinate to God. As the anti-god in charge of the underworld, he is effectively treated as a lesser deity. In *Matthew* 12:24, Satan is referred to as "the ruler of demons." Similarities to even older supernatural belief systems are apparent in Roman Catholicism (as well as certain other branches of Christianity). Going back to animism, the Holy Ghost is usually illustrated in the form of a dove. Going back to ancestorism, God is referred to as "Our Father" in The Lord's Prayer. Finally, going back to matristicism, the mother of Jesus is revered at a level even above sainthood. Many Catholic churches have a separate alter dedicated to the Virgin Mary (e.g., see upper left-hand figure of next page). An extreme veneration of Mary, which grew substantially in Medieval time, is sometimes referred to as the *cult of Mary*.[9,20]

Inside the Roman Catholic *Techny Towers Chapel of the Holy Spirit* in Northbrook, Illinois. Built in the early twentieth century, it contains beautiful sculptural renderings of some of Roman Catholicism's most highly revered icons. Although the crucified Christ (not shown) is the central figure, dozens of other revered figures are represented around the church, including (*top left*) a large alter for the Virgin Mary, mother of Jesus; (*top right*) a large backlit window of the Holy Spirit as a dove above the main entrance; and (*bottom*) a dozen or so eight- or nine-foot-tall statues of important saints, four of which are shown here.

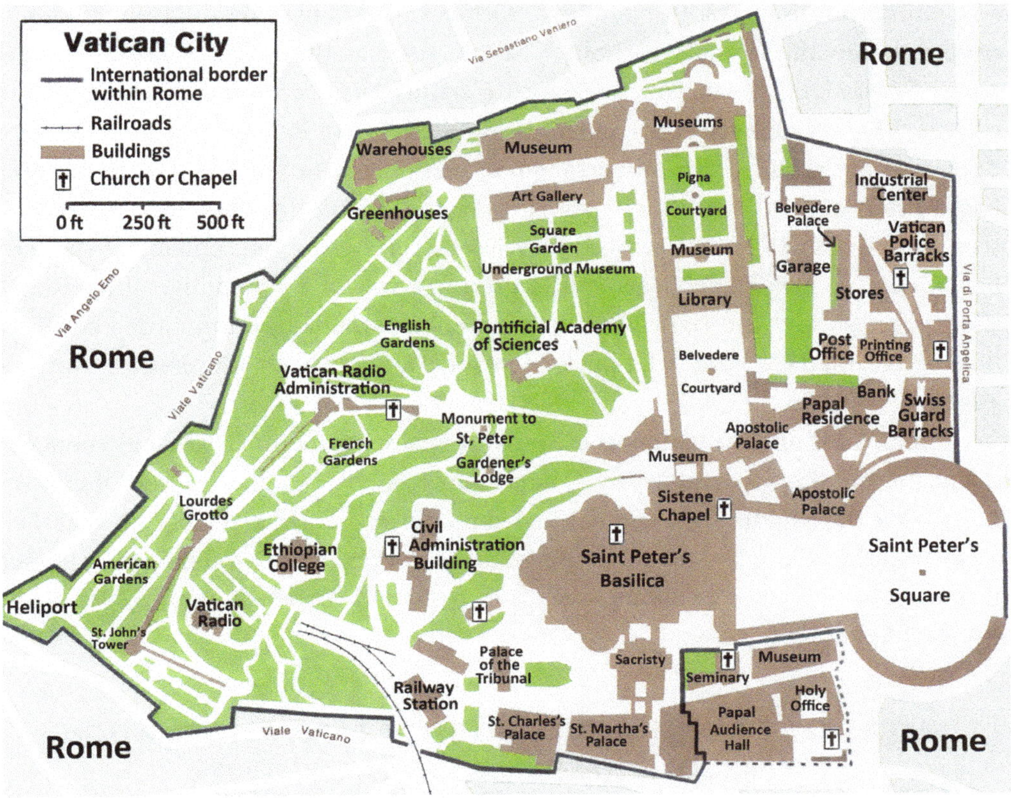

Vatican City, the world's smallest sovereign state, is 100 percent Catholic and ruled by the Roman Catholic pope. It is Roman Catholicism's one remaining theocracy. It has an area of 110 acres with a population of about one thousand people, according to the U.S. Central Intelligence Agency, and is surrounded by the city of Rome. The Vatican is home to Saint Peter's Basilica, said to be the burial site of Saint Peter and the world's largest church building, with a capacity of sixty thousand people. (*Top*) Map of the city. (*Bottom*) View of Saint Peter's Square facing the domed Basilica in the distance.

Today, Roman Catholicism accounts for more than half of all Christians.[29] Although Roman Catholicism strives to project a strong uniformity throughout its membership, elements of diversification have evolved within it that involve slight variations in tradition, mission, or vows. Some of the resulting variants are regional, influenced by earlier indigenous beliefs, such as the *Church of Our Lady of Guadalupe* discussed in chapter 5. Other variants include different religious orders within the Roman Catholic Church, such as the *Augustinians, Benedictines, Dominicans, Franciscans, Jesuits, Carthusians, Carmelites, Salesians,* and others. These groups are not currently in strong conflict with each other, nor does any one group claim universal superiority over another. Each order simply has a different set of traditional priorities (e.g., the way they feel they can best be true to their faith). However, that may have been how some schisms of the past began. It also may be laying a subtle groundwork for further diversification in the future, should the right sort of social environment develop. Some conflicts between Roman Catholic religious orders have already been addressed by various authors.[39,40]

Roman Catholicism claims the Roman papacy as one of its core elements of oversight and doctrinal authority. The papacy has been a lightning rod of controversy since the Middle Ages and has fueled numerous schisms. As discussed earlier, Roman Catholicism has had good popes and bad popes. Controversial popes helped fuel one of the biggest religious schisms in history, *The Reformation,* which is the subject of chapter 9.

PAPAL REJECTIONISM: FUELING THE *PROTO-REFORMATION* TO BECOME *REFORMATION CHRISTIANITY*

The Roman papacy continued to be a major factor in fueling discontent within Western Christianity well after the East-West Schism of 1054 CE, particularly during the late Middle Ages (thirteenth through sixteenth centuries). This schism was the result of the office being led by a succession of powerful and controversial popes in concert with the Holy Roman Empire. Papal indiscretion and struggles for power, such as those mentioned earlier in this chapter, continued to create unrest. The residence of the pope moved to Avignon in the Kingdom of Arles (part of the Holy Roman Empire located in what is now France) from 1309 to 1377. After it was moved back to Rome in 1378, a *papal schism* ensued when Avignon continued to appoint its own popes until 1417. At times, as many as three men simultaneously claimed to be the true pope. The Church in Rome eventually declared the later Avignon popes as "antipopes"; but for a time, the rival claims to the papacy contributed to the already damaged reputation of the office. Furthermore, continued abuses of the papacy fueled an ever-growing dissatisfaction with the clerical hierarchy. The series of bad popes mentioned earlier leading up to the East-West Schism, continued in the West as major motivators for further change and reformation The following extreme examples of papal abuse led to the backlash of *Reformation Christianity* in the early sixteenth century.

Pope Urban VI was the Archbishop of Bari (southern Italy) who was elected pope from 1378 to 1389. He was the last pope elected who was not from the College of Cardinals and the first pope after the official end of the Avignon Papacy. He had a tyrannical and extremely violent nature. To an adviser who questioned the pope's authority to excommunicate someone for a mild misdemeanor, he replied, "I can do anything, anything!"[41] The cardinals at the time thought he was mad. After a group of cardinals failed in their attempt to depose him, he had them seized, tortured, and put to death. He reportedly complained that he did not hear enough screaming when the cardinals were being tortured.[25] In protest and opposition to Pope Urban, a number of cardinals moved to Anagni, Italy, where they elected the first anti-pope Clement VII (formerly Robert of Geneva). This action began the previously mentioned *papal schism* in which two or three men simultaneously claimed to be the true pope.[1,26]

Pope Alexander VI was the Roman pope from 1492 to 1503. He was born as Rodrigo de Borja. Before becoming pope, he was the second richest man in the College of Cardinals, a group that at the time included the richest men in Europe.[25,26] He was known for his notorious sale of indulgences, even while he was a cardinal. These special "passes into heaven" were issued by the Church and could be had for a price. They could be bought in advance for one's self or for a deceased relative. It did not much matter what the offenses in life had been. When someone protested Cardinal Borja's massive sale of indulgences for even the most heinous crimes—one of which included pardoning a father for murdering his daughter—Borja responded, "It is not God's wish that a sinner should die, but that he should live—and pay."[42] He bribed his way to the papacy, buying a majority of votes from other cardinals. He became one of the most controversial of the Renaissance popes, in part because he acknowledged fathering several children with a number of mistresses. He lavished vast amounts of money on his mistresses and children, and he attempted to use his power to gain the most advantageous conditions for his family. He became a symbol of papal decadence. Subsequently, his italicized Valencian surname, *Borgia*, became a symbol for nepotism and corruption.

Finally, *Pope Leo X* was the Roman pope from 1513 to 1521, a period overlapping the great split in the Church as a result of the Christian Reformation. He was a member of the Medici family, the bourgeois family that ruled Florence and Tuscany, Italy.[26] Leo was one of the many Renaissance popes who acted more like a ruling prince than an ecclesiastical leader. He is famous for saying "God has given us the papacy—let us enjoy it."[43] Leo was also known as a spendthrift who depleted the papal treasury in the process of lavishly rebuilding and decorating Saint Peter's Basilica, using the world's best and most expensive artists, including Raphael and Leonardo da Vinci. He is another pope who was closely associated with granting indulgences. He led the Church of Rome to use these indulgences to raise money for grandiose reconstruction and unprecedented decoration of Saint Peter's Basilica. Leo let Johan Tetzel, who was a Dominican monk and the Grand Inquisitor of Heresy to Poland, lead the campaign. Tetzel urged followers: "Won't you part with even a farthing to buy this letter? It won't bring you money but rather a divine and immortal soul, whole and secure in the Kingdom of Heaven."[44] One of his common pitches was "As soon as the coin in the coffer rings, the soul from out of the fire springs."[45]

By the late Middle Ages, the power of Western Christianity and its papal authority in Rome had grown immensely. Widespread dissatisfaction with the central church powers grew to the point at which cries were building for reformation. This was not so much a problem in the Eastern and Middle Eastern branches of Christianity that had been weakened by Muslim conquest and their own decentralization.[10] The problem was centered in Western Christianity and the Holy Roman Empire that had continued to grow in power and spread throughout Europe without serious challenge for nearly a thousand years. By late Medieval time, people were becoming increasingly resentful of the vast wealth and land accumulated by the central church in Rome. Some people were becoming wary of what they saw as superstitious practices being adopted by the Church, such as increasing veneration of saints and relics (e.g., bones of saints, nails supposedly from the cross, and objects supposedly touched by Jesus, his mother Mary, or various saints). Many of the relics were obvious fakes or pieces of doubtful authenticity. Veneration of saints increased to seemingly absurd levels by the beginning of the sixteenth century. Frederick the Wise, Prince of Saxony (and later, a protector of Martin Luther), had more than five thousand relics on display in his castle. The viewing of each was said to provide one hundred days' remission from purgatory. A grand total reduction of more than five hundred thousand days of purgatory could be obtained from seeing the entire collection.[37]

All of this contributed to the general movement of *Papal Rejectionism* and *Reformation Christianity*. In the beginning, a few false starts faded to obscurity, but not without leaving their influence on the later Reformation. Some are mentioned in the next section (e.g., John Wycliffe with his followers, the *Lollards*, and Jan Hus with his followers the *Hussites*) because of their strong influence on Martin Luther and the branches of what became *Reformation Christianity*.

LOLLARDS, *HUSSITES*, AND THE *PROTO-REFORMATION*

In the 1370s, an Oxford philosopher named John Wycliffe (1320–1384), later dubbed "the Morning Star of Reformation" began to criticize the Church in Rome as a "false Church." He rejected the authority of the pope as well as the privileges of the priesthood.[20,33] His ideas eventually inspired a movement called the *Lollardism* (also known as *Lollardy*), meaning "ones who mumble."[46] The name was originally intended as a pejorative term by those disputing Wycliffe's teachings. Wycliffe urged people to go back to the Traditional Christian Bible themselves, and to dismiss many newer rituals and ceremonies condoned by the church hierarchy in Rome that he considered distortions of Christ's word. He engendered great controversy by stressing the importance of civil power within the Church, and by scandalizing the pope and higher-ranking clergy as corupt.[46] He also took particular issue with the doctrine of *Transubstantiation* (the doctrine that the sacrament of Holy Communion turned the bread and wine into the actual body and blood of Jesus). He preached that this was only superstition and that scripture intended the ceremony to be symbolic. He also rejected the sale of indulgences, the need for confession, praying to the saints and the Virgin Mary, praying for the dead, holy water, and a variety of other Roman Catholic traditions.[46] The clergy in Rome reacted strongly against Wycliffe, and in 1377, Pope Gregory XI accused Wycliffe of heresy. Only Wycliffe's connections to English royals and noblemen saved him from being burned at the stake. In 1438, however, Church authorities went to his gravesite, dug up his bones, ceremoniously burned them for heresy, and threw the ashes into the River Swift.[23]

One of the major clashes that Wycliffe had with the church in Rome was his efforts to translate the Bible into vernacular English so that common people could read and interpret its words themselves.[20] His followers produced the first complete translation of the Bible (Old and New Testament) from Latin to Middle English. Copies were rare and printed one at a time by hand because the printing press had not yet been introduced to the Western World. The few copies that existed were widely distributed and the vernacular translation brought the scripture directly to a broader base of people.

Wycliffe's Lollardism was, above all, a religion of vernacular scripture.[46] The rise of Lollardism and the translation of the Bible to a language easily understandable to the common people of England did not go over well with the clerical hierarchy of Rome. They saw easy access of sacred scripture to laypeople as undermining clerical authority. The Catholic Church had long sought to keep the Bible untranslated lest people develop their own ideas of how to interpret it and the text should become a source of schism (**Note 8k**).[48] In 1407, twenty-three years after Wycliffe's death, all existing versions of the Bible translated by the Lollards into English were officially banned by the Church hierarchy. The Church also would not sanction any more English translations of the Bible until over a century later.[1] Lollardism as a distinct belief system did not survive for long. Exactly what Lollardism was and what it became in the sixteenth century remains controversial.[46] However, many of its ideas and principles appear to have been incorporated into emerging branches of Reformation Christianity.[20]

In the early fifteenth century, a Roman Catholic Priest named Jan Hus (1373–1415), who was influenced by (or at least in agreement with) the writings of Wycliffe, founded another early reform movement known as the *Hussite Movement*. Hus, like Wycliffe, was critical of what he saw as excess in Rome and corruption of the papacy, such as the commercial sale of indulgences for church profit. In his 1413 treatise, *De Ecclesia*, which was translated into English in 1915, Hus is quoted as saying "the pope is not the bond of Christianity, and nowhere does he represent the sacrament as a bond of unity." Instead, "wherever two or three are gathered in Christ's name, there is the Church."[49] He also preached that the pope did not have the power to exclude people from heaven through excommunication.[33] This all proved fatal for Hus. He was excommunicated, imprisoned, and burned at the stake for heresy on July 6, 1415. Martin Luther would later praise Hus as a martyr and forerunner of Protestant Reformation, saying, "We are all Hussites without knowing it."[50]

Shortly after Hus was executed, ritualistic differences arose among the Hussites resulting in a schism over how to deliver the Sacrament of Holy Communion to the laity, and a few more minor issues. This led to the differentiation of several Hussite subbranches, including the moderate (close to Catholic) *Ultraquists* (also known as *Calixtinists*) and the more radical *Taborites*, which included its own subbranches, such as the *Orebites* (also known as the *Hradecti*) and the *Orphans* (also known as the *Sirotci*).[51,52] The pope proclaimed five consecutive brutal Crusades against the Hussites between 1419 and 1431, which became known as the Hussite wars (or Bohemian wars). In the end, the Hussites survived the Crusades. However, in 1432, the Ultraquists switched sides to join Catholics (as battle allies, not as Roman Catholic themselves) in a new war against the Taborites. In 1434, the Taborites were thoroughly defeated at the Battle of Lipany by the Ultraquists and effectively disappeared from history. In 1457, surviving Hussites founded the *Unitas Fratrum* (also known as the *Bohemian Brethren*), which later became known as the *Moravian* Church.

Controversy over papal authority and over Rome's tight authoritarian control on Western Christianity continued. The Church of Rome began to be seen by many Christians as inflexible, dogmatic, and wrong-headed. Proto-Reformers following in the steps of Wycliffe and Hus were convinced that the Roman Church had fallen away from the one founded by Jesus. Some even believed that the papacy had become associated with the Antichrist. The environment was ripe for schismatic change. Pope Leo X (discussed earlier in this chapter) continued to add tension within Western Christianity by his opulent spending habits, his sale of indulgences, and his failing to take the developing Protestant Reformation seriously. He eventually excommunicated another reformist named Martin Luther in 1521, which fueled the formal emergence of *Reformation Christianity*, which is the subject of the next chapter.

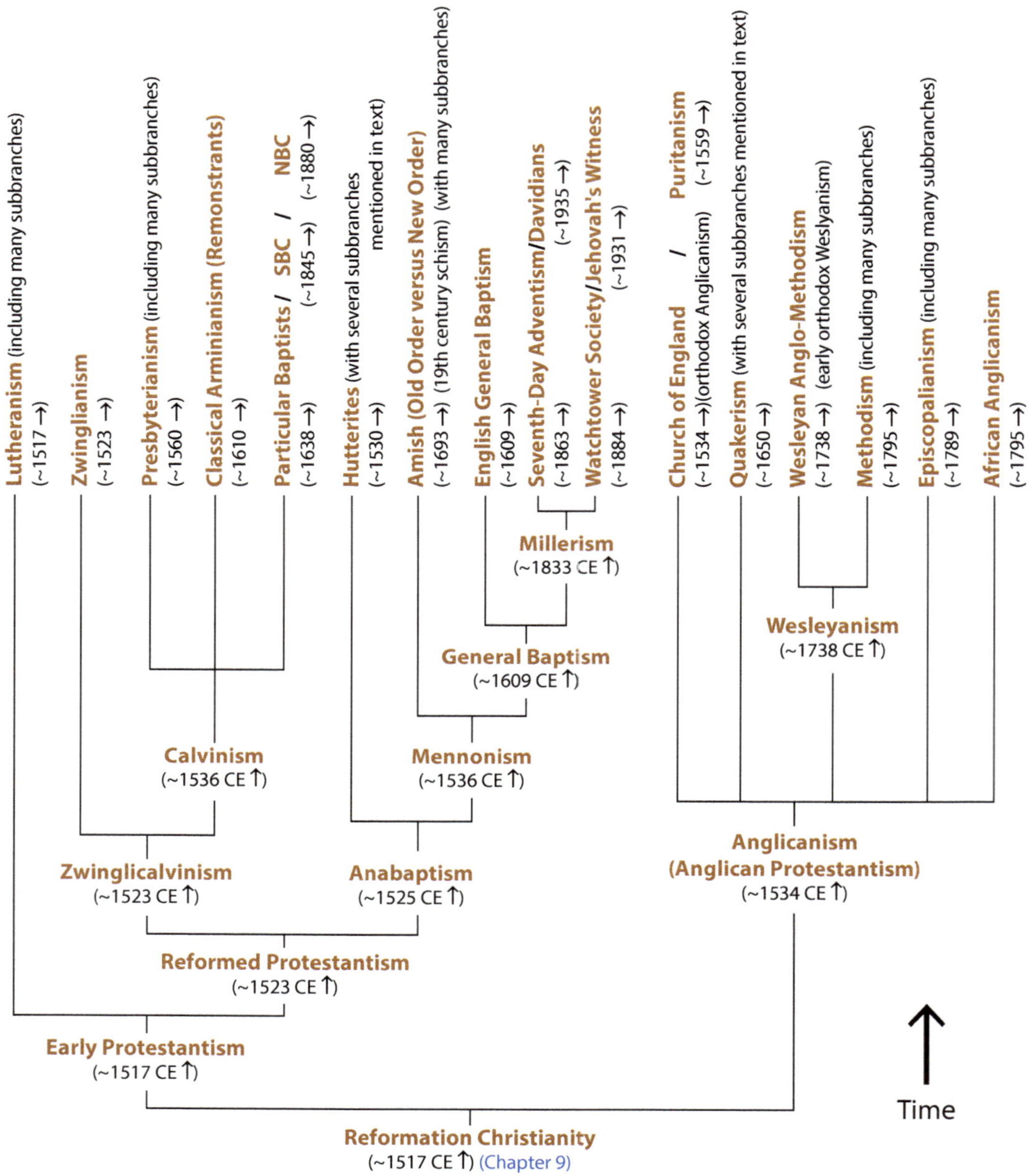
Lutheranism (including many subbranches)
(~1517 →)
Zwinglianism
(~1523 →)
Presbyterianism (including many subbranches)
(~1560 →)
Classical Arminianism (Remonstrants)
(~1610 →)
Particular Baptists / SBC / NBC
(~1638 →) (~1845 →) (~1880 →)
Hutterites (with several subbranches mentioned in text)
(~1530 →)
Amish (Old Order versus New Order)
(~1693 →) (19th century schism) (with many subbranches)
English General Baptism
(~1609 →)
Seventh-Day Adventism/Davidians
(~1863 →) (~1935 →)
Watchtower Society/Jehovah's Witness
(~1884 →) (~1931 →)
Church of England / Puritanism
(~1534 →)(orthodox Anglicanism) (~1559 →)
Quakerism (with several subbranches mentioned in text)
(~1650 →)
Wesleyan Anglo-Methodism
(~1738 →) (early orthodox Weslyanism)
Methodism (including many subbranches)
(~1795 →)
Episcopalianism (including many subbranches)
(~1789 →)
African Anglicanism
(~1795 →)
Millerism
(~1833 CE ↑)
Wesleyanism
(~1738 CE ↑)
General Baptism
(~1609 CE ↑)
Calvinism
(~1536 CE ↑)
Mennonism
(~1536 CE ↑)
Zwinglicalvinism
(~1523 CE ↑)
Anabaptism
(~1525 CE ↑)
Anglicanism
(Anglican Protestantism)
(~1534 CE ↑)
Reformed Protestantism
(~1523 CE ↑)
Early Protestantism
(~1517 CE ↑)
Time
Reformation Christianity
(~1517 CE ↑) (Chapter 9)

9

REFORMATION CHRISTIANITY

It is evident that a pope living contrary to Christ, like any other perverted person, is called by common consent, antichrist

—JAN HUSS, 1413

Reformation Christianity is a collective term for the Christian reform movements that became codified in Western Europe during the early sixteenth century. Revolution was growing against corruption within the Roman papacy and an increasing propagation of various doctrinal elements (e.g., saints, relics, and various rituals seen by some as superstition). The movement sought to return to earlier models of Traditional Christianity that it saw as having been corrupted by the Medieval church of Rome.[1]The Reformation was also a result of rising nationalism in cities, such as Germany, Switzerland, and England.[2] Although the roots of Reformation Christianity began with the Proto-Reformation in the fourteenth and fifteenth centuries, with movements such as those of the Lollards and the Hussites discussed in chapter 8, it did not formally establish itself until the early sixteenth century (phylogenetic trees on pages 322 and 366).

Reformation Christianity arrived in two major forms. One was *Early Protestantism* in the form of *Lutheranism*, beginning in Germany around 1517. The other was *Anglicanism* (*Anglican Protestantism*), which began as the *Church of England*, and grew out of the Reformation in 1534. Early Protestantism (after the Latin word *protestari*, meaning "to protest") began as a backlash against what Roman Catholicism had become, while Anglicanism first saw itself as a bridge between Roman Catholicism and Early Protestantism.[3] The emergence of Reformation Christianity resulted in an explosive diversification of new religious branches over the next several centuries. It also set the stage for more than a century of wars between *Reform Christians* and a *Counter Reformation* movement of Roman Catholicism (discussed further in **Note 9A**).

Timeline for certain events, people, and beliefs of early Reformation Christianity

Episcopalianism (an American offshoot of *Anglican Protestantism*) is codified and formally separates from the *Church of England*, rejecting the supremacy of the English monarchy →	~1789 CE
English colonists (the Pilgrims) bring *Protestantism* to the New World, beginning in the New England region with *Puritanism,* and followed a few years later by *Quakerism* (both offshoots of *Anglican Protestantism*) →	~1620 CE
King James Version of the Bible was published in English (authorized by King James I of England) and became one of the single greatest influences on the English-speaking world →	~1611 CE
First copy of the Bible published in English (called "The Great Bible" and authorized by King Henry VIII). A decade later it it was used to help construct the Anglican *Book of Common Prayer* used to this day by the *Church of England* →	~1539 CE
Anglican Protestantism (*Anglicanism*) is formally instituted by King Henry VIII and the *Church of England*. The *Act of Supremacy* makes the King of England the Supreme Head of the *Church of England* →	~1534 CE
Over the centuries, *Protestantism* diversifies into many different branches, beginning with a schism between *Reformed Protestantism* and *Lutheranism* →	~1523 CE ↑
Protestant Reformation formally instituted by Martin Luther in Germany with what became *Lutheranism* →	~1517 CE
Invention of the *Gutenberg printing press*, making consistant and widespread dissemination of Europe's religious literature easier (e.g., *Gutenburg Bible* printed in Latin) →	~1436 CE
Proto-Reformation (including the *Lollards and Hussites*) was initiated by John Wycliff and Jan Hus. Fueled by rejection of papal authority, it moved toward *Reformation Christianity* →	-14th and 15th centuries CE

EARLY PROTESTANTISM AND ITS BREAK FROM THE ROMAN CATHOLIC CHURCH AND THE PAPACY

The formalization of *Early Protestantism* and *Reformation Christianity* began in Germany. It rejected the authority of the Roman papacy as well as many Roman Catholic traditions (e.g., mandatory celibacy of priests and the sale of indulgences, discussed in chapter 8). Christ was still the central figure with the devil as his adversary. However, the Virgin Mary and the saints were no longer revered at the level that they were in Roman Catholicism.[4] Most of Early Protestantism also dismissed five sacraments that the Roman Catholic Church had added to Traditional Christianity by the early fifteenth century, including *Confession and Penance* (together forming a single sacrament), *Confirmation, Matrimony, Holy Orders,* and *Extreme Unction* (anointing the sick and dying). However, it upheld the two sacraments of earlier Christianity (*Baptism* and *Holy Communion*).

Early Protestantism encouraged individual followers to read the Bible rather than simply relying on interpretations from the papacy and clergy. The overall reduction of church ritual and elimination of priestly vestments desacralized the ministry and softened the distinction between the clergy and the laity. A significant advantage that early Protestants had going for them in reaching common people was the invention of the Gutenberg printing press. This technology was lacking in earlier attempts of Christian reform. By Luther's time, several printing presses throughout Europe could produce identical pamphlets and vernacular Bibles to be more widely distributed to a public that was gradually becoming more literate. This advancement was the beginning of an early "information age" that marked the end of the Middle Ages as well as the end of Roman Catholicism's monopolistic hold on the Western Christian Church.

The conflict between the Roman Church and Early Protestantism began as one of long-lasting, bitter dispute that would last for centuries. Remnants of it still exist today. I remember one of the nuns in the Roman Catholic grade school I attended during the early 1960s telling us that Roman Catholics were the chosen ones and that our Protestant friends would not see heaven unless they converted to Roman Catholicism before they died. Not that they would necessarily go to hell. The implication was that they would go someplace in between, not unlike babies who died before being baptized who supposedly went to a place called "limbo," or people who had died before being forgiven of venial sins who supposedly went to purgatory. In the late twentieth century, during the Second Vatican Council (1962–1965), the Roman Church made substantial official efforts to heal old wounds with other Christian denominations.[5] The Council no longer referred to Protestants as heretics, and church officials began referring to Protestants simply as "other Christians."

However, even such seemingly minor efforts of conciliation contributed to further schism within the church (discussed in **Note 9B**).

Once Early Protestantism took hold in Western Christianity, it did not stop with Lutheranism. Innovation and schism continued, resulting in the evolution of many denominations. Early in its history, Protestantism diverged into two major subgroups: *Lutheranism* (the beginnings of *Early Protestantism*) and *Reformed Protestantism*, which branched off only a few years later.

THE RISE OF LUTHERANISM WITH LUTHER THE "EVANGELICAL"

The pressure for reform within Western Christianity grew to a major breaking point in the early sixteenth century. A German priest and professor of biblical studies, Martin Luther (1483–1546) followed many of the tenets of Jan Hus and the Hussites.[4,5] However, Luther's influence went beyond that of his ideological predecessors. He set Reformation Christianity on a course to become one of the world's major religious branches. He most famously shunned indulgences, opposing the idea that any spiritual influence could be bought and sold. He, much like Wycliffe of the Proto-Reformation discussed in chapter 8, placed reading the Bible above the authority of the papacy. He also taught that humans were saved by their faith in God's grace, not simply by the good works they do.[2]

According to some historians, Luther nailed a list of points for debate called *Ninety-Five Theses* on the door of All Saints Catholic Church in Wittenberg in 1517.[4] This was the same church where he sometimes preached and where he was buried after his death in 1546. The posting of this list is often taken as the official start of Lutheranism. Luther, however, disliked the term Lutheranism. He preferred the term "Evangelical."

The Roman Church interpreted Luther's Ninety-Five Theses as an act of rebellion and gave a series of heavy-handed responses. While Luther's focus was on Paul the Apostle's teachings about grace and salvation, the Roman Church's response focused on the infallible authority of the Roman Church and its papacy.[6] Repeated confrontations fueled the image of Luther as a reincarnation of the executed rebel, Jan Hus. In 1521, Pope Leo X excommunicated Luther as a heretic. In defiance, Luther publicly burned the papal decree excommunicating him, cheered on by the townsfolk to whom he had become a hero.[6] Lutheranism took hold and spread rapidly, beginning in the German-speaking region of the Holy Roman Empire.

In addition to his early acts of defiance, Luther blatantly rejected clerical celibacy and the monastic way of life. He married a former nun with whom he had six children.[4] He escaped being burned at the stake for heresy largely because Prince Frederick III

Martin Luther (1483–1546), founder of Lutheranism and one of the founders of the Protestant Reformation. Here he is seen in a painting by artist Ferdinand Pauwels (1830–1904) at the gate of the castle church in Wittenberg where he is said to have posted his ninety-five theses in protest over the practices of the Catholic Church. This event is recognized by many historians as the formal start of the Reformation.

of Saxony hid him in Wartburg castle. During his period of refuge, Luther translated the Bible into German, which later greatly influenced German religious practices, helped shape language, and increased literacy. Luther maintained some conservative principles of Western Christianity, such as the interdependence with secular authority and the idea that the bread of the Eucharist is the actual body of Christ.

Once Lutheranism had spread throughout much of Germany, it then spread into the Baltic States. It came to the United States with German and Scandinavian immigrants, where it became part of a new Protestant Empire. Ideological diversification of Lutheranism began shortly after Luther's death, resulting in several rival branches. The most prominent of these earliest branches were the *Gnesiolutherans* and the *Philippists*.[7] The following centuries saw a complex evolutionary history of diversification that cannot be adequately covered here (nor has it broadly been covered elsewhere that I could find). Lutheranism continues to evolve with new denominations still forming in the early twenty-first century (e.g., the *Evangelical Lutheran Diocese of North America* and the *Association of Confessional Lutheran Churches*). According to the *lutheranworld* website, today, there are more than seventy-seven million Lutherans in one hundred and forty nine putative denominations around the world.[8] However, many of those denominations are so far distinguished mainly by nationalistic or regional boundaries rather than by ideological characteristics. A comprehensive phylogenetic study of the many Lutheran subbranches based on ideological diversification over time has yet to be done.

Not long after Lutheranism emerged in mainland Europe, reform continued to grow within Early Protestantism. This led to another schism and the development of *Reformed Protestantism*.

REFORMED PROTESTANTISM: REFORMING FASTER

Reformed Protestantism split from Lutheranism in the early sixteenth century. The schism was over endlessly controversial doctrinal issues, such as the meaning of the Eucharist (Holy Communion), baptism, and the use of religious imagery.[6] The resulting conflict, sometimes over seemingly minor issues to those outside the fray, often led to hardline divisions and even violent bloodshed. The detailed historical relationships among the many branches of Reformed Protestantism are poorly resolved and in need of further comparative work. I provisionally divide Reformed Protestantism into two major Reformed Protestant groups based on historical records of early relationship and ideological influence. The first group I call *Zwinglicalvinism* (the Reformed Protestantism subgroup that emerged under the influence of Ulrich Zwingli and John Calvin). The second group is *Anabaptism* (also known as *Radical Reform Protestantism*).

REFORMED PROTESTANTISM GROUP 1: *ZWINGLICALVINISM*

I use the term *Zwinglicalvinism* for a branch of Reformed Protestantism that emerged in the early sixteenth century. Ulrich Zwingli and John Calvin were both active Protestant reformers in the Swiss cities of Zurich and Geneva during the early 1500s. They were both set on purging much of Roman Catholicism from Western Christianity. Zwingli and Calvin, just as Luther before them, still attempted to work with the European secular magistrates (state authorities, such as princes, magistrates, city councils, and kings). They believed government to have been ordained by God unless they were found to have violated God's law. For that reason, their type of Reformed Protestantism is sometimes referred to as *Magisterial Reformed Protestantism*.[9] Early Zwinglicalvinism also believed in the concept of predestination, which will be explained in the next section.

As the name suggests, I provisionally identify two major subbranches of *Zwinglicalvinism*: *Zwinglianism* and *Calvinism*.

ZWINGLIANISM: PONDERING THE PARADOX OF FREE WILL

Zwinglianism follows the teachings of Huldrych (or Ulrich) Zwingli (1458–1531) a parish priest and army chaplain who became one of the most important Swiss leaders of the Protestant Reformation.[5,6] His followers became known as "Zwinglians," which was initially a derogatory term used by Roman Catholic officials who considered followers of Zwingli to be heretics. According to Zwingli, his theology was more consistent with biblical scripture than that of the pope or the Church Fathers (the church authorities charged with defending Roman Catholic doctrine and identifying heresy). He began teaching his Reformed version of Protestantism in Zurich in the early 1500s. In 1523, Zwingli attended a large conference in Zurich where he presented his "Sixty-Seven Articles" to describe his ideas of Protestant Reformation in detail. The articles also contained diatribes against the Roman Catholics, the Lutherans, and the Anabaptists. I date the beginning of Zwinglianism in that year.

Zwingli, like most other Reform Protestants, believed that the bread and wine were only symbolic of Christ, much like the beliefs of the earlier reformer John Wycliffe who was presented in the last chapter. Zwingli also had new ideas about infant baptism. Although he supported the Lutheran and Roman Catholic idea of baptizing infants, he taught that its purpose was to welcome children into the church and a pledge to God rather than a magical washing away of *original sin*. Luther's response to that was bitter, telling his followers that they would be better off getting married and having their children baptized in Roman Catholic churches rather than among Zwinglians.[6]

Like Martin Luther, Zwingli rebelled against the corruption of the papacy (although Luther usually overshadowed him in this regard). He also rebelled against numerous Roman Catholic practices, such as mandatory celibacy for the clergy, veneration of the saints, fasting during Lent, and the power of papal excommunication. He not only rejected iconoclastic imagery, such as statues and crucifixes, he also rejected alters, incense, and even music. Following Wycliffe, he pointed out that the Roman Catholics had intentionally hidden the original second commandment against "graven images"[10] (discussed in chapter 8). His stand against music constituted a major shift from Roman Catholic and Lutheran services. As a result, Zwinglian church services became dull and less engaging, driving people to rebel. In 1598, the ban on music was lifted so people could once again sing their hymns and psalms. Zwinglianism was similar to Lutheranism and other Zwinglicalvinist branches in its close relationship with the state authority. Zwingli saw government as ordained by God, so Christians were obliged to obey civil authorities unless those authorities acted against God's law.

Zwingli saw the practice of infant baptism as an important means to welcome children to the Church. He strongly defended the practice against the Anabaptists who considered it as unnecessary and illegitimate. Anabaptism (to be discussed later in this chapter) was another even more radical reform version of Protestantism arising in Switzerland and in conflict with Zwinglianism. Followers of Zwingli were the first to persecute Anabaptists in a long and violent campaign that would last for centuries.

Zwingli believed that salvation was something that was predetermined for the "elect." The Zwinglian idea of predestination was something that Calvinism would expand on. Salvation was not something that could come from the development of faith or belief. Faith was only a symptom of those who had already been elected. The believer believed because he had already been predestined for eternal salvation. This was how Zwinglians and other predestination believers grappled with the so-called paradox of free will. They reason that God's omniscience is incompatible with the belief that human free will had the power to change their destiny. If God knows all, then he must already know who is predestined for salvation even before they are born.

Although Zwinglianism and Lutheranism agreed on some issues, on several issues, they could not. For example, Luther believed that the Eucharist *became* the body of Christ (like the Roman Catholics), where Zwingli believed that it only *represented* the body of Christ. Also, as noted, Zwinglians more strictly forbade religious imagery. The bitterness between Luther and Zwingli became intense. Luther saw Zwingli as more in error than even the Roman Catholic pope.[6] The Roman Church and the Lutherans both considered Zwingli a heretic. In 1531, religious war broke out between Roman Catholics and the Protestants in Switzerland. Zwingli, who served as a chaplain, died on the battlefield that same year at the relatively young age of forty-seven. The Roman Catholic victors desecrated Zwingli's body. They cut it into quarters, mixed it with dung, and burned it in posthumous execution for heresy.[11]

Zwinglianism was never very successful on its own, even though Zwingli was one of the great Protestant Reformers. Zwingli was succeeded by Heinrich Bullinger (1504–1557) who, as chief minister in Zurich, did much to spread Zwingli's ideas within the country.[5] Outside of Switzerland, no church counts Zwingli or Bullinger as its founder. Although Zwingli and Bullinger began the Reform movement in Switzerland, their roles were eventually eclipsed by John Calvin and *Calvinism*.

CALVINISM: ANOTHER BRANCH PROMOTING A DOCTRINE OF DOUBLE PREDESTINATION

Calvinism was a further development of *Zwinglicalvinism* that arose in the sixteenth century. It was founded by French theologian and lawyer, John Calvin (1509–1564).[1,6] Building on the work of Zwingli and Bullinger, Calvin's contributions made

Geneva a center for the expansion of *Reformed Protestantism*. Calvin eventually had a much broader impact than even Martin Luther did on the growth of *Early Protestantism*. In the early years after his death, Luther's ideas were mainly confined to Germany.[2] Calvinism, in contrast, was established as an international religion by the end of the sixteenth century. It ranged from Europe to the new-world colonies. Calvinism came to be seen as a progressive new wave of the Reformation, particularly in places like the American colonies. Even people who no longer believed in God could retain a secular notion of Calvinist predestination (the ultimate sense of entitlement). They could envision themselves as the nation "predestined" for success, whose flag and ideals had a semidivine purpose.[2] The name *Calvinism* was not the idea of John Calvin. It was instead the invention of Lutheran critics of Calvin as a derogatory term. Calvin later responded after hearing the name: "They can find no more horrific insult . . . , than the term, Calvinism". "From where this bitter hatred against me comes is clear enough."[12]

Calvin, like Zwingli, accepted ideas such as trinitarianism and infant baptism. He believed that human nature was almost irresistibly drawn toward evil. Also, like Zwingli, he was aniconistic, encouraging removal of all statues and pictures from churches and adoption of congregational singing that had no organ accompaniment.[4] However, where Luther, Zwingli, and Roman Catholicism all shared a high veneration of Mary, mother of Jesus, Calvin did not.

Calvin accepted Zwingli's idea of *double predestination*. The doctrine of double predestination was never the leading axiom of John Calvin's thought,[5] but it grew in significance over time. After Calvin's death, it became a foundational characteristic distinguishing Calvinism from rival Protestant groups.[1,2] Double predestination took the idea of God's power to its logical end: if God is all-powerful and all-knowing, then God has already decreed who will be saved (the "elect") and who will be damned at birth.[1,6] One's deeds in life had no effect on one's salvation or damnation; they were merely a sign of what God has already decreed. Following John Calvin's death, Calvinists were taught to assume they were among the elect, because any doubt about this was a self-fulfilling prophesy. Although they could not *change* their eternal destiny, they could demonstrate through their actions that they were already *among* the elect. The *Five Points of Calvinistic Predestination* that developed (known by the acronym TULIP) were later challenged with the competing ideology of *Classical Arminianism* and their counterpoints to TULIP. These counterpoints became known as the *Five Articles of Remonstrants* (listed in **Note 9C**). Arminianism gave more power to human free will and felt that anyone could potentially achieve salvation through his or her acts in life.

Calvinism instructed its followers to be solid citizens and not rebel against authority.[1] It was the magistrate authorities who were to be obeyed, not the papal authorities of Rome. Because the pope had obstructed Reform efforts within the

Christian Church, Calvin reasoned that the pope was the Antichrist, and that Protestants were the true catholics (in the most general sense of the word).[6] Calvin broke with the Roman Catholic Church sometime around 1530.

Calvin and the collaborative Swiss authorities could be ruthless at times. In 1553, Calvin helped convict scientist and theologian Michael Servetus of heresy. Servetus had rejected trinitarianism, considering it a form of polytheism. He instead supported the stricter monotheistic concept of Jewish scriptures.[2] (His antitrinitarian writings would later influence the rise of *Unitarianism* in Poland and Transylvania.) Servetus also rejected the practice of infant baptism, which he called an invention of the devil. Calvin found the writings of Servetus to be blasphemies and helped the Geneva City Council convict him of heresy. Servetus was subsequently sentenced to be burned at the stake.[2] Although Calvin requested the more "merciful execution" of beheading instead, that request was denied and Servetus was burned.

Calvinism formed an ideological base for the development of *Presbyterianism, Classical Arminianism*, and *Particular Baptists*. Its best-known direct descendent is *Presbyterianism*.

PRESBYTERIANISM: SCOTTISH REFORM

Although Scottish Presbyterians trace certain elements of their religion back to the sixth century, the formal establishment of the Presbyterian Church as a distinct denomination began around 1560 in Scotland. At that time, the Parliament adopted the *Scots Confession* as the creed of the Scottish Kingdom, as well as the *First Book of Discipline*, which outlined important doctrinal issues for the church. The latter book also established regulators for church government including ten ecclesiastical districts known as "Presbyteries." The founder of Presbyterianism was John Knox (1513–1572), an ex–Roman Catholic priest who left the Church and became leader of the Reform Protestant movement in Scotland.[5] Although he had many early influences, including Roman Catholicism and Anglicanism, the most important influence on him was Calvinism. Knox preached against the veneration of the Virgin Mary, purgatory, the Mass, and prayers for the dead. Like Calvin, he compared the pope to the Antichrist. In 1551, Knox became a royal chaplain to King Edward VI, as Protestant influence was growing in England.

When Mary Tudor ascended to the throne of England and Ireland in 1553, she violently attempted to reestablish Roman Catholicism as the state religion (discussed again later in this chapter). Knox was forced to resign his position as Royal Chaplin to escape the hostile environment created by Queen Mary. England was no longer a safe place for Protestant preachers and Knox fled to Europe. He eventually ended up in Geneva where he studied with John Calvin.[4] After Queen Mary died in 1558, her efforts to move the region toward Roman Catholicism were reversed. Knox returned to Scotland, and collaborating with Scottish nobility, he eventually instituted a new branch of Calvinism: the Presbyterian Church of Scotland (the roots of early Presbyterianism)[5]. Presbyterianism spread to colonial America in 1644 to establish the First Presbyterian Church in Hempstead, New York. Although the mother church is in Scotland, today Presbyterianism has evolved into many branches around the world, far too numerous to mention.[5]

Other branches besides Presbyterianism were rooted in, but diverged from, Calvinism. One that splintered from it over doctrinal issues was *Classical Arminianism*. Its members preferred to be known as the *Remonstrants.*

CLASSICAL ARMINIANISM (*REMONSTRANTS*): PUSHBACK AGAINST THE DOCTRINE OF PREDESTINATION

Classical Arminianism (also known as the *Remonstrants*) was first established in 1610 with the *Five Articles of Remonstrance* mentioned earlier. These articles formed a theological statement signed by forty-five ministers of mostly Dutch Protestants following the ideas of Jacobus Arminius (1510–1609).[5] Classical Arminianism began in the Netherlands within Reformed Protestantism as part of a theological debate countering Calvinistic predestination (**Note 9C**). Jacobus Arminius was a Dutch theologian who died a year before the religion named after him was established. Arminius was a student of Theodore Beza, Calvin's successor. Arminius was trained in the Calvinistic school, although he came to reject certain early Calvinistic ideas.[6]

Classical Arminianism can be distinguished from other branches of Calvinism in two major ways.[5,13] First, Classical Arminians believe that humans are redeemable and have the freedom to pursue redemption through faith in God. Everyone has a chance at salvation, and God's decision about who is destined for salvation is based on an individual's acts in life. Where most Calvinists believe that they owe their faith to their predestined election, Arminians believe that they owe their election to maintaining their faith in life. Second, major Classical Arminianism holds that Jesus died for all. Humans can accept or reject God's call to grace and salvation. No "predetermined elect" is unable to fall from grace and guaranteed salvation at birth. Classical Arminianism's rejection of predestination is unusual for Zwinglicalvinistic belief systems, but because Classical Arminianism's founder was trained as a Calvinist, I consider the absence of predestination beliefs in the branch to be a *secondary loss* or *character reversal* (the concepts of secondary loss and character reversal are discussed in **Note 7C**).

Arminius's followers, choosing not to call themselves by their leader's name, called themselves *Remonstrants*. Nevertheless, as is often the case for denominational names, the moniker that stuck (*Arminianism*) was the one that its enemies used.[6] Arminianism had far-reaching influence. It was later influential to John Wesley and the development of the *Weslyanism* branch of *Anglicanism*,[5] as well as Thomas Hewley's branch of *General Baptism* (discussed later in this chapter). Another early offshoot of Calvinism that emerged in the 1630s was *Particular Baptism*.

PARTICULAR BAPTISTS: CALVINISM WITH ADULT BAPTISM

Particular Baptists (also collectively known as *Reform Baptists* or *Calvinistic Baptists*) evolved in early seventeenth-century England with broad roots in Calvinism. The Particular Baptist doctrine on baptism (substituting adult baptism for infant baptism) was influenced by the Anabaptists (discussed later in this chapter as the sister group to Zwinglicalvinism; see the tree on page 366).[14] However, the strongest influence for Particular Baptism came from the basic principles of Calvinistic theology, which is why I present it as a subbranch of Calvinism on my phylogenetic tree. Like the Presbyterians, the Particular Baptists believe that Christ died only for the elect who were predetermined for salvation before birth. Also like the Presbyterians, they believe in unconditional election and predestination (that God unconditionally elected those destined for salvation before their birth).[5] *Particular Baptists*, as defined here, should not be confused with *General Baptists*. General Baptists appear to follow a Mennonite theology, whereas Particular Baptists follow more of a Calvinist theology. Even though Particular Baptists and General Baptists share the word "Baptist" as part of their name, they are thought to be only distantly related.[14] The relationships of these and other Reformed Protestant groups are in need of further comparative analysis to test and possibly revise the pattern shown in this hypothetical tree.

Many historians place the origin of the Particular Baptists in 1638, when Minister John Spilsbery (1593–1668) helped found the first Calvinistic Baptist Church in London, England.[14,15] Spilsbery was a member of the London Separatist Church, which had split from Anglicanism. Like many early Protestants, he believed there were many Antichrists, including the popes of Rome. He preached Calvinistic principles, including *believer's baptism* (adult baptism). He saw infant baptism as part of the Antichrist's system,[16] and he did not support the idea that unbaptized infants who die are rejected from salvation.

Over time, Particular Baptism spread around the world, and diverged into many subbranches, the largest of which was the *Southern Baptist* branch (officially known as the *Southern Baptist Convention* or *SBC*).

SBC VERSUS *NBC*: CONFLICT AND SCHISM WITHIN *PARTICULAR BAPTISM*

The *Southern Baptist Convention* (*SBC*) is a branch of Particular Baptism that evolved in the United States. It began over the issue of slavery.[5] By the beginning of the nineteenth century, *Particular Baptism* was growing in both the northern and southern regions of the United States. Particular Baptist preachers in the south coordinated their evangelistic efforts with the concerns of southern society. The powers of that society had become dependent on slavery for its agricultural economy, and they generally saw Blacks as inferior beings to whites. They perceived their right to own slaves, and for slave owners to become missionaries.[17] Some even interpreted the Bible as supporting slavery and Black inferiority.[18]

As antislavery sentiment grew in the north, tension began to build within Particular Baptist parishes in the south over issues such as whether or not slaveholders could serve as missionaries. Particular Baptist parishes in the north did not want to endorse slavery. Such issues led to the *Particular Baptist Schism* of 1845 that separated the *Southern Baptist Convention* (*SBC*) from other Particular Baptists. Its first church opened in Augusta Georgia in 1845. After the Civil War ended and the Confederacy was defeated in 1865, racial tensions continued to escalate in the South. White SBC members defended their earlier racist views resulting in the development of organizations such as the Ku Klux Klan. In response, Black Southern Baptists eventually formed their own branches of Particular Baptism, culminating with the *National Baptist Convention* (*NBC*). The NBC was founded in 1895 by Baptist pastor and freed slave, Elias Camp Morris (1855-1922). It became the largest predominately African American Christian denomination in the United States, growing to more than 3.5 million members by 2010.[19]

Up through the 1960s, the SBC was one of the last bastions of racial segregation.[20] White Baptist pastors and most SBC members fought against integration and for white supremacy. By the early 1970s, however, moderate elements started to grow within the SBC. Televangelism was a medium that developed predominately within the SBC. Well-known evangelistic Baptist preachers such as the conservative-moderate Jerry Falwell (1933–2007) and the more liberal-moderate Billy Graham (1918–2018) reached hundreds of millions of listeners worldwide at their peak. Finally, in 1995, the SBC adopted a resolution renouncing its racist history and apologizing for its former defense of slavery, segregation, and white supremacy.[21] It was the denomination's first formal acknowledgment of its early racist history, which was later followed by a number of other such admissions.[18] By the late twentieth and early twentyfirst century, the SBC had evolved from an organization of strict racial segregation to one of the more ethnically diverse Protestant denominations in North America. Today, it is the world's largest Baptist denomination with more than sixteen million members in the United States alone.[19]

Presbyterians, Classical Arminians, and *Particular Baptists* (including *SBC* and *NBC*) were all rooted in Calvinism to varying degrees. Calvinism and Zwinglianism together made up one of the major branches of Reformed Protestantism: *Zwinglicalvinism* (see the phylogenetic tree on page 366). The other major branch of Reformed Protestantism to emerge in the early sixteenth century was *Anabaptism.*

REFORMED PROTESTANTISM GROUP 2: *ANABAPTISM*

Anabaptism is the sister group to *Zwinglicalvinism* (tree on page 366). It rejected earlier ideas of magisterial collaboration (secular authority within the church) and favored a clear separation of church and state. It also rejected *infant baptism* in favor of *believer's baptism,* postponing baptism until the recipient was old enough to earnestly profess their faith. It arose in a time of great social turmoil. In the early 1500s, large areas of central Europe were convulsed in revolt against church leaders and their magistrate collaborators (the royal princes).[7] Some denominations were vying to assert dominance, whereas others were simply struggling for survival. The early revolts were brutally crushed, which added to the passion of the revolutionaries. Luther, who loathed the resulting disorder, sided with the magistrate in this case. Lutheranism joined Zwinglianism, Calvinism, and even Roman Catholicism in the brutal efforts to stamp out Anabaptist denominations, because they all deemed Anabaptists to be heretics.

The practice of believer's baptism was known as "rebaptism" in the case of converted Roman Catholic who had already been baptized as infants. As with many religions that diverge from the mainstream, the name Anabaptism was first used as a pejorative term invented by those who persecuted this group. During the early history of Anabaptism, members were violently persecuted. Their different interpretations of scripture put them at odds with both the State Church and the magistrate. The early story of Anabaptists is a particularly horrific one, even as Medieval religious history goes. They were first persecuted in 1525 by Roman Catholics who executed them simply as "Protestants."[22] (The religious battles of the Reformation between Protestants and Catholics were still flaring at that time.) Soon afterward, other Protestant groups joined in on the persecution of Anabaptists because of their conflicting views on baptism.

The early Anabaptists of Zurich were called the *Swiss Brethren* in Zurich. The group was founded by three leaders: Felix Manz, Conrad Grebel, and George Blaurock.[5,6] They broke with Zwinglicalvinism because they believed it was not reforming far enough or fast enough. This is why some people refer to Anabaptism as

Radical Reformed Protestantism.[5] The Anabaptist rejection of infant baptism was unacceptable for Zwingli and others, and the response to the Swiss Brethren was swift and harsh.

Felix Manz (1498–1527), was the first of the three founders of Swiss Brethren to be executed. He became one of Anabaptism's better-known martyrs. The Zwinglian Zurich Council condemned him to death for the offense of "rebaptizing." On January 5, 1527, he was taken by boat onto Lake Zurich. With his hands bound and pulled behind his knees, he was thrown into the lake and drowned. Most early Anabaptists were unwilling to fight for their lives because of their pacifist views.

In the early sixteenth century, another founder of the Swiss Brethren branch of Anabaptism, Conrad Grebel (1498–1526), wrote "True Christian believers are sheep among wolves, sheep for the slaughter . . . Neither do they use worldly sword or war, since all killing has ceased with them."[23] Grebel died at an early age of the plague in 1526, which probably saved him from a fate similar to that of Manz.

George Blaurock (1491–1529) was the third founder of the Swiss Brethren to die. He worked closely with Felix Manz until 1527, when Manz was executed. On the same day that Manz was executed, Blaurock was beaten and expelled from Zurich. He moved to Tyrol, Austria, where he built a successful Anabaptist ministry, but eventually authorities there arrested him for his heretical Anabaptist evangelizing. He was convicted of heresy and burned at the stake in 1529.[24]

Another early leader of Anabaptism was Hans Hut (1490–1527), a German bookbinder and fiery evangelist.[22] He was rebaptized in 1526, the same year that the town of Zurich introduced the death penalty for Anabaptists, and he became a zealous missionary in Southern Germany and Austria.[24] Hut was a pacifist who emphasized the communal sharing of material goods among the Anabaptist community. Even the raising of children was done communally. Hut was convinced that the Second Coming of Christ and the apocalypse would occur in 1528.[24] He would not live long enough to see that he was wrong. In 1527, he was arrested, accused of heresy for his Anabaptist beliefs. He was severely tortured and died later in his cell by asphyxiation before he could be convicted and executed. Not to be outdone by fate, the trial authorities wheeled his dead body into the judgment hall, finished their trial, and found him guilty of heresy. They sentenced his dead body to "death" by fire, and burned it to ashes.[22] By 1530, at least one thousand Anabaptists are believed to have been executed for heresy.[24]

In the midst of severe persecution, Anabaptism was further formalized with the seven articles of the *Schleitheim Confession of Faith* (the SCF).[25] This document represented the basic tenets of what we think of as Anabaptism today. It was approved at a meeting of early Swiss Anabaptists in 1527 led by Michael Sattler, an ex–Roman Catholic monk.[22] Response from other branches of Western Christianity to the formal institutionalization of Anabaptism was both swift and brutal. Roman Catholics, Lutherans, and Zwinglicalvinists all demonized Anabaptists as arch-heretics. A few months after the SCF meeting, Sattler was arrested. He was charged with various "heretical" acts that included rejection of infant baptism, dishonoring the saints, and marrying (as a supposed celibate monk). He was tried by a group of Roman Catholic theologians and judges. They convicted him and sentenced him to death. Sattler was brutally tortured, his tongue was cut out, and he was burnt alive. His wife and thirteen of his associates from the SCF meeting were also executed, either by beheading or by drowning. The violent response to Anabaptism seems all the more brutal because Anabaptism was specified in its SCF document as a strictly pacifistic religion. The sixth article of the SCF forbids all forms of violence by its members.[25]

The first and most controversial article of Anabaptism's SCF was its official rejection of infant baptism. The article specifies that baptism is reserved only for one who "desires to walk in the resurrection of Jesus Christ."[26] Infants, therefore, were not to be baptized since they have no conscious knowledge of Christ. Anabaptists believe that infant baptism was never specified in the Bible, and that baptism is only for those who were old enough to understand the significance of the ritual. (After all, even Jesus was an adult when he was baptized.) They therefore rejected infant baptism as illegitimate. This departure from the way baptismal rituals had evolved elsewhere within Western Christianity was what brought such a violent response against the Anabaptists. The Anabaptist "rebaptism" of adults was interpreted as a form of heresy punishable by execution. Anabaptists continued to be severely persecuted by Lutherans, Zwinglicalvinists (particularly Zwinglians), and Roman Catholics during the sixteenth and seventeenth centuries. During that time, thousands of them were put to death.[5,6]

The most common manner to execute Anabaptists, when they were not burnt at the stake, was by drowning. This was called the "third baptism" of Anabaptists by their persecutors. According to Holy Roman Emperor King Ferdinand I, it was "the best antidote for Anabaptism." Ferdinand was a staunch Roman Catholic who was against all forms of Protestantism, but he had a special hatred for the Anabaptists. He organized groups of Anabaptist hunters called *Tauferjager*, and he appointed special judges whose job it was to convict Anabaptists of heresy. The punishment was always the same: death.[22] Even those who recanted their Anabaptist beliefs under torture were killed, although instead of being burnt alive or drowned, they were granted the more merciful form of execution: beheading.

Although early Anabaptism struggled against fierce persecution, it managed to survive. By the sixteenth and seventeenth centuries, the basic religious branch had given rise to a number of subbranches. Some of the best known included the *Hutterites*, *Mennonism*, *General Baptism*, and the *Amish*.

HUTTERITES AND HUTTER THE HATMAKER

Hutterism began in the Tyrol region of Austria. One of the Swiss Brethren founders, George Blaurock, brought Anabaptist ideas to Tyrol in 1527. He heavily influenced a hatmaker named Jacob Hutter (1500–1536). Hutter (not to be confused with the Anabaptist Hans Hut, mentioned earlier) carried on as pastor in Tyrol after Blaurock was burnt at the stake in 1529.[24] Hutter developed his own evangelistic branch of Anabaptism between 1530 and 1535, which eventually was named after him. The Hutterites are extreme pacifists, and they forbid members from taking part in military activities, police activities, or even paying taxes for a war.[24] They developed an extreme form of communal society-building on earlier Anabaptist ideals of pacifism and socialism, stressing denial of private property. Like other Anabaptists, they rejected infant baptism.

Although Hutterites and other Anabaptists of the sixteenth to eighteenth centuries had radical social beliefs for their time, including nonviolence against competing religions, and separation of church and state, they also retained certain traditional views of the time about women.[27] They believed that God dictated general male superiority and female submission. One of the early key figures of the Hutterites and prolific writer of Hutterite doctrine was Peter Riedemann (1506–1556). He is considered to be the second founder of the Hutterites. While in prison, Riedemann wrote the great *Rechenschaft* (*Confessions of Faith*) summarizing the basic beliefs and practices of the Hutterites, first published in 1540. Regarding women he wrote, "We say, first, that since woman was taken from man, and not man from woman, man hath lordship but woman weakness, humility and submission, therefore she should be under yoke of man and obedient to him, even as the woman was commanded by God when he said to her 'The man shall be thy lord.' "[27] Such sentiment was particularly common among early branches of Abrahamic Monotheism.

The Hutterites were persecuted by Roman Catholics, Lutherans, and Zwinglicalvinists mainly because of their Anabaptist practices, such as the rejection of infant baptism. By 1535, influence from Roman Catholicism's King Ferdinand I, the Holy Roman Emperor, had reached Tyrol. The Hutterites went into hiding for their lives.[24] A bounty was placed on Hutter's head by Roman Catholic authorities. On November 29, 1535, he and his wife were captured. He was severely tortured to give up the names of other Hutterites, but he refused to reveal them. Two months later, his captors brutally executed him in public. First, they put him in a bath of freezing water and then they moved him to a sweltering room where they beat him severely. Later, they poured brandy over his lacerated body and set the alcohol aflame in the town square of Innsbruck, Austria.[24] His wife somehow managed to escape, but she was recaptured two years later and executed by drowning.

During the late sixteenth and early seventeenth centuries, the Hutterites went through successive periods of growth and persecution. Then in 1621, bubonic plague killed one out of every three remaining Hutterites. They almost became extinct by the early eighteenth century, down to only a few hundred individuals. They eventually evolved into two distinct branches that migrated to North America in 1874.[24] These were the *Schmiedeleut* and the *Dariusleut* branches. Later, in 1877 South Dakota, another branch emerged known as the *Lehrerleut*. By the mid-1990s, the population of Hutterites recovered to around thirty-five thousand individuals living mostly in western Canada and the Upper Great Plains of the United States.[24] They still avoid military service and taxes, but because Hutterite colony members who work within their closed community do not receive a salary, they are not subject to state or federal income taxes. They also do not pay Social Security taxes or collect Social Security benefits.

In a "better late than never" move, representatives from the Catholic and Protestant churches met with Hutterite representatives to work toward reconciliation in 2007.[28] They also held a small ceremony to commemorate the meeting in the town of Innsbruck, where Hutter had been executed 470 years earlier.

MENNONISM AND ITS FOUNDER MENNO ON THE RUN

Mennonites were another group that evolved from basic Anabaptism. Their founder was Menno Simons (1496–1561). Menno was a former Roman Catholic priest who became troubled by questions surrounding the Roman Catholic doctrine of Transubstantiation (the belief that the bread and wine of mass literally turned into the actual body and blood of Christ). His search of scriptures also left him doubting that infant baptism was prescribed anywhere in the Bible.[22] In 1536, he finally rejected the Roman Catholic Church and his clerical appointment, and he cast his lot with the Anabaptists. A year after becoming an Anabaptist, he became one of its leaders. By 1544, the term Mennonite (after Menno) became the name used for Dutch Anabaptists. Menno was excommunicated and frequently charged with heresy by the Roman Catholic Church, but he always managed to avoid capture and execution. Menno was one of the few early Anabaptist leaders who managed to survive persecution long enough to die a natural death.

The sixteenth-century Mennonites were mostly farmers who adhered to the *Schleitheim Articles of Faith* that came with Swiss Anabaptism. Persecution of Swiss Mennonites continued, although by the late sixteenth century, the death penalty was no being longer applied. Instead, other severe measures were imposed, such as imprisonment or being taken to sea as galley slaves. Swiss Mennonites were not allowed to be buried in community cemeteries. Most Mennonites were branded on their heads as undesirables and shunned from society. Because of their segregation, few were able to get a formal education.

Mennonism gave rise to the majority of later Anabaptist branches, most of which lived socially segregated from the rest of society. Over time, Mennonites were well known for their commitment to pacifism. Two of the better-known Mennonite subgroups were the *General Baptists* and the *Amish*.

AMISH: THE SWISS BRETHREN RETURN TO SIMPLICITY IN DAY-TO-DAY LIVING

In the late sixteenth century, a schism among the Swiss Mennonites (most of who preferred to continue using the old name "Swiss Brethren") resulted in the rise of the *Amish*. Jacob Ammann (1640–between 1712 and 1730) was the Swiss Mennonite who founded the new religious branch,[22] although he denied that he was trying to start a new branch of religion. He turned Anabaptist social segregation from being a hardship into being a fundamental core of a new religious belief system. His opponents coined the name "Amish" after Ammann, and it was originally used as a derogatory term. Ammann was born in 1644 and disappeared from historical records after 1712. He is thought to have died sometime between 1712 and 1730. Ammann had little or no formal education and could not even write his full name. He signed his name with a "mark" (sometimes initials) and had other people read and write correspondence for him. He promoted a simple life for his followers as separate from the non-Amish world. The Amish developed as a closed society. He opposed men shaving their beards. He also shunned wearing "prideful" clothing and associating or dining with individuals who had been excommunicated from the congregation. The Amish doctrine continued to revolve around the Anabaptist *Schleitheim Confession of Faith*. Ammann also considered lying to be grounds for excommunication.

Even today, the Amish are known to be a closed society that emphasizes communal cooperation. They live simply, and many dress plainly and shun certain technology, such as power-line electricity, telephones, and automobiles. They shun labor-saving technologies for fear that it might make members less dependent on the community. They also shun photographs because they might cultivate vanity. They are mostly rural people who value manual labor, humility, and honesty. They do not buy commercial insurance or participate in the federal Social Security program. As with many other Anabaptists, they practice nonresistance and will not perform any type of military service, law enforcement activity, or political action. They receive adult baptism and believe in having large families. The first Amish came to North America around 1720.[22] Over time, the Amish evolved into at least forty subbranches based on minor doctrinal disputes.[29] Some authors cite a schism in the nineteenth century that divided the Amish into "Old Order Amish" and a more progressive "New Order Amish." The progressive branches modified the strict dress code, allowed the use of certain technologies forbidden by the Old Order, and changed various worship rituals.[22] Another major group with roots in Mennonism is *General Baptism.*

GENERAL BAPTISM: MENNONITE BAPTISM WITH ARMINIAN INFLUENCE

The group commonly known as *General Baptism* has different roots than the Calvinist-grounded *Particular Baptists* discussed earlier.[30] The General Baptism branch of Radical Reformed Protestantism developed under heavy influence from Anabaptism, even before the Amish. It was later influenced by the antipredestination elements of Arminianism. As opposed to the Calvinism's Particular Baptists and Presbyterians who believe that Christ died only for the predestined elect, the General Baptists were a Mennonite subgroup that followed the Arminian view that Christ died for everybody. Only those with faith in Christ would be saved, but his death was for the sake of all people. General Baptism (as defined here) was cofounded by John Smyth and Thomas Helwys in Amsterdam around 1609, with John Smyth as its pastor.

John Smyth (1570–1612) was originally ordained as an Anglican priest in England in 1594. He grew frustrated with the Anglican Church and much of the Western Christian Church in general. By 1606, he had renounced Anglicanism, becoming a Puritan and an English Separatist. Still not satisfied, and fleeing persecution in England, he led a group of forty followers to Amsterdam in 1608 to study scripture.[22] There, he was heavily influenced by the Mennonites, and he came to reject infant baptism based on his interpretations of scripture that infants who died without being baptized would not be damned. He instead embraced adult baptism (believer's baptism) and was convinced that for baptism to be valid, it had to follow a mature confession of faith. He first baptized himself, and then others, to begin the General Baptist movement. At first, General Baptists were baptized with water poured or sprinkled over them. Eventually the practice evolved into one of total immersion (immersing the entire body under water). The name *Baptist* was a derogatory term used by Calvinists who thought the practice of total immersion was strange. They were later identified as *General Baptists* to avoid being confused with the Calvinist *Particular Baptists.*

Another characteristic of Smyth that later General Baptists followed closely was his refusal to use a translation of the Bible for worship. Instead, he translated spontaneously from biblical languages. After Smyth left the group, cofounder Thomas Helwys (1575–1616) took over leadership of the General Baptists. He went back to England and started its first congregation there. He wrote a treatise on religious freedom and its independence from the monarchy that he unwisely sent to King James I. The king imprisoned him for endorsing the separation of church and state, and Helwys died in his prison cell in 1616.

General Baptist beliefs are wide ranging because each church congregation is autonomous. They believe that for something to become a General Baptist principle, it must be explicitly stated in the Bible. However, different interpretations of biblical verses have often been the source of schism and diversification. General Baptism appears to have given rise to a number of major hyperapocalyptic lineages in England (*English General Baptism*) and in the United States (*Millerism*).

MILLERISM: PREPARING FOR THE APOCALYPSE

Millerism arose in North America out of the evangelistic preaching of Baptist William Miller (1782–1849). Although apocalyptic belief systems were nothing new, dating back at least as far as the 12th century BCE with Zoroastrianism (chapter 6), the hyperdevelopment of end-of-times eschatology is often associated with Miller. In 1833, he asserted that he knew the date of the coming apocalypse and the Second Coming of Christ. This was a time when the earth would be destroyed and the elect would be raptured into heaven. Miller claimed to have deciphered the date of this event from coded information in the King James Bible.[6] At first, he predicted that the end would come sometime between March 21, 1843, and March 21, 1844. He built a huge following, delivering hundreds of lecturers to groups as large as 1,800 people at a time.[31] He had a cadre of assisting preachers also spreading his message.

When March 1844 passed by without the world ending, Miller was not discouraged. He revised the date to October 22, 1844. The eager anticipation of Miller's followers returned. Farmers abandoned their crops, stores had clearance sales to get ready, and people gave away their possessions. Millerites were eagerly waiting to be uplifted to heaven, and personal possessions would be unnecessary if the world was about to end. Newspapers told of how some people were getting ready, like one woman who strapped herself to a trunk containing her entire wardrobe, so not to leave without her best clothes. Another woman was reported to have jumped out of her third-floor apartment window, expecting to fly to heaven as part of the rapture.[31] As the October date passed without the end of times, expectations and morale of the Millerites began to plummet. This period became known as the *Great Disappointment*.

The Millerites not only had their expectations shattered when the apocalypse failed to arrive but also had to deal with public criticism and sometimes even violent retribution. Several of their churches were burned or vandalized. People that had given up all of their possessions in anticipation of the rapture now had to rebuild and put their lives back together. After the Great Disappointment, some abandoned the faith altogether, whereas others followed new subgroups evolving from Millerism with the belief that the exact timing of the apocalypse had simply been miscalculated. Some of the most successful subbranches to emerge from Millerism include the *Seventh-Day Adventists* and the *Jehovah's Witnesses.*

SEVENTH-DAY ADVENTISM

Seventh-Day Adventism (also referred to as the *SDA*) began in 1863 as one of the groups of disenchanted Millerites seeking new meaning after the Great Disappointment (failure of the apocalypse to come in 1844). It was founded by Joseph Bates (1792–1872) and James White (1821–1881) and his wife Ellen White (1827–1915). Ellen White claimed to have received more than two thousand visions, many directly from God. Although she did not specifically refer to herself as a prophet, early Adventists viewed her visions as a product of biblical prophecy. She was the most influential founder of the SDA. She was extremely prolific, publishing twenty-six books and thousands of journal articles.[22] Her books have been translated into more than 160 languages. Today, the SDA has more than twenty million adult followers.[32]

Seventh-Day Adventism differs from most other branches of Christianity in its observance of the Hebrew *seventh-day Sabbath* (Saturday) instead of the traditional Christian *first-day Sabbath* (Sunday) as the holy day for worship and abstinence from secular work.[33] They believe that the shift to Sunday by early Traditional Christians from the earlier Hebrew tradition was due to pagan influence from polytheistic sun god worshippers (e.g., the god Sol Invictus discussed in chapters 7 and 8). Seventh-Day Adventists avoid non-Kosher meats, alcohol, tobacco, caffeine, body piercing, and tattoos. They also discourage dancing and listening to rock music. The SDA followers do not believe in hell, but instead believe that after death the wicked simply cease to exist forever. They also do not believe in the soul, but instead believe that those judged worthy will be resurrected in mind, body, and spirit. Until then, the dead will have no consciousness. Seventh-Day Adventists, like other Millerite derivatives, still believe that the Second Coming (Advent) of Jesus Christ is imminent. They believe that it will include the *apocalypse* (the *end of times*) followed by the *rapture* (when all the chosen will ascend to salvation). After the number of false starts in Millerism, however, most SDA leaders have pulled back on predicting the exact day of the event.

Like all major branches of Organized Religion, the SDA went through numerous schisms and developmental changes. This evolution was sometimes the result of intense competition for leadership within a branch. Other times, it was the result of the apocalypse not materializing on a date predicted by one branch founder or another. Such events were the drivers of schism and branch diversification over time. Some extreme sects that split from the SDA were the result of the social isolation of small communities led by charismatic zealots. One such sect included the Davidians, which later largely transferred into the *Branch Davidians*. They and their sympathizers came under the influence of another self-professed prophet who renamed himself David Koresh. A few years later, he brought his followers into violent conflict with federal authorities in the United States, resulting in monumental tragedy. Charismatic, self-proclaimed messiahs can be a powerfully influential force on segments human society.

THE *DAVIDIANS*, THE *BRANCH DAVIDIANS*, AND THE *KORESHIANS*

The *Davidians* were another apocalyptic branch of the SDA. Although elements of it had first begun to develop in Los Angeles in 1929, it was formally founded in 1935 by Victor Houteff (1885-1955).[34] He established it in Texas within the confines of a large remote compound of several hundred acres, where many of its followers lived. As the outskirts of the nearby town of Waco expanded toward the compound, it moved to a seventy-eight-acre plot farther from town to maintain its isolation from outside society. The Houteffs named the compound Mount Carmel Center, after the biblical Mount Carmel in northern Israel. After Victor Houteff's death in 1955, his wife Florence took charge of the branch. She not only took charge but also claimed to have had a vision that the apocalypse would occur on April 22, 1959.[34] After that did not occur, the Davidians, which had grown to more than nine hundred members at its peak, began to dissolve as a group. Some were absorbed by the newly developing *Branch Davidians*.

The *Branch Davidian SDA* (*Branch Davidians* or *The Branch* for short) was a group that had originally been founded in 1955 by Benjamin Roden (1902–1978), a few years before the Davidian apocalypse had failed to materialize.[34] Many Davidians, who had become disenchanted with Florence Houteff after her failed prophesy, began to see Benjamin Roden as their new prophet. In most respects, the Branch Davidians were the "new Davidians." Roden even laid claim to the Mount Carmel compound as the home of the Branch Davidians.

Benjamin Roden was a converted Jew who claimed to be King David's successor. He and his wife Lois Roden (1916–1959) governed the Mount Carmel compound as a tightly knit apocalyptic group (as their predecessors the Davidians had done before them). They continued to keep the residents of the compound socially isolated from general society. In 1977, Lois claimed to have had a vision of the Holy

Spirit claiming to be a female entity. Lois became the coprophet of the Branch Davidians until her husband's death in 1978, after which she became its sole leader. In 1983, her leadership was challenged by a man named Vernon Howell, who later renamed himself as David Koresh (1959–1993).

Vernon Howell was born to a fourteen-year-old single mother, and he never met his father.[35] He had a difficult childhood that included periods of physical and sexual abuse. His mother abandoned him to his grandparents at the age of four, although he was reunited with her several years later. He had various learning disabilities and dropped out of high school in his junior year. Howell eventually joined his mother's church, the SDA, and became deeply in interested in their ideas about the coming apocalypse. He believed that the SDA church had fallen away from its original mission. He also questioned why the SDA had no current prophet. Like other Davidian branches that questioned the direction of the church, Howell was rejected from mainstream SDA.

Howell moved to Waco, Texas, in 1981 and joined the Branch Davidians, where he eventually had an affair with Lois (he was in his twenties and she in her late sixties at the time). He claimed that God had chosen him to father a child with her who would be "The Chosen One." The child never materialized. Then, in 1983, he claimed the gift of prophesy and vied for leadership of the Branch Davidians. A schism eventually developed within the group, with some following Howell as the leader, and others following Lois Roden's son George, who was also fighting for leadership of the cult. George Roden's group managed to drive Howell's group from Mount Carmel for a period.

After Lois Roden died in 1986, Howell again moved to claim leadership of the Branch Davidians and Mount Carmel. In 1987, he drove George Roden and Roden's followers from the Mount Carmel compound in a fierce gun battle, after which he moved his followers back to the compound. He claimed that the world had already entered the period of tribulation preceding Armageddon, and that the apocalypse would one day begin at Mount Carmel.

In 1990, Howell legally renamed himself David Koresh and proclaimed himself to be a messiah and the earth's final prophet before the end of times.[35,36] He chose the first name David because he believed he was a descendant of the great Israelite King David. He chose the last name Koresh after the biblical name for Cyrus the Great (600–530 BCE) who freed the Jews from Babylonian captivity. Shortly after changing his name, his followers began to use his name *Koreshians* for their sect to distinguish themselves from the rest of the Branch Davidians.

The details of this complex story leading to what became known as the "Waco siege" are widely available in books, reports, and news archives.[35,36,37,38] I provide this example only to show how the persuasive religious charisma of an individual can sometimes move large numbers of people, even in the modern-day United States. It also exemplifies the potential danger of combining religious tribalism with social isolation. Koresh's charismatic approach charmed the hundreds of followers at the Waco compound, which he eventually renamed "Ranch Apocalypse." Koresh not only believed he was a messiah, but he also believed he was messianic husband to all of the female members of the Koreshian community.[35] He saw his mission as fathering a new community of divine offspring and through them, a "House of David" army.

The Koreshians remained largely isolated from society at the Mount Carmel compound. The women of the compound birthed at least a dozen children for which he was the father. Some of his many wives were as young as twelve years old. In preparation for his prophesized end of times, he stockpiled large amounts of arms and explosives for what he saw as the coming holy battle. By the early 1990s, those who had left the cult were worried about what they began to see happening at Mount Carmel. One defector named Marc Breault began a campaign to expose Koresh as a false prophet.[35] He portrayed Koresh as a religious megalomaniac, polygamist, and child molester. In 1992, he suggested to Texas authorities that the Koreshians were planning to commit mass-suicide similar to the one that had happened fourteen years previously at Jonestown with Jim Jones and his *People's Temple* (the Jonestown massacre described in **Note 9D**). His campaign was ultimately successful in engaging the authorities, as well as a great deal of media attention.

In 1993, federal and state agencies tried to serve a warrant at the ranch compound for alleged sexual abuse and illegal firearms violations. As a result, an intense firefight and standoff broke out that went on for two months with much media attention. Eventually, the FBI launched a tear gas assault using more than two thousand tear gas canisters. Fire engulfed the compound, ending in the deaths of eighty Koreshians and four federal agents.[37] In addition to Koresh, the Koreshians killed during the Mount Carmel siege included fifty-eight adults, and twenty-one children between the ages of one and fifteen. For many people, this was indeed an apocalypse. The government claimed that the Koreshians fired first to start the firefight and that they set the fires in the compound themselves. The few surviving Koreshians claimed that the government had fired first to start the firefight and that the government's tear gas assault started the fires. That controversy was never conclusively settled. Some people believe that the FBI exploited the media throughout the episode.[35] To make matters worse, the tragedy did not end with Waco. Two years after that deadly clash, Timothy McVeigh, motivated by revenge against the federal government for the tragedy at Mount Carmel, blew up the federal building in Oklahoma City. The attack killed 168 people and wounded 680, making it the deadliest act of domestic terrorism in U.S. history.

David Koresh remains one of several, well-documented examples of self-proclaimed prophets in modern time, as well as the potential of charismatic proselytism to go in a bad direction. Others examples include Jim Jones and the People's Temple sect (**Note 9D**), and Marshall Applewhite and the Heaven's Gate sect (**Note 9F**) Controversy still exists over how to historically frame the tragedy at Waco with the Koreshians. Some see it as a justified intervention to try and save children and avert a mass suicide. Others see it as the government's single most significant breach of American religious freedom in the twentieth century.[35] Whatever the truth is, all agree that it was a monumental tragedy. Today, a few Koreshians still survive, as well as a new subgroup of Branch Davidians called *The Branch, The Lord Our Righteousness*.[39] Some of these adherents are waiting for David Koresh to be resurrected.

Understanding where to draw the line between religious tolerance and predatory behavior in the name of religion in a globally acceptable way is a problem that human society has yet to master. Fortunately, the history of Koreshianism was not at all typical for mainstream branches of the SDA, which today includes more than twenty-one million followers. Another offshoot of Millerism that evolved after Millerism's Great Disappointment was *The Watchtower Society.*

THE WATCHTOWER SOCIETY AND *JEHOVAH'S WITNESSES*

Another major religious belief system to branch from Millerism after its Great Disappointment (no-show apocalypse) was *The Watchtower Society*, later known as *Jehovah's Witnesses*.[6] Like Millerism, its original founder attempted to predict the exact time of the apocalypse, and when it did not appear at the predicted time, followers were greatly disappointed. This branch began with Charles Taze Russell (1852–1916), who published a monthly religious journal called *Zion's Watchtower and Herald of Christ's Presence*. His group, founded in 1884, was called the *Zion's Watchtower Tract Society* (or simply *The Watchtower Society*).[40]

The Watchtower Society was another apocalyptic group of Christians who believed that destruction of the present world was imminent.[4] They believed that the Second Coming of Christ was the only solution for all of the problems faced by humanity. They did not believe in the divinity of Jesus and adopted the anti-trinitarian views of Arianism. They believed that only God (Jehovah, the *Universal Sovereign*) should be worshipped. Jesus was considered to be an intermediary between God and humanity. He was appointed by God as the surrogate king and judge of God's kingdom, but Jesus was not believed to be divine. Therefore, The

Watchtower Society adherents did not celebrate Christmas, Easter, or many other religious holidays, which they believed to be pagan rituals. They did not condone religious symbols, nor did they believe in hell.

Charles Russell also predicted the date of the apocalypse, which he was convinced would occur in 1914; to his surprise (and disappointment), that did not happen. He later emended his prediction to say that Armageddon *began* in 1914, with the beginning of World War I, and its exact duration could not be predicted. He also stated that because Armageddon had begun, "millions now living will never die."[40] He later made additional predictions that never came to fruition. Subsequently, large numbers of people defected from the religion, and later church officials moved away from predicting exact dates for eschatological (end of times) events.

Russell died in 1916, and in 1917, Joseph Rutherford (1869–1942) was elected as the new leader. The Watchtower Society strongly criticized Christian involvement in World War I, thinking it was the start of Armageddon. As a result, its directors were jailed for sedition under the Espionage Act of 1918 and other members were subjected to mob violence.[40] In 1931, Rutherford changed the name of the group he led to *Jehovah's Witnesses* to distinguish it from other branches that also had diverged from the mainstream *Watchtower Society.*

The Jehovah's Witnesses are best known for their door-to-door preaching; distribution of religious pamphlets, such as *The Watchtower* and *Awake!*; and various other evangelistic activities. Other than proselytizing, they limit their contact with people from outside their faith, because they believe that they are the chosen people and that the rest of the world is controlled by Satan.[40] Jehovah's Witness members refuse military service or loyalty to the military objectives of any country.[4] They do not vote and do not salute any national flag. They avoid nontheocratic social activity and generally avoid universities and institutions of advanced education. They are also famous for refusing blood transfusions for themselves or family members, even to save a life. They do this because of their interpretation of biblical scripture forbidding the ingestion of blood. They generally are not concerned that refusing a transfusion may kill them or their relatives. They believe that the end of the world is coming soon anyway and that dying now will merely ensure their entry to the next world after resurrection.[40] The Jehovah's Witnesses have more than eight million followers worldwide.

As discussed in this chapter so far, the subbranches of *Early Protestantism* are numerous and diverse, making it the largest of the two major branches on my phylogenetic tree of Reformation Christianity (page 366). The other major branch on that tree is *Anglicanism,* also called *Anglican Protestantism.*

ANGLICANISM: THE *CHURCH OF ENGLAND* AND THE KING'S DIVORCE FROM ROMAN CATHOLICISM

After *Early Protestantism* had evolved in Germany and other parts of mainland Europe as a protest movement against Roman Catholicism, *Anglicanism* (*Anglican Protestantism*) evolved in England as an attempted compromise between Early Protestantism and the Roman Catholicism.[3,4,5,41] The mother church of Anglicanism was the *Church of England*.[4] Unlike the Early Protestant Reformation in Europe, the early Anglican Reformation of England was first focused on the needs of the king, and only secondarily on those of the lay piety.[5] The growth and diversification of Anglicanism eventually extended well beyond the British Isles, particularly with the colonization of the Americas. In the United States after the Revolutionary War, Anglicanism abandoned allegiance to the king of England and continued to evolve. Today, the many known derivatives of Anglicanism are in need of much further study to better understand their detailed diversity and complex historical interrelationships. I review Anglicanism only within the broadest context.

Anglicanism split from Roman Catholicism in the early sixteenth century under King Henry VIII (1491–1547). Centuries before, the English branch of Roman Catholicism had already declared its authority regarding religious matters within the English border with clause 1 of the Magna Carta, in 1215. The original intent of the clause (which referred to the English Church as *Anglicana ecclesia*) was to prevent the king from interfering with church matters. However, the attraction between Organized Religion and monarchical power is inevitably a strong one. Over the next few centuries, magisterial influence grew to have fundamental influence over the English Church, as the church moved closer to the monarchy. In 1521, Pope Leo X conferred the title "Defender of the Faith" to Henry VIII for his treatise defending the papacy and attacking Martin Luther. A little more than a decade later, the cordial relationship between the king and the pope abruptly soured.

A schism splitting the Anglican Church from Roman Catholicism came in 1534 with King Henry VIII's *Act of Supremacy*. This act specified the king and successive monarchs as the Supreme Head of the Anglican Church (*Anglicanism*) and declared the pope was no longer the leader of England's Christians. King Henry's main motivation for this move was to get an annulment of his marriage to Catherine of Aragon so he could marry Anne Boleyn. The Roman pope at the time was not cooperating with Henry's wish for an annulment, so Henry simply gave himself the authority in the easiest way he knew how. He replaced the pope's authority with his own, and he ordered the archbishop of Canterbury (senior bishop for the Anglican Church) to grant the annulment. This allowed Henry to marry Anne (who was incidentally beheaded only a few years later for questionable charges of adultery and incest to make room for Henry's next wife).[6] King Henry's Act of Supremacy proclamation severed the ties between the Church of England and Rome. Following Henry's contrived annulment and grab for Anglican religious power, he was excommunicated by Pope Paul III in 1538.

Theologically, Anglicanism began as a confusing ideological mix of Roman Catholicism with Early Protestantism. A series of defining documents were published over time attempting to better codify the movement, beginning with the *Ten Articles* in 1536.[41] Reflecting the Catholic influence on Anglicanism, the Ten Articles supported infant baptism (as well as adult baptism), confession, penance, absolution, and trinitarianism. It also supported praying to saints and the Virgin Mary as well as the observance of most Catholic holy days, as long as "superstition" was avoided (lending a certain element of subjectivity regarding the term "superstition"). The observance of various Catholic rites and ceremonies, such as clerical vestments, sprinkling of holy water, and giving of ashes on Ash Wednesday, was also permitted. Reflecting the Early Protestant (anti-Catholic) influence on Anglicanism was the rejection of kneeling before iconic images, papal indulgences, or any recognition of papal authority. The Ten Articles went through a number of revisions over the years, culminating with the mildly Calvinistic *Thirty-Nine Articles* in 1571. The Thirty-Nine Articles were eventually incorporated into the *Book of Common Prayer*, the major prayer book of the Anglican Church.

One of Anglicanism's early strengths was that it produced several highly influential books in English that were increasingly available thanks to the introduction of the Gutenberg printing press. These included the first authorized edition of the Christian Bible in English published in 1539. It was called *The Great Bible*. It was authorized by King Henry to be read aloud at English church services. Another important book was the previously mentioned *Book of Common Prayer* first printed in 1549. This was the first prayer book published in vernacular English, and versions of it are still used today. Probably the best-known book of early Anglicanism is the King James Bible, first published in 1611. Named for its sponsor, King James I of England, it became tremendously influential on the still-emerging English language.[4] The King James Bible standardized English phraseology and sentence structure and substantially increased the literacy rate of the English-speaking world (doing for it what Luther's German translation of the Bible had done for the German-speaking world). In addition, more than four centuries after its first printing, it remains the most famous Bible translation in history and one of the most printed books ever made.

Mary Tudor, daughter of King Henry VIII and Anne Boleyn, became queen in 1553. She tried to reverse the English Reformation and restore Roman Catholicism to England. The effort entailed a violent campaign that earned her the nickname of "Bloody Mary" by her Protestant opponents. During her five-year reign, she had

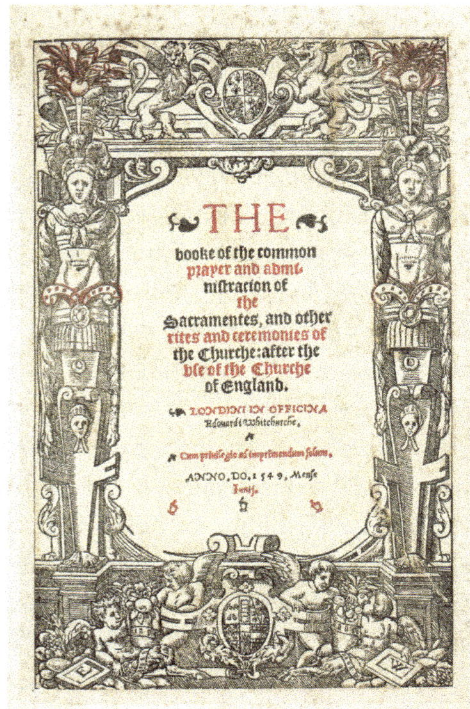

THE
booke of the common
prayer and admi-
nistracion of
the
Sacramentes, and other
rites and ceremonies of
the Churche: after the
vse of the Churche
of England.

LONDINI IN OFFICINA
Edouardi Whitchurche.
Cum privilegio ad imprimendum solum.
ANNO. DO. 1549. Mense
Iunij.

The development of Anglicanism and the Church of England. (*Left*) King Henry VIII, who founded the Church of England (early Anglicanism) in 1531 when he appointed himself as its first supreme head. Portrait by Hans Holbein, King's painter for Henry VIII. Currently in the collection of the Thyssen-Bornemisza Museum in Madrid, Spain. (*Right*) Cover page from the first edition of the *Book of Common Prayer*, published in 1549. This was the first prayer book published in vernacular English, and it was highly influential in the spread of Anglicanism.

more than 280 Protestants burned at the stake under Roman Catholic heresy laws. Those executed included some of Anglicanism's leading Reformers.[6] More than eight hundred Anglican Protestants chose exile instead, escaping with their lives.[42] However, the effort to revert the English Church to Roman Catholicism never stuck. Once Mary's reign ended with her death in 1558, the Church returned to its development as a branch of Anglicanism.

PURITANISM: THE RESURGENCE OF ANTI-CATHOLICISM WITHIN THE CHURCH OF ENGLAND

After Queen Mary's attempt to bring the Church of England back to Roman Catholicism failed, a movement called *Puritanism* arose within the Church of England to purify the English Church from its "Roman Catholic" influence under Queen Mary.

Early Puritanism included many people who had returned to England after exile under the Roman Catholic queen.[5] By the seventeenth century, the Puritans and the Church of England grew closer to each other, even using each other's churches.[5] However, some ideological differences remained. Puritanism was influenced by the Calvinist views on *predestination* (believing that they were God's elect, predestined for salvation before they were even born), which brought them into conflict with the rest of the English Church. Its urgent apocalyptic spirituality "inspired anxiety and a harsh intolerance of those who were not among the elect."[57] Conflicting religious ideologies and traditions between Puritans and other branches of Anglicanism figured in as part of the English Civil War of 1642, which is sometimes referred to as the Puritan Revolution.[5,6]

Migrations of Puritans to the American colonies of New England (the Pilgrims) began during the early seventeenth century forming a sort of colonial theocracy. Harvard University, the oldest institution of higher learning in the United States, was founded in 1636 to accommodate the anticipated need for Puritan clergy.[43] By then, more than seventeen thousand Anglican Protestants had relocated to the area. Strong efforts to convert the Native American population from their native supernatural belief systems began early on in the Puritan era of the New World. In 1648, an order was issued by the Anglican colony stating that "no Indians shall at any time powaw, or preform outward worship to their false gods, or to the devil in any part of our jurisdiction."[44] Economic and political competition with Spain and France also contributed to the competition for souls. There was a drive to convert Native Americans to Anglican Protestantism before they could be converted to Roman Catholicism.[45]

On a dark note, the Puritans are associated with the Salem witch trials of 1692–1693 in Massachusetts. These trials resulted in nineteen people being executed for witchcraft by hanging, another five people convicted as witches dying in prison, and one person being pressed to death for not pleading. This proved to be a historic turning point after which colonial New England began to veer away from its theocratic nature, and Puritanism went into decline.[46] The Puritan era ended by the mid-eighteenth century.

Over time, many branches of Anglicanism came to reject the influence of Calvinistic predestination. One of the first of these anti-Calvinist branches was *Quakerism*.

QUAKERISM: THE QUAKING "SOCIETY OF FRIENDS"

Quakerism, formerly known as the *Society of Friends*, evolved in the mid-seventeenth century and today is represented by several hundred thousand members around the world. The founder of Quakerism was George Fox (1624–1691), who was born in the strongly Puritan village of Leicestershire, England.[4,47] He had no formal education but learned how to read and write. He became an apprentice for a shoemaker and pursued simplicity in life with a deep interest in religion. By 1647, he began preaching and attracted a small following. He maintained an opposition to iconic religious art as well as the use of music in church services. He taught that it was more important for someone to access his or her "Inner Light" of God rather than to depend on a central church structure. He rejected the Calvinistic Puritan idea of predestination. He was convinced that it was possible to have a direct experience of Christ through nothing more than personal study of the Christian Bible. Despite poor health, to spread his teachings, he traveled widely through England, Ireland, the West Indies, and eventually the American colonies.[5]

Quakers have no ordained clergy, and church services are conducted largely in silence without ritual, but any Quaker who feels so inclined, can stand up and preach.[4] Although their origins were closely related to Puritanism, they rejected the Puritan tendency to transform religious beliefs into authoritarian theology.[48] Quakers are extreme pacifists who prohibit all forms of violence against other human beings. They are also teetotalers (they avoid all alcoholic beverages) and plain dressers. Like the Anabaptists and various other groups of Abrahamic Monotheism, they refuse to swear oaths. Many are known for their use of "thee" (rather than "you") when referring to an ordinary person in normal conversation. Quakers rejected water baptism of any kind (adult or infant) in favor of inward transformation, or Inner Light, seeing water baptism as a meaningless, superstitious ritual. The name "Quaker" came from George Fox's belief that people should "quake at the Word of the Lord."

Quakerism expanded to Colonial America not long after it developed in England. In Massachusetts, Quakers were persecuted by the Church of England beginning in 1656. They were branded as heretics because of their preaching about the Inner Light. As a result, many of them, including George Fox, were imprisoned for blasphemy. Others were banished, deported, or executed with their books burned and property confiscated. In 1660, English Quaker Mary Dyer was hanged in Boston for repeatedly defying a Puritan law banning Quakers from the colony.[49] She was one of four Quakers who were executed in what became known as the Boston Martyrs. In 1661, King Charles II forbade the execution of any more Quakers in Massachusetts, and Quakers were integrated into the colonies. As the Society of Friends grew in number, internal schisms eventually resulted in the diversification of several subbranches not shown on my phylogenetic tree. Some of these subbranches are discussed next.[48]

The *Hicksites* formed a liberal branch founded by Elias Hicks (1748–1830). Hicks was a traveling Quaker minister from Long Island, New York, who objected to an increasingly centralized organization and growing oligarchic leadership within

Quakerism. He emphasized social reform and an unorthodox doctrine. He was less bible-centered than most Quakers. He did not believe in an external devil and considered obedience to an individual's Inner Light to be the sole rule of faith.

Quakerism also has a number of more conservative branches. Two of the most prominent ones are the *Gurneyites* and the *Wilburites*. The *Gurneyites* formed the more evangelical group of the two, and it was much more bible-centered than the Hicksites, following Joseph John Gurney (1788–1847). Gurney was a banker and an evangelical Quaker minister who emphasized scriptural authority and advocated working closely with other Christian groups. He moved away from the Hicksite focus on Inner Light, believing the Hicksites did not give proper due to the Bible. In contrast, the *Wilburites* were more traditional Quakers devoted to individual spiritual inspiration who followed John Wilbur (1774–1856). Wilbur was a Quaker minister from Rhode Island who thought that the Gurneyites had gone too far in their total rejection of the Inner Light preached by the Hicksites. He took a somewhat intermediate position, believing in the importance of the Inner Light as well as in the importance of the Christian Bible. He was disowned (effectively excommunicated) by the bulk of the Quakers, who were mostly Gurneyites. His followers eventually became known as the Wilburites.

One of the more unusual and least successful branches to develop out of Quakerism was *Shakerism.*

SHAKERISM: THE CELIBATE PACIFISTS

Shakerism evolved from Quakerism in the mid-eighteenth century. Shakers were originally known as the "Shaking Quakers" because of their trembling ritualistic dancing at services. Their frenetic spiritual expressions and thousands of songs are what distinguished them from the more subdued Quakers of the mid-eighteenth century.[4] Like the Quakers, they are extreme pacifists and lead a life of simplicity. At one time, they lived in communities primarily dependent on farming. "Mother" Ann Lee (1736–1784), who came from England to New York State, founded Shakerism. She claimed that she has received a revelation that she was the female counterpart of Christ.[5] The Shakers were never destined for success, in part because they preached complete celibacy. They believed that Adam's sin was sex; therefore, Shakers prohibited both sex and marriage. Children were added to the religion only through adoption, so adults could be added only through conversion from another religion. That alone would explain why Shakerism never had more than a few thousand adherents, even at their peak. They eventually all but disappeared. History of Religion scholar, Stephen Taysom, estimates there were only three remaining Shakers in the world by 2020.[50]

In addition to Quakerism, the Anglican Church also gave rise to *Wesleyanism* (as shown on the tree on page 366).

WESLEYANISM (*WESLEYAN ANGLO-METHODISM* AND *METHODISM*)

Wesleyanism (also known as *Wesleyan-Arminian* theology) evolved from early Anglicanism, but it became most successful in the American colonies. Its founder was John Wesley (1703–1791), an Oxford University instructor known to be a logical thinker. He was also an Anglican priest, like his father before him.[6] He started the "Holy Club" society at the university. During his lifetime, he witnessed the growth of his followers from four to more than one hundred thousand. He never formally separated from the Anglican Church.[5,6] For that reason, the movement before his death is termed *Wesleyan Anglo-Methodism* (the early orthodox form of Wesleyanism).

The formalized distinction between "Methodism" and the more orthodox branches of Welseyanism developed several years after Wesley's death. In 1795, the *Plan of Pacification* was signed at the Methodist Annual Conference, which allowed Methodist officials to appoint clergy who could administer sacraments in their own chapels. By officially allowing these preachers to administer sacraments separately from the Church of England, Wesley's Methodist belief system transformed into a genuinely dissenting denomination.[51] This further separated Methodism from the Church of England, and it is where I date formalized Methodism on my phylogenetic tree (page 366).

Wesley was deeply committed to the Church of England, but he was also influenced by Classical Arminian Christianity.[5] He traveled the world, from the American colony of Georgia to many regions across the British Isles to spread his theological message. He traveled hundreds of thousands of miles on horseback or by ship during his lifetime as a missionary. He famously said, "I look upon all the world as my parish."[52] Wesley rejected the Calvinistic and Puritan concepts of predestination and theological determination.[5,6] He preferred the Classical Arminian belief in the importance of free will, human effort, and faith to achieve salvation. He accepted the practice of baptism (adult or infant), although he did not clearly mandate it. He saw it as symbolic rather than as a way to cleanse away original sin. Wesley also oversaw the building of an infrastructure that included a headquarters in London and Bristol and, more important, included preaching houses in open-air locations across rural areas to serve a multitude of enthusiastic followers. He was evangelistic in his approach, and part of his mission was helping the poor and common people, creating orphanages, soup kitchens, hospitals, and schools. The Industrial Revolution, of the late eighteenth century, was causing swift, major shifts of population centers and resulted in many rural areas that lacked church facilities. A social gap between mainstream Anglicanism and the common people was widening, with priests set apart from parishioners by their status as "gentlemen."[53]

Wesley came up with a remedy for the resulting shortage of church buildings and accessible clergy in outlying regions throughout Europe. He began preaching to crowds in open-air locations across the continent as revivalist Jesuits did in Roman Catholic Europe. This remedy helped close the growing elitist gap between the clergy and the rural people. His colleague George Whitefield (1714–1770), who would later be credited as a cofounder of Methodism, joined him in these efforts. Excited crowds who were not used to such direct personal attention enthusiastically received his efforts. They were often gripped with mass emotion. They reportedly reacted wildly with laughter, weeping, and rolling on the ground. His influence continued to expand. By 1739, Wesley began personally approving local lay preachers to fill a shortage of clergy. The Anglican Church officials did not officially ordain the men he appointed, and Wesley was seen as disregarding church rules concerning who had the authority to preach. Consequently, he and his followers faced considerable hostility from mainstream Anglicanism and its hierarchy.[6] Nevertheless, Wesley continued to appoint lay preachers, which eventually became one of the key factors differentiating Methodism from the Church of England.

John Wesley endorsed the practice of singing hymns in his services and offered a large number of possible selections from which to choose. His prolific brother Charles produced around nine thousand different songs.[6] The noisy and disruptive behavior of Wesley's followers worried Anglican Church authorities. Even as the divide between Wesley's followers and the English Church widened, he still refused any official separation. In 1758, he published an article entitled *Reasons Against a Separation from the Church of England*. He chose to appreciate the traditions he had in common with the rest of Anglicanism, rather than dwell on the differences. It was not until four years after his death that his followers (the Wesleyanists), together with his colleague George Whitefield, formally split from mainstream Wesleyanism (Wesleyan Anglo-Methodism) as independent "Methodists." However, Methodism retained most of Wesleyanism at its core.

METHODISM: LEARNING TO AGREE TO DISAGREE

The founding of *Methodism* is often collectively credited to three people: John Wesley, his brother Charles, and George Whitefield. The earliest and most fundamental influence was from John Wesley, although as mentioned, he always resisted any separation from the Church of England. He also strongly disagreed with Whitefield's support of Calvinistic predestination.[5] Wesley's ideas about rejecting a belief in predestination prevailed in the end. Although Wesley and Whitefield often debated each other vigorously, Wesley retained his relationship with Whitefield. Wesley is credited with coining the phrase "agree to disagree" as a manner of holding his own convictions while remaining in contact with those with whom he disagreed (a glimmer of increased religious tolerance emerging in Western Christianity).

Like *Wesleyan Anglo-Methodism*, early formalized *Methodism* was an evangelistic product of the eighteenth-century revivalist movement that brought preachers from the large towns to the small villages. The Methodist camp meetings and rural preaching events were often like festivals, fueled with music, alcohol, and loose constraints.[53] They served as effective recruiting tools. Popular Methodist preachers were believed to have powers to cure the sick and to ward off demons. They preached about God's grace in the common language of the people, and some even spoke of supernatural entities like fairies and witches in daily life.

Methodism took off in both Europe and the newly formed United States. Early Methodism emphasized connections with the less fortunate more than many other branches of Traditional Christianity. It was one of the first to make inroads with Americans who were enslaved. By 1800, about a fifth of all American Methodists were enslaved people.[6] Because of the strong segregation that existed, even with free men, African Americans created their own branches of Methodism. One was the *African Methodist Episcopal Church* (which grew to 2.5 million members by 2006), and another was the *African Methodist Episcopal Zion Church* (which grew to 1.4 million members by 2006).[5] The *Methodist Episcopal Church* (the largest and oldest Methodist Church in the United States) divided in 1844 over the issue of slavery. Much of the northern and southern branches attempted to reunite in 1939 to form the *United Methodist Church*. By 2006, Methodism had sixty million followers in one hundred countries,[5] with more than one hundred subbranches, many of which were the result of minor doctrinal schisms and syncretizations. Many of the subbranches are currently classified according to region rather than their historical relationships. The detailed diversity and historical diversification of Methodism needs broad phylogenetic study before it can be expressed in a tree or more complex network form. One of the better-known branches that evolved from Methodism in the late nineteenth century is the *Salvation Army*, whose presence is seen across the United States today.

SALVATION ARMY: A MILITARIZED CHURCH OF PRAGMATISM

The *Salvation Army* is a familiar name to most people in the United States because of their many thrift stores and charity shops around the country. These shops sell donated goods at a low cost to the public and use the profits to fund shelters for the homeless, rehabilitation facilities for addicts, and humanitarian relief efforts. The denomination evolved out of Methodism. It was cofounded by William Booth (1829–1912) and his wife Catherine (1829–1890) in London, England. It evolved from a pragmatic evangelistic welfare effort of Wesleyan revivalism into a distinct evangelistic church. However, its development as a formal religion did not fully coalesce until the late twentieth century.[54]

In 1849, William Booth was a pawnbroker and Wesleyan Methodist lay preacher in London's East End.[55] He became convinced that his denomination was in need of reform and converted to the *Methodist Reform Church* in 1851. After that denomination would not support the time that he needed to accomplish his evangelistic goals, he converted to yet another branch of Methodism, the *Methodist New Connexion*. He married Catherine Mumford, a Methodist, in 1855. She also had a strong Methodist upbringing and had read the Christian Bible eight times in its entirety by age twelve. She became a partner in her husband's work, advocated for women in the clergy, and became a powerful preacher in her own right. By 1862, they had seven thousand converts, but a schism was developing between their group and the mainstream Methodist Church. Together, the Booths developed an innovative organization that they christened as the Christian Mission in 1865. Later, in 1878, they changed the name to the *Salvation Army*. Its Arminian-Wesleyan doctrinal convictions are embodied in its *Orders and Regulations for the Salvation Army*, which was first published the same year in London.[5]

The core doctrinal convictions of the Salvation Army reflect the Methodist background of the Booths with certain modifications. It agrees with the Methodist views on free will and its rejection of Calvinistic predestination. It also incorporates rousing music and revivalist services to engage rural populations. However, the Salvation Army abandoned the practice of baptism and the Eucharist altogether as unnecessary, much like the Quakers did. It tends to play down the clerical hierarchy that exists within much of Western Christianity. William Booth stated that "the Holy Ghost is just as willing and just as able to speak through a washerwoman as through a bishop."[56]

The Salvation Army is a highly pragmatic institution, heavily focused on charity. It was organized along military lines, because Booth originally considered it an "Army of Salvation." The head of the church is a general. Its clerical organization is a network of colonels, majors, captains, lieutenants, and other military ranks meant to incorporate military-like efficiency into their organization's efforts to address pragmatic human needs. The model was successful in expanding its global reach. Today, the Salvation Army has spread to 121 countries,[55] and its membership ranges between 1.5 and 2 million members worldwide according to various Salvation Army websites. Another major offshoot of Anglicanism to develop was *Episcopalianism*.

EPISCOPALIANISM: THE AMERICANIZATION OF ANGLICANISM

During the seventeenth and eighteenth centuries, many followers of Anglicanism relocated to the American colonies, particularly to the New England area. As the governing body of the American colonies moved away from the English rule, the American branch of Anglican Protestantism moved away from the Church of England and its allegiance to the king as its head. On July 4, 1776, the Continental Congress formally adopted the Declaration of Independence, separating the colonies from England. On the same day, the Congress and several colonial states passed laws making prayers to the king of England to be acts of treason.[58] In 1784, the Anglican clergy of Connecticut elected Samuel Seabury as the first Bishop to support a ministry outside of the British Isles.[5] By 1789, American *Episcopalianism* was codified, with its own revised version of the Book of Common Prayer. It reflected an Americanized version of Anglicanism, free of any reverence toward the king of England or to the British Parliament. It began as the first major branch of Anglican Protestantism outside of the British Isles. According to the Episcopal Church website, in 2021 the Episcopal Church had about 1.74 million members, most of whom live in the United States .[59]

Another branch of Anglicanism, which Christianized the southern Cape of Africa, is *African Anglicanism.*

AFRICAN ANGLICANISM: THE RISE AND FALL OF APARTHEID

African Anglicanism began with Anglican clergy who accompanied British troops to occupy the Cape Colony of present-day South Africa in 1795. The British were there to fight for territory against the earlier Dutch settlers, sometimes referred to as the "Afrikaners." By the end of the eighteenth century, the three major cultural groups in the region were the Afrikaners, the British, and the indigenous Africans. In the early 1820s, Anglican missionaries began arriving in the Cape from an organization called the *Society for the Propagation of the Gospel* (a group that was also active in the American colonies) to convert indigenous Africans to Anglicanism. The adherents of African Anglicanism steadily increased.

South Africa's Afrikaners attained independence from British rule in 1910, after which they dominated South African politics until 1994. Shortly after gaining their independence, the Afrikaners changed their fear of being controlled by the British to fear of being controlled by indigenous Africans.[60] As a result, in 1948, the Afrikaners began to institute a harsh system of institutionalized racial and cultural oppression called apartheid. This sparked significant international and domestic opposition that grew into some of the most influential global social movements of the twentieth century. One of the leaders of that movement in South Africa was a native South African and Anglican cleric named Desmond Tutu (1931–2021).

Tutu was born in South Africa to a Motswana family. He was originally baptized in a Methodist Church, but he and his family later converted to Anglicanism. He married a Roman Catholic woman named Nomalizo Leah Shenxane in a Roman Catholic Church with a Roman Catholic ceremony. His multireligious experience throughout his life probably helped develop his nondiscriminating view of religions. He became an ordained Anglican priest in 1961. He became the bishop of Johannesburg in 1985, and finally was named archbishop in 1986.[5] Tutu's approach to Anglicanism has since been characterized as "Anglo-Catholic" in nature.[61]

Tutu became a well-known force in the late twentieth century against apartheid, which was not repealed until 1991. In 1995, he became chair of the Truth and Reconciliation Commission of South Africa, an organization authorized by Nelson Mandela, to facilitate the move away from apartheid and investigate gross human rights violations that were perpetuated by it. He consistently promoted nonviolence, cultural inclusiveness, and recognition of indigenous African culture. Although he was an Anglican Christian cleric, he never denied the legitimacy of other faiths, and he endorsed a policy of inclusiveness rather than one of exclusionism. He authored books on this topic, such as *God Is Not a Christian* (2011), and he won the Nobel Peace Prize in 1984, as well as numerous other honors. African Anglicanism today has more than three million followers in South Africa. It is but one of many branches that have existed within Anglican Protestantism, only a few of which have been discussed in this chapter.

Together, Early Protestantism and Anglican Protestantism accounted for the bulk of Reformation Christianity, one of the most explosive examples of religious diversification in the history of Abrahamic Monotheism (see the tree on page 366). The historical relationships of other Christ-revering, post-Reformation groups within Christianity remain unclear.

POST-REFORMATION CHRISTIAN FAITHS OF UNCERTAIN HISTORICAL RELATIONSHIP

I was unable to clearly place several Christ-revering religions within my phylogenetic trees of Christianity because of their highly syncretic nature and insufficient knowledge of their origins. Their founders were surely influenced by preexisting religions. However, little historical information is available about how their ideas were originally formed and shaped over time. Nevertheless, some major, putative groups are worth mentioning. Perhaps further study of them will better resolve their relationships in the future. Some are clearly non-monophyletic and need to be divided into their unrelated parts before those components can be accurately incorporated into the phylogenetic tree of all religions. Some of the better-known groups whose relationships are not fully resolved include *Unitarianism*, *Mormonism*, *Unificationism*, *Evangelicalism*, and *Pentecostalism*.

Unitarianism began with a strict "Unitarian" doctrine declaring that God is one and indivisible (i.e., not part of a Holy Trinity).[5] (The name "Unitarianism" is capitalized here when it is used as a proper name identifying a particular branch of religion, and it is uncapitalized when it is used as an adjective.) Unitarianism is sometimes considered to be a late formalized branch of Reformation Christianity. However, in its strictest ideological sense, it has a long and complex history. It dates back well before the Reformation, at least to the fourth century CE in the form of *Arianism* (discussed in chapter 8). Arianism was largely overcome by trinitarian *Nicene Christianity* later in the fourth century, which was declared as the state religion of the Roman Empire (chapter 8). Conflict between unitarians and trinitarians continued to flare up in a variety of forms for centuries after that.

Unitarian ideology began to see a formal resurgence during the sixteenth and seventeenth centuries in the form of *Socianism*, led by humanist and theologian Lelio Sozzini (1525–1562).[5] As a more modern formalized movement bearing the unitarian name, *Unitarianism* began to establish itself as a specific Christian movement in England by 1774 and the United States by 1782.[5] One of the formalized U.S. branches developed into the *American Unitarian Association* (the AUA) in 1825.

In 1961, AUA officially merged with the *Universalist Church of America* (founded in 1866) to form the *Unitarian Universalist Association* (the UUA). Although this branch evolved from a largely Protestant Unitarian tradition, the UUA professes to be a *universalist* tradition. It claims to be a "non-creedal, non-doctrinal religion which affirms the individual's freedom of belief" open to many different religious beliefs.[62] Although the name "Unitarian" is still included as part of its name, after merging with the Universalist Church, they were no longer strict Unitarians, or even strict Christians. They became open to all other peaceful, ethical belief systems and moved away from strict denominationalism. Today, the UUA claims to welcome the beliefs of polytheists, trinitarians, Jews, unitarian Christians, Buddhists, Hindus, and Muslims. They take a somewhat agnostic approach of "not having all the answers."[62] It appears to be one of the most inclusive religious branches today, focusing mostly on shared humanistic teachings rather than a specific, exclusivist creed. Members of the UUA place more emphasis on a search for the truth and meaning in this life than on specific deities and salvation in an afterlife. The Seven Principles of the UUA are an affirmation of the following: (1) The inherent worth and dignity of every person. (2) Justice, equality, and compassion in human relations. (3) The acceptance of one another. (4) Free responsible search for truth and meaning. (5) The use of a democratic process in congregations and society. (6) The goal of a world community. (7) Respect for the interdependent web of all existence. The UUA also professes to periodically review their principles to maintain their relevance to the modern world. Each of the more than one thousand congregations in the United States and Canada is self-governing.[5] The UUA is not the only branch of religion to adopt a strongly universalist approach. Another religion to have done so independently is the Baha'i religion, which has its early roots in Shia Islam (discussed in chapter 11).

Because of its complex syncretic history, its continuously changing nature over time, and its possible nonmonophyly, I was unable to place a putative group containing Unitarianism, the AUA, and the later UUA accurately within my phylogenetic trees as a distinct branch. All three of these groups appear to have roots in Christianity. This group is in need of further study to more accurately place it (or more likely independent subcomponents of it) within my trees.

Mormonism is also known as the main branch of the *Church of Jesus Christ of Latter-Day Saints* (the LDS). Although it belongs somewhere on the phylogenetic tree of Abrahamic Monotheism, and probably within Christianity, it is a syncretized collection of ideologies with pieces taken from Jewish Christianity to Reformation Christianity. Its precise placement on the phylogenetic tree is therefore unclear. The founder of Mormonism was Joseph Smith (1805–1844), who was born in upstate New York.[4,5] His followers considered him to be a prophet comparable to Moses. They believed Mormonism to be the restoration of authentic Christianity, which had been lost.[6] Such movements that profess to be reverting back to Apostolic Christianity are sometimes referred to as forms of *Restorationism*. Restorationism is a characteristic that has occurred independently in many different branches of religion and does not represent a single group of monophyletic origin.

As a boy, Smith was troubled by the conflicts and fighting among different Christian denominations. When he was fourteen, he claimed to have had a vision in which God and Jesus told him that none of the Christian denominations was correct. Furthermore, they told him that "true Christianity" had disappeared long ago when the last of the early apostles died. At age seventeen, he supposedly had another vision directing him to find some mysterious golden tablets and assorted articles with which to translate the tablets. With this, he composed the sacred book of the LDS called the *Book of Mormon*, which he published seven years later as the means to restore authentic Christianity through Mormonism.[6] The Book of Mormon creatively draws the United States into biblical history. Perhaps another way to put it is that it Americanizes the Bible. It teaches that the people who built the tower of Babel (*Genesis* 11:1–9) eventually settled in America. It also teaches that in 600 BCE, the Israelites came to America as two warring factions, the Nephites and the Lamanites, and that Jesus came to America after his resurrection to preach. The Lamanites went on to become the people we know as pre-Columbian indigenous Americans. Today, more than one hundred million copies of the Book of Mormon have been distributed around the world.[4]

Smith also had other visions. One instructed him to take the message of Mormonism to other regions of the country. He and his followers began their missionary work in Ohio, Missouri, and Illinois, where they often received vigorous opposition to the Mormon message and practices. They faced persecution and often were run out of town for their unorthodox beliefs, such as polygamy. In Ohio, Smith and a colleague were beaten, tarred and feathered, and left for dead. Smith was often in conflict with state government authorities, but his evangelistic skill was formidable. Charisma is a powerful force, and it seems that plenty of people are always waiting for someone to give them meaningful direction. Smith's followers eventually grew into a small army. He was even ready to declare his candidacy for the 1844 presidential election.[6] Before he could make his declaration, however, he and his brother were arrested and jailed on a charge of inciting a riot. The charge was later upgraded to treason. A few days later, Joseph and his brother were shot and killed by a mob that broke into the jail.[4,6]

After Smith's death in 1844, the Mormons split into two branches over a number of issues, including who should be Smith's successor, and polygamy.[5] The larger group moved to what would become the Utah Territory with a new leader, Brigham Young (1801–1877), who had been one of Joseph Smith's long-standing liutenants.[6] Young founded Salt Lake City in Utah, which remains the headquarters of the LDS to this day. Where Joseph Smith had married thirty wives, Brigham Young outdid him by having fifty-five wives. With sixteen of them, he had fifty-six children (one way to increase your congregation!) By 1890, most of Mormonism had rejected polygamy, but some secluded Mormon families continue to practice it even to the present day.[6] For much of its history, Mormonism was a segregationist religion. It was not until 1978 that African American men were allowed to join the Mormon priesthood. Mormons avoid tobacco, alcohol, coffee, tea, and even chocolate.

Baptism of the dead is a particular Mormon practice that allows Mormons to baptize dead ancestors by proxy. This has generated an extreme interest in genealogy, and for this reason, the Mormons maintain the largest source of genealogical records in the world.[4] The records are kept within an environmentally controlled vault in Utah, built hundreds of feet deep within a granite mountain. The highly secure facility is meant to survive even a nuclear bomb strike, and contains billions of images, microfilm, and digital records of Mormon genealogy. Members of the eclectic group of people who have been baptized by proxy in the Mormon Church include Elvis Presley, Pope John Paul II, Anne Frank, and Adolf Hitler.

The theological origin of Mormonism is not clear. Even though Joseph Smith may have had some early interest in Methodism at a young age, he was not happy with any particular existing branch of Traditional Christianity. Although the Book of Mormon recognizes a trinitarian view of God, the resurrection of Jesus, and other Christian beliefs, Mormonism clearly has a syncretized origin within Reformation Christianity. It was clearly influenced by Smith's limited knowledge of history and the world, and his involvement in the early westward movement of the American frontier. It is a prime example of how founders are influenced by other aspects of their culture when constructing the details of their religion. Most Mormons consider the president of the LDS to be a prophet of God. Mormonism is in need of deeper comparative study before it can be more confidently positioned within my phylogenetic trees of Reform Christianity. Today, the Mormon Church has more than fourteen million members with about a dozen subbranches and about sixty thousand missionaries.[4]

Unificationism was founded by Reverend Sun Myung Moon (1920–2012) in South Korea in 1954. Its followers are sometimes colloquially referred to as "Moonies" after their founder, but many members consider that name to be derogatory. To tens of thousands of Unificationist members around the world, Moon is The Messiah returned to earth.[35] He is one of the most recognizable religious leaders to have materialized in the late twentieth century. He has addressed the U.S. Congress twice, and his wife delivered a speech to the United Nations.[35]

Moon was born in North Korea. His religious activities brought him into conflict with the Communist government, and in 1947, he was convicted of spying for the South Koreans. For that, he was sentenced to five years in a labor camp. He escaped to South Korea during the Korean War. Moon's family was Christian, and was influenced by Buddhism and Confucianism.[4] He was raised as a Presbyterian, but he was excommunicated for heresy shortly before establishing his own church in 1954. Moon claimed to have had visions communicating with other religious figures, including Moses and even Buddha.[5] He came to call himself the Jesus of the Second Coming.[4]

Moon claimed that the Christian Bible was not the "truth" but rather an approximation. As the newly self-proclaimed messiah, he preached that his own book *The Divine Principle* was God's revelation to him. Unificationist doctrine includes the belief that the fall of man from God's grace was the result of biblical Eve having a "spiritual affair" with Satan before having a sexual relationship with Adam. The result was that their children and their future descendants were all born with "defective natures."[5] Reverend Moon frequently performed massive wedding ceremonies involving hundreds of couples at a time. Children of couples married at such ceremonies were said to be born without the curse of "defective natures."

Moon and his organization were well supported (spiritually and monetarily) by the membership. In 1971, he moved to upstate New York, where he purchased an $850,000 home. His financial success also brought him some problems. In 1982, he was convicted in U.S. federal court of conspiracy and filing false income tax returns. He was sentenced to eighteen months in a federal penitentiary, of which he served only thirteen months because of good behavior. While he was in prison, Unification members equated his suffering to that of Jesus. Several more-traditional Christian leaders (e.g., Southern Baptist televangelist, Jerry Falwell and United Methodist minister, Joseph Lowery) also signed petitions in defense of Moon, fearing increased government taxation of church activity (including their own). Today, Unification members number between one and three million, most of which are in Korea. Where this branch (or its components) fit within the trees of Abrahamic Monotheism remains unresolved.

A number of other vernacular categories within Reformation Christianity also technically are not single denominations, because they are based on characteristics that evolved independently many times. These names are more like adjectives (characteristics) than nouns (independent group names) and include *evangelicalism* and *pentecostalism*.

Evangelicalism (also called *Pietism*) is a series of distantly related Protestant branches that all claim a "born again" experience (a form of personal conversion to seeing the Bible as the sole authority in matters of faith[5]). They are often the product of "revivalist" or "Great Awakening" movements. However, that experience is an additive characteristic to a variety of different core denominations and lineages with different histories and theological relationships. Just as birds, bats, and butterflies each evolved the ability of flight independently rather than inheriting it from a single ancestor, many evangelistic branches evolved their evangelic characteristic independently. Thus, the general term *evangelism* is seen here as an adjective or characteristic rather than a denominational noun. Some of the groups today which self-identify as evangelistic include *Evangelical Lutheranism* (sometimes known simply as *Evangelicalism*), *Evangelical Presbyterians*, *Evangelical Methodists*, *Evangelical Baptists*, *Evangelical Quakers* (*Evangelical Friends*), and many others. According to the Pew Research Center, groups that today self-identify as evangelistic include an estimated 4.1% of the global population.[63]

Some evangelistic groups are apocalyptic and focus on the afterlife at the expense of the present life. *Dispensationalist Christian Zionists* sometimes referred to as the "Armageddon Lobby" eagerly anticipate the "end of days" because they believe it is divinely ordained and will pave the way for the rapture (ascent of the faithfully righteous into paradise without dying).[64,65] They believe that this must begin in the Middle East. Consequently, they are not concerned that some of their lobbying efforts with the U.S. federal government have effectively raised tension between Arabs and Jews in the Middle East. Their enthusiastic anticipation of the apocalypse could one day become a self-fulfilling prophecy. It is another prime example of the dangerous direction that even some modern religions can take (Sam Harris comments on the Armageddon Lobby in **Note 9E**).

Pentecostalism is a characteristic present in a variety of different groups, including Presbyterians, Methodists, and Baptists. It arose as a charismatic movement that "pentecostalized" several different branches independently, making pentecostalism non-monophyletic (thus, it does not fit on a phylogenetic tree as a single branch). Like general evangelism, general term *pentecostalism* is treated here as an adjective or characteristic, rather than as a denominational noun. It may be better represented as a number of independently derived groups, such as *Pentecostal Presbyterianism* (as an offshoot of Presbyterianism), *Pentecostal Methodists* (as an offshoot of Methodism), and others. It warrants much further study to incorporate its components into an evolutionary context and a phylogenetic tree. That is not to demean its membership. It is meant only to encourage efforts to clarify the historical relationships involved.

Charles Parham (1873–1929) founded a formal version of pentecostalism in 1900.[5,33] Although he was trained as a Methodist preacher, his ideas diverged from mainstream belief where he began preaching an evangelistic belief system and incorporating elements of other movements into his preaching. One of his students, William J. Seymour (1870–1922), sparked the four-year-long *Azusa Street Revival* in 1906, which was characterized by people speaking in tongues, making prophetic predictions, and having other spiritual experiences. The event is credited for boosting the widespread growth of pentecostalism across the United States in the early twentieth century.[66]

Like other forms of evangelical Protestantism, pentecostalism adheres to the inerrancy of the Bible and the necessity of accepting Jesus Christ as Savior. The movement takes its name from an incident described in the Book of Acts at the Jewish feast of the Pentecost. The Holy Spirit descended on the apostles and they began "speaking in other tongues" (also called *glossolalia*) so that the many listeners gathered in Jerusalem could all hear them speaking in every language represented in the crowds.[6] Although most pentecostal groups are trinitarian, at least one is not: *Oneness Pentecostalism*. Some pentecostalists see a need for believer's baptism, but most do not. Some pentecostalists are Wesleyan derivatives, whereas others are not. It is a complex conglomeration of different religious belief systems.

Random spontaneity is a common element of pentecostalism. Probably the best-known example of this is "speaking in tongues" (or at least the belief that members who do this are in direct communication with God, and that the Holy Spirit is directing this behavior).[4] Speaking in tongues generally involves loudly vocalizing unintelligible utterances or phrases thought to be in a language of which the utterer supposedly has no knowledge. Adherents sometimes start into such rants in the middle of an evangelistic church service. Other spontaneous practices include being "slain in spirit" (a form of prostration in which a person falls backward as if fainting), "dancing in spirit" (where a person spontaneously jumps up with their eyes closed and starts dancing), and "holy laughter" (where a person spontaneously begins uncontrollably laughing). Such spontaneous and charismatic forms of emotional response are thought of as gifts brought by the presence of the Holy Spirit.[4,6]

Beginning in early twentieth-century Appalachia, some pentecostalist preachers added the handling of venomous snakes during church services, sometimes with deadly results. They also would drink poison thinking it would not hurt them. They see these acts as obedience to God, inspired by passages from the Bible (*Mark* 16:18): "They will pick up snakes with their hands; and when they drink deadly poison, it will not hurt them at all." Their concern is more with the afterlife than with the present life, and they see snake handling in their worship tradition as assuring eternal salvation.

Pentecostalism began as an extreme form of evangelism, and its leaders were pioneers in using mass media to spread its message. The swashbuckling Aimee McPherson (1890–1944) was an early pentecostal celebrity. Sister Aimee, as she was known, dropped bundles of "handbills for God" from airplanes in the 1920s and presided over the first-known church radio station.[6] In her time, she was the most publicized Protestant evangelist, using her radio station to tap into the growing appeal of popular entertainment in North America. Oral Roberts (1918–2009) was one of the first televangelists. His charismatic pioneering efforts began in 1947 with radio, and he moved to television in 1954. He became to television what Aimee McPherson had been to radio.[6] He was so successful that in 1996 he founded his own Christian television network, GEB America, with twenty-four-hour programing. Later, other popular pentecostal televangelists followed, such as Jimmy Swaggart and Jim and Tammy Fay Baker.

Another famous, and much darker, pentecostalist derivative was led by the Reverend Jim Jones and his *Peoples Temple* branch in Jonestown. Jones was famously responsible for the mass-suicide and murder of more than nine hundred of his followers, including three hundred younger children in 1978. They were all convinced by Jones that this was their path to paradise (for details, see **Note 9D**). It is another example of the profound power that charismatic individuals can have over large numbers of people. A challenge for us as a species is to learn how to channel such power for the collective good and learn how better to deal with charismatic proselytism gone bad. Fortunately, most Pentecostal branches are not so extreme (or deadly).

By 2006, pentecostalism had more than 130 million adherents worldwide,[5] although some authors put the number much higher.[66] Part of the problem in calculating the size of pentecostalism appears to be in clearly identifying its ideological parameters as a distinct religion, rather than simply a characteristic shared by portions of several different religions. As stated by a number of History of Religion specialists, there are many "diverging opinions" on the size, diversity, and definition of pentecostalism.[66]

FINAL THOUGHTS ON THE EVOLUTION OF *TRADITIONAL CHRISTIANITY*

Traditional Christianity was instrumental in the rise of Western civilization, just as the rise of Western empires was instrumental in the evolution of Traditional Christianity. It is by far the largest branch to evolve from Abrahamic Monotheism, and by 2010, it included more than 32 percent of all religious people.[67] This account is highly simplified and focused on primary branches. Much comparative work is needed to understand its more detailed diversity, history of diversification, and phylogenetic interrelationships.

Other major Christ-revering religions also were in competition with early Traditional Christianity during its first 1,500 years. Some of them did not fare well, eventually fading to extinction. Many of these ill-fated branches are classified in a group called *Biblical Demiurgism*, the subject of chapter 10.

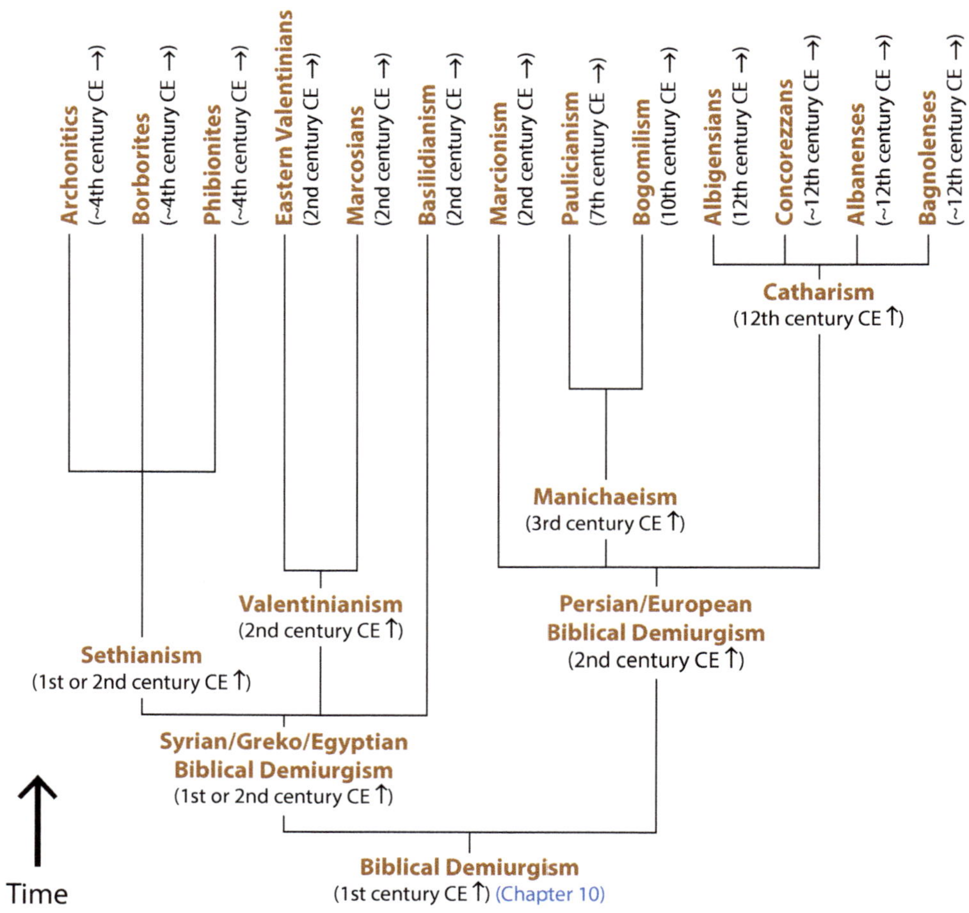

Archonitics (~4th century CE →)
Borborites (~4th century CE →)
Phibionites (~4th century CE →)
Eastern Valentinians (2nd century CE →)
Marcosians (2nd century CE →)
Basilidianism (2nd century CE →)
Marcionism (2nd century CE →)
Paulicianism (7th century CE →)
Bogomilism (10th century CE →)
Albigensians (12th century CE →)
Concorezzans (~12th century CE →)
Albanenses (~12th century CE →)
Bagnolenses (~12th century CE →)
Catharism (12th century CE ↑)
Manichaeism (3rd century CE ↑)
Valentinianism (2nd century CE ↑)
Persian/European Biblical Demiurgism (2nd century CE ↑)
Sethianism (1st or 2nd century CE ↑)
Syrian/Greko/Egyptian Biblical Demiurgism (1st or 2nd century CE ↑)
Biblical Demiurgism (1st century CE ↑) (Chapter 10)
Time

10

BIBLICAL DEMIURGISM

A SUBGROUP OF "GNOSTICISM"

The only good is knowledge, and the only evil is ignorance.

—SOCRATES

During the first few centuries CE, many different Christ-revering beliefs evolved in close association with each other. There The newly developing authoritative scripture (New Testament) was being interpreted in conflicting ways. Each religious belief system was claiming to be the true path to salvation with Jesus as a savior, prophet, messiah, angel, or even an aspect of God. It was a fierce and often violent competition for the hearts and minds of the emerging Western World. Some groups became dominant, others struggled as minor branches, and yet others drifted to extinction. Traditional Christianity (chapter 8) teamed up with the Roman Empire by the fourth century CE and became extremely successful. Several competing Christ-revering branches were deemed heresy by Traditional Christian heresiologists (the Church officials also known as the Church Fathers who were charged with identifying heretics, among other things). These defeated belief systems faded either to obscurity or to extinction over time, after years of persecution and intercultural conflict. The term "gnosticism" was later used for a hodgepodge assortment of particular religious belief systems that generally revered Christ in one way or another, but were rejected by the heresiologists. This catchall "group of convenience" is clearly non-monophyletic and could not be placed on a phylogenetic tree as a single group. In this chapter, however, the focus is on what appears to be a monophyletic subset of that group: *Biblical Demiurgism*.

In 1996, comparative religion scholar Michael Allen Williams recognized the problem of non-monophyly with "gnosticism" as a category of religion (non-monophyly defined in **Note 1L**). To rectify this, he distinguished a specific subset of gnostic subbranches that he called Biblical Demiurgical belief systems (referred

Timeline for certain events, people, and struggles of Biblical Demiurgism

Discovery of the ancient *Nag Hamnadi Library* in Egypt, containing the first major collection of early texts about Biblical Demiurgism written *by* Biblical Demiurgists →	–1945
Revival of Biblical Demiurgism in the form of *"Catharism,"* best best known for its prolonged persecution by the Catholic Church. It was driven to extinction by the 14th century →	–12th century CE
Traditional Christian heresiologists, Iraneaus and Hippolytus, write their biased interpretations of *Biblical Demiurgism,* which remained the main record of that faith until excavation of the ancient Nag Hamnadi library in 1945 (see above) →	–2nd and 3rd centuries CE
Manichaeism was founded by a Persian preacher named Mani (212–276 CE). It thrived between the 3rd and 7th century, but faded to near extinction by the 15th century, persecuted by both Traditional Christianity and Zoroastrianism →	–3rd century CE
Basilidianism, was founded by Basilides (117–138 CE). It was soon a major early branch of Christ-revering religion, but was deemed heresy by leaders of Traditional Christianity and faded toward extinction after two centuries of prominence →	–2nd century CE
Marcionism founded by Marcion of Sinope (85–160 CE), who was branded a heretic by Traditional Christianity. It becomes largely extinct by the 5th century, partly due to persecution, and partly because of its requirement of celibacy →	–2nd century CE
Valentianism develops, founded by the theologian, Valentinius (100–180 CE). Labeled as heresy by Traditional Christianity's heresiologists, it becomes largely extinct by the 7th century →	–early 2nd century CE
Sethianism develops as one of the earliest branches of Biblical Demiurgism, but becomes largely extinct by Middle Ages →	–1st century CE
Biblical Demiurgism develops in parallel with Traditional Christianity as another version of Christ-revering religion →	–1st century CE

to here as *Biblical Demiurgism*).[1] Unlike gnosticism, Biblical Demiurgism can be more legitimately hypothesized as a monophyletic branch, based on certain shared characteristics hypothesized to be uniquely derived. It is therefore represented in that context on my provisional phylogenetic trees of religions. Some "gnostic" sects mentioned in the literature, but not included on my trees, may have existed only in the minds of the heresiologists (e.g., *Cainites, Simonians, Nicolaitans*).[1] Other gnostic subgroups (e.g., the *Mandeans*, a surviving gnostic group in southern Iraq) are in need of further study before they can be placed on a phylogenetic tree with confidence.

Biblical Demiurgism survived only marginally beyond the sixth century CE. It experienced a brief resurgence in the twelfth century by a branch provisionally called *Catharism* (a controversial name questioned by some authors).[2] However, the Cathars were brutally crushed by militant crusades of Roman Catholicism during the fourteenth century. Biblical Demiurgism was a highly complex group of religions that are still poorly known. My presentation is somewhat selective, simplified, and incomplete in its coverage. It is a group that warrants much further phylogenetic study.

BIBLICAL DEMIURGISM: A GOOD GOD VERSUS IGNORANT GOD AND THE REDEEMING POWER OF GNOSIS

Biblical Demiurgism originated in the Middle East in the first or second century CE, evolving in parallel with Traditional Christianity, Nazarenes, and Ebionites[3] (see the phylogenetic tree on page 286). This early Christian branch did not accept the Traditional Christian idea of the good and all-knowing Creator God or of a trinitarian set of godly aspects. Instead, it envisioned God as two separate aspects (a form of *binitarianism*). The first of these was the all-good *Supreme God*, generally referred to as the *Monad, Bythos*, or *The One*. He was considered the perfect yet unknowable Father. He existed within a higher, perfect spiritual realm of the *Pleroma*. He was mysterious and was not directly responsible for creating anything. However, he emanated (mysteriously manifested) a series of thirty personified aspects of his divine nature called *Aeons* (spheres of being), ranging from *Nous* (Mind) to *Sophia* (Wisdom) (e.g., Aeons of one branch are listed in **Note 10A**). The Monad did not exist in any sense that the human mind could understand, and he was more of a spiritual essence than an anthropomorphized deity.[4] Like emanations and servants of the Supreme Being that often are depicted as beings of light, the Aeons were similar to Judeo-Christian angels.

The second god of Biblical Demiurgism was the imperfect *creator god* or *god of the Old Testament* called the *Demiurge* (also called *Yaldabaoth* in some branches, or identified as *Satan* in some Cathar sects). It was an antagonistic entity to the Supreme God. The concept of a Demiurge as an imperfect creator god is thought to have been influenced by the Greek philosopher, Plato.[5] Because the concept of a Demiurge was a secondary characteristic adopted by a Christ-revering tradition, Biblical Demiurgism is included here as a derivative of Abrahamic Monotheism and Jesu-venerationism (tree on page 286). The Demiurge was believed to have been accidently created by the Aeon Sophia (Wisdom). The Demiurge did the dirty work of creating the imperfect material universe in its entirety. He was believed by Biblical Demiurgists not only to be imperfect but also to be lacking in *gnosis* (Greek for knowledge). He did not exist in the Pleroma and was unaware of the Supreme Monad above him.[1] Biblical Demiurgists reasoned that a perfect God could not have created a world so full of pain and evil and that the hands-on creation of that world must have been the work of a lesser being.[4]

The Demiurge was a single entity in some branches of Biblical Demiurgism. Other branches had a demiurgic collective of dark creator archangels, archons, or other evil emanations with a head demiurge. The Demiurge was sometimes a vengeful, jealous deity who perpetrated ignorance and cruelty. He was thought by many Biblical Demiurgists to be the equivalent of Yahweh in the Old Testament. He demanded to be worshipped by people as the only god.[1] He commanded the Israelites (ancestors to the early people of Abrahamic Monotheism) to be his servants in return for having created them (as in the Torah and Old Testament Book of *Isaiah* 44:21). Although some branches of Biblical Demiurgism portrayed the Demiurge as evil, others considered him to be simply ignorant and misguided. Those that considered the Demiurge merely ignorant attributed evil to one of his subsequent creations: the devil. Many groups symbolized the Demiurge as having a lion's head and a serpent's body.

The Monad blamed the Demiurge for the imperfections, ignorance, and perversions of the material world, as well as for the fall of Aeons from the Pleroma.[4,6] Some branches of Biblical Demiurgism saw Christ (not necessarily synonymous with Jesus) as the spiritual emissary of the Monad sent to repair the damage done by the Demiurge. Other branches saw Christ and the Holy Spirit as creations of the Aeon *Nous* (Mind), with Christ tasked with reminding the Aeons that the Monad was incomprehensible, and with the Holy Spirit charged with bringing harmony to the Aeons and teaching them to give thanks. Eventually, the Aeons were thought to have pooled their qualities to create Jesus as the Second Christ.[1] They also created an escort of angels for him. The different branches of Biblical Demiurgism feature many variations of these themes. The distinction of God as two different aspects with vastly different levels of gnosis and one of the two aspects being an imperfect, ignorant creator (the Demiurge) are the two most general defining characteristics of Biblical Demiurgism.[1]

The lion-faced serpent symbol commonly used as a symbol of *Yaldabaoth*, the Demiurge in Biblical Demiurgism. Image by Bernard de Montfaucon (1655–1741), French Benedictine monk and early archaeologist.

Biblical Demiurgism and Traditional Christianity had reciprocal effects on each other during their early development. In their earliest days, they shared many traditions, including a particular reverence for Paul the Apostle.[5,7,8] Although the early leaders of both Biblical Demiurgism and Traditional Christianity claimed priority as the core of early Christianity, and both claimed a connection to the apostles of Jesus Christ, no conclusive evidence shows which branch was an earlier movement of Jesu-venerationism.

Before 1945, our knowledge of Biblical Demiurgism was generally a biased, second-hand accounting. During that time, the only surviving historical descriptions of its subgroups came from writings of the *heresiologists*—the clerics and theologians charged with finding heresy and promoting the superiority of Traditional Christianity. The core beliefs of the Biblical Demiurgists concerning Christ and the imperfect creator god put them near the top of Traditional Christianity's list of heretics. Some of the most critical texts concerning Biblical Demiurgism included *Against Heresies* by the principle heresiologist, Irenaeus (130–202 CE); *Refutation Against All Heresies* by an anonymous heresiologist, questionably thought by some historians to have been a pupil of Irenaeus named Hippolytus (170–235 CE);[9] and *"Medicine Chest" of Antidotes for a List of Eighty Heresiological Afflictions* by the later heresiologist Epiphanius (315–403 CE).[1,6] The Traditional Christian Church later canonized Irenaeus, Hippolytus, and Epiphanius as saints.[1] Most texts written by the Biblical Demiurgists to describe their faith mysteriously disappeared. They were either hidden or destroyed by Traditional Christian Church officials to effectively bury or rewrite history. Others were simply no longer copied by the copyists (mostly Traditional Christian volunteers focused on reproducing parts of the Bible). As a result, most of the only surviving accounts and interpretations of Biblical Demiurgism through the mid-twentieth century were colored with the negative bias and subjectivity of the heresiologists. Historically, the winners of ideological conquest and subjugation have been the ones to write and record history.

Our knowledge of Biblical Demiurgism and the earliest development of Christian doctrine entered a new phase with the discovery of an archaeological site in Nag Hammadi Egypt in 1945.[5] This site was one of the most important archaeological finds of the twentieth century. It contained an extensive library of original documents written on papyrus in Coptic Egyptian about the gnostic branches of Jesus-revering branches of religion from the second and third centuries CE. This ancient library had been buried in the fourth century after the Nicene Christian Church condemned public use of such noncanonical books. Its discovery led to a wealth of information about the early formation of Christian belief systems that is still being evaluated today.

Early competition among the different branches that venerated Jesus Christ at one level or another was intense. Adherents of Biblical Demiurgism eventually saw the clerical hierarchy of the Traditional Christian Church all the way up to the pope as being corrupt and evil. In response, Traditional Christian heresiologists stigmatized the followers of Biblical Demiurgism as being either severely misled, or “the first-born of Satan” (Irenaeus III: 3:4).[10] Many followers of early Biblical Demiurgism appear to have considered Christ to be a divine Savior, composed only of spirit. Many believed the Christ spirit entered the man Jesus at his baptism and conveyed to him the knowledge to liberate the “sparks” of divine light that are buried within people. At the crucifixion, they believed that the spirit briefly left Jesus to suffer his death alone, which is why Jesus cried out, “My God, my God, why have you forsaken me?” (Matthew 27:46) After Jesus died, the Christ spirit came back marking the resurrection.[10,11]

Biblical Demiurgism shared a number of other basic characteristics with the rest of Abrahamic Monotheism, although in modified form. Abraham, Noah, Moses, Jesus, and other important figures of Abrahamic Monotheism were important figures in Biblical Demiurgism as well. However, they generally held different mythological roles there than in other branches of Abrahamic Monotheism.[1,12,13] Biblical Demiurgism rejected most of the Old Testament, preferring to concentrate on books in the New Testament that had preserved the spirit of Jesus Christ. Biblical Demiurgism eliminated the Jewish and Traditional Christian reverence for the creator god

(Yahweh). They recognized an enhanced dualistic nature of God. Although most Biblical Demiurgists considered themselves to be monotheists, most envisioned the Demiurge as an ignorant, lesser aspect of supernatural power. The binitarian view of God in Biblical Demiurgism (the Monad and the Demiurge) was similar to the trinitarian Christian view in Roman Catholicism (the Father, Son and Holy Spirit), in that they were both forms of soft polytheism (or soft monotheism) a doctrine discussed previously in chapter 6.

One of the Monad's Aeons of particular importance to most of Biblical Demiurgism is Sophia. Sophia not only was seen as analogous to the human soul but also as the feminine aspect or emanation of the Monad. She was sometimes represented as a dove, much like the Roman Catholic Holy Ghost. She fell out of favor with the Supreme God after she accidently created both the Demiurge and matter itself (this is a long and complicated story with numerous versions).[1] However, she also had an especially good side as the spark of divine light that resides in all people and awaits awakening. Reviving this spark was the road to salvation. The spiritual Christ was sent by the Supreme God to guide Sophia and humanity back from the evil, materialistic world to the spiritual light.

In Biblical Demiurgism, gnosis is of primal importance and the key to salvation for people. This is not unlike Socrates' sentiment that "the only good is knowledge, and the only evil is ignorance." Biblical Demiurgism proposes that it is ignorance and unconsciousness of our inner spirit that is responsible for evil, not sin or guilt. Gnosis is the remedy, and Jesus Christ delivered this gnosis to humanity. This special knowledge instructs that one must reject the moral evil of the Demiurge to embrace the Monad and the spiritual world. Biblical Demiurgism deals less with the problem of sin, and more with that of ignorance. To Biblical Demiurgists, it is all about knowledge and giving people the gnosis to rescue themselves from the physical world and return to spirituality. In most branches of Biblical Demiurgism, gnosis was emphasized even more than faith.

Biblical Demiurgism flourished in the Mediterranean region as an early form of Christianity in the second century CE, but by the late third century, it began to decline as the Traditional Christian Church continued to condemn it as heresy. It declined further in the fourth and fifth centuries, as Traditional Christianity was embraced by the Roman Empire. Little of Biblical Demiurgism survived beyond the sixth century. It became overwhelmed by persecution from both the Eastern and the Western Christian Church and its collaborative monarchs. In twelfth- and thirteenth-century Europe, a revival of Biblical Demiurgism took the form of *Catharism*. The Cathars were once again persecuted and severely beaten back, this time by the Catholic Crusaders. By the mid-fourteenth century, they had been hunted to virtual extinction in one of history's most remarkable examples of genocide.

Biblical Demiurgism had a diverse and complex early history. Early major branches that thrived during the first few centuries included *Syrian/Greco/Egyptian* Biblical Demiurgism (*Sethianism, Valentinianism,* and *Basilidianism*) and *Persian/European* Biblical Demiurgism (*Manichaeism, Marcionism,* and *Catharism*). Most or all of these branches in turn, split into subdivisions of their own over time. Unfortunately, most of our knowledge about the branches of Biblical Demiurgism, their detailed characteristics, and their complex histories is fragmentary and consists almost exclusively of second- and third-hand summaries from negatively biased sources. As Traditional Christianity achieved dominance among the early Christ-revering branches, it lost or modified much of the recorded history of Biblical Demiurgism. Consequently, the ideological diversity of Biblical Demiurgism and the detailed interrelationships of its subgroups are far from resolved on my phylogenetic trees. The simplified tree and discussion provide only minimal information about the diversity of supernatural belief systems that existed within Biblical Demiurgism.

Biblical Demiurgism, as currently constructed, generally falls into two major groups: *Syrian/Greco/Egyptian BD* and *Persian/European BD* (see the phylogenetic tree on page 414).

SYRIAN/GRECO/EGYPTIAN BIBLICAL DEMIURGISM AND THE INFLUENCE OF PLATO

The Syrian/Greco/Egyptian branches of Biblical Demiurgism included its earliest branches. They were strongly influenced by Egyptian history, the rise of Traditional Christianity, and the Hellenistic philosophy of Plato (*Platonism*). The Supreme God is seen as a transcendent being that operates through divine intermediaries in the form of divine aspects and Aeons. By creating the world, the Demiurge transferred matter, which is a receptacle of evil, into the divine soul of the world where it continues to operate as a source of evil. Although the demiurge thinks he is creating matter solely on his or her own, the overall process was believed to be ultimately under the control of divine providence.

I include the following three branches of Syrian/Greco/Egyptian Biblical Demiurgism: *Sethianism, Valentinianism,* and *Basilidianism.*

SETHIANISM AND THE THIRD SON OF ADAM AND EVE

Sethianism was one of the earliest branches of Biblical Demiurgism from ancient Persia, or possibly Egypt, for which we have contemporary textual evidence (**Note 10B**). Its name comes from the practice of venerating Seth, son of the biblical Adam and Eve.[1] Unlike his older brothers Cain and Abel, Seth was said to possess gnosis from the spiritual world. Its early roots penetrated Second Temple Judaism of the first century BCE.[14] It was Christianized in the first century CE and evolved along the eastern half of the Mediterranean region in parallel with Traditional Christianity. By the second century, it became increasingly estranged from Traditional Christianity as its own doctrine took shape. By the early third century, Sethianism and Traditional Christianity were in heated competition for those who revered Christ. Sethianism was deemed heresy by Traditional Christian heresiologists. According to History of Religion scholar, John D. Turner (1938–2019), by the fourth century, Sethianism had split into a number of subbranches based on developing doctrinal differences. These included the *Archonitics*, the *Borborites*, and the *Phibionites*.[1,14] Sethianism eventually lost the competition with other early branches of Abrahamic Monotheism, and by the late Middle Ages, it was largely extinct.

Sethians believed that the Supreme God of Pleroma, the Monad, emanated the angel-like Aeons as a series of paired male and female beings that represented the various attributes of the Monad. They ranged from *Nous* (Mind) and *Aletheia* (Truth) to *Theletos* (Desired) and *Sophia* (Wisdom). As in other branches of Biblical Demiurgism, the creator god (Demiurge) of the material world was an ignorant, lesser god than the Monad. His name among Sethians was Yaldabaoth (or Ialdabaoth).[14,15] He led a gang of minor deities called *archons* (equivalent to the fallen angels of Judeo-Christianity) to help him control the realm of darkness.

As with Yahweh in the *Genesis* story of the Old Testament, the Demiurge first created Adam and then Eve. They were placed in the Garden of Eden with trees bearing edible fruit. From that point, the Sethian version of the Garden of Eden story diverges sharply from that of Traditional Christianity. The Demiurge proclaimed himself the only God and demanded both Adam and Eve venerate him as such. The Garden, created by Demiurge, was a false paradise whose trees bore fruits of sin, lust, ignorance, confinement, and death, with one exception: the tree of knowledge. Once Adam and Eve tasted the fruit from the tree of knowledge, they had a positive revelation.[1,15] The Sethian version of the Garden of Eden story ends with Adam and Eve escaping from ignorance and false paradise rather than the Traditional Christian/traditional Jewish version where they receive divine punishment for eating the fruit.

Sethian interpretation and portrayal of Jesus Christ came to be radically different from that of early Traditional Christianity. In its later phase, the Christ of Sethianism comes as an agent of Pleroma (the higher realm) and as the high priest of the Monad to bring light into the evil, ignorant, material world. To many Sethians, Christ was the spiritual aspect of Seth.[14] Some of the earliest accounts of Sethian mythology come from writings by the early Christian chief heresiologist, Irenaeus, from the second century, who was first to officially pronounced Sethianism as heresy. Other ancient accounts appear in the *Secret Book of John the Apostle* (also called the *Apocryphon of John*). This so-called Lost Gospel describes Jesus appearing to John after the crucifixion to give him secret knowledge (gnosis) about creation, the spiritual world, and the road to salvation. It is one of the earliest documentations of gnosis as being the key to salvation in Sethianism. Another Lost Gospel often attributed to Sethianism was discovered in the late 1970s in an Egyptian tomb.[16,17] It is a copy of the *Gospel of Judas*, dated to the third century CE and is thought to have been written in the second century. In it, the apostle Judas is portrayed as a hero rather than the one who betrayed Jesus. In Traditional Christianity's New Testament Gospels, Judas betrays Jesus for "30 pieces of silver," identifying him with a kiss in front of Roman soldiers. He is later overcome with guilt and commits suicide. In stark contrast, the Gospel of Judas gives a completely different account. As the story goes, Jesus asks Judas, who is his friend, to betray him so he could complete his mission on earth. He singles Judas out for special status among his disciples. In a key passage, Jesus tells Judas, "You will be greater than them all. For you will sacrifice the man who carries me about" (Gospel of Judas 56) This is a prime example of the doctrinal diversity of Christ-revering belief systems that existed in the first few centuries CE.[17]

By the Middle Ages, Sethianism was largely extinct. Early in its history, it influenced other Syrian/Greco/Egyptian branches of Biblical Demiurgism, such as *Valentinianism*.

VALENTINIANISM: GNOSIS, NOT FAITH, AS THE KEY TO SALVATION

Valentinianism was the largest branch of Biblical Demiurgism, founded in the early second century CE by the theologian Valentinus (100–180 CE). He was a native of Egypt who was educated in Greece. He was said to have been an eloquent speaker and a prolific writer, although all original texts were lost or destroyed. What we know about Valentinus and his writing comes mainly from fragmentary "quotes" transmitted by critics, such as the heresiologist Irenaeus, who were focused on refuting Valentinus as a heretic.[10,18] Valentinus was originally a prominent member of the early Traditional Christian community in Rome who, according to some historians, broke with the Church after being passed over for the position of bishop.[6,19] According to some historians, Valentinianism differentiated into Eastern ("Italian") and Western ("Marcosian") branches early in its development, and each of those subgroups subsequently divided into subsects. The taxonomy and historical diversification of Valentinianism remains to be analyzed and clarified in a more rigorous way.

Valentinians saw themselves as successors of Paul the Apostle through a succession of teachers.[8] Valentinus was instructed by an early Biblical Demiurgic thinker named Theudas, who was a student of Paul. He is also thought to have been a contemporary of Basilides, founder of Basilidianism (discussed in the next section). Valentinianism spread throughout Rome and from Northwest Africa to Egypt through Asia Minor and Syria during the second through fourth centuries. Early Traditional Christian leaders considered it to be threat because of its beliefs.

In Valentinianism, much like Sethianism, the Aeons were coupled as male-female pairs (**Note 10A**). According to Irenaeus,[10] the Primal Father Aeon of Valentinianism, also known as "The One," was *Bythos* (the *Monad* equivalent of other branches of Biblical Demiurgism). In the beginning, Bythos existed within the Pleroma in silence and contemplation. After ages passed, Bythos projected other celestial Aeons through the process of emanation (mysterious manifestation). A few of the Aeons are known by different names among the different branches of Biblical Demiurgism, while others (like Sophia) was more universally known by the same name. Sophia was filled with curiosity and passion, which eventually led to her fall from Pleroma. The most significant aspect of her sin was giving rise to matter and ultimately to the Demiurge.[20]

The Demiurge of Valentinianism, like that of other branches of Biblical Demiurgism, is equivalent to the creator god of the Old Testament, Yahweh. In most subgroups of Valentinianism, however, he is not as malevolent as he is portrayed in Sethianism. He is simply ignorant and misguided.[1] The Valentinians, like other branches of Biblical Demiurgism, believed that the attainment of gnosis (knowledge), rather than faith, was the key to salvation. The ultimate goal of this gnosis for all men and women was to reunite with their spiritual counterparts in the Pleroma. Valentinus was reported to have identified three general human natures.[20] The first are the *pneumatici* (spiritual): those who will return to the grossness of matter and ultimately be consumed by fire. The second are the *psychici* (psychical): those who will enter a middle state of neither Pleroma (heaven) nor matter (hell), together with the Demiurge as their master. The third are the *hylici* (material): those who will be freed from the influence of the Demiurge and be reunited with their spiritual counterparts in the Pleroma. The three outcomes are similar to the concepts of hell, purgatory, and heaven found in other branches of Abrahamic Monotheism.

Valentinianism was an early competitor with Traditional Christianity for the followers of Christ, but by the seventh century, it had all but disappeared. Because of the scarcity of surviving original historical documentation, we cannot say much about the numerous Valentinian belief systems, nor can we clearly separate them all. The details of Valentinianism as reported by Irenaeus[10] are complicated and include a confusing mix of different subgroups. What I have presented is both condensed and highly simplified. Another contemporary branch of Syrian/Greco/Egyptian Biblical Demiurgism was *Basilidianism*.

BASILIDIANISM: A POORLY KNOWN BRANCH OF BIBLICAL DEMIURGISM BELIEVING IN REINCARNATION

Basilidianism is a branch of Biblical Demiurgism that arose in the second century CE and persisted into the fourth century. Founded by a religious teacher from Alexandria named Basilides (117–138 CE),[21] it was contemporary with Valentinianism but not as popular. Historical details of Basilidianism and its founder from the ancient heresiologists Irenaeus and Hippolytus are contradictory. I follow the reports of Irenaeus, as do many historians. Most of the original writings of Basilides have been lost, except for various fragments preserved to supplement the writings of heresiologists criticizing Basilides as a heretic. Basilides claimed to have been taught by Glaucus, a disciple of the apostle Peter. Like other Biblical Demiurgic branches, Basilidianism equated the material world with evil. He set the ultimate goal of Basilidians to achieve gnosis and ascend back to the spiritual world. Basilidianism taught that salvation comes through knowledge (gnosis), although unlike Valentinianism, Basilidianism did not make a strong distinction between knowledge and faith.[21]

Basilidianism taught that in the beginning, the Supreme, unknowable God inhabited the Pleroma, as in most branches of Biblical Demiurgism.[21] Eventually, a pentad of Aeons emanated from the Supreme God, beginning with the first-born *Nous* (Mind). Then came *Logos* (Word) and *Phronensis* (Prudence). Finally, the last two members of the pentad emanated were *Dynamis* (Power) and the ever-present female figure of Biblical Demiurgism, *Sophia* (Wisdom). From this point, the story markedly diverges from other belief systems of Biblical Demiurgism. In one version, Sophia and Dynamis gave birth to a set of Creator Angels that made the highest level of heaven outside the Pleroma. This set of Creator Angels also made a second set of Creator Angels who then made the second level of heaven. The second set of Creator Angels made a third set of Creator Angels who made a third level of heaven, and so on until there were 365 levels of heaven, each with its own set of Creator Angels. The 365th level of heaven is the one we see when we look up into the sky. Each of the 365 levels of heaven is said to be ruled by an archon (a type of entity created by the Demiurge). However, there are several different versions of this story.

In Basilidianism, Christ is the spirit of the Supreme God's first-born Aeon, *Nous* (Mind).[21] He was sent to the material world to liberate people's spirits from the power of the Demiurge and to facilitate their salvation to the spiritual world. The Christ spirit entered the body of Jesus at his baptism. The body of Jesus endured suffering at the crucifixion, but not the spirit of Christ, which ascended safely into the Pleroma. Basilides ascribed unspecified sinfulness even to the human Jesus, for only the Supreme God was considered "wholly good."

Basilidianism accepted reincarnation after death, to give some souls enough time to achieve gnosis and eventual salvation. Basilides believed that the souls that appeared to be punished for no reason were being punished because they had sinned in a previous life. Basilidianism did not believe in hell, and instead it believed in reincarnation until the soul got it right (much like some branches of Asian Cyclicism). This concept did not sit well with Traditional Christian clerics who complained that by eliminating the possibility of hell after death, Basilides deprived humans of a necessary salutary fear.

Many details of Basilidianism remain unclear, but it shared enough similarities with the other branches of Biblical Demiurgism to belong in this group. Much more would be known today about the history of Basilides and his followers had it not been for the ecclesiastical establishment of the Traditional Christian Church and the efforts of the heresiologist chroniclers, who were responsible for the loss or alteration of most original records. Nevertheless, enough is known about Basilidianism to recognize that it was a major early branch of Christ-revering religion for more than two centuries. Some authors consider Basilidianism to be a sect of Valentianism.[1]

In addition to the *Syrian/Greco/Egyptian* branches of Biblical Demiurgism, I also include Persian/European branches (see the tree on page 414).

PERSIAN/EUROPEAN BIBLICAL DEMIURGISM

The *Persian/European* branch of Biblical Demiurgism arose at least by the second century CE. Like the Syrian/Egyptian branch, it was an outgrowth of early Christianity. Unlike the Syrian/Greco/Egyptian branch, it was more heavily influenced by Zoroastrianism and Buddhism. The Persian/European branch of Biblical Demiurgism included the subbranches of *Marcionism*, *Manichaeism*, and *Catharism*.

MARCIONISM: AN EARLY SOURCE OF NEW TESTAMENT DOCUMENTS

Marcionism was an early branch of Biblical Demiurgism that originated in the Roman Empire during the second century CE after the teachings of *Marcion of Sinope* (85–160 CE). Marcion was the son of an early Christian bishop. Whatever Marcion wrote did not survive to the present day, and his teachings have been reconstructed based on writings of heresiologists and other enemies (discussed further in **Note 10C**).[22] According to those records, Marcion was highly influenced by the teachings of Paul the Apostle, whom he considered to be the only faithful apostle of Jesus Christ.[23] The apostolic letters of Paul were reportedly the most sacred documents of Marcionism. Marcion challenged the early Traditional Christian Church and its strict use of the Old Testament. He was guided by one overpowering idea: God, the benevolent father of Christ, was not Yahweh, the malevolent creator god and Demiurge.[1,22] Marcion was said to be the first to publish a comprehensive list of apostolic writings (mostly by Paul) and gospels that contained no Old Testament. This ultimately led to the completion of the New Testament by the Traditional Christian Church.[1,23] Before Marcion, the Church had not appreciated its own heritage of apostolic writings.

Marcion considered his proto-New Testament as completely superseding the Old Testament. He was so convinced of the need for change in a church that would not listen to him that he started his own branch of religion. Marcionism had the clear characteristics of Biblical Demiurgism. However, Marcion considered it not only to be *a* Christian branch; he considered it the *only true* branch of Christianity.[1] He preached that the father of Christ was the Supreme, Unknowable God of pure spirit (the Monad equivalent) who existed in the spiritual world. He was the good and merciful God of the New Testament who no one had heard of until Christ came to deliver knowledge of him. Marcion considered the creator god of the Old Testament (Yahweh) to be the lesser god of the material world: the evil Demiurge. The material world of the Demiurge was considered to be inherently evil. As with most ideas that challenged the early Traditional Christian Church of the time, Marcion's efforts were not met with warm ecclesiastic reception. He was excommunicated by his father and branded a heretic by Irenaeus. Ironically, Marcion was the first writer of religious doctrine to set forth a formal New Testament (i.e., the collection of Christian writings that became the core sacred scripture of mainstream Traditional Christianity).[1]

In Marcionism, Jesus came to save the material world by delivering knowledge of the Supreme God of mercy, as opposed to the Demiurge. Jesus was not The Messiah who fulfilled the prophecies of Jewish scripture.[1] He was instead a prophet with a message of love for the poor and humble, healing them of their afflictions.[1] Marcion contrasted the God of Jesus who proposed to "love your enemies" with the creator god of the Old Testament who said to kill them. In Marcionism, faith (the Greek *pistis*) was valued as much as knowledge (the Greek *gnosis*). Marcion preached that Christ came into the world as a spirit, entering the body of Jesus after his birth and departing the body of Jesus before he died, without experiencing the human suffering like the physical body of Jesus. Marcion could not conceive of Christ being carried in a female womb or of Mary going through the ordeal of childbirth. According to the heresiologists, Marcion found the whole idea of conception and childbirth to be disgusting and unacceptable.[1,23] He also found it hard to accept that the divine Christ would inhabit a body, as he put it, "full of dung."[1]

From its beginning, Marcionism was on course for extinction because of its mandatory requirement of universal celibacy. The only way to sustain membership was with a constant flow of converts. Although the physical presence of traditional Marcionism died out in the fifth century, some of its elements influenced the later rise of Catharism in the twelfth century, as did another early Persian offshoot of Biblical Demiurgism called *Manichaeism*.

MANICHAEISM: A SERIOUS RIVAL OF TRADITIONAL CHRISTIANITY THAT IS NEARLY EXTINCT TODAY

Manichaeism was a major religious movement that thrived between the third and seventh centuries and faded to extinction by the fifteenth century, except for a few remnants in Southern China that survived into modern time.[24] It was founded by a Persian preacher named Mani (216–276 CE), who born in the southern Babylonian region of Mesopotamia.[1] Although Mani came from a family in an early Judeo-Christian tradition known as the *Elcesaites*, Mani's teachings represented a move into Biblical Demiurgism. Mani received revelations at the ages of twelve and twenty-four. He considered himself to be the final successor in a series of prophets or patriarchs that included Buddha, Zoroaster, and Jesus.[6,25] His predecessor Jesus came to reveal to humans the divine light of their soul as well as its painful captivity by the material body. As with several other branches of Biblical Demiurgism, Mani believed that the divine spirit of Christ did not enter Jesus until he was baptized. Mani declared himself to be an "apostle of Jesus Christ" and an emissary for a universal message. Hoping to prevent corruption of his message and to promote doctrinal consistency, he personally recorded his teachings in writing to give them canonical status during his lifetime.[26]

Manichaeism barely fits the parameters of Biblical Demiurgism, but I believe that it shares enough of its characteristics to be hypothetically included. Much of my highly simplified description of Manichaeism is based on writings by the German-born-and-educated, American Jewish philosopher, Hans Jonas, and a few recent sources.[25,27] In Manichaeism, the unnamed Supreme God within the world of light (a modified version of the Pleroma) is the all-good "Father of Greatness." He is not the creator god. He has Aeon-like entities (called evocations), including a female "Mother of Life" with her son the "Original Man," and a second creation called the "Living Spirit" who had five sons. The Demiurgic creator component of Manichaeism is a collective involving many aspects. The Mother of Life, the Living Spirit, and the five sons created most of the material universe, including ten heavens and eight earths, using the bodies of evil beings from the World of Darkness. As with other branches, Manichaeism saw matter and the material world as evil

(*Left*) The four primary prophets of Manichaeism. From left to right: Mani, Zoroaster, The Buddha, and Jesus. Scroll painting from the thirteenth or fourteenth century CE (Zsuzanna, 2015). (*Right*) Portrait of Jesus on a Manichaean temple from the tenth century CE. Xinijiang, China. Jesus was revered as the penultimate prophet of Manichaeism by the Manichaeans.

and not as the creation of the good God of the spiritual realm. Adam and Eve were created by fallen angels or Demiurge-like demonic archons. Although humans are believed to be temporarily trapped in the material world, they each have a spark from the spiritual world within their soul waiting to be discovered. The knowledge (gnosis) that this spark comes from the world of light is the only way to salvation. The spark of light is believed to be present not just in humans but elsewhere, including in flora such as trees and certain fruits such as melons. Manichaean ascetics (the *elect*) consumed melons and believed that through their burps and farts they released the sacred light particles to ascend to the "ships" of the moon and sun, and then to the Milky Way and the realm of light.[27]

The world of darkness (the opposing realm to the world of light) had an evil King of Darkness (also called Ahriman, Satan, or the devil) who opposed the Father of Greatness. They battled each other through the human soul, which contained both light and dark elements. In Manichaeism, the Father of Greatness was not omnipotent, and the battle between good and evil was unending. Everything in Manichaeism broke down into sharp principles of light and dark, love and hate, and good and evil.

Manichaeism had two tiers: the *elect* and the *auditors* (also called the *hearers*). The elect were committed to a missionary life of poverty and celibacy. The main burden of Manichean ethics was on them. They were forbidden to eat meat or even to harvest plants, considering all of these activities to be forms of murder (sounding somewhat like the extreme asceticism of the Jains discussed in chapter 4). The elect were cared for, in part, by the auditors. They survived only because the auditors were willing to incur the sin of food harvesting and preparation in exchange for release from that sin by the elect who ate the food.[6] This clearly symbiotic arrangement extended from the physical world to the spiritual. The auditors also could have wives or consorts, presumably to enable the production of future Manicheans. Manichaeism incorporated the teachings of Jesus, Zoroaster, and Buddha, with those of Mani.

At first, Mani won the support of many high-ranking political figures in the Persian Empire, but this support did not last. After failing to win the favor of the next generation of royalty and because of building conflicts with Zoroastrian priests, Mani was imprisoned and brutally executed by Emperor Bahram I in 276 CE.[6] The emperor was a strict follower of Zoroastrianism and ordered the killing of many Manicheans during his reign. He also made an example of Mani. He had him flayed, crucified, and then decapitated, with his body on public display as a warning to any who might follow his teachings.[28]

Over the ensuing years, the followers of Manichaeism continued to be widely persecuted by church and state. In 296 CE, Roman Emperor Diocletian, also known for his persecution of Traditional Christians, decreed all Manichean leaders to be burnt alive in an effort to reassert Roman Polytheism. He also ordered the destruction of all Manichean scriptures. Nearly a century later, in 382 CE, Theodosius I issued a decree of death for all Manichean monks, shortly before he declared Nicene Christianity to be the state religion of the Roman Empire.[29] In 780, the Islamic leader and third Abbasid caliph, Abu Abdallah Muhammad ibn Abdallah al-Mansur, began a campaign in Baghdad to put to death any Manicheans who would not renounce their faith. In 843, Emperor Wuzong of Tang issued a death decree for all Manichean clerics in China as part of his campaign to promote Taoism as the state religion.[30] The establishment of the Inquisition in 1232 brought further persecution from the Western Christian Church. It seems almost every competing religion was out to eliminate Manichaeism.

In spite of its widespread persecution in Persia and many other parts of the world, at its height, Manichaeism was widespread. It was even made the state religion of the Turkish Uyghur (Uigur) Empire for a time during the late eighth century.[6] Its churches sprang up as far west as the Roman Empire and as far east as China. It briefly rivaled Traditional Christianity as the dominant religion in many parts of the world that were replacing earlier polytheistic belief systems. Between the seventh and tenth century CE, many historians believe that Manichaeism diversified into, or at least influenced the development of, a few smaller, poorly known subgroups, such as *Paulicianism* and *Bogomilism*.[31,32] Followers of both groups were deemed heretics by Traditional Christianity. Most historical records of these subbranches were destroyed or distorted by the heresiologists of Traditional Christianity. Today, the origin, composition, and monophyly of these subgroups is debated.[27] Manichaeism lasted longer in the east than in the west and has persisted in parts of south China even into modern time.[24]

The *Paulician* branch of Manichaeism flourished between the seventh and ninth centuries in Armenia and the Eastern Byzantine Empire. However, it was eventually stamped out by competing branches of Traditional Christianity. The identity of the branch's namesake is disputed.[32] Most believe it to have been the third-century Bishop of Antioch, Paul of Samosata (200–275 CE), whereas others believe it to have been Paul the Apostle. The founder of Paulicianism was Constantine-Silvanus (620–684 CE), and his successor was Simeon-Titus (?–690 CE). Like other branches of Biblical Demiurgism, Paulicianism was demiurgic, believing in a lesser creator god as the lord of the material world who is separate from the unknowable Supreme God. Unlike some branches of Biblical Demiurgism who believed that the Demiurge was simply ignorant, the Paulicians believed that he was also evil.[32] They believed that Christ came down to Earth to emancipate humans from the material world. They considered themselves to be the "good Christians" and considered Traditional Christians to be "Romanists."[33] During the late seventh century, Paulicianism was severely persecuted by the Byzantine emperors Constantine III and Justinian II. Sometimes they allied themselves with Muslims against the forces of Traditional Christianity.[6] Paulician founder Constantine-Silvanus was stoned to death in 684 CE, and his successor Simeon-Titus was burned to death in 690 CE by Traditional Christian authorities. Paulicians are thought to have influenced the rise of another subgroup of Manichaeism, *Bogomilism*.

Bogomilism arose in the Balkan area in the tenth century and flourished there until the fifteenth century.[34] Its founder was a Bulgarian priest named Bogomil, who we know very little about. Some later leaders of Bogomilism were burned at the stake for heresy.[34] Bogomilism evolved as a merger of doctrines from Paulicianism and earlier elements of Biblical Demiurgism. Like the Paulicians, they believed the Demiurge to be evil.[6] The Bogomils were strongly ascetic, rejecting sex, marriage, meat eating, and worldly possessions.[6] They believed that the soul must be freed from evil and thus the body.

Bogomilism, Paulicianism, and other forms of Persian/European Biblical Demiurgism were widely persecuted by supporters of Traditional Christianity and Islam. Manichaeism was influential on the later development of another major branch of Biblical Demiurgism that developed in the twelfth century called *Catharism*.

CATHARISM: VANQUISHED BY THE CATHOLIC CRUSADES OF MEDIEVAL ROME

Catharism was a Middle Ages revival of Biblical Demiurgism in Southern Europe. Like other branches of Biblical Demiurgism, its history is difficult to reconstruct because most of the surviving historical writings about it came from people who deemed it heretical and focused on trying to erase it from history. It became clearly identifiable by the mid-twelfth century, although it was strongly influenced by the earlier branches, such as Manicheanism and Marcionism.[35]

Catharism flourished through the thirteenth and fourteenth centuries, and its followers are known as Cathars. During those two centuries, Catharism evolved into a number of subbranches, including the *Albigensians* (which some historians use as a term for all of early Catharism), the *Concorezzans*, the *Albanenses*, and the *Bagnolenses*.[36] These sects differed in the way they interpreted the nature of the good God, the Demiurge, and certain sacramental rituals.[36,37] Cathars saw themselves as the only true Christians and denied the title to the Church of Rome because they considered the papacy to be spiritually corrupt.[6,38] They saw the regal buildings and opulent trappings of the Western Christian Church as immoral. The Cathars held their services in fields, barns, and homes. They did not recognize any rigid clerical hierarchy. They led lives of poverty and simplicity, and they rejected all forms of killing, whether it was killing animals for food, killing people in battle, or capital punishment. They also rejected the *tithes* (obligatory church taxes of the day). The traditional Western Christian Church, now in the age of its Crusades, saw these beliefs as a threat not only to their authority but also to their revenue stream. The Christian Church leadership in Rome pronounced the Cathars as heretics in need of conversion or elimination. Pope Innocent III launched the Albigensian Crusades in 1209 CE to eliminate Catharism from Southern France. The conflict did not end well for the Cathars, and they became best known for their prolonged and brutal persecution by the Catholic Church of Rome, leading to their virtual extinction.

Catharism, like other branches of Biblical Demiurgism, believed that there was a good Supreme God, responsible for the spiritual realm. He was a being of pure spirit unsullied by the taint of matter. The Cathars believed this to be the God of the New Testament. Like their earlier predecessors the Marcionites and the Manichaeans, the Cathars completely rejected the Old Testament in favor of the New Testament. Where the Marcionites held the apostolic letters of Paul as most sacred, and the Manichaeans followed a mix of Zoroastrian-Buddhist-Early Christian influence and teachings by Mani, the most sacred text to the Cathars was the apostolic Gospel of John. Cathar religious beliefs included the basic tenets of Biblical Demiurgism, including an evil creator god who was the primary deity described in the Old Testament. He was the Demiurge who claimed to be the one and only god, even though he was a false and cruel god who created the material world. Some Cathars equated him with Satan. Like other branches of Biblical Demiurgism, all matter was evil, particularly the human body.[38] The spirit of each person was trapped within his or her human body while they remained in the polluted material world of the Demiurge.

Catharism had two levels of personal spiritualism. Followers started out at the level *Credentes* (also called *believers*). The higher level was that of the *Perfect* (plural form *Perfecti*). Perfecti at any one point in time were relatively few in number. Female Perfecti were known as *Parfaites*. The higher level was the shortest route to the Pleroma, and it required the believer to commit to self-denial of the material world and go through a ritual called *Consolamentum*.[35] After becoming a Perfect or Parfait, one was expected to surrender all their possessions to the community, vest themselves in a black robe with a cord belt, and undertake a life devoted to purity, preaching, and charitable work. Because everything in the creator god's material world was evil, it all had to be given up to the degree possible.[39] The Perfecti were responsible for helping others find the road that led from the dark material world to the realm of light. They led ascetic lives, not allowing themselves any food that was the product of animal sexual reproduction ("coition" as they called it), with the occasional exception of fish. They were required to avoid all sexual contact because it could produce more bodies in which the Demiurge could imprison souls. Many believers would wait to receive Consolamentum until they were on their deathbed, when the strict obligations of Perfecti purity would be relatively short-lived. To the Catholic Church of the time, the Cather perfects were simply "heretic perfects."[40]

Catharism saw human spirits as the spirits of angels trapped within materialistic creations of the Demiurge. As an evolutionary parallelism or convergence with Dharmic religions, they believed that people would reincarnate after each death, until they achieved salvation by committing to self-denial of the material world as a perfect. Only then would they reunite with the spiritual world. Cathars did not believe in hell or purgatory (another heretical trait according to Western Christians of the time). For the Cathars, this world was the only hell, and there was nothing worse to fear after death than reincarnation back to this world. (Life for common people in Medieval Europe was apparently no picnic.) For Cathar Credentes, sexual relations were permitted, but sex for the purpose of procreation was seen as immoral because it continued the chain of reincarnation and suffering in the material world. Cathars did not rely on an established clergy. Even the Perfecti were only leading by example. They had no equivalent to the pope or clerical hierarchy. They believed that it was the responsibility of the individual to develop a personal inner-relationship with God. As in other branches of Biblical Demiurgism, individual gnosis (knowledge) was the key to salvation. Cathars were antisacramental as well as anticlerical. They rejected infant baptism, confirmation, extreme unction, and the Mass.[39] They also rejected the Old Testament in its entirety, believing its God to be the evil Demiurgic creator god.

Cathars did not accept the Western Christian principle of trinitarianism. They considered Jesus to be a human aspect of an angel and his human sufferings and death to be only an illusion. They even rejected the iconic Christian imagery of the cross. Early Cathars considered gender relatively meaningless, because a man could be reincarnated as a woman or vice versa. As a result, Cathars saw women equally suited to be spiritual leaders, which undermined the rigid male-dominated culture of contemporary Western Christianity.[38] Mary Magdalene had special significance in Catharism, where she had even greater importance than Saint Peter who was the patriarch of the Roman Christian Church and its papacy.[41] In its later days, Catharism became more misogynistic, incorporating a belief that a person's last incarnation had to be as a man to break the cycle of reincarnation, which eventually excluded women from being Perfecti.

Competitive conflict between the Cathars and Traditional Christianity escalated during the twelfth century and into the early thirteenth century until reaching a breaking point in 1209. At that time, Pope Innocent III declared a deadly crusade against the Cathars (previously noted) that lasted more than twenty years and became known as the Albigensian Crusade.[39] Some of the attacks were barbaric, especially under the command of the papal legate, Arnaud Amaury. The most famous was the massacre of the town of Beziers, where the crusaders killed nearly every man, woman, and child on July 22, 1209. Reportedly, Amaury was asked by one of his troops before the final siege how to distinguish the Western Christians from the Cathars. His reply was "Kill them all! God will know his own."[42] The doors of the Church of Mary Magdalene were broken down and the refugees were slaughtered. Prisoners were dragged behind horses and used for target practice. The entire population was slaughtered and the city was looted and burned to the ground. In the aftermath, Amaury wrote to Pope Innocent III, "Today your Holiness, twenty thousand heretics were put to the sword, regardless of rank, age, or sex."[43] Only a year later, another commander of the Western Christian crusaders and ruthless "suppressor of heresy" committed a memorable atrocity landing in the history books. As a prelude to sacking the Cathar stronghold of Lastours, Simon de Montfort, fifth earl of Leicester, cut off the noses and upper lips of one hundred Cathar prisoners and then blinded them, leaving one man with one of his eyes to lead the group into Lastours ahead of Montfort and his men.[38] This was extraordinarily brutal, even by Medieval Crusader standards.

From the mid-twelfth century onward, the Cathars were hunted down by the Western Christian crusaders. Their religious texts were systematically destroyed and their traditions lost to history because they were no longer transcribed. Cathars who refused to recant were hanged or burnt at the stake (and they usually did refuse to recant, in part because lying was against Cathar doctrine). Those that did recant were forced to sew yellow crosses on their clothing worn in public.[38] In 1244, another symbolically important massacre took place at the Cathar fortress of Montsegur, where the troops of the archbishop of Narbonne burned alive more than two hundred Cathar Perfecti in an enormous pyre at the foot of the castle.[44] The last known Cathar Perfect was executed by the Inquisition in Languedoc France in 1321.[38] By the late fourteenth century, Catharism had effectively been driven to extinction.

FINAL THOUGHTS ON THE EMERGENCE OF TRADITIONAL CHRISTIANITY AND BIBLICAL DEMIURGISM

The early centuries of the Common Era were cauldrons of development and competition for many early monotheistic religions. A particularly violent struggle for domination ensued among the early branches that revered Jesus Christ as a major figure, which eventually drove some belief systems, such as Biblical Demiurgism, to extinction. It seems ironic that differences between religions based on the New Testament's "Prince of Peace" (*Isaiah* 9:6) were used so many times through history as justification for violent human conquest, forcible indoctrination, and murder. Biblical Demiurgism was also a victim of historical obfuscation, because most records of it were written by the heresiologists of Traditional Christianity. Hopefully, one day more substantial records of this belief system and its subbranches will be discovered that were written by the Biblical Demiurgists themselves.

Several centuries after the first appearance of Traditional Christianity and Biblical Demiurgism, another major branch of Jesu-venerationism evolved. However, they revered Jesus as the penultimate (twenty-fourth) prophet rather than a divinity. They placed Muhammad as the twenty-fifth and final prophet. This branch was *Islam*, which is the subject of the next chapter.

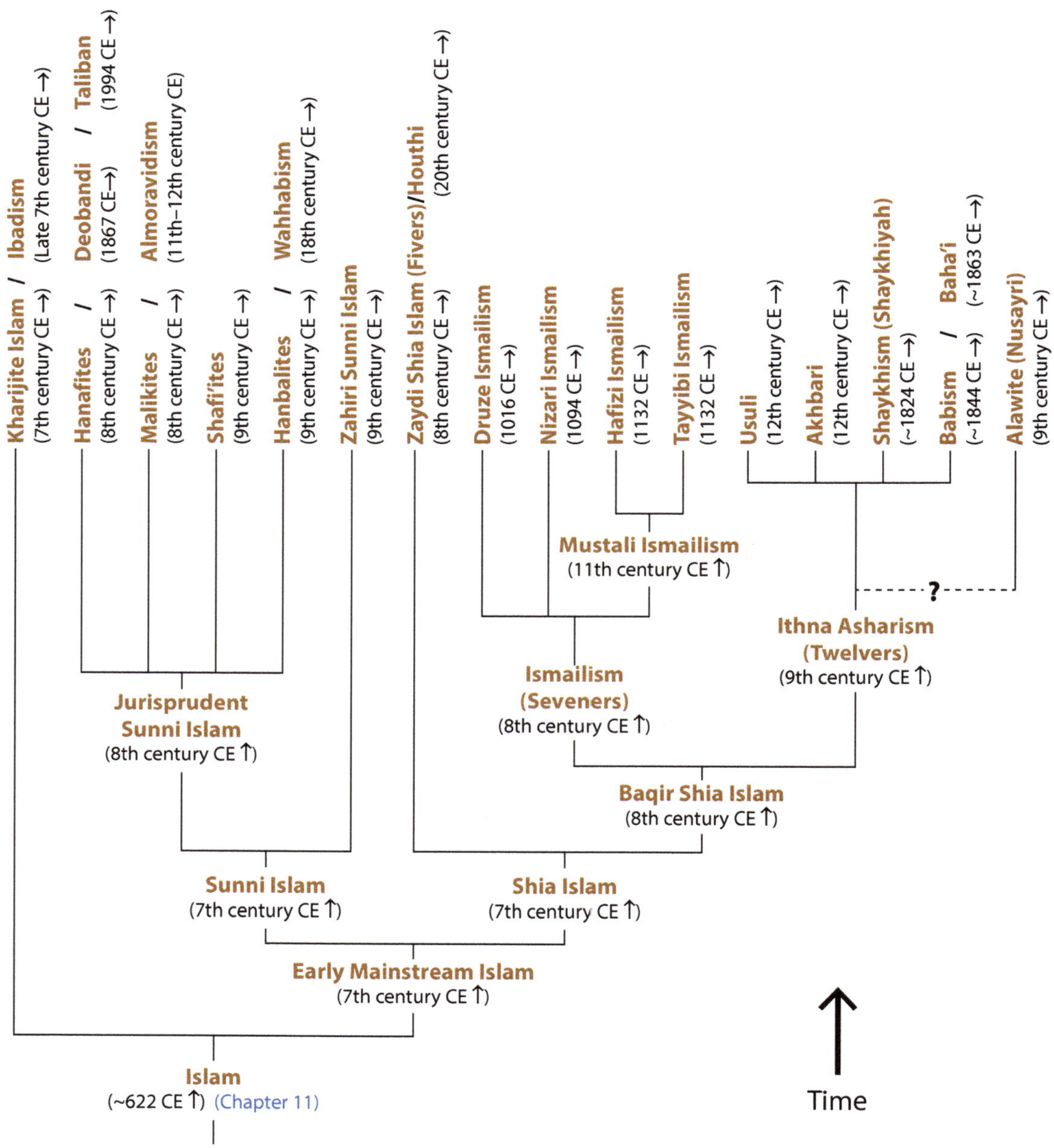

Kharijite Islam (7th century CE →) / Ibadism (Late 7th century CE →)
Hanafites (8th century CE →) / Deobandi (1867 CE →) / Taliban (1994 CE →)
Malikites (8th century CE →) / Almoravidism (11th–12th century CE)
Shafi'ites (9th century CE →)
Hanbalites (9th century CE →) / Wahhabism (18th century CE →)
Zahiri Sunni Islam (9th century CE →)
Zaydi Shia Islam (Fivers)/Houthi (8th century CE →) (20th century CE →)
Druze Ismailism (1016 CE →)
Nizari Ismailism (1094 CE →)
Hafizi Ismailism (1132 CE →)
Tayyibi Ismailism (1132 CE →)
Usuli (12th century CE →)
Akhbari (12th century CE →)
Shaykhism (Shaykhiyah) (~1824 CE →)
Babism (~1844 CE →) / Baha'i (~1863 CE →)
Alawite (Nusayri) (9th century CE →)
Mustali Ismailism (11th century CE ↑)
?
Ithna Asharism (Twelvers) (9th century CE ↑)
Ismailism (Seveners) (8th century CE ↑)
Jurisprudent Sunni Islam (8th century CE ↑)
Baqir Shia Islam (8th century CE ↑)
Sunni Islam (7th century CE ↑)
Shia Islam (7th century CE ↑)
Early Mainstream Islam (7th century CE ↑)
Islam (~622 CE ↑) (Chapter 11)
Time

11

ISLAM

Islam is a way of living rather than a way of thinking.

—ALIJA IZETBEGOVIC

Islam (the Muslim faith) coalesced as a branch of Abrahamic Monotheism in the seventh century CE, centuries after the appearance of Rabbinic Judaism, Traditional Christianity, and Biblical Demiurgism. It began as a syncretization of elements from the Hebrew Tanakh and the Christian New Testament, within an Arab cultural context. Like Christianity, it shares many characteristics with Second Temple Judaism because of its historical and ideological roots. Also, like Traditional Christianity, Biblical Demiurgism, and most of the other branches of Jesu-venerationism, its followers venerate Jesus. However, Muslims venerate him only as the penultimate prophet, rather than as a trinitarian aspect or son of God. Unique to Islam is its most iconic inspirational figure: Muhammad ibn Abdallah, a charismatic son of a merchant family from Mecca (modern-day Saudi Arabia). Muslims see him as the final prophet of Abrahamic Monotheism (succeeding Jesus). In Islam he has the title of "Seal of the Prophets," indicating that his message supersedes that of all earlier prophets.[1]

Muhammad claimed to have received revelations over the course of twenty-three years from God (known as *Allah* among Muslims). These revelations were said to have been delivered to him in the Middle Eastern regions of Mecca and Medina by the archangel Gabriel. Although he was illiterate, scribes recorded and memorized these revelations. Eventually these records were compiled into Islam's most sacred text, the *Quran*, several years after Muhammad's death. Muhammad perceived his revelations as divine intervention to correct original Abrahamic truths that had been corrupted over time by conflicting Zoroastrian, Judaic, and Christian belief systems. Followers of Islam call themselves *Muslims*, meaning "those who submit to God."

Timeline for certain events, people, and beliefs of early Islam

Christian crusaders of the *Reconquista* overthrow the last Muslim dynasty in Spain, soon after which they expel all Muslims (and Sephardi Jews) from the Iberian peninsula →	~1492 CE
Islamic Ottoman Empire overthrows Constantinople and the Eastern Roman Empire, shifting the political center of *Islam* to that city →	~1453 CE
Mongol siege of Baghdad ends the Abbasid Caliphate and the Golden Age of Islam, shifting the political center of *Islam* to Egypt →	~1258 CE
The Crusades begin as a series of "Holy Wars" between Christians and Muslims for control of the holy lands in the Middle East that were considered sacred to both groups →	~1095–1291 CE
Control of Muslim world shifts from Damascus (Syria) to Bagdad (Iraq) where it remained for over 500 years, beginning the *Golden Age of Islam* →	~750 CE ↑
Islamic conquest of Spain giving rule of it to Muslim monarchies for much of the next 780 years →	~711 CE ↑
Rashidun Caliphate conquers Persian Empire and changes the state religion there from *Zoroastrianism* to *Islam* →	~651 CE
Quran completed in written form →	~650 CE
Death of Muhammad and beginning of schism and ideological diversification within *Islam* →	~632 CE
After building an army, Muhammad returns to Mecca, taking control of the city and institutionalizing *Islam* for its impending expansion throughout the Arab world →	~630 CE
Muhammad and some early followers flee persecution in Mecca and emigrate to Yatribe (later named Medina) →	~622 CE
Muhammad experiences first revelation in a cave →	~610 CE
Muhammad born in the town of Mecca (in the present country of Saudi Arabia) →	~570 CE

Pre-Islamic Arabia was a place of nomadic tribalism in a desert environment. Muslims refer to this period as the *Jahiliyyad* (time of ignorance). The tribes were in a constant state of warfare. Parts of their culture were influenced by other regions of the Fertile Crescent, through trade, conquest, missionary activity, and other northern tribal connections.[2] Early Arab culture included populations that practiced forms of Zoroastrianism, Pharisaic Judaism, Traditional Christianity, and even Biblical Demiurgism. Others, including the Quraysh tribe that Muhammad was born into, had polytheistic beliefs with diverse pantheons of gods, goddesses, and Jinni (genies).[3] Their early pantheon was headed by a High God named *al-Lah* (Arabic for "the God").[1,4,5]

As Islam developed, "the God" became "The One and Only God," and the name *al-Lah* was subsequently consolidated as *Allah*.[4] Islam's early evolution (like that of Christianity) was tied to tribalistic conflict and imperial conquest within the Arab community, although during various periods in its later history it would also become a victim of Western imperialism (discussed later in this chapter). Its early success was tied to the charismatic proselytism and organizational efforts of Muhammad, who skillfully managed to bring nearly all of the tribes of Arabia into a single united community. He was a charismatic man who had become deeply familiar with the doctrines of the other major branches of Abrahamic Monotheism. He ultimately concluded that Allah was identical to the God worshipped by Judeo-Christianity, even though those religions disagreed about the significance of Jesus and the nature of God. Muhammad came to believe that God had become misrepresented in other Abrahamic religions and that the revelations he received from the archangel Gabriel would set the record straight.[5]

MUHAMMAD AND THE EARLY RISE OF ISLAM

It is difficult to decipher the life history of Muhammad. The Quran was not intended to provide a biographical account of him. Most written records of Muhammad's life are faith-based recollections produced more than a century after his death.[6] These include the *hadith* (referring to words or assorted teachings and traditions of Muhammad compiled several generations after his death) and a few documents compiled by historians of the late eighth and ninth centuries with limited biographical information. As with the core inspirational figures of so many other ancient religions, it is a great challenge to separate the mythological Muhammad from the historical one. He left no personal writings about himself. According to Muslim belief, Muhammad was illiterate.[6] Images and artistic recreations of him were historically forbidden. In addition, no known significant archaeological evidence proves his existence. Perhaps archaeological evidence exists, but non-Muslim historians are forbidden to enter the two holiest city centers of Islam where he is said to have lived. The first is Mecca, where Muhammad was born and lived his early life. The second is Medina, where he lived his last decade.

According to consensus, Muhammad was born in 570 CE in the town of Mecca,[1,3] a town featuring a local shrine with a sacred black stone.[4] This stone was a black meteorite roughly twelve inches in diameter that fell in Mecca during ancient times. It was first believed to have been sent from heaven by the gods, and a black cube-shaped shrine was built around it called the Kaaba (or Kabah). At the time of Muhammad's birth, hundreds of images featured tribal gods and goddesses within the Kaaba together with the sacred stone.[3] Yearly pilgrimages to Mecca were already common at the time. An annual four-month truce among the warring Arabic tribes even allowed for these sacred visits.

As a young man, Muhammad was a caravan worker. From his early travels, he became familiar with the major traditions of Rabbinic Judaism and Traditional Christianity that had entered the Arabian Peninsula.[4] This familiarity is apparent by the preoccupation of the Quran with those two branches of Abrahamic Monotheism. By the end of the sixth century, Muhammad had seen his tribe, the Quraysh, become rich. His tribe had been extremely successful in making Mecca an international center of trade and the wealthiest settlement in Arabia.[5] Individualism began to replace the communal spirit and competition for personal wealth began to replace concern for the weaker members of society. The different clans of Quraysh fought one another for their share of Mecca's riches in a ruthless form of capitalism. Muhammad began to believe that the Quraysh were making a religion out of money.[5] He became convinced that unless the tribe overcame its greed and went back to its egalitarian principles, the social structure would tear itself apart.

In 610 CE, when Muhammad was around forty years of age, he claimed to have experienced visions of the Angel Gabriel, offering him the first of many revelations he would experience over the following two decades.[1,4] His experience began in a cave in the hills surrounding Mecca where he often went to pray and meditate.[3] He interpreted these revelations as the words of God, and he came to believe himself to be divinely chosen to deliver these words to the world. As he preached to people over the years, he acquired a growing band of followers. The verses and lessons he transmitted to scribes over the years eventually formed Islam's most holy book, the *Quran*.[1]

The Quran preached that Islam was the natural successor to *Pharisaic Judaism* and *Traditional Christianity* (in the sense that those names are used on my phylogenetic tree on page 286). Where much of the New Testament was about Jesus, the Quran was not about Muhammad. Instead, it was presented as a message from God to the people, with Muhammad serving as only an intermediary and prophet.

Sixteenth-century painting entitled *The Ascent of Muhammad to Heaven* by artist Sultan Muhammad (1470–1555). The prophet Muhammad, accompanied by angels, is riding the steed Buraq on his ascent to heaven. Although the Quran and hadith do not explicitly describe Buraq as having a human face (i.e., as a therianthrope), Near East and Persian art usually portray him as such. Following Islamic tradition, the face of the prophet Muhammad was not depicted by the artist. In the collection of the British Library, London.

The charismatic power of Muhammad's words were a combination of his talented delivery and the exceptional beauty of the verses when portrayed in the Arabic language. He is said to have had the communication skills of a talented poet in his sermons.[6] According to the Quran, Muhammad was the last in a succession of twenty-five holy prophets, beginning with Adam (page 445).[7] These prophets include many iconic figures shared with other branches of Abrahamic Monotheism, such as Abraham, Noah, and Moses. Jesus is considered by Muslims to be the twenty-fourth prophet directly before Muhammad. Muhammad did not consider himself divine. He thought of himself as a messenger relaying the words of God. Where the Christian creed proclaims that Jesus Christ is divine as the Son of God, Islam proclaims neither Muhammad nor Jesus as a divinity. They are revered only as prophets and spiritual leaders, and Jesus is considered to be the Messiah (referred to as such in the Quran). The Islamic creed focuses on the unitarian principle of "I believe there is no god but *the* God."

Islam follows the Hebrew prohibition against idolatry and the making of images. In Islam, this policy of aniconography extends particularly to images of God, Muhammad, Islamic prophets, and certain relatives of Muhammad. In paintings that include Muhammad with angels and other religious beings, the face of Muhammad is usually shown as veiled or is left as a blank space (see the figure on page 443). Acceptable Islamic iconography consists mostly of non-anthropomorphic line art, floral, and geometric design, using calligraphy, depictions of artistic patterns, and architectural features.

Many practices and beliefs of Islam have roots in earlier branches of Abrahamic Monotheism. Like early Judaic traditions, Islam attaches great importance to fasting. Like Traditional Christianity, they recognize the importance of Christ and his mother Mary. Islam also includes many characters from the Torah and the New Testament, including Adam and Eve, Noah, Abraham, John the Baptist, Jesus as Prophet, his mother as the Virgin Mary, and some of the apostles.[8] Muslims also have saints (called *Wali*) chosen by God and endowed with exceptional gifts, including the ability to work miracles. Islam, like most other branches of Abrahamic Monotheism, also believes in angels and archangels. Muhammad was originally closer to the Judeo-Christian belief system than he was later in life. Historians contend that he originally instructed his followers to pray facing Jerusalem.[4] He altered the direction of prayer to face Mecca only later, after Judeo-Christian-Muslim relations no longer seemed promising.

Muhammad's earliest following started small, beginning with his wife Khadijah (555–619 CE). There was also his cousin and son-in-law, Ali b. Abi Talib, known simply as Ali (600–661 CE). Finally there was his close friend and father in-law Abu Bakr (573–634 CE). Khadijah was Muhammad's first of at least a dozen wives,

The twenty-five consecutive prophets of Islam (according to the Quran)

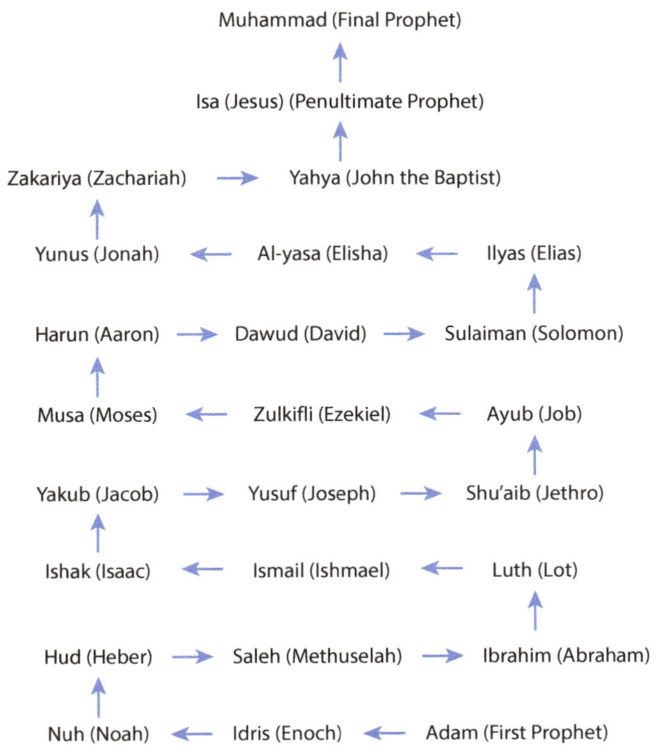

although during her lifetime she was his only wife. They married when he was twenty-five and she was about forty. Together they had three sons and three daughters. All three of their sons died in childhood, causing later problems for those hoping for a heredity-based succession to Muhammad.[3] Khadijah was the first person to believe in Muhammad's prophethood.[9] She encouraged him to accept his experiences as true communications from God.

As Muhammad began to proclaim his revelations more widely, he started to antagonize the polytheistic Quraysh who still revered multiple deities but were now being asked to destroy all statues and images. He also antagonized the local business people for advocating the prohibition of exorbitant interest on loans. The opposition against Muhammad in Mecca grew; but his uncle, Abu Talib, an influential Quraysh clan leader and father of Muhammad's cousin Ali, protected him.[3,6] After Talib died in 619 CE, persecution of Muhammad and his followers by the Quraysh intensified, and early Islamic ideology began to be oppressed. Plots were made against Muhammad's life. During the year 622 CE, Muhammad and some of his followers emigrated more than two hundred miles north of Mecca to the Arabian city of Yathrib (renamed Medina later that same year). This historic flight, called the *Hegira* (also called *Hijra*), marks the beginning of the Islamic era. It also marks the beginning of the Muslim calendar: year AH 1 (*anno Hegirae*, Latin for "in the year of the Hegira").

Over time, Muhammad was able to convert some of the larger Arab tribes of Yathrib to Islam, although his Islamic message conflicted with some of the smaller Jewish Arab tribes. In spite of these early pockets of resistance, Muhammad's power grew. He began to entertain the idea of defining his religious message in terms of law (jurisprudence). Because Judeo-Christianity had laws, he reasoned that Islam should have laws of its own. This marked the earliest conception of an Islamic legal system that would later be formally institutionalized. This legal system would become one of Islam's greatest innovations as well as a source of schism among various subbranches of Sunni Islam.[2] Muhammad was eventually able to banish or execute many Islamic dissenters, gaining control of Yathrib. The city's name was eventually changed to Medina (short for "City of the Prophet") in his honor.

By 629 CE, Muhammad had built a large army. He used this army to return to Mecca and face the polytheistic faction of the Quraysh tribe. By 630 CE, he conquered that city for Islam with minimal bloodshed. Once he took control of Mecca, he destroyed the polytheistic images in the Kaaba and institutionalized Islam as its sanctioned religion. The Quraysh eventually came to embrace Islam. Muhammad had come to see himself as the instrument of God's will. Mecca, Medina, and Jerusalem became the most sacred cities of Islam. After his victory in Mecca, Muhammad returned to Medina, where he spent the rest of his days, until he died in 632 CE from an unknown illness. By the time of his death, he had managed to bring nearly all the tribes of Arabia into a single united community, or *ummah*. Millions of Muslims visit his burial site in Medina each year after their pilgrimage to Mecca and the sacred site of Kaaba. In Islamic Mecca, the nature of the pilgrimages to Kaaba changed radically from its polytheistic heritage. People now came to worship only one God: Allah. Although the numerous images to other gods in the Kaaba are gone today, the black meteorite is still there. It is treated with reverence as a Muslim relic dating back to Adam and Eve. According to one story, Muhammad had kissed the stone, and today, many Muslim visitors try to kiss the stone, emulating the kiss it received from the Islamic prophet.

Taking after other major groups of Abrahamic Monotheism, Islam's rapid growth and institutionalism eventually became a story of power struggles and conquest. Muhammad had originally set out to create a compassionate society in the name of Allah. In the process of winning the hearts and minds of the Arab world, however, he was eventually transformed into a warrior prophet. During the last decade of his

The Kaaba at Mecca, Saudi Arabia (the dark, cube-like building at the center of the enormous crowd). It houses the famous "black stone," which is thought to be a meteorite and also thought by some to have been kissed by the prophet Muhammad.

life, he commanded armies united in the idea of Holy War. He fought many battles. By the time of his death, Muhammad had forcefully united most of Arabia as the heart of Islam. Muslim conquests continued after Muhammad's death, leading to the creation of *caliphates* (Muslim governing bodies or dynasties) occupying vast geographical areas with Islam as the state religion. Following the death of Muhammad, the caliphates, and empires of Muslim states were often in conflict. The ideas of holy war (*public jihadism*) and martyrdom (*shahada*) continued. Islam, from its earliest start, survived and expanded with the help of military conquest in one form or another.[4] It is often described by non-Muslims as a faith established by the sword rather divine miracles.[4] However, most other major branches of Abrahamic Monotheism have also had a history of violent conquest during their rise to prominence. The Jewish Torah and Christian Old Testament both contain vignettes describing even Abraham as a military commander (*Genesis* 14).

Early in its history, Muslims took over much of the Middle Eastern world where the other major Abrahamic religions (including Rabbinic Judaism and Traditional Christianity) had begun. This was part of what drove the center of Christianity westward to Rome. During the transition, even the historical records of Christianity were expunged from the Middle East. Although Christianity flourished in parts of Arabia before Islam, no trace of a Christian Bible in Arabic from that time is known to exist today.[4] In Arabia, the ultimate sacred text became that of Islam: the *Quran*.

THE QURAN: DOCTRINAL CORE OF ISLAM

The Quran is said to have originated from Muhammad. For decades, it was preserved only through oral transmission by Muhammad's disciples, who claimed that they heard him speak the verses. To Muhammad, the Quran was a series of miraculous revelations that he would bring to people. Eventually, scribes wrote down parts of it on papyrus, leather, palm leaves, stones, and even sheep bones. Unlike the Jewish Torah or the Christian Old Testament, which according to biblical accounts was revealed to Moses all at once on Mount Sinai, the Quran is believed by Muslims to have been revealed to Muhammad bit-by-bit, over a period of twenty-three years. This began in the fortieth year of Muhammad's life and ended just before his death. These revelations were called *Ayah* (sign of God). In early Islam, religious leaders had to memorize all parts of the Quran, which included thousands of verses and had to recite it to multiple scribes at a time as the only means of "mass production." It was not until around twenty years after Muhammad's death (650 CE) that the third Rashidun caliph, Uthman ibn Affan, is said to have collated and edited the various fragments into a single book. This was the first complete, standardized text of the Quran to be produced.[2,3,6]

A few historians (although a minority) contend that much of the Quran before Uthman's work was influenced by earlier Syrian Christian liturgies (Oriental Orthodoxy, *sensu* chapter 8). They also believe that it reflected an earlier period of greater Christian influence on Arab culture.[10]They cite that later, when the caliph Uthman produced his edition of the Quran, he ordered all earlier manuscripts to be destroyed. Some historians suggest that this destruction obscured many Syriac antecedents of the Quran.[10]

The name Quran means "recitation" and its poetic verses are repetitive in places and have a rhyme and rhythm meant to be recited. In Islamic centers, verses are chanted aloud with hypnotic resonance by followers during daily prayers. They are heard on Muslim radio, particularly during Ramadan (the fasting month for Muslims). The Quran is not like the Torah or the Christian Bible, from which rabbis, priests, or ministers read passages to the laity. The Quran is a book meant to be recited by all Muslims. The repetition of its phrases is both meaningful and comforting to Muslims who have recited them aloud in daily prayers since childhood. The moral theme of the Quran is emphasized in its *Five Pillars of Islam*, outlining the fundamental core tenets of the faith.[3] These basic rituals are obligatory for all Muslims who are physically able to perform them, and they are as follows.

1. *Shahadah* (Islamic Creed). "There is no God but Allah, and Muhammad is his messenger." For Muslims this is the first thing whispered into a newborn infant's ear, it is recited daily in prayers and it is whispered into a person's ear again on their deathbed. It is inscribed in Arabic on people's doors and on the inside of mosque domes.
2. *Salat* or *Solat* (prayer). Formal prayers to Allah proclaiming his greatness are said five times a day: before dawn, midday, midafternoon, sunset, and nighttime. Those who pray face Mecca. Muhammad originally specified facing Jerusalem. He later changed this to facing Mecca.
3. *Zakat* (charity to the poor). Believers donate a certain portion of their wealth to the poor as a form of social justice. This is similar to alms-giving of many other Organized Religions.
4. *Sawm* or *Saum* (fasting during Ramadan). Muslims fast during the ninth month of the Muslim calendar. Devout Muslims avoid all food, liquid, tobacco, and sex, from dawn to dusk. It is seen as a purification ritual. The end of Ramadan is celebrated with a feast.
5. *Hajj* or *Haj* (pilgrimage to Mecca). This journey is expected of all Muslims at least once in their lifetime, unless prevented by poverty or sickness. It is seen as a way to embrace the entire Muslim community and draw closer to Allah.

The Quran prohibits the consumption of pork (much like traditional Judaism). It also prohibits consumption of wine, and most Muslim scholars interpret this as a prohibition against all alcohol. It also prohibits homosexuality, gambling, and charging interest on loans.

The Quran also introduced the concept of *jihad*, a word meaning "struggle" in Arabic. Islam has two types of jihad.[1] The first is *individual jihad* (also called *greater jihad*): the personal daily struggle to live virtuously.[3] It is the warfare within oneself against evil and temptation. It is considered the most important type of jihad. The second type is *public jihad* (also called *lesser jihad*): the defense of Islam or a Muslim country against aggression. This can take the form of the "pen and tongue" (verbal defense), or in some cases, it can take the form of regulated violence. Public jihad is generally considered obligatory for a segment of the Muslim community, much like the military service in many other countries. It can become violent when it involves what Muslims perceive as justified warfare. This is the type of jihad most familiar to Western society because of its connection to well-publicized acts of violence around the world by extremist Islamic branches, such as ISIL (the Islamic

State of Iraq and the Levant), also known as ISIS (the Islamic State of Iraq and Syria) and al-Qaeda. Public jihad is regulated and to be used only in defense. The killing of noncombatants is prohibited.[1] As with many religions, extremist groups often ignore ethical standards for what they perceive as larger issues, and the fatalities of non-combatants is unavoidable. Also, the broadest interpretation of what constitutes "defense," can be highly subjective. One of the challenges today for some devout Muslims is accepting moral and religious differences in the modern world while being faithful to the positive ideals of jihad and Islamic law.

Some parts of the Quran are seen by non-Muslims as brutal because of certain proscribed punishments, such as hand amputation for stealing (*Quran* 5:38). Like all ancient holy books, the 6,236 verses of the Quran were written down and compiled by people from a different time and cultural context. The few violent phrases of the Quran are sometimes used to portray all of Islam as violent. However, the Quran primarily contains messages of peace and tolerance. As in all major religions, the use of scripture is often a matter of selective interpretation that people in power can use to justify an agenda. Islam, like other major Organized Religions, has many branches, the vast majority of which are nonviolent.

The Quran is the center of Islamic religion and has been translated into more than one hundred languages, although Muslims feel that it loses something important in translated form. Quranic Arabic is considered to be a divine language. Therefore, translated versions are often accompanied by the Arabic text. Another sacred text of Islam is the *hadith* (plural form *ahadith*). The hadith contains numerous reports about the words, deeds, and behavior of Muhammad. For most Muslims, it is the source of religious law and moral guidance. Parts of it were orally transmitted over several generations before they were finally written down in the ninth century CE. The significance of the hadith is considered second only to that of the Quran, but it is of a different nature. Where the Quran has very few verses pertaining to Islamic law, the hadith focuses on *Sharia* (Islamic canonical laws about how Muslims should live) and acceptable Muslim behavior.[9] Over time, differences developed among Muslims in how the hadith should be interpreted and applied. This helped fuel schism and the diversification of Islamic branches.

THE EARLY EVOLUTION (DIVERSIFICATION) OF ISLAM

The early struggles of Muslims were not only against repression from outside powers; they were also within the Muslim community. The ultimate result was ideological diversification within Islam. As long as Muhammad lived, he was the uncontested ruler and judge of the Muslim community.[11] However, after his death in 632 CE, an internal power struggle developed over who his successor would be.[2]

This struggle fueled a number of early schisms. Many, or most, of the conflicts involved full-fledged military campaigns. With Islam, there was no holding back passionate ideological diversification, as with all major Organized Religions. The complex environment for diversification was enhanced by the spread of Islam to many different countries and cultures. The mix of religion with other aspects of culture created schisms and syncretizations. I focus on general religious aspects of major Islamic branches, but I acknowledge that many Muslim people are better described as "cultural Muslims." Although they might be born into a particular branch of Islam, they might not necessarily conform to every ideological detail of the religious branch that they identify with.

One of the first religious belief systems to differentiate from the rest of Islam was a small group called the *Kharijites.*

KHARIJITE ISLAM: EARLY ISLAMIC REBELS

The *Kharijites* (also called *Khawarij*) comprised the first major branch to diverge from Early Mainstream Islam. The name Khawarij literally means "seceder." Their succession as an independent movement is generally set in 657 or 658 CE.[12] They were associated with the earliest Islamic militants. The Kharijites believed that Muslims who commit certain sins were infidels and not true Muslims. They believed these infidels should be hunted down and killed.[13] Sins considered as deserving of such punishment could be something as simple as refusal to accept Kharijite authority and its views on how to select the leader of the Muslim world. They also believed that fights between conflicting Muslim schools over certain issues had to be fought to the finish without arbitration, until one side repented or was killed. Kharijites believed that no single leader was infallible, and that Muslims had a duty to revolt against any ruler who deviated from the Kharijite interpretation of Islam. Early Kharijites condoned political assassination as a means of enforcement, if necessary. They possessed a strong desire for martyrdom and dying for the sake of Allah. For that reason, the Kharijites are often seen as political rebels and forerunners of contemporary Islamic militant movements.[14] They have sometimes been compared to extremist groups such as al-Qaeda and ISIL (the Islamic State of Iraq and the Levant, also known as ISIS, the Islamic State of Iraq and Syria).[15]

The detailed origin of the Kharijites is unclear, but Muslim historians connect it to Muhammad's cousin and son-in-law, Ali, and a rebellion within his army.[3,14] During an Islamic civil war and the battle of Siffin, Ali agreed to arbitration talks with an opposing general rather than continue a lengthy and bloody battle to the death.[3] Some of his troops saw this as a capital sin. In their eyes, Ali should have battled to the death of one side or the other, because they believed "judgement belongs to God alone," and that the victor preferred by Allah always prevails. Tensions escalated until Ali and the rest of his army were forced to fight the rebels, defeating them in 658 CE.[2,3] The surviving rebels later became known as the Kharijites.[3] The Kharijites never forgave Ali, and eventually a Kharijite assassin murdered him in 661 while he was praying in a mosque.

After Ali's murder, most other early Muslims saw the Kharijites as evil, because Ali had been highly regarded by most of Early Mainstream Islam and he was related to the Prophet Muhammad himself. The Kharijites continued to be a source of insurrection against both the caliphate (the Islamic state under caliph leadership) and the imamate (the Islamic state under imam leadership). This insurrection lasted for more than a hundred years, until the late eighth century when they ceased to be a military threat. By that time, they had evolved into around 20 smaller subgroups,[16] most of which are extinct today.

The Kharijites never constituted more than a small percentage of Muslims. They were mostly a tribal society of Bedouin nomads who resented the centralization of power in the developing Islamic state. They believed that the supreme leader of Islam did not have to be from Muhammad's tribe (the Quraysh) or from his family line. They believed instead that the caliph should be elected by the Muslim community based solely on merit, and that he could potentially be anyone, even a non-Arab or black slave.[13] They largely disappeared after the tenth century. A few Kharijite sub branches still survive today living in scattered rural and tribal areas. Surviving Kharijites are mostly *Ibadis* (*Ibadism* on my phylogenetic tree).[1] They make up the majority of the population of Oman and they have a presence in North Africa and Zanzibar.[1] The clergy-led Ibadite community elects an imam based on merit and piety. The imam holds absolute ruling authority over the community, but that community can remove him if he wavers from Ibadi principles and law. The Ibadis were not completely wedded to the necessity of having a single leader for all Muslims. It there is no one fit for the job, Ibadi Kharijites believe that Muslim communities can rule themselves. The Ibadis are more moderate than their earlier Kharijite ancestors were. They no longer consider all non-Kharijite Muslims as heretics or infidels, and today they reject political assassination as a weapon.[1] The Ibadis have survived into the twenty-first century with fewer than one million followers.[9] Even though they were never a very large Islamic branch, they influenced other branches, such as the Sunni Wahhabi.

EARLY MAINSTREAM ISLAM AND THE RISING SCHISM BETWEEN THE SUNNI AND SHIA

Shortly after the schism with Kharijite Islam, another major schism formed within Early Mainstream Islam. The conflict was over the line of succession after the death of Muhammad. Muhammad had left no clear instructions about who his successor should be.[2]

Much like the Kharijites, the majority of Mainstream Islam, (which came to be known as the *Sunni*), did not think the leader of Islam had to be descended from Muhammad's immediate family. They believed that the leader (whom they would call *caliph*) should be chosen by a consensus of his peers, based on his being the strongest and most capable person in the Quraysh community to govern.[3] They chose Abu Bakr, one of Muhammad's close companions and father-in-law, to be Muhammad's successor.

Alternatively, a smaller opposing group, which came to be known as the *Shia*, believed that only God could choose the leader (whom they would call *imam*). They believed that because God had chosen Muhammad, successors to Muhammad had to be descendants from his immediate family. By sharing his bloodline, Muhammad's descendants were also thought to have the blessings of Allah. Therefore, the Shias believed that Muhammad's cousin Ali, was the rightful successor to Muhammad, rather than someone outside of Muhammad's bloodline. They rejected the first three leaders (caliphs) chosen by the Sunni because they were outside Muhammad's bloodline. Conversely, the Sunni rejected the Shias choice of Ali.

Although the Sunni did not consider Ali as the first caliph in 632 CE, they chose him as the fourth beginning in 656 CE. As a result, for a very brief time the Sunni and Shia views on contemporaneous leadership of Islam were aligned (i.e.,the Sunni caliph and the Shia imam were the same person). However, the alignment of the two groups did not last long. A further controversy developed over who deserved to be venerated in Islam. All Muslims agreed with the veneration (although not worship) of the Prophet Muhammad. However, Shias also venerated Ali, whom the Sunni saw as a mere human being. Anyone who advocated for Ali as the most elevated Muslim after Muhammad was declared by the Sunni authorities to be Shia (from *shi at Ali* meaning "the party of Ali").[17] Many Sunnis eventually came to see Shias as heretics, and tensions grew between the two branches.[3] Sunni extremists sometimes refer to Shias in a derogatory sense as "polytheists" because of the reverence that the Shias show for Ali. A 2012 study by the Pew Research Center found that even today, in many Islamic countries, at least 40 percent of Sunnis do not accept Shias as Muslims.[18]

Another major difference between Sunni Islam and Shia Islam is that Sunnis consider the *Companions* (those believed to have interacted with, heard, or seen the Prophet Muhammad) as authorities on the conduct of Muhammad. They are regarded as immune from major sin and beyond criticism. In contrast, the Shias see the Companions as guilty of sin and morally culpable for blocking Ali, the rightful successor to Muhammad, from being the first caliph.[9] Over time, the Sunnis were much more successful in growing their numbers than the Shias. Today, *Sunni Islam* represents between 85 percent and 90 percent of Muslims, and the *Shia Islam* represents only between 10 percent and 15 percent.[3]

SUNNI ISLAM AND ITS EARLY EXPANSION

The name *Sunni* comes from the word *Sunnah*, referring to "practice" or "tradition" based on the exemplar behavior of the Prophet Muhammad. It is a denominational name that was not used with any regularity until the ninth century CE.[18] The *hadith* is a sacred record of Sunnah in literary or verbal form. It is the way that knowledge of Sunnah has been transmitted from generation to generation. Although the hadith helps guide all Muslims, *Sunni* Muslims stress its importance more than *Shia* Muslims.[10] The hadith is the second most sacred source for Sunni jurisprudence (school of legal thought) after the Quran. Only in the absence of answers from the Quran and hadith will most Sunni scholars resort to Islamic community consensus (*ijma*) or analogical reasoning (*qiyas*) to interpret divine law. The Sunni put social and political authority in the hands of their leader, the *caliph*. However, they reserve religious authority for the broader community of religious legal specialists. The caliph is not seen as infallible or a source of revelation in Sunni Islam. He is elected, and he can even be deposed if there is a valid reason to do so. It was the *caliphate* (the succession of caliphs who lead the Islamic state) that did the most to expand Islam during the first millennium CE (see the figure on next page).

THE EARLY ISLAMIC CALIPHATES: RASHIDUN (632–661 CE) AND UMAYYAD (661–750 CE)

The Sunni caliphate is a politico-religious office. The role of the caliph is to expand the realm of Islam, enforce Islamic law, and act as the general supervisor of the government.[1,10] The first four Sunni caliphs after the death of Muhammad were collectively referred to as the *Rashidun Caliphate* (632–661 CE) or "Rightly Guided Caliphs," and they were elected by consensus of Muslim community elders. Their capital was initially located in Medina (modern-day Saudi Arabia) from 632 to 656. It moved to Kufa (in modern-day Iraq) from 656 to 661. It was the Rashidun Caliphate that conquered the Persian Empire and changed the state religion of that region from Zoroastrianism to Islam (from chapter 6).

The first Rashidun caliph elected by the Sunni was Muhammad's companion, Abu Bakr (573–634 CE). He believed that leadership of Muslims belonged to Muhammad's Meccan companions rather than to Muhammad's family, and he became known for launching a number of military attacks against those who abandoned

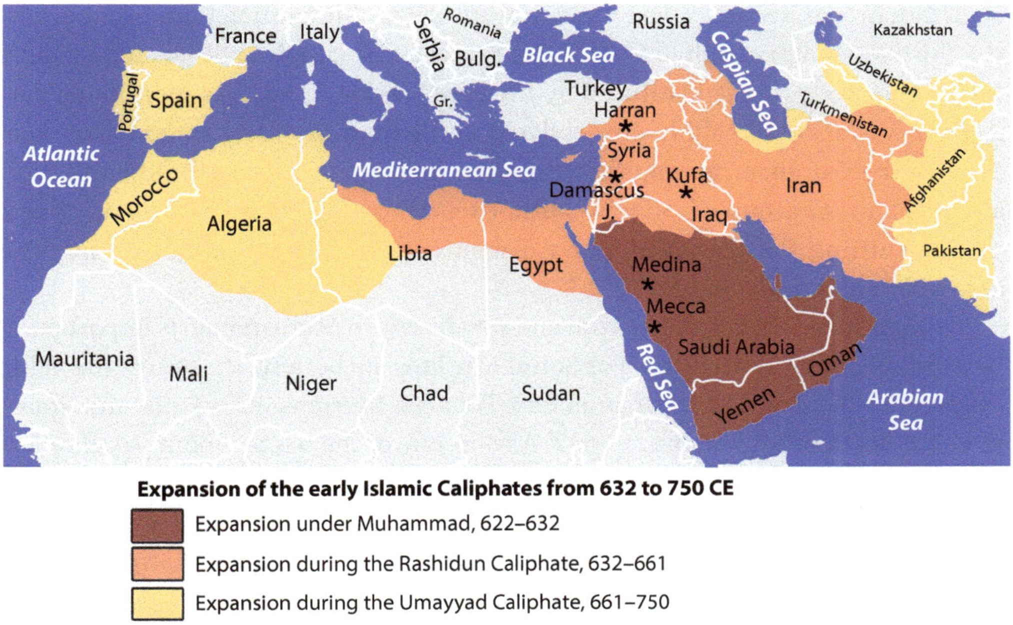

Map showing expansion of the Islamic Caliphates during the seventh and eighth centuries CE, along with the various early capital cities. Superimposed on modern national borders for geographic context.

Islam outside of Medina as well as Muslims who failed to recognize his status as caliph.[9] Abu Bakr accelerated the spread of Islam across Arabia during his reign. He eventually died of natural causes in 634 CE. Before dying, he had chosen his successor, Umar ibn al-Khattab (584–644 CE).

Umar was the second Rashidun caliph and a member of the Quraysh tribe. Originally, he was an enemy of Islam, born in Mecca and set on defending the traditional polytheistic religions of Arabia. Then, about four years before the Hegira (Muhammad's flight from Mecca to Yathrib), Umar had a sudden conversion to Islam. He, like Abu Bakr, became a companion of Muhammad and was a great military leader who organized the Islamic conquests. He eventually became the main early architect of the Arab Islamic Empire.[1] He transformed the Arab army into an elite military class whose leaders ruled the regions they conquered.[9] Muslims, at the time, were not as concerned with proselytism as early Christians were.[19] In the conquered territories, Judeo-Christians, Zoroastrians, and other non-Muslims who did not convert to Islam were required to pay a pole tax in exchange for keeping their non-Islamic religion, but they were not required to convert to Islam. This helped support the Muslim populations. At the height of his power, Umar was assassinated in Medina. His successor was Uthman ibn Affan (576–656 CE).

Uthman was second cousin and son-in-law to Muhammad, and he became the third Rashidun caliph in 644 CE. He was the first caliph to be elected through the committee process instituted by his predecessor, Umar. As mentioned earlier, he was also the caliph who collated and edited the early fragmentary sections of the Quran, arranging them into the first single, standardized book. This angered an elite group of religious authorities who previously had been in charge of transmitting the Quran. Popular discontent led to a mutiny of troops who assassinated him in 656 CE.

As mentioned earlier, Ali, cousin and son-in-law to Muhammad, is important to both Sunni and Shia Muslims. For Sunni Muslims, he became the fourth Rashidun caliph after the death of Uthman in 644. For Shia Muslims, he was the first imam after the death of Muhammad in 632. Ali was one of the ten to whom paradise was promised by Muhammad himself.[1] Ali was another fierce warrior who fought many battles for Islam and within it. He presided over the first Islamic civil war, initiating the final overt split between Sunni and Shia Islam. In 661 CE, a Kharijite assassin murdered him while he was praying in a mosque.

Following Ali's murder, battles over succession and assassinations continued, most notably including Ali's second son Husayn (626–680 CE). Husayn was killed and beheaded at the battle of Karbala by Sunni forces. Husayn's family and troops were also slaughtered, martyred in the eyes of the Shias. The battle further hardened a growing schism between Shia and Sunni Islam, and enhanced the Shiite sense of righteousness and persecution. Today, Husayn is revered as a Shia martyr, and his gravesite in Karbala is a major holy place for Shiite pilgrimage.[1,3] Husayn became a symbol of political resistance, revolution, and spiritual destiny in the centuries that followed. (The Shias and Husayn will be revisited later in this chapter).

After the end of the Rashidun Caliphate in 661 CE, the *Umayyad Caliphate* began a dynastic succession of fourteen caliphs who would reign until 750. This caliphate was founded in 661 by Muawiyya I (602–680 CE), who was a late convert to Islam but had eventually become one of Muhammad's scribes. It was Muawiyya who changed the character of the caliphate to one of heredity monarchy.[1,3] He moved the capital to Damascus, Syria, and continued the rapid military expansion of the Islamic Empire. At its greatest extent (during the eighth century), the Umayyad Caliphate came to encompass more than four million square miles on three continents (Africa, Europe, and Asia). The fourteenth and last Umayyad caliph, Marwan II (688–750 CE), moved the capital to Harran, Turkey, where it lasted from 744 to 750 CE. The Rashidun and Umayyad Caliphates represented the most highly successful historical collaboration between Islam and imperial power. This collaboration also resulted in the largest empire the world had yet seen.

THE RISE AND FALL OF ISLAM'S GOLDEN AGE: THE ABBASID CALIPHATE (751–1258 CE)

The Islamic conquests and expansion during the Rashidun and Umayyad Caliphates were extremely successful. At its peak, Islam came closer to uniting all of humankind under a common theocracy than any other religion.[10] By the eighth century, it was one of history's largest empire in terms of area and proportion of world population, extending well beyond Arabia across much of Northern Africa, Southeast Asia, and even parts of Europe. Its rise was remarkably swift, more rapid than even the rise of the Roman Empire.[19] The Umayyad Caliphate had even taken control of Spain by the eighth century CE.

In 750 CE, theocratic control of Islam shifted from Damascus (Syria) to Iraq, where it remained for more than five hundred years during a period of great cultural evolution with the Abbasid Caliphate. Like the Umayyad Caliphate, the Abbasid Caliphate was a heredity dynasty. Their name is derived from their ancestor, al-Abbas, who was Muhammad's uncle.[1] Where the previous Rashidun and Umayyad Caliphates had focused on military expansion and capturing territory, the Abbasid Caliphate focused on intellectual expansion and archiving knowledge. This came to be known as the *Golden Age of Islam*.[3,9,16] During that time, Muslim rulers relaxed the prohibition of images (aniconography) and proceeded with an ambitious program to assimilate the knowledge and cultures of the many civilizations that previous Islamic caliphates had conquered. The Abbasid Caliphate heavily patronized its scholars and intellectuals to translate classic works of antiquity from other cultures that might otherwise have been lost. They translated the great works from Greek, Egyptian, and other languages into Arabic. During this period, the Islamic world absorbed a broad range of subjects, including logic, philosophy, art, medicine, legal theory, science, and mathematics.[10] Among other things, it established algebra as a distinct discipline. Baghdad and Damascus became world centers of learning and scientific inquiry during the early Middle Ages. During the same time, Western Europe was in decline (referred to by some historians as the Dark Ages because it left little in the way of cultural or literary records).

One result of this Golden Age of Islam, according to legend, was the formation of one of the world's greatest libraries of knowledge at the time, called the *House of Wisdom* (*Bayt al-Hikmah*), centered in Baghdad.[20,21] Much of what Western civilization knows of Greek philosophy and literature may have come from these Arabic translations that were later translated into Latin.[19] The precise breadth and depth of the library is unclear and debated among scholars, because little of its original physical contents survived to the present day. The Mongols, led by the grandson of Genghis Khan, supposedly burned it to the ground during the Siege of Baghdad in

1258 CE. Most knowledge of the library is derived from the writings of contemporary scholars of the time.

By the early thirteenth century, the Islamic Empire had become too widespread to be politically ruled by a single leader. It gradually fragmented and went into decline. The Mongol siege of Baghdad brought the Abbasid Caliphate, as well as the Golden Age of Islam, to an end. The Mongolian siege was one of territorial conquest, not religious proselytism. Unlike most major empires whose conquests were intertwined with one particular religion or another, such as the Roman Empire's entwinement with the expansion of Nicene Christianity or the early caliphate's entwinement with the expansion of Islam, the early Mongol Empire took a more pragmatic approach to religious and cultural differences. They believed that religion was a personal concept not subject to secular law or interference. Mongol leaders generally practiced a form of shamanistic animism. Their repertoire of intimidating military campaigns, however, was tempered with protection of certain religious traditions including Buddhism, Islam, Taoism, and Christianity.[19] This helped the Mongols eventually create an empire during the thirteenth century that was even larger than that of the Abbasid Caliphate, although it fragmented quickly after reaching its peak.[19]

THE SIXTEENTH-CENTURY RISE OF THE OTTOMAN CALIPHATE (1517–1924 CE)

After their fall in 1258 CE, members of the Islamic Abbasid dynasty remained only as titular caliphs in Cairo until the Ottoman Turkish Conquest of Egypt in 1517. After that, a succession of Turkish Ottoman sultans was recognized as the Ottoman Caliphate until 1924, when the secular Republic of Turkey was established and the caliphate was officially abolished.[3] Islam has continued to have a strong presence in the world to the present day, and some Islamic branches even hope to one day reestablish the caliphate (mentioned later in this chapter).

THE EVOLUTION OF SUNNI ISLAM

Like most of Islam, the Sunni do not separate political life from religious life. The early codification of Sunni religious diversification was intertwined with codification of their political laws. Over time, new branches formed as diverging schools of Islamic jurisprudence (different Islamic schools, each guided by their own legal philosophy). These branches evolved in local communities (Mecca, Yathrib/Medina, Kufa, Syria) around local theologians and jurists called *faqih* (trusted experts in Islamic legal theory discussed later in this chapter). Where disputes about the succession of Islamic leadership were the major drivers of schism among Shia Muslims (discussed later in this chapter), the main cause of schism among Sunni Muslims generally had to do with disputes about the relative importance of various legal factors in the interpretation of Islamic doctrine.

Many of the earliest Sunni branches were short-lived. Extinct branches not shown on my tree include the *Awzai* (eighth-eleventh centuries CE), the *Laythi* (eighth century CE), the *Thawri* (eighth century CE), the *Mutazilites* (eighth–tenth centuries CE), and the *Jariri* (ninth–eleventh centuries CE). Eventually four early schools formed in the eighth and ninth centuries CE that became major Sunni branches of today. These include the *Hanafites*, the *Hanbalites*, the *Malikites*, and the *Shafi'ites*. These four groups are collectively grouped as *Jurisprudent Sunni Islam* (see the tree on page 438). In the nineteenth century, the *Salafi reform movement* appeared among Sunni Muslims as a response to European imperialism. It did not evolve from any particular Sunni branch and therefore is not included on my phylogenetic tree. It is a generalization rather than a specific religious branch, somewhat like "evangelism" or "orthodoxy" is to Christianity. As a revivalist movement, it affected multiple branches of Islam, particularly the Hanbali school.

JURISPRUDENT SUNNI ISLAM: DOCTRINAL SCHOOLS OF ISLAMIC LAW

Jurisprudent Sunni Islam includes several branches or schools (also called *madhhab*), each of which was named after a type of ancient Islamic theologian and legal expert called a *faqih* (plural *fuqaha*). The early fuqaha lived in the eighth and ninth centuries CE, and were the accepted authorities on Islamic religious law. Each faqih was thought by its particular school to have had the *right of independent judgment* for interpreting divine Islamic law (Sharia law). They constituted a major segment of the religious elite and were guardians of the community and its religion. The teachings of the early fuqaha inspired the rise of the Sunni schools of jurisprudence, beginning in the mid-ninth century and solidifying by the tenth century.[2,22] This fueled the early diversification of Sunni Islam. The four most significant fuqaha whose later schools persisted to the present day were: Abu Hanifa of Kufa of Iraq, who inspired the *Hanafi* school; Malik bin Anas of Medina who inspired the *Maliki* school; Muhammad al-Shafi'i of Medina who inspired the *Shafi'i* school; and Ahmad bin Hanbal of Baghdad who inspired the *Hanbali* school.

All Islam follows the Quran, but the Quran is relatively unconcerned with rule making, and it is inexplicit about certain details of tradition. Jurisprudent Sunni Islam evolved out of various attempts of the fuqaha jurists to more precisely interpret divine Islamic law. Many early schools of Jurisprudent Sunni Islam became extinct within the first few centuries.[17] By the eleventh century, most educated

Sunni belonged to one of the four large schools mentioned in the last paragraph. These schools were each legally formulated using a variety of particular characteristics, including the approved use or nonuse of the following: personal opinion and free reasoning (*ray*), consensus of the Islamic community (*ijma*), deduction by analogic reasoning (*qiyas*), judicial discretion by suitably qualified legal scholars or "juristic discretion" (*istihsan*), and decisions with the public interest in mind (*istislah*). Differences in how these various factors were prioritized or excluded is mainly what differentiated the various schools of Jurisprudent Sunni Islam from each other ideologically. There are slight variations in the ways some of these terms are defined by different authors. Further study of these characteristics and others could possibly result in better resolution of the Sunni tree of ideological diversity and interrelationships, but I leave that to specialists in the field.

Each school of Jurisprudent Sunni Islam has its own method of attempting to understand divine Islamic law. The body of rulings was known as *fiqh*. Although all Muslims see Sharia law as part of Islamic tradition, the Sunni schools of jurisprudence accept human effort to understand it as fallible and subject to variation.[9] The idea of legal schools in Islam was extremely successful. By the end of the eleventh century, Sunni jurists became established as Islam's foremost authorities.[23] They founded branches that ensured the continued Sunni dominance of Islam. The system they devised excluded other claimants to religious authority, including the caliphs, imams, and various theologians of the time.

Although much work is yet needed to understand the detailed early evolution of Sunni Islam and its numerous books of law and doctrinal interpretation, I simplify the story by beginning with the four schools of *Jurisprudent Sunni Islam* that have been studied by historians in most detail. I begin with the Hanafites (see the tree on page 438).

HANAFI SUNNI ISLAM AND ITS SUBBRANCHES FROM THE *DEOBANDI* TO THE *TALIBAN*

The *Hanafi* Sunni follow a school of legal thought named for Abu Hanifa (699–767 CE). He was a merchant and faqih. He was a native of Kufa, Iraq, but was of Persian descent.[9] He stressed the idea of an international Muslim community based on the Quran and Sunnah. He was eventually offered the post of chief judge of state by the Abbasid monarch and caliph, Al-Mansur, which he turned down. He was later imprisoned for his beliefs and for refusing this post, and he eventually died in prison.[1]

The Hanafi school was later formally developed in Iraq (then the intellectual capital of the Muslim world), primarily by two of Abu Hanifa's disciples: Abu Yusuf (735–798 CE) and Muhammad al-Shaybani (740–805 CE).[1,9] They declared that the

Quran and hadith were insufficient for certain Muslim legal issues. Thus, analogic reasoning (*qiyas*); personal opinion (*ray*); Islamic community consensus (*ijma*); and juristic discretion (*istihsan*) should be legitimate considerations in formulating Islamic law. This was the first Sunni school to formulate contract rules for taxation and business transactions involving resale for profit. The Hanafi school is the most widespread Sunni school today, prevailing in much of the Middle East, India, Pakistan, and has presence in China, and even parts of the former Soviet Union.[9]

A number of well-known subbranches evolved from this school, including the *Deobandi* in British-occupied India. The British attempted to rule the Indian subcontinent during the mid-nineteenth century, and one of their tactics to weaken the indigenous population was to heighten tensions between Hindus and Muslims in the region. The Deobandi movement rose in 1867, as a Hanafite revivalist movement, to preserve Islamic teachings that Muslims thought were being corrupted.[9] The Deobandi represent a sternly religious movement. It opposes education in business and modern science, arguing that education should be entirely religious.[3] As part of an Indian independence movement, the Deobandi (like the Hindus) advocated a form of Indian nationalism that worked against the British occupation. The Islamic Deobandi were also against partitioning of the Indian subcontinent over religious differences. In the end, their efforts for unity failed. In 1947, India split into the Dominion of Pakistan (later named the Islamic Republic of Pakistan, which is 97% Islamic) and the Dominion of India (later named the Republic of India, which is 85% Hindu)[3].

Today the Deobandi form a significant part of the conservative Muslim population of Pakistan, a country where blasphemy is a capital offense still enforced in the twenty-first century (Section 295-C of the Pakistan penal code). The country remains in violent conflict with India. Conflict over Muslim-Hindu religious differences famously sparked the assassination of Mohandas Gandhi, among others. The partitioning of India and Pakistan resulted in a rapid, massive population relocation of millions of Muslims from India to Pakistan, and millions of Hindus and Sikhs from Pakistan to India. The migrations resulted in the loss of hundreds of thousands of lives and violent religious-based persecution. Religious-based tensions have continued to grow between the two regions to this day. India named its most powerful nuclear missiles after the Hindu Vedic god of fire, *Agni*.[24,25] In response, Pakistan named its powerful nuclear missiles after Muhammad of Ghor, the Sunni ruler of the Ghurid Empire who led the Muslim conquest of the Indian subcontinent in the late twelfth and early thirteenth century.

One of the groups that evolved from the ranks of the Deobandi in the late twentieth century was the militant *Taliban* of Afghanistan.[9] This fundamentalist movement emerged out of the devastating Afghan civil war that began in 1989. In 1994, they established the Islamic Republic of Afghanistan, which was supported and recognized only by Saudi Arabia, Pakistan, and the United Arab Emirates. The Taliban represent one of the strictest branches of Islam, mandating segregation of sexes, full veiling in the burqa for women, and prohibition of women driving, working outside the house, or even seeking an education. The regime was denounced by international human rights organizations for its refusal to allow the practice or presence of any religion besides Islam in the country and for its treatment of women. The Taliban also provided shelter and training grounds for Osama bin Laden and his associates. They were forcibly removed from power after the September 11, 2001, attacks in the United States on the World Trade Center and the Pentagon. The Taliban regained control of Afghanistan in August 2021, after capturing the capitol city, Kabul. Currently, Afghanistan is 88 percent Sunni and 12 percent Shia Islam.

The majority of Hanafi Sunni Islam today is peaceful, and it includes about one-third of all Muslims.[9] Another major school of Jurisprudent Sunni Islam is the *Maliki school.*

THE *MALIKITES*: ALLOWING THE CONSENSUS OF MEDINA JURISTS TO INFLUENCE RULINGS

The *Maliki* Sunni follow a school of legal thought developed by the theologian and Islamic legal specialist, Malik ibn Anas (711–795 CE). He was born and lived in Medina, leaving only to perform pilgrimages to Mecca.[1,9] His teachings reflect the scholarship of eighth-century Medina, from which he did not deviate. As a youth, he memorized the Quran, and he supported the use of holy relics in seeking religious blessings. He was author of one of the earliest surviving Muslim law books, the *Muwatta.* It contains doctrines of law and religion, including sayings attributed to Muhammad's wives, relatives, and other associates.

Distinguishing features of the Maliki school include its reliance on the *Companions* of Medina (followers of Muhammad who had personal contact with him) as an authoritative source of law. This school was favored by the Umayyad Caliphate because of their Medinese roots.[2] In addition to the hadith, the Maliki school emphasizes the legal rulings of the first four caliphs (the *Rashidun*), especially Umar. It also values consideration of public interest (*istislah*) as a guide to interpretation of Islamic law, even allowing it to overrule a deduction from the Quran and Sunnah. Like the Hanafites, the Malikites also value personal opinion (*ray*), juristic discretion (*istihsan*), Islamic community consensus (*ijma*), and analogical reasoning (*qiyas*) to influence Islamic law. The Malaki school is closely related to the Hanafi school, but does not assign as much weight to analogy (qiyas).

Some authors consider Malikites to be intermediate between the Hanafites and the Shafi'ites.[1]

One of the offshoots of the Maliki school was the *Almoravid Islam*. The Almoravids began in North Africa in the eleventh century as a strict disciplinary branch of Sunni Islam focused on the Quran, hadith, and the Maliki school of law.[26] They were militaristic Saharan desert ascetics who shunned lavish decorations, palaces, and monuments. They closed taverns, destroyed musical instruments, and enforced Islamic law in the distribution of war booty. They were also fierce warriors. Their successful military conquests eventually spread Almoravidism into Morocco and Spain. Their successes also sparked a determination among Christian forces for the reconquest of Spain. After the twelfth century, the Almoravids were largely extinct as a religious branch, although other Maliki branches survive. The Maliki school is today found predominately in northern and western Africa (sometimes known as the Muslim West) and to a lesser degree in areas across the Middle East.

Another major school of Jurisprudent Sunni Islam is *Shafi'I Sunni Islam*.

THE *SHAFI'ITES*: CLARIFICATION OF ISLAMIC LAW

The *Shafi'ites* follow a school of legal thought developed by Abu Abd Allah Muhammad al-Shafi'i (767–820 CE) (also referred to as "al-Shafi'i" for short). Al-Shafi'i was born in Gaza and died in Egypt. He received his theological education in Mecca, Medina, and Iraq. Although a faqih, he began as a student of another faqih, Malik ibn Anas.[1,9] Shafi'i eventually diverged from Malik's teachings and was first to categorize the theoretical bases of Islamic law and clarify the differences among the early schools of Jurisprudent Sunni Islam. For this reason, he is known as "the architect of Islamic law." The Shafi'ite school was later founded by his students and followers.[1]

Following the Maliki school, Al-Shafi'i attempted to establish rules that would provide a middle ground between conservative traditionalism (e.g., the Hanbali school) and innovation (e.g., Hanafi school). The Shafi'i school supports the use of Islamic community consensus (*ijma*) and analogical reasoning (*qiyas*). It rejects the use of jurist discretion (*istihsan*) as a source of Islamic law because it is too much like human legislation of Islamic law. It also rejects the use of personal opinion (*ray*). The Shafi'ites are prominent in Egypt, Palestine, Jordan, Ethiopia, Somalia, Yemen, Kurdistan, and various other regions of Africa, the Caucasus, and the Middle East. They also have a significant presence in Syria, Lebanon, Iran, Yemen, Iraq, Pakistan, India, and Indonesia.[9]

Another major branch of jurisprudence Islam is the Hanabali.

THE *HANBALITES*: CONSERVATIVE IN RITUAL, LIBERAL IN COMMERCE

The *Hanbali* Sunni follow a school of legal thought first developed by Ahmad ibn Hanbal of Baghdad (780–855 CE).[1,9] Ibn Hanbal established no formal school of his own, but some of his disciples systematized his pronouncements on legal issues. Hanbali Sunni Islam is the official school of Saudi Arabia, Qatar, Kuwait, and the United Arab Emirates. It also has followings in Palestine, Syria, Bahrain, Oman, Yemen, and Iraq. The Hanbali school is the most conservative of the Jurisprudent Sunni schools in ritualistic matters, but most liberal in commercial matters. It is characterized by its dogmatic nature, recognizing no source other than the Quran and the Sunnah in Islamic law. It rejects blind adherence to the opinions of scholars. Unlike the Hanafites and the Malikites, but similar to the Shafi'ites, the Hanbalites reject jurist discretion (*istihsan*) and personal opinion (*ray*). Unlike the Hanafites, Malikites, and Shafi'ites, the early Hanbalites also rejected community consensus (*ijma*) and analogic reasoning (*qiyas*), as sound basis from which to derive Islamic law. However, later Hanbalites readopted qiyas to accommodate a trend in mainstream Islam toward less conservative tradition.[2]

The Hanbali school gave rise to the ultra-conservative Wahhabi movement in the eighteenth century[1] which is today often associated with the Salafi movement.[9,51] This puritanical Muslim movement is named after the preacher and activist Muhammad ibn Abd al-Wahhab (1703–1792 CE) who started it in Najd, on the Arabian Peninsula. His father was a Hanbali jurist. The general goal of Wahhabism began as a social and moral reconstruction of traditional Muslim society in response to what was perceived as a moral decline of the Muslim community in Arabia. The Wahhabi were extremists in dealing with Muslims who did not subscribe to their theology.[9] They exercised violent opposition to the Muslim mystical cult of saints as well as to shrine and tomb visitations, even to that of Muhammad. Somewhat resembling the New England Puritans practice who of not celebrating Christmas, the Wahhabis do not celebrate Muhammad's birthday.[7,51] They introduced the *mutawwi'un* ("enforcers of obedience") as a kind of religious police to monitor public and private conformity to their idea of appropriate Islamic tradition.[1] They sacked the Shia shrines in Najaf and Karbala in 1801–1802, killing many Shias in the process. They have continued to destroy many sites associated with the early rise of Islam, such as the graves of the Companions of Muhammad, because of their prohibition against veneration of sacred objects and places. An early collaboration with the Saud family in the late eighteenth century eventually led to an important Wahhabi presence in what is today Saudi Arabia. (The relationship between Wahhabism and Saudi Arabia is briefly discussed in **Note 11A**). The Hanbali school is becoming increasingly influential today, in part because of their successful proselytizing efforts in Africa and the East.[1]

These four branches of *Jurisprudent Sunni Islam*, along with their various derivatives, collectively form the sister group to *Zahiri Sunni Islam* on my phylogenetic tree (page 438).

ZAHIRI SUNNI ISLAM: ORTHODOX SUNNI OUTSIDE THE SCHOOLS OF LAW

Many theologians also include the largely extinct *Zahiri Islam* as an early Sunni branch, or even as a fifth school of Sunni jurisprudence. However, the Zahiris were a highly conservative group that largely rejected the use of legal opinions to derive Islamic laws. They claimed that Muslim life should be derived only from the Quran and Sunna. For this, they are sometimes referred to as the literalist school.[2] Therefore, I include it as *Zahiri Sunni Islam*, sister group of Jurisprudent Sunni Islam (tree on page 438).

Zahiri Sunni Islam was founded in the ninth century CE based on the teachings of Dawud al-Zahiri (815–883 CE).[2,29] Al-Zahiri's father was a scribe and follower of the Hanafi school.[27] Al-Zahiri was originally trained in the principles of *Shafi'i* Islam, but he later branched off with new principles of his own. Very little is known about his life. Zahiri Islam followed close literal interpretation of the Quran and Sunnah to govern Muslim life.[1] They held the conservative approach that the Quran did not contain figurative or metaphorical expression open to subjective interpretation. Unlike most of Jurisprudent Sunni Islam (other than Hanbalites), it rejected the use of analogical deduction (*qiyas*) in interpreting Islamic law. For example, the Quran specifically prohibits the use of wine (*Quran* 5:90). Jurisprudent Sunni Islam extrapolates this, by analogy, to mean the prohibition of any intoxicating substance. In contrast, Zahiri Sunni Islam rejects any such analogy, recognizing a more literal interpretation of the Quran as prohibiting only alcohol. Like Hanbali Islam, Zahiri Islam also rejects the use of personal opinion (*ray*) and jurist discretion (*istihsan*) in formulating Islamic law. Unlike the branches of Jurisprudent Sunni Islam, Zahiri Sunni Islam did not seek to establish its own formal school of law.[27] In the words of one Islamic scholar, Zahiri Sunni Islam showed "some tendencies in legal thought" but "diverged from mainstream jurisprudence."[28]

Zahiri Sunni Islam flourished in the Iberian Peninsula (Spain and Portugal), but it was also present in Syria, Iraq, Egypt, and elsewhere. Over the course of its history, a number of minor subbranches evolved, including one that considered the Virgin Mary to be a prophet.[25] Zahirihism reached its peak in the eleventh century CE, after which it began to decline. By the fourteenth century, it was largely extinct.[9]

With the Reconquista (Christian reconquest of territory lost earlier to Muslim rule) and loss of the Iberian Peninsula back to Christian rule by 1492, many Zahiri records were lost, making historical study a challenge (as with many early Islamic branches).

SUNNI EFFORTS TOWARD PAN-ISLAMIC REUNIFICATION

Sunni efforts to reunite Islamic branches into a Pan-Islamic state have existed since the early days of Islam.[9] These efforts have been strongly influenced by the Wahhabi and Deobandi branches, as well as by Afghani cultural traditions.[9] In the mid-nineteenth century, it was discussed as a possible way to save the Sunni Ottoman (Turkish) Empire from fragmentation. By the twentieth century, this movement was used to marshal Pan-Islamic support against foreign aggression and imperialism. Sometimes the fear of a common enemy is the easiest way to bring people together (consider the ancient proverb, "the enemy of my enemy is my friend"). An unfortunate side effect is that it has attracted a number of well-known extremist groups to violence. Their militant actions are rejected by the majority of Muslims. Even though such groups constitute a minority of Sunni Muslims, they have done much to equate Islam with violence in the eyes of the non-Muslim world. The most famous of these fringe groups in the West today are al-Qaeda (1986–present), the Taliban (1994–present), and ISIL (1999–present).

Just as early Jews and Traditional Christians sometimes used the Torah or Old Testament to justify imperialism and violence against unbelievers and heretics, Sunni militant groups, such as al-Qaeda, have used the Quran to justify ideologically motivated nationalism. Although the bulk of the Quran is nonviolent, does not promote violence, and actively extolls peace (e.g., Quran 8:61), it has a few passages that Muslim extremists have taken out of context (and out of ancient times) to justify violence against what it perceives as enemies of Islam (*Quran* 8:12 and 8:60). This is not a universal product of Islam. It is instead partly a politicized form of Islam trying to recreate the caliphate, and partly a strike against what some Islamic groups see as the "evil" west. It is not just about religion; it is also about power and control. Sunni extremism is no greater than the extremism found in early Western Christianity (chapters 8 and 9). It is also partly in response to what they see as a history of Western imperialism in the Middle East by Great Britain, France, and the United States through most of the twentieth and twenty-first centuries.[29] It is part of a lethal cycle of conflict that has grown between the Arab world and the Western world for centuries, fueled by ideological righteousness, nationalism, fear, and endless cycles of retaliation.

One of the best-known recent effects of this cycle to affect the United States in recent history involved the destruction of the World Trade Center in New York

City and part of the Pentagon in Washington, D.C., together known as the September 11 terrorist attack. In 1998, Osama Bin Laden, leader of the Sunni militant group al-Qaeda, issued a fatwa that decreed the killing of Americans and their allies everywhere, both civilian and military, as a duty of every Muslim.[30,31] The main justification that he gave for this included the continued occupation of Islamic lands by the United States (primarily on the Arabian Peninsula), its massive aggression toward the Islamic State (e.g., the bombing of Iraq by the United States and the United Kingdom in 1998), and its continued support of the Jewish State of Israel.[32]

On September 11, 2001, nineteen al-Qaeda Sunni Muslims hijacked four commercial airliners filled with passengers and flew three of them into the buildings. The fourth never made it to its intended target and crashed into a field. This attack resulted in the loss of more than three thousand lives and more than six thousand wounded. Partly motivated by the September 11 terrorist attack, as well as by some suspected Iraqi weapons of mass destruction, which were never found, the United States invaded Iraq in 2003. After three years in Iraq, the invasion cost more than a hundred thousand lives, mostly Iraqi noncombatants (discussed further in chapter 12).

For al-Qaeda, the war with America was partly a religious war (or at least religion was used as a way to justify the violence). Bin Laden's handwritten instructions to the September 11 hijackers read, "Continue to recite the Quran. Purify your heart and clean it from all earthly matters. The time of fun and waste is gone. The time of judgement has arrived. . . . You will be entering paradise. You will be entering the happiest life, everlasting life."[33] The al-Qaeda leader's promise of guaranteed entry into heaven for dying in the fight against unbelievers and "Western immorality" is reminiscent of the eleventh-century popes of Western Christianity who, during the Crusades, promised that to fight and die for Roman Catholicism against its enemies, guaranteed immediate entry into heaven (as discussed in chapter 8). The promise of eternal paradise is a powerful motivator for the faithful. Major acts of violence have also been committed in the name of Islam by the Taliban and ISIL. These Muslim extremist groups did not materialize without reason, however, and they are politicized forms of religion. Major acts of violence have been committed in the name of every major religion at one time or another (discussed further in chapter 12). Such violence is not a direct product of religion. Instead, it is an example of the way that individuals and groups can leverage religion to moralize aggression in the name of rightful expansionism, preemptive defense, or justified revenge. It is also an example of the horrendously inadequate intercultural communication and understanding that persists in the modern world.

Groups such as al-Qaeda often see themselves as acting "defensively" in response to outside aggression against Muslim territory and tradition. The Quran tells them: "Whoso commits aggression against you, do you commit aggression against him"

(*Quran* 2:194). In that sense, the Quran is not unlike the Torah and Old Testament's sentiment of an "eye for an eye" (*Deuteronomy* 9:21). Al-Qaeda began as a response to the Soviet invasion of Afghanistan.[34] Once the war with the Soviets ended, al-Qaeda continued fighting Islamist causes and eventually turned the focus of its efforts toward other countries with major military campaigns in the Middle East, such as the United States. They came to see the United States as an evil enemy of Islam seeking to overthrow Muslim countries.

Like all major religious documents, the interpretation of its holy scripture is in the eye of the beholder. A person can selectively use passages to defend nearly any kind of action, even though the majority of the Quran has nothing to do with violence. Expressing a more peaceful interpretation of the Quran, even the twenty-first century Roman Catholic Pope Francis wrote: "Authentic Islam and the proper reading of the Quran are opposed to every form of violence."[35] In spite of a few extremist groups, the majority of Sunni Muslims are peaceful and recognize the necessity for increased tolerance of cultural diversity in a multicultural world.[3]

About 86 percent of Muslims today consider themselves to be Sunni,[36] although many Sunnis identify with one particular subbranch or another. The smaller sister group to Sunni Islam is *Shia* Islam (see the phylogenetic tree on page 438).

SHIA ISLAM: ONLY GOD CAN ELECT A LEADER, SO HE MUST BE A DESCENDENT OF MUHAMMAD

Shia Islam and its diverse branches believe that the rightful successor to the Prophet Muhammad must be a descendant from his family. They believe that the first rightful successor was Muhammad's cousin, Ali ibn Ahi Talib ("Ali"). Most Shias believe that Muhammad had designated Ali as his immediate successor on several occasions, including on his deathbed, a claim disputed by the Sunni. The Sunni rejected Ali and elected a succession of three supreme leaders before him that were not descendants of Muhammad's family (the caliphs Abu Bakr, Umar, and Uthman) before finally electing Ali as their fourth caliph. This created the early roots of schism between Sunni and Shia Muslims. The Shias also believe that upon Ali's death, the supreme leadership of Islam should have passed to his son Husayn; but before Husayn could assume that position, Sunni forces killed him. The Shias see Husayn's burial site in Karbala (in present-day Iraq) as a major holy place and pilgrimage center.

The Shias defined *eligibility* for supreme Muslim leadership more narrowly than either the Kharijites or the Sunni. They believe that the leader of Islam must be a blood relative descended from the immediate family of Muhammad through his daughter Fatima Zahra or his cousin Ali. Unlike the Sunni who use the term "imam" for positions of much less significance (e.g., merely a jurist or a leader of prayer in a mosque), the Shias see the imam as having social, political, and religious authority. Also unlike the Sunni, most Shias (other than the Zaydis) see their supreme leader as sinless and infallible.[1,11] Disobeying a rightfully appointed imam is equivalent to disobeying Muhammad.

In contrast to the Jurisprudent Sunni Muslims, whose early internal schisms were based on differing doctrinal parameters and interpretation of Islamic law, early schisms within Shia Islam resulted from differences on who should occupy the position of imam (successor to Muhammad). Most Shia believe that there have been a total of twelve successors or imams, beginning with Ali and ending with the twelfth born in 874 CE who has been in a hidden form of spiritual existence ever since. This branch is known as *Ithna Ashari* (sometimes referred to as the *Twelvers*). Smaller Shia branches disagree with the Twelvers about the number and identity of imams, including the *Ismailis* (also called *Seveners*), who disagree with the Twelvers on about the identity of the seventh Imam; and the *Zaydi* (known as the *Fivers*), who disagree with the Twelvers about the identity of the fifth imam (tree on page 438).

As in the early development of Sunni Islam, the early development of Shia Islam included a violent competition among Muslim warlords, which was particularly dangerous for its supreme leaders. Nearly all of these leaders were murdered over the course of history. According to Shia writers, the five imams of the Fivers, the seven imams of the Seveners, and eleven of the twelve imams of the Twelvers were all assassinated (mostly by poisoning) between the seventh and ninth centuries. This was done at the order of Sunni caliphs. The one Twelver to theoretically have escaped assassination is claimed to have been the twelfth "Hidden Imam," Muhammad al-Mahdi.[11] He disappeared from known human contact while still a child, and the Twelvers believe that he will reappear at the end of time with Jesus to fight the Antichrist. Shiite writers note some similarity between stories of Muhammad al-Mahdi and stories of the Traditional Christian Jesus.[11]

Dozens of early Shiite branches (sects) evolved during the seventh through tenth centuries CE, but little is known about their comparative characteristics and historical interrelationships.[37] Most became extinct within the first few centuries of their existence. By the early 1500s, Shia Islam had become the state religion of Iran while much of the rest of the Muslim world had become Sunni under the Ottoman state. Today, Iran remains the center of Shia Islam, but Iraq, which is the burial place of Husayn, Ali, and four other imams, is also a major spiritual center of Shia Islam. Although Shias account for only 10 percent to 15 percent of Muslims today, they are the majority in certain oil-rich lands of the Middle East, adding to their power and influence.

Shia Islam can be divided into three major branches: the *Zaydi*, the *Ismaili*, and by far the largest branch, the *Ithna Ashari*. Where the schisms among Sunni subbranches were largely driven by different interpretations of Islamic law (jurisprudence), the major schisms and diversifications among Shia subbranches continued to be largely over issues of leadership and succession.

ZAYDI SHIA ISLAM: *FIVERS* AND THEIR INTERPRETATION OF THE FIFTH IMAM

Zaydi Shia Islam (also known as *Zaydi*, *Zaidi*, or *Fivers*) split from the rest of Shia Islam by recognizing a different fifth imam (hence the name "Fivers").[1,9,38] They chose *Zayd ibn Ali* (695–740 CE), grandson of Muhammad's cousin Ali, as the fifth imam. They favored him because of his activist revolutionary position against the Sunni Umayyad dynasty. The rest of Shia Islam (termed *Baqir Shia Islam*) determined Zayd's brother Muhammad al-Baqir (677–733 CE) to be fifth imam. The Zaydi were the first to become independent of the rest of Shia Islam, eventually founding state dynasties in Tabaristan (northern Iran) in 864 CE and Yemen in 893 CE. Unlike the Baqir Shia, the Zaydi do not view the imam as a supernaturally endowed, infallible person representing God on earth.[9] The only necessary qualification for imam in Zaydi Shia Islam is descent from Muhammad's cousin Ali, Muhammad's daughter Fatima, or Muhammad's grandsons, al-Hasan or al-Husayn. The imam must also have an absence of physical imperfections, possess personal piety, and be willing to take up the sword for Islam and fight against corrupt rulers. The Zaydi retain some Sunni traits that are not shared with the rest of the Shias, including its own school of law and acceptance of Abu Bakr and Umar as legitimate caliphs.

Where they exist, today's Zaydi subgroups are minority populations. The more violent of these subgroups fetch headlines: Since 2004, Zaydi Shia fighters in Yemen from the Houthi branch (also spelled Huthi) have been uprising against the Sunni (Salafi) majority of the country, in part to promote the Zaydi religious identity in the country where the Sunni majority was shutting down Zaydi schools.[39] It also contains a core of anti-western rhetoric. One of their early slogans was "Allah is Great. Death to America. Death to Israel. Curse on the Jews. Victory to Islam."[40] The ongoing fighting has resulted in many casualties in Yemen, and by 2015, the conflict between the Sunni and the Houthi evolved into a full-fledged civil war.

The extant sister group to Zaydi Shia Islam, which includes most of Shia Islam today, is the *Baqir Shia Islam* branch (tree on page 438).

BAQIR SHIA ISLAM: EARLY CONFLICT WITH THE SUNNI

The branch of *Baqir Shia Islam* is defined based on its followers recognizing Muhammad al-Baqir as the fifth imam.[9] He was poisoned by a Sunni caliph and buried in

Medina under the dome in al-Baqi, where two other imams were also buried. The grave was demolished by the Wahhabi Sunni Muslims in 1806, later rebuilt by Shias, and demolished again by the Wahhabi in 1926. Among other things, Muhammad al-Baqir supported the Islamic concept of *taqiyya* (permitting the denial or lying of religious belief and practice in the face of persecution). This was important to early Shia Muslims because Sunni Muslims so commonly persecuted them. Unlike the Zaydi Shia, Baqir Shia Islam considers their imams to be infallible.

Baqir Shia Islam has continued to grow, although it has never caught up with the Sunni branch in membership. Eventually, a schism developed within Baqir Shia Islam between the *Ismailis* (*Seveners*) and *Ithna Ashari* (Twelvers) over imam succession (tree on page 438).

ISMAILISM (*SEVENERS*): CONTROVERSY OVER THE SEVENTH IMAM

The schism between the *Ismailism* (*Seveners*) and the *Ithna Ashari* (*Twelvers*) occurred in the eighth century CE over who would be recognized as the seventh imam and over the number of imams that would succeed Muhammad.[1,9] The Seveners believe that the seventh imam should have been Ismail ibn Jafar (719 CE–762 CE) the first-born son of their sixth imam, Ja'far al Sadiq (702 CE–765 CE). The situation became complicated when Ismail died several years before his father did, which ultimately resulted in a major schism. The Twelvers (discussed in the next section) thought that because the first-born son of the sixth imam had died, that his oldest *surviving* son should be known as the seventh imam. The Seveners thought that the right to succeed the sixth imam had been passed to Ismail even before his father died. Therefore, the Seveners believe that Ismail had *posthumously* became the seventh imam (and should be known as such). The Seveners believe that their prophets come in cycles of seven.[9] The Seveners are also characterized by believing Islam to have seven pillars of faith rather than the five of other Islamic branches.[1] They add *Taharah* (purity) and *Jihad* (struggle). Details of the pillars vary somewhat among the different Ismali branches. Over the years, the Seveners have evolved into their own complex phylogenetic tree, which includes three major branches: *Druze Ismailism*, *Nizari Ismailism*, and the *Mustali Ismailism*. Many other branches, not included here, have evolved from this group.[9]

The *Druze* were an early offshoot of Ismailism that began in early eleventh-century Egypt and Syria, and contained syncretized elements of Asian Cyclicism and Linear Monotheism. Hamza ibn Ali (985–1021 CE) founded the branch, although it is named after Muhammad ad-Darazi, an early leader of the faith who presented himself as a new prophet. Ad-Darazi was labeled a heretic in 1016 CE. In 1018, he was executed for his teachings. The Druze are distinctive within Islam in believing in reincarnation and in forbidding polygamy and concubinage.[1,9] In their version of reincarnation, a Druze can be reincarnated only as another Druze, and after a series of successive reincarnations, the soul may one day be united with the Cosmic Mind for ultimate happiness (the Druze version of heaven). Until then, hell is a series of successive reincarnations resulting in distance from God. Because of their strict adherence to monotheism, they self-identify as "unitarians," not to be confused with the nontrinitarian Christian groups that go by the same name. Many Muslims consider the Druze to no longer be a school of Islam, even though it historically evolved from it. They were persecuted as heretics beginning in 1021 CE, and by 1043, their missionary activities ended. Conversions were no longer accepted, and the faith became cloaked in secrecy, concealing doctrines and practices from the uninitiated.[1] This prevented the understanding of its origins, doctrines, and practices until relatively recent times. Even today, they are targeted by ISIL and al-Qaeda for persecution. Today, Druze populations survive in Lebanon, Syria, and Israel with around a half-million members.[1]

The *Nizari* and the *Mustali* diverged from the rest of mainstream Ismailism in 1094 CE over a leadership issue, as with so many other schisms within Islam. The Nizari Ismalis recognized Abu Mansur al-Nizar, surnamed al-Mustapha al-dinillah (Nizari for short) as the Ismail leader, even though he was an infant. In contrast, the Mustali recognized Nazari's brother al-Mustali as the Ismail leader.[9] Each branch has evolved and diversified into a number of subbranches.

The Nizari Ismailis were particularly passionate in their opposition to Sunni rule, and their tactics of violence and terror during the first several centuries of their existence earned them the epithet of "Assassins" by the Christian Crusaders.[38] This image changed dramatically toward efforts of a more peaceful nature by the nineteenth century. At that time, the forty-sixth Nizari Ismaili imam fled to India to establish a line of imams with the title *Aga Khan*. The current and fourth Aga Khan, who considers himself to be the descendant of Ali and Fatima (daughter of Muhammad), is Harvard-educated Shah Karim al-Hussayni (1957–). He has focused on cultivating the spiritual and cultural lives of the more than fifteen million Nizari Ismaili and improving their quality of life. He has also created a number of institutions focused on eliminating global poverty, advancing the rights of women, promoting Islamic art, and fostering pluralistic values in society.[38] Most Nizari are today found in Africa and West Asia.[9]

The Mustali Ismailis differ from the Nizari and Druze primarily with regard to who they accepted as their leader (as explained earlier). They originated in Egypt, later moving their center to Yemen and also spreading to Western India through missionaries in the eleventh century. Mustali Ismailism split into the *Hafizi* and the

Tayyibi by the early twelfth century CE, over differences in whom they accepted as the legitimate Imam.[41]

Today, several million Ismailis are scattered as minorities across more than twenty-five countries of Asia, the Middle East, Africa, Europe, and North America.[9] They have a strong presence in India and East Africa. The sister group to Ismailism (Seveners) on my phylogenetic tree is the *Ithna Asharism* (the Twelvers) (phylogenetic tree on page 438).

ITHNA ASHARISM (*TWELVERS*): ELEVEN PAST IMAMS AND THE AWAITED ONE

Ithna Asharism (also known as the *Ithna Ashariyah*) is the branch of Shia Islam recognizing twelve particular imams; the last of which they believe has been in spiritual hiding for more than 1,200 years. He is known as Muhammad al-Mahdi (also called Muhammad al-Muntzar, the "Awaited One" or "the Mahdi"). He is not mentioned anywhere in the Quran. He is a concept that was developed by the Twelver Shia.

Most Shias believe that at the end of time, the twelfth imam, *Imam al-Mahdi*, will return as a messianic figure, together with *Isa* (Jesus Christ) to defeat the Antichrist in an apocalyptic battle and bring an era of peace and justice to the world.[38] They believe that he was born in the late ninth century CE and became imam at the age of five after the Sunni assassination of his father, the eleventh imam.[1] The Mahdi was taken into seclusion and was seen only by senior figures and supposedly answered questions only through deputies (*wakil*). Eventually, it was announced that he would remain hidden until God gave him permission to manifest himself again as the twelfth imam to take on the Antichrist. He became known as the *Hidden Imam*, suspended in a form of spiritual existence known as occultation, where he was to remain, patiently waiting for the end of times. There are many variations of this story, depending on which Shia school or historian tells it. The core idea of the Twelvers is that after the first twelve imams, the line of legitimate imams ended. Since the disappearance of the twelfth imam, the ruler of the Shia state has not been a true imam, but rather the most learned jurist or jurists serving as the twelfth imam's representative. Since the late nineteenth century, these jurists have been called *ayatollahs* (meaning "signs of God").[11] Unlike most other branches of Islam, the Twelver Shias have a clerical hierarchy whose apex consists of religious leaders acknowledged by their followers as ayatollahs because of their reputation for knowledge and piety.[38]

Little is known about the branching history of early Twelver Islam until the twelfth century CE, even though proto-Twelver Islam is thought to have existed at least as far back as the ninth century CE. Today they predominate in Iran and

Southern Iraq. One of the better-known and largest branches of Ithna Asharism is the *Usuli*.

USULI: SHIA JURISPRUDENCE AND THE IRANIAN ISLAMIC REVOLUTION

The *Usuli* branch of Twelver Islam is a Shia branch that was influenced by the Sunni idea of organizing itself as a school of jurisprudence.[1,9] The name comes from the Muslim term *Usul al-fiqh*, meaning "principles of jurisprudence." It was founded by Aqa Muhammad Baqir ibn Muhammad Akmal (1702–1792 CE). Usulism relies on a series of rational processes and independent reasoning (*ijtihad*) by special religious authorities called *ulama*. The ulama were the guardians, transmitters, and interpreters of Islamic law and tradition. Some of them declared themselves representatives of the Hidden Imam, the Mahdi. Usuli requires strict obedience to the rulings of the ulama by followers. Such obedience allowed jurists to radically influence Iranian history. The tradition fueled the political activism of the Shias in the nineteenth and twentieth centuries, and was a precursor to the Iranian Islamic revolution of 1978 and 1979.[9]

The Usuli use independent reasoning to arrive at novel and unprecedented decisions. Their interpretations of Islamic law are in contrast to another contemporaneous Shia branch, the Akhbari, who required precedence for all decisions about Islamic law rather than independent reasoning. Early Usuli leaders were extremely aggressive, declaring the Akhbari (discussed in the next section) as infidels (*kafirs*) and expelling them to lives of marginal existence in southern Iraq. They employed religious police (*mirghadabs* or "executers of wrath") to enforce their views, a precursor to the *Islamic Revolutionary Guards* of Iran.[1]

One of the most widely known Usuli was a grand ayatollah from Iran by the name of Ruhollah Khomeini (1902–1989). He was known in the West simply as the Ayatollah Khomeini. He was the figurehead for the 1979–1980 Iranian Islamic Revolution that overthrew the U.S.-backed authoritarian shah of Iran. The revolutionaries rebelled against the shah's authoritarian efforts to secularize Iranian society. The ayatollah championed the idea that the government of Iran should be in the hands of Muslim jurists, not the increasingly Westernized shah. The revolution of Khomeini and his followers eventually succeeded, and in 1979, the shah fled the country into exile in Egypt. Khomeini installed what he referred to as "God's government" in Iran.[42] The revolution transformed the country into the Islamic Republic of Iran, whose state religion is Shia Islam. Khomeini called for further Islamic revolutions across the Muslim world to combat secularism, including growing secularism within its neighbor, Iraq, whose nationalist leader was Saddam Hussein.

Iraq feared that the Iranian revolution would lead Iraq's Shiite majority to rebel against its more secular Baathist government. In an attempt to avert this, Iraq invaded Iran in 1980, beginning a brutal war that lasted from 1980 to 1988. The United States and the Soviet Union (among others) provided political support to Iraq, while Iran remained relatively isolated during the conflict. It was one of the bloodiest conflicts of the century, resulting in an estimated 680,000 dead, and another one-and-a-half million wounded and maimed.[43] It aroused world condemnation for its use of indiscriminate ballistic-missile attacks on cities by both sides and heavy use of chemical weapons (mostly by Iraq). Fighting between the two countries ended in August 1988, with both countries having sustained heavy losses but neither side able to claim victory.

Ayatollah Khomeini was named *Time* magazine's 1979 *Man of the Year* for his international influence and leadership during the revolution. To the Shiite Twelvers, he was one of the most charismatic leaders in modern time and was the first and only Iranian cleric to be addressed as imam. His governing resulted in strict enforcement of Shia norms and harsh measures against those considered to be deviant threats to his ideals, such as the *Shaykhis*, *Babis*, and *Baha'i*.[9] Khomeini continued to call upon Muslims of the Persian Gulf and throughout the world to rise up against secular, pro-Western government. He referred to the United States as the "Great Satan" and the former Soviet Union as the "Lesser Satan." His fatwa calling for the assassination of Salman Rushdie, author of *The Satanic Verses* (discussed later in this chapter) was widely accepted among Muslims. His death in 1989 led to widespread outpouring of grief throughout the Shia Islamic world.

The Usuli represent the overwhelming majority of Twelver Shias today. One of its main Twelver opponents are the Akhbari.

AKHBARI: REJECTING THE USULI AS EVIL SUNNI INFLUENCE

The *Akhbari* are a twelfth-century branch of Twelver Shia Islam that strongly opposed the Usuli. They saw the Usuli school of jurisprudence as an evil Sunni influence.[23] Akhbari see the uncertainty and conflicting opinions among the jurists of such schools as an unnecessary and evil defect. They prefer the single authoritative opinion by a central imam. Most Akhbari believe that no one has the authority to set new religious rules until the return of the Hidden Imam, the Mahdi. Therefore, they also reject the authority of the Usuli *ulama* (Usuli scholars, preachers, and judges). Today, the Akhbari movement is a minority Shia group found mainly in Southern Iraq and Bahrain. The third major Twelver branch on my tree (page 438) is *Shaykhism*.

SHAYKHISM: PERSECUTED AS HERETICS BY THE USULI

Shaykhism (the *Shaykhiyah*) is a branch of Twelver Shia Islam named after Shaykh Ahmad al-Ahsai (1753–1826 CE) who founded the religion in 1824 CE. He was influenced by the Akhbari Twelver school.[9] Although he was born in Bahrain and died in Saudi Arabia, he spent fifteen years in Iran where he acquired many followers. Part of Shaykhis doctrine teaches that at the end of the world, resurrection will be spiritual rather than physical. He reasoned that the physical body would be of no use in the next world. He also claimed to have received this and other information as mystical visions direct from the infallible imams who had chosen him as the spokesperson for the Hidden Imam.[1,9] The Iranian Usuli considered this heresy, and al-Ahsai was condemned by the traditional Usuli ulama (religious authorities). He was exiled to Mecca, where he eventually died. His teachings were carried on by his appointed successor, Sayyd Kazim Rashti (1793-1843 CE) in Karbala, who organized Shaykhism more formally. Following the Islamic Revolution in Iran in 1979, the Shaykhi were persecuted and their leader moved the Shaykhi branch headquarters to Iraq.[9] They still maintain minority followings in Iran, Iraq, Syria, Pakistan, Kuwait, and Saudi Arabia. After the death of Kazim in the mid-nineteenth century, many Shaykhis converted to *Babism* or *Baha'i*.[1]

BABISM AND THE MERCHANT BAB CLAIMING TO BE THE HIDDEN IMAM

Babism, also known as the *Bayani* branch of Twelver Shia Islam, evolved from earlier Shaykhi roots. It originated as a militant Iranian movement led by a merchant from southern Iraq named Sayyid Ali Muhammad Shirazi (1819–1850 CE). In 1844, after a series of revelatory visions, he claimed to be the *Bab* (gate) to inborn knowledge of the Quran, and the connection to the *Hidden Imam* of Twelver Shia Islam.[9] At first, he attracted mainly Shaykhi followers, but soon he gained a much wider footing, becoming well established in many parts of Iran. Mainstream Muslim leaders attempted to stem the growth of Babism, which drew a growing militancy from the Babis.[1] Bab preached an intensified version of Islamic law and militancy in preparation for what he believed was going to be the reappearance of the Hidden Imam, the Mahdi. He believed the date of this reappearance would be in 1845 or 1846.[9] The eventual nonappearance of the Madhi did not dissuade the majority of Bab's followers. In 1846, the Bab was arrested by order of the Iranian government and imprisoned in an effort to contain his influence. This only made him more impassioned. He now claimed that he was the Hidden Imam in person, designated by divine authority. Babism's perceived interference and disrespect of Quranic law caused the Islamic state to act decisively and violently, accusing him of blasphemy against Islam.[44] The Islamic state attacked and killed most of Babism's leadership along with thousands of its followers. It executed the thirty-year-old Bab in 1850 by firing squad in a public square.[1,9] Most of Bab's prophetic career had been spent under house arrest or in Iranian prisons. After his death, Babism became a suppressed movement and most of its followers either disappeared or became influential to development of the *Baha'i* movement.

BAHA'I: BROAD IDEOLOGICAL INCLUSIVENESS AND REINTERPRETATION OF SHIA ISLAM

The *Baha'i* religion evolved from Babism in the mid-nineteenth century, after most surviving Babis had turned to a new leader named Mirza Husayn Ali Nuri (1817–1892).[44] Ali Nuri was born in Tehran to parents who were prominent members of the Muslim community. Although brought up as a traditional Muslim, he became an early and longtime follower of Babism and its leader, the Bab. More than a decade after the Bab's execution, Ali Nuri changed his name to Baha'u'llah, meaning "the glory of God," proclaiming himself to be the new "Promised One" that Bab had prophesized. He claimed to have experienced divine visions during the period from 1853 to 1863 appraising him of his mission to establish the Baha'i faith.[9] He was condemned as a heretic and exiled by the Ottoman Caliphate to the prison city of Acre (also spelled Akko), in modern-day Palestine, where he eventually died in 1892 and was buried.[1]

Although the Baha'u'llah founded the Baha'i faith, the Bab remained as its most iconic figure. The Bab's birth, death, and declaration anniversaries are considered Baha'i Holy Days. The Baha'i consider Bab to be a manifestation of God, and they ascribe miracles to him, such as surviving the first attempt by a firing squad to kill him (although the second attempt did the job). As the Baha'i faith developed, it deviated even further from mainstream Shia Islam than either Shaykhism or Babism and adopted a more humanistic, universalist approach. The Baha'i believe in progressive revelation through a continuum of divine manifestations of God.[9] The manifestations have included the core iconic figures of many major Organized Religions through history, such as Zoroaster, Abraham, Moses, Buddha, Jesus, Muhammad, Bab, and for the present age, Baha'u'llah.[1]

With this continuity of divine messengers comes the Baha'i directive to accept all of the world's major religions as valid. It is a highly inclusive belief system, much like other universalist practices, such as those of the *Unitarian Universalists* (mentioned in chapter 9). The Baha'i see all major religions as unified in purpose but varied in social implementation and practice. The Baha'i faith also aspires to reject

The *Baha'i* Temple in Wilmette, Illinois. This is the oldest surviving Baha'i House of Worship. Its construction began in 1912 and was completed in 1953. It is open as a place of worship for people of any faith. Like a few other small branches of religion (e.g., the *Cao Dai* discussed in chapter 4 and the *Unitarian Universalist Church* discussed in chapter 9), the Baha'i attempt to create an inclusive, nonviolent ideological system accepting of all religious faiths as valid.

prejudice and condone universal education. It promotes harmony between science and religion, equality of men and women, and a spiritual basis for society. Baha'i holds a basic belief that all religions ultimately come from the same God and teach the same essential truths. Adherents attribute the differences among religions to historical, geographic, linguistic, and social factors. There is no priesthood, and church administration rests with elected councils (Spiritual Assemblies) with the *Universal House of Justice* in Israel having supreme authority.

Membership seats in the Universal House of Justice are limited to males, providing an exception to the Baha'i principle of equality between the sexes. Baha'i documents do not explain why, but Baha'i authorities claim that the reason will be revealed with time. Although it is one of the world's smaller Organized Religions, Baha'i is one of the most geographically widespread religions, with followers in nearly two hundred countries. In 2010, there were just over 7.3 million adherents, with the largest membership in India.[36] Although the religion is historically rooted in Twelver Shiism through Babism and Shaykhism, most Muslims consider Baha'i to be heretical and exclude it from Islam.[9] However, many Baha'i simply distinguish themselves as a new interpretation or descendent of Shia Islam.

SYNCRETIZED TWELVER GROUPS OF UNCERTAIN RELATIONSHIP

The *Alawites* (also called *Alawis* or *Nusairi*) are a highly secretive Shia school of thought located in northwestern Syria, northern Lebanon, and southern Turkey.[1,9,11] The school was founded in the ninth century by Muhammad ibn Nusayr, a follower of the eleventh Shia imam. The Alawites were often persecuted early in their history: first by the Christian Crusaders in the late eleventh century and then by the Ottoman Caliphate in the sixteenth century.[11] The Ottoman Empire oppressed them and forced many to convert to Sunni Islam.

The precise relationship of the Alawite branch within Shia Islam is unknown for two major reasons: they are highly secretive about much of their religion, and the Alawite branch appears to be a complicated syncretization of elements from Shia Islam and Christianity.[11] It is unclear whether they began as Twelvers influenced by Christianity, or as Christians influenced by Twelver Islam. As with the Shia Twelver branch, they believe in the twelve imams. They consider the Quran to be one of their holy books and consider Ali to have been the rightful leader of Islam after Muhammad. However, in addition to celebrating the Muslim holidays, they also celebrate Christian holidays, such as Christmas. They also recognize the biblical patriarch Saint Peter as the embodiment of true Islam.[11] They have a revered trinity that includes God (Allah), Ali, and Muhammad as three aspects of one person.[11] This is highly controversial within Islam and in contrast to the Quran, which states "do not speak of a 'Trinity'; desist - it is better for you. Indeed, Allah is but one God." (Quran 4:171-173) The Alawites believe that God appeared on earth seven times, and they deify Ali as his last manifestation. The Alawites also believe in reincarnation, and they allow social drinking of alcohol. Many Shias consider Alawites as heretics and exclude them from Shia Islam. They are considered *ghulat* (a Muslim term for those considered to be deviant by orthodox Shia standards).[11] The name ghulat is an Arabic term meaning, "to exaggerate or exceed the proper bounds." Of all major Islamic branches, the Alawites appear to be closest to the Twelvers, but this belief system warrants further study to determine its precise historical origin and to determine whether it is a monophyletic group. The Alawites have more than one million followers today.[1] Since the 1980s, they have become extremely powerful in Syrian politics. The Assad family is Alawite and has ruled Syria since 1971.

Many other poorly known religious belief systems were founded in the thirteenth to fifteenth centuries CE that appear to be syncretizations of elements from

Twelver Shia Islam with non-Shia Islam, or even non-Islamic groups. Some of these belief systems include *Alevism, Safavism,* and Bektasism (not shown on my phylogenetic tree).[11,45] They often are secretive and combine unexpected combinations of theological characteristics. Some may not even constitute single groups at all. Like the Twelvers, most of these groups recognize twelve imams, and they hold the first (Ali) with special reverence. They also follow many of the Shia dietary taboos. However, most generally do not make the pilgrimage to Mecca nor do they pray openly five times a day or acknowledge the Quran as God's word.[45] These groups are in need of much further study to determine their historical origins and position within the phylogenetic tree of Islam.

NON-MONOPHYLETIC GROUPS AND GROUPS OF QUESTIONABLE IDEOLOGICAL ORIGIN, WITH TIES TO ISLAM

Sufism is not a specific denomination or branch of Islam in the evolutionary sense. Instead, it is a generic name for the intense mystical expression of Islamic devotion present in both Sunni and Shia Islam. Some forms of Sufism even show cross influence between Islam and Hinduism.[1,18] Much like evangelism within Christianity, Sufism is an intermittent characteristic that does not uniquely define a single branch of the Islamic phylogenetic tree. It is sometimes referred to as "Islamic mysticism." One of the best-known practices within Sufism is *fana*: the inducement of a state of ecstasy.[1] In this state, Sufis claim they sense a loss of self where all that remains is God. This has been controversial within Islam. Al-Hallaj (858–922 CE) was a Sunni and an early major figure of Sufism. He memorized the entire Quran before he was twelve years old. For many years, he addressed himself to popular audiences, some who reported his performing miracles. He would sometimes fall into trances in which he claimed to be in the presence of God. While in the state of fana, he proclaimed of God, "I am He whom I love and He whom I love is I. We are two spirits dwelling in the same body."[32] Islamic authorities eventually found him guilty of blasphemy, and he was brutally executed in 922 CE. Sufis continue to come into conflict with traditional Islamic authorities because of their veneration of both living and dead, and a variety of other practices in conflict with more orthodox Muslims.[3] Sufistic elements can be found in most every major branch of Islam today, although they are not universally accepted within those branches.

The *Islamic Brotherhood,* also known as the *Muslim Brotherhood,* is another movement that is not a distinct branch.[1,9] Instead, it is a Sunni revivalist group (one of the Salafi movements) containing members from all four major Sunni branches. Although it is a Sunni group, it is not concerned with intergroup differences among the major Sunni branches. It was founded in 1928 in Egypt by Hasan al-Banna

(628–680 CE) who was later assassinated. The movement originally formed as a pan-Islamic political group whose objective was to free Egypt from British colonialist occupation, which began in 1882 and did not end until 1956. Today, the Brotherhood's aim is a politicized form of Islam that aims to reestablish the caliphate (which was abolished in 1924 by the nationalist government of Turkey) and work for Muslimization of society. Although it claims to be primarily a peaceful, democratic organization, methods of the Muslim Brotherhood have ranged from activism, economic institution building, and political accommodation to militancy, antiregime assassinations, and violence. Splinter groups include *Hamas*, which is currently fighting to establish an Islamic state in Palestine.

Ahmadiyya, with more than ten million followers today, is based in Pakistan and has followers around the world. Self-proclaimed prophet Mirza Ghulam Ahmad (1835–1908), founded Ahmadiyya in Punjab (British-controlled India) in 1889. This branch believes that Jesus did not die on the cross, but rather survived and died in India at the age of 120. They believe that his tomb is in Srinagar, the Indian-controlled part of Kashmir. Because Ahmadis accept Mirza Ghulam Ahmad as a prophet (denying Muhammad of being the last of the prophets), the majority of Muslims consider it heresy to practice, preach, or propagate Ahmadiyya in the name of Islam or to identify their places of worship as mosques. The Ahmadis, however, consider themselves to be the one true Islam (a familiar story of religious schism). Pakistan prohibits Ahmadiyya, and the Ahmadis are severely persecuted there. Some Islamic countries have killed hundreds of Ahmadis for their religious beliefs. The current spiritual leader of the Ahmadis lives in London. He was forced into exile by the Pakistani government.[9] When applying for a Pakistani passport, Pakistanis are required to declare that Mirza Ghulam Ahmad was an impostor prophet and his followers are non-Muslims. In contrast, Ahmadiyya Islam advocates nonviolence, shares many traditional beliefs of Islam, and accepts the Quran as its holy text.

The *Nation of Islam* (NOI), also known as *The Lost-Found Nation of Islam*, is an eclectic movement whose primary derivation from Islam is questionable.[3,46] It has elements of Islam, and claims a particular relationship to Sunni Islam, but its founders quote the Judo-Christian Bible more than the Quran, and a later movement leader (Louis Farrakhan) even claimed influence from Scientology.[46] Mainstream Muslims consider it to have selectively adopted certain Islamic beliefs, but not to be truly Islamic.[47] The NOI began in the 20th century as a racially motivated movement exclusively for African Americans. Part of its mission is to instill pride to African Americans by instilling virtues of thrift, hard work, education, self-defense, and political activism. It was also a black separatist organization rejecting integration of black and white races. It was co-founded by Saudi Arabian immigrant, Wallace Fard

(1977–1934), together with his successor, Elijah Muhammad (1897–1975) who was son of a Baptist preacher.[1,3] Fard, became known as Prophet Fard, and eventually as *Allah on earth* to many followers. He preached about "*the deceptive character of the white man and the glorious history of the black race.*"[5] According to historians, he and Elijah Muhammad preached that black Americans descended from the Tribe of Shabaz, an "Asian Black Nation," 66 trillion years ago. Then around 6,600 years ago, white men were created by a rogue black scientist named Yakub. His creations were "devils" made to plague the black race.[46] On judgement day, white people and Christianity (identified as "white man's religion") will be destroyed, and a black nation will rise again. The first centers of worship for the Nation of Islam were located in Detroit Michigan and Chicago Illinois. Besides Fard and Muhammad, the Nation of Islam came to include a number of other inspiring orators such as Malcolm X (1925–1965) and Louis Farrakhan (born in 1933). In the 1960's, Malcolm X became disillusioned with the Nation of Islam and its leader Elijah Muhammad. He renounced the movement and left it in 1964. The next year he was assassinated by three Nation of Islam members. After Malcolm's death, Louis Farrakhan became a major spokesperson for the Nation of Islam. Farrakhan is an admitted black supremist and antisemitic (referring at times to Jews as "satanic"), but he speaks with skill to his followers and is an influential politician.[1] In 2007, Nation of Islam had around 50,000 members (according to Wikipedia).

The evolution of Quran-based religion has gone in many different directions over time. Much of it remains poorly understood or feared, because of the actions of fundamentalists. However, religious ideologies should not be confused with the way extremists have used them for violent and unscrupulous purposes.

CHALLENGES OF ISLAMIC SENSITIVITY AND EXTREMISM IN THE MODERN WORLD

Agnostic studies of Islam are much less common than agnostic studies of other branches of Abrahamic Monotheism. Christians and Rabbinic Jews today live largely in secular societies. Most Muslims do not. Questioning the Bible or maintaining a secular approach to the discussion of religion is a protected right in most of Western civilization. People in certain Muslim countries do not enjoy such rights. It is against Islamic religious law to examine Muhammad in a secular, critical manner or even to draw his face.[6] A verse from the Quran states, "So when the sacred months have passed away, then slay the idolaters wherever you find them" (*Quran* 9:5).

With extremists, it can be extremely dangerous to disrespect Muslim doctrine or Muhammad, even for those outside of Islamic countries. Examples of such danger have populated the headlines of recent decades. With his 1988 book *The Satanic*

Verses, Salman Rushdie won the Whitbread Award for best novel of the year, and it has sold more than one million copies. However, Islamic spiritual leaders thought it was a case of blasphemy, because it contained a fictional version of the Prophet Muhammad in an irreverent light. In 1989, Ayatollah Khomeini, Usuli Shia Islamic leader in Iran, issued a *fatwa* (point of Islamic law given by a recognized authority) officially accusing Rushdie of blasphemy and ordering his assassination. Rushdie's name was added to a "hit list" in the al-Qaeda magazine, *Inspire*. Iranian groups placed a bounty on his head worth nearly three million dollars (which was raised to $3.3 million in 2012). Attempts on Rushdie's life and others associated with the publication of his book have resulted in several deaths over the years. Muslim assassins that have been killed in their efforts to assassinate Rushdie or publishing associates are revered as Islamic martyrs. On Friday, August 12, 2022, more than three decades after the fatwa was originally declared, an assassination attempt was made on Rushdie (then seventy-five years old) in upstate New York. It was made by a man named Hadi Matar. Rushdie was about to give a public presentation at the Chautauqua Institution, ironically on the importance of America giving asylum to exiled writers. Matar attacked him on stage with a knife, gravely injuring him before police stopped the attack. A preliminary review of Matar's social media indicated that he had sympathies for Shia extremist groups (reported the same day by Joshua Goodman of NBC News in New York).[48]

Another famous target of Islamic fundamentalists was the French satirical newspaper, *Charlie Hebdo*, in Paris. The newspaper regularly mocked Rabbinic Judaism, Traditional Christianity, and Islam in the context of unfolding world events (always a dangerous proposition). They published some irreverent pictures of Muhammad in 2012, following a number of violent attacks on U.S. embassies in the Middle East. In response, the Charlie Hebdo editor, Stéphane Charbonnier, was accused of blasphemy by al-Qaeda and placed on their hit list in *Inspire*. Two years later, al-Qaeda gunmen attacked the office of Charlie Hebdo, killing seventeen people, including Charbonnier, five cartoonists, and a number of other staff members. The cases of Salman Rushdie and Charlie Hebdo represent only two of several examples of Islamic extremist retribution for what they consider to be disrespectful depictions of Muhammad.

Much like early Traditional Christian heresiologists, Muslim extremists sometimes go further than targeting the present day events in their efforts against idolatry. They occasionally also target historical records of pre-Muslim faiths to eradicate what they interpret as heresy, as though wiping out the knowledge of history can change what has already happened. Important Mesopotamian and Roman archaeological sites and artifacts have been destroyed by Islamic fundamentalists in the twenty-first century because they were deemed idolatrous (see the figure on page 185).[49,50] This seems misguided in its attempt to erase records of history, diversity, and cultural relationships. Ignorance of the past is not a recipe for a better future.

Uncompromising ideas of cultural domination need to change in Islam as well as in all branches of Abrahamic Monotheism (e.g., Traditional Christianity). Improved understanding, respect for cultural diversity, and intercultural cooperation are necessary for the sake of human survival. This includes not only understanding the beliefs and traditions outside of one's culture but also acknowledging historical mistakes of one's religious institution and traditions. In a world with nearly eight billion people and climbing, the current levels of chauvinistic intolerance, cyclic aggression, and violent retribution are unsustainable.

ISLAM'S PLACE ON THE PHYLOGENETIC TREE OF ORGANIZED RELIGION

Islam is the fastest-growing religion today, and a fitting last chapter on the major branches of Organized Religion. I have but scratched the surface. I have highlighted the Middle Eastern branches of Islam because most of the mainstream media and literature on Islam is Middle Eastern–centric. However, the vast majority of Muslims today live outside the Middle East in Africa and the Asia Pacific region (Turkey to Indonesia).[36] A complete assessment of the modern branches of Islam is well beyond the scope of this book. It is extremely complex, and some of its branches have become more like theocratic political entities than specific religions. At times, such studies can involve elements of personal risk, particularly in the Middle East. I hope that with continued study of Islam in a respectful comparative context, its broader understanding by Western society will improve. Finding more effective ways of communication and understanding between Islamic culture, Western culture, and Eastern culture, is clearly a matter of growing importance in the modern world (**Note 11B**).

Within the scope of history, Islam is but one of many branches of Abrahamic Monotheism, just as Abrahamic Monotheism is one of several branches of Linear Monotheism, and Linear Monotheism is one of many branches of Organized Religion. These groups within groups within still-larger groups of ideological belief systems illustrate a complex pattern of historical relationships. Every religion can benefit from a better understanding of its historical interconnections and shared interests with the rest of the world's religious belief systems. In the next and final chapter, I summarize major conclusions about the evolution of religions and human society, ending with a look toward the future.

CONCLUDING SECTION

FOUR HISTORICAL TRENDS IN THE EVOLUTION OF RELIGIONS AND CONSIDERATIONS FOR THE FUTURE OF HUMANITY

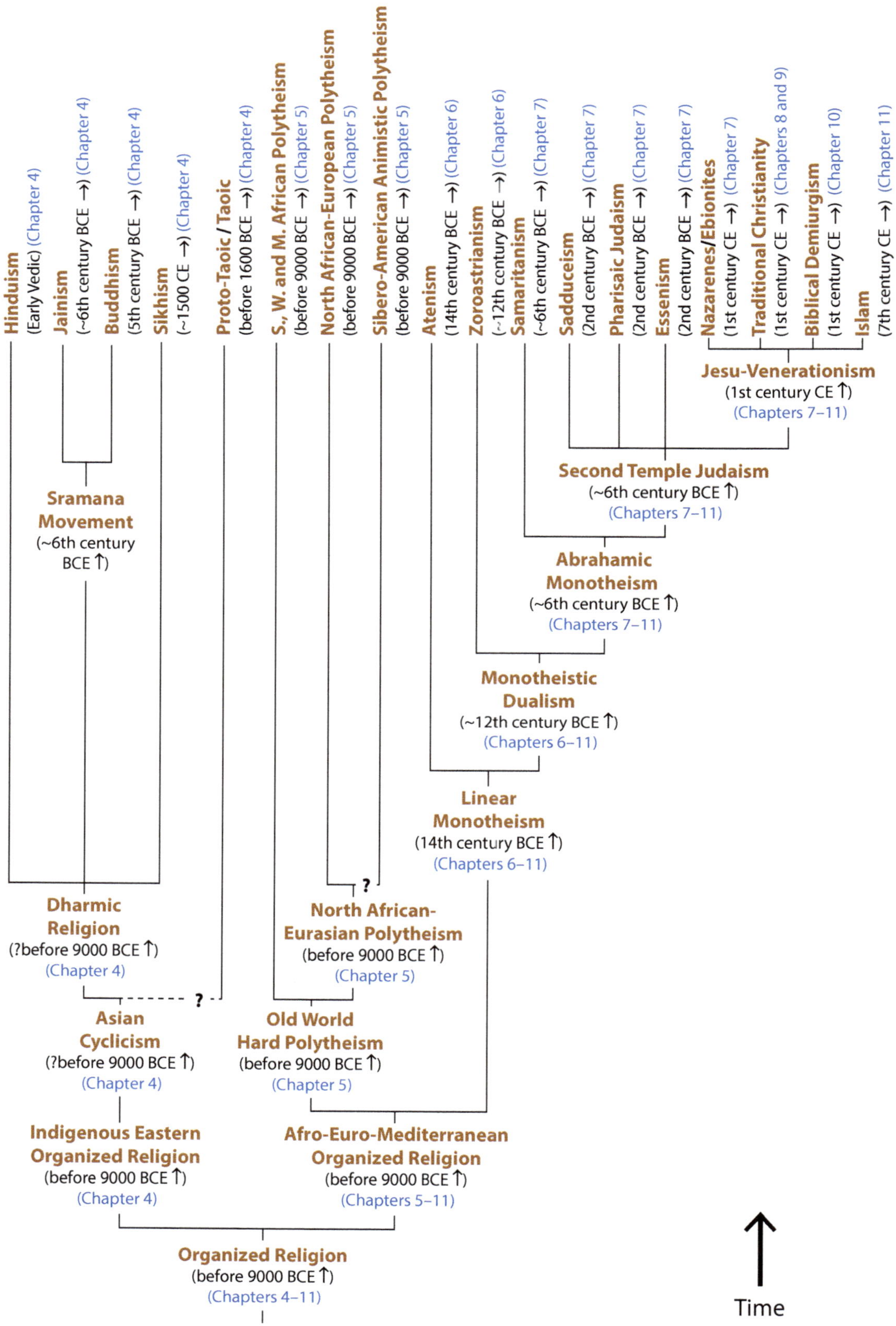
Hinduism
(Early Vedic) (Chapter 4)
Jainism
(~6th century BCE →) (Chapter 4)
Buddhism
(5th century BCE →) (Chapter 4)
Sikhism
(~1500 CE →) (Chapter 4)
Proto-Taoic / Taoic
(before 1600 BCE →) (Chapter 4)
S., W. and M. African Polytheism
(before 9000 BCE →) (Chapter 5)
North African-European Polytheism
(before 9000 BCE →) (Chapter 5)
Sibero-American Animistic Polytheism
(before 9000 BCE →) (Chapter 5)
Atenism
(14th century BCE →) (Chapter 6)
Zoroastrianism
(~12th century BCE →) (Chapter 6)
Samaritanism
(~6th century BCE →) (Chapter 7)
Sadduceism
(2nd century BCE →) (Chapter 7)
Pharisaic Judaism
(2nd century BCE →) (Chapter 7)
Essenism
(2nd century BCE →) (Chapter 7)
Nazarenes/Ebionites
(1st century CE →) (Chapter 7)
Traditional Christianity
(1st century CE →) (Chapters 8 and 9)
Biblical Demiurgism
(1st century CE →) (Chapter 10)
Islam
(7th century CE →) (Chapter 11)
Jesu-Venerationism
(1st century CE ↑)
(Chapters 7–11)
Second Temple Judaism
(~6th century BCE ↑)
(Chapters 7–11)
Sramana
Movement
(~6th century
BCE ↑)
Abrahamic
Monotheism
(~6th century BCE ↑)
(Chapters 7–11)
Monotheistic
Dualism
(~12th century BCE ↑)
(Chapters 6–11)
Linear
Monotheism
(14th century BCE ↑)
(Chapters 6–11)
?
North African-
Eurasian Polytheism
(before 9000 BCE ↑)
(Chapter 5)
Dharmic
Religion
(?before 9000 BCE ↑)
(Chapter 4)
?
Asian
Cyclicism
(?before 9000 BCE ↑)
(Chapter 4)
Old World
Hard Polytheism
(before 9000 BCE ↑)
(Chapter 5)
Indigenous Eastern
Organized Religion
(before 9000 BCE ↑)
(Chapter 4)
Afro-Euro-Mediterranean
Organized Religion
(before 9000 BCE ↑)
(Chapters 5–11)
Organized Religion
(before 9000 BCE ↑)
(Chapters 4–11)
Time

12

ORGANIZED RELIGIONS AND THE EVOLUTION OF HUMAN SOCIETY

Learn from yesterday, live for today, hope for tomorrow.

—ALBERT EINSTEIN

The process of evolution results in the continual diversification of groups over time. This is true whether those groups are biological or cultural. In this book, my focus has been to apply modern evolutionary theory and philosophy to one of the most impactful aspects of human culture: religion. I use historical records compiled by many historians and anthropologists, together with a wide variety of shared ideological characteristics, to reconstruct hypotheses of relationship among major religions of the world. My focus is primarily on what I term *Organized Religions* (formally codified or institutionalized forms of religion).

The phylogenetic tree on page 486 is a hypothesis summarizing how most known major branches of Organized Religion, in their original ideological form, are interrelated. Several branches are greatly expanded into phylogenetic trees of their own in previous chapters of this book (as indicated on the tree in blue text). There are, of course, many gaps and missing elements in the trees of this book, such as the earliest forms of Asian Cyclicism or Old World Hard Polytheism practiced by hunter-gatherer societies (religious belief systems that left little in the way of contemporaneous records). Also missing are most early indigenous supernatural belief systems of Africa, North America, and much of the South Pacific region (all with very few surviving contemporaneous records). These relatively unexplored groups, plus the various non-monophyletic groups (the so-called Chinese folk religions), are in need of much further study before they (or their monophyletic subgroups) can be confidently added within one of the larger trees. Also, I reiterate that many of the dates on my hypothetical trees are approximations based on limited records. However, as discussed in chapter 1, missing information is a challenging reality for any scientific study of deep history (i.e., of macroevolution), whether that study is of biological groups (e.g., birds, mammals, or flowering plants) or aspects of culture (e.g., languages, empires, or religions). Historical hypotheses such as the interconnected trees in this book are not meant as the final word or unchangeable truth. Instead, they are meant to be tested, expanded, enhanced, and revised, as comparative research and new discoveries continue. That is a critical part of the scientific process. Also, I have only begun to look at the major diversification of religions over time. I hope that historians will continue to investigate the more detailed diversification of terminal branches proposed within this book, as well as the detailed complexities within unresolved major non-monophyletic groups (e.g., "Chinese folk religion" *sensu* Johnson and Grimm[1]).

RELIGIONS IN THE TWENTIETH AND TWENTY-FIRST CENTURIES

The influence and representation of particular religions have significantly changed over the course of history, and they continue to change in modern time. The broader cultural affiliation of many religions has also shifted. For example, today most Muslims are not Arabs, and most Christians are located in Africa, Asia, and the Americas.[1] A map on page 489 shows the largest major religion for each country in 2010 (although each of those countries includes at least some members of many other religions).

The chart on page 490 shows the major beliefs to which people self-identified as a percent of world population in 2010. Today, however, I reiterate that each of the major groups on the chart has hundreds of different subbranches. There are probably at least ten thousand different religions that still exist today as distinct belief systems.[1] Although Christianity is the largest major religion today, Islam is the second-largest as well as the fastest-growing major religion.[1] Between 1910 and 2010, Christianity, as a percentage of the world's population, declined slightly from 34.8 percent to 32.8 percent. During the same period, Islam increased from 12.6 percent to 22.5 percent.[1] Other changes between 1910 and 2010 for religions as a percentage of world population include *increases* for Hinduism (12.7 percent to 13.8 percent), agnostics (0.2 percent to 9.8 percent), and atheists (less than 0.1 percent to 2.0 percent); and *decreases* for Buddhism (7.9 percent to 7.2 percent) and traditional Judaism (0.8 percent to 0.2 percent).[1]

Largest religion in each country, 2010*

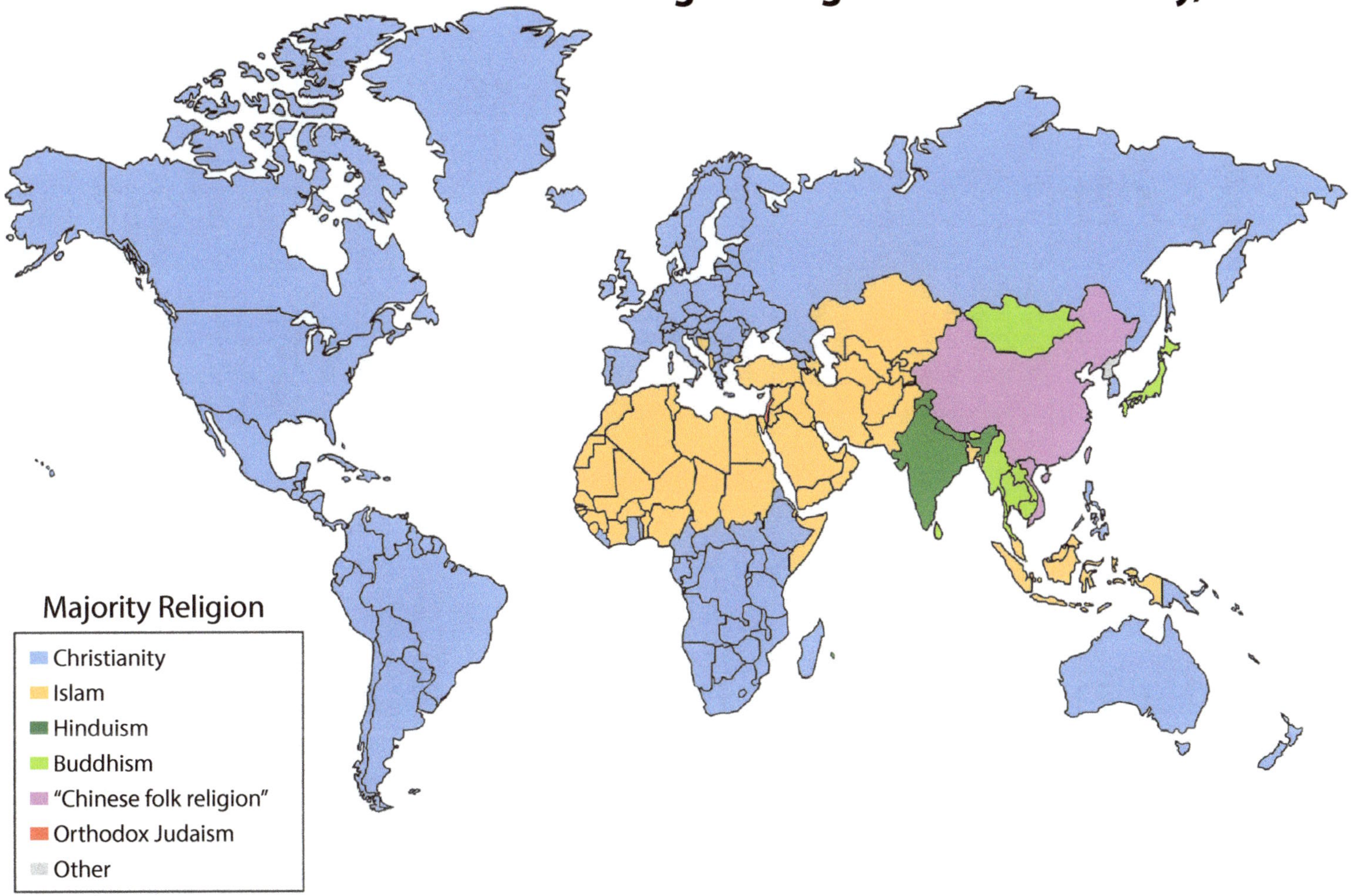

Beliefs to which people self-identified in 2010, as a percent of world population

Branches of World Religion		Percent of World Population in 2010
Christianity		**32.780%**
Roman Catholicism	[51.7% of Christians]	
Eastern/Oriental Orthodox	[12.2% of Christians]	
Protestantism	[18.6% of Christians]	
Anglicanism	[3.8% of Christians]	
Other	[13.7% of Christians]	
Islam		**22.531%**
Sunni	[85.6% of Muslims]	
Shia	[12.9% of Muslims]	
Other	[1.5% of Muslims]	
Asian Cyclicism		**21.612%**
Hinduism	[59.8% of Asian Cyclicism]	13.8%
Vashnaivism	*(67.6% of Hindus)*	
Shaivism	*(26.6% of Hindus)*	
Shaktism	*(3.2% of Hindus)*	
Other	*(2.6% of Hindus)*	
Buddhism	[36.9% of Asian Cyclicism]	7.2%
Mahayana	*(53.2% of Buddhists)*	
Theravada	*(35.8% of Buddhists)*	
Other	*(11.0% of Buddhists)*	
Sikhism	[0.9% of Asian Cyclicism]	0.3%
Taoic Religion	[2.0% of Asian Cyclicism]	0.2%
Taoism	*(47.0% of Taoics)*	
Confucianism/Neoconfucianism	*(36.7% of Taoics)*	
Shintoism	*(16.3% of Taoics)*	
Jainism	[0.4% of Asian Cyclicism]	0.1%
"Traditional Judaism" **		**0.214%**
Zoroastrianism		**0.003%**
Other religions yet to be comparatively analyzed* **		**11.061%**
Agnosticism		**9.817%**
Atheism		**1.982%**

*Based on 2010 world population of 6,895,889,000.

** "traditional Judaism" (*Abrahamic Monotheism* excluding what is referred to here as *Jesu-venerationism*. It today consists almost exclusively of *Rabbinic Judaism* (as defined here).

***"Chinese folk religion" plus many smaller traditions (e.g., various polytheistic branches).

GENERAL CONCLUSIONS ABOUT THE HISTORY OF ORGANIZED RELIGION

The study presented in this book leads to several general conclusions about the history of Organized Religion, including the following four points:

- *First Point.* Most or all known ideological branches of Organized Religion are historically related to each other at one level or another.
- *Second Point.* Branches of Organized Religion that were symbiotically intertwined with ruling power during their development had a selective advantage over those branches that did not (e.g., Nicene Christianity with the Roman Empire, Sunni Islam with the Medieval Caliphates, Hinduism with the Gupta dynasty, and Buddhism with the Maurya Empire under Ashoka and the Tibetan Empire under King Songtsen Gampo).
- *Third Point.* The prominence of women in Organized Religion generally declined with the rise of male-dominated chiefdoms, monarchies, clergies, and governments.
- *Fourth Point.* Religions have had a profound impact on the evolutionary fitness and success of humanity (as expounded on later in this chapter). In today's complex global society, having a broader and better historical understanding of religious diversity, as well as of certain other aspects of culture, may be necessary for the continued survival of our species. Cultural diversification and environmental change are both inevitable parts of history. A growing challenge for humanity is learning how to adapt to such changes and to develop globally cooperative solutions to increasingly complex problems.

In the following sections, I offer some final thoughts on each of these four points.

FIRST POINT: *ALL MAJOR ORGANIZED RELIGIONS ARE HISTORICALLY INTERRELATED AT SOME LEVEL*

As discussed in the preface and first two chapters, by focusing on historical records of relationship and shared characteristics rather than differences, we can construct unambiguous hypotheses of historical relationship among Organized Religions. We can graphically illustrate those hypotheses with phylogenetic trees. The "supertree" figure at the beginning of this chapter illustrates how all of the phylogenetic trees throughout this book fit together. I reiterate three qualifications for my phylogenetic trees here.

The first qualification of my phylogenetic tree is that the order from left to right on my trees is not a qualitative ranking; it simply indicates a hypothesis of historical relationships. As stated in the preface, and elsewhere throughout this book, the phylogenetic tree illustrates a hypothetical pattern of relationship, based on a combination of shared characteristics and historical records. It is not a ladder of improvement or progress. Higher branches on the phylogenetic tree are not universally better or worse than lower branches. Instead, the tree is a retrospective map of interrelationships that theoretically did not follow a predetermined path. The pattern of branching was largely shaped by the ever-changing, multiple environments in which those branches existed.

The second qualification of my phylogenetic trees is that the relative number of subbranches is not indicative of relative historical diversity. It is instead largely an artifact of the imbalanced availability of historical information for broad studies such as this. The branches of Abrahamic Monotheism, for example, are much more numerous and presented in much more detail on my trees than certain other major branches of religion (e.g., South, West, and Central African Polytheism). This is largely due to the relative abundance of written, understandable records available for Abrahamic Monotheism, particularly in Western society. Written records of diversity were admittedly often selectively produced and preserved by the powers and cultures that eventually came to dominate over the course of history.

The third qualification for my phylogenetic trees is that they are all only hypotheses and thus are open to challenge and revision. These trees illustrate the existence of historical relationships based on special similarities and interpretations of historical records that were available for this study. Some anthropologists prefer to look at many religions as completely independent inventions not historically connected to any other religion. Some may have indeed been independently derived (particularly early localized belief systems that developed before populations began mixing and transmitting ideas). If we accept, however, that cultural ideologies, much like biological organisms, inevitably diversify over time, then the comparative search for patterns of relationship with phylogenetic trees is a logical place to start. This is particularly true for what I call Organized Religions. Many of the branches of these trees are clearly the result of shared historical origins. Fundamentalists of many Organized Religions resist the notion of comparing ideological tenets of their religion to those of other religions. They often refuse to see that elements of their faith could be descended from, or ancestral to, any other. (Young Earth Creationists similarly take a stand against recognizing historical interrelationships among biological species.[2]) However, if you follow most branches of Organized Religion back in time before their various schisms and later modifications, they converge on certain elements of basic ideology. This convergence is part of what I referred to as the *law of cultural connectivity* in chapter 1.

The idea of relationships among religions is not to imply that all religions are the same. The differences among branches are what give us distinguishable units to compare for historical analysis. Supernatural belief systems are like genetic patterns of DNA in that they never stop evolving for long. Diversification and change account for the natural and inevitable course of history. Over time, some branches remain similar to their ancestral roots, whereas others dramatically diverge from them. Some imitate elements from other branches or syncretize with them, whereas others add innovative new elements to their belief system.

As part of the evolutionary process of diversification, successful religious belief systems (defined as those that have survived and expanded) have been the ones to adapt to the cultural contexts, needs, and ethical standards of the host cultures and environments in which they existed. The ancient followers of Mesopotamian Polytheism were concerned mainly with their existence in the present life rather than any afterlife. Their concerns were with various elements of nature, growing enough food for the community to survive, raising children, and waging war. Their solution was to pray to the gods for rain, bountiful crops, protection from floods, the birth and survival of children, and success in battle. In contrast, most Buddhists perceived the primary challenge as suffering in the present life and their continuing cycle of reincarnation. Their solution became a search for personal enlightenment to achieve nirvana and freedom from endless cycles of death, suffering, and rebirth leading to the same. Yet another major group, Abrahamic Monotheism, was originally concerned with the predominance of sin and injustice in this life and the risk of eternal damnation. Their solution became blind (unquestioning) faith, active proselytism, and certain duties in this life to attain eternal rewards in heaven.

Organized Religions turn to supernaturalism in various ways for solutions to problems and needs. Its altruistic elements made critical contributions to the development of human society (summarized in the discussion of the fourth point). Unfortunately, the devil is in the details. Conflicting interpretations of doctrinal details among different branches of religion have all too often resulted in violent aggression on a massive scale. The various sacred scriptures (e.g., the Vedas, the Sutras, the Avesta, the Torah, the New Testament, and the Quran) have such a broad selection of stories and interpretations that cultures can selectively emphasize any part of them, ranging from peace and love to tribalistic conquest and genocide. It is up to the people who interpret these documents to use them for good or bad.

SECOND POINT: *ON THE SYMBIOTIC DEVELOPMENT BETWEEN ORGANIZED RELIGIONS AND EMPIRES*

In the Old World, as well as in the New World, many branches of Organized Religion willingly aligned themselves with rulers of kingdoms, empires, dynasties, and caliphates. This served to strengthen popular allegiance to those ruling entities. In return, the ruling entities or political regimes strategically promoted those religions, providing them with a selective evolutionary advantage for growth. It was clearly a symbiotic relationship, and an example of what evolutionists refer to as *coevolution* (when two entities develop in response to each other's evolutionary changes[2]). The early development of a ruling class often reflected the acceptance of law-giving deities. Law-giving deities and their associated commandments provided organizational and ethical standards for control of civilizations. Some social anthropologists contend that religion helped provide the original foundation necessary for large-scale urban civilization to develop or that it at least helped sustain the stability of societies once they began to develop. Religion provided a clear incentive for people to work for altruistic causes and to support public services whose benefits might otherwise have been too abstract for people to appreciate. It provided a degree of order and jurisprudence to crowded societies. It has often been used by imperial power to moralize the use of violent aggression to enlarge the domain of the state. Whether it was religion or urban civilization that came first is a "chicken or the egg" type of question (further discussed in **Note 12A**).

EMPERORS AS GODS

Monarchs and governments of the ancient world often used their ties to hard polytheism to strengthen their power and influence. Some claimed that their authority was given to them by the gods themselves.[3] In the fourth century BCE, the ancient Greek philosopher Aristotle wrote: "A tyrant must put on the appearance of uncommon devotion to religion. Subjects are less apprehensive of illegal treatment from a ruler whom they consider god-fearing and pious. On the other hand, they do less easily move against him, believing that he has the gods on his side."[4] The ultimate early connection between rulers and religion eventually reached a pinnacle with the induction of monarchs into the godly pantheon. Naram-Sin had himself deified as "the god of Agade" to consolidate his rule over Mesopotamia during the period of 2254 BCE to 2218 BCE.[5] Many Egyptian pharaohs from the fourth through first millennium BCE were considered living gods or "god-kings." Deification of rulers helped maintain order in Greco-Roman civilizations. Roman Emperor Julius Caesar was officially deified by decree of the Roman Senate in 42 BCE (chapter 5).

After Caesar, more than a dozen other succeeding Roman emperors were also deified. In the New World, the Incan emperor, referred to as *Sapa Inca*, was thought by the Inca to be a divine being who could mediate with other gods of the state pantheon and ask for their assistance.[6] The belief in divine rulership persisted into the twentieth century. Until the end of World War II, the Japanese emperors were regarded as having divine ancestry by the state religion of Shintoism (chapter 4).

THE INTERTWINED GROWTH OF RELIGION AND EMPIRES

Many rulers did not claim their own divinity, but nevertheless supported the growth of a particular religion with their imperial powers. Identification with a religion often enhanced devotion toward a ruler as well as to support a spirit of nationalism. Promoting state-sponsored religions often helped fuel the massive expansion of empires by way of faith-driven allegiance and sometimes by conquest. It was a quid-pro-quo arrangement, and the success of those empires often meant the success of the religions that they sponsored and leveraged for credibility.

It is a strategy that has persisted into modern times. Many examples of Organized Religion coevolving with empires have been discussed in this book. Dharmic Religion grew together with the Gupta dynasty (chapter 4). The expansion of Egyptian Polytheism was intimately intertwined with the early growth of the Egyptian Empire (chapter 5). The expansion of Buddhism grew together with the expansion of the Maurya Empire under Emperor Ashoka (chapter 4).

The collaborative relationship between Organized Religion and imperial power was particularly strong in Linear Monotheism, beginning with *Atenism* in the mid-fourteenth century BCE and *Zoroastrianism* by the sixth century BCE (both discussed in chapter 6). The shift of the Roman Empire from Old World Hard Polytheism to Traditional Christianity, during the fourth century CE, enhanced the connection between religion and imperial rulers. Roman emperors, from Constantine I and beyond, and various other emperors, kings, and rulers, spent great effort in organizing, instituting, and spreading Christianity while fighting and suppressing other forms of religion.

The most dramatic early expansion of Islam occurred with the Rashidun and Umayyad Caliphates during the seventh and eighth centuries CE (chapter 11). That politico-religious collaboration between Islam and Arab rulers resulted in an immense expansion of Islam and Arab territory. By the eighth century, the Islamic Caliphate was not only one of the largest empires in the world, it had also become one of the world's major religions in what became known as the *Islamic Golden Age* (chapter 11). Centuries later, clashes with the Mongol- and Christian-supported empires resulted in the fall from its Golden Age and the further rise of Traditional

Christianity as the world's largest religion. Nevertheless, Islam remains the world's second-largest major religion, after Christianity. Several Islamic theocracies still exist in the Arab world (e.g., Iran, Afghanistan, Pakistan, Mauritania, Yemen, and Sudan).

Some of the invasive expansions of Christianity through imperial conquest were massively global in scope, in places such as Africa, the South Pacific, and particularly the Americas (chapters 5 and 8). During the fifteenth through nineteenth centuries, old-world Christian conquests of European empires into the Americas eventually overwhelmed and all but erased many indigenous cultures throughout the entire New World (chapter 5). Missionary activity was a key step in the cultural reformation of the Americas, and one of history's largest examples of cultural colonialism. Over just a few centuries, Western European conquest efforts advanced to the point at which the European invaders eventually proclaimed themselves to be the new "Americans" and referred to indigenous Americans as "Indians." The Indigenous religions of Native Americans were rapidly overtaken as part of the process.

The original peoples of the New World were clearly at a disadvantage when it came to early conflict with the Europeans.[7] They lacked steel, guns, horses, and the wheel. They were also highly vulnerable to old-world disease pathogens that the Europeans brought with them. Collectively, it all enabled comparatively small invasion forces from Europe to conquer American civilizations thousands of times their size. It culturally reshaped half the world in the process. As old-world conquest prevailed over rulers in the New World, Western monotheism prevailed over the indigenous polytheistic belief systems there. Conservative Christians simply claimed that God was on our side, However, sometimes the events of history are shaped by a combination of superior resources, nationalistic alliance, careful timing, and good luck. One wonders what the cultures of the world and their respective historical records would be like today if the New World had successfully colonized the Old World.

Another advantage Western Europeans of the conquest had was literacy. Our knowledge of deep history is shaped more by the people who record and interpret it, than by the people who actually live it. Most pre-Columbian cultures of the New World had not developed a written language to document their history. The earliest written records we have of these cultures were made by the Spaniard Christian invaders. It is impossible to know how much of that record is politicized Old World bias, distorting the true nature of indigenous belief systems. The effect of conquest has often been not only to change things for the future but also to prejudice records of the past.

The intimate connection between national governance, intercultural conflict, and Organized Religion has continued into modern times. Although in the Western world, Middle Eastern Islamic countries are most often used as examples (chapter 11), the connection is also common in predominantly Christian countries. In 1954, the U.S. Congress, at the urging of Presbyterian President Dwight D. Eisenhower, added the words "under God" to the U.S. pledge of allegiance (referring to the monotheistic Abrahamic God). This was an effective form of social bonding for Western democracy during the Cold War. It helped portray the Soviet Union as "Godless." At least one later Christian U.S. president was reported to have decided to go to war based on personal advice from his God (discussed in the fourth point of this chapter). Even Adolf Hitler, baptized as a Roman Catholic, claimed to be "fighting for the work of the Lord" in his wartime efforts (chapter 7).

In 2022, Russian President Vladimir Putin, who currently claims to be a Russian Orthodox Christian, invoked his religion to justify his brutal invasion of Ukraine as a Holy War.[8,9,10] The current head of the Russian Orthodox Church, Patriarch Kirill (age seventy-five), has openly supported Putin's imperial and spiritual claim on Ukraine, endorsing the Russian aggression as a struggle against "sinful Western culture."[11] He referred to Putin "a miracle of God."[12] In response, the current head of the Ukrainian Orthodox Church, Metropolitan Epiphanius (age forty-three), likened Putin to both the Antichrist and Adolf Hitler. He also referred to the Russian invasion as a "fratricidal war."[11] The rift between the Russian Orthodox Church and the Ukrainian Orthodox Church represents a major modern schism within Eastern Orthodoxy that is intertwined with brutal military conflict between Russia and Ukraine. The fault does not lie with religion itself. Instead, it lies with the way in which people of authority and influence have *used* religion as a tool of manipulation to justify power struggles, violent tribalistic conflict, and ambitions of conquest.

THIRD POINT: *THE LEADERSHIP ROLE OF WOMEN IN RELIGION HAS GENERALLY DECLINED THROUGH HISTORY*

The overall status of women in Organized Religion has historically declined, in general. This gradually evolved in the pantheons, in clerical leaderships, and in general social empowerment. In the modern world, feminist movements have attempted to push back against this in many religions. However, the status given women in most Organized Religions remains lower than that given to men.

In the Upper Paleolithic, women appear to have been highly revered, based on the predominance of feminine-oriented iconography known as the Venus figurines, between 37,000 BCE and 11,000 BCE (chapter 3). It seems intuitive that in the Paleolithic, when the largest social groups were family-centered bands of thirty to forty people, women would be revered for their unique ability to propagate the clan. Those groups that valued that ability were the ones that survived, providing a cultural selection for matristicism in the days before urban civilization and institutionalized social protections. Those groups that did not support those with the ability to propagate the clan programmed their own extinction.

Eventually, larger scale communities began to develop with chiefdoms and states. Those centers generally were protected and expanded with the help of male-dominated hunting groups and eventually the formation of male-dominated war parties or armies.[13] As a result, a paternal social structure evolved that was usually led by a chief, king, pharaoh, or emperor. Along with the stratified organization of civil control, early forms of polytheism began to take shape in the Old World with a hierarchically organized pantheon of both goddesses and a growing proportion of male gods.

Initially, the supreme deities of the pantheons often still included earth goddesses or mother goddesses (chapter 5). The Mesopotamian goddess *Ninhursag* was not only Earth Mother, she also was thought to be mother of the rest of the gods in the pantheon other than the father *An*.[3] Mesopotamian priestesses held equal status to the priests. The Egyptian goddess *Mut* was called the "Mother of the Gods" and "Earth Mother." The Greek *Gaia* was the earth goddess as well as mother to the gods of the first supreme pantheon (the Titans). The Roman goddess *Tellus* (*Terra*) was the first earth goddess of the Roman Pantheon and goddess of human fecundity. Many scholars consider her to be the equivalent of the Greek goddess Gaia.[14] In Greco-Roman Polytheism, priestesses attended the female gods, and priests attended the male gods.[15] A central male-dominated church hierarchy did not yet exist.

Increasingly, the continued development of male-dominated monarchies and empires influenced the development of Organized Religions associated with them. Many Greek and Roman goddesses were gradually replaced with male gods, and the pantheons became increasingly male dominated. The male gods came to rule the Greco-Roman pantheons. Zeus became head of the Greek pantheon, and Jupiter became the head of the Roman pantheon. The church hierarchies became clearly male dominated.

Asian Cyclicism also saw elements of gender bias develop throughout history. Although some of its branches are still female-centric, with the veneration of female cosmic energy personified as deities (e.g., Shaktism), many branches did not allow women to officiate ceremonies. In its extreme, some branches of Hinduism have had a sporadic history of *suttee* (or *sati*). This practice first appeared in the fifth and sixth century CE, became more widespread by the twelfth century, and is occasionally present to this day.[16] It involves the burning alive of widows on their husband's funeral pyre (although usually voluntarily). It was thought to be sinful for a woman to survive her dead husband.[17] Consequently, they would "join" their deceased spouse by burning to death either directly on the spousal funeral pyre or by a secondary fire ignited with embers from that pyre. It was not until 1829, that suttee became outlawed as homicide in India (by the British government). Nevertheless, some Hindus still support this tradition,[18] and Hindu traditionalists still campaign to have the 1829 ruling against it overturned. Illegal instances of suttee have occurred even in the late twentieth century.[16]

Some branches of Jainism (Digambara) eventually believed that women must be reborn as men before they can achieve spiritual liberation (chapter 4). Some Buddhist dogma taught that women were impure, having a more sinful karma than men.[19] Their menstrual cycles were considered to be spiritually polluting. Buddhist and Confucian belief systems largely supported laws that took away women's rights to inherit property or divorce their husbands. Confucian tradition argued that women should first obey their fathers, then obey their husbands, and finally obey their sons.[19]

During the earliest history of Linear Monotheism (Atenism), women still had nearly equivalent status with men. Atenism worshiped a single animistic deity, *Aten*, with neither masculine nor feminine characteristics. It was represented as a sun disc with downward-pointing rays (page 264).[20] Its dominant earthly representatives included both the pharaoh (Amenhotep IV) and his queen (Nefertiti), both of whom were recognized as having semidivine origins. Amenhotep and Nefertiti were the supreme priest and priestess (chapter 6). Atenism collapsed by about a decade after the pharaoh's death. After that, the general status and role of women in Organized Religion continued to erode.

Paternalization of Linear Monotheism increased with the development of Zoroastrianism, where the sole deity, *Ahura Mazda*, was clearly male. Zoroastrian doctrine was composed and interpreted only by male priests.[21] The Zoroastrian clerics, referred to as *Mobad* or *Magi*, were always male. In early Zoroastrianism (e.g., Sasanian society), a woman's role and influence within the church was generally less than that of a male, a characteristic that would later influence Judaism, Christianity, and Islam.

After Zoroastrianism, came Abrahamic Monotheism's worship of one God clearly depicted with masculine characteristics. Its iconic characters are almost exclusively men (Abraham, Moses, Noah, Jesus, the Twelve Apostles, Paul of Tarsus). Although Mary, mother of Jesus, is the most widely venerated woman in Christianity, she is virtually unknown for anything other than being the mother of Jesus (discussed further in **Note 12B**). The gospels (at least those that went into the New Testament) were all written by men. The largest branches of Abrahamic Monotheism eventually were managed almost exclusively by men (chapters 7–11). The fundamental scriptural texts of the major Abrahamic religions (e.g., the Torah, Talmud, Christian Bible, and the Quran) are diverse in their content with regard to the status of women. However, for every verse describing Jesus, Muhammad, or other

Abrahamic icons defending equal status of women, other verses command women to be subservient to their husbands.

It is good to remember that whether or not these documents were originally divinely inspired, their interpretation is the work of people (almost exclusively men), not gods. Many passages from the sacred books of Abrahamic scripture are blatantly discriminatory against women. The Jewish Talmud says that Jewish males should include a prayer each morning thanking God for not making them a gentile, woman, or slave (Talmud Menahoth 43b-44a). Most biblical stories of conflict on the battlefield involve iconic male figures and warriors fighting in the name of the God Yahweh or El. These sources occasionally treat women as booty of conquest. The Book of Numbers portrays Moses as commanding the Israelites to "execute the Lord's vengeance" on the Midian people in retaliation for the Midianite women supposedly having seduced a number of Israelites to the Canaanite god Baal, the onetime ideological competitor of Yahweh. After conquering the region and killing all of the adult Midian males, Moses told the troops to keep all the captured young girls who were still virgin for themselves and to kill the rest (*Numbers* 31:17–18).

Passages from the Christian New Testament are sometimes used to justify exclusion of women from prominent roles in public worship. For example, 1 *Timothy* 2:12–14 says, "I do not permit a woman to teach or assume authority over a man; she must be quiet. For Adam was created first, then Eve. And Adam was not the one deceived; it was the woman who was deceived and became a sinner." To this day, much of Traditional Christianity excludes women from clerical positions of leadership. Roman Catholicism has never allowed a woman to serve as a pope or a bishop (discussed further in **Note 12C**). In 2007, the Vatican issued a General Decree that any bishop who attempted to ordain a woman as a priest, as well as the woman involved, would be punished with excommunication.[22] As of 2022, Canon law of the Catholic Church still forbids women from being ordained as priests or bishops.

Islam began on better footing for women, with at least one key female character. In early Islam, Muhammad's wife Khadijah played a fundamental role in its development (chapter 11). According to the Sunni narrative, she was the first person to convert to Islam. She was one of the first to actively spread Muhammad's message. Her encouragement helped Muhammad believe in his mission to spread Islam. This led to an equality of the sexes for a while. Muslim women could accumulate wealth, and they had equal rights of inheritance and divorce. Early branches of Judaism or Traditional Christianity did not have comparable equality of the sexes at the time.

Islam was eventually transformed by men who reinterpreted sacred text (the Quran) in a way that made women more subservient to men. For example, the Quran does not clearly dictate wearing the veil (hijab) for women. However, in many Islamic regions (e.g., Muslim extremist regimes in Afghanistan, Iran, Iraq, and the Indonesian province of Aceh), the veil is required by law for Muslim women. Not wearing it, or even wearing it improperly, can result in fines, lashes, or imprisonment (discussed further in **Note 12D**). Muslim women have had less influence on their own lives than their male counterparts. For example, Muslim women in Saudi Arabia, the largest country in the Middle East, did not have the right to vote until 2012 and could not legally drive a car until 2019. Some Islamic women in the late twentieth century began to rebel. Meena Keshwar Kamal (1956–1987) was a lawyer, a feminist and a women's rights activist in Afghanistan and Pakistan who famously said, "Afghan women are sleeping lions, when awoken, they can play a wonderful role in any social revolution."[23] She fought against violence toward Islamic women. She founded the *Revolutionary Association of the Women of Afghanistan* to support Afghan women, and launched the feminist magazine *Payam-e-zan*. She was assassinated at the age of thirty by two Muslim fundamentalists who were thought to have possibly been agents of the Afghan government.

Although the modern world has seen efforts from various feminists and feminist movements to improve gender equality, the relationship between religion and gender remains complex. In general, men have clearly come to occupy a dominant place in most branches of Organized Religion, unless they are homosexual (discussed further in **Note 12E**). However, the current status of women in Organized Religion continues to be challenged by feminists today, arguing for more equality with men. In Hindu cultures, for example, women have been speaking out about such issues as gaining equal education rights and equal land ownership rights, while rejecting sex-selective abortion favoring males and speaking against rare extremes such as suttee mentioned on page 498. There have also been a number of feminist movements within various Abrahamic branches (e.g., Heterodox Judaism in the nineteenth century discussed in chapter 7 and Christian feminism which continues to be addressed by theologians such as Ann Loads[24]). Also, the subservient status of women in Muslim countries continues to be challenged by Muslim feminists, often at their own peril (e.g., see the example in the last paragraph about Meena Keshwar Kamal).

FOURTH POINT: *ON THE IMPACT OF RELIGIONS ON THE EVOLUTIONARY FITNESS OF OUR SPECIES: GOOD AND BAD, PAST AND PRESENT*

The tribalistic element of religion has historically been a two-edged sword. It played a fundamental role in the development of early human society. It is one of the most complex and enduring systems of power and influence in human history. Emile Durkheim (1858–1917), one of the principal architects of modern social science, concluded that much of the early development of human society, and its major institutions, evolved with the help of Organized Religion. The early evolution of

Organized Religion may have enhanced certain ethical codes, such as community altruism, respect for one's ancestors, and laws against murder and theft. Many Organized Religions teach the "golden rule," promoting kindness, mercy, and charity. Whether people are more charitable because of heartfelt generosity, or only because they fear consequences in the afterlife for not being charitable, either way it can enhance social cohesiveness. Organized Religion also inspired much of humanity's greatest works of art, music, and literature. It inspired increased literacy rates among early civilizations and was instrumental in forming the earliest institutions of higher education (discussed further in **Note 12F**).

Organized Religion was also closely associated with the idea of hospitals to treat the sick. In Greece, during the fifth century BCE, healing temples called *asclepions* were dedicated to the god Asklepios (also spelled Asclepius). Asclepius was the Greek god of medicine and healing, and he was also son of Apollo, the national god of Greece. The asclepions were the precursors to hospitals.[15,25] The early Greek physician Hippocrates (460–370 BCE), known as the "Father of Medicine," is said to have received his medical training in one. He is thought to be author of the Hippocratic Oath, historically taken by medical school graduates. In its original form, the oath required physicians to swear by a number of healing gods, and to ethical standards, some of which remain paramount today. After the legalization of Christianity in fourth-century Rome, public hospitals began to be supported by Christian groups, because care for the sick and needy was clearly spelled out in the New Testament. Later, Islam and other religions also built hospitals. During the Islamic Golden Age (the eighth to thirteenth centuries) the Islamic world was instrumental in protecting, elaborating, and reintroducing Greek medical knowledge to Europe, after Europe's decline during the early Middle Ages.[26] Buddhism also made major contributions to the early development of medical care, as documented with the work of historians such as Pierce Salguero and others.[27,28,29] Nearly all medical facilities were associated with one Organized Religion or another until the twentieth century. The association of religious faith with the healing process remains significant. A 1995 study carried out at the Dartmouth-Hitchcock Medical Center found that those with religious faith were three times as likely to survive heart surgery as those without such faith.[30]

Religion often plays a significant role in helping people personally cope with stress, fear, loneliness, and trauma. It enhances group cohesion and brings special comfort in many people's everyday lives. For example, candlelight vigils with prayer after traumatic mass-shootings often bring comfort and a sense of connection to communities of people. Religious historian and ex–Roman Catholic nun, Karen Armstrong, explains religion as an inevitable effort by most humans to address feelings of emptiness they could not otherwise endure.[31] Many people are personally comforted by their private faith and express it in a nonintrusive, peaceful way. My father, who was Roman Catholic, was a good example. He worked extremely hard to provide for his family and went to church each Sunday. He never tried to force his beliefs on anyone else, even in his own family. His way of looking at religion brought him some comfort in his life, which was often a hard one.

It is clear that religion, particularly Organized Religion, has often played a positive, even imperative, role in the development of what we think of today as human society. However, as indicated so often in previous chapters, religion also has a darker side in the way it has been used by nations, rulers, and cultures to moralize large-scale cycles of violence and other unethical behaviors. It is that darker side that needs to be addressed for the sake of human survival in the modern age. The blind, unquestioning faith that sometimes comes with Organized Religion has often been used by people in power to justify tribalistic hypernationalism, aggressive expansionism, and cultural subjugation. Abrahamic religions, in particular, have had a long history of forceful proselytism and promoting violent conquest, graphically illustrated by the Western conquest of the New World (chapter 5). Religious differences have often been the source of in-group and out-group hostilities. In the words of the History of Religion scholar, J. Z. Smith, in his 1982 book *Imagining Religion*, "Religion is not nice; it has been responsible for more death and suffering than any other human activity."[32] Historically, the two largest Organized Religions today (e.g., Traditional Christianity and Islam) became that way, in part, by collaborating with aggressive empires and state-sponsored conquest (as discussed in the second point). They subsequently became the prevailing religions in most countries of the world today (page 489), and they are often in deep cultural conflict with each other. It is not strictly the religious belief system that is the problem. Instead, it is the way certain people use their political power and authority to facilitate a violent or otherwise unethical agenda claiming God or sacred scripture is on their side.

Religion-based conflict often begins with intolerance of ideological diversity and a profound sense of self-righteousness. It often oversimplifies complex social issues. This often leads to escalating conflict, with opposing belief systems each claiming God is on their side. Eventually both sides justify violence in the name of expansionism, revenge, and growing polarization. Religious nationalism, lethal "retaliation" for past violence, and eventually "preemptive self-defense" add to a never-ending cycle of Holy Wars. Compromise and forgiveness for past transgressions become idealistic principles that remain out of reach. This starts a no-win spiral of violently competitive tribalism and deep-seated conflict. As noted by many History of Religion specialists, most long-running and seemingly insoluble conflicts in the world today have been enhanced by conflicting religious belief systems.[17]

The long list of religiously enhanced conflict, even in the modern world, includes the following: Northern Ireland (Protestants versus Roman Catholics), the Balkans

(Eastern Orthodox Christianity versus Roman Catholicism versus Muslims), Kashmir (Muslims versus Hindus), Khalistan (Sikhs versus Hindus), Pakistan (Muslim vs Hindu), Sri Lanka (Buddhists versus Hindus), Indonesia (Muslim vs Christian), the Philippines (Roman Catholics versus Muslims), Israel and Palestine (Muslims versus traditional Judaism), Iran and Iraq (Shia versus Sunni Islam), Sudan (Muslims versus Christians and animists), and Nigeria and Ethiopia (Muslim versus Christian) and many more. As I began working on this book, it was reported that Islamic terrorist groups were calling for attacks on the Roman Catholic Pope and for Roman Catholic icons to be destroyed.[33] At the same time, it was reported that Islamic mosques in the United States were being bombed and vandalized at a rate of nine per month.[34] These conflicts reflect a fundamental problem that will not be resolved simply with chauvinistic force.

Today in the West, Muslims are frequently singled out as the primary sponsors of terrorism based on a minority of politicized extremist groups (chapter 11). Islam as a whole is often unfairly blamed by the West for the acts of a few. That is partly a result of subjectively filtering events through a Western (usually Christian) cultural context and media platform. Much of Islamic militancy today is in response to Western imperialism in the Middle East through much of the early twentieth century. The aggressive posture of Islamic extremists is enhanced by a deep-seated mistrust they have of the West for various reasons, one being a feeling of betrayal after the end of World War I regarding what they interpret as a broken promise of an Arab homeland.[35] The Middle East has also been a region exploited by Western interests for its oil reserves. Islam certainly has had its own history of major conquest, particularly in Persia and North Africa. However, they have historically often let local cultural traditions (including non-Islamic religions) continue.

Islamic militancy is often a responsive action for what it perceives to be deep-seated threats from the West. Muslim extremists have committed acts of extreme violence in the name of Islam, to be sure. However, extremists using religion to justify their actions have existed in *all* major Organized Religions. Placing exclusive blame for terrorism on a specific religion is a product of cultural chauvinism. Religious ideology is not the cause of harm. It is instead the people who use that ideology to justify violence or other unscrupulous purposes that do the damage. Many historical examples of such abuse are given throughout this book.

Organized Religion is also often used to justify forceful actions by cultures who themselves have been repeated victims of prejudice and genocide. For example, some Jewish fundamentalists who have settled in the Occupied Territories of the West Bank and the Gaza Strip have avowed their intention of driving out the Arab inhabitants, using force if necessary.[31] They claim Godly support. Their inspirational leaders have included Orthodox Judaism Rabbi Meir Kahane, who is quoted proclaiming, "There are not several messages in Judaism. There is only one. And this message is to do what God wants. Sometimes he wants us to go to war, sometimes he wants us to live in peace. . . . But there is only one message: God wanted us to come to this country [Israel] and create a Jewish state."[36] Kahane was assassinated in New York in 1990 by an Egyptian-born Muslim and U.S. citizen named El Sayyid Nosair.

Traditional Christianity also still has its share of fundamentalists inciting violence. For example, the *News-Sentinel*, a newspaper from Fort Wayne Indiana, reported that Christian politician Randall Terry gave a speech in Indiana where he said "I want you to just let a wave of intolerance wash over you. I want you to let a wave of hatred wash over you. Yes, hate is good . . . Our goal is a Christian Nation. We have a biblical duty to conquer this country. We don't want equal time. We don't want pluralism."[37] His inciteful tone is hardly different from that of other extremists bent on chauvinistic conquest in the name of one religion or another. Although his views have been clearly extreme, Terry ran as a Republican for the U.S. Congress in 1998 and for the U.S. Senate in 2006. In 2011, he made a run for the presidency of the United States as a Democrat, winning 18 percent of the vote in the Oklahoma primary of 2012 (more than twenty thousand votes) before withdrawing from the race.[38] After that, he abandoned both the Democratic and Republican parties and declared himself an Independent.

In 2003, Methodist President George W. Bush commanded the U.S. Armed Forces to invade Iraq, partly in response to the September 11 terrorist attacks, and partly because he believed that the country's leader, Saddam Hussein, was hiding weapons of mass destruction (discussed in chapter 11). Bush made the claim about hidden weapons even though some U.S. intelligence analysts and Cabinet members advised him there was no good evidence for it. It was later reported by the BBC and other media that Bush announced in a meeting with Palestinian ministers that he made the decision to preemptively invade Iraq because God told or inspired him to do it.[39,40,41] Although the context of his statements is disputed, it is not difficult to imagine how this report was received in the Islamic world.

Some of the president's top advisers who supported the invasion of Iraq were also influenced by their religious beliefs. Born-again Christian General William G. Boykin was the U.S. Undersecretary of Defense for Intelligence for President Bush from 2002 to 2007. He was reported to have claimed that George Bush (who had won the electoral vote but lost the popular vote in the 2000 national election) was in the White House because "God put him there."[42,43,44] Boykin referred to Muslims as idol worshippers (even though they are even more strictly against idolatry than Christians; chapter 11). He reportedly referred to the United States as a Judeo-Christian Nation whose God was bigger than the God of Islam (even though the God of Islam and the God of Christianity are technically the same Abrahamic God).[42,43,45] Boykin reportedly claimed that in the war against Iraq, "the enemy was a guy called Satan."[42] Boykin later apologized for his remarks, but the damage was done.

The decisions of people in power often have powerful unintended consequences. That is part of the burden of leadership. Between 2003 and 2006, the war in Iraq resulted in more than a hundred thousand people killed, most of whom were civilian noncombatant Muslims.[46] No weapons of mass destruction were ever found. The invasion greatly enhanced the rise of ISIS (the Islamic State of Syria), also known as ISIL (the Islamic State of the Levant).[47] This group is a Muslim fundamentalist organization that continues to fight Western interests around the world. They claim to be a worldwide Islamic Caliphate fighting in the name of God (Allah, in Islam). They continue to paint the United States as the "Great Satan" or "Iblis" (the primary devil of Islam). It is yet another vicious cycle of major world conflict in which each opponent contends that God is on their side and condones their violent actions. But no God makes the decision to commit such violence in this world; instead, people do, and sometimes claim to do it in the name of, or with the blessings of, their God.

Fundamentalists who are thoroughly convinced that they are fighting in the name of God are often the most intolerant and dangerous people of all. One group's holy warrior is the other group's terrorist (and vice versa). Each side of such conflicts often claims the other is aligned with the Satan.[45] It is a passionate cycle inflamed by conflicting beliefs, a sense of persecution, and the human propensity for violence and retribution. It is also a highly successful tool to passionately rile up the masses. It often seems impossible to forgive past transgressions, with a constant war of words between each group claiming to be God's chosen people. Does it come down to a contest based on proselytizing salesmanship and power? Is it really "might that makes right?" If so, then it is a sad commentary on the state of human ethics, with bleak prospects for the future of our species.

Mohandas Gandhi went to great lengths to try to break down chauvinistic barriers between Islam, Traditional Christianity, Hindu, Buddhist, Sikh, Jains, and others (chapter 4). While recognizing that religious belief systems were distinctively different, he claimed that the nonviolent parts of all religions served a common purpose. He spent much of his adult life trying to persuade people to stop the fighting and killing in the name of religious nationalism. Sadly, his advocacy for peaceful spiritual inclusivism led to his assassination by an upper-caste Hindu Nationalist. Other recent efforts to create more inclusive or pluralistic belief systems include the *Unitarian Universalist Association* (chapter 9) and the *Baha'i* (chapter 11). Neither of those branches has ever come close to rivaling the size of Roman Catholicism, Sunni Islam, or many other more exclusivist belief systems.

Organized Religion is largely an attempt by humans to understand and explain supernatural matters (e.g., events that cannot yet be explained or influenced by natural causes). It is ironic that so many religious fundamentalists admit that God is unknowable, yet then create a multitude of specific details about God as "truth" that cannot be questioned. Each Organized Religion codifies its own unique combination of beliefs, rules, and doctrines. Must we pray once a day, five times a day, or more? What do we say, and which holy entity or entities do we pray to? Which ethical parameters do we conform to, and why? Many fundamentalists effectively treat their religion as though *it* were God. Slight deviation from subjectively determined detail can be deemed worthy of severe punishment ranging from excommunication to execution. Violence-of-God verses sporadically embedded in sacred texts, such as the Torah, Bible, or Quran, are often selected out to justify human violence in the name of God or in the pursuit of "God's objectives."[48] However, the scribes, editors, and assemblers of these sacred documents were human, not divine. Clearly, we cannot objectively know which of the many conflicting views and interpretations of scripture are correct.

Organized Religion has too often been used by ethnic, state, or church authorities to moralize violent actions in long-standing conflicts. Again, Organized Religions are the constructs of people and organizations, and not of the gods, prophets, or other entities that people venerate. Organized Religion continues to evolve long after the founding iconic figure is gone. Jesus, Buddha, Muhammad, and other iconic figures who reportedly reinforced many better parts of human nature during their lifetime, should not be held responsible for all the diverse later belief systems that inevitably evolved in their name. Even if one or more of these figures represented some sort of link to a higher power, the later evolution of the religious institutions that they inspired were the constructs of people, usually men. Cultural evolution eventually modifies the original intent of iconic teachings in many different, often conflicting, directions and traditions. Each resulting group claims to be the true path to salvation and to know the true intent of their religion's iconic inspirational figures. History clearly suggests that we cannot prevent the evolutionary diversification of supernatural belief systems or the interpretation of sacred founders and icons over time. Furthermore, there is no way to know, if they were to reappear today, what kind of Christian Christ would be, what kind of Buddhist The Buddha would be, or what kind of Muslim Muhammad would be.

The concept of "faith" is most often used by competing religious belief systems to justify why their beliefs are true and the others are not (or even why other groups are "heretical"). However, if someone uses faith to justify harm or irresponsibility toward nonviolent unbelievers, it is not really a morally defensible belief system. In that context, I agree with Richard Dawkins that there is something evil about it.[49] In the extreme, "blind faith" (unquestioning faith without objective reason, or even unquestioning faith in spite of evidence to the contrary) often gives people unquestionable confidence in their own righteousness, sometimes for behaviors that are immoral and detrimental to society as a whole. At times, this type of faith has provided powerful people with the self-righteous resolve to subjugate, enslave, or persecute massive populations of people who did not share a monarch, king, or pope's perspective (chapters 5 and 7). Blind faith has also strengthened the resolve of some extremists to kill themselves and others in the name of their particular beliefs, while demonizing the targets of their aggression.

Another problem with blind faith is that it is sometimes used as an excuse to neglect ethical responsibilities and the broadest altruistic action for the common good of humanity. In that sense, it is allowing our species to drift toward extinction. For example, in 2009, during a House Subcommittee meeting on Energy and Environment by the U.S. Congress, Illinois Representative John Shimkus opened with the claim that we should not try to fight global warming, because of what the Bible tells us. After quoting *Genesis* 8:21–22 and *Matthew* 24:31, he stated, "The earth will end only when God declares it is time to be over. Man will not destroy the earth."[50] In other words, the congressman suggests that global warming is not something we should even try to manage, because the Bible tells us not to worry about it. While global warming may not have been a problem in the age when the New Testament was composed, it is surely a problem today. Another example of outdated thinking is the Catholic Church leadership's stance against contraception as a form of birth control. In a world spiraling toward multiple calamities resulting from overpopulation, to outlaw voluntary contraception seems both immoral and tragically thoughtless toward future generations (discussed further in **Note 12G**). It seems that the connection between faith and morality is a subjective one. Organized Religion has been used far too many times through history to justify immoral inactions as showing virtuous faith in God. If personal morality is subjective and partisan, then it does not necessarily serve a broader ethical purpose. This is but one of the problems with proselytism (**Note 12H**).

Another troubling aspect of some of the most powerful Organized Religions (particularly branches of Traditional Christianity and Islam) has been the effort to erase or distort cultural history other than their own. Even the Catholic Crusades against Muslims are still defended by certain Catholics as "defensive and just."[51] In the words of Peter Turchin and his colleagues of the *Seshat Project* (**Note 12I**), "If we want to live in better—more peaceful, wealthy, and just—societies, we need to understand the major evolutionary transition (or transitions) that occurred in our past and why they resulted in such divergent outcomes in the present."[52] Such understanding should be based in unbiased historical analysis, not tainted efforts to favor one religion or another. As noted by the renowned religion scholar, Stephen Prothero, "To reckon with the world as it is, we need religious literacy."[53] We need to honestly separate and understand those elements of religion that have helped our species, from those that continually undermine its stability. Those religions that forget or ignore the mistakes of their own histories are doomed to repeat them.

True tolerance of different belief systems with peaceful intent includes tolerance of atheism as well. According to Johnson and Grim's landmark study of global religions, based on a summary of over 2500 surveys and population registers, it was estimated that about 2 percent of the world's population self-identified as atheists in 2010 (table on page 490).[1] Atheists, as a group, are no less moral or violent than religious people. It is true that particular atheists have acted immorally or violently (e.g., Lenin, Mao). However, as discussed throughout this book, even more individuals claiming to be religious have acted immorally and violently. There are countless examples of people committing atrocities in the name of one Organized Religion or another, ranging from burning heretics at the stake and genocide, to human sacrifice, suicide bombers, and organizations supporting the path toward Armageddon in anticipation of the rapture. Conversely, who commits atrocities in the name of atheism? (Discussed further in **Note 12J**.)

Notable atheists also see the problems in Organized Religion and have promoted intolerance of religious pluralism as a solution. Sam Harris, for example, attacks "the very ideal of religious tolerance" as "one of the principal forces driving us toward the abyss."[54] I agree with Harris about the numerous problems of blind religious faith being used to justify acts of violence. I also agree that the very survival of our species is in danger from taboos against the questioning details of one's own religious beliefs. However, I disagree with Harris's notion that the best, most pragmatic strategy for the future is to push for universal *intolerance* of all religions. It is the element of intolerance (e.g., violently enforced exclusivism) in religion that is its main problem today. Trading one sort of intolerance for another is not the answer, especially for an aspect of culture so deeply embedded within the human species.

Religion has helped many people endure stresses of the world that would otherwise be unbearable. Religion was also a fundamental element in the early structural development and regional stability of human society. The proclivity to harbor some sort of supernatural belief is deeply ingrained in the human psyche, and the suggestion to completely eliminate it is not a pragmatic solution to the problems of society. It is unrealistic to believe that we could universally wipe out the human instinct toward all supernatural beliefs even if we wanted to. So how do we better satisfy the human inclination for supernatural beliefs than with the more peaceful elements of religion? With hypernationalism, hypercapitalism, or hyper-partisan politics? I do not think so.

The most likely solution, no matter how idealistic it may sound, lies with the collective broadening of tolerance for the nonviolent and altruistic aspects of all religions. We should, of course, reject practices such as human sacrifice, competitive

Holy Wars, inhumanely prejudicial caste systems, and the demonization and persecution of nonbelievers. However, we need a more tolerant way of allowing different cultures to express and maintain the peaceful aspects of their particular belief systems. This is a complex problem that needs deeper philosophical discussion and more collaborative pragmatic solutions. It also needs a broad consensus to acknowledge, but somehow move beyond, past atrocities and prejudice. Hopefully, our species has the cognitive ability to build a more broadly altruistic, inclusive, and beneficially tolerant social environment. (The tricky issue of tolerance is discussed further in **Note 12K**.)

Finding ways to decrease the violent tendencies of human competitiveness and chauvinistic coercion, while accepting the peaceful elements of diverse personal belief systems, will not be possible without a broader, less biased understanding of cultural history and diversity. We must try to better understand the major evolutionary relationships, transitions, and behaviors of "other" belief systems that have occurred through human history. We must also honestly confront our own belief system's history (the good and the bad) to learn lessons from it. Every major Organized Religion has been used at some point by nationalist groups or ruling authorities to justify violent conflict and other forms of immoral behavior. A more honest understanding of cultural history, and the will to become more pluralistic in the way we collectively move forward, is critical for the future of our species. Such tolerance may decrease the chances of chauvinistic violence and ignorance becoming sources of our eventual extinction as a species.

Broad tolerance of religious diversity (*pluralism*) is much easier said than done, and it is resisted by many theologians today.[55] The attraction of *exclusivism* (acceptance of only a single religion as true and rejection of all others as heretical) or of *selective inclusivism* (acceptance of a single religion as true, but also accepting that certain aspects of truth can be found in other religions) are exceedingly strong to most theologians, particularly those advocating for particular branches of Abrahamic Monotheism.[56] In the words of Lutheran religious scholar, Martin Marty: "Nowhere does [pluralism] and its implicit world view prosper in open competition with anti-tolerant and proselytizing groups." He continues, "If pluralism is to survive . . . it is likely to demand conviction and commitment, and with them, more reflective philosophical supports."[57] True pluralism will require a much deeper understanding of the history and diversity of religion. In the words of theologian Jerald Gort: "Learning both to know and understand people of other faiths and their religions is absolutely requisite to the restoration and preservation of peaceful coexistence in the intolerant and often violent pluralistic societies in which the vast numbers of people throughout the world live today."[58] General ignorance of the "religious other," as well as ignorance of history, are often at the heart of cultural, and even nationalistic, conflict.

RELIGION'S ROLE IN THE FUTURE SURVIVAL OF THE HUMAN SPECIES

As discussed in chapter 2, evolutionary theory holds that each species' survival is tied to its ability to adapt to its ever-changing environment. For the human species, that environment has radically changed since the earliest days of civilization (as defined in **Note 3C**), and it has continued to change and diversify at an even faster pace through the late twentieth and early twenty-first century. In the words of Edward O. Wilson, "the problem with humanity today is that we have paleolithic emotions, Medieval institutions, and god-like technology."[59] Understanding today's cultural diversity, as well as its unbiased history, will be necessary prerequisites for us to adapt to the future. Many cultural traits that were selective advantages for the success of hunter-gatherer clans, kingdoms, and early empires, during Paleolithic through Medieval times, are now selective disadvantageous for the future survival of our species. This is due to a number of factors, particularly population growth, technological advancement, and the overall complexity of the modern global environment. We must learn to collectively adapt to a plurality of social, as well as physical, environments on this planet, or we risk our extinction as a species.

The human population has reached eight billion people, most of whom hold one sort of supernatural belief or another. Modern technology has added increasingly efficient means of global interconnectivity among most of those people, beginning with radio, television, and finally the internet. The vast interconnectivity of social media that has evolved today, including the widespread use of personal smartphones and various other devices, together with the use of complex algorithms targeting human emotional vulnerability, represent a vastly new social paradigm. It has a virtually unregulated potential to influence, proselytize, and selectively exploit the passionate emotions of millions, or even hundreds of millions of people at a time. It is as though the social intellect of our species today is being rewired and increasingly manipulated with this technology. The environment in which our species must adapt to or face extinction, is fundamentally changing. The most complex social issues of our species today often seem to be influenced more by short bursts of widely dispersed, deeply persuasive manipulation, than by the more in-depth collaborative study that these issues deserve. Far too often, large groups of people are united mainly by identifying another cultural group as a common enemy. Taking advantage of ingrained tribalism still appears to be one of the easiest, most-frequently used ways to build a passionately devoted coalition. This problem will only become more of a challenge as AI technology (artificial intelligence) becomes more sophisticated in its integration with search engines and other influential aspects of social media.

Today's technology also includes an unprecedented assortment of increasingly powerful weapons of destruction, ranging from bombs to biotechnology to cybertechnology. These innovations all have the potential to cause catastrophic devastation on a global scale. Such tools of broad influence and potential destruction are available not only to nations but also to smaller cultural subgroups and extremists that sometimes fall back on ethnic or religious differences to justify, or even glorify, violent conflict. These issues need to be addressed globally, in a way that brings all cultural entities to the table. They need to be addressed in more thoughtful, broadly inclusive ways, rather than using extreme, provocative messages and fearmongering to incite deep-seated intercultural rivalries and tribalistic sensitivities. Past violence and certain imperialistic or colonialist behaviors should provide lessons of what not to do in the future, rather than continuing the seemingly never-ending cycles of violent and vengeful vendettas. Although religion is not the only aspect of culture that has exacerbated tribalistic extremism today, it is certainly one of them. In the broadest scope of religious history and its role in the justification of social tension, no major branch is without guilt. However, I would like to believe that humans are capable of the fundamental change needed to focus more on the common good of our species in the modern world.

In addition to challenges of violent conflict, a number of other growing problems require globally collaborative solutions. These include climate change, pollution, pandemics, regional poverty, and overpopulation. Religious differences, empowered by tribalistic narrow-mindedness, contribute to a lack of global cooperation in addressing such critical issues. Personal religious conviction is not an ethical excuse to ignore global responsibility.

Tribalism has historically been a double-edged sword. At first, it may have strengthened early altruistic characteristics and group cohesiveness necessary for the development of organized communities and major civilizations. However, it eventually also fueled subjectively biased attitudes that have often divided human societies in harmful ways. Such attitudes have often led to violent conflict and discrimination as well as collaborative inaction on broader global problems. Today, the socially divisive aspect of tribalism is setting us on a dangerous course, and creating an increasingly challenging environment for the long-term survival of our species. Specific subbranches of Organized Religion have often been used to justify violent intercultural conflict as well as inaction on global issues in need of fixing. We need to start thinking of our "tribe" as all of humanity and to expand our bubble of empathy accordingly. We need to collaboratively address universal problems requiring globally collaborative solutions. Ultimately, the survival of our species through the twenty-first century will depend on finding innovative ways to modify and manage the divisive aspect of our ingrained tribalistic instincts, some of which have a religious core.

Organized Religion is but one of several aspects of culture that has led to violent tribalistic conflict, conquest, and failure to collectively address global issues. Other factors contributing to these problems (often deeply entangled with religion) include uncompromising partisan politics, nationalism, racism, and an overly aggressive quest for superiority. Some religious apologists argue that religion is not the problem and that it is the other divisive social factors that are at fault, as if to summarily excuse the role Organized Religion has often played in intercultural tension.[60] Although many complex reasons exist for violent conflict and lack of social cohesion today, we now need to honestly address *all* aspects of human culture that contribute to social discord, including religious conflict. Religion is one of the most socially ingrained and broadly held aspects of human culture contributing to social discord. Therefore, it represents a good (possibly even a necessary) starting point for broad discussion. If we could emphasize its positive moral and ethical elements (e.g., humanistic altruism, honesty, social bonding, and peace of mind) and work to eliminate the harmful ones (e.g., bigotry, forceful proselytism, institutionalized sexism, environmental irresponsibility, and various forms of religiously justified violence), religion might even provide a selective *advantage* for the continued survival of our species through the twenty-first century. So far, humans appear to be far from mastering such a social transformation.

One step toward such a change may require some sort of inclusive, United Nations–like world council with representation from all major belief systems (all major religions plus atheistic belief systems). Its mission would *not* be to create or decide on a single belief system. It would instead be to find pragmatic ways for different belief systems to peacefully coexist and work toward common ethical goals. Most religions include deep ethical components of one kind or another. Past attempts to establish such a council have had limited success because the extreme sensitivities, regional customs, and nationalistic complexities involved (discussed further in **Note 12L**).

If there is a legitimate reason to discriminate, it should be against those who use exclusivist, chauvinistic doctrine to justify violence, racism, conquest, or other types of clearly unscrupulous behavior. It should also be used against those who use forceful proselytism against outsiders that they demonize as heretics. Outsider and heretic, after all, are only relative terms used for the opposing side of an ideological battle. As noted, we should also be critical of those who use religion as an excuse *not* to collectively address complex issues of global significance, such as climate change, global pandemics, food insecurity, and extreme poverty.

As I have said throughout this book, my work here is preliminary, broadly focused, and from the perspective of an evolutionary systematist. My intent is not to proselytize for any particular religion, nor to suggest that every one of them is false. Each person has their own way of dealing with the profound meaning of their personal existence. I believe that we need a broader, less biased understanding of world religions, as well as of other deeply ingrained aspects of human culture that have been used at times to unite us and at other times to violently divide us. Such knowledge may help enable the continued survival of our species into the future.

We need to collectively acknowledge our past and present intercultural conflicts, while finding ways to peacefully move beyond them in the modern world. Although religious chauvinism is not the only cause of social strife, it has certainly been one of them. We must adopt a broader intercultural perspective to more effectively focus on matters of common human interest. We need to end the repeated cycles of oppressive indoctrination, violent intolerance, and deadly retaliation, that far too often have been justified in the name of one religious ideology or another. Culturally, we may be products of our past, but that doesn't mean we have to be prisoners of it. The rapidly growing problems facing our species today cannot be solved by violently intolerant intercultural competition. We must diffuse the extremes of cultural intolerance, as well as extremes of ideological conquest, vindictive retribution, and hypernationalism. Those traits today serve to increasingly divide us and collectively weaken our species' ability to adapt to the changing global environment. Going forward, intolerance should be reserved for traditions that are violently oppressive, aggressive, and divisive. We should learn to accept, and better work with, the inevitably diverse nature of human society. If we, as a species, can collectively adopt such adjustments to our social behavior, perhaps then we can collectively adapt to the increasingly complex, and ever-changing, challenges of the future. Not doing so may ultimately drive us to extinction. In the words of late-twentieth-century French biologist and social evolutionist, Rene Dubos: "Human diversity makes tolerance more than a virtue; it makes it a requirement for survival."[61]

Amen to that.

ACKNOWLEDGMENTS

I thank the many people who read and commented on parts of this book or on certain ideas contained within it. My wife Dianne read and commented on the manuscript in its entirety (reading some sections several times). I can never thank her enough for that as well as for her unrelenting moral support. Others who read versions of the entire manuscript and provided useful commentary include Elizabeth Grande, Mark Alvey, Ben Heaney, Anne Trubek, Michael Rosenberg, and Charles Stanish. Those who provided useful discussion and comments on various parts of or points in the manuscript include Jim Philips, Scott Lucas, Michael Allen Williams, Gary Feinman, Ryan Williams, Scott Lidgard, Peter Turchin, Laurie Zoloth, Peter Flügel, Vishruth Venkat, and Paul Harrison. I, of course, take responsibility for all mistakes contained within this book, which will hopefully be corrected by cultural comparativists, cultural particularists, and other historians. I thank the Negaunee Foundation for their generous support of my research over the years. Finally, I thank the Field Museum in Chicago for its continued support of interdisciplinary research. Sometimes, it takes a syncretization of philosophies or perspectives to move things in a newly productive direction.

GLOSSARY OF SELECT TERMS

Many of the terms defined below are used in a variety of very different ways by other authors. For the sake of clarity, I define them here the way they are used in this book.

ADHERENTS: followers of a particular ideology or teacher.

AFTERLIFE: some sort of existence after death (e.g., heaven, hell, purgatory, hades, the netherworld, the underworld, the kingdom of the dead, Elysium, reincarnation, nirvana, becoming one with a spiritual force, etc.).

AGAMAS: a name for certain, post-Vedic religious *scriptures* within many Dharmic Religions.

AGNOSTICISM: a form of accepted uncertainty. A belief that the ultimate reality of any particular spiritual entity or religion is objectively unknown and probably unknowable. It is effectively more pluralistic than exclusivistic. Also see **Note PrA**.

ALTRUISM: behavior of an individual that benefits another at his or her own expense.

ANAGENESIS (ANAGENETIC CHANGE): the gradual evolution of a group as a single, nonbranching lineage.

ANCESTORISM: the general belief that the spirits of the dead continue to dwell in the natural world and have the power to influence the fates and fortunes of the living.

ANIMISM: the belief in a spiritual essence or soul in living things, nonliving things, and even in forces of nature. I use the term in the nonderogatory sense of Graham Harvey's "New Animism" discussed in chapter 3 (Harvey, 2006).

ANTHROPOMORPHIC SUPERNATURALISM: the attribution of human traits to supernatural elements.

APOCALYPTICISM: the belief that the end of the world is imminent, often including a great battle, final judgment, and reward of the good.

APOMORPHY: a homologous character that is uniquely derived for a particular group of focus (e.g., mammary glands in mammals; or the recognition of Abraham as a *patriarch* in Abrahamic Monotheism). A uniquely shared apomorphy between two or more groups is also referred to as a *synapomorphy*.

ASCETIC: severe self-discipline and abstention from all forms of indulgence, typically for religious reasons.

ASPECT: a particular feature of something (e.g., a particular feature of natural history or cultural history, as shown on page 25). With regard to a *deity*, it can also mean a particular form of a supreme *deity* (e.g., the Sri and Kali aspects of the goddess Shakti in Hinduism; or the Father, Son, and Holy Ghost aspects of Yahweh in Roman Catholicism).

ATHEISM: a denial of the existence of one or more supernatural beings. It is a characteristic and not a specific religion in of itself (e.g., **Note 3A**). Some religions (as defined here) are atheistic in that, although they do not believe in supernatural beings, they nevertheless hold some codified elements of supernatural belief (e.g., reincarnation). The ancient Greeks used an early form of the term atheist (*atheos*) as a derogatory term for the impious. Later Greek literature used the term for *lacking the belief in the gods* (Whitmarsh, 2021). By the fifth century CE, Chalcedonian Christians appropriated the term and changed its meaning to refer to someone who did not believe in the Abrahamic God.

AVATAR: earthly manifestations of a *deity*.

AVESTA: the primary collection of religious texts for Zoroastrianism.

BCE (BEFORE COMMON ERA OR BEFORE CHRISTIAN ERA): CE (Common Era) and BCE (Before Common Era) are the world's most widely used calendar era abbreviations, and they are secularized alternatives to the AD (Anno Domini) and BC (Before Christ) notations invented by the sixth-century Eastern Orthodox monk Dionysius Exiguus (470–544 CE).

BELIEF SYSTEM: a systematic set of beliefs that fit together to explain the universe and a person's place in it.

BIBLE: a collection of religious texts and *scriptures* that is sacred within Abrahamic Monotheism.

BINITARIANISM: the belief that a supreme God has two divine aspects (e.g., the Monad and Demiurge of Biblical Demiurgism).

BISHOP: an ordained clergy member who is entrusted with a position of authority over a number of churches or a diocese.

BLASPHEMY: the offense of speaking or acting sacrilegiously about a god, gods, or sacred things.

BRANCHING DIAGRAM: a *cladogram* or *phylogenetic tree* suggesting historical interrelationships among *comparands* (see comparand).

BUDDHA: The enlightened one. The term *buddha* (lower case) is often used for anyone who has attained nirvana. The term "The Buddha" (upper case) is used for Siddhartha Gautama, the founder of Buddhism.

CALIPH: the civil and religious leader of a Muslim state usually considered to be a representative of Allah on earth. He is also considered to be a political-religious successor to the Prophet Muhammad.

CALIPHATE: a political-religious state under the leadership of a caliph (chief Muslim ruler).

CANON: a collection of authoritative *scripture* accepted as genuine, usually pertaining to general law, rule, and principle.

CE (COMMON ERA): see *BCE*.

CHARACTER: a diagnostic feature or trait of an individual or a group.

CATHOLIC: (with a small c) used here as a vernacular and general characteristic of Nicene Christianity. Not to be confused with Roman Catholicism, a more specific, formalized catholic branch. Also see **Note 8D**.

CHIEF GOD: the highest, mightiest god of a *pantheon*; typically, the *patriarch* of the *pantheon*.

CHRISTIANIZATION: the conversion of a society to Christianity, or infusion of Christian characteristics on other belief systems or cultures.

CHRONICLER: a person who writes accounts of important or historical events.

CIVILIZATION: see **Note 3C**.

CLADISTICS: a method of *classification* of *comparands* according to the proportion of measurable characteristics that they have in common.

CLADOGENESIS (CLADOGENETIC CHANGE): the evolutionary splitting of an ancestral group into two descendant subgroups.

CLADOGRAM: a branching diagram that illustrates group interrelationships based on the most parsimonious summary of shared characters among comparable entities (i.e., among *comparands*). Also see **Note 1B**.

CLASSIFICATION: a hierarchical organization of groups based on special shared characters that are hypothesized to be indicative of some sort of relationship. It is hierarchical only in the sense of being a system of "nested groups" (groups, within groups, within even larger groups) based on special diagnostic similarities defining each group. The initial classification results in putative groups or "kinds" for comparison; phylogenetic comparison tests the historical integrity and unique reality of those groups.

CLERGY: designated religious leaders, often by means of ordination.

CODIFICATION: the action and process of arranging rules and laws according to a system.

COEVOLUTION: the influence of closely associated entities on each other in their evolution.

COMPARANDS: putative groups of comparison, or typified "kinds," in comparative studies.

COMPARATIVISM: looking at different typified "kinds" (*comparands*) of something collectively to better understand their natures (e.g., their similarities, differences, interrelationships, and histories).

CONQUEST: the subjugation of a people or culture by military force.

CONVERGENCE: where two different groups evolve a similar characteristic independently. A similarity between two different groups that is not due to inheritance from a unique common ancestor (e.g., the possession of flight wings in flies, birds, and bats).

CREATOR GOD (or **CREATOR DEITY**)**:** a god responsible for the creation of the earth, the universe, and its inhabitants.

CRUSADES: campaigns and wars sanctioned by the Western Christian Church against non-Christian regions of the world, particularly during Medieval time.

CULTURAL SELECTION: see *selection*.

CULTURE: the behavior, institutions, language, customs, beliefs, arts, skills, and achievements of a particular social group.

DALAI LAMA: foremost spiritual leader of the Geluk or "Yellow Hat" branch of Tibetan Buddhism (currently the most dominant of the four main schools of Tibetan Buddhism). Also, served as head of the Tibetan government from 1642 to 1705 and from 1750 to 1951.

DATA MATRIX: a collection of presence versus absence data arranged in vertical columns of comparative groups (*comparands*) and horizontal rows of hypothetical characters to decipher group interrelationships (see the figure on page 17) used to construct a *cladogram.*

DEITY: the essential nature of a god, goddess, or divinity.

DENOMINATION: a religious subgroup that shares a particular combination of ideological characteristics. In practice, today there has been a general move away from traditional denominationalism by many religious groups (e.g., the Unitarian Universalist Association mentioned in chapter 9, the Bahai mentioned in chapter 11). There has often been a move away from denominationalism in practice, indicated by a modified way in which many people practice the particular religion to which they self-identify. However, the subject of this book is on the evolutionary interrelationships of core religious ideologies, which were historically denominational.

DHARMA: religious law, justice, and righteousness in most of Dharmic Religion.

DIAGNOSE: to recognize a category of something based on a unique combination of characteristics; to identify the nature of something.

DIASPORA: dispersal of people from their original homeland. A scattered population whose origin lies in a different geographic place.

DIVINE TRIAD: three recognized aspects of a divine entity, such as a *Holy Trinity* of Christianity, the *Trimurti* of Hinduism, the *Three Pure Ones* of Taoism, the *Thebian Triad* of Egyptian Polytheism, and many others.

DIVINITY: a religious being, like a god, goddess, or angel.

DOCTRINE: a body of principles or codified set of beliefs.

DOUBLE PREDESTINATION: the idea that God not only chooses some people to be saved but also creates some people who will be damned. The difference between predestination and double predestination is philosophically confusing and somewhat controversial (e.g., the difference between "not saved" and "damned").

DUALISM (IN RELIGION): the conceptual division into two contrasting or opposing aspects or principles. Dualism can be non-conflicting and cyclic as in various branches of *Asian Cyclicism* (e.g., breathing in and out, yin and yang, day and night) or it can be conflicting as in *Dualistic Monotheism* (e.g., good vs evil, heaven vs hell, God vs Satan).

ECUMENICAL PATRIARCH: (also known as the bishop of Constantinople). Known as "first among equals" among the bishops of the Eastern Orthodox Church, and the spiritual leader of Eastern Orthodoxy.

ENVIRONMENT: the surrounding conditions in which something exists and operates.

ESCHATOLOGY: the study of matters concerning the ultimate destiny of humankind and the world.

EUGENICS: a racially or socially biased set of beliefs and practices attempting to improve the genetic quality of a human population, historically by excluding (or even seeking to decrease the population of) people and groups judged to be inferior and promoting those judged to be superior.

EVANGELISM: the zealous advocacy of a belief, cause, or *scripture.*

EVOLUTION: diversification and change over time. In biological evolution, components of *evolutionary process theory* include genotypes, phenotypes, species, populations, and various other elements. In cultural evolution, components of *evolutionary process theory* include memotypes, ideologies, denominations, populations, and other elements.

EVOLUTIONARY PROCESS THEORY: a theory used to explain *patterns* of diversity as being due to common (shared) ancestry; evolutionary processes used to explain the natural order of *cladograms* and data-based *branching diagrams*. Also see **Note 1A**.

EVOLUTIONARY PROGRESSIONISM: a theory that evolution is a universal ladder of progress. As a general theory, it is rejected in *The Evolution of Religions*. Also see **Note 2A**.

EVOLUTIONARY PROGRESSIVISM: an obsolete theory that organisms have an innate tendency to evolve in a definite direction toward a particular goal due to an internal driving force.

EXCLUSIVISM: the action or policy of being exclusive. In religion, this generally refers to the chauvinistic practice of accepting only one's particular religious ideology as true and excluding all others as heretical or ignorant.

FATWA: a point of Islamic law given by a recognized Muslim authority.

FITNESS: the ability to adapt to a given environment and reproduce successive generations. Also see **Note 2D**

FLOOD MYTH: *myths* in which a great flood destroys most or all of the human population, often a population that has been disobedient or disappointing to a *deity*.

FOUNDER: one who first establishes something such as a belief or an institution.

FUNDAMENTALISM: strict, literal interpretation of religious *scripture*.

GENE: a biological element of heredity that is transferred from one generation to the next, which determines certain characteristics of the offspring. The functional unit of biological inheritance. Analogous to the "meme" of cultural inheritance.

GENERAL POLYTHEISM: the belief in multiple deities or aspects. *Hard polytheism* is the belief in multiple gods or goddesses; *soft polytheism* is the belief in multiple aspects of a single *deity* Sometimes, soft polytheism is referred to as soft monotheism..

GENOTYPE: the complete set of genes possessed by an individual.

GOD, GODDESS, OR GOD: see *deity*.

GROUP: a collective of individuals sharing some similarity or historical connection.

GURU GRANTH SAHIB: a collection of religious texts and *scriptures* that is sacred within Sikhism.

HEBREW BIBLE: the canonical collection of scriptures known as the Tanakh in traditional Judaism and as the Old Testament in Christianity.

HELLENISM: greek culture beginning around the fourth century BCE and ending around the fourth century CE (Bowker, 2006).

HENOTHEISM: worship of a single, supreme god while not denying the existence or possible existence of other lower deities.

HERETIC: one who practices *heresy*.

HERESY: a belief or practice that is profoundly contrary to orthodox religious *doctrine*.

HERESIOLOGIST: a person who writes about heresy, particularly from the Medieval perspective of Christianity or Islam.

HETERODOX: belief or opinion at variance with the official or traditional (*orthodox*) position.

HIGH PRIEST: a chief cleric of a religious belief.

HIGHER TAXONOMIC GROUP, OR "HIGHER TAXON": a taxonomic group that is relatively greater in size and breadth than another in terms of subgroup membership.

HOLY LAND: the areas of the Middle East mentioned in the Abrahamic Bible; roughly corresponding to the modern State of Israel, the Palestinian territories, western Jordan, and parts of southern Lebanon and southwestern Syria.

HOLY TRINITY: see *divine triad* and *trinitarianism.*

HOMOLOGY: a homologous characteristic is one that is possessed by two or more groups and is thought to be derived (inherited) from a *unique common ancestor.* See *apomorphy* and *plesiomorphy.*

HOMOPLASY: a general term for certain nonsynapomorphic similarities (e.g., due to convergence, independent deviation, or evolutionary reversal).

HYPOTHESIS: a testable supposition or proposed explanation made on the basis of limited evidence as a point for further investigation (e.g., something proposed as a *pattern* of *shared derived characters* among different groups of comparison). Also see **Note 2G**.

ICON: a symbolic object regarded as worthy of veneration.

IDEOLOGY: a system of ideas.

IDOLATRY: the sin of idol worship; the veneration of a human-made object instead of the transcendent God.

IMAM: to the Shia, the Imam has supreme political and religious authority. He is thought to be one of the hereditary successors of Ali (cousin, son-in-law, and Companion of the Islamic Prophet Muhammad). To the Sunni, the Imam title is only a title given to a leader of worship in a mosque, and not necessarily with any hereditary connection to Ali.

INCLUSIVISM: with regard to religion, this term generally refers to the acceptance of only one particular religious ideology being "true", but accepting that there are also certain elements of truth that can be found in other religions (definition of Hick, 2016).

INDIGENOUS: native or original, something occurring naturally in its place of origin.

INQUISITION (ALSO KNOWN AS THE CATHOLIC INQUISITION): a Roman Catholic tribunal formally established in 1232 under Pope Gregory IX, for the suppression of heresy and the punishment of heretics. It was most prominent in northern Italy and southern France, and it was notorious for its use of torture. The "Spanish Inquisition" was a separate nationalized part of the institution established in 1478, and endorsed by Pope Sixtus IV in 1483 (Bowker, 2006:272). Also see **Note 8H**.

ISLAMIZATION (ALSO CALLED MUSLIMIZATION): the conversion of a society to Islam or infusion of Islamic characteristics on other belief systems or cultures.

INTERRELATIONSHIPS: the way three or more things are interrelated.

JIHAD: meaning "struggle" in Arabic, with two forms: "individual jihad", and "greater jihad" (discussed further on page 449).

JURISPRUDENCE: a legal system of philosophy of law.

KARMA: the sum of a person's actions in this and previous states of existence, viewed as deciding their fate in future existences.

LINEAGE: a continuous series or string of ancestor-descendant populations over time

LINEAR MONOTHEISM: a form of *monotheism* that sees time as linear (one birth, one life, one death, and finally eternal existence in the afterlife). See *monotheism.*

MACROEVOLUTION: a vague and subjective term, referring to major evolutionary diversification. It generally refers to inherited change significant enough to result in speciation (in biological evolution) or social schism (in culture). As opposed to much finer variation levels referred to as *microevolution.*

MARTYR: one who suffers death on behalf of their faith.

MATRISTICISM: matristicism, also known as the *Cult of the Mother Goddess* by some authors, focuses on maternalism and human fertility as being sacred.

MEDIEVAL TIME: the so-called Middle Ages, which was the period in European history ranging from the collapse of the Western Roman Empire in the fifth century CE to the beginning of the Renaissance in the fourteenth or fifteenth century.

MEME: an element of culture that is transferred from one generation to the next by non-genetic means, particularly by teaching, example, and imitation. The functional unit of cultural inheritance. Also see **Note 2C**

MESSIAH: in Abrahamic religions, a messiah is a savior or messenger sent by God.

MICROEVOLUTION: a vague term, referring to very slight change within a group (relative to *macroevolution*).

MONARCHICAL: referring to a *monarchy.*

MONARCHY: a political system based on the sovereignty of a single ruler such as a pharaoh, king, or emperor.

MONASTICS: *monks, nuns,* or others living under special religious vows.

MONK: a member of a religious community of men typically living under vows of poverty, chastity, and obedience.

MONOPHYLETIC GROUPS: groups whose members show special uniquely shared similarities that are theoretically derived from a *unique common ancestor.* Also see **Note 1L**.

MONOPHYLY: the group condition of sharing a *unique common ancestor* and including all known descendants of that common ancestor.

MONOTHEISM: belief in one god, and one God only (e.g., Aten, Ahura Mazda, El, Yahweh, Jehova, Yaldabaoth, Allah, Ik Onkar, El Shaddai, God, G_d).

MUSLIM: a follower of the religion Islam.

MUSLIMIZATION: see *Islamization.*

MYTH: a traditional story involving supernatural beings or events.

NATURAL SELECTION: see *selection.*

NETWORK: a complex system of interconnected things or groups.

NEW TESTAMENT: the second division of the Christian biblical canon.

NEW WORLD: the Americas; the majority of the world's Western Hemisphere. See *Old World.*

NIRVANA: the final escape from samsara (the cycles of rebirth) in Buddhism.

NOMENCLATURE: the system of names and naming.

NON-MONOPHYLETIC GROUPS OF CONVENIENCE: provisional groups of *comparands* that are admittedly not *monophyletic* and are used only on occasion as a category of convenience for the purpose of discussion (e.g., "things that fly," "invertebrates," "gnosticism," "Judaism," excluding Jesu-venerationsim).

NUN: a member of a religious community of women, especially a cloistered one, living under vows of poverty, chastity, and obedience.

OLD TESTAMENT: a Christian reorganization of the Hebrew *Tanakh.*

OLD WORLD: Africa, Europe, and Asia; generally, a term for those parts of the world other than the Americas. The term originated in Europe during the late sixteenth century, shortly after Europe became aware of the Americas.

ORGANIZED RELIGION: a formally codified or institutionalized form of religion. It generally involves specific *doctrine, authority, rituals, myths, practice, sacred elements*, and *taboo elements*.

ORTHODOX: belief or opinion conforming to the original or traditional position. The departure from orthodoxy is *heterodoxy*.

PAGAN: a derogatory (and nonmonophyletic) term for non-Abrahamic religions, coined in the fourth century CE.

PALEOLITHIC TIME: a period of prehistory beginning with the origin of stone tools by close human relatives around three million years ago until the end of the Pleistocene 11,650 years ago. Also known as the Stone Age.

PANTHEISM: a belief that all souls and components of the universe are collectively synonymous with God; the belief that everything in the universe is divine.

PANTHEON: the group and organization of deities or great spirits recognized by a particular *polytheistic* religion.

PARALLELISM: another term (in addition to convergence) for a case in which two different groups evolve a similar characteristic independently. A similarity between two different groups that is not due to inheritance from a unique common ancestor (e.g., the possession of flight wings in flies and birds).

PARSIMONY: choosing the simplest explanation that fits all the available evidence. The principle of parsimony refers to explaining something with the fewest possible assumptions. In the biological sciences, the most parsimonious ordering of shared characteristics is the main driver of phylogenetic analysis (the reconstruction of historical diversification). In the social sciences (and unlike in the biological sciences), additional information in the form of many contemporaneous historical records (e.g., humanly observed schisms, founders, and other factors resulting in diversification) can overrule the simple parsimonious assessment of characters. Also see **Note 1N**.

PARTICULARISM: the exclusive focus on a particular subgroup rather than comparing it with others. This method is sometimes attributed to the social scientist Franz Boas of the late nineteenth and early twentieth century. See, in contrast, *comparativism*. Also see **Note 3D**.

PATRIARCH: the male head of a group or movement.

PATTERN (PHYLOGENETIC PATTERN): statistically nonrandom order of *comparands* based on shared characteristics and historical records.

PHYLETIC GRADUALISM: gradual evolutionary change over a relatively long period of time (opposite of *punctuated equilibrium*).

PHYLOGENETIC: relating to the diversification of subgroups within groups over time.

PHYLOGENETIC TREE: effectively a map of historical interconnection, often with general estimates of time included. It is a *cladogram* interpreted in an evolutionary context.

PHYLOGENY: the classification of monophyletic groups as a system of nested sets.

PLATONISM: a school of early Greek philosophy beginning with Plato (fifth or fourth century BCE), affirming the existence of an ideal realm of intellect that recognizes the existence of abstract objects (e.g., numbers, sets, truth values). This realm is thought to be distinct from the sensible external world of everyday life and concrete objects (e.g., plants, human beings, planets) and to exist outside of (independent of) the internal world of human consciousness. Platonism holds that ultimate truth lies within this ideal realm. In the third century CE, Roman philosopher Plotinus (204–270 CE) added certain mystic elements to Platonism to found *Neoplatonism*, which was highly influential to Western culture during Late Antiquity, the Middle Ages, and the Renaissance. Although some authors consider Platonism to be an ancient form of religion (Turchin et al., 2022), in *The Evolution of Religions* it is considered to be only a general philosophy that was independently adopted by various groups of religions.

PLESIOMORPHY: a homologous character that is *not* uniquely derived for the particular group of focus but is instead uniquely derived for a larger group to which the group of focus belongs (e.g., the presence of vertebrae in birds, or the recognition of Abraham as a *patriarch* in Roman Catholicism). As a shared character, plesiomorphy is also referred to as *symplesiomorphy*.

PLURALISM: where two or more forms of something coexist peacefully. Religious philosopher, John Hick defines religious pluralism as an acceptance that many different beliefs facilitate the transformation of human culture from complete self-centerdness to recognition of a greater peaceful purpose or "reality-centerdness" (Hick, 2016). Many recent authors who reject pluralism in favor of exclusivism are consciously or unconsciously advocating for one particular religion or another (Donavan, 1993). As used here, advocacy of pluralistic tolerance of diversity does not contend that "all religions are true," that "all religions are the same," or that "none of them are true." It simply contends that, in the long run, tolerance of peaceful diversity and finding globally collaborative solutions to globally impactful challenges are the best courses of action for the future survival of our species. The present study is a pluralistic one that is focused on the historical interrelationships of many different ideologies rather than any sort of qualitative ranking or preference. Comparativism is a necessary component of pluralism, because in order to be tolerant of diversity, one first must attempt to identify and understand it. See, in contrast, *exclusivism*.

POLYTHEISM: a supernatural belief system involving a *pantheon* of more than one *deity* or *great spirit*. Sometimes divided into "hard polytheism" (each deity or great spirit distinct and individually named and worshipped) and "soft polytheism" (one deity or great spirit with multiple aspects, such as *trinitarianism* and *binitarianism*). Also see **Note 8I**.

POPE (also known as the bishop of Rome): Head of the Roman Catholic Church, and theocratic head of state for Vatican City. Within Roman Catholicism, the title "pope" has been used exclusively for the bishop of Rome since the eleventh century, but was more widely used before that time. The title is also used more broadly for priests within Eastern Orthodoxy (e.g., within Greek Orthodoxy). Also see **Note 8B**.

PRE-COLUMBIAN: the pre-colonial period of the Americas; referring to the time prior to the first arrival of the Spanish and the landing of Christopher Columbus.

PREDESTINATION: the belief that, because God is all-knowing and all-powerful, a human being's ultimate reward in the afterlife was already decreed by God and unchangeable during his or her lifetime. See *double predestination*.

PRIEST: a person with the authority to perform certain religious rites, rituals, and practices within a religious tradition.

PROPHET: a person regarded as an inspired teacher or proclaimer of the will of God; one who speaks on behalf of God.

PROSELYTIZATION: the policy of attempting to actively convert a people's religious beliefs, often forcefully. Also see **Note 3G** and **Note 12H**. Proselytism is often an element of religious *exclusivism*.

PUNCTUATED EQUILIBRIUM: relatively abrupt evolutionary change over a relatively short period of time. See, in contrast, *phyletic gradualism*.

QURAN: the central religious text of Islam, believed by Muslims to be a revelation from God.

RABBI: a Jewish scholar or spiritual leader, especially one who studies or teaches Jewish law.

RECONQUISTA: a series of military campaigns in Medieval Spain and Portugal (eighth to fifteenth centuries) by Christian states to recapture territory from the Muslims.

RELIGION: a system of shared beliefs and practices connecting the physical world to the supernatural or sacred. It often involves specific *doctrine*, *rituals*, *myths*, practice, sacred elements, and *taboo* elements, all in the form of *Organized Religion*. I use the term *religion* interchangeably with the term *supernaturalism*. It is an anthropological term, not a theological one (Smith, 1998:269). As defined in *The Evolution of Religions*, religion is not necessarily synonymous with theism. Some forms of religion are atheistic in that they do not believe in spiritual beings, but still hold to certain supernatural beliefs. Also see **Note 1H**.

RETICULATE: net-like complexity.

RITUAL: a solum ceremony consisting of actions performed according to a prescribed order.

ROMANIZATION: the conversion of a non-Roman culture to one showing Roman influence.

SACRED: dedicated to a religious purpose and worthy of veneration or reverence. Also see **Note 3K**.

SAINTS: exemplary religious individuals who are venerated and invoked in prayer.

SAMPRADAYAS: denominations of Dharmic Religion.

SCHISM: a split or division that develops between subgroups of a religious belief system over the growth of strong ideological differences. A primary factor in cultural differentiation over time.

SCRIPTURE: sacred writings of a religion.

SECONDARY CHARACTER LOSS: where the absence of a character in a *lineage* is the result of losing it rather than never having had it.

SECT: an inexact term, usually used to indicate a small branch of religion separated from a more established church.

SECULAR: worldly, not sacred; nonreligious.

SECULARIZATION OF RELIGIONS: the general trend away from core supernatural belief that has occurred independently in factions of many different religions over time (e.g., Buddhism, Judaism, and Christianity discussed in chapters 1, 4, 7, 9, and 11).

SELECTION (NATURAL OR CULTURAL): the process through which populations or ideologies are thought to change, in response to the environment in which they exist. In general, as the environment changes, populations or ideologies either adapt, change, or become extinct.

SHAMAN: a person regarded as having access to, and influence in, the world of good and evil spirits, generally for the good of the tribe or group to which the person belongs.

SISTER GROUP (SISTER BRANCHES): the two most closely related branches on a *phylogenetic tree* or *cladogram*.

SOCIAL SCIENCES: scientific study of society and how various aspects of culture developed.

SPIRITUALISM: a doctrine that the spirit exists as distinct from the material world.

STATE RELIGION: the religion officially endorsed by a sovereign state.

STOICISM: a school of Hellenistic philosophy founded by Zeno of Citium in Athens (334–262 BCE). It asserts that the practice of virtue, in accordance with nature, is both necessary and sufficient to achieve *endaimonia* (happiness and good spirit). Although some authors consider stoicism to be an ancient form of religion (Turchin et al., 2022), in *The Evolution of Religions,* it is considered only to be a general philosophy that was independently adopted by various groups of religions.

SUPERNATURALISM: a traditional belief in a realm or spiritual force beyond the physical world. Also see **Note 3A**.

SUTRAS: a genre of ancient and Medieval Indian texts found in Hinduism, Buddhism, and Jainism.

SYMBIOTIC: relating to different entities developing a mutually beneficial relationship.

SYMPLESIOMORPHY: a *shared plesiomorphy* (sometimes referred to as a *shared primitive character*). See *plesiomorphy* and *primitive.*

SYNAPOMORPHY: a *shared apomorphy* (also referred to as a *shared derived character*) thought to indicate close relationship. See *apomorphy*. Also see **Note 1L**.

SYNCRETISM: the combination of different forms of belief or practice.

SYSTEMATICS: the branch of science that deals with *classification* and *nomenclature*; *taxonomy.*

TABOO: a strong social prohibition.

TALMUD: The classic rabbinic discussions of the ancient code of Jewish law.

TANAKH: The Jewish name for the Hebrew Bible. Used to make the Old Testament in Christianity.

TAO-TE CHING: a collection of religious texts and *scriptures* that is sacred within Taoism.

TAXA: named representative samples of diversity (singular is *taxon*).

TAXON: a group of things seen by taxonomists to form a unit, based on special shared similarities among the members of that group.

TAXONOMY: the *classification* of *taxa,* and the basic principles involved in their recognition and hierarchical organization (hierarchical in the sense of different levels of inclusiveness, or groups within groups). In the words of History of Religion scholar, Jonathan Z. Smith (Smith, 2004:175), "Scholarly labor is a disciplined exaggeration in the direction of knowledge; taxonomy is a valuable tool in achieving that necessary distortion."

TERMINAL GROUPS (OR TERMINAL TAXA): the smallest branches or subgroups in a classification; a relative term referring to the subgroups at the very tips of a phylogenetic tree or cladogram.

THEISM: the belief in supernatural beings.

THEOCRACY: a system of government believed to be under divine guidance, led by a person who is believed to be under divine guidance; a form of government that does not have a separation between church and state.

THEOLOGY: reflection on the nature of religious belief or supernatural powers.

THEORETICAL COMMON ANCESTOR: a theorized uniquely shared ancestor to explain groups that share statistically significant similarities (apomorphies) with each other.

THEORY: a system of ideas intended to explain something based on general principles independent of the thing to be explained (e.g., Darwin's theory of evolutionary process to explain hypothetical *patterns* of diversity). Also see **Note 2G**.

THERIANTHROPE: a mythical being that is part human and part nonhuman animal.

TORAH: the compilation of the first five books of the Hebrew Bible or Tanakh; namely, the books of Genesis, Exodus, Leviticus, Numbers, and Deuteronomy.

TOTEMISM: a system of belief in which humans have kinship or a mystical relationship with a spirit-being, such as that of an animal, plant, or sacred object. The word "totem," derived from an Ojibwa word meaning "he is a relative of mine," usually refers to an animal figure that is revered for its symbolic meaning and is often considered to be related by blood to a family or clan.

TRADITION: the transmission of beliefs and practices from generation to generation.

TREES: see *phylogenetic tree.*

TRIBALISM: the condition of creating an exclusive group or community beyond immediate family ties, based on shared beliefs, shared history, shared governance, shared patriotism, or genealogical relationship; differences in beliefs and practices that sometimes lead to discriminatory behavior or attitudes toward out-groups, based on in-group loyalty or in-group history.

TRINITARIANISM (AS AN ADJECTIVE): the belief that a Supreme God has three distinct aspects (e.g., the Father, Son, and Holy Ghost of Roman Catholicism). This divine triad is known in most of Christianity as the Holy Trinity.

TRIPITAKA (ALSO TIPITAKA): a comprehensive collection of religious texts and *scriptures* that is sacred within Buddhism. Only the compilation of the Theravada school, written in Pali, survives in its entirety.

UNIQUE COMMON ANCESTOR (UNIQUELY SHARED ANCESTOR AMONG DIFFERENT DERIVATIVES): the most recent form from which all *monophyletic group* members or branches are descended.

UPANISHADS: late Vedic Sanskrit texts of Hindu philosophy that supplied the basis of later Hindu philosophy.

UNITARIANISM (AS AN ADJECTIVE): a belief that asserts the unity of a single God with no multiple aspects (e.g., rejection of *trinitarianism* and *binitarianism*).

VEDAS: a collection of religious texts and *scripture* originating in India that is sacred to Hinduism.

VENERATE: to regard with reverential respect or devotion.

WESTERN WORLD (also known as **WESTERN CIVILIZATION**, **WESTERN CULTURE**, or simply **THE WEST**): a common name for various regions, nations, and states that share certain cultural characteristics inherited from the Mediterranean region of ancient Greece and Rome. Over the millennia, "the West" expanded to include most of Europe, North America, and Oceania (Australia plus certain nearby Pacific regions). The division between the Western world and the Eastern world was fueled, in part, by the East-West Schism in 1054 CE, between the Roman Catholic Church and the Eastern Orthodox Church.

YANG: the aspect of reality that expresses itself in speech, heat, and light.

YIN: the aspect of the universe that expresses itself as silence, coolness, rest, and darkness.

NOTES

PREFACE: A MODERN EVOLUTIONARY APPROACH TO HISTORY OF RELIGION STUDIES

PrA. Agnosticism as an objective, impartial perspective for the comparative study of religions. Agnosticism (a form of accepted uncertainty) is not the same as atheism (the unequivocal rejection of the existence of God, gods, or spiritual beings). It is simply a philosophy that the ultimate, specific reality of any particular spiritual entity or religion is objectively unknown and probably unknowable. The agnostic approach to History of Religion studies is not a value judgment about people of particular faiths or about atheists. It is instead a nondiscriminatory search for patterns of ideological relationships among different belief systems over time. However, agnosticism is a perspective that is far from being universally embraced. Agnostics are sometimes criticized by atheists as being too timid in their refusal to take a stand about the existence or the nonexistence of God (Dawkins, 2006). Alternatively, in the eyes of fundamentalists and many highly religious people, agnostics are considered to be atheists, heretics, or lost souls in need of saving because they do not pick a particular religion. When it comes to the issue of religion, we face an underlying pressure to take sides. Nevertheless, agnosticism is a logical and objective context in which to empirically investigate the comparative properties and broad history of religions. According to Johnson and Grim (2013), agnosticism increased from 0.2 percent of the global population in 1910, to 9.8 percent of the population by 2010.

Greek philosopher Protagoras (490–420 BCE) was one of the earliest proponents of agnostic thinking, stating: "As to the gods, I have no means of knowing either that they do exist or that they do not exist. For there are many obstacles to knowing: the obscurity of the matter and the fact that man's life is short" (Broadie, 2003:80). The famous Austrian-British philosopher of science Karl Popper (1902–1994) was another supporter of agnosticism with its unpretentious element of uncertainty. He once stated, "I don't know whether God exists or not . . . but agnosticism—to admit we don't know and to search—is alright" (Zerin, 1998:46). Agnosticism is simply the honest acknowledgment of an element of uncertainty about specific belief systems. This uncertainty is deeply buried in some people, unacknowledged by many, and honestly confronted by others. In the words of Mark Twain, "Religion consists of a set of things which the average man thinks he believes and wishes he was certain of" (University of Cambridge Correspondence Project; accessed March 1, 2023).

Many famous evolutionary biologists throughout history have also been agnostics, ranging from "Darwin's Bulldog" T. H. Huxley (1825–1895) to late-twentieth-century evolutionist Stephen J. Gould (1941–2002). Even Darwin was an admitted agnostic in his later years. He saw no *necessary* conflict between religions and evolutionism, except in the case of strict biblical literalism and fundamentalism. In 1879, while working on his autobiography and in a letter to evangelical minister John Fordyce, Darwin wrote: "I think that generally (and more and more as I grow older), but not always, that an agnostic would be the more correct description of my state of mind" (Darwin, 1879).

Organized Religion is a faith-based construct of human effort to understand the spiritual world, and there have been tens of thousands of such efforts over time. Interpretative details of Organized Religion vary from culture to culture, region to region, or even generation to generation, and these differences have often sparked violent conflict. Perhaps the Biblical Demiurgists were right in their belief in a Supreme God that is not meant to be understood by mortal beings (chapter 10). Clearly, not all conflicting interpretations of the supernatural can be correct, nor can we objectively know which one, if any, is. Therefore, there is no reason to persecute any supernatural belief system, or lack thereof, that does no harm to others.

We cannot scientifically prove the existence or the nonexistence of a God, gods, or other supernatural entities, nor can we prove the historic accuracy or reality of many ancient myths. That is not a matter for science. However, the job of developing classifications and hypothetical patterns of historical relationship among different kinds of belief systems clearly *is* a matter for science (and particularly for cultural historians). Agnosticism offers a necessary degree of impartiality to keep broad studies of comparative religion as objective as possible. It effectively also supports a more pluralist (rather than exclusivist) environment within human society, by presenting a less partisan survey of known cultural diversity and history.

PrB. Further discussion about the use of trees to illustrate phylogenetic relationships. Phylogenetic trees are ideal metaphorical tools to clearly illustrate hypotheses of interrelationship. Most of them are continually updated, revised, and expanded as science and continued study refine analyses and add new data (new characters, new historical records, and new subgroups). Phylogenetic trees are also continuing works in progress. They are hypotheses that are never meant to be final words of authority, and they are always open to challenge, revision, clarification, and expansion.

Although phylogenetic trees are commonly used in evolutionary biology, some religious historians believe that trees are misleading in the historical study of religions.

For example, theologian William H. Harrison expresses two major concerns (Harrison, 2014:163): First, he states that trees are problematic because they indicate "a branch always comes out of one other branch rather than several." Second, he states that trees mistakenly show branches as "discrete entities, meaning that they do not all meld together." I believe that these two objections are the result of a misunderstanding of what modern phylogenetic trees represent.

Regarding Harrison's first point: the branch proximity on a modern phylogenetic tree does *not* mean that a branch necessarily evolved directly out of the branch next to it. Trees, such as the ones in this book, are meant as simplified diagrams of *relative* relationship among named (and typified) groups within a given study. The closest branches on a phylogenetic tree (i.e., sister groups) do not necessarily represent direct ancestor-descendant pairings. Instead, they hypothesize the most closely related forms among those groups being analyzed. There could have historically been many missing intermediate groups (missing stages of transitional ideology) between two sister branches on the tree.

Regarding Harrison's second concern about trees not showing melded branches, I would make two points. First, some putative examples of what Harrison interprets as melding together (presumed fusions, reticulations, or other forms of complete syncretization) may instead be oversimplifications of much more complex historical processes of branching differentiation. The evolution of a particular branch of religious ideology may include several imitated details from other distantly related branches, and vice versa, without complete fusion among those branches. Syncretization may be partial, at best, including only certain elements of each religious ideology. Such similarities, in many cases, could be considered examples of parallelism or character convergence in two different branches, rather than the complete melding of two branches. Another point I would make is that even if a study needed to illustrate an actual complete fusion (complete and absolute melding together of two branches rather than partial reciprocal imitation), such an event could easily be shown on a tree by bringing branch tips together, or by turning parts of the tree into more complex networks (see **Note 2E**). I agree with Harrison that partial syncretizations (or at least cross-influences) of particular elements among different religions have been common through history. However, I believe that instances of complete and absolute fusion of branches of religion have been relatively rare, and some putative examples of such complete syncretizations are oversimplifications of historical process and diversification.

PrC. Method of evolutionary analysis. The method of evolutionary (phylogenetic) analysis, as used here and in most modern studies, can be summarized with three basic steps.

First, you identify ("diagnose") typified samples of diversity ("kinds" or "comparands") for comparative analysis. These may include different species of organisms, dialects of languages, denominations of religion, etc. Here, I clarify that it is religious ideological variation I am analyzing, and not types of people or other aspects of culture. Even a single individual person can switch between vastly different ideological lineages within their lifetime. The focus of this book is on the historical relationships among different religious ideologies.

Second, you organize ("classify") the different comparands into hierarchically inclusive, hypothetical patterns or "cladograms" (i.e., groups within larger groups, within even larger groups) based on different levels of shared, hypothetically derived characteristics. High levels of congruent, uniquely shared group characteristics can be hypothesized to have been uniquely derived (uniquely inherited from a common ancestor). If contemporaneous records of historical ancestry are available (as is often the case for aspects of human culture in particular), those records should be incorporated into the overall organizational pattern. With regard to aspects of human culture, historical records of imitation and radical shifts may overrule parsimony in the cladistic reconstruction of phylogeny, indicating that certain shared characteristics are parallelisms or examples of convergence, rather than uniquely derived. Remember that each organizational pattern produced by phylogenetic analysis (the phylogenetic classification or tree) is only a hypothesis of relationship that is subject to revision and expansion with further study.

Third, you look for possible historical process explanations (e.g., mutation, schism, selection, extinction, inheritance, imitation) that could be used to theorize how parsimony-based patterns of organization came to be (see **Note 1A**).

1 RELIGIONS, CLASSIFICATION, AND PHYLOGENETIC PATTERN

1A. ***Pattern analysis* versus *process theory* in the study of evolution.** In the study of deep evolutionary history, the search for, and reconstruction of, hypothetical *patterns* (e.g., cladogram or phylogenetic tree) is both independent of, and a prerequisite to, theoretical *process explanations*. Process explanations include *mutation, anagenesis, cladogenesis, natural selection, sexual selection, artificial selection, Lamarkianism, orthogenesis, speciation,* and other mechanisms used to theoretically explain the origins of historical patterns. Even the concept of evolution is a process explanation. In the words of systematic biologists Nelson and Platnick (1981:35) regarding the character-based study of evolutionary relationships, "A *process* is that which is the cause of *pattern*. No more, no less." In general, if there is neither a hypothetical pattern of group organization (e.g., taxonomic framework, cladogram, or tree), nor contemporaneous written records of diversification (records that sometimes exists in the case of cultural evolution), then empirically based evolutionary analysis is not possible (also see **Note PrC**).

1B. ***Cladograms* versus *phylogenetic trees*, and the trees in this book.** *Cladograms* are branching diagrams of relative relationship based on nested-group hierarchies of shared derived characteristics—that is, shared-character-based groups within larger shared-character-based groups, within even larger shared-character-based groups. Cladograms do not require specific dates, time ranges, or specific ancestors; however, in evolutionary studies, they indicate hypothetical sister groups theorized to be the result of shared ancestry. A cladogram is a tree-like type of classification that is simply the most parsimonious summary of comparative empirical data about the groups being studied (pages 19 and 20). (*Parsimony* is discussed in **Note 1N**.)

Phylogenetic trees are cladograms with the addition of specific dates, specific ancestors, or other such genealogical information. Although phylogenetic trees add details of evolutionary history to cladograms, they are not generally meant to show historical variation in its finest detail (e.g., they are not usually meant to show the most detailed genetic variations of individual biological organisms or the most detailed ideological variations of individual beliefs). An important function of the tree diagram is to distill something intuitively understandable out of something that is infinitely complex. Phylogenetic trees are hypotheses that are used by generalists to represent broad macroevolutionary patterns of interrelationship and historical diversification. Although such trees generally illustrate more organizational detail (more splitting events) than either simple classification systems or strict cladograms, they generally do not illustrate as much organizational detail as *reticulate evolutionary networks* (see **Note 2E** for explanation of reticulate networks). They also generally do not illustrate hybridizations, mergers, or extreme syncretic blends of groups that are better illustrated with net-like reticulate networks.

All the trees in this book are phylogenetic trees rather than strict cladograms, because I have combined time-based information from contemporaneous (or near contemporaneous) written records together with the cladistic patterns. Such a hybrid system is ideal for the study of Organized Religion because of the relative recency of so much of the major diversification (unlike the case with biological speciation). Also, I note that I am a generalist with regard to History of Religion studies. I leave it to specialists focused on particular branches of Organized Religion to build more detailed character matrices and trees to test, modify, and expand the trees in this book.

1C. Phylogenetic work on other aspects of culture. Another aspect of culture that has been investigated by a number of authors using phylogenetic techniques is language. The field of evolutionary linguistics is a robust one (e.g., Gray and Atkinson, 2003; Bouckaert et al., 2012). Although these relatively recent attempts have been met with much controversy (Heggarty, 2006; Mallory, 2020), controversy is part of any attempted paradigm shift within the scientific community. One premise for phylogenetic linguistics is that the entity that evolves (the entity that diversifies over time) is cultural, not biological.

1D. Aristotle's Ladder of Nature (Scala Naturae). Expounding on the earlier concept of immutable essence by Plato (427–348 BCE), Aristotle developed the idea that there are "natural kinds" (taxonomic groups) diagnosed on the basis of fixed properties.

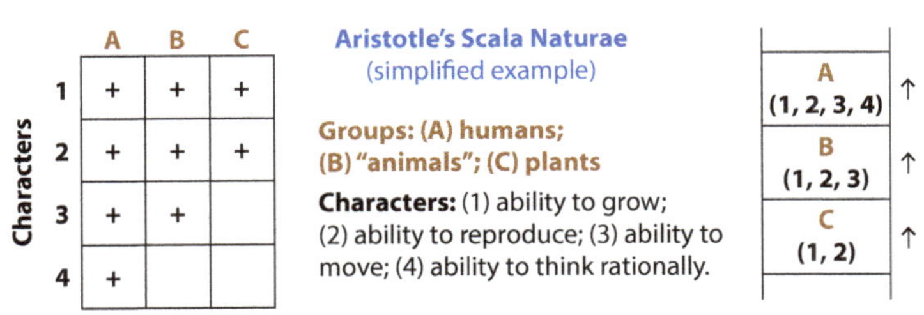

Phylogenetic reinterpretation
(of Aristotle's ladder of characters)

Groups: (A) humans; (B) "animals"; (C) plants

Characters: (1) ability to grow; (2) ability to reproduce; (3) ability to move; (4) ability to think rationally.

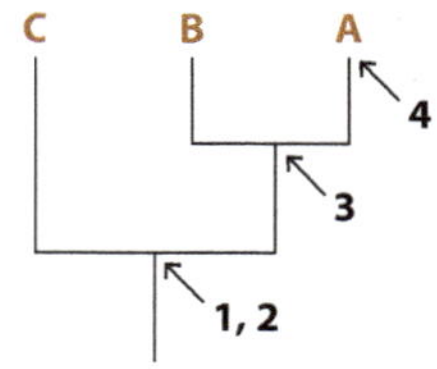

This can be extrapolated as an early (pre-evolutionary) form of classification for all living things based on shared characteristics among group members. Aristotle's group characters were very general (e.g., "ability to grow and reproduce," "ability to move," and "ability to think rationally"). His natural kinds were arranged in a progressive series beginning with inanimate nature (at the bottom of the chain), to plants, to invertebrate animals, to a series of vertebrate animals (excluding humans), and finally to humans (at the top of the biological chain). It was not only a classification but was also a qualitative ladder of "natural kinds," expressed as the *Great Chain of Being*, or the *Ladder of Nature* (*Scala Naturae*) (Pavlinov, 2021:26: Lovejoy, 1936: 58-59).

Aristotle's ladder-like qualitative ranking of groups may have later influenced the misdirected "progressive ladder approach" of early evolutionists. He believed that Special Creationism, rather than Natural Selection, explained the origin of classifiable patterns. The groups in Aristotle's system did not evolve, diversify, or ever become extinct. They were instead believed to be disconnected, independent entities that had existed from the beginning of time and were expected to exist infinitely into the future (Hull, 1988: 82). Aristotle's natural kinds were simply different categories of being, based on their cumulative number of subjective "attributes." However, Aristotle's system was also an early beginning of hypothetical, character-based analysis for the differentiation of "natural kinds." The identification of hypothetically natural kinds as comparands is the first step in any modern, phylogenetic analysis (Note PrC). Note that Aristotle's data (figure on previous page) could also have been used to construct a tree rather than a ladder as in the image at the top of this page.

1E. International Codes of Nomenclature. Major philosophical misunderstandings can easily arise when different things have the same name, or the same thing has different names. In biology today internationally recognized codes of nomenclature (rulebooks) govern the creation and use of official taxonomic names. For certain phylogenetic group names to be internationally recognized as valid, they must follow official rules outlined in these legalistic volumes. (Note: This does not apply to *vernacular* names that are intended only as nonphylogenetic group names of convenience, such as "fishes," "invertebrates," or "flying animals.") The net effect is an element of consistency for comparative studies of biological taxa. Some of the better-known volumes are the ICZN (the International Code of Zoological Nomenclature), the ICN (the International Code of Nomenclature for Algae, Fungi, and Plants), and the ICNP (the International Code of Nomenclature for Bacteria and Archaea). Comparative cultural studies could benefit from a similar set of guidelines to ensure that group names are used in a more consistent way. No doubt, getting some sort of international consensus for the naming of "cultural taxa" would be a daunting and controversial task. However, if successful, it might eventually simplify and improve intercultural discussions about the social history of our species. Striving for objective, intercultural consistency in terminology is a necessary part of meaningful cross-cultural communication and understanding.

1F. When did the cognitive ability for early supernaturalism evolve? The answer to that question will always be speculative, because human thoughts are not preserved as fossils. This explains the vague designation of Paleolithic for the approximate dates of origin for supernaturalism, animism, ancestorism, anthropomorphic supernaturalism, matristicism, and general polytheism on my phylogenetic tree on page 44. The most basic cognitive abilities of modern *Homo sapiens* may have developed very early in the evolution of the species, perhaps even from their first appearance more than three hundred thousand years ago. This early development is indicated by the suggestions by some social scientists and biologists that even nonhuman mammals possess significant levels of cognition (e.g., Heyes and Huber, 2000). Some historians suggest a somewhat later development of full cognitivism seventy thousand years ago that they refer to as the *Cognitive Revolution* in which humans gained the ability to speak and create myths (Harari, 2014:3). This date too is highly speculative.

Early elements of animism, ancestorism, matristicism, and general polytheism may have developed shortly after the development of cognitivism. Theoretically, having a deeper understanding of nature, the wisdom of ancestors, and the importance of motherhood would have all been selective advantages for survival of the human species. Animism, anthropomorphic supernaturalism, ancestorism, matristicism, and general polytheism became some of the fundamental underlying characteristics of the later emerging branches of supernaturalism (e.g., tree on page 44). Although eventually some more formalized sects of animism, ancestorism, and matristicism that developed over time could legitimately be classified as Organized Religions, most of that part of supernaturalism's diversification is beyond the scope of this book and is left to specialists in those fields to better illuminate.

1G. Religions as evolving entities. The point of religions functioning as evolving entities is graphically made by the well-known biologist Edward O. Wilson, who writes: "Religions are analogous to superorganisms. They have a lifecycle. They are born, they grow, they compete, they reproduce, and in the fullness of time, most die. In each phase they represent the human organisms that nourish them" (Wilson,1998:280).

1H. Should someone even try to define "religion"? This is (somewhat ironically) a highly controversial question in History of Religion studies. Some authors argue against trying to define this term at all, and do not even believe "religion" to be a distinct aspect of human culture (Fitzgerald, 2007). I disagree. Culture is extremely complex and includes many interacting aspects. However, in order to understand it in detail, we sometimes need to dissect it. The term "religion" is simply a term (or systematic tool) used by some scientists and historians who wish to look at this aspect of culture in a broadly comparative and historical context. It is not a theological term; it is an anthropological one (Smith, 1998:269). It may be that the word "religion" is a Western invention, but so is the formalized concept of systematic, comparative study. In the field of biology, the word "species" is also a Western invention that has often been controversial (e.g., Mishler, 2021). However, the comparative study of biological species has resulted in countless hypotheses of historical relationship helping us decipher the history of life. There is no universally accepted way of defining the term "religion," and there might not ever be one. The most important part of defining a diagnostic term for a group, in any comprehensive, systematic study, is for the author of the study to clearly define how *they* consistently use the term *in their particular analysis* (more on this in **Note 1I**). I have done that here with no expectation that this definition of religion will be universally accepted.

Specific religions, as with most aspects of culture, "do not exist apart from the individual minds that constitute them and the environments that constrain them" (Atran, 2002:10). Similarly, biological taxa, as with most aspects of nature, do not exist apart from the human minds that define them or the environments that confront them. In both cases, the identification of comparative "kinds" (religions or biological species) is necessary to better understand the history, diversity, and interrelationships within a given aspect of culture or nature. In the case of religious diversity, theoretical *memes* (as the term is used here) are heritable elements that translate into traditions (ideological elements that are subject to cultural selection and change over time) from generation to generation (**Note 2C**). In the case of biological diversity, *genes* are heritable elements that translate into phenotypes (which are subject to biological selection and change over time) from generation to generation.

1I. **The importance of a consistent context for building phylogenetic hypotheses of historical diversification.** A Western scientist may perform their character analysis of religious belief systems within one cultural context, while the Eastern scientist of Comparative Religion may perform their character analysis of religious belief systems within a completely different one. One person's classification of religions, or list of significant characteristics, may be completely different from another's. One person may also use a group name one way, where another may use it in a completely different way, because of the lack of nomenclatural rules in the comparative study of culture. These can all be obstacles to the broad analytical study of any aspect of culture (e.g., religion). Given that challenge, and in lieu of a group naming convention for cultural groups, there is at least a degree of methodological uniformity by having all characters and groups of religion broadly evaluated through the same subjective filter, or the same set of eyes (mine, in this case). It at least allows for meaningful comparisons *within a given study*. In the words of History of Religion scholar Jonathan Z. Smith, "'Religion' is not a native term; it is [an anthropological] term created by scholars for their intellectual purposes and is theirs to define. It is a second-order, generic concept that plays the same role in establishing a disciplinary horizon that a concept such as 'language' plays in Linguistics, or 'culture' plays in anthropology. There can be no disciplined study of religion without such a horizon" (Smith, 1998:269). I hope that a more consistent method of naming cultural groups will one day be developed within the social sciences. That might help facilitate more meaningful comparisons among different studies by different authors.

In addition, a better consensus on the relative ranking of groups within groups would also be a useful organizational tool for History of Religion studies. Such ranking would have nothing to do with qualitative levels of "importance." It would indicate only the relative generality or inclusiveness of each group in a given study. With biological studies, it is understood that *genera* include subgroups called *species*, *families* include subgroups called *genera*, *orders* include subgroups called *families*, *classes* include subgroups called *orders*, and *kingdoms* include subgroups called *classes*. There is no comparably systematic way of organizing and naming religious belief systems. Names such as *branch, denomination, order, school, tradition, congregation*, and *sect* are used indiscriminately and interchangeably without any formal consensus on relative group-rank level or naming criteria. Such inconsistency can add a great deal of confusion when comparing the study of one historian to another. Again, a level of contextual consistency is necessary for comparative work to be meaningful and understandable.

1J. The defining characteristics and naming criteria for groups of comparison in the social sciences. As discussed in **Note 1I**, there are no consistent ways of organized naming (e.g., Linnaean equivalent), or formalized parameters, for the identification of cultural groups. In particular, the identification of religious groups (or "taxa" equivalents) varies in several ways. Sometimes religious belief systems are named and defined based on their iconic figures (e.g., Christianity, Lutheranism, Buddhism, Zoroastrianism, and Confucianism). Sometimes they are based on reverence to a specific God (e.g., Atenism, Shaktism, Shaivism, and Islam). Other times, religious belief systems are based on specific cultures (e.g., Inca Polytheism or Aztec Polytheism), or specific regions (e.g., Egyptian Polytheism, Greek Polytheism, and Roman Polytheism). The highly simplified norm for identifying branches of Old World Hard Polytheism and New World Hard Polytheism here, in particular, is generally by geographic region (with a few minor exceptions, e.g., **Note 5G**). I take this approach because of a combination of factors reflecting our poor, fragmentary knowledge about such ancient belief systems. First, much of hard polytheism is so ancient that it developed and eventually became extinct entirely during preliterate time, with few or no surviving contemporaneous records of its history. Consequently, we have little detail about most of these belief systems. Hard polytheism appears to have begun more than ten thousand years ago, and much of it (unlike Asian Cyclicism and Abrahamic Monotheism) became extinct more than a millennium ago. Second, records of many of these ancient belief systems were intentionally destroyed by invading conquests wishing to infuse their own religions into conquered indigenous populations (e.g., in the post-Columbian New World). Third, some of these belief systems apparently never self-identified as distinct denominations. Fourth, there is an inherent sensitivity by many indigenous cultures to being placed in a systematic Westernized context, based on past transgressions (particularly in the case of pre-colonial New World Animistic Polytheism). Fifth, the diversity of hard polytheistic branches was so extreme, and changed so frequently with the expansions, contractions, and reorganizations of pantheons over time, that, in my opinion, we are not even close to understanding the complex evolution of those belief systems, especially considering the added complications from the previous four factors.

Consequently, the naming convention used here for groups of *Old World Hard Polytheism* (chapter 5) is even more provisional than that used for other major groups of *Organized Religion* in this book (chapters 6–11). Here I simply capitalize the word Polytheism where I use it as a name for a specific branch, and leave it uncapitalized where it simply refers to the general characteristic of recognizing multiple deities. Hopefully, we will one day better understand the historical diversity of Old World and New World Hard Polytheism based on defining characteristics of its subgroups and a detailed comparative analysis of historical records. Such studies may require greater intercultural cooperation and less intercultural competition and mistrust. It should also be remembered that composition and defined limits of groups for evolutionary analyses (cultural or biological) are only hypothetical, just as the branching diagrams of relationship that they suggest are. A better understanding of these belief systems will be an ongoing process of progressive studies.

The etymology of names specifying geographic regions (e.g., Asian Cyclicism and Old World Hard Polytheism) are not meant to imply that those belief systems remained restricted to those regions. This use indicates only where they are thought to have originated.

1K. Discernable, discrete boundaries among taxa for classification. The reality of boundaries between taxa (groups of comparison) is hypothetical. In the evolutionary context of *phyletic gradualism* (the gradual, even continuum of phenotypic change over time within an entire lineage), many such groups are readily identifiable as discrete groups because of *gaps* in our knowledge of transitional forms as illustrated on page 14 (e.g., gaps due to "missing links" or transitional forms). Missing links are theorized to be extremely numerous in both the fossil record and the archaeological record. Within the *biological sciences*, it is estimated that more than 99.95 percent of all "species" that ever existed on this planet are extinct and left no fossil record (Grande, 2017:162). Similarly, within the *cultural sciences* the majority of small early religious belief systems that ever existed on this planet are extinct and have no known archaeological record.

1L. Monophyletic groups (i.e., phylogenetic groups) based on uniquely derived characters (synapomorphies) versus nonmonophyletic groups (phylogenetically artificial "groups of convenience") not indicative of evolutionary relationships. One of the main jobs of modern evolutionary science is the search for *monophyletic groups*, also called *phylogenetic groups*. In terms of evolution, a monophyletic group is one whose members (e.g., species or religious ideologies) are all more closely related to each other than they are to any other groups. This similarity is due to all of its members being theoretically evolved from a uniquely shared common ancestor. Monophyletic groups also include *all* descendant branches. For example, the vernacular group name "fish" does not represent a monophyletic group, because it does not include a major group descended from one of its branches (tetrapods, which descended from sarcopterygian fishes several hundred million years ago). Monophyletic groups are analytically determined by using what are hypothesized to be uniquely derived, shared characteristics, thought in theory to be derived from a common (shared) ancestor (e.g., page 17). A group hierarchy of monophyletic groups is used to reconstruct hypothetical maps of evolutionary diversification.

Alternatively, scientists occasionally use nonmonophyletic vernacular names in a nonevolutionary context. These can be either paraphyletic (groups that theoretically evolved from two distantly related ancestors independently) or polyphyletic (groups that evolved independently from three or more distantly related ancestors). Nonmonophyletic group names are used only in a vernacular sense and they have no phylogenetic significance. In a phylogenetic sense, they are not even true taxa, and they are sometimes called "groups of convenience" by systematists. A graphic example of a nonmonophyletic group of convenience for evolutionary biologists would be "flying animals" (mentioned in **Note 1N**) or "fish" (mentioned in this note).

Another type of nonmonophyletic group is a group of convenience defined only by what it lacks. A commonly used group of convenience in biology for much of the twentieth century was the Invertebrata, which was used for animals that lack a vertebral column. Nonmonophyletic names are rarely used today in by evolutionary biologists. When they are used, it is understood that they used only in an informal, vernacular sense and are not meant to reflect historical relationship or group monophyly.

Nonmonophyletic groups of convenience used for religious ideologies include *traditional Judaism* in its commonly used vernacular that excludes Christianity (discussed in chapter 7). Largely for ethnic historical reasons, the group name is often used that way because many people choose to self-identify in that more restrictive sense. However, linking religions together based on their historical relationships results in traditional Judaism being nonmonophyletic. On my phylogenetic tree, the Ebionites (commonly thought of as Jewish) and what is sometimes called Jewish Christianity (a name not on my tree) are historically more closely related to Traditional Christianity, Biblical Demiurgism, and Islam than they are to Sadduceism or Pharisaic Judaism (branches commonly thought of as traditional Judaism). To be a monophyletic religious group, Judaism would have to include what is called Jesu-venerationism here (page 286), because Jesu-venerationism was a descendant of a Judaic religion. Even Jesus was a practicing Jew (page 319). To be a strictly monophyletic religious ideology, Judaism would have to be equivalent to Abrahamic Monotheism on the phylogenetic tree on page 286. Jesu-venerationism would be one of its major subbranches. Alternatively, subgroups of Judaism could be ideologically monophyletic if they could be diagnosed by uniquely derived characteristics (e.g., Pharisaic Judaism on the same tree).

Another early example of a nonmonophyletic use of a group name for religion is the way early Roman Christians used the name "pagan" (discussed in chapter 5). They generally used the name as a collective derogatory term for most every non-Abrahamic religion of the time. Occasionally that term is still used that way even today by fundamentalists for non-Abrahamic sects.

Yet another example of a nonmonophyletic group of convenience is "Chinese folk religion" (discussed in chapter 4 and **Note 4L**). This is not a real group within an evolutionary context, and its component subgroups remain to be sufficiently characterized. It is defined only by ethnicity and geography. Before it can be included on phylogenetic trees in this book, its distinct parts must be more clearly defined and classified within the larger framework of worldwide comparison. Different parts of Chinese folk religion belong to different phylogenetic trees within this book (e.g., some parts with Old World Hard Polytheism, some parts with Asian Cyclicism). Other nonmonophyletic group names include gnosticism (chapter 10) and sufism (chapter 11).

I also use a few nonmonophyletic group names of convenience to express some general characteristics in the earliest development of religion. These names are uncapitalized at the base of my phylogenetic trees on page 44. They represent elements that first developed during a period in which we have very little sociocultural information about humans. They are used only to indicate certain widespread characteristics of religion that probably developed independently more than once and developed well before the major branches of Organized Religion did.

1M. Separating *synapomorphy* (similarity theoretically due to unique common ancestry) from *homoplasy* or *symplesiomorphy* (similarity not due to unique common ancestry). Many social anthropologists contend that similarity due to independent derivations (homoplasy) is much more common for aspects of culture (e.g., branches of religions) than for aspects of nature (e.g., species of plants and animals), causing a problem for applying phylogenetic analysis to culture. However, distinguishing synapomorphies from homoplasies, is a challenging problem for evolutionary biologists as well. Many biological characters that were at one time thought to be traceable to a single ancestral origin were later statistically determined to have been independently derived from two or more different groups (e.g., parallelisms, convergent characters, or character reversals) (Futuyma, 2013). Homoplasy at the most general levels is not at all uncommon among biological species, particularly when widely different lineages are convergently adapting to the same environment (e.g., the independent loss of eyes in various cave-dwelling species of fishes, amphibians, crustaceans, spiders, and insects). In the end, it is mainly statistical analysis and congruence with other characters that theoretically separate the synapomorphies (uniquely derived characters from a single ancestral source) from the homoplasies (characters independently derived from different ancestral lineages) (discussed further in **Note 1N**). Even if homoplasies are more common in studies of cultural evolution than in studies of biological evolution, social scientists also have a major advantage for their studies of evolutionary history that biologists lack. They have many contemporaneously written records of schisms, historical interrelationships, and other observed phases of phylogenetic diversification. In fact, the majority of today's known Organized Religions evolved within the period of human literacy. In contrast, virtually all biological speciation occurred well before the period of human literacy.

With regard to religious belief systems, homoplasy would be most likely for the oldest, most elemental characteristics of supernaturalism. Such characteristics are indicated with uncapitalized names in two of my phylogenetic trees and in the text (e.g., animism, anthropomorphic supernaturalism, ancestorism, matristicism, and general polytheism). Each of these group characteristics are likely to have developed independently in different isolated cultures because these characteristics are of such a general nature, and because they developed well before there was sufficient interchange among organized groups of humans to allow intercultural transmission. As indicated in the first chapter, once intercultural influence began, due to trade, warfare, and other social interactions, religious belief systems would have had the opportunity to be freely transmitted from one culture to another and probably were. After all, it would only take a single immigrant or visitor to transmit such ideas; much like it takes only the spread of a single genetic mutation to begin a biological lineage.

The intent of this book is to use evolutionary theory as the default explanation for the diversification of Organized Religion, just as biologists do for the diversification of species. Certain details may be proved wrong in some cases, but that risk is always the case with scientific theory. And once again, studies of cultural evolution have the added benefit of contemporaneous written historical records that contribute to the robustness and precision of phylogenetic hypotheses. Contemporaneous written records of historical diversification generally have priority over cladistic evidence in the study of cultural evolution.

1N. The use of *parsimony* to hypothesize different types of character similarity. The *principle of parsimony* refers to explaining something with the fewest possible assumptions. It is a logical starting point for any historical analysis. Wherever the strongest evidence points statistically, that is the most parsimonious hypotheses. It is similar to the concept of Occam's razor, which says that when two competing hypotheses explain the data equally well, *choose the simpler one.* It is sometimes jokingly referred to as the KISS ("keep it simple, stupid") principle. In terms of classification, parsimony is the most efficient organization of characteristics and hypothetical groups of kinds (e.g., the cladogram of shapes on page 20 and the parsimonious classification on the page opposing page 3).

With biological organisms, I generally use a parsimonious summary of character data for cladogram and phylogenetic tree construction (sometimes referred to as the *Maximum Parsimony* method). For religious ideologies (and also applicable to other aspects of culture), I supplement (and sometimes overrule) such character data with contemporaneous written records of diversification (i.e., observed process explanation of pattern) where such records exist. The addition of time elements and hypothesis of evolutionary relationship to explain cladistic pattern makes my trees phylogenetic, rather than strict cladograms. As mentioned elsewhere in this book, the existence of so many contemporaneous (or even near-contemporaneous) written records of evolutionary process represents a distinct advantage that cultural evolutionary studies have over biological evolutionary studies. Many major schisms within Organized Religion were documented with written records during contemporaneous or near-contemporaneous time.

Maximum Parsimony methods by themselves do not give time-scaled, relative branch lengths on the resulting tree, nor do they show certain microevolutionary complexities (e.g., reticulate patterns from extreme syncretization, recombinations). They give only the relative order of branching and hierarchical group organization based on special shared characteristics. However, such methods are a good place to start because they involve minimal assumptions about unobserved process details (which can always be added later). Alternatives to Maximum Parsimony methods include probabilistic methods of analysis that can assume certain evolutionary rates or branch lengths. These include *maximum likelihood* and *Bayesian inference*; these two analytical techniques are generally used by evolutionists who are more focused on microevolutionary detail and reticulate patterns of relationship (e.g., molecular data or other forms of sub-specific population data). Maximum Parsimony is still preferred by many evolutionists for constructing trees of general group relationship (Sober, 1988; Brower, 2018).

A primary aim of cladistics, and of the cladistic study of historical relationships, is to distinguish similarities that are hypothetically indicative of unique relationship (*synapomorphies*) from those that are not (*plesiomorphies* and *homoplasies*). Synapomorphies are special shared characteristics that have been statistically determined to be uniquely derived for the group of focus. For example, hair and mammary glands are synapomorphies for mammals (the group Mammalia). In an evolutionary context (i.e., under a theory that observed diversity is the product of past evolutionary diversification), synapomorphies are thought to be the historical result of inheritance from a common ancestral group. All putative similarities need congruence with other characters to statistically test their legitimacy as synapomorphies. Each synapomorphy is only hypothetically indicative of ancestral origin. Classifications with a cladistic core tend to be the most empirically supported and relatively stable over time. In contrast to *synapomorphies*, other types of character similarities are not thought to be indicative of relationship or monophyletic groups (monophyly is discussed in **Note 1L**). These include *plesiomorphies* and *homoplasies*.

Plesiomorphies (also sometimes called *symplesiomorphies*) are effectively shared characteristics that are synapomorphies for a group larger than the focus of analysis. For example, although hair and mammary glands are synapomorphies for Mammalia (as previously mentioned), those features are symplesiomorphies for a group containing only primates and squirrels. The terms synapomorphy and plesiomorphy are relative terms dependent on the level of group context.

A good example of a *homoplasy* (also discussed in **Note 1M**) is flight among animals. The group "flying animals" is a nonmonophyletic group. There is a group of archosaurs that fly (e.g., birds), there is a group of mammals that fly (bats), and there are several different groups of insects that fly (e.g., butterflies). Based on a preponderance of other characters, no one today would contend that birds, bats, and butterflies had a unique common ancestor shared by no other organisms. Therefore, we identify the analogous characteristic of "flight" as homoplasy (i.e., independently evolved in different groups) rather than a synapomorphy (i.e., a trait uniquely inherited from a theoretical common ancestor). However, the only way we were able to identify general flight as homoplasy (rather than synapomorphy) is with an overriding number of other characters tying the butterfly to other insects (e.g., six legs, exoskeleton, two pairs of scaled wings), the bat to other mammals (e.g., mammary glands, fur, internal bone structures), and the sparrow to other birds (e.g., feathers, internal bone structures). In addition, the bird wing, bat wing, and butterfly wing are each made differently. The only logical conclusion is that flying is not a single synapomorphy, and that bird flight, bat flight, and butterfly flight are each independently evolved synapomorphies within their particular subgroup. We separate synapomorphies from homoplasies by using statistical analysis of character distribution and the principle of parsimony. Distinguishing whether certain character similarities are synapomorphies or are homoplasies is often more difficult than this example of flight. Again, various forms of statistical analysis can be used to facilitate this task.

As in biology, we can use character analysis and parsimony in historical studies of Organized Religion to distinguish synapomorphies from homoplasies. For example, Pharisaic Judaism, Traditional Christianity, and Islam share uniquely derived characteristics from their sacred books of revelation (the Torah and the Old Testament) that are absent in Hinduism. Based on a combination of character congruence and historical records, many of these characteristics are considered to be synapomorphies. Evolutionary theory proposes that the reason those religions uniquely share these and other characteristics is that they are more closely related to each other than any of them are to Hinduism. Homoplasies also commonly occur among religious belief systems. For example, the concept of reincarnation that is a synapomorphy among Dharmic religions later occurred independently within Catharism (see chapter 10). Soft polytheism also appears to have evolved independently several times. It evolved once within Asian Cyclicism, another time within the *Amun-Re* sect of Egyptian Polytheism, another time within the trinitarian branches of Traditional Christianity, and so on. It is only with statistical analysis of many characters that we find these instances of soft polytheism to be cases of homoplasy (independently derived or convergent characteristics) rather than synapomorphy (similarities theoretically derived from a theoretical common ancestor). Similarly, general monotheism has evolved independently more than once within Organized Religion (e.g., in Sikhism and certain Hindu subgroups in chapter 4 and again in Linear Monotheism in chapters 6–11).

2 EXPLAINING HYPOTHETICAL PATTERNS WITH EVOLUTIONARY PROCESS THEORY

2A. Haeckel's use of evolutionary progressionism. Although some recent authors have attempted to defend Haeckel as not intending to establish a discriminatory racial praxis (Levit and Hossfeld, 2020), his work was deeply racist by today's standards. His ideas of evolutionary progressionism were morally offensive, and they damaged the practice of comparativism within the social sciences for over a century. To his credit, he discovered, described, and named thousands of new species of plants and animals. He also coined many biological terms used today, such as *ecology*, *phylum*, *heterochrony*, and *phylogeny*. To his discredit, he promoted racism and helped lay the groundwork for a number of major discriminatory movements ranging from Social Darwinism and eugenics, to core elements of Nazism.

Based on biological characteristics, Haeckel divided living humans into "twelve species of men" containing thirty-six races (Haeckel, 1883:309. Haeckel's classification of living human races as a map of diversity was not the problem. Such an understanding of relationships could potentially be important for various types of medical research on genetically linked diseases or resistance to such diseases. Using racial differences in the study of optimal nutritional requirements also has potential health benefits to certain races. An understanding of racial differences and similarities is necessary to fully understand specific cultural histories, racial sensitivities, and even the existence of specific racial inequalities.

The ethical problem with Haeckel's classification of human races was in his interpretation of phylogenetic trees as progressive ladders rather than simple maps of relationship. He considered some human races as universally superior while considering others as inferior and fated for extinction. He considered the group he called the "Mediterranean" (his species *Homo Mediterraneus*) as superior to the other eleven living human species. *Homo Mediterraneus* included the Indo-Germanian, Caucasian, Semite, and Basque races. According to Haeckel, *Homo Mediterraneus* was "the most highly developed and perfect" of the human species. He claimed that "In bodily as well as mental qualities, no other human species can equal the Mediterranean" (Haeckel, 1883:321). He further claimed that *within Homo Mediterraneus*, the Indo-Germanian race "far surpassed all other races of men in mental development" (setting the stage for the basic idea of the "master race"). Note, however, that unlike the later Nazi version of race superiority, Haeckel also ranked Semites (including Jews) within the top group next to the Indo-Germanians, which became one reason that the anti-Semitic Nazis ultimately found him unsuitable for their purposes (Levit and Hossfeld, 2020).

At the bottom end of Haeckel's progressionist ladder of human races, he ranked the "Papuan," "Hottentot," "Kaffre," and "Negro" species as being composed of "the lowest stage of mankind" (Haeckel, 1880). He named these four as *Homo Papua*, *Homo Hottentottus*, *Homo Cafer*, and *Homo Niger*. Each of the four species was assigned between two and four racial subgroups. He wrote that these four species are collectively "at a much lower stage of development and more like apes than most other human species" (Haeckel, 1880). He also wrote that they are "incapable of a true inner culture and of a higher mental development, even under the favorable conditions of adaptation now offered to them in the United States of North America" (Haeckel, 1883:307). He further claimed that none of these "lower" species "has ever had an important history" and that those species would "sooner or later completely succumb in the struggle for existence to the Mediterranean races [*Homo Mediterraneus*]" (Haeckel, 1883:325). His was a Western cultural bias of the time, and a prime example of what is today sometimes referred to as "scientific racism."

2B. "Vertical" versus "horizontal" transmission of traits in evolutionary theory. In biological evolution, most genetic information is passed "vertically" from parent to offspring. This type of transmission is known as *vertical gene transfer*. In rare instances, however, genetic information is also transferred "horizontally" between contemporaneous, but different, species, mainly by way of the following: (1) hybridization, (2) parasite to host transfer, (3) predator to prey transfer, or (4) virus infection. This is known *horizontal gene transfer* (also called *lateral gene transfer*). In the evolution of biological organisms, horizontal transfer is particularly important among species of bacteria (Futuyma, 2013:43).

Much like biological evolution, cultural evolution involves both vertical and horizontal types of trait transmission. In cultural evolution, the traits are ideological rather than biological, and horizontal transmission among religious belief systems appears to be more common than it is among biological species. Some scientists contend that horizontal transfer is a serious problem for phylogenetic studies, whereas others argue that it is not (e.g., Currie, Greenhill, and Mace, 2010). The perceived problem of syncretized merging or blending is often an exaggerated oversimplification of detail for the evolution of religious traditions. For example, religion *C*, defined as a blended form of religions *A* plus *B*, may turn out to be two independent (unblended or unmerged) branches: one branch *A'* (primarily *A* with later influence from *B*) and another branch *B'* (primarily *B* with later influence from *A*). In that sense, sometimes what superficially appears to be complete syncretization, hybridization, or merging is instead a more complex situation involving cross-imitation of certain elements between two distinct branches (also see **Note PrB**). In addition, restricting evolutionary analysis to only a single aspect of culture (e.g., religion) rather than to multiple aspects of culture as a homogenous entity, often results in less overlap between groups (discussed on page 24).

2C. **Memes.** I use the *meme* of Dawkins (1976, 2006), which was further developed by Blackmore (1999), hypothetically as the cultural analogue of the gene in biology. The meme can be thought of as the smallest identifiable element of cultural tradition that is replicated and passed down through successive ideological generations. Just as a collection of genes (the genotype), passed down from one generation to another, enables the replication of physical traits that are more adaptable or less adaptable within a given biological environment, a collection of memes (the memotype), passed down from one ideological generation to another, theoretically enables the replication of cultural traits that are more adaptable or less adaptable within a given cultural environment. Where the gene is a physical biological entity, the meme is largely an ideological cultural one. Memes are generally replicated and passed down by way of educational transmission or behavioral imitation, and they do not have to follow biological lines. They deal primarily with the relationships of ideological factors rather than with the people that possess them, for they can change significantly even within an individual during his or her lifetime.

The concept of meme/gene analogy is controversial (e.g., Atran, 2002), particularly with social scientists who are hostile toward theories of cultural evolution (Lewens, 2018), or Christian authors who are responding to the Dawkins's (2006) book *The God Delusion* (McGrath and McGrath, 2007). Fracchia and Lewontin (2005:18) write "unlike genes, memes are not entities with an existence independent of theory. They are a mental construct whose only defined property is to fill the gap with an elaborate metaphor." I agree that the terms *meme* and *memotype* are only placeholders until we can better understand the details of social evolutionary processes. However, I consider that sufficient for the time being. Placeholders are often used in science for unobserved but theoretically predicted phenomena (e.g., *dark matter* in modern physics). Darwin, at the time of *Origin of Species* (1859), did not yet know the nature of the replicating element for biological characteristics (i.e., genes). However, that did not stop him from proposing his concepts of *gemmules* and *pangenesis* as placeholders to serve that function of a theoretical mechanism for heredity before the development of genetics

(Eldredge, 2005). *Pangenesis* proposed that each part of the body continually emitted small organic particles called *gemmules* that aggregated in the gonads to transmit heritable information to the sperm and ovum. The details of transmission were modified once the science of genetics developed. Just as we propose that a biological genotype leads to a phenotype that is theoretically subject to natural selection, we can propose that a cultural "memotype" leads to an ideological variant that is theoretically subject to cultural selection. Perhaps, one day, a more precise mechanism and terminology for cultural inheritance will be defined, as it eventually was for biological inheritance.

2D. **Fitness.** The idea of *evolutionary fitness* is that certain physical variations develop over time within a population that affect its ability ("fitness") to adapt to its environment and produce offspring. Although the term "fitness" is often credited to Darwin in his *Origin of Species*, he only mentioned it once as an informal name meaning "a state of being well adapted" (Costa, 2013). It was not until five years later that the English social anthropologist and biologist Herbert Spencer coined the phrase "survival of the fittest,"

2E. **Evolutionary reticulate networks.** Evolutionary *reticulate networks* showing reticulations and recombinations are sometimes used by specialists to represent the most complex microevolutionary processes and patterns (Gontier, 2015). In *evolutionary biology*, microevolutionary processes include minor genetic mutation, hybridization, lateral gene transfer, gene flow, genetic drift, and other detailed developments that collectively result in net-like recombination patterns of merging and diverging lines, such as the highly simplified diagram below leading to forms A, B, and C.

To express the broadest aspects of evolutionary pattern in a simplified, intuitive way, biologists usually summarize relationships using trees of species or other groups (samples of diversity) arranged hierarchically within larger and larger (i.e., increasingly inclusive) taxa.

In the *evolutionary study of religions*, microevolutionary process and patterns often include slight changes, such as minor ideological innovations, hybridizations, reticulations, extreme syncretizations, cultural drift, and other detailed developments that also collectively result in reticulated, network-like recombination patterns, much like those described for biological organisms (Holden and Shennan, 2005). Some authors, including the famous evolutionist Stephen J. Gould, suggest that reticulate network-like amalgamations of lineages are more common in the evolution of culture than they are in evolutionary biology, making human culture more difficult to represent with tree-like topologies (Gould, 2010). However, with that statement Gould understates the true complexity of biological evolution with its hundreds of billions of different genotype lineages. At the microevolutionary level, the net-like amalgamation is present in biological evolution, much as it is in cultural evolution.

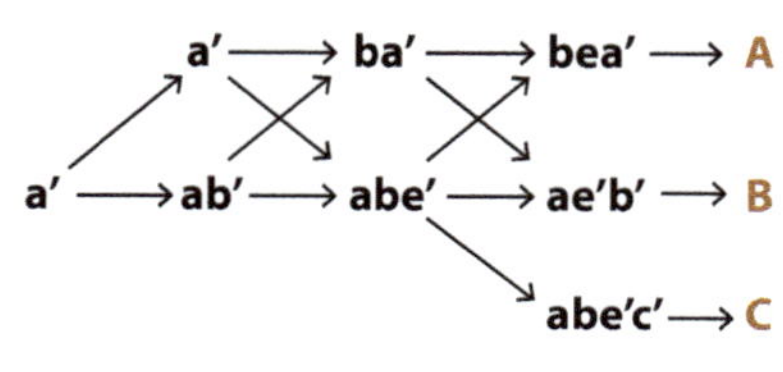

In addition, confusing mixes of cultural characteristics are sometimes collectively oversimplified as "fusions" or "reticulations." It is true that there are some diverse syncretizations of religious ideologies. However, these branches are rarely exact fifty–fifty mixes, and they rarely represent complete merging of ideologies. Furthermore, they do not materialize out of nothing. Generally, one initial religious tradition was later modified by influences from another distantly related belief system. For example, Sufism (discussed in chapter 11) is often described as a blend (extreme syncretization) between Islam and Hinduism (e.g., Doniger, 2009). The "fusion" or "blend" descriptors that are sometimes used may be only oversimplifications masking a more complex history of diversifications and evolutionary convergence, as a result of one branch imitating certain elements of another branch. Sufism, for example, may actually include some primarily Islamic branches that were partly influenced by Hinduism as well as some primarily Hindu branches that were partly influenced by Islam. Collectively, Sufism as commonly understood may be a non-monophyletic group (*monophyly* and *non-monophyly* discussed in **Note 1L**). Similarly, many polytheistic branches syncretized with Christianity over time, in one form or another, as discussed in chapter 5 (e.g., syncretization of Roman Catholicism with Yoruba Polytheism to form *Vodun*, and the syncretization of the Incan goddess Pacha Mama with the Virgin Mary of Roman Catholicism). Such simplifications are often the result of inadequate information to reveal even more complex patterns of differentiation and coevolution. The detailed written records and character details of many syncretized groups are in need of further systematic study to better determine their patterns of diversification and evolutionary histories.

In the present study, the complexity of cultural evolution is simplified in two major ways. First, the focus is on one individual aspect of culture (i.e., religious traditions in this case) rather than viewing culture as though it were a homogenous entity. Second, I focus on broader, generalized (and necessarily typified) samples of religious belief systems (somewhat analogous to typical *phenotypes* or *species* in biology), rather than the countless microevolutionary variations, syncretizations, and convergences of religious traditions that exist (analogous to biological *genotypes*, and detailed intraspecific variations such as eye color). Consequently, I rely primarily on phylogenetic trees (rather than networks of reticulations and recombinations) to illustrate the broadest relationships. I nevertheless recognize that reticulate networks may reflect many smaller details of the evolutionary history not indicated by the trees. Such detail is better covered by specialists or cultural "particularists."

2F. Diversification, entropy, and the sum of evolution. While *Natural Selection* (also termed *Environmental Selection* by some authors) explains the survival or the extinction of variants, it does not explain the continual appearance of variation over time. Some evolutionary biologists (e.g., Brooks and Wiley, 1988), in an attempt to connect evolution to the laws of physics, have proposed evolution as a form of entropy. However, the inclusion of selection as part of entropy may be an overreach. The second law of thermodynamics specifies that entropy (also known as disorder) is always irreversibly increasing over time. My philosophical view is that entropy can explain the inevitable and continual *origin of differentiation* (*irreversible mutations*). However, the sum of evolution is more complex, including two distinct stages: (1) entropy (the origin of differentiation), and (2) selection (i.e., physical, biological, or cultural factors that determine the success or failure of each different variant).

2G. Differences between hypothesis and theory. Currently, no firm demarcation exists where a hypothesis ends and a theory begins. In general, a *theory* (e.g., the theory of evolution by natural selection or the theory of gravity) is of a broader nature than a hypothesis. It is a principle, indicated by a hypothetical pattern, to explain a broader phenomenon. A *hypothesis* (e.g., a specific cladogram or phylogenetic tree for all primate species, or for Abrahamic religions) often focuses on details of larger theories (e.g., a hypothesis of relationship within the theory of evolution). A hypothesis is more frequently tested, falsified, or replaced with alternate conclusions than a theory is. Both theories and hypotheses seek to explain or interpret phenomena. Both are critical parts of the scientific method.

3 EARLY SUPERNATURALISM AND THE DEVELOPMENT OF ORGANIZED RELIGION

3A. Traditions not considered here to be forms of *supernaturalism* (religion). Although some writers refer to *Atheism* (or *New Atheism*) as a stealth religion (e.g., Wilson, 2007; Wallace, 2009; Prothero, 2011), I do not include it as a religion because it is defined only by what it lacks. Atheism has been noted historically as a philosophy going back at least as far as the ancient Greeks. Ancient Roman polytheists used the term for early monotheists because the monotheists rejected all but one God from the Roman Pantheon (chapter 5). Atheism also became prominent after the so-called Age of Enlightenment in the eighteenth to nineteenth centuries. However, it is not a natural religious group because within a religious context, it is defined only by what it lacks (i.e., lacking belief in a deity). There are many unrelated reasons that an individual could lack such beliefs, such as never having been exposed to religion in his or her life, or having rejected one or more of a number of particular religions. Groups defined only by the absence of characteristics are artificial, nonmonophyletic as groups (also discussed in **Note 1L**). Even a rock lacks a religious belief (as far as we know), but most people would not consider a rock to be an atheist. I also do not consider communism, nationalism, consumerism, or capitalism to be religions, although some authors do (Harari, 2014; Harrison, 2014). Other groups excluded from my definition of religion include the following.

Secular Humanism (also called *Humanism*) is a philosophy that the U.S. Supreme Court decided in 1961 should receive the same sort of first amendment protection that religions enjoy (*Torcaso vs. Watkins, 367 U.S. 488, 1961*). Secular Humanists focus on secular ethics and moral values from the perspective of atheism. Some religion historians consider Humanism to be a religion (e.g., Armstrong, 1993). However, because Secular Humanists define themselves as strictly secular and reject all forms of supernaturalism, they are not included as a religion.

Some organizations claiming to be religions today clearly began as something other than a supernatural belief system. One example involves the popular *Star Wars* series created by American filmmaker, George Lucas (1944–present). When Lucas first set out to write *Star Wars*, he said he wanted to distill the essence of most of the world's religions to create the *Star Wars* universe. He created a good "Light Side" and a bad "Dark Side" realm, and powerful forces of good and evil that were in constant conflict. The Jedi were the apostle-like saviors of humanity who channeled the great unknowable "Force." The movie saga, including a series of nine movies released between 1977 and 2019, went on to become the second most successful movie franchise of all time. In 2007, a religion called *Jediism* was founded by people extrapolating the philosophy of the Star Wars movie characters into a religion. Its professed mission was stated as "to reaffirm and restore the knowledge of the True Inner Self . . . and invite all to personal faith in the Light Side of the Force. . . . We believe in the path of the true Jedi Knights, and with it the code of Jediism" (Grant and Mansell, 2008:249). In the 2001 census for England and Wales, 0.7 percent of the population claimed Jediism as their religion, surpassing Rabbinic Judaism, Buddhism, and Sikhism (Ethnicity and Religion, "Jedi." Census 2001. National Report for England and Wales [accessed October 10, 2019]). Although much of this early support for Jediism is suspected to have begun as a hoax, some followers took it seriously enough to officially launch it as a religion in 2007 and created an extensive website with its own online doctrine. Today you can read the stories of many apparently serious followers online. The IRS even gave Jediism ("The Temple of the Jedi Order") tax-exempt status in 2015. (They are legally prevented from passing judgments on the merits of any asserted religious belief, according to *U.S. vs Ballard*, 322 U.S. 78, 1943.) It is hard to predict where the efforts of Jediism may evolve, but perhaps centuries from now, some part of it will have morphed into a major religious belief system or have syncretized with some existing religion to create yet another religion.

Another organization that began as something other than a religion, but today claims to be a religion, is *Scientology*. The core of this organization was not originally founded as a religion. It started out claiming to be a nonreligious school of psychotherapy (Bowker, 2006:256,521). Its founder, L. Ron Hubbard (1911–1986), began as a writer of science fiction and fantasy stories. His science fiction book *Battlefield Earth* sold more than six million copies and was later made into a Hollywood movie starring John Travolta in 2000. In 1950, Hubbard published the book *Dianetics: The Modern Science of Mental Health*. A review in the journal *Scientific American* said that Hubbard's book contained "more promises and less evidence per page than any publication since the invention of printing" (Rabi, 1951:57). Nevertheless, the highly charismatic Hubbard developed a substantial cult following, emphasizing his tradition as a psychotherapeutically oriented religion (Lewis, 2012:133). It is sometimes disputed whether the movement is a religion or a form of psychotherapy (Bowker, 2006:256). Hubbard made quite a bit of money with the concept. Although at first, he was hostile to the idea of religion in general, he eventually decided to morph his Dianetics into a "religion" he called Scientology. Making it a religion was a way to keep revenue tax free. He is often quoted as having said, "Writing for a penny a word is ridiculous. If a man really wants to make a million dollars, the best way would be to start his own religion" (Wakefield, 2009:21; Lewis, 2012:135). Hubbard died a very wealthy man in 1986, on his sixty-acre ranch near Creston, California. The base of Scientology remains in California, where celebrities are still recruited to endorse it. Hubbard understood early on that a bit of charisma goes a long way. He created "Project Celebrity" to recruit celebrities who would then go to the public to recruit for Scientology. Modern-day celebrities who have endorsed it include U.S. recording star Beck and Hollywood celebrities Tom Cruise and John Travolta. Their involvement with Scientology has been a frequent story on the front page of the supermarket tabloids. Scientology has a hierarchical, rather than congregational, organization, with secret details revealed only to advanced members. Although it may be an "ology" of some sort, it did not start out as a religion. It appears to have become one of sorts, but many of its details are poorly known to outsiders. It is difficult to track where some of its derivatives have gone, particularly in traditions that are secretive about much of their internal doctrine. Humans have a proclivity for supernatural enhancement of formerly secular traditions and for secrecy about privilaged elements of ritual.

Perhaps one of the most exotic twenty-first-century ideas now claiming to be a religion is one created for future advances with artificial intelligence (AI) called *Way of the Future* (WOTF for short). Anthony Levandowski, an engineer from Silicon Valley, claimed to be its "prophet." He said that his "religion" sought to build working relationships within the AI industry and to grow its membership beginning with "laypersons who are interested in the worship of a Godhead based on AI" (Harris, 2017). Levandowski claimed that the eventual arrival of AI "a billion times smarter than the smartest human" was inevitable. He further claimed, "What is going to be created will effectively be a god." He believed that humans are only in charge of the planet now because they are smarter than the other animals. However, when the transition to smarter AI comes in the future, he claimed there would be a "transition as to who is actually in charge." It sounds very much like the meat of a science fiction story, but Levandowski and his handful of followers appeared to be convinced the transition is coming. Levandowski claimed that his aim was to promote the use of divine AI for the "betterment of society" and to effectively ensure that the super-smart computers understand our better nature. Other technology leaders such as Bill Gates, Elon Musk, and the late physicist Stephen Hawking, have agreed that super-intelligent AI is on the way and warned that its appearance may eventually spell trouble for humans. Elon Musk was famously quoted saying, "With artificial intelligence we are summoning the demon" (Harris, 2017).

The IRS granted Levandowski's church a tax-exempt status in August of 2017. In spite of this decision by the U.S. federal government, the WOTF is not included as a religion because it is not a supernatural belief system. It is instead a belief in (and fear of) certain advances in future technology. However, if it persists over time, it could one day incorporate supernatural belief elements that make it a religion. Again, there seems to be no limit to the cultural creativity of the human species. Levandowski eventually closed his branch of the WOTF and donated the church's funds, totaling just over $175,000 to the NAACP Legal Defense Fund (Williams, 2021). Ideological vestiges of this "movement" remain alive on social media (e.g., Twitter). Once a new belief system is ignited, it is often beyond the ability of any single individual to extinguish it completely.

3B. **The invention of writing and literacy.** Sumerian (Mesopotamian) cuneiform script is considered to be the world's earliest-known writing system. It is first known as inscription on clay tablets from around 3400 BCE (Black and Green, 1992). These tablets are from archaeological sites in the ancient Mesopotamian civilization of Sumer, located in what is today southern Iraq. Although at first, the tablet script was purely pictographic (a form of "protoliteracy"), eventually it began to contain syllabic elements showing the structure of the Sumerian language (Collins, 2021; Walker, 1987).

In the New World, writing is thought to have developed independently and much later. One of its oldest-known occurrences is from the Olmec civilization of Mesoamerica, somewhere between 1200 BCE and 600 BCE) (Woods, Emberling, and Teeter, 2010).

3C. Defining "civilization." The term "civilization" is used here to mean *a society characterized by urban development, social stratification, symbolic systems of communication* (e.g., *writing* or *art*), and *a perceived independence from the natural environment.* Among anthropologists, the term has been controversial because of its early misuse in a colonialist, racist sense. Early cultural anthropologists (e.g., Tylor, Spencer, and Haeckel) used it as a qualitative term to contrast what they deemed to be advanced Western European culture with the so-called lower, savage, or barbaric cultures thought to be predestined to extinction. It is not used here in that way, nor is it used here to distinguish "higher" social groups from "lower ones".

3D. Franz Boas and the rise of historical particularism in the social sciences. Franz Boas (1858–1942) was a German born pioneer of modern anthropology, and in 1894, he was the first curator at the institution where I have spent my career in Chicago (the Field Museum). He is known as the "Father of American Anthropology" (Grande, 2017b:264). He developed the concept of *Cultural Relativism* as a backlash against the colonialist attitudes of early cultural anthropologists. The Boas concept asserted that each culture must be understood within its own cultural context rather than be judged by the criteria of a different culture. Boas admirably rejected the racist concept of *Unilineal Evolution,* in which all "primitive" cultures were thought to "progress" in a stage-based evolutionary progression toward "advanced" Western European society, which was thought to represent the pinnacle of social evolution. Boas rejected the generalized distinction between "primitive" and "advanced" culture altogether. For example, he rejected the idea held by contemporary anthropologists that the Inuit people he studied in the northern Arctic were socially unfit primitives (at a time when the word "primitive" meant universally inferior to many who used the term). He argued that the Inuit should not be qualitatively judged within a biased Western standard. Within the context of the northern Arctic environment, the Inuit people were actually more advanced than Western Europeans. In other words, a person's cultural fitness is not absolute, but rather it is relative to the particular environment in which a person lives (much like the concept of fitness for species in evolutionary biology). Boas did much to combat racism in his time. He was a vocal archrival of Madison Grant and the rise of the racist eugenics movement in the United States (discussed in chapter 2).

Boas went even further in his backlash against cultural progressive evolutionism. He is thought by many scholars to have pushed anthropology away from comparativism in general and was often branded as a historical "particularist" (Harris, 1968). He discouraged cultural systemization and the use of the comparative method for cultures because of the way such practices had been misapplied in a racist fashion. The push in anthropology toward historical particularism seems to have resulted in a focus more about differences than similarities. This set up a controversy between comparativists and particularists in cultural anthropology that exists to this day, as well as a divide between anthropologists and archaeologists (Lincoln, 2018; Parkinson, 2017; also discussed in chapter 2). Although Boas accepted that different cultures had a number of fundamental traits in common, early in his career, he chose not see these shared characteristics as indicative of historical relationship. Instead, he believed them to be independently developed from different sources in parallel, resulting from "laws . . . which govern the growth of human culture" (Boas, 1896:5). Later in his career, he backed away from the hope of being able to discover such laws (Harris, 1968). He also became less of a particularist, accepting a level of comparativism that excluded the idea of a progressive cultural ladder.

Boas' early form of historical particularism was more than just a stance against racism. To his credit, it was also a statement about the importance of personal fieldwork and empiricism (emphasizing the role of personal observation rather than subjective conjecture). Little was known about non-Western cultures at the time. Boas opposed the sort of amateur "armchair anthropology" based on little or no evidence that was often done in his day, involving little or no immersive field experience. He favored the detailed, hands-on personal study of cultures as isolated entities. However, in the words of anthropologist Gary Feinman, "Human populations construct their cultural practices through connectivity with others, and not in isolation" (Feinman, 2019:38). There is no denying the importance of particularists who focus on individual cultures. Similarly, there is also no denying the importance of their immersive fieldwork to help gain a deeper understanding of those cultures. However, if we are also interested in the broader historical significance of specific belief systems, we cannot abandon also including comparative work of a broader nature, particularly if it incorporates the careful work of specialists (particularists). It is only through broadly comparative, group-level studies of these belief systems that we can identify shared characteristics giving us a deeper picture of historical relationships. Focused studies of specific, regional groups are the necessary first steps, but to take that information to the next level, there must also be periodic synthetic studies of a broadly comparative nature.

Boas became more of a comparativist and cultural evolutionist later in his career, after he had been unable to find support for his theory of universal laws governing the development of society. In a 1908 publication simply titled "Anthropology," he wrote: "Anthropological research leads us to two fundamental questions: Why are the tribes [cultures] of the world different, and how have the present differences developed?" (Boas, 1908:7). These questions require a comparative approach, especially because the last part of his question seeks information about the deep history of culture. First, we must focus on a particular aspect of culture because culture is not a homogenous entity (discussed in chapter 2). Second, we must identify the different samples of diversity for that particular aspect of comparison (discussed in chapter 1). Third, we look for shared characteristics among the units of comparison to provide evidence of historical interrelationships.

Boas's first doctoral student at Columbia University was Alfred Kroeber (1876–1960). Like Boas, Kroeber was influential in the development of cultural anthropology as a discipline in the United States, and he too came out against the general classification and comparison of cultures. In 1915, he delivered an unambiguous credo against using an evolutionary perspective with the study of human culture. Kroeber added further to the bias against comparativism that continued to grow within anthropology. This bias predominated the discipline of social anthropology until the 1930s, when some social scientists began to drift back toward the comparative method (Harris, 1968). Nevertheless, the conflict between historical particularism and historical comparativism still exists today (as discussed in chapter 2).

It is important to remember that the unethical part of colonialism was not its attempt to use comparativism to understand non-Western cultures. The ethical problem was its biased subjectivity and underlying motive to demonstrate Western superiority.

3E. Is the study of history a science or an art? It is sometimes debated whether the discipline of history is an art, a science, or both. Some historians consider it primarily an art because their work focuses on the literary, artistic aspect. They often build imaginative scenarios around archaeological objects or aspects of modern culture to create engaging stories in cases in which little empirical support exists. Other historians consider the discipline as primarily a science because they focus on using more empirical data and analytical methods such as comparativism to formulate testable, unambiguous hypotheses. I prefer the latter definition, with the acknowledgment that no empirical work is theory-free (Popper, 1959). Religion, in particular, has long been deemed the exclusive province of the humanities. After almost a century, scientific interest in religion is once again on the rise (Norenzayan and Gervais, 2012). However, the taboos that have sometimes kept the study of religion out of the broader scientific realm still exist.

3F. Branch position on a cladogram or phylogenetic tree. In modern evolutionary studies, the cladogram or phylogenetic tree indicates the relative relationships among groups, in a nonqualitative sense. Being a lower branch on a cladogram or phylogenetic tree is not a pejorative or qualitative assessment. For example, the phylogenetic tree at the beginning of chapter 3 has collapsed all the complex branches of animism into a single branch and has expanded the anthropomorphic supernaturalism branch into many subbranches. If instead we had represented many branches of animism on the phylogenetic tree and collapsed all the branches of anthropomorphic supernaturalism into a single branch, then anthropomorphic supernaturalism (including all monotheistic religions) would appear as the basal (or lower) branch to animism. It is all relative and simply a matter of contextual scale.

3G. Defining proselytism, and changing ideas concerning it. Although the earlier use of the term proselytization sometimes simply referred to the act of evangelism (spreading the "word" by preaching scripture), the term eventually developed a more pejorative meaning because of the violent and negative way it was often imposed. Today, proselytism most often refers to conversion efforts that fail to respect the prospective convert's freedom and dignity and are sometimes employed in a forceful manner. That is how the term is used here. At the time I wrote this, even the current Roman Catholic

Pope, Pope Francis, had been quoted as calling proselytism a "grave sin" (Serratelli, 2018). He has referred to the Catholic Crusades as "shameful" (Prestigiacomo, 2019). In 2022, the pope traveled to Canada to apologize for the Catholic Church's complicity in the "catastrophic" history of Residential Schools by catholic missionaries (**Note 5C**), admitting that it destroyed indigenous cultures, severed families, and marginalized generations of indigenous people (Winfield and Smith, 2022).

3H. Decimating conquest of the New World by the Old World. The late pre-Columbian (e.g., 1491) population of indigenous peoples of the Americas is estimated to have been as high as between 75 and 145 million people (at the time, larger than Europe and Russia combined), with 25 million people in Mexico (Sullivan, 2002), and 8 to 18 million people living north of Mexico, mostly in what is now the United States (Stannard, 1992). Although precise population numbers are controversial, without doubt, over the late fifteenth through nineteenth centuries, indigenous American populations were decimated with the arrival and conquests of Europeans. First, the Spaniards brought a number of diseases with them that were fatal to indigenous Americans. These diseases, previously absent from the Americas, included smallpox, influenza, and pneumonic plague. This spread of disease was paired with a growing expansion of European conquest, which eventually included enslavement and outright massacres of indigenous peoples in an ever-increasing quest for territory, resources, and cultural appropriation. The Europeans also had advanced weapon technology including steel (swords and armor) and guns, which the indigenous Americans lacked (Diamond, 1997).

Many citizens of the United States do not think of their country as one with a history of imperial conquest. However, shortly after gaining independence, the United States picked up where the Spaniards, French, and British left off as the primary force of imperial conquest in the New World. With the Louisiana Purchase of 1803, Thomas Jefferson purchased most of the American territory west of the Mississippi. However, he purchased this from France, not from the indigenous nations that lived there. The heart of the purchase agreement was the exclusive right of conquest within that territory, to the exclusion of other colonial powers. In 1823, the U.S. Supreme Court declared Native North Americans to be "an inferior race of people, without the privileges of citizens." In 1831, Chief Justice John Marshall described Native American nations as "domestic dependent nations" whose relation to the United States was one of "award to his guardian" (Burbank and Cooper, 2010:263). In 1871, the U.S. Congress declared "no [future] Indian nation or tribe within the territory of the United States shall be recognized as an independent nation, tribe, or power with whom the United States may contract by treaty" (Burbank and Cooper, 2010:266).

By the end of the nineteenth century, the population of indigenous people from North America is estimated to have been reduced by more than 90 percent from early fifteenth-century levels. The cumulative decimation of indigenous Americans by European conquest between the fifteenth and nineteenth centuries has been labeled by some historians as the largest act of genocide in history (Stannard, 1992).

3I. Eco-animism today in popular Western culture. Modern Western culture has a cultural attraction to certain eco-animistic principles. In *The Lion King*, the concept of animism is a major theme. Although the characters are all talking animals, they deal with a variety of ethical and spiritual issues. In a key scene, the lion Mufasa explain to his son Simba that there is a balance to the universe in which each thing has a part. In one instance, he says, "When we die, our bodies become grass, and the antelope eats the grass, so we are all connected in the great circle of life." *The Lion King* became the world's highest grossing film during the year it was released, in 1994. Similarly, the movie *Avatar* relays a message that trees and all living things have souls. It also conveys that the forest is suffused with spiritual essence and is inseparably connected to all life, including people. Avatar became North America's highest grossing film of all time shortly after its release in 2009. Although it is a stretch to characterize either of these movies as touting religion, the eco-animistic overtones are clear.

3J. The Virgin Mary as a Mother Goddess. A belief system called *Collyridianism* is thought by some historians to have taken the worship of the Virgin Mary to the extreme in pre-Islamic Arabia (Chrystal, 1908). They are thought to have worshipped Mary as a true Goddess who took the place of the Holy Spirit in Christian trinitarianism. Collyridianism was branded as heresy by the early Christian Church father Epiphanius in the fourth century CE and disappeared soon after that.

3K. Sacred texts of Organized Religion. The sacred texts of Organized Religion (e.g., the Egyptian Book of the Dead, the Avesta, the Vedas, the Tanakh, Old Testament, New Testament, and the Quran) are historical only in the sense of being a record of what some people believed or proselytized at the time they were written. They are mainly records of doctrinal beliefs over the course of human literacy. They are not unembellished life histories of prophets or messiahs. Even though sacred scriptural myths surrounding iconic individuals and prophets of the most successful Organized Religions are often taken as historical fact by their adherents, most of these sacred stories are not objective depictions of history. One of the challenges for some historians is to distinguish between verifiable historical facts and faith-based mythological constructs. Faith supplies answers to people seeking relief from uncertainty. Its professed truth often requires blind acceptance of certain elements not identifiable by the senses or verified with evidence. The challenging, often thankless, and sometimes even dangerous charge of objective historians is to ask hard questions about highly sensitive matters, make hypotheses based on unbiased evidence, and maintain a scientific perspective. In this book, I am primarily interested in the evolution of the belief systems themselves rather than in trying to verify or contradict the personal histories of the entities being venerated. Historical science is not a contest of biased righteousness. It is instead an objective effort to discover and record the true history of past events.

4 INDIGENOUS EASTERN ORGANIZED RELIGION AND ASIAN CYCLICISM

4A. Major Hindu deities. A few of the many gods and goddesses of Vedic and Hindu religions are those listed in this note (Molloy, 2018; Bowker, 2006). Some Hindu sects focused on (or at least emphasized) only one of these or on multiple aspects of one of them. Other sects revered many deities at a time. Some resemble gods in the pantheons of Old World Hard Polytheism. Also, sometimes one of them was considered as the Supreme God of the pantheon for various regional populations or sects, much like the case in some regional branches of Roman and Greek Polytheism.

Brahma	god of creation and part of a Hindu divine triad with *Vishnu* and *Shiva.*
Vishnu	god of preservation, goodness, and mercy.
Shiva/Rudra	god or gods of destruction and re-creation, sometimes as different aspects or gods.
Kirshna	god of love, compassion, tenderness, and protection; most celebrated avatar of *Vishnu.*
Rama	royal manifestation of *Vishnu.*
Vasudeva	one of the earliest-known manifestations of *Vishnu* attributed to Vaishnavism.
Shakti	divine goddess of female power, represented by many aspects ranging from the beautiful *Lakshmi* to the terrible *Kali.*
Durga	goddess of perseverance; slayer of demons
Parvati	goddess of power, energy, and nourishment. Consort of *Shiva.*
Lakshmi	goddess of success, love, fertility, and beauty. Consort of *Vishnu.*
Saraswati	goddess of music and culture; consort of *Brahma*. In some Hindu branches she is the first consort of *Vishnu.*
Kali	goddess of destruction and recreation.
Radha	goddess of love, avatar of *Lakshmi*, and consort or feminine aspect of *Krishna*.
Sita	incarnation of *Lakshmi*, consort of *Rama*, and patron of agriculture.
Ganesha	elephant-headed son of *Shiva* and *Parvati*. God of success.
Indra	Supreme God of the Indo-central Asians, often represented with thunderbolts.
Savitri	daughter of the sun and mother of the Vedas
Agni	god of fire and mediator between humankind and the heavenly realm.

There are two divine triads in Hinduism. One is the masculine Trimurti (*Brahma*, *Vishnu*, and *Shiva*). The other is the feminine Tridevi (*Parvati*, *Lakshmi*, and Saraswati).

4B. The Ten Yamas: The *Ten Yamas* are codified ethical rules of Hinduism that advocate for non-violence, truthfulness, not coveting another's property, chastity, kindness toward all life, straightforwardness, patience, non-possessiveness, measured diet, and the cleansing of body and mind (Aiyar, 1914). The Yamas were first mentioned in the Rig Veda, and later expanded on in the Sandilya Upanishad.

4C. Sacred texts of Hinduism. The sacred scripture of Hinduism is not consolidated in a single text like the Jewish Torah, the Christian New Testament, or the Muslim Quran.

Instead, it is a lengthy complex of many different ancient documents. The most sacred texts of Hinduism are the four *Vedas*. The word Veda is Sanskrit for "knowledge" or "wisdom." The four Vedas include the *Rig Veda*, the *Yajur Veda*, the *Sama Veda*, and the *Atharva Veda* (Doniger, 2009; Molloy, 2018). The *Rig Veda* (Knowledge of Verses) includes more than one thousand chants to various gods. The *Yajur Veda* (Knowledge of Sacrifice) contains recitations for use during sacrifice. The *Sama Veda* (Knowledge of Songs) contains musical elaborations of Vedic chants. The *Atharva Veda* (Knowledge of the Fire Priest) consists of charms, magic spells, and prayers of protection against sickness and harm.

Each Veda has three or four subdivisions, including the *Samhitas* (mantras and benedictions), the *Aranyakas* (text on rituals, ceremonies, sacrifices, and symbolic-sacrifices), the *Brahmanas* (commentaries on rituals, ceremonies, and sacrifices), and the *Upanishads* (texts discussing meditation, philosophy, and spiritual knowledge). The Upanishads (the last chapter and appendix for each Veda), were composed around 500 BCE (Doniger, 2009). These sections were written in a more reader-friendly style for a broader base of people than the previous sections. It explains spiritual Vedic tradition and philosophy in a way meant to be understood by any class of people, not just priests and ascetics (Molloy, 2018).

The Vedas are considered by many Hindus to be timeless knowledge never to have been authored, edited, or redacted by humans. The texts were not written down until after they had been orally handed down over many centuries. The tradition of oral transmission continues to this day by Brahmin high priests. The oldest of the four Vedas is the Rig Veda. It is reported by most scholars to have been composed around 1500 BCE, although many Hindus believe that date to be much earlier. Hinduism scholar, Wendy Doniger, estimates the earliest composition of the Rig Veda to have been between 1700 BCE and 1500 BCE (Doniger, 2009).

Important post-Vedic texts that helped spread Hinduism to a broader base of people include the *Epics* (epic tales demonstrating the critical importance of dharma) and the *Puranas* (stories of the gods, heroes, and kings). The Epics follow the practice of the ancient Sumerians of composing poetic stories that were focused on the power of the gods. These stories help create ideals for people to follow. The best-known of the Epics are the *Mahabharata* and the *Ramayana*. These two volumes eventually became more influential among many modern Hindus than even the Vedas. The dates of origins given for such works are highly speculative estimates because these stories were passed down through countless generations of oral transmission before they were ever transcribed in dateable form (Doniger, 2009). The *Mahabharata Epic*, composed somewhere between 400 BCE to 400 CE, is remarkable for its longevity as a popular element of Hindu tradition and for its monumental length. At roughly one hundred thousand verses and 1.8 million words, it dwarfs the Bible, the *Iliad*, and the *Odyssey* combined (Prothero, 2011). It contains one of the best-known Hindu texts, the *Bhagavad Gita*. This is a philosophical text said to have been spoken by the god Krishna to a spiritual son of the god Indra named Prince Arjuna (Doniger, 2009). It focuses mainly on the search for serenity and calmness in a world that was rapidly changing through urbanization during the fifth through first centuries BCE. It contends that change is the law of the universe. The *Ramayana Epic,* completed around 200 BCE to 200 CE, is not as long as the Mahabharata but still twice as long as the New Testament. It is revered not only by Hindus but also by Jains, Buddhists, and even Muslims. It has a central theme of love, longing, and fidelity to the ethics of marriage. The Ramayana Epic is one of the four most influential books ever composed, along with the Bible, the Quran, and the *Analects* of Confucius.

The *Puranas,* composed around 350 CE to 750 CE (Doniger, 2009), are major post-Vedic texts that include a historical narrative of the universe from creation to destruction. They discuss the genealogies of the deities, heroes, and kings important to most branches of Hinduism. They were written not only in Sanskrit, but in other languages as well, making them accessible to a much broader base of people than older texts. The Puranas fueled the widespread Vedic edification of lower castes and women as well as upper-caste men (Bowker, 2006). Most of the rituals developed in the Puranas do not require the mediation of a Brahmin priest, which eliminated the stranglehold of the priests on Vedic ritual (Doniger, 2009). Although many Hindus have disputed Puranic religiosity, these texts remain as some of the most popular in Hinduism today (Bowker, 2006).

Both the Epics and the Puranas were part of the period of great ideological consolidation and development within Hinduism called the *Hindu Synthesis,* which occurred around 500 BCE to 467 CE (Hiltebeitel, 2013). These and other documents helped clarify the obscurities of the Vedas, further bringing Hinduism to a broader base of people.

4D. Religious affiliation in India in 2011. In 2011, the religion with which people in India self-identified, according to a survey by the Pew Research Foundation, found the following percentages (reported by Kramer, 2021): Hinduism, 79.8 percent; Islam, 14.2 percent; Christian, 2.3 percent; Sikh, 1.7 percent; Buddhist, 0.7 percent; Jain, 0.4 percent; and all others combined at less than 1 percent.

4E. Examples of Vaishnavism classifications from the literature. A few of the major groups of Vaishnavism mentioned in the literature (Bowker, 2006; Srinivasan, 2011; Flood, 2003) include *Sri Vaishnavism* (tenth century CE), which split into the *Thenkalai* and the *Vadakalai* during the sixteenth century; *Madhava Vaishnavism* (also known as *Bhagavata Vaishnavism*), which evolved around the thirteenth century CE; and *Pushti Vaishnavism,* which evolved around the fifteenth century CE. Each of these branches eventually evolved into its own set of subbranches. Some have become monotheistic, whereas others continue to recognize multiple deities or godly aspects. In a somewhat different context, the *New World Encyclopedia* (entry for Vaishnavism, accessed online March 29, 2021) divides Vaishnavism into two major Groups: the *Bhagavatas* and the *Pancharatas.* They list six major subgroups (sampradayas) of the Bhagavatas group, including *Lakshmisampradaya* (eleventh century), the *Sanakadisampradaya* (thirteenth century), the *Brahmasampradaya* (thirteenth century), *Rudrasampradaya* (fifteenth century), *Gaudiya Vaishnavism* (fifteenth century), and the *Ramanandi movement* (fourteenth century).

Systematic classification of the numerous subgroups of Vaishnavism is extremely complex, and its organization is not consistent or broadly inclusive. Although these subbranches are not included on my phylogenetic trees, I mention a few (although by no means all) examples of recognized subgroups and classifications. As with Shaivism, listing all of the existing classifications of Vaishnavism is an extremely complex problem I will not approach here.

4F. Implausible biblical stories. Followers of Traditional Christianity and Rabbinic Judaism should not be too quick in mocking the implausibility of ancient Jain scriptural text. The Old Testament of the Abrahamic Bible has its giants and embellishments, too. For example, it describes the Amorite people as "the height of the Cedars" (*Amos* 2:9–10). The Amorites were said to be descendants of Noah (*Genesis* 10:15–16), a man who lived to be 950 years old (*Genesis* 9:28–29). With no previous engineering experience Noah and his family built a mega-supertanker out of logs that was large enough to house two of every species on earth, along with enough food to feed the animals and his family for five months of float time (*Genesis* 8:3–5). All scripture composed in the first few millennia BCE contain such stories. If we are to grant metaphorical license and sensationalism to some religions (e.g., Abrahamic religions), we should do so for all.

4G. Phylogenetic tree for the schools of Buddhism. I constructed my phylogenetic trees and denominational classifications of Buddhism mainly after the work of Buswell and Lopez (2014), which is based on the works of many others. That work contains one of the clearest broad classifications of Buddhist schools that I could find. However, there are different interpretations of how these schools are interrelated or distinguished from each other. One such alternate example is the work of Baruah (2000), in which some of the same names are defined or grouped differently. A major part of the challenge is the paucity of firsthand information about many of the extinct schools. Another problem is the absence of broadly comparative studies that have clearly summarized characterizations of the many Buddhist schools in a phylogenetic context. My work here barely scratches the surface.

One putative school of Buddhism that I omitted from my trees, which is frequently discussed as a branch of Mahasamghika in the literature, is called *Caitya* (sometimes spelled *Caitika* or *Caityaka*). I did not include it on my phylogenetic trees because I could not find sufficient information to clearly distinguish it as a uniquely formalized branch of Mahasamghika. It is vaguely diagnosed, mainly by the practice of worshipping caitya (Buddhist shrines and sanctuaries).

4H. "Noble" teachings of Buddhism. Many of Buddhism's teachings are bundled in groups of four or eight, perhaps as an aid to memorization (Molloy, 2018). The eight-spoked wheel is the ancient symbol of Buddhism. The *Four Noble Truths* and the *Noble Eightfold Path* are often used to summarize the essence of Buddhism. The Four Noble Truths are as follows: (1) Suffering exists in life. (2) The cause of suffering is desire and craving for things to be a certain way. (3) Suffering can be ended by detaching one's self from craving and desire. (4) The way to stop craving and to achieve release from suffering and the limitations of this world is to follow the Noble Eightfold Path. The Nobel Eightfold Path entails the following: (1) Correct understanding (recognizing the impermanence of life, the mechanism of desire, and the cause of suffering). (2) Correct intention (that one's motives are pure and not tainted by selfish desires or emotions). (3) Correct speech (speaking honestly, kindly, and in positive ways). (4) Correct action (not hurting other people or organisms that can feel hurt). (5) Correct work (not permitting one's job to hurt others). (6) Correct effort (striving for improvement). (7) Correct meditation (using meditation and focused awareness to contemplate the nature of reality more deeply. (8) Correct contemplation (cultivating states of blissful inner peace).

Buddhists also have *Five Precepts*, which state basic moral obligations of lay Buddhists. These prohibit killing, stealing, sexual misconduct, lying, and intoxicants (Salguero, 2020).

4I. The concept of "anatman" in Buddhism. A common perception in Buddhism is that there is no subsistent self or soul underlying the reoccurring human presence. As figuratively explained by religious scholar John Bowker (2006), the *anatman* of Buddhism (also called the *anatta*) does not have a Hindu-like *atman* (permanent, individual self, spirit, or soul) inside the human body that, like a bus driver, gets out at the end of one life only to climb into the next reincarnation. Instead, the human appearance is constituted by an aggregate of five independent elements called *skandha*. These include (1) *rupa* (material composition), (2) *vedana* (sensing), (3) *samjna* (perception), (4) *samskara* (mental formations producing character), and (5) *vijnana* (consciousness). These elements are in a constant process of change and do not constitute a self. The rebirth concept in Buddhism asserts that rebirth does not necessarily mean coming back as another human being. The *anatman* in Buddhism refers specifically to the rejection of the *atman* of Hinduism and Sikhism.

4J. The Canonical Councils of Buddhism. The Buddhist Councils were somewhat analogous to the Ecumenical Councils of Western Christianity (**Note 8C**). Like the Christian councils, the Buddhist councils were large canonical gatherings of the church community organized to better define belief systems and settle doctrinal disputes. Also like the Christian councils, they were often sponsored by the ruling political monarchy (kings or emperors) and often resulted in schisms within the religion. Briefly, the six Buddhist councils are listed in the following bullet points (after Buswell and Lopez, 2014; Baruah, 2000; Bowker, 2006; Harvey, 2013). Although the exact dates for the early councils are controversial, the first three councils convened on the Indian subcontinent are thought to have been particularly important in the shaping the development of early Buddhism. The dates of the first two councils were later reconstructions based on Buddhist texts and the hypothesized date of The Buddha's death, which is also controversial (discussed in the text).

- The *First Buddhist Council* is thought by many authors to have been the *Council of Rajagriha* held around 486 BCE, shortly after the Buddha's death. It was convened at a cave near Rajagriha (today's Rajgir, India) under the patronage of King Ajatshatru, leader of the Haryanka dynasty of East India. It was attended by about five hundred senior monks with the objective of establishing Buddhist canon (authoritative scripture) and preserving the Buddha's teachings before they became corrupted.
- The *Second Buddhist Council* is thought by many to have been the *Council of Vaisali*, possibly held in 386 BCE. It was held in the ancient city of Vaisali (present-day Bahir, India) and convened under the patronage of King Kalasoka, leader of the Magadha region of ancient India. It was attended by about seven hundred Buddhist

monks to discuss monastic practices, including ten points of controversy ranging from dietary restrictions to whether or not monks should be allowed to handle money. It is thought by several authors to have resulted in the Great Schism in Buddhism (also known as the Basic Schism), eventually resulting in the split between *Sthaviranikaya* and *Mahasamghika* (phylogenetic tree on page 110).

- The *Third Buddhist Council* was the *Council of Pataliputra* held in 250 BCE. It was held in the ancient city of Pataliputra, capital of the Maurya Empire on the Indian subcontinent. It was said to have been attended by about one thousand monks. It was convened under the patronage of Ashoka, emperor of the Maurya Empire. Under his authority, a number of false monks were defrocked and Buddhist missionaries were dispatched to foreign lands. Some authors believe the Great Schism occurred as a result of this council rather than the second (Bowker, 2006). In addition, this council is thought to have led to a number of other schisms mentioned in the text.
- The *Fourth Buddhist Council* was the *Council of Alu Vihara*, held in Sri Lanka around 25 BCE under the patronage of King Vattagamani Abhaya. Until that time, the Buddhist canon had been maintained only through generations of oral transmission by a specific line of monks. Fearing that famine and social discord might result in loss of those monks as well as the sacred scripture with them, the king convened five hundred monks and inscribed the canon onto palm leaves as the earliest-known written Buddhist scripture (Buswell and Lopez, 2014).
- The *Fifth Buddhist Council* is thought by some historians to have been the *Council of Kundalavana*, convened in Kashmir around 72 AD. It was held under the patronage of King Kanishka, ruler of the Kushan Empire. Some authors report that there was a schism during the council that led to the formation of Mahayana Buddhism (Bowker, 2006). Other authors believe that this "council" was only a later reconstruction based on myth (Buswell and Lopez, 2014).
- The *Sixth Buddhist Council* was the Theravada *Council of Mandalay*, convened in Burma in 1868 by King Mindon Min. He summoned 2,400 monks to revise the Pali Canon (the Tipitaka). In 1871, the revised Pali Canon from the conference was inscribed on 729 stone slabs.
- The *Seventh Buddhist Council* was the Theravada *Council of Kaba Aye*. It was convened in Yangon (formerly Rangoon), Burma, from 1954 to 1956, to commemorate the 2,500th anniversary of The Buddha's death. It was sponsored by the Burmese government under Prime Minister U Nu and included 2,500 monks to edit and recite the Pali Canon.

4K. Earliest evolution of Chinese Religion, and the lack of original records. The prospect of understanding the roots, diversity, and relationships of the early Chinese branches of religion is intimidating. Until the fifth century BCE, "China" as we know it today did not exist. Instead, more than one hundred different independent states were constantly at war with each other. Each state had an independent history, and most of them had their own religious traditions. The collective assortment of philosophies and religious belief systems that flourished among them was later referred to as the *Hundred Schools of Thought*. As the area that would one day become China gradually consolidated into just seven independent states by around 476 BCE, the interstate wars intensified. Conquest and consolidation continued until around 221 BCE, when the first Chinese Imperial dynasty, the Qin dynasty, brought about the first unified China. Its first emperor was Qin Shi Huang, who abolished the older feudal system of the region, built a network of highways for his troops, and started the Great Wall. His mission to unify the geographic regions of China included efforts to consolidate the hearts and minds of the Chinese people. The emperor was infamous for many things, including his attempt to standardize the religions of China. His efforts resulted in the loss of many early cultural records. The loss of historical documents from the so-called Hundred Schools of Thought obscured our knowledge of many early Chinese religious belief systems, except for the Taoic and Buddhist branches that received significant monarchical patronage over the following centuries.

Even in the twentieth century, political efforts toward consolidation of culture in China increasingly obscured the earlier history of early Chinese Religions. The Cultural Revolution in China, beginning in 1966, slowed opportunities for Western anthropologists and archaeologists to conduct research in China on its religious history (or other aspects of culture as well). Even such work by in-country Chinese Nationals was curtailed. Under Mao Zedong from 1966 to 1976, thousands of monasteries, churches, and mosques were destroyed. Religious literature was deemed "anticommunist" and censored, and Chinese intellectuals were persecuted. A cult-like admiration of Mao (*Maoism*) developed with its own semireligious overtones (Cheek, 1997). The proclivity of humans to deify new prophets or individuals in position of great power seems instinctual.

The historian Arthur Waldron stated that "communism was, in effect, a religion for its early Chinese converts: more than a sociological analysis, it was a revelation and a prophecy that engaged their entire beings and was expounded in sacred texts" (Waldron, 1998:325). Some philosophers see Maoism as an attempt to combine socialism with Confucianism (Cohen, 2001). The attempts of the Cultural Revolution to ban certain religions was largely unsuccessful, and eventually the government softened on this (at least on paper) when the Constitution of the People's Republic of China added the "freedom of religion" article 36 in 1978. Article 36 states: "Citizens of the People's Republic of China enjoy freedom of religious belief. No state organ, public organization or individual may compel citizens to believe in, or not to believe in, any religion; nor may they discriminate against citizens who believe in, or do not believe in, any religion. The state protects normal religious activities. No one may make use of religion to engage in activities that disrupt public order, impair the health of citizens or interfere with the educational system of the state. Religious bodies and religious affairs are not subject to any foreign domination."

In 2005, another policy called the "Intangible Cultural Heritage" was implemented to protect native Chinese religions (Liang, 2016). By then, however, many smaller indigenous religions in China as well as the historical records of their existence had been lost. Since 2012, efforts to "Sinicize religion" have stepped up to make religion in China "more Chinese" with stronger socialist core values. The effect still seems to be as much about obscuring historical information as clarifying it. In the end, the effect of trying to stamp out or control religion has led only to the evolution of new religions and an unfortunate loss of information about older ones. Many of these poorly known branches of early Chinese religion were eventually grouped together in what became known as "Chinese folk religion" (see **Note 4L**).

4L. "Chinese folk religion": a non-monophyletic group of convenience in need of further analysis. *Chinese folk religion* (as mentioned in **Note 1L**) is a poorly understood "group" of religions (hence, the quotation marks). As currently defined, it is a hodgepodge collection of supernatural belief systems whose interrelationships and monophyletic components remain largely unknown. Overall, it appears to be a non-monophyletic group. I did not come close to understanding its detailed composition. It is ambiguously defined only as a collection of "traditional religions" of the Chinese and Taiwanese people. Because the primary definition of this group is based on geography rather than religious characteristics, its evolutionary context and subgroup classifications remain obscure. It is in need of much more detailed comparative study, and its subgroups are not yet sufficiently defined and classified to be included on my phylogenetic trees. I will say a few words about it for general historical context.

Much of what is called "Chinese folk religion" originated during or before the Shang dynasty (1600–1046 BCE). Shang religion contained early elements of ancestor worship as well as god-king spirits who remain an unseen force after death (Ohnuki-Tierney, 1991). The Shang religion was a form of ancestorism in which royal ancestors eventually became deified. The Zhou religion, which took hold after overthrowing the Shang, was rooted in agriculture and polytheism, with a number of branches that had pantheons of gods or their aspects (Libbrecht, 2007).

"Chinese folk religion," as currently defined in a vernacular sense, has more than a half billion followers today that could potentially be sorted into hundreds of distinct subbranches. Some of these subbranches may fit into the phylogenetic tree of Taoic Religion, whereas others might be better classified within the branches of Dharmic Religion, ancestorism, Old World Hard Polytheism, or other groups of Organized Religion. Chinese folk religion varies from province to province and even from village to village (Fan and Chen, 2014). Since the 1970s, several Chinese scholars have proposed simply referring to it as a unified "Chinese Religion," effectively nationalizing whatever combination of religions exists now and ignoring most of their evolutionary context. This may have certain social or political benefits, but it also obscures the historical roots of many modern religions in China. Hopefully, scholars will eventually sort out the detailed diversity and historical relationships of belief systems provisionally included within this group.

5 AFRO-EURO-MEDITERRANEAN ORGANIZED RELIGION, BEGINNING WITH OLD WORLD HARD POLYTHEISM

5A. Early similarities between the Old World and the New World supernatural belief systems suggest a Late Paleolithic origin for hard polytheism. Archaeological evidence indicates that hard polytheism was already present in both the Old World

(Europe, Asia, and Africa) and the New World (North America, Central America, and South America) by at least the fourth millennium BCE, which suggests that the concept of hard polytheism may be much older than that in both regions. Elements of hard polytheism most likely developed first in the Old World because of the much older human presence there. From around seventy to eleven thousand years before present, there were long periods of expansive land connection between Asia and North America called Beringia (page 209). Large-scale population migration from the Old World to the Americas would have been possible over the land bridge, and along a continuous coastline with small boats, for thousands of years. This corridor was disrupted in 9000 BCE when the land bridge submerged, separating the Old World from the Americas by the Pacific Ocean. Mass population migration from Eurasia to the Americas was relatively cut off until post-Columbian time when large groups of humans could once again cross from the Old World to the New World using large ships. Archaeological evidence of hard polytheism on both sides of the Pacific dating to the fourth millennium BCE suggest that hard polytheism may have originated and spread before the loss of the Beringian land connection. Of course, it is impossible to know whether or not hard polytheism evolved independently in the New World and the Old World. However, early trans-Pacific ideological similarities make independent derivation seem less likely.

The earliest origin of hard polytheism remains unclear. Hopefully, it will become better understood with new archaeological discoveries. However, the archaeological record is intermittent at best, and it will always be an incomplete record of history.

5B. The true diversity of early hard polytheism: still (and perhaps will always be) a mystery. Although today, hard polytheism constitutes a relatively small percentage of practicing religions, it once included the majority of the world's branches of Organized Religion. I place only a few better-known examples on my phylogenetic tree. Some of the major branches of hard polytheism omitted include unnamed belief systems with origins in Baltic, Basque, Finnic, Georgian, Aboriginal Germanic, Slavic, Vainakh, and aboriginal East Asian regions. There are many books about specific branches of polytheism, and I only scratch the surface. Even the best-known ancient branches (e.g., Greek Polytheism, Roman Polytheism) are highly simplified on my phylogenetic tree. There may have been hundreds of subbranches within each of these, and the characteristics given are meant only to representative the most general examples.

The ancient branches of hard polytheism were some of the most numerous and complex. However, their detailed diversity is poorly known and inadequately classified. We do not know how the followers of most of these supernatural belief systems referred to themselves. There are probably thousands of polytheistic belief systems whose names and traditions were lost or never clearly distinguished. In contrast to other chapters of this book, most branch tips of the phylogenetic trees for this chapter (pages 158 and 208) are named after the cultures (e.g., Inca or Aztec) or geographic regions (e.g., Greek or Roman) in which these supernatural belief systems primarily existed. Naming conventions (described in chapter 1 for comparative biology) are sorely lacking in the comparative study of religions, particularly within early major groups of Organized Religion, such as Old World Hard Polytheism. We do not have the names for the hundreds of different sects of Greek and Roman Polytheism in ancient times. This naming is in contrast to the names of religions at the tips of other chapter phylogenetic trees, such as Sikhism (chapter 4), Samaritanism (chapter 6), Roman Catholicism (chapter 8), and Zahiri Sunni Islam (chapter 11). Those names are more closely tied to the ideological elements of the belief systems that they represent, and they more clearly transcend cultural and geographic borders. The vague classification of subgroups within hard polytheism is also due to the fact that the early roots of these subgroups were so early in preliterate time. Detailed records of most hard polytheistic subgroups, and how they may have self-identified, were lost in translation by later Traditional Christian and Islamic chroniclers who considered it all "pagan" history not worth clearly differentiating. Many historical details of hard polytheism became victims of forceful proselytism and reframing by the winners of conquest. Much detailed comparative work is needed to catalogue and classify the true diversity and evolutionary history of hard polytheism groups that existed in prehistoric and early historic time (also see **Note 1J**).

5C. The "Residential Schools" imposed on Native American people. Efforts to eradicate indigenous American religions in the New World increased over time, with Christian missionaries continuing to use forceful means of proselytism. During the late seventeenth through twentieth centuries, large-scale "conversion" efforts were cooperative ventures between the government and the church. Government officials often used Western Christianity as a tool of cultural assimilation, while the Church often used governmental powers to enforce proselytization efforts. Particularly during the nineteenth and twentieth centuries, Native American children were forcibly removed from their families into government-operated Western Christian boarding schools called the Residential Schools (also referred to as "American Indian Boarding Schools"), where they were strongly encouraged to accept Western Christian beliefs and values in place of their native traditions. At the same time, the U.S. and Canadian governments also suppressed the practice of many Indigenous traditions. For more than eighty years, there were federal laws banning ceremonies such as sun dances and sweat lodges (Rhodes, 1991). This forcible conversion and assimilation continued until the 1970s (Hall, 2013; Lajimodiere, 2019; Boxer, 2009). The primary objective was "civilizing" or assimilating Native American children into European American culture. Similar Christianization efforts were also ongoing in Latin America, the Arctic, and the Pacific (Smith, 2009).

The Residential Schools were not officially discontinued in the United States until 1978 with the passage of the American Indian Religious Freedom Act. It is ironic that a country claiming to have been founded on a principle of religious freedom could be so harsh with religions other than those sanctioned by the government. The Canadian Prime Minister Stephen Harper issued a formal apology for the residential schools in 2008, calling them a sad chapter in Canadian history. In 2009, the 111th U.S. Congress finally issued a formal apology to the "native peoples" of the United States for its "long history of official depredations and ill-conceived policies" regarding indigenous peoples of North America (S. J. Resolution 14 of the 111th U. S. Congress). By the twenty-first century, even the Roman Catholic pope, Pope Francis, apologized for the

Church's "grave sin" of violent proselytism and the Catholic Church's complicity in the "catastrophic" Residential Schools (**Note 3G**). By the twenty-first century, membership in traditional Native American religions had all but disappeared. In the United States, it had fallen to as few as nine thousand people, and most knowledge about its former diversity had been lost (Wybraniec and Finke, 2004).

5D. Child sacrifice in early Abrahamic religions. In addition to the biblical story of Abraham ready to kill his son as a sacrifice to his God, described in Genesis, other biblical examples of potential human sacrifice exist. In the Book of Judges, chapter 11:30-39, an Israelite judge and military leader named Jephthah is said to have made a bargain with Yahweh. If Yahweh would grant him victory over the Ammonites in an upcoming battle, he would sacrifice "whatsoever cometh forth of the doors of my house to meet me, when I return." He then defeated the Ammonites in a great slaughter and returned home victorious. The first person to come out of his house to meet him was his daughter and only child. According to scripture, he let her go into the mountains to bewail her virginity for two months. Upon her return, he killed her and burnt her body as a sacrifice.

5E. Records of Mayan Polytheism. Mayan Polytheism is one of the better-known pre-Columbian Organized Religions from Mesoamerica, although most of its surviving written documentation reflects Spanish Christian (Roman Catholic) influence. One of its fundamental documents is the *Popol Vuh*, which was composed in the language of the Quiche Maya (inhabitants of the southern highlands of what is now Guatemala) well before the Spanish Conquest. It was translated into Spanish by a Christian monk named Francisco Ximenez at the beginning of the eighteenth century, after which the original Quiche manuscript disappeared (Christenson, 2007). Most, if not all, pre-conquest Mayan spiritual texts going back at least to the second century CE were destroyed by the Spanish during the Conquest to "protect" the Maya from their traditional religion and accelerate their conversion to Christianity (Christenson, 2007). The original copies of the seventeenth- and eighteenth century books of *Chilam Balam* on the Yucatec Maya culture of the Yucatan Peninsula, like those of the Popol Vuh, were lost and are survived only by postconquest translations.

5F. Controversy over the origin of Andean urbanization. Some of the earliest-known major civilizations in South America (and in the New World, for that matter) are indicated by archaeological sites in the *Fortaleza*, *Pativilca*, *Huaura*, and *Supe* River Valley regions of Peru. These are sometimes collectively referred to as the Norte Chico urban complex (Mann, 2005b). Parts of it were contemporary with the building of the Egyptian pyramids and the rise of the earliest empires in the Old World. In the New World, it predated the Mesoamerican Olmec civilization by more than a millennium.

The Norte Chico complex is thought to represent the cradle of Andean civilization. Its discovery and excavation began in earnest with the work of Peruvian archaeologist Ruth Shady Solis in the mid-1990s. She focused on a site called Caral as well as a number of other sites, all in the Supe Valley. Caral was determined to have been inhabited by more than three thousand people between the twenty-sixth and twentieth centuries BCE. It became the best studied of all the Norte Chico sites, largely due to the work of Shady and her colleagues. However, in the twenty-first century, an intense controversy blew up between two international groups of archaeologists working in the region, including Shady and other archaeologists from Peru and certain archaeologists in the United States. This became yet another example of the extreme sensitivities often involved with the study of cultural history. Former Peruvian and non-Peruvian colleagues who once worked together in the region eventually turned against each other with public charges of plagiarism, academic imperialism, intellectual territorialism, and shoddy work. The controversy became referred to in *Archaeology* as the "Monumental Feud" (Atwood, 2005) and in *Discover* as the "Showdown at the O.K. Caral" (Miller, 2005). Among the hotly disputed scientific issues were disagreements about whether the urban sites had arisen from a farming culture inland or from a fishing culture along the coast. Other disagreements concerned the age of certain artifacts and archaeological sites. There was even passionate dispute over what to call the civilization, "Caral-Supe" or "Norte Chico." (Colleagues warned me that using either name could be seen by some as "taking sides," so I use both). The abundance of controversy is covered in Miller (2005), and I do not cover it in detail because the focus of this book is on the evolution and broad interrelationships of Organized Religions. My intent is not to take sides, and not to act in an imperialistic or a culturally provincial way. I use the name Norte Chico as the more general name, because it contains both the Supe River Valley and the Pativilca River Valley. It is therefore less likely to be confused with a more restricted use (i.e., only the Supe River Valley). The most significant religious object reported from the Norte Chico region that is relevant to the subject of this book is a staff god image etched on a gourd from the Pativilca River Valley, not the Supe River Valley (page 242).

According to radiocarbon tests, the Norte Chico gourd with the image was from between 2280 BCE and 2180 BCE. Haas, Creamer, and Ruiz (2003) argue that, based on this artifact, the principal Andean spiritual tradition of having a staff god originated in Norte Chico. This was more than a millennium before the Chavin civilization is thought to have existed, and the staff god tradition endured for at least four thousand years through many other Andean civilizations and supernatural belief systems. Other anthropologists argue that although the gourd was four thousand years old, the image of the staff god may have been etched into it much later, because it was discovered in strata only 1,100 to 700 years old (Mann, 2005b). There is no definitive way to tell which scenario is correct.

5G. *Yaya-Mama* as an early Andean "religious" belief system of the Titicaca Basin. The late Andean archaeologist of Central Michigan University, Karen Mohr Chavez (1941–2001), described what she called the *Yaya-Mama Religious* belief system as a shared belief among several cultures of the Titicaca Basin of Peru and Bolivia. It was identified based on a number of archaeological elements, most of which are not clearly religious in nature, such as architecture and an artistic style. She characterized Yaya-Mama with the following features: "(1) temple-storage centers such as at Chiripa, (2) Yaya-Mama style stone sculpture having supernatural images associated with the temples, (3) ritual paraphernalia and ceramic trumpets and ceremonial burners, and (4) supernatural iconography including heads having rayed appendages and a vertically divided eye." (Chavez, 1988:17)

Yaya-Mama was named primarily after its associated style of stone sculpture. This style was characterized by (1) being in the form of slabs or "stelae" with four corners rather than rounded, carved on all four sides; (2) motifs predominately consisting of heads or faces with projecting appendages, undulating serpents, anthropomorphic figures, checkered crosses, four-legged animals in profile, frogs, rings, and forked serpent tongues; and (3) the composition of sculptural motifs using opposition and symmetry (Chavez, 1988; Stanish, 2003). These sculptures have no readable inscriptions. Chavez also characterized Yaya-Mama by having ritual paraphernalia made in a particular way (e.g., ceramic trumpets made with grass molds) and buildings constructed in a particular way that included a sunken central floor. Chavez interpreted these buildings as temples. Chavez also observed the presence of supernatural iconography and motifs with rayed heads, but such characters are far from unique to the Titicaca Basin area and are not of much use in diagnosing Yaya-Mama as a distinct belief system.

The current diagnosis of Yaya-Mama as a distinct religion of the Titicaca Basin poses two major challenges. First, some of the characteristics listed by Chavez (e.g., artistic style, production methods for ritualistic objects, and building architecture) are not religious ideologies. They may well have nothing to do with religion. Second, even those characteristics that are of a religious nature (e.g., supernatural images and images with rayed heads) need broader comparison with other South American iconography to understand the scope of group membership that they encompass. Those characteristics are also found in forms of South American Polytheism outside of the Titicaca region and even outside of Mesoamerica. Additional work is needed to verify the significance of Yaya-Mama as a distinct religious tradition and to determine its broader ideological relationships.

6 LINEAR MONOTHEISM

6A. Could monotheism have come before polytheistic belief systems? Wilhelm Schmidt (1868–1954) was a Roman Catholic priest and linguist in Europe who became interested in the evolution of religion. He proposed a theory that polytheism (as well as all other religions) evolved from monotheism. In his 1912 book on *The Origin of the Idea of God*, he proposed that originally, the earliest humans began with the notion of a single god, usually a sky god, who created the world and governed human affairs from afar (Armstrong, 1993). This theory has never been widely accepted, and what meager evidence exists about the earliest notions of supernaturalism suggests otherwise (i.e., that it was early monotheism that evolved from polytheism). But who knows for certain? The earliest notions of supernaturalism happened far too long ago for absolute certainty. Thus, we rely on hypotheses and comparative analysis, and the consensus today is that polytheism came before monotheism.

6B. Hebrew alphabet, and its spelling of God's name. The Hebrew alphabet does not contain vowels. The original Hebrew name for God was *YHWH* (Known as the Tetragrammaton). Traditionally, the name was considered too sacred to speak aloud, and it was therefore read by early Abrahamic Monotheists as *Adoni* (my Lord) or *Hashem* (the name). Because the letter J represents the Hebrew equivalent of the letter Y and the letter W the equivalent of the letter V, YHWH (= JHVH) is also sometimes translated with added vowels as *Jehovah*. Thus, *Yahweh* and *Jehova* are different transliterations of the same Hebrew word.

6C. Early life stories of Moses and Sargon. The story of Moses's birth and adoption is similar to a much earlier story from Mesopotamian Polytheism about Sargon of Akkad, also known as Sargon the Great (Westenholz, 1997). Sargon's birth and early childhood was first transcribed in a Neo-Assyrian text from the seventh century BCE based on oral transmissions that went even further back in time. Some historians place his alleged birth around the twenty-fourth century BCE in the Mesopotamian town of Azupiranu. Sargon was the illegitimate son of a high priestess of the goddess Inanna. Being an illegitimate son of a priestess put his life in danger. To save him, his birth mother left him in a reed basket in the river. A member of the Mesopotamian royal court (the Sumerian king's gardener) discovered the basket and took Sargon into his custody. He raised him under the protection of royalty, after which Sargon eventually became an iconic figure and historic leader of the Mesopotamians. He was regarded as a semisacred figure by subsequent Mesopotamian empires for more than 1,500 years after his death. It is difficult to ignore the similarities between the early developmental story of Sargon and that of Moses. Both were also believers in divine guidance. Where Moses venerated the God Yahweh as his spiritual guide, Sargon venerated the Goddess Inanna from the Mesopotamian pantheon. The Abrahamic story of Moses may have been influenced by this earlier story (Lewis, 1980).

6D. The darker side of biblical Moses. Biblical accounts of Moses portray him as having quite a vengeful temper, which fit with much of the overall tone of the Old Testament. He was not a great role model for modern moralists. As pointed out by the prominent ethologist and outspoken atheist, Richard Dawkins, "modern morality, wherever else it comes from, does not come from the Bible" (Dawkins, 2006:279). The example with the golden calf was only one of several showing the intolerant, violent side of Moses's nature. He justified such actions by simply saying that he was following the orders of Yahweh. Another example of Moses's darker side involved the Midianites.

The Midian war is described in the Old Testament book of *Numbers*. In the decades following their departure from Egypt, Moses had organized a significant army of Israelites set on conquest. At some point of the biblical account of the Exodus, Moses tells the Israelites that he was commanded by Yahweh to exterminate the cities and warriors of the Midianites, a league of tribes descended from Midian, son of Abraham. Moses explains that this was necessary because the Midianites had led the people of Israel to sin. The battle ensued, and the Israelite army of Moses was successful in their efforts to destroy the cities and wipe out the warriors. When the troops returned from battle, Moses went down to meet them. When he heard that the army had spared most of the women and children of Midian, he was incensed. The biblical passage reads as follows:

> Moses and Eleazar the priest, and all the princes of the congregation, went forth to meet them without the camp; and Moses was wroth with the officers of the host, with the captains over thousands, and captains over hundreds, which came for the battle. Moses said

to them, "Have ye saved all the women alive?" Behold, these caused the children of Israel, through the counsel of Balaam, to commit trespass against the Lord in the matter of Peor, and there was a plague among the congregation of the Lord. Now therefore, kill every male among the little ones, and kill every woman that hath known a man by lying with him; but all the women-children that have not known a man by lying with him [virgins], keep alive for yourselves. (*Numbers* 31:13–18)

6E. The Egyptian negative confessions. The Forty-Two Negative Confessions from the Egyptian Book of the Dead are as follows (from the Papyrus of the Theban scribe Ani, as translated in Goelet et al., 2015:plates 31-A and 31-B):

1. I have not done wrong.
2. I have not robbed.
3. I have not stolen.
4. I have not slain people.
5. I have not destroyed food offerings.
6. I have not reduced measures.
7. I have not stolen the god's property.
8. I have not told lies.
9. I have not stolen food.
10. I have not been sullen.
11. I have not fornicated with the fornicator.
12. I have not caused (anyone) to weep.
13. I have not dissembled.
14. I have not transgressed.
15. I have not done grain profiteering.
16. I have not robbed a parcel of land.
17. I have not discussed secrets.
18. I have brought no lawsuits.
19. I have not disputed at all about property.
20. I have not had intercourse with a married woman.
21. I have not had intercourse with a married woman.(Repeats the previous affirmation with a different god).
22. I have not wrongly copulated.
23. I have not struck terror.
24. I have not transgressed (the Law).
25. I have not been hot-tempered.
26. I have not been neglectful of truthful words.
27. I have not cursed.
28. I have not been violent.
29. I have not confounded the truth.
30. I have not been impatient.
31. I have not discussed (pried into matters).
32. I have not been garrulous about matters.

33. I have not done wrong; I have not done evil.
34. I have not disputed the King (blasphemed against the King).
35. I have not waded in (stopped the flow of) water.
36. I have not raised my voice (spoken arrogantly, or in anger).
37. I have not cursed (or blasphemed) a god.
38. I have not made "extollings" (bragged).
39. I have not harmed the bread ration of the gods.
40. I have not stolen the khenef-cakes from the Blessed (spirits of the dead).
41. I have not stolen Hefnu-cakes of youth, (nor) have I fettered the god of my town.
42. I have not slain sacred cattle.

7 THE EARLY DIVERSIFICATION OF ABRAHAMIC MONOTHEISM

7A. Evolutionary sequence in culture and biology. Some anthropologists state that Christianity is not a later stage of Judaism (Atran, 2002). Although in the most typological sense they are different, in evolutionary science, we can nevertheless find evidence of relative relationship among these groups reflecting their history. We accept that many clearly diagnosed stages are missing in any evolutionary lineage or evolutionary network with a deep history, whether we are looking at culture or at biological species (see the figure on page 14). Therefore, we generally look for patterns of relationship among the kinds we do see, rather than look for specific ancestors. A species of bird is not a later stage of a species of a crocodile. However, any species of bird is thought to be more closely related to any species of crocodile than it is to any species of mushroom. Similarly, Judaism and Christianity are thought to be more closely related to each other than either is to Hinduism.

7B. The face of Jesus in the Shroud of Turin. The *Shroud of Turin* is a length of linen cloth with a negative image of a man's face. Some believe it to be the burial shroud of Jesus, and therefore believe the image to be the actual face of Jesus (Taylor and Bar-Yosef, 2016). Today, this shroud is generally considered a fake. The first mention of the cloth was not until the fourteenth century (Collinge, 2012), making its existence in the time of Jesus highly unlikely given that it would have been mentioned by at least one of the early church fathers. French bishop Pierre d'Arcis stated in 1389 that the shroud was a fake produced in his time by a confessed artist. In 1988, laboratories in three different countries performed carbon-14 dating tests on small squares of the cloth, and all yielded dates between 1260 CE and 1390 CE (Collinge, 2012). Not even the current Roman Catholic pope, Pope Francis, attests to its authenticity, asserting only that it is an "icon of man" (Allen, 2013).

7C. Secondary losses (character reversals) and radical character changes that mask historical connections. My inclusion of *Traditional Christianity* and the rest of Jesu-venerationism within *Second Temple Judaism* on page 286 may make some people uncomfortable. However, phylogenetic trees are maps of historical connections, not summaries of exclusionary preferences or levels of ideological difference. Radical modification or even character reversal of certain aspects within one branch does not erase its historical (or ancestral) connections to another. I am related to my great-grandfather whether or not I wish to be, and no matter how different I may be from him today. Such connections, even when clouded by the addition or loss of characters over time, remain discernable through other lines of evidence.

In *biology*, for example, snakes are included in the superclass Tetrapoda (animals with four limbs) even though most do not possess any limbs. The absence of limbs in snakes does not erase their historical connection to (and inclusion in) the larger group Tetrapoda. Much evidence from other shared characteristics and from embryological observations shows that snakes evolved from ancestors that had four limbs. The absence of legs in modern snakes is not because its lineage never had them. It is instead due to a *secondary loss* (or *character reversal*) early within the lineage leading to them.

In *Organized Religion*, for a similar example, the characteristic of Traditional Christianity to be no longer awaiting a prophesized Savior, does not erase its historical connection to the larger group Second Temple Judaism (who are still awaiting such a Savior). Much evidence supports a historical relationship between the two groups, including a multitude of *shared characteristics* (e.g., the Christian Old Testament and the Hebrew Tanakh being the same sacred text) and *historical records* (e.g., records clearly indicating that Traditional Christianity evolved from older religious belief systems still anticipating a coming messiah).

7D. Paul's biblical writings are the earliest-known documents about Jesus, but they are not biographical. Paul the Apostle (commonly known as Saint Paul or by his Jewish name Saul of Tarsus) was an important early author of Traditional Christianity. His texts, thought to have been written between 50 CE and 62 CE (the *Epistles* of the New Testament), include the oldest texts to be incorporated into what became the New Testament. However, the *Pauline Epistles* were not biographical about Jesus. They took the form of letters written to churches and individuals about proper church order, beliefs, and problem solving (Molloy, 2018; Marty, 1959).

7E. The *Russian Skoptsy*. The Russian Skoptsy (meaning castrate) is a splinter group from the Russian Orthodox church. They advocate breast and genital mutilation in women and the removal of the penis and testes in men because they believe that restores their bodies to the original pre-fall condition of Adam and Eve (Dunbar, 2022). They also believe that Jesus Christ was a castrate. The tradition was founded in the 1760s by Kondratiy Ivanovich Selivanov (?–1832) who at one point claimed to be the reincarnation of Christ. Russian Skoptsky reached a peak of as many as one hundred thousand members by the early twentieth century and then faded to obscurity. A few members have persisted even into the twenty-first century.

8 TRADITIONAL CHRISTIANITY

8A. Name confusion regarding *Apostolic Christianity*. The *Apostolic Christianity* branch, as used here, should not be confused with the much later Protestant branch that self-identifies as *the Apostolic Christian Church* (ACC). The ACC arose in the nineteenth and twentieth century from the Anabaptist group of religions. Here (see the phylogenetic tree on page 322), Apostolic Christianity refers to the earliest belief systems of *Traditional Christianity* during the life of the original twelve apostles and certain other key proselytizers, such as Paul of Tarsus (5 BC–67 CE) and James, brother of Jesus (?–67 or 69 CE). The name, *Apostolic Christianity*, represents a provisional, period-based group designation that will hopefully either be reinforced with ideological characterizations from future studies, or be replaced with a more clearly substantiated classification.

8B. The title of Pope. The supreme clerical authority of the Traditional Christian Church during the first millennium was usually referred to as the *bishop of Rome*. The term *pope* (after the Greek word *pappas*, meaning father) came into common use for the bishop of Rome after the tenth century. Many historians, as well as the Catholic Church, retroactively extend the title of pope to all supreme authorities of the Church (bishops of Rome) going back into the first century. For the sake of simplification, I follow historians that use the term *pope* for all supreme authorities of the Roman Catholic and pre-Reformation Christian Churches, and the term *papacy* for the administrative structure over which that authority presides.

Although Roman Catholicism can name every bishop of Rome in an unbroken line extending from the apostle Peter in the first century to the current-day Pope Francis, most of the popes from the first few centuries are relatively obscure figures. Pope Miltiades (Bishop of Rome from 311 to 314) was pope during the period when Christianity was first legalized in Rome. It was the relationship between him and Emperor Constantine that began the collaborative link between the papacy and the imperial monarchy. Pope Miltiades also received a large palace from Constantine (the Lateran Palace), which became the papal seat and residence for the next one thousand years.

8C. The early Ecumenical Councils of Western Christianity. The seven earliest Ecumenical Councils are thought by many to be the most important because they are recognized as ecumenical ("universal Christian") by both *Western Christianity* (those early Christian branches most closely associated with the Roman Empire and the papacy) and the *Eastern Orthodox* Church. Each of the seven councils was convoked by a Roman emperor who enforced the decisions of these councils within the State Church of the Roman Empire (Christianity). In return, most of these emperors were eventually glorified as saints by various Christian branches, particularly the Oriental Orthodox and Eastern Orthodox Churches. None of the Council meetings were held in the City of Rome. All were held in the eastern half of the Roman Empire, referred to by historians since the sixteenth century as the Byzantine Empire. The council decisions formally institutionalized branches of Traditional Christianity. These decisions also led to several schisms (see the phylogenetic tree on page 322). Later, branches of Reformation Christianity rejected the concept of Ecumenical Councils establishing doctrine for the entire Christian faith. Briefly, the seven Ecumenical Councils are as follows (mostly after MacCulloch, 2010 and Ehrman, 2018).

- The *first Ecumenical Council* was the *Council of Nicaea* in 325 CE. It was convened by the Roman Emperor *Constantine I*, emperor from 306 CE to 337 CE. It was held in Nicaea, a city in the ancient Roman province of Bithynia (today known as the

Turkish province of Bursa). More than three hundred Christian bishops attended from around Christendom. It produced the first uniform Christian doctrine, called the Nicene Creed (see text). The theologian Arius (256 CE–336 CE) was condemned at the conference, and *Arianism* was rejected from *Nicene Christianity* as the first Ecumenical Council defined it. It not only formalized early doctrine for Traditional Christianity, it also increasingly drew the Roman Empire into the process of Christianization.

- The *second Ecumenical Council* was the *Council of Constantinople I* in 381 CE. It was convened by the Roman Emperor *Theodosius I*, emperor from 379 CE to 395 CE. It was held in Constantinople (present-day Istanbul, Turkey), which at the time was the capital city of the Eastern Roman Empire. More than 150 bishops attended, and they expanded and amended the Nicene Creed developed by the first Ecumenical Council. The second Ecumenical Council also reiterated the trinitarian concept of the Holy Ghost with the Father (God) and the Son (Jesus). Macedonius was a bishop of Constantinople from 342 CE to 346 CE who taught that the trinitarian Holy Ghost was not a person but simply a power of God. He was condemned as a heretic at this council meeting for his "subversive" teachings, and *Macedonianism* was rejected from *Amended Nicene Christianity* as defined by the Second Council.
- The *third Ecumenical Council* was the *Council of Ephesus* in 431 CE. It was convened by *Theodosius II*, emperor of the Eastern Roman Empire from 402 CE to 450 CE. It was held in Ephesus (present-day Izmir Province, Turkey). It was attended by more than 250 bishops. It focused on establishing Mary, Mother of Jesus, as "Birth giver of God" rather than "Birth giver of Jesus." Nestorius (386–450 CE), Archbishop of Constantinople, proposed that Mary was simply the birth giver of Jesus and that the divine nature of Jesus did not enter Jesus until after his birth. The controversy that ensued resulted in Nestorius being labeled as a heretic and *Nestorianism* being rejected from *Ephesus Christianity* as defined by the Council.
- The *fourth Ecumenical Council* was the *Council of Chalcedon* in 451 CE. It was convened by *Marcian*, emperor of the Eastern Roman Empire from 450 CE to 457 CE. It was held in Chalcedon (a town just east of Constantinople, today called Kadikoy in Turkey). It was presided over by the Patriarch of Alexandria and more than five hundred bishops attended. It focused on establishing that Jesus was truly God and truly man, without mixing the two natures. The controversy that ensued resulted in a split between *Chalcedonian Christianity* and *Oriental Orthodoxy*.
- The *fifth Ecumenical Council* was the *Council of Constantinople II* in 553 CE. It was convened by *Justinian I*, emperor of the Eastern Roman Empire from 527 CE to 565 CE. It was held in Constantinople like the second Ecumenical Council. It was attended by 152 bishops. Like the previous council, it continued to focus on explaining the two natures of Jesus (man and God). It condemned the teachings of the early Christian scholar Origen (184–253 CE) on the preexistence of the soul. It also reaffirmed its stance against Nestorianism.
- The *sixth Ecumenical Council* was the *Council of Constantinople III* in 680 CE and 681 CE. It was convened by *Constantine IV*, emperor of the Eastern Roman Empire from 668 CE to 685 CE. It was attended by 170 bishops and was again held

in Constantinople. Its asserted that Christ had a separate will for each of his two natures (human and divine) and condemned the teachings of the Monothelites as heresy. Monothelitism teaches that Jesus Christ had two natures but only one will. For their part in their support of Monothelitism, Pope Honorius of Rome, three Patriarchs of Constantinople (Sergius I, Paul II, and Peter), and one patriarch of Alexandria (Cyrus) were all anathematized as heretics.

- The *seventh Ecumenical Council* was the *Council of Nicaea II* in 787. It was convened by *Constantine VI*, emperor of the Eastern Roman Empire from 780 CE to 797 CE. It was held in Nicaea like the first Council meeting and was attended by 370 bishops. This council meeting eased the restrictions against the veneration of icons, although it still forbid their worship as objects. This effectively changed the second commandment in early Western Christianity from how it was portrayed in the Torah.

8D. Name confusion: catholic versus Roman Catholicism. Today the terms "catholic" and "Roman Catholic" are often used as interchangeable names, adding confusion to History of Religion studies. The histories and original meanings of the two terms are quite different.

The simple term *catholic* is, by far, the older of the two names (from the Greek word *Katholike* meaning "universal"). It was first used by Christian Bishop Ignatius of Antioch (35–108 CE?) to refer to general or "universal" Christianity (Ehrman, 2003). It was a term initially claimed by all of Christianity. Although the term *catholic* is still claimed by many different Christian branches, today it is sometimes also used as a vernacular for *Roman Catholicism*. I use the term (with a small *c*) as an interchangeable vernacular for Nicene Christianity of the fourth century CE, where the first formal uniform doctrine of Traditional Christianity was assembled (the *Nicene Creed*). Because of the confusion surrounding the term "catholic," I rarely use it here. Instead, I use the more specific, formal terms for the Traditional Christian branches that recognize papal authority (e.g., *Nicene Christianity, Ephesus Christianity, Chalcedonian Christianity, Western Christianity*, and *Roman Catholicism*) (see the phylogenetic tree on page 322).

The term *Roman Catholicism* also has been used by different authors in a variety of ways. Some use it interchangeably with Traditional Christianity going back to the first century, some use it interchangeably with Western Christianity after the East-West Schism of 1054, and some use it in yet other ways. However, it was the Proto-Reformation movement beginning in the fourteenth century, and the later Reformation movement that first clearly distinguished Roman Catholicism from other branches of Traditional Christianity by defining it as an adversary. The term "Roman Catholic" was first used by Anglican Protestants of England as a slur against the Rome-based Christian contingent, insinuating that it had a traitorous allegiance to a foreign authority: the Roman pope (Burbank and Cooper, 2010). The Roman Catholic Church preferred (and still prefers) the more abbreviated name, "The Catholic Church." Reform Christians essentially resented the attempt by the Roman Church to monopolize the term Catholic, which they believed referred to universal Christianity. The response from the Rome-based Christians ("Roman Catholics") was to claim that theirs was the branch leading most directly to Jesus Christ so their use of the term was justified. The Vatican

Archives (1993) still do not use the term Roman Catholic, but simply use the term "Catholic." To clarify terminology for the purposes of this book, I use the term *Roman Catholicism* to refer to the part of Western Christianity that retained allegiance to the Roman papacy after the Proto-Reform and Reform movements beginning in the late fourteenth century (see the phylogenetic tree on page 322). I also use the term "Nicene Christianity" rather than "Catholic" to denote Pre-Reform branches of Traditional Christianity that formalized after the first Ecumenical Council (**Note 8C**).

Greater nomenclatural convention is needed for comparative studies of religious belief systems. It does not matter so much what those names are. What does matter is consistency in the way in which names are applied to groups being compared. In comparative biology, this is so critical that a book of international rules governs the creation of valid names for species, genera, and families of animals (ICZN, 1999; also mentioned in **Note 1E**). No such book of rules (or anything close to it) exists for the naming of religions. Lack of nomenclatural consistency adds much difficulty to broad comparative studies, and the inconsistent use of names has added much confusion about the historical diversification of religions.

8E. The Edict of Thessalonica. The following translation of the Edit of Thessalonica from Latin to English dates from 380 CE:

> EMPERORS GRATIAN, VALENTINIAN AND THEODOSIUS AUGUSTI. EDICT TO THE PEOPLE OF CONSTANTINOPLE. *It is our desire that all the various nations which are subject to our Clemency and Moderation, should continue to profess that religion which was delivered to the Romans by the divine Apostle Peter, as it has been preserved by faithful tradition, and which is now professed by the Pontiff Damasus and by Peter, Bishop of Alexandria, a man of apostolic holiness. According to the apostolic teaching and the doctrine of the Gospel, let us believe in the one deity of the Father, the Son and the Holy Spirit, in equal majesty and in a holy Trinity. We authorize the followers of this law to assume the title of Catholic Christians; but as for the others, since, in our judgment they are foolish madmen, we decree that they shall be branded with the ignominious name of heretics, and shall not presume to give to their conventicles the name of churches. They will suffer in the first place the chastisement of the divine condemnation and in the second the punishment of our authority which in accordance with the will of Heaven we shall decide to inflict.*
>
> GIVEN IN THESSALONICA ON THE THIRD DAY FROM THE CALENDS OF MARCH, DURING THE FIFTH CONSULATE OF GRATIAN AUGUSTUS AND FIRST OF THEODOSIUS AUGUSTUS
>
> — Codex Theodosians, xvi.1.2

8F. The origin and consolidation of Italy. Italy, as a formal country with its capital in Rome, did not take its current shape until the nineteenth century. After the fall of the Roman Empire in 476 CE, a number of urban centers on the Italian peninsula continued to flourish. By the eighth century, the peninsula began to see the rise of autonomous city-states that would persist and fragment the region over the next millennium, and the united country of Italy as we see today did not exist. The papacy's influence in Rome during this period was dominant in both religious and secular matters. Then, in

1848, the city-states of the peninsula began to consolidate as a unified country, ending with the formation of the Kingdom of Italy in 1861. The city of Rome remained under tight control of the papacy until 1870 when Italian troops took control of it and moved the capital of Italy from Florence to Rome. The pope refused to include the papacy within the newly united Italy and took refuge in the Vatican (a small theocratic enclave within the city of Rome). In 1929, the Lateran Treaty was signed by Mussolini (dictator of Italy) and Pope Pius XI (leader of the Roman Catholic Church). This established Vatican City as an independent country or "city-state", and the papacy thereby maintained autonomy and rule over a small territory it could call its own.

8G. What was the "Holy Roman Empire?" The *Holy Roman Empire* (800–1806 CE) and the *Roman Empire* (27 BCE–476 CE) were two very different entities. The earlier of the two, the *Roman Empire*, began with the crowning of its first emperor, Caesar Augustus (adopted son of Julius Caesar) in 27 BCE (Burbank and Cooper, 2010). Its capital was the city of Rome. The Roman Empire had replaced the earlier Roman Republic (509–27 BCE) which had replaced the Roman Kingdom before it (753–509 BCE). As discussed in chapter 8, in 395 CE, the Roman Empire formally split into eastern and western halves, each with its own imperial court (effectively its own capital). This was the result of Constantine moving the capital of the Roman Empire east to Byzantium (later renamed as Constantinople) for strategic reasons. The eastern half (*Byzantine Empire*) included modern-day Asia Minor and the Balkans Peninsula, and it persisted until 1453 CE after which it was conquered by the Islamic Ottoman Empire. The western half of the Roman Empire (page 343) persisted only until 476 CE (Molloy, 2018). The Western Empire's fall was the result of invasion from various Germanic tribes. The victorious invaders divided the region among the Franks (including future French), Lombards (including future Italians), Goths (including future Spaniards), and other clans. After the fall of the united Empire, Chalcedonian Christianity began to collaborate with various kings, monarchs, and emperors, particularly with the Franks.

The *Holy Roman Empire* had a very different history than the Roman Empire. In 800 CE, Pope Leo III revived the tradition of a Western Empire by crowning the Frankish King Charlemagne as Holy Emperor. This event gave informal rise to what became known as the Holy Roman Empire, a multicultural complex, including the Kingdoms of Germany, Italy, Burgundy, Bohemia, as well as other territories. Charlemagne and the Roman papacy forged a close and mutually beneficial association between church and state that would last a millennium. This enabled Latin Christianity's supremacy as the dominant Christian power of the world. For a while, between the sixteenth and eighteenth centuries, the name changed to the *Holy Roman Empire of the German Nation* to reflect the growing importance of the German Imperial States. The Holy Roman Empire name was sardonically summarized by the French historian and poet, Voltaire, as being an agglomeration "neither holy, nor Roman, nor an Empire" (Renna, 2015:60). The Holy Roman Empire persisted in one form or another until the last of the Holy Roman Emperors was overthrown during the Napoleonic wars in 1806. However, the power of the Roman Catholic Church would continue to flourish and live on to the present day (MacCulloch, 2010).

8H. **The Catholic Inquisitions.** The Roman Catholic Inquisition was a tribunal that was formally established in 1232 under Pope Gregory IX. Its primary purpose was for the suppression of heresy and the punishment of heretics. It was most prominent in northern Italy and southern France, and it was notorious for its use of torture and execution, especially early in its history. However, the secular courts in those days were equally, if not more, involved in the use of torture (Monter, 1990). The "Spanish Inquisition" was a separate nationalized part of the institution established in 1478, and endorsed by Pope Sixtus IV in 1483 (Bowker, 2006). It blurred the lines between sin and crime. The lead inquisitor of the Spanish Inquisition was called the Grand Inquisitor, or Inquisitor General, the first of which was Tomas de Torquemada from 1480 to 1498. Under his watch, as many as two thousand people were burned at the stake for heresy (Bowker, 2006). Many times that number were tortured or imprisoned in the name of the inquisition. In 1542, the Papal Inquisition became an established part of the papal government that was used to combat Protestantism. Its inquisitors were selected primarily among the Dominican order, partly because of the presumed detachment that order had from the rest of the world. Although the inquisition was most commonly used as a tool of violent proselytism and to combat competing religious belief systems (e.g., Biblical Demiurgism, Rabbinic Judaism, and Islam), it was also occasionally used to punish early scientists whose ideas conflicted with those of the Catholic Church. In 1632, the inquisition famously convicted the early astronomer, Galileo Galilei, of heresy for supporting the Copernican notion that the earth orbited the sun. This was in contrast to the more commonly held notion of the Church at the time, that the sun orbited the earth. As a result, Galileo was put under house arrest for the last eight years of his life. It was not until 1992 that the Vatican formally and publicly pardoned him (Cowell, 1992).

The Catholic Inquisition was formally abolished in Portugal in 1821 and in Spain in 1834 (Bowker, 2006). The Vatican, through Pope John Paul II, formally apologized for the abuses of the inquisition in the year 2000.

8I. **"Soft monotheism."** Some authors call the soft polytheism of Christian trinitarianism "soft monotheism" (e.g., Prothero, 2011:68). Both involve the practice of recognizing multiple aspects of a single God.

8J. **Ritualistic trinitarianism and veneration of Mary in Roman Catholicism.** The Rosary is a common trinitarian prayer ritual that emphasizes the veneration of Mary, mother of Jesus (also known as the *Dominican Rosary*). When I was an early adolescent, saying the Rosary was a common prayer ritual for Roman Catholic families. My Roman Catholic grandmother, who was very religious, recited it every night. The Rosary uses a string of beads with a small crucifix at the end to count the number and specific order of prayers to be recited. The ritual strongly promotes both *trinitarianism* and the *veneration of the Virgin Mary*. In addition to reciting series of certain one-line "mysteries" about Jesus that are meant for meditation, the Rosary consists of sixty-eight official prayers. These include the *Sign of the Cross* (a trinitarian dedication), once at the beginning and once at the end. They also include *The Apostles Creed* (a trinitarian prayer), six *Our Fathers* (also known as The Lord's Prayer), six *Glory Be's* (a trinitarian proclamation), and fifty-three *Hail Mary's* (a prayer venerating the Virgin Mary). Some Roman Catholic traditions believe that the Rosary was given to Saint

Dominic (1170–1221 CE) through a vision he received from the Virgin Mary. Dominic was the founder of the Dominican order and the patron saint of astronomers.

In its broadest characterization, the concept of a divine triad, is widespread among Organized Religions (Griffiths, 1996). Some examples illustrated in this book include the Hindu *Trimurti* (page 77), the Mahayana Buddhist *Trikaya* (page 127), the Taoic *Three Pure Ones* (page 146), the Egyptian *Theban Triad* (page 191), and the Roman Catholic *Holy Trinity* (pages 313 and 340). Many others are not illustrated here (e.g., the Celtic *Three Morrigna* and the Hindu *Tridevi*, both mentioned in the text). Although some theologians object to such comparisons on the grounds that adherents understand those triads according to detailed differences (for objections to comparisons between the Mahayana Trikaya and the Christian Trinity, see Habito, 1986), I am focused on the similarities that do exist. All triads, as I recognize them, generally involve three aspects, (or "modes of being") of a revered entity. The number three (as well as the number twelve) has widespread symbolic importance among many Organized Religions through history, often reflecting a sense of completeness.

8K. Medieval church decrees against translation of the Christian Bible into vernacular language for the laity. Clerical authorities of the Roman Catholic Church sought to keep the bible untranslated in order to prevent people from developing their own interpretation of scripture as a possible source of schism. Two examples of this include the following.

The council of Toulouse in 1229 CE, called by the Bishop of Toulouse, issued the following decree: "*We prohibit also that the laity should be permitted to have the books of the Old and the New Testament; unless anyone from the motives of devotion should wish to have the Psalter or the Breviary for divine offices or the hours of the blessed Virgin; but we most strictly forbid their having any translation of these books*" (Peters, 1980:194–95).

The council of Tarragona ruled in 1234 CE that "*No one may possess the books of the Old and New Testaments in the Romance language, and if anyone possesses them he must turn them over to the local bishop within eight days after the promulgation of this decree, so that they may be burned*" (Lortsch, 1910:14).

9 REFORMATION CHRISTIANITY

9A. The *Counter Reformation* movement. The Reformation movement eventually led to a Counter Reformation movement of revival and reform within the Roman Catholic Church during the sixteenth and seventeenth centuries. It reacted to Protestant challenges by reaffirming traditional doctrines on sin and sacraments and loyalty to the Roman papacy. It formally established the Jesuit order (a specialized male order of certain priests and brothers within Roman Catholicism). Jesuit founder, Ignatius of Loyola (1491–1556), was pledged allegiance to the Roman Catholic papacy with his famous quote, "What I see as white, I will believe to be black if the hierarchical Church thus determines it" (Prothero, 2011:78). The Counter Reformation movement had the effect of hardening the schism between Roman Catholicism and Protestantism.

It fueled more than a century of "Holy Wars" that did not end until the mid-seventeenth century (Bowker, 2006). Although the complete story of the Counter Reformation is not covered in detail here, its history is described elsewhere (e.g., Dickens, 1969).

9B. Relations between Roman Catholicism and Protestantism in the twentieth and twenty-first century. Roman Catholicism's stance against Protestantism has softened considerably over the past several decades, particularly since the Second Vatican Council of 1962–1965. That Council sought to bring the Church more firmly into the modern world and to recognize Protestants simply as being other Christians rather than heretics. However, even this seemingly minor move toward reconciliation contributed to a schism within Roman Catholicism. Some conservative branches, such as the *Sedevacantists*, rejected the recommendations of the Second Council, believing that salvation is attainable only through traditional ("true") Roman Catholicism (Collinge, 2012). Sedevacantists believe that the Vatican has become flawed since the Second Vatican Council and that the Vatican has had no true pope since the death of Pope Pius XII in 1958. They see mainstream Roman Catholic Church's embrace of modernism as heresy. Today, Sedevacantism has tens of thousands or possibly even the hundreds of thousands of followers (Collinge, 2012). This schism between mainstream Roman Catholicism and Sedevacantism is yet another example of how religion never stops evolving, even in the modern world.

In 2015, Pope Francis again attempted to soften the stance of Roman Catholicism against Protestantism (as well as against other Christian religions) by declaring that the divisions among Christian faiths to be the work of the devil (Wooden, 2015). Admitting that some conservatives would take his comments "as controversial, or even heretical", he stated that Christian branches "must shorten the distance between us to strengthen our bonds of brotherhood." He followed that "it won't be theologians who bring about unity among us." "Theologians may help us . . . but if we hope that theologians will agree with one another, we will reach unity the day after judgement day" (Vatican archives, Phoenix, 23, May). In 2016, he extended his pluralistic message to include Muslims and Hindus (Matharu, 2016). The pluralistic, non-orthodox ideas of Pope Francis have been strongly resisted by many conservative Roman Catholic theologians and senior members of the Church, including many in the College of Cardinals, who accuse him of flirting with heresy (Brown, 2017). The history of Roman Catholicism with other early branches of Western Christianity is reviewed further in chapters 8 and 9.

9C. Calvinistic theology versus Arminian theology. The *Five points of Calvinistic Predestination* are often summarized with the Acronym TULIP. These points were summarized more than fifty years after John Calvin's death by the Synod of Dort (1618–1619) in response to the rise of Arminianism. The TULIP acronym for these points did not come into use until beginning in the early twentieth century. The TULIP points of Calvinism are as follows:

T = *Total Depravity*. Original Sin has affected all of humankind. People are so irredeemably affected by it that they are incapable of choosing between good and evil. Therefore, God must *predestine* a person's spiritual fate between salvation and damnation.

U = *Unconditional Election.* Because human beings can only choose evil, God does not base his election of someone for salvation on the acts of that person during his or her life. A person is powerless to affect their fate. God has chosen (elected) some people for salvation without any conditions, leaving others to be damned.

L = *Limited Atonement.* Jesus died only for the predestined elect. God's forgiveness is limited to those he has elected.

I = *Irresistible Grace.* When God calls his elect into salvation, they cannot resist. Although the gospel message is offered to all people, God extends a unique internal call to the elect through the Holy Spirit. Election cannot be refused because God has decreed it.

P = *Preservation of the Saints.* If you are one of the elect, you cannot lose your election to salvation. The elect are eternally secure.

The Arminian theology is in opposition to Calvinist theology. *The Five Counterpoints of Arminianism* (or the *Five Articles of Remonstrants*) are as follows:

1 = *Deprivation* (counter to Calvinist point **T**). Humans are redeemable and can be transformed by God's grace. God's prevenient grace allows humanity the freedom of will for redemption.

2 = *Conditional Election* (counter to Calvinist point **U**). God calls all of humanity to exercise their freedom of will to be righteous by his grace, as a condition of election for salvation.

3 = *Unlimited Atonement* (counter to Calvinist point **L**). Jesus died for all of humanity, and God offers the possibility of atonement for all.

4 = *Irresistible Grace* (counter to Calvinist point **I**). God's grace is freely offered to all; however, humans have the freedom to resist His grace.

5 = *Assurance and Security* (counter to Calvinist **P**). Election for salvation is dependent upon continued faithfulness and not guaranteed to a predestined elect.

9D. The story of Jim Jones and the Jonestown "massacre." The story of David Koresh and the Waco tragedy is discussed in this chapter. An abbreviated story of Jim Jones and the Jonestown massacre (also referred to as the "Jonestown mass-murder" or the "Jonestown mass-suicide" by various sources) follows.

In 1956, the Reverend Jim Jones (1931–1978) was officially ordained by the *Assemblies of God* branch of Pentecostalism. His evangelistic style and charisma led to an ever-increasing following. In 1964, he was ordained by the *Disciples of Christ* church, and then in 1965, he incorporated the *People of the Temple, Disciples of Christ Church* (Smith, 1982). Later the name of his church was shortened to the *People's Temple.* By 1972, Jones's evangelistic reach had expanded through much of northern California, and he had a half-hour radio show. In 1973, he began building an "agriculture mission" in Guyana. By 1974, at least one newspaper declared Jones' *People's Temple* to be the largest Protestant congregation in northern California (Smith, 1982). In 1975, Jones was named one of the one hundred most outstanding clergymen in the United States. He was an active member of the presidential campaign for Jimmy Carter and was invited to the inauguration at the White House (Smith, 1982). He received *Humanitarian of the Year* awards, first in 1976 from the *Los Angeles Herald*, and then in 1977 from the Martin Luther King Foundation.

In 1977, Jones and more than nine hundred of his followers moved to the agricultural mission in Guyana, which was renamed "Jonestown" (Bowker, 2006; Wade, 2009; Smith, 1982). They practiced what they called "Apostolic Socialism," resulting in a guarded compound in which no one was allowed to leave. Reports of human rights abuse began to come from Jonestown as well as hints of a planned mass suicide. In response, a delegation from the United States led by U.S. Representative Leo Ryan flew down to investigate what was going on. Jones saw this as the arrival of demonic agents. Ryan and others were gunned down and killed while attempting to board a return flight with defectors from the People's Temple. Then on November 18, 1978, Jones and his inner circle committed mass murder-suicide including Jones and more than nine hundred of his followers. Around three hundred of those who died were infants and small children that were injected with the poison against their will. Even the dogs, livestock, and fishponds were poisoned. The suicides were famously accomplished by drinking cyanide poisoned "Kool-Aid" (actually the English knock-off, *Flavor-Aid*). They became the source of the phrase "drinking the Kool-Aid," which became a vernacular expression for someone who believes in a doomed idea because of perceived high reward in return. The event was one that Jones and his followers had rehearsed for a number of times before carrying it out, referring to it as the "White Night" (Smith, 1982). A tape recording of incident was later discovered. On it, chants from Jones and some of followers are heard uttering phrases, such as "Let's get gone," "This world is not our home," and "Step to the other side." It was the collective voice of a group convinced they were on the path to paradise. It was also another example of how charismatic proselytizing figures can profoundly mold the will of followers, and still do today.

9E. Sam Harris's comment on the Armageddon Lobby. Sam Harris captures the essence of the problem with the Armageddon Lobby in his 2006 book, *Letter to a Christian Nation*, where he writes:

> "It is, therefore, not an exaggeration to say that if the city of New York were suddenly replaced by a ball of fire, some significant percentage of the American population would see a silver lining in the subsequent mushroom cloud, as it would suggest to them that the best thing that is ever going to happen was about to happen: the return of Christ. It should be blindingly obvious that beliefs of that sort will do little to help us create a durable future for ourselves—socially, economically, environmentally, or geopolitically. Imagine the consequences if any significant component of the U.S. government actually believed that the world was about to end and that its ending would be glorious. The fact that a significant percentage of the American population believes this, purely on the basis of religious dogma, should be considered a moral and intellectual emergency." (Harris, 2006:xii)

9F. **The Heaven's Gate movement** (most of this note is based on Chryssides, 2011, and Zeller, 2014). This cult was founded in 1974 by Bonnie Nettles (1927–1985) and Marshall Applewhite (1931–1997). To followers of the movement, Nettles became known as Ti, and Applewhite became known as Do. They were collectively known as the UFO Two.

Nettles was a nurse with an interest in astrology and Biblical prophecy. She had been raised as a Baptist, but she was not particularly fervent to that tradition. She authored an astrology column in a local newspaper. Applewhite was the son of a popular Presbyterian Minister and he was very religious as a child. He entered a Presbyterian seminary for a while, but later he left it, deciding instead to pursue a career in music. He became the music director of a Presbyterian church in North Carolina for a while, after which he moved between various religious music and teaching jobs. Along the way, he continued to study the Judeo-Christian bible, and he also developed a strong interest in science fiction (e.g., Robert Heinlein, Arthur C. Clarke, and UFO's). Nettles and Applewhite met in 1972, shortly after which they forged a spiritual bond that developed into the Heaven's Gate movement. Nettles had made an astrological chart for Applewhite early in their relationship suggesting that they had known each other in a previous lifetime. After that they worked closely together. They studied passages from the Old Testament and the New Testament, using it to build and defend their religious movement. They eventually became convinced that they were the two witnesses mentioned as prophets in the New Testament's Book of Revelation 11:3 (Wade, 2009). Their theology continued to develop as a mixture of Christianity, New Age, astrology, and UFO mythology. They came to believe that after their deaths, they and their followers would all be resurrected as immortal aliens on a spaceship that would take them to heaven. They acquired a number of faithful followers along the way. By 1976, the movement stopped recruiting and became more reclusive and behaviorally rigid. Applewhite banned sexual activity among members, promoting castration as a method of abstinence.

In 1985, Nettles died, leaving Applewhite to assume control as the charismatic leader of the tightly-knit movement. He had come to see himself as the current representative of Christ on earth. In March of 1997, Applewhite and his 38 active followers, between the ages of 26 and 72, committed suicide in a well-coordinated, ritualistic fashion to facilitate their resurrection, transformation, and voyage to heaven. They did so in a rented upscale villa in San Diego County, California, that they referred to as "The Monastery." They used a poisonous mixture of barbiturates and alcohol, to which they added asphyxiation under plastic bags. The dead were all found wearing armband patches reading "Heaven's Gate Away Team." It was the largest-known mass-suicide ever to have occurred in the United States. However, the participants considered what they were doing not to be suicide, but rather a religious act taking them to the "next evolutionary level." Just before they ritualistically killed themselves, they left a message on their website (heavens gate.com). In it they explained that they had timed their suicide with the approach of the Hale-Bopp comment, which they interpreted as a spiritual sign to end their time on earth. Their last update to their website stated "We are happily prepared to leave this world and go with Ti's crew."

10 BIBLICAL DEMIURGISM: A SUBGROUP OF "GNOSTICISM"

10A. The thirty Aeons of Valentinianism (Mead, 1890). The triacontad of Aeons in the Pleroma, in Valentinianism includes the Ogdad (with four male-female pairs of Aeons), the Decad (with five male-female pairs of Aeons), and the Dodecad (with six male-female pairs of Aeons). The Ogdad Aeons include *Bythos* (Depth or Profundity, and the Primal Father) and *Sige* or *Ennoia* (Silence, Idea); *Nous* (Mind) and *Aletheia* (Truth); *Logos* (Word) and *Zoe* (Life); and *Anthropos* (Man) and *Ekklesia* (Church). The Decad Aeons include *Bythios* (Deep) and *Mixis* (Mingling); *Ageratos* (Unageing) and *Henosis* (Union); *Autophues* (Self-existent) and *Hedone* (Bliss); *Akinetos* (Immovable) and *Synkrasis* (Blending); and *Monongenes* (Only-begotten) and *Makaria* (Happiness). The Dodecad Aeons include *Parakletos* (Advocate or Comforter) and *Pistis* (Faith); *Patrikos* (Paternal) and *Elpis* (Hope); *Metrikos* (Maternal) and *Agape* (Love); *Ainos* (Everlasting) and *Synesis* (Understanding); *Ekklesiastikos* (Ecclesiastical) and *Makariotes* (Happiness, Bliss); and *Theletos* (Desired, Longed-for) and *Sophia* (Wisdom). Some branches of Biblical Demiurgism have somewhat different Aeon pantheons.

10B. Contemporaneous texts on Sethianism. According Turner (2001), several treatises of the Nag Hammadi Library fit the designation "Sethian Gnostic." They are Coptic translations of earlier Greek texts. These documents are thought to have been written in the second and third centuries CE.

10C. A heresiologist's version of Marcionism. The primary source for Marcionism is a lengthy tome called *Adversus Marcionem* (meaning "Against Marcion"). It was written by the early Christian Church Father and heresiologist, Tertullian (170–230 CE). Tertullian was an early harsh critic of Biblical Demiurgism and of Marcion in particular. Tertullian was also the early author of the Christian doctrine of trinitarianism (Marian, 2012), which became formally adopted by Traditional Christianity in the fourth century CE (see the figure on page 340). He was also the father of Latin Christianity (the practice of employing Latin for liturgical rites). He also had some extremely harsh things to say about women (see chapter 12). Unlike his contemporaneous heresiologists, Irenaeus and Hippolytus, the Christian Church never canonized Tertullian as a saint, because he eventually converted to what the church determined to be a heretical sect: Montanism. Montanists (according to heresiologists) believed that a second prophet named Montanus received revelation from the Holy Spirit in the second century CE that cleared up some of the ambiguities of older Christian scripture. They also believed that the end of the world was imminent (Bowker, 2006; Armstrong, 1993). They believed that the revelations of Montanus could even supersede the authority of Jesus and Paul the Apostle. Their followers were told to lead severely ascetic and celibate lives and that martyrdom was the only sure path to God. Thus, they were enthusiastic martyrs (Bowker, 2006). Our understanding of Montanistic tradition may be flawed because it is based on heresiologist writings by those who considered Montanism a form of heresy.

11 ISLAM

11A. The relationship between Wahhabism and Saudi Arabia. The Wahhabi began what proved to be a durable alliance with the Saud family, beginning with Muhammad ibn Saud in 1747. The power of this politico-religious theocracy waxed and waned over the next 150 years, eventually leading to the proclamation of the Kingdom of Saudi Arabia in 1932. The kingdom's power was further enhanced in the 1960s and 1970s as it was financed with the rich petroleum reserves of the region. The Wahhabi assert their ultraconservative, puritanical brand of Islam in Saudi Arabia with a force of "religious police" (also called "enforcers of obedience") who monitor public and private conformity to Islam (Bowker, 2006:633). They have traditionally enforced strict codes of behavior, such as observance of Islamic ritual and gender segregation. Because Wahhabism forbids the veneration of shrines, tombs, and objects, an aspect of their religious enforcement has been the intentional destruction of archaeological records. Preservationists have estimated that as many as 95 percent of the historic sites around Mecca and Medina have been razed under Saudi rule (Tesch, 2018). The Wahhabi movement is an extreme offshoot branch from what was already a conservative branch of Sunni Islam. In spite of its early ties to the Hanbali School of law, Wahhabism has long been rejected by many Hanbali Muslims.

11B. The Coexist Foundation. An early twenty-first-century effort to increase understanding among Islamic, Jewish, and Christian faiths was the creation of the *Coexist Foundation* in 2005. The original organization, founded by Middle Eastern philanthropist Mohammed Abdul Latif Jameel (according to Cagney and Ross, 2013) had a series of directors from the United Kingdom, Saudi Arabia, Bangladesh, and Israel, until it was officially dissolved in 2016. This effort initially sought to enhance tolerance of cultural diversity and social cohesion within Abrahamic Monotheism. The original organization was not primarily concerned with religions and cultures outside of that group, although later iterations by the same name have attempted to expand its scope. There are a number of later organizations with the same or similar names whose connections with the original foundation are somewhat confusing. A number of later commercial enterprises also usurped the "Coexist" logo for a variety of enterprises, ranging from the sale of bumper stickers to rock concerts, further adding to the confusion concerning the organization's history. For discussion of broader pluralistic initiatives regarding religion (i.e., beyond Abrahamic Monotheism), see **Note 12L.**

12 ORGANIZED RELIGIONS AND THE EVOLUTION OF HUMAN SOCIETY

12A. Which came first: Organized Religion or urban civilization? With regard to the question of whether urban civilization or Organized Religion came first, anthropologists are on each side of the issue. Most propose that *urban civilization gave rise to Organized Religion*. Many of those anthropologists tie the origin of urban civilization to the rise of agriculture (farming and animal husbandry) rather than to religion (Turchin et al., 2015). The Agricultural Revolution arose independently in different parts of the world. Today, around 90 percent of the calories that humans consume come from species that our ancestors domesticated between around 9500 BCE to 3500 BCE. This includes wheat, corn, rice, potatoes, barley, millet, and more (Hariri, 2014. Various animal species were also domesticated for food during this time, including dogs, goats, and cattle. Once humans could produce and store enough food to feed those beyond their direct families and kin groups, much larger urban centers became possible. The populations of these centers increased at an exponentially greater rate than neighboring regions with only hunter-gatherer clans for food production. Most of the earliest known urban centers in the Old World appear to have developed between 9,000 and 3,500 years ago in major river valleys and mountain regions of the world, because those regions were the most conducive to farming. These included the Indus River Valley and its precursors, the Fertile Crescent of Mesopotamia and Egypt, and the major river valleys of China, including the Yellow, the Yangtze, and the Liao (chapter 4). In the New World, the oldest-known civilization was the Norte Chico civilization in the river valley and coastal regions of Peru that flourished between around 5,500 and 3,800 years ago (chapter 4). They had a well-developed system of agriculture and fishing to feed their population.

Alternatively, a few anthropologists suggest that *Organized Religion gave rise to urban civilization*. They sometimes point to the ancient archaeological site of Gobekli Tepe as evidence (chapter 3). The site is more than eleven thousand years old and indicates some of the earliest elements of what may have been Organized Religion, with an organized society large enough to build an enormous structure thought to be a temple. The structure includes hundreds of massive pillars, some of which weigh up to twenty tons each and many of which have engravings. This construction would have required a relatively sophisticated social organization to build. Nothing indicates that this civilization had yet developed large-scale agriculture or an organized social class system, and it has been suggested that it was an organized form of religion that drew people to the site to form one of the earliest examples of human civilization. This theory suggests that the development of Organized Religion helped maintain a level of communal cooperation necessary to maintain civil order. However, without written records, these hypotheses remain highly speculative.

Whichever came first, Organized Religion or political influence from within urban civilization, one thing is clear. In the words of historian Fredrick H. Russell, "religion and politics often aid each other in their respective evolutions" (1975:1). Historically, the most successful political systems have acted hand-in-hand with the most successful religious systems (and vice versa), even in the case of states claiming to be secular entities.

12B. Mary, mother of Jesus, in the Bible and in Traditional Christianity. The one woman most widely venerated in Christianity is Mary, mother of Jesus. However, her thoughts and teachings are completely absent from the New Testament. Her importance is as a mother. She was eventually portrayed as the eternal "Virgin Mother" of Jesus by the male authors of early Traditional Christianity, even though she is reported to have given birth to several children over her lifetime (*Matthew* 12:46–47; *John* 2:12). According to *Matthew* 1:18–25, Mary was pregnant when she met Joseph. It seems her female sexuality was thought of as something to be hidden by much of the church.

12C. Pope Joan. There is a legend of one female pope, later referred to as Pope Joan (Bowker, 2006). As the legend goes, around the year 1100, a woman in male disguise succeeded into the papacy. After two years, the ruse ended when she gave birth to a child. She mysteriously died immediately afterward. The tale is first attributed to a Dominican chronicler of the mid-thirteenth century named Jean Pierier of Mailly (also known as Jean de Mailly). Although it was widely believed through the Middle Ages, today it is considered by most historians to be fictitious.

12D. The wearing of the veil (*hijab*) by Islamic women. Wearing of the veil is also voluntary by some Muslim women. For example, as a part of the Iranian revolution in the late 1970s, a number of Iranian women began to don the veil once again, even though they had previously abandoned the practice (Horsley, 2003). Many women did this mainly in symbolic protest against what they saw as an invasion of Western imperial culture perpetrated by the shah, Mohammad Reza Pahlavi. The shah had made the abandonment of the hijab a central issue of Westernizing reform in Iran. He decreed it to be illegal, even ordering police to use force in removing veils from women in public. His decree became seen as an example of Western imperialism (Horsley, 2003). In response, some women began to wear it in protest. They saw it as not so different from head coverings worn by followers of other religious belief systems (e.g., the wearing of the kippah or yarmulke by Orthodox Jews, or head coverings worn by various Christian branches). However, after the overthrow of the shah, many of those women were unprepared for the way conservative *ulama* (guardians of Islamic tradition) would strictly impose the veil, even for those who did not wish to wear it. Not wearing it became a criminal offense for women, punishable by fines or imprisonment. As I write this, there is a general uprising in Iran of women against this after the "morality police" of the government arrested a woman named Mahsa Amini for wearing her hijab improperly, and she died while in their custody (al Talei, Bazoobandi, and Khorami, 2022). As with so many cultural traditions tied to religion, the hijab has a complex history. The ethical problem is not with women who willingly choose to wear the veil. The problem is with men and theocratic governments who force women to wear it against their will under the threat of extreme harm to them if they do not. The issue can also be philosophically complex. In France today, the hijab is prohibited in public schools for reasons of church/state separation, but in Sweden it is allowed in the name of religious liberty (Prothero, 2011:56).

Although the requirement of head covering for women is strictest in Islam, head coverings for women have also been elements of many other Abrahamic religions (although to a lesser degree). According to *halacha* (Jewish law based on the Talmud) married Orthodox Jewish women are expected to cover their hair when in the presence of men other than their husbands or close family. Head coverings for Catholics are prescribed in the New Testament (*Corinthians* 11:4–5), which says that a wife who prays or prophesies with her head uncovered dishonors her head. I remember well when growing up and attending Roman Catholic church services with my parents, all women were required to wear head coverings in church. Head coverings for women are also prescribed for Oriental Orthodoxy, Eastern Orthodoxy, and even certain branches of Reformation Christianity (e.g., the Amish).

12E. Exceptions to the dominance of men in Organized Religion. Homosexual men are a notable exception to male dominance in some Organized Religions. Although homosexuality was a common element of Greek and Roman Polytheism (Wade, 2009) and it was accepted as a third nature of human sexuality in the Kama-sutra text of Hinduism (Doniger, 2009), it was never a dominant element of those religions. With the rise of Abrahamic Monotheism, an antihomosexual doctrinal bias materialized. In the early development of its monotheistic branches, the penalty for homosexuality was death by stoning (as proscribed in the Torah and Old Testament; *Leviticus* 20:13). Although tolerance for homosexuality has generally improved since those times, the prejudice continues in fundamentalist branches. A number of Muslim countries that have Sharia-based criminal laws, still consider homosexuality to be a capital offense, although most Muslim countries no longer enforce the death penalty.

12F. Religion's role in the early development of literacy and higher education. The development of literacy was integrally linked to the spread of major Organized Religions (Sharot, 2001). The earliest preserved writings east of the Indus River are those from the Shang dynasty of ancient China in the second millennium BCE. They consist of questions written on cattle bones and turtle shells submitted to the gods (Wright, 2009). The earliest-known writing system from Western Eurasia, birthplace of Western civilization, was from Mesopotamia on clay tablets from the fourth and third millennia BCE. These included an ode to the religious temple and to the gods represented within it (discussed here in chapter 5).

The early expansion of literacy among general populations of the world was often advanced by the need to read scripture (e.g., Torah, Old Testament, New Testament). Historians Botticini and Eckstein (2012:2), for example, write that Second Temple Judaism was "a religion whose main norm required every Jewish man to read and to study the Torah in Hebrew and to send his sons from the age of 6 or 7 to primary school or synagogue to learn to do so." The Anglican King James Bible of the early seventeenth century fueled the early expansion of literacy in the English-speaking world, just as Martin Luther's sixteenth-century translation of the Bible fueled the advance of literacy in the German-speaking world (chapter 9).

The early development of a higher education system was another thing linked to the development of Organized Religion. The priests of ancient Egyptian and Babylonian Polytheism established higher learning opportunities for the elite as early as the second millennium BCE (Rudy, 1984). The growth of Christianity in the Eastern Roman Empire, as well as the early spread of Islam, in the mid- to late first millennium CE, played important roles in the diffusion of higher learning, although at first it was still informal and focused primarily on the privileged class. In the Middle Ages, the earliest formalized universities with regular teaching staff, regular courses, and certificates of accomplishment arose. These early institutions are referred to by historians as Medieval universities. These developed from earlier church agencies such as monasteries and cathedrals. They were supported by various members of the papacy or Holy Roman Empire. In the United States, the training of Christian clergy was a founding mission for many of its oldest universities (e.g., Harvard, Yale, Princeton, Dartmouth, Rutgers, and the College of William and Mary).

12G. Population growth and Organized Religion in the modern world. Efforts toward a more broadly altruistic society must include a more responsible approach to the complex global issue of population growth in the modern world. During the years when Christianity was first developing, overpopulation was less problematic. The push to procreate was even strategic. The planet still had plenty of room for more bodies, or at least the human species had not yet exceeded the earth's carrying capacity (the maximum population size that the environment can sustain indefinitely). The Roman Catholic doctrine that maximized population growth in its earliest days probably helped it overtake many competing religions over time (chapter 8). Today, Roman Catholicism contains half of all the world's Christians. However, even now in the twenty-first century, the Roman Catholic Church has never revised its restriction against voluntary contraception. This attitude shows a certain level of indifference toward the sufferings and challenges of the less fortunate segments of human society. The Roman Catholic pressure to defund public support of contraception affects primarily the poor. As former Anglican priest and professor of religious studies, Tom Harpur, suggests, the Roman Catholic leadership's attitude about this fosters a belief that "suffering is good for you, that you must resign yourself and, above all, comply and conform" (Harpur, 2004:46). In practice, many or most who today claim their religion to be Roman Catholicism ignore the Church ruling against birth control (Lacey and Oakley, 2011).

The attitude of the Roman Catholic ministry against any form of birth control is partly an artifact of ancient principles meant for an ancient time when far fewer people were on the planet. However, in the twenty-first century the human population is climbing above global carrying capacity (Ehrlich and Ehrlich, 2004). With a population tripling in size since 1950, surpassing eight billion people in 2022, and anticipated to reach ten billion by 2055 (Cilluffo and Ruiz, 2019), the default control against overpopulation has primarily become war, genocide, starvation, and disease. What God would want those to be the only forms of population control? Christian fundamentalists often respond, "You must have faith," or "The future is in God's hands." That response is yet another instance of blind faith in institutionalized church dogma being used to justify cultural and ethical irresponsibility. Faith in a church ministry should not be used as a catchall excuse for irresponsible action or inaction on complex social issues. There is even a self-defeating logic in the moral argument of Christian fundamentalists. According to several studies (e.g., Miller and Valente, 2016) making access to contraception more difficult or expensive adds significantly to the level of abortions, presumably a worse alternative in the eyes of Christian fundamentalists. Finally, I note that the supposedly celibate Roman Catholic clergy, as a whole, has little credibility in dictating sexual norms for the laity. This has become increasingly clear with the numerous and widespread reports of sexual abuse of minors by Catholic priests, nuns, bishops, and even cardinals over the last several decades, as well as the extensive early church efforts to cover up those abuses (Blasi and Oviedo, 2020; Hidalgo, 2007; Renzetti and Yocum, 2013).

12H. The difficult philosophical question of proselytism. Proselytizing efforts have sometimes amounted to little more than cultural colonialization. Some charitable aspects of missionary work that support societies in need of medical, educational, or other critical resources, are admirable and serve a noble purpose. However, overzealous and coercive proselytism poses ethical and philosophical questions best summed up with the following quote by American author Annie Dillard: "An Inuit hunter asked a local missionary priest: "'If I did not know about [the Christian] God and sin, would I go to hell?' 'No' said the priest, 'not if you did not know'. 'Then why' asked the Inuit earnestly, 'did you tell me?'"(Dillard, 1974:123). It is hard to justify pressuring a stable culture whose beliefs are peaceful and nonintrusive to convert to something foreign and less compatible with their own cultural environment. This is made worse by intimidating threats of eternal suffering for noncompliance and with violently coerced proselytism. Historically, missionaries and soldiers of religious conquest sometimes attempted to justify their proselytizing efforts as being for the target culture's "own good," even though the invaders may not have understood the cultural context of the indigenous society they were trying to forcefully change. In extreme cases, such as in the sixteenth- through nineteenth-century Americas, this eventually resulted in colonialist acts of aggression ranging from cultural ethnocide to genocide (chapters 3 and 5).

12I. Seshat. *The Seshat: Global History Databank* is an international scientific research project founded in 2011 by University of Connecticut evolutionary biologist Peter Turchin and colleagues. It gathers data on many aspects of cultural evolution, including religion, governance, social competition, economics, warfare, language, and technology. It combines this information into a single large database in the hope of deciphering how modern societies evolved from ancestral ones and to identify the sort of long-term dynamics that affect the course of human history (Turchin et al. 2015).

12J. The case of communism and atheism. Some authors bring up "communism" as an example of violence in the name of atheism, particularly with regard to governments such as the old Soviet Union (McGrath, 2005). However, it was not atheism that drove this violence. People were not burned at the stake, drowned, or tortured for failing to adhere to atheistic principles. They may have been killed for broader political reasons or ethnic prejudice, but not exclusively for having religious beliefs. Atheism is not a problem any more than theism is, in this regard. The Communist State's legislative campaign against religion in the Soviet Union (which never made religion a capital offense) was only a small part of a much larger effort by the Soviet State to exert comprehensive political control over its people. The Soviet Union and Leninism may have been a hostile, oppressive, and cruel government in many ways, but atheism was never the driver of that violence. The institutionalized abuse of an ideology does not necessarily discredit the ideology itself (McGrath, 2005).

Karl Marx (1818–1883) is often cited as a founding father of Communist atheism. It is true that Marx is famous for calling religion "the opiate of the masses." However, he saw both a good side and a bad side to religion. On the bad side, he saw religion as a symptom of sickness in society. On the good side, he saw religion as giving people the hope of a life after death that was free of suffering. Marx believed that the need for religion would disappear on its own once society improved and matured sufficiently (Molloy, 2018). Neo-Marxists see most religions as masking political and economic inequality and, in effect, allowing leaders and privileged members of society to manipulate people into accepting class divisions (Stanish, 2017). Similarly, Chinese Communism is also blamed for vast persecution of religious believers in East Asia. Admittedly, Buddhism and other religions were severely repressed when Communist Party governments took over a number of countries there, including Mongolia in 1921, China in 1949, and Tibet in 1959. The Dalai Lama went into exile, and at least a million Tibetans were thought to have died in the ensuing persecution of the 1950s and 1960s (Molloy, 2018). This was largely influenced by *Maoism* becoming so reverently portrayed that it almost became a type of religion itself. Mao's quotations were often treated as holy writ. When Mao's successor, Deng Xiaping, took over in 1978, religious persecution was relaxed and five major religions were officially endorsed: Buddhism, Taoism, Roman Catholicism, Protestantism, and parts of Islam (Prothero, 2011).

Today, the People's Republic of China officially claims to protect freedom of certain branches of religion (also mentioned **Note 4K**), although it severely suppresses and rigidly controls some of the other branches, such as the Islamic tradition of the Uighur people in Xinjiang Province. There, according to the Council on Foreign Relations, more than a million Muslims are detained in "reeducation camps" because the Chinese government has come to fear them as the core of an extremist separatist movement (Maizland, 2020). Chinese Communist Party officials claim that there have been a number of terrorist attacks in China by Uighur Muslims and that the reeducation camps have prevented further terrorist violence. Critics of the Chinese policy claim it amounts to nothing short of genocide. Whatever the case, it is not atheism that is driving this inhumane treatment, it is political fear of cultural diversity, and efforts by people in government to control which beliefs survive and which ones do not.

12K. The *Paradox of tolerance*. One of the more complex philosophical issues concerning the issue of tolerance is how to pragmatically and strategically increase it in a diverse, interconnected world. The famous Austrian-British philosopher, Karl Popper (1902–1994), was a strong advocate of tolerance as a general humanistic trait. However, he pointed out a problem with "unlimited tolerance" using what he called the *Paradox of Toleration* (Popper, 1945). That paradox reasoned that unlimited tolerance could ultimately lead to the *disappearance* of tolerance. If unlimited tolerance was extended even to the violently intolerant, and if the tolerant were unwilling to defend a tolerant society against an onslaught of intolerance, then the tolerant and the climate of tolerance would ultimately be destroyed. Therefore, Popper contended we should claim the right to suppress intolerant actions that infringe on the peaceful liberties and safety of other people. He reasoned that this would be the only way to maximize and maintain an environment of tolerance. With regard to religious beliefs, we should suppress those who use violence to forcefully proselytize an exclusive ideology. Where to draw the line against a particular form of intolerance is consequently a complex and potentially subjective issue. It is an issue that could be addressed by a broadly collaborative world council (e.g., see **Note 12L**). Ideally, tolerance should require no lack of commitment to one's peaceful beliefs, but also should support the rights of others with different peaceful beliefs. Perhaps then, the net effect of religions could be a more positive one, helping to break through geopolitical, nationalistic barriers. Even countries in political conflict often share common religions that could provide international connections for compromise and collaboration.

12L. The challenges and prospects for an effective world council of religions. There has been, and still is, much resistance to religious pluralism, because of various tribalistic factors including nationalism, religious fundamentalism, and various other cultural differences. Noted psychologist, Sigmund Freud, once noted, "Cruelty and intolerance to those who do not belong to it are natural to every religion" (Freud, 1922:98). Formal efforts to build tolerance of religious diversity over the last century and a half have been met with much resistance. Efforts toward pluralism have often been met with violent hostility (e.g., Gandhi's fate discussed in chapters 4 and 12). The current Pope of Roman Catholicism, Pope Francis, recently made efforts to develop a more pluralistic climate within Christianity (e.g., Catholicism, Lutherans, Oriental Orthodoxy) as well as including certain non-Christian religions (Muslims and Hindus), stating "We have different cultures and religions, but we are brothers and we want to live in peace" (Wooden, 2015; Matharu, 2016). However (as mentioned in Note 9B) his pluralistic agenda, along with his other progressive ideas, were met with hostility from many conservative senior members of the Roman Catholic Church who accused him of flirting with heresy (Brown 2017).

Although pluralistic initiatives have made seemingly little progress against religious-based conflict around the world, efforts continue even by a few relatively small religious movements (e.g., the UUA church mentioned in chapter 9, and the Bahai movement mentioned in chapter 11). In addition, a few other limited efforts continue, such as the original Coexist Foundation (see **Note 11B**) or a Christian-Muslim effort that developed in Lagos, Nigeria, in the 1970s that goes by the name of *Chrislam* (Marloes, 2016). Building a more effective climate of tolerance may require some sort of inclusive world council or parliament that is more than simply a new "universal religion." It may require a forum for *different* religions to find ways to peacefully coexist and negotiate common ground. Its diverse membership would try to better understand each other's cultural context and differences, as well as common interests, to collectively achieve common goals of global significance. This would admittedly be a significant challenge.

An early attempt to form a world council of religions was in 1893, at the World's Fair in Chicago. It was called the *World's Parliament of Religions* (Seager, 1995), and it set out to bring the major religions of the world together to find ethical elements in common. It set out to discuss religious diversity without proselytism and included representatives from Hinduism, Jainism, Buddhism, Confucianism, Shintoism, Rabbinic Judaism, Traditional Christianity, Reform Christianity, and Islam. The group's attempt to establish a World's Council of Religions was coolly received (Bowker, 2006), and it was often ideologically competitive, favoring some faiths over others (Seager, 1995). Christian Protestants dominated the number of speakers and attendees. However, the event did establish that interest was growing in expanding the breadth of religious pluralism. The council reconvened again one hundred years later, once again in Chicago. By that time, other supernatural belief systems had been added, such as Sikhism and certain sects of animism. Around eight thousand people from around the world attended. This time, a Roman Catholic cardinal was there, and the keynote address was given by Mahayana Buddhism's fourteenth Dalai Lama. The group agreed to meet more frequently after that. They met in 1999 in South Africa, 2004 in Spain, 2009 in Australia, 2014 in Utah, 2018 in Canada, and 2021 online because of the Covid-19 pandemic. What has yet been lacking is sufficient support from, and broad participation by, ruling national powers. Nationalism is far too often exclusively entwined with specific religions, rather than supporting the idea of pluralism. Historically, the easiest route for ruling entities to build patriotic support from their subjects has often been to selectively combine elements of a particular religion with nationalistic objectives (e.g., see the "Second Point" in chapter 12, page 494).

Another attempt to form a world council of religion was a little-known global grassroots interreligious network called the *United Religions Initiative* (URI), founded by a retired Episcopal bishop named William E. Swing. A formal charter for the movement was signed in 2000 (URI, 2000). In it, the organizers pledged that the movement would be a bridge-building organization between the major existing religions, rather than creating a new religion. The URI charter rejected forceful proselytization of any particular religion. It stated that the purpose of the movement was "to end religiously motivated violence and to create cultures of peace, justice and healing for the Earth and all living beings." It pledged to act within sound ecological practices to protect and preserve the earth for both present and future generations. Several prominent religious leaders have endorsed the URI goal of increased religious tolerance, including the Roman Catholic Pope Francis (United Nations News, 2021). The group even holds consultative status with the United Nations Economic and Social Council (ECOSOC). However, some Christian fundamentalists resist the URI. Using fear as a tactic to support evangelistic selectivity, they claim pluralistic initiatives are paving the way for the Antichrist. Some Christian fundamentalists warn that "the only time that the world should be united is when Christ returns to reign" (Penn, 2004:213). Such strongly evangelistic messaging impedes any broad advancement of religious pluralism.

Another challenge to pluralism is that some elements of fundamentalism and religious tribalism are broadly and deeply ingrained. For example, a recent survey conducted by the Pew Research Center suggested that the majority of people in Egypt (who are mostly Sunni Muslims) believe that converting from Islam to a different religion should be punishable by death (Sahgal and Grim, 2013). One wonders how such a violent sentiment became so broadly disseminated throughout that country. Such deeply seated fundamentalist issues will be a challenge for any major pluralistic initiatives on a global level. Also, there is the question of where to draw the line regarding toleration (see the *Paradox of Toleration* in **Note 12K**).

I do not claim to have a detailed plan for building a better climate of tolerance, nor do I know precisely how to form an effective global council of religions to facilitate such an atmosphere. That is the collective responsibility of ethical religious authorities and world leaders. However, there must be better ways of dealing with today's global problems than religious aggression, ideological chauvinism, or waiting for the biblical "end of times." Efforts toward building a successful council or parliament of world religions, with equitable representation and participation from all major religions, will hopefully continue for the sake of addressing one of humanities most deeply ingrained challenges: how to peacefully coexist in an environment of unavoidable cultural diversity and continual diversification.

CHAPTER CITATIONS

PREFACE

1 Neville, 2018; 2 Smith, 1998:269; 3 Wilson, 1998:294.

CHAPTER 1

1 Lincoln, 2018; 2 Dunbar, 2022; 3 Freiberger, 2019; 4 Grande and Rieppel, 1994; 5 Feinman, 2000; 6 Johnson and Grim, 2013:41; 7 Winzeler, 2008; 8 de Waal, 2013; 9 Allan, 2005; 10 Bowker, 2006; 11 Wade, 2009; 12 Pew Research Center, 2014; 13 Alper, 2008; 14 Hamer, 2004; 15 Wallace, 1966; 16 Wilson, 1998; 17 Nelson and Platnick, 1981; 18 Smith, 2004:175; 19 Linnaeus, 1735; 20 Darwin, 1859; 21 Futuyma, 2013:22, G-7; 22 ICZN, 1999; 23 Smith, 2004:175; 24 Prothero, 2011:xxxiii; 25 Antonio, 2017; 26 Martin, 2008; 27 Boyarin, 2017; 28 Talal, 1993; 29 Brittain, 2005; 30Smith, 1982:xi; 31 Muller, 1873:68; 32 Harris, 1968; 33 Josephson-Storm, 2017; 34 Mishler, 2021; 35 Richards, 2010; 36 Singh, 2012; 37 Phalet et al., 2013; 38 Nicholson, 2013; 39 Roscoe, 2006; 40 Saler, 2001:268; 41 Sharot, 2001:10; 42 Mora et al., 2011; 43 Smith, 1982; 44 Harvey, 2006:xii; 45 Eldredge, 2005.

CHAPTER 2

1 Terrell and Hart, 2002; 2 Molloy, 2018; 3 Feinman, 2000:8; 4 Asad, 1993; 5 Grande, 1994; 6 Grande, 1985; 7 Hofstadter, 1955; 8 Futuyma, 2013; 9 Darwin, 1859; 10 Darwin, 1871; 11 Crook, 2007; 12 Haeckel, 1883; 13 Archibald, 2014:203–206; 14 Jackson and Weidman, 2004; 15 Levit and Hossfeld, 2020; 16 Gould, 1977; 17 Eldredge, 2005; 18 Spiro, 2009:357; 19 Anthony, 2007; 20 Smith, 2004; 21 Freiberger, 2019; 22 Lincoln, 2018:4; 23 Smith, 1971; 24 Doniger, 1996:532; 25 Gould, 2011; 26 Merriam-Webster, 2012; 27 Roscoe, 2006;

28 Eliade, 1978; 29 Eliade, 1982; 30 Eliade, 1985; 31 Paden, 2016; 32 Peterson et al., 2019; 33 Popper, 1959; 34 Doniger, 2000:66; 35 Richerson and Henrich, 2012; 36 Turchin, 2009; 37 Wilson, 2003; 38 Grande, 2017a; 39 Grantham, 2004; 40 Parkinson, 2017; 41 Eldredge and Gould, 1972; 42 Gnuse, 1997; 43 Richerson and Boyd, 2005:237.; 44 Cliquet and Avramov, 2018.

CHAPTER 3

1 Harvey, 2006; 2 Harrod, 2014; 3 Smith, 1998:269; 4 Smith, 1982:xi; 5 Tylor, 1871; 6 Rossano, 2006:347; 7 Dunbar, 2022:xv; 8 Smith, 1962:131; 9 Williams and Nash, 2016:133; 10 Stanish, 2017:88; 11 Nickerson, 1998; 12 Baumeister and Vohs, 2007; 13 Smith, 2004; 14 Wade, 2009; 15 Schopf, 2002; 16 Futuyma, 2013; 17 Clark, 2010; 18 Goodall, 1986; 19 Nishida et al., 1999; 20 Joordens et al., 2015; 21 Harari, 2014; 22 Brumm et al., 2021; 23 Aubert et al., 2018; 24 Henshilwood et al., 2018; 25 Armstrong, 1993; 26 Frick and Lautenshlager, 2007; 27 Molloy, 2018:12; 28 Durkheim, 1915 [1964]:62; 29 Winzeler, 2008; 30 Mounier and Lahr, 2019; 31 Cliquet and Avramov, 2018; 32 Clark et al., 2003; 33 Trinkaus et al., 2014; 34 Saliba, 1976; 35 Neville, 2018; 36 Callaway, 2017; 36 Freiberger, 2019:1; 37 Bowker, 2006; 38 Ingold, 2000; 39 Willerslev and Ulturgasheva, 2012; 40 Hoppal, 1987; 41 Rasmussen, 1921; 42 Gostin and Chong, 1954; 43 Taylor, 2009:10; 44 Kind et al., 2014; 45 Spencer, 1898:440; 46 Pettitt, 2011; 47 Mayer, Vandermeersch, and Bar-Yosef, 2009; 48 Lieberman, 1991; 49 Yang, An, and Turner, 2005; 50 Steadman, Palmer, and Tilley, 1996; 51 Stearns, 2005; 52 Beck, 2000; 53 Benigni, 2013; 54 Adovasio, Soffer, and Page, 2007; 55 Singh, 2008; 56 Rountree, 2001; 57 Culotta, 2009; 58 Motz, 1997; 59 James and Dillon, 2012; 60 Spretnak, 2011; 61 Chrystal, 1908; 62 Jones and Palffy, 2013; 63 Wright, 2009; 64 Schmidt, 2010:quote on page 256; 65 Peters and Schmidt, 2004; 66 Gresky, Haelm, and Clare, 2017; 67 Feinman, 2016; 68 Gray and Watts, 2017; 69 Norenzayan and Gervais, 2012; 70 Norenzayan and Sharif, 2008; 71 Turchin et al., 2022; 72 Norenzayan, 2013; 73 Doniger, 2009.

CHAPTER 4

1 Hiltebeitel, 2013; 2 Armstrong, 1993; 3 Johnson and Grim, 2013; 4 Morgan, 2001; 5 Doniger, 2009; 6 Singh, n.d.; 7 Shrama, 2001; 8 Dundas, 2002; 9 Bowker, 2006; 10 Molloy, 2018; 11 Possehl, 2002; 12 Anthony, 2007; 13 Feuerstein, Kak, and Frawley, 1995; 14 Kenoyer et al., 1983; 15 Chattopadhyaya, 2016; 16 Wesler, 2012; 17 Kramer, 2021; 18 Srinivasan, 1975; 19 Marshall, 1931; 20 Flood, 1996; 21 Puri, 2014; 22 Gethin, 1998; 23 Fenech and McLeod, 2014; 24 Poussin, 1917; 25 Mallinson and Singleton, 2017; 26 Jacobsen, 2012 ; 27 Cort, 2001; 28 Dandekar, 2005; 29 Johnson, 2009; 30 Salguero, 2017; 31 Srinivasan, 2011; 32 Macdonell, 1995; 33 Reddy, 1999; 34 Jacobs, 2010; 35 Radhakrishnan and Moore, 1957; 36 Lipner, 1994; 37 Bloch, Keppens, and Hegde, 2010; 38 Doniger, 2017; 39 Bailey, 2017; 40 Redden, 2016; 41 Prothero, 2011; 42 Jones and Palffy, 2013; 43 Griffiths, 1996; 44 Bayly, 1999; 45 Boesch, 2002; 46 *National Geographic*, 2003; 47 Smita, 1999; 48 Muller, 1867:300; 49 Singh, 2004:44; 50 Gandhi, 1949; 51 Kamath, 2007:195; 52 Godse, 1948; 53 Chakravarti, 1994; 54 Marshall, 1931; 55 Woodard, 2010; 56 Doniger, 1980; 57 Flood, 1996; 58 Lorenzen,

2005; 59 Melton and Bauman, 2010; 60 Atapur, 2002; 61 Lang, 2009; 62 Keyt, 1960; 63 Krishan, 1986; 64 Zimmer, 1969; 65 Yang, 1961:27; 66 Babb, 2015; 67 Flügel, 2006; 68 Doniger, 2009; 69 Sangave, 2001; 70 Jacobi, 1884; 71 von Glasenapp, 1925:16; 72 Uttal, 2004; 73 Sangave, 1980; 74 Singh, 2008; 75 Jain, 1975; 76 Flügel, 2005; 77 Salguero, 2020; 78 Salguero, 2022; 79 Damascene, 1914 [1967]; 80 Buswell and Lopez, 2014; 81 Eliade, 1982; 82 Laumakis, 2008; 83 Harrison, 2014; 84 Boyer, 2001; 85 Fowler, 1999; 86 Nikam and McKeon, 1959; 87 Nakamura, 1980; 88 Sharot, 2001; 89 Baruah, 2000; 90 Shizuka, 1997; 91 Van Norden and Jones, 2019; 92 Harvey, 2013; 93 Bareau, 1955; 94 Hirakawa, 1990; 95 Williams, 2008; 96 Harrison, 2018; 97 Akira, 1993; 98 Xing, 2004:66; 99 Liu, 2010; 100 Drewes, 2015; 101 Fowler, 2005; 102 Ford, 2017:16, 267; 103 Prentiss, 2014; 104 Mandair, 2013; 105 Puri, 2003; 106 McLeod, 2009; 107 Nirbhai, 1990; 108 Singh and Fenech, 2014; 109 Oberoi, 1994; 110 Chen, 2013:127; 111 Min, 2012; 112 Goldin, 2005; 113 Sybesma, 2015; 114 Juergensmeyer, 2005; 115 Robinet, 1997; 116 Jha, 2014; 117 Keightley, 2004; 118 Wang, 2012; 119 Markham and Ruparell, 2001; 120 Tsu, 1972; 121 Yao, 2000; 122 Lee, 2017; 123 Partin, 2005; 124 Kirkland, 2004; 125 British Taoist Association, n.d.; 126 Eskildsen, 2004; 127 Bing, 2012; 128 Chan, 2005; 129 Li, 2008; 130 Littlejohn, 2010; 131 Goldin, 2011; 132 Hofstede and Bond, 1988; 133 Nobutako et al., 2003; 134 Toshio, 1981; 135 Breen and Teeuwen, 2010; 136 Hardacre, 2017; 137 Philippi, 1969; 138 Boyd and Williams, 2005; 139 Ashkenazi, 2003; 140 Philippi, 1969; 141 Aston, 1956; 142 Holt, 2000.

CHAPTER 5

1 Callaway, 2017; 2 Bourgeon, 2021; 3 Wright, 2009; 4 Tylor, 1871; 5 Uttal, 2004:138; 6 Davies, 2011; 7 Brown, 1999:625; 8 Boyer, 2001; 9 Parrinder, 1967; 10 Lewis-Williams, 1995; 11 Horton, 1993:161; 12 Antonio, 2017; 13 Lovejoy, 2012; 14 Bowker, 2006; 15 Love, 2012; 16 Molloy, 2018; 17 Prothero, 2011; 18 Abimbola, 2006; 19 Alana, 2004; 20 Mbiti, 1969; 21 Castañeda, 2007; 22 Desmangles, 1992; 23 Childs, 2003; 24 Hall, 1997:169; 25 Howorth, 1880; 26 Forsyth, 1991; 27 Doniger, 2009:689; 28 Anthony, 2007; 29 Mallory and Adams, 2006; 30 Gray and Atkinson, 2003; 31 Ryder and Nicholls, 2011; 32 Haak et al., 2015; 33 Day, 2000; 34 Commarosano, 2018; 35 Eliade, 1982; 36 Hutton, 1991; 37 Ross, 1974; 38 MacNullen, 1986; 39 Berger, 1985; 40 Schultz, 2000; 41 MacKillop, 2004; 42 Rhys, 1888; 43 Hull, 1898; 44 Bede, 731 [1990]; 45 Cunliffe, 2008; 46 Wood, 2000; 47 Cunliffe, 1997:309; 48 DeWitt, 1938; 49 Rives, 1995; 50 Monaghan, 2004; 51 Lindow, 2001; 52 Abram, 2011; 53 Andrén, 2011; 54 Andrén, 2014; 55 Sturluson, 1220 [1987]; 56 Andren et al., 2006; 57 Dubois, 1999; 58 Acker and Larrington, 2015; 59 Turville-Petre, 1975; 60 Winroth, 2014; 61 Black and Green, 1992; 62 Bottero, 2001; 63 Jacobson, 1976; 64 Kramer, 1963; 65 Black, Cunningham, and Robson, 2006; 66 Collins, 2021; 67 Kramer, 1956; 68 Gnuse, 1997; 69 Collins, 2007; 70 Smith, 2002; 71 George, 1999; 72 Choksi, 2014; 73 Hoffman, 1979; 74 Wilkinson, 2003; 75 Raof, 1990; 76 Frankfort, 2000; 77 Hart, 1997; 78 Allen, 2015; 79 Goelet et al., 2015; 80 Hart, 2005; 81 Redford, 2001; 82 Taylor, 2001; 83 Van Dijk, 2003; 84 Baines and Malek, 2000; 85 Morford, Lenardon, and Sham, 2011; 86 Bengtson, 2009; 87 Kelder, 2010; 88 Evans, 1921; 89 Evans, 1925; 90 Garland, 2013; 91 Walbank, 1982; 92 Durant, 1939:23; 93 Lattimore, 1959; 94 March, 2014; 95 Ahuvia, 2012; 96 Burkert, 1985; 97 Puhvel, 1987;

98 Graves, 1960; 99 Rüpke, 2007; 100 Stoddart, 2009; 101 Cazanove, 2007; 102 Stone, 2019; 103 Futuyma, 2013; 104 O'Brien et al., 2014; 105 Morritt, 2011; 106 Terberger, Zhilin, and Savchenko, 2021; 107 Zhilin et al., 2018; 108 Ravenstein, 1861; 109 Halpin, 1981; 110 Willerslev and Ulturgasheva, 2012; 111 Willerslev, 2007; 112 Howorth, 1880; 113 Forsyth, 1994; 114 Raff and Bolnick, 2014; 115 Krasinski and Yesner, 2008; 116 Bennett et al., 2021; 117 Rasmussen et al., 2010; 118 Stern, 2013; 119 Mann, 2013; 120 Callaway, 2015; 121 Hagan, 2021; 122 Sullivan, 2002; 123 Sahagun, 1540–1585 [1950–1982]; 124 Gray and Watts, 2017; 125 Carrasco, 2002; 126 Wiley et al., 2016; 127 Christenson, 2007; 128 Jones and Palffy, 2013; 129 Pool, 2007; 130 Spores and Balkansky, 2013; 131 Diehl, 2004; 132 Martinez et al., 2006; 133 Coe, 1968; 134 Belli, 2010; 135 Markman and Markman, 1989; 136 Thompson, 1970; 137 Garza, 2002; 138 Joyce et al., 1991; 139 Sharer and Traxler, 2006; 140 Bergh, 2012; 141 Blanton et al., 1993; 142 Lind, 2015; 143 Florescano, 1999; 144 Terraciano, 2001; 145 Jansen and Jiminez, 2011; 146 Phillips, 2015; 147 Sweeny, 2008; 148 Brinton, 1887; 149 Hassig, 2003; 150 Harner, 1977; 151 Wade, 2018; 152 Smith, Wharton, and Olson, 2003; 153 Restall, 2003; 154 Ricard, 1966; 155 Mann, 2005b; 156 Diamond, 1997:37; 157 Wright, 2013; 158 Lau, 2008; 159 Stanish, 2017; 160 Shady, 2014; 161 Shady Solis, Haas, and Creamer, 2001; 162 Haas, Creamer, and Ruiz , 2004; 163 Mann, 2005a; 164 McEwan, 2006; 165 Haas, Creamer, and Ruiz, 2003; 166 Bischof, 2008:111; 167 Burger, 2008; 168 Moseley, 2001; 169 Silverman and Proulx, 2002; 170 Proulx, 2008; 171 Towsend, 1985; 172 Quilter, 2011; 173 Popson, 1982; 174 Chavez, 1988; 175 Keys, 1999; 176 Kolata, 1993; 177 Stanish, 2003; 178 Protzen and Nair, 2000; 179 Isbell, 2008; 180 Zuidema, 2002; 181 Janusek, 2004; 182 Williams and Nash, 2016; 183 Verano, 2008; 184 Rossano, 2006; 185 Schreiber, 2005; 186 Moore to and Mackey, 2008; 187 Romey, 2018; 188 Davies, 2007; 189 Covey, 2008; 190 Toohey, 2013; 191 Urton, 1999; 192 Kubler, 1945: (quote on page 422); 193 Diamond, 1997:69–71; 194 Raghavan, 2014.

CHAPTER 6

1 Hart, 2005; 2 Wright, 2009; 3 Reeves, 2001; 4 Teeter, 2011; 5 Harpur, 2004; 6 Bowker, 2006; 7 Boyce, 1979; 8 Nigosian, 1993; 9 Rose, 2011; 10 Uttal, 2004; 11 Molloy, 2018; 12 Harpur, 2004:329; 13 Skjaervo, 2011; 14 Hintze, 2014; 15 Martin, 2005; 16 Boyce, 1984; 17 MacCulloch, 2010; 18 Randeria, 1993; 19 Modi, 1928; 20 Boyer, 2001:213; 21 Srivastave, 2001; 22 Johnson and Grim, 2013; 23 Gnuse, 1997; 24 Doniger, 2009; 25 Smith, 2001; 26 Prothero, 2011; 27 Kramer, 1963; 28Armstrong, 1993; 29 Dever, 2003; 30 Smith, 2002; 31 Day, 2000; 32 Betz, 2000; 33 Wade, 2009; 34 Brown, 2017; 35 Tigay, 1982; 36 Levenson, 2012; 37 McNutt, 1999; 38 Neusner, 2004; 39 Freud, 1939; 40 Wright, 2009:431.; 41 Nelson-Pallmeyer, 2003.

CHAPTER 7

1 Johnson and Grim, 2013; 2 Kramer, 1963; 3 Israel Central Bureau of Statistics, 2010; 4 Boyarin, 1994:244; 5 Dever, 2003; 6 Wade, 2009; 7Armstrong, 1993; 8 Wright, 2009; 9 Smith, 2002; 10 Molloy, 2018; 11 Day, 2000; 12 Rogerson, 2003; 13 Gnuse, 1997; 14 Brown,

2017; 15 Pummer, 2010; 16 Grabbe, 2010; 17 Bowker, 2006; 18 Wigoder, Skolnik, and Himelstein, 2002; 19 Neusner, 2004; 20 Betz, 2000; 21 Nelson-Pallmeyer, 2003; 22 Berquist, 2007; 23 Prothero, 2011; 24 Uttal, 2004; 25 Dawkins, 2006:269; 26 Peters, 2009; 27 Shulamit, 1994; 28 Horsley, 2003; 29 Botticini and Eckstein, 2012; 30 Neusner and Avery-Peck, 2004; 31 Greenspahn, 2011; 32 Flavius, 1800; 33 Omer, 2015; 34 Freidson, 2016; 35 Kaplan, 1995; 36 Chazan, 1997; 37 Levy, 1981; 38 MacCulloch, 2010; 39 Riley-Smith, 1991; 40 Reston, 2001:xiii; 41 Ben-Sasson et al., 2007; 42 Shirer, 1960; 43 Hitler, 1999:171; 44 Hitler, 1999:65; 45 Isabella I and Ferdinand II, 1492; 46 Angel, 1991; 47 Lowensten, 1994; 48 Dundas, 2002; 49 Planning and Construction Law 5655-1965 of the Israeli Knesset; 50 Wine, 1995; 51 Kaplan, 1934; 52 Harpur, 2004; 53 Bond, 2012; 54 Spong, 2007; 55 Case, 1914; 56 Marian, 2012; 57 Pritz, 1988; 58 Werblowsky and Wigoder, 1997; 59 Eliade, 1982; 60 Marty, 1959; 61 Ehrman, 2014; 62 Burbank and Cooper, 2010; 63 Twain, 1935:328; 64 https://www.guttmacher.org/article/2012/02/guttmacher-statistic-catholic-womens-contraceptive-use [accessed 12/22]; 65 Pummer, 2015.

CHAPTER 8

1 MacCulloch, 2010; 2 Ehrman, 2003; 3 Marty, 1959; 4 Ehrman, 1993; 5 Stark, 1997; 6 Spong, 2007; 7 Case, 1914; 8 Faust, 2006; 9 Armstrong, 1993; 10 Molloy, 2018; 11 Harpur, 2004; 12 Pagels, 1992; 13 Ehrman, 2018; 14 Wade, 2009; 15 Ehrman, 2006a; 16 Wypustek, 2013:19; 17 Dawkins, 2006:396; 18 Burbank and Cooper, 2010; 19 Burbank and Cooper, 2010:62; 20 Bowker, 2006; 21 Schaferdiek, 2014; 22 Hanson, 1988; 23 Hudson and Kenny, 2004; 24 Pluralism Project of Harvard University, 2022; 25 Chamberlin, 2003; 26 Kelly and Walsh, 2010; 27 McGukin, 2020; 28 Pew Research Center, 2014; 29 Johnson and Grim, 2013; 30 Libreria Editrice Vaticana, 2019:paragraph 882; 31 Bond, 2012; 32 Peters, 1998; 33 Dyck, 1993; 34 Levi, 1981; 35 Armstrong, 1994:197; 36 Carroll, 2000; 37 Sokal, 2019; 38 Vatican Archives, 1993; 39 McGrath, 2010; 40 Steinkerchner, 2020; 41 Chamberlin, 2003:143; 42 Chamberlin, 2003:167; 43 Chamberlin, 2003:248; 44 MacCulloch, 2010:609.; 45 Hendrix, 2015:58; 46 Hildebrand, 1996; 47 Sharot, 2001; 48 Wade, 2009:256; 49 Huss, 1915:xii, xiii; 50 Dyck, 1993:26; 51 Grant, 2015; 52 Fudge, 1998.

CHAPTER 9

1 Wilson, 2003; 2 Armstrong, 1993; 3 Prothero, 2011; 4 Molloy, 2018; 5 Bowker, 2006; 6 MacCulloch, 2010; 7 Vainio, 2008; 8 lutheranworld.org/member-churches; 9 MacCulloch, 2010:622; 10 MacCulloch, 2010:618-619, 443-444; 11 Jackson, 1901; 12 Gordon, 2009:330; 13 Stanglin and McCall, 2012; 14 Belyea, 2007; 15 Brackney, 2009; 16 Spilsbery, 1652; 17 Hankins, 2002:241; 18 Mohler, et al., 2018; 19 Lindner, 2011; 20 Hankins, 2002:242; 21 Priest and Priest, 2007; 22 Dyck, 1993; 23 Broc, 1972:60; 24 Hostetler, 1997; 25 Kraybill, 2010:184; 26 Dyck:1993:435; 27 Harrison, 1992; 28 Hutterian Brethren Website, n.d.; 29 Petrovich, 2017; 30 McBeth, 1987; 31 Aveni, 2016; 32 Fortin and Moon, 2013; 33 Jones and Palffy, 2013; 34 Pitts, 1995.; 35 Cowen and Bromley, 2008; 36 Breaualt and King, 1993; 37 Bromley and Silver, 1995; 38 Teasley, 1999; 39 Burnett, 2013; 40 Rogerson, 1969;

41 Marshall, 2017; 42 Waller, 2006; 43 Bremer, 2009; 44 Massachusetts, 1648:29; 45 Knapp, 1998; 46 Burr, 1914; 47 Nickalls, 1952; 48 Ryan, 2009; 49 Rogers, 1896; 50 Taysom, 2011:196; 51 Webb, 2013; 52 Web, 2013:20; 53 Sharot, 2001; 54 Taylor, 2014:100-102; 55 Taylor, 2015; 56 Taylor, 2014:75; 57 Armstrong, 1993:284; 58 Hein and Shattuck, 2004; 59 https://extranet.generalconvention.org/staff/files/download/32040; 60 Mungazi, 2000; 61 Allen, 2006; 62 Unitarian Universalist Association website; 63 Johnson and Grimm, 2013:16; 64 Weber, 2004; 65 Haija, 2006; 66 Anderson et al., 2010; 67 Johnson and Grim, 2013;

CHAPTER 10

1 Williams, 1996; 2 Sennis, 2016; 3 Uttal, 2004; 4 Armstrong, 1993; 5 MacCulloch, 2010; 6 Bowker, 2006; 7 Reynolds, 1993; 8 Pagels, 1992; 9 Litwa, 2016; 10 Irenaeus, 180; 11 Ehrman, 2003; 12 van den Broek, 2013; 13 DeConick, 2010; 14 Turner, 2001; 15 Rasimus, 2009; 16 Ehrman, 2006b; 17 Lovgren, 2006; 18 Wilson, 1958; 19 Roukema, 1998; 20 Bousset, 1911; 21 Pearson, 2008; 22 Lieu, 2015; 23 Bruce, 1988; 24 Lieu, 1980; 25 Jonas, 1958; 26 McKenzie, 1979; 27 Garsoian, 1967; 28 Al-Biruni, 1000; 29 Melton, 2014; 30 Lieu, 1998; 31 Bihaliji-Merin and Benac, 1963; 32 Paulicianism, 2020; 33 Moosa, 1987; 34 Bogomil, 2020; 35 Lambert, 1998; 36 Roach, 2005; 37 Moore, 2012; 38 Martin, 2005; 39 Leglu et al., 2014; 40 Bruschi and Biller, 2003; 41 O'Shea, 2000; 42 Martin, 2005:13.; 43 Sibly and Sibly, 2003:128; 44 Sumption, 1999;

CHAPTER 11

1 Bowker, 2006; 2 Hallaq, 2005; 3 Molloy, 2018; 4 MacCulloch, 2010; 5 Armstrong, 1993; 6 Peters, 2009; 7 Kadir, 2017; 8 Levinson, 2012; 9 Esposito, 2003; 10 Hodgson, 1974; 11 Moosa, 1987; 12 Hoffman, 2012; 13 Kennedy, 2016; 14 Kenny, 2006; 15 Mohammed and Kayode, 2021; 16 Martin, 2016; 17 Lucas, 2011; 18 Pew Research Center, 2012; 19 Burbank and Cooper, 2010; 20 Algeriani and Mohadi, 2017; 21 Al-Khalili, 2011; 22 Melchert, 2016; 23 Stewart, 1998; 24 Doniger, 2009; 25 Sonbal, 2005; 26 Messier, 2010; 27 Osman, 2014; 28 Hallaq, 2016:257; 29 Fieldhouse, 2006; 30 Nelson-Pallmeyer, 2003; 31 Kibble, 2002; 32 Lewis, 1998; 33 Kibble, 2002:34–35; 34 Faisal, 2005; 35 Pope Francis, 2013; 36 Johnson and Grim, 2013; 37 Al-Nawbakhti, 2007; 38 Esposito, 2011; 39 Human Rights Watch, 2008; 40 Bowker, 2006:26; 41 Daftary, 2007; 42 Moin, 1999; 43 Razoux, 2015; 44 Uttal, 2004; 45 Prochazka-Eisl, 2016; 46 Saunders, 2017; 47 Clark, 2006; 48 Goodman, 2022; 49 Turku, 2018; 50 Cuno, 2015; 51 Prothero, 2011:53–54.

CHAPTER 12

1 Johnson and Grim, 2013; 2 Futuyma, 2013; 3 Kramer, 1963; 4 Aristotle; 5 Collins, 2021; 6 McEwan, 2006; 7 Diamond, 1997; 8 Elie, 2022; 9 Jenkins, 2022; 10 Thames, 2022; 11 Cunningham, 2022; 12 Khodarkovsky, 2022; 13 Adams, 1983; 14 March, 1998; 15 Garland, 2013; 16 Johnson, 2009; 17 Bowker, 2006; 18 Doniger, 2009; 19 Martinez, 1994; 20 Teeter, 2011;

21 Rose, 2011; 22 Vatican, 2007.; 23 Wikipedia entry for *Meena Keshwar Kamal*, accessed on May 14, 2023; 24 Loads, 1990; 25 Risse, 1995; 26 Martin, 2016; 27 Salguero, 2017; 28 Salguero, 2020; 29 Salguero, 2022; 30 Goleman, 1995; 31 Armstrong, 1993; 32 Smith, 1982:110; 33 Maza, 2017; 34 Colman, 2017; 35 Fieldhouse, 2006; 36 Armstrong, 1993:391; 37 Caylor, 1993; 38 Dinan, 2012; 39 Cornwell, 2005; 40 Percy, 2005; 41 Record, 2010; 42 Richter, 2003; 43 Graham, 2003; 44 Utter and True, 2004; 45 Zakaria, 2003; 46 Burnham et al., 2006; 47 Gerges, 2016; 48 Nelson-Pallmeyer, 2003; 49 Dawkins, 1997; 50 Werbach, 2010; 51 Slade, 2019; 52 Turchin et al. 2015:79; 53 Prothero, 2011; 54 Harris, 2004:15; 55 Donovan, 1993; 56 Hick, 2016; 57 Marty, 1988:163; 58 Gort, 2006:2; 59 Ratcliffe, 2016; 60 McGrath and McGrath, 2007; 61 Dubos, 1981:115;

REFERENCES

Abimbola, K. 2006. *Yoruba Culture: A Philosophical Account.* Birmingham, UK: Iroko Academic Publishers.

Abram, C. 2011. *Myths of Pagan North: The Gods of the Norsemen.* New York: Continuum.

Acker, P., and C. Larrington, eds. 2015. *The Poetic Edda: Essays on Old Norse Mythology.* London: Routledge.

Adams, D. 1983. "Why There Are So Few Women Warriors." *Behavior Science Research* 18 (3): 196–212.

Adovasio, J. M., O. Soffer, and J. Page. 2007. *The Invisible Sex: Uncovering the True Roles of Women in Prehistory.* New York: Smithsonian Books/Harper Collins.

Ahuvia, K. 2102. *Homer: a Guide for the Perplexed.* London, Bloomsbury.

Aiyar, K. N. 1914. [Translation of] Thirty minor Upanishads. Vasanta Press.

Akira, H. 1993. *A History of Indian Buddhism.* Translated and edited by Paul Groner. Delhi: Motilal Banarsidass.

Al-Biruni, M. 1000. *Chronology of the Ancient Nations* 1879 English translation by E. Sachau. London: W. H. Allen.

Al-Khalili, J. 2011. *The House of Wisdom. How Arabic Science Saved Knowledge and Gave Us the Renaissance.* New York: Penguin.

Al-Nawbakhti, A. 2007. *Shia Sects.* Translation of an earlier work by tenth-century Islamic author A. Kadhim. London: ICAS Press.

al Talei, Bazoobandi, and Khorami. 2022. "Hijab in Iran: From Religious to Political Symbol." Carnegie Endowment for International Peace, October 13, 2022. https://carnegieendowment.org/sada/88152.

Alana, O. 2004. "Traditional Religion." In *Understanding Yoruba Life and Culture,* ed. N. Lawal, M. Sadiku, and A. Dopamu, 65–80. Trenton, NJ: Africa World Press.

Algeriani, A., and M. Mohadi. 2017. "The House of Wisdom (Bayt al-Hikmah) and Its Civilizational Impact on Islamic Libraries: A Historical Perspective." *Mediterranean Journal of Social Science* 8 (5): 179–187.

Allan, K. 2005. *Explorations in Classical Sociological Theory: Seeing the Social World*. Thousand Oaks, CA: Pine Forge.

Allen, J. 2006. *Rabble-Rouser for Peace: The Authorized Biography of Desmond Tutu*. London: Rider.

Allen, J. 2013. "Pope Francis and the Shroud of Turin." *National Catholic Reporter*, April 1, 2013.

Allen, J. 2015. *The Ancient Egyptian Pyramid Texts*. 2nd ed. Atlanta, GA: SBC Press.

Almere, K. 1998. *Time and Sacrifice in the Aztec Cosmos*. Bloomington: Indiana University Press.

Alper, M. 2008. *The "God" Part of the Brain: A Scientific Interpretation of Human Spirituality and God*. Naperville, IL: Sourcebooks.

Anderson, A., M. Bergunder, A. Droogers, and C. van der Laan. 2010. "Introduction." In *Studying Global Pentecostalism: Theories and Methods*, 1–9. Berkeley: University of California Press.

Andrén, A. 2011. "Old Norse and Germanic Religion." In *The Oxford Handbook of the Archaeology of Ritual and Religion*, ed. T. Insoll, 846–862. Oxford: Oxford University Press.

Andrén, A. 2014. *Tracing Old Norse Cosmology: The World Tree, Middle Earth, and the Sun in Archaeological Perspectives*. Lund, Sweden: Nordic Academic Press.

Andrén, A., K Jennbert, and C Audvere. "Old Norse Religions. Some Problems and Prospectives. Origins, Changes, and Interactions" In *Old Norse Religion in Long-Term Prospects*, ed. A. Andrén, K. Jennbert, and C Audvere, 11–15. Sweden: Nordic Academic Press.

Angel, M. 1991. *Voices in Exile: A Study in Sephardic Intellectual History*. Hoboken, NJ: KTAV.

Anthony, D. 2007. *The Horse, the Wheel, and Language: How Bronze-Age Riders from the Eurasian Steppes Shaped the Modern World*. Princeton, NJ: Princeton University Press.

Antonio, E. 2017. "Indigenous African Traditions as Models for Theorizing Religion." In *Religion, Theory, Critique: Classic and Contemporary Approaches and Methodologies*, ed. R. King, 147–154. New York: Columbia University Press.

Archibald, J. 2014. *Aristotle's Ladder, Darwin's tree. The Evolution of Visual Metaphors for Biological Order*. Columbia University Press.

Aristotle. Fourth century BCE. *Politics*. Book 5, 1314b.39.

Armstrong, K. 1993. *A History of God: The 4,000-Year Quest of Judaism, Christianity, and Islam*. Westminster, MD: Random House.

Asad, T. 1993. *Genealogies of Religion: Discipline and Reasons of Power in Christianity and Islam*. Baltimore, MD: Johns Hopkins University Press.

Ashkenazi, M. 2003. *Handbook of Japanese Mythology*. Santa Barbara, CA: ABC-CLIO, Inc.

Aston, W., trans. 1956. *Nihongi: Chronicles of Japan from the Earliest Times CE 697*. London: Allen and Unwin.

Atapur, A. 2002. "Killing for 'Mother Kali.'" *Time* (Asian Edition), September 13, 2002.

Atran, S. 2002. *In Gods We Trust: The Evolutionary Landscape of Religion*. Oxford: Oxford University Press.

Atwood, R. 2005. "A Monumental Feud." *Archaeology* 58 (4): 18.

Aubert, M., et al. 2018. "Paleolithic Cave Art in Borneo." *Nature* 564: 254–257.

Aveni, A. 2016. *Apocalyptic Anxiety*. Boulder: University Press of Colorado.

Babb, L. 2015. *Understanding Jainism*. London: Dunedin.

Bailey, G. 2017. "Hinduism Contextualized." In *Hinduism in India*, ed. G. Bailey. Los Angeles, 7–26: Sage.

Bareau, A. 1955. *Les Sectes bouddhiques de Petit Vehicule*. Ho Chi Minh City, Vietnam: Saigon Ecole francaise d'Extreme Orient.

Baruah, B. 2000. *Buddhist Sects and Sectarianism*. New Delhi: Sarup.

Baumeister, R., and K. Vohs. 2007. "Confirmation Bias." In *Encyclopedia of Social Psychology*. London: Sage.

Bhattacharyya, N. 1996. *History of the Sakta Religion*. 2nd ed. First edition, 1974. New Delhi: Munshiram Manoharial.

Baines, J., and J. Malek. 2000. *Cultural Atlas of Ancient Egypt*. Rev. ed. Original edition, 1980. New York: Checkmark.

Baruah, B. 2000. *Buddhist Sects and Sectarianism*. New Delhi: Sarup.

Bayly, S. 1999. *Caste, Society and Politics in India from the Eighteenth Century to the Modern Age*. Cambridge: Cambridge University Press.

Beck, M. 2000. "Female Figurines in the European Upper Paleolithic. Politics and Bias in Archaeological Interpretation." In *Reading the Body. Representation and Remains in the Archaeological Record*, ed. A. Rautman, 202–214. Philadelphia: University of Pennsylvania Press.

Bede. 731. *The Ecclesiastical History of the English People*. Translation of revised edition from 1990. London: Penguin.

Belli, F. 2010. *The First Maya Civilization. Ritual Power Before the Classic Period*. London: Routeledge.

Belyea, G. 2007. "Origins of the Particular Baptists." *Themelios* 32 (3): 40–67.

Bengtson, H. 2009. *Griechische Geschichte*. Munich, Germany: C. H. Beck.

Benigni, H., ed. 2013. *The Mythology of Venus: Ancient Calendars and Archaeoastronomy*. Lanham, MD: University Press of America.

Bennett, M., et al. 2021. "Evidence of Humans in North America During the Last Glacial Maximum." *Science* 373: 1528–1531.

Ben-Sasson, H., S. Stern, R. Weltsch, M. Berenbaum, J. Levinger, and C. Yahil. 2007. "Germany." In *Encyclopedia Judaica*, ed. M. Berenbaum and F. Skolnik. Detroit, MI: Macmillan Reference.

Bergh, S. 2012. *Wari: Lord of the Ancient Andes*. London: Thames and Hudson.

Berger, P. 1985. *The Goddess Obscured: Transformation of the Grain Protectress from Goddess to Saint*. Boston, MA: Beacon.

Berquist, J. 2007. *Approaching Yehud: New Approaches to the Study of the Persian Period*. Atlanta, GA: Society of Biblical Literature.

Betz, A. G. 2000. "Monotheism." In *Eerdmans Dictionary of the Bible*, ed. D. Freedman and A. Meyer. Grand Rapids, MI: Eerdmans.

Bihalji-Merin, O., and A. Benac. 1963. *Bogomil Sculpture. Essays by Oto Bihalji-Merin and Alojz Benac*. New York: Harcourt Brace.

Bing, C. 2012. "The Revival of the Longmen Branch of the Quanzen School in the Qing Dynasty." In *Taoism*, ed. M. Zhongjan, 225–251. Boston, MA: Brill.

Bischof, H. 2008. "Context and Contents of Early Chavin Art." In *Chavin: Art Architecture, and Culture*, ed. W. Conklin and J. Quilter. Cotsen Institute of Archaeology, Monograph 61, 107–142. Los Angeles: UCLA Cotsen Institute of Archaeology Press.

Black, J., G. Cunningham, and E. Robson. 2006. *The Literature of Ancient Sumer*. Oxford: Oxford University Press.

Black, J., and A. Green. 1992. *Gods, Demons and Symbols of Ancient Mesopotamia: An Illustrated Dictionary*. Austin: University of Texas Press.

Blackham, M. 1996. "Further Investigations as to the Relationship of Samarran and Ubaid Ceramic Assemblages." *Iraq*. 58: 1–15.

Blackmore, S. 1999. *The Meme Machine*. Oxford: Oxford University Press.

Blanton, R., S. Kowalewski, G. Feinman, and L. Finsten. 1993. *Ancient Mesoamerica: A Comparison of Change in Three Regions*. 2nd ed. Cambridge: Cambridge University Press.

Blasi, A., and L. Oviedo, eds. 2020. *The Abuse of Minors in the Catholic Church*. London: Routledge.

Bloch, E., M. Keppens, and R. Hegde, eds. 2010. *Rethinking Religion in India. The Colonial Construction of Hinduism*. London: Routledge.

Boas, F. 1896. "The Limitations of the Comparative Method of Anthropology." *Science* IV: 901–908.

Boas, F. 1908. *Anthropology*. New York: Columbia University Press.

Boesche, R. 2002. *The First Great Political Realist*. Lanham, MD: Lexington.

"Bogomil." 1998. Britannica Online Encyclopedia. https://www.britannica.com/topic/Bogomils.

Bond, H. 2012. *The Historical Jesus: A Guide for the Perplexed*. London: Bloomsbury.

Bottero, J. 2001. *Religion in Ancient Mesopotamia*. Chicago: University of Chicago Press.

Botticini, M., and Z. Eckstein. 2012. *The Chosen Few: How Education Shaped Jewish History*. Princeton, NJ: Princeton University Press.

Bouckaert, R., et al. 2012. "Mapping the Origin and Expansion of the Indo-European Language." *Science* 337 (6097): 957–960.

Bourgeon, L. 2021. "Revisiting the Mammoth Bone Modification from Bluefish Caves (YT, Canada)." *Journal of Archaeological Science: Reports* 37: 1–10.

Bousset, W. 1911. "Valentinus and the Valentinians." In *Encyclopedia Britannica*, 11th ed., H. Chisholm, 852–857. Cambridge: Cambridge University Press.

Bowker, J. 2006. *The Concise Oxford Dictionary of World Religions*. Oxford: Oxford University Press.

Bowles, S., and H. Gintis. 2011. *A Cooperative Species: Human Reciprocity and Its Evolution*. Princeton, NJ: Princeton University Press.

Boxer, A. 2009. "Native Americans and Federal Government." *History Review* 64 (September).

Boyarin, D. 1994. *A Radical Jew: Paul and the Politics of Identity*. Berkeley: University of California Press.

Boyarin, D. 2017. "Nominalist 'Judaism' and the Late-Ancient Invention of Religion." In *Religion, Theory, Critique: Classic and Contemporary Approaches and Methodologies*, ed. R. King, 23–40. New York: Columbia University Press.

Boyce, M. 1979. *Zoroastrians: Their Religious Beliefs and Practices*. London: Routledge.

Boyce, M. 1984. *Textural Sources for the Study of Religion, Zoroastrianism*. Chicago: University of Chicago Press.

Boyd, J., and R. Williams. 2005. "Japanese Shinto: An Interpretation of a Priestly Perspective." *Philosophy East and West* 55 (1): 33–63.

Boyer, P. 2001. *Religion Explained: The Evolutionary Origins of Religious Thought*. New York: Basic Books.

Brackney, W. 2009. *Historical Dictionary of the Baptists*. Lanham, MD: Scarecrow.

Braun, W., and R. McCutcheon, eds. 2018. *Reading J. Z. Smith. Interviews and Essay*. New York: Oxford University Press.

Breault, M., and M. King. 1993. *Inside the Cult. A Member's Chilling Exclusive Account of Madness and Depravity in David Koresh's Compound*. New York: Signet.

Breen, J., and M. Teeuwen 2010. *A New History of Shinto*. Hoboken, NJ: Wiley-Blackwell.

Bremer, F. 2009. *Puritanism: A Very Short Introduction*. New York: Oxford University Press.

Brinton, D. 1887. "Were the Toltecs an Historical Nationality?" *Proceedings of the American Philosophical Society* 24 (126): 229–241.

British Taoist Association. n.d. Website. http://www.taoists.co.uk/main_files/quanzhen.htm Accessed May 20, 2022.

Brittain, C. 2005. "The 'Secular' as a Tragic Category: On Talal Asad, Religion, and Representation." *Method and Study of Religion* 17 (2): 149–165.

Broadie, S. 2003. "The Sophists and Socrates." In *The Cambridge Companion to Greek and Roman Philosophy*, ed. D. Sedley, 73–97. New York: Cambridge University Press.

Broc, P. 1972. *Pacifism in Europe to 1914*. New Jersey: Princeton Legacy Library.

van den Broek, R. 2013. *Gnostic Religion in Antiquity*. New York: Cambridge University Press.

Bromley, D., and E. Silver. 1995. "The Davidian Tradition." In *Armageddon at Waco*, ed. S. Wright, 43–74. Chicago: University of Chicago Press.

Brooks, D., and E. Wiley. 1988. *Evolution as Entropy*. 2nd ed. Chicago: University of Chicago Press.

Brower, A. 2018. "Statistical Consistency and Phylogenetic Inference: A Brief Review." *Cladistics* 34 (5): 562–567.

Brown, A. 2017. "The War Against Pope Francis." *The Guardian*. October 27, 2017.

Brown, G. and P. Richardson. 2013. *Applying Evolutionary Theory to Human Behavior: Past Differences and Current Debates*. New York: Springer Science and Business Media.

Brown, N. 2012. *Song of the Vikings: Snorri and the Making of Norse Myths*. New York: Palgrave Macmillan.

Brown, P. 1999. In *Late Antiquity: A Guide to the Postclassical World*, G. Bowersock, P. Brown, and O. Grabar. Cambridge, MA: Belknap Press of Harvard University Press.

Brown, W. 2017. "Ancient Israelite and Judean Religion." World History Encyclopedia. https://www.worldhistory.org/.

Bruce, F. 1988. *The Canon of Scripture*. Downers Grove, IL: InterVarsity.

Brumm, A., A. Oktaviana, B. Burhan, B. Hakim, R. Lebe, J. Zhao, P. Sulistyarto, M. Ririmasse, S. Adhityatama, I. Sumantri, and M Aubert. 2021. "Oldest Cave Art Found in Sulawesi." *Science Advances* 7 (3). https://www.science.org/doi/10.1126/sciadv.abd4648.

Bruschi, C., and P. Biller, eds. 2003. *Texts and Repression of Medieval Heresy*. Rochester, NY: York Medieval Press.

Burbank, J., and F. Cooper. 2010. *Empires in World History*. Princeton, NJ: Princeton University Press.

Burger, R. 2008. "Chavin de Huantar and Its Sphere of Influence." In *The Handbook of South American Archaeology*, ed. H. Silverman and W. Isbell, 681–704. New York: Springer.

Burkert, W. 1985. *Greek Religion*. Cambridge, MA: Harvard University Press.

Burnett, J. 2013. "Two Decades Later, Some Branch Davidians Still Believe." *NPR Weekend Edition*, April 20.

Burnham, G., S. Doocy, E. Dzeng, R. Lafta, L. Roberts. 2006. *The Human Cost of the War in Iraq. A Mortality Study 2002–2006*. Joint study published by The Bloomberg School of Public Health at John Hopkins University, Maryland; The Center for International Studies at Massachusetts Institute of Technology at the University of Massachusetts; and the School of Medicine at Mustansiriya University in Iraq. http://web.mit.edu/humancostiraq/reports/human-cost-war-101106.pdf.

Burr, G. 1914. *Narratives of the Witchcraft Cases, 1648–1706*. New York: Scribner's.

Buswell, R., and D. Lopez. 2014. *The Princeton Dictionary of Buddhism*. Princeton, NJ: Princeton University Press.

Cagney, P., and B. Ross. 2013. *Global Fundraising: How the World Is Changing the Rules of Philanthropy*. Hoboken, NJ: Wiley.

Callaway, E. 2015. "North America's Oldest Mummy Returned to US Tribe After Genome Sequencing." *Nature* 540: 178–179.

Callaway, E. 2017. "Oldest *Homo sapiens* Fossil Claim Rewrites Our Species' History." nature.com. https://www.nature.com/articles/nature.2017.22109.pdf.

Carrasco, D. 2002. "Aztec Religion: Sacred Cities, Sacred Actions." In *Native Religions and Cultures of Central and South America. Anthropology of the Sacred*, ed. L. Sullivan, 9–32. New York: Continuum International.

Carroll, R. 2000. "Pope Says Sorry for Sins of Church." *World News*, March 13, 2000.

Case, S. 1914. *The Evolution of Early Christianity*. Chicago: University of Chicago Press.

Castañeda, A. 2007. "The African Diaspora in Mexico: Santería, Tourism, and Representations of the State." In *The African Diaspora and the Study of Religion*, ed. T. Trost, 131–150. New York: Palgrave Macmillan.

Caylor, B. 1993. "Terry Preaches Theocratic Rule 'No More Mr. Nice Christian' is the Pro-Life Activist's Theme for the 90's." News Sentinel, Fort Wayne, August 16, 1993.

Cazanove, O. 2007. "Etruria." In *Ancient Religions*, S. Johnston, 220–224. Cambridge, MA: Belknap Press of Harvard University Press.

Chakravarti, M. 1994. *The Concept of Rudra-Śiva Through The Ages*. 2nd. Rev. ed. Delhi: Motilal Banarsidass.

Chamberlin, R. 2003. *The Bad Popes*. Gloucestershire, UK: Sutton.

Chan, R. 1989. *The Olmec: Mother Culture of Mesoamerica*. New York: Rizzoli.

Chan, K. 2005. "Religion in China in the Twenty-First Century: Some Scenarios." *Religion, State and Society* 33 (2): 87–119.

Chattopadhyaya, I. 2016. "Living Tradition: A Study of Prehistoric Rock-Paintings and Indigenous Art from District Sonbhadra, Southern Uttar Pradesh, India." *Athens Journal of Humanities and Arts* 3 (4): 251–270.

Chavez, K. 1988. "The Significance of Chiripa in Lake Titicaca Basin Developments." *Expedition* 30 (3): 17–26.

Chazan, R. 1997. *In the Year 1096: The First Crusade and the Jews*. Philadelphia, PA: Jewish Publication Society.

Cheek, T. 1997. *Propaganda and Culture in Mao's China: Deng Tuo and the Intelligentsia*. Oxford: Oxford University Press.

Chen, Y. 2013. *Confucianism as Religion: Controversies and Consequences*. Leiden, Netherlands: Brill.

Childs, G. 2003. *An Introduction to African Languages*. Philadelphia, PA: John Benjamin.

Choksi, M. 2014. "Ancient Mesopotamian Beliefs in the Afterlife." *World History Encyclopedia*. Last modified June 20, 2014. https://www.worldhistory.org/article/701/ancient-mesopotamian-beliefs-in-the-afterlife/.

Christenson, A. 2007. *Popol Vuhi: The Sacred Book of the Maya*. Norman: University of Oklahoma Press.

Chryssides, G, 2011. *Heaven's Gate: Postmodernity and Popular Culture in a Suicide Group*. Great Britain: Applegate Publishing.

Chrystal, J. 1908. *Authoritative Christianity. The Third World Council . . . Which Was Held A.D. 431, at Ephesus in Asia Volume 3*. Jersey City, NJ: James Chrystal.

Cilluffo, A., and N. Ruiz. 2019. "World's Population Is Projected to Nearly Stop by the End of the Century." *Pew Research Center Fact Tank*. https://www.pewresearch.org/short-reads/2019/06/17/worlds-population-is-projected-to-nearly-stop-growing-by-the-end-of-the-century.

Clark, E. 2010. "The Problem of Biological Individuality." *Biological Theory* 5 (4): 312–325.

Clark, J., et al. 2003. "Stratigraphic, Chronological and Behavioral Contents of Pleistocene *Homo sapiens* from Middle Awash, Ethiopia." *Nature* 423 (6941): 747–752.

Clark, P. 2006. "Nation of Islam" In, *Encyclopedia of New Religious Movements*, P. Clark, ed., 426–429. London: Routledge.

Cliquet, R., and D. Avramov. 2018. *Evolution Science and Ethics in the Third Millennium*. Cham, Switzerland: Springer.

Coe, M. 1968. *America's First Civilization: Discovering the Olmec*. New York: American Heritage.

Coe, M. 1980. *The Maya*. 2nd ed. New York: Thames and Hudson.

Cohen, M. 2001. *Political Philosophy from Plato to Mao*. London: Pluto.

Coleman, N. 2017. "On Average, 9 Mosques Have Been Targeted Every Month This Year." CNN, August 7, 2017. https://www.cnn.com/2017/03/20/us/mosques-targeted-2017-trnd/index.html.

Collinge, W. 2012. *Historical Dictionary of Catholicism*. Lanham, MD: Scarecrow.

Collins, J. 2007. "Israel." In *Ancient Religions*, ed. S Johnston, 181–188. Cambridge: Belknap Press.

Collins, P. 2021. *The Sumerians: Lost Civilizations*. London: Reaktion.

Commarosano, M. 2018. *Hittite Local Cults*. Atlanta, GA: Society of Biblical Literature.

Compleat Aberree. 1954. "Three Churches Are Given Charters in New Jersey." *Compleat Aberree* 1 (1): 4.

Cordy-Collins, A. 1992. "Archaism or Tradition? The Decapitation Theme in Cupisnique and Moche Iconography." *Latin American Antiquity* 3: 206–220.

Cornwell, R. 2005. "Bush: God Told Me to Invade Iraq." *The Independent* (UK), October 7, 2005.

Cort, J. 2001. *Jains in the World: Religious Values and Ideology in India.* New York: Oxford University Press.

Costa, J. 2103. *Hamiltonian Inclusive Fitness: A Fitter Fitness Concept. Biology Letters* 9: 21030335.

Cousins, L. 2001. "On the Vibhajjavādins." *Buddhist Studies Review* 18 (2): 131–182.

Covey, R. 2008. "The Inca Empire." In *The Handbook of South American Archaeology*, ed. H. Silverman and W. Isbell, 809–830. New York: Springer.

Cowell, A. 1992. "After 350 Years, Vatican Says Galileo Was Right: It Moves." *New York Times*, October 31, 1992.

Cowen, D., and D. Bromley. 2008. *Cults and New Religions.* Chichester, UK: Wiley-Blackwell.

Crook, P. 2007. *Darwin's Coat-tails: Essays on Social Darwinism.* New York: Peter Lang.

Culotta, E. 2009. "On the Origin of Religion." *Science* 326: 784–787.

Cunliffe, B. 1997. *The Ancient Celts.* Oxford: Oxford University Press.

Cunliffe, B. 2008. "A Race Apart: Insularity and Connectivity." *Proceedings of the Prehistoric Society* 75: 55–64.

Cunningham, E. 2022. "How Russia's War in Ukraine Is Dividing the Orthodox Christian World." *Washington Post*, April 24, 2022.

Cunningham, L. 2009. *An Introduction to Catholicism.* Cambridge: Cambridge University Press.

Cuno, J. 2015. "ISIS and the Threat to Our Cultural Heritage: What Can the World Do? A Five Point Proposal." The Gerson Lectures Foundation, Groningen, Netherlands.

Currie, T., S. Greenhill, and R. Mace. 2010. "Is Horizontal Transmission Really a Problem for Phylogenetic Comparative Methods? A Simulation Study Using Continuous Cultural Traits." *Philosophical Transactions of the Royal Society of London: Biological Sciences* 365: 3903–3912.

Daftary, F. 2007. *The Isma'ilis: Their History and Doctrines.* 2nd ed. Cambridge: Cambridge University Press.

Damascene, J. 1914. *Barlam and Ioasaph.* Translated by G. R. Woodward. Reprint with new forward, 1967. Cambridge, MA: Harvard University Press.

Dandekar, R. 2005. "Vaisnavism." In *Encyclopedia of Religion Online*, ed. L Jones, 9498-9501. GaleeBooks.link.gale.com/apps/doc/CX3424503238/GVRL?u=chic_rbw&sid=bookmark-GVRL&xid=964c59eb. Accessed 22 May 2023.

Darwin, C. 1859. *The Origin of Species.* London: John Murray.

Darwin, C. 1871. *The Decent of Man and Selection in Relation to Sex.* London: John Murray.

Darwin, C. 1879. "Letter 12041—Darwin, C. R. to Fordyce, John, 7 May, 1879." *Darwin Correspondence Project.* darwinproject.ac.uk/letter/DCP-LETT-12041.xml.

Davies, N. 2007. *The Incas.* Boulder: University Press of Colorado.

Davies, O. 2011. *Paganism: A Very Short Introduction.* New York: Oxford University Press.

Davies, P. 2010. "Urban Religion and Rural Religion." In *Religious Diversity in Ancient Israel and Judah*, ed. F. Stawakopoulou and J. Barton, 104–117. New York: Continuum International.

Dawkins, R. 1976. *The Selfish Gene.* Oxford: Oxford University Press.

Dawkins, R. 1998. *Is Science a Religion? Humanist in Canada* 31 (4): 6–9, 38–39.

Dawkins, R. 2006. *The God Delusion*. New York: Mariner.

Day, J. 2000. *Yahweh and the Gods and Goddesses of Canaan*. London: Sheffield Academic Press.

DeConick, A. 2010. "After the Gospel of Judas: Reassessing What We Have Known To Be True About Cain and Judas." In *In Search of Truth. Augustine Manichaeism, and Other Gnosticism*, ed. J. van der Berg et al, 627–661. Leiden, Netherlands: Brill.

Desmangles, L. 1992. *The Faces of the Gods: Vodou and Roman Catholicism in Haiti*. Chapel Hill: University of North Carolina Press.

Dever, W. 2003. *Who Were the Early Israelites and Where Did They Come From?* Grand Rapids, MI: Eerdmans.

de Waal, F. 2013. *The Bonobo and the Atheist: In Search of Humanism Among the Primates*. New York: W. W. Norton.

DeWitt, N. 1938. "The Druids and Romanization" *Transactions and Proceedings of the American Philological Association* 69: 319–332.

Diamond, J. 1997. *Guns, Germs, and Steel*. New York: W. W. Norton.

Dickens, A. 1969. *The Counter Reformation*. New York: Harcourt, Brace & World.

Diehl, R. 2004. *The Olmecs: America's First Civilization*. London: Thames and Hudson.

Dillard, A. 1974. *Pilgrim at Tinker Creek*. New York: HarperCollins.

Dillehay, T., et al. 2015. "New Archaeological Evidence for an Early Human Presence at Monte Verde, Chile." *PLoS ONE* 10 (12): e0145471.

Dinan, S. 2012. "Obama Renomination Won't Be Unanimous." *Washington Times*, March 7, 2012.

Doniger, W. 1980. "Dionysus and Shiva: Parallel Patterns in Two Pairs of Myths." *History of Religion* 20: 81–111.

Doniger, W. 1996. "Myths and Methods in the Dark." *Journal of Religion* 76: 531–547.

Doniger, W. 2000. "Post-Modern and -Colonial—Structural Comparisons." In *A Magic Still Dwells: Comparative Religion in the Modern Age*, ed. C. Kimberely, C. Patton, and B. Ray, 63–74. Berkeley: University of California Berkeley Press.

Doniger, W. 2009. *The Hindus: An Alternative History*. London: Penguin.

Doniger, W. 2017. "Hinduism and Its Complicated History with Cows (and People Who Eat Them)." *The Conversation*, July 16, 2017.

Donovan, P. 1993. "The Intolerance of Religious Pluralism." *Religious Studies* 29 (2): 217–229.

Drewes, D. 2015. "Oral Texts in Indian Mahayana." *Indo-Iranian Journal* 58 (2): 132–133.

Dubois, T. 1999. *Nordic Religions in the Viking Age*. Philadelphia: University of Pennsylvania Press.

Dubos, R. 1981. *Celebrations of Life*. New York: McGraw-Hill.

Dunbar, R. 2022. *How Religion Evolved and Why It Endures*. Oxford: Oxford University Press.

Dundas, A. 2002. *The Shabbat Elevator and other Sabbath Subterfuges: An Unorthodox Essay on Circumventing Custom and Jewish Character*. Lanham, MD: Rowman & Littlefield.

Dundas, P. 2002. *The Jains*. 2nd ed. New York: Routledge.

Durant, W. 1939. *The Story of Civilization Part II*. New York: Simon & Schuster.

Durkheim, D. 1915. *Elementary Forms of Religious Life*. English translation from French, 1964. London: Allen and Unwin.

Dyck, C. 1993. *An Introduction to Mennonite History*. 3rd ed. Scottdale, PA: Herald Press.

Ehrlich, P., and A. Ehrlich. 2004. *One with Nineveh: Politics, Consumption, and the Human Future*. Washington, DC: Island Press.

Ehrman, B. 1993. *The Orthodox Corruption of Scripture. The Effect of Early Christological Controversies on the Text of the New Testament*. Oxford: Oxford University Press.

Ehrman, B. 2003. *Lost Christianities: Christian Scriptures and the Battles over Authentication*. Oxford: Oxford University Press.

Ehrman, B. 2006a. *Studies in the Textural Criticism of the New Testament*. Leiden, Netherlands: Brill.

Ehrman, B. 2006b. *The Lost Gospel of Judas Iscariot*. New York: Oxford University Press.

Ehrman, B. 2014. *How Jesus Became God: The Exaltation of a Jewish Preacher from Galilee*. New York: Harper One.

Ehrman, B. 2018. *The Triumph of Christianity. How a Forbidden Religion Swept the World*. New York: Simon and Schuster.

Eldredge, N. 2005. *Darwin. Discovering the Tree of Life*. New York: W.W. Norton.

Eldredge, N. and S. Gould. 1972. "Punctuated Equilibria: An Alternative to Phyletic Gradualism." In *Models in Paleobiology*, ed. T. Schopf. San Francisco, CA, 82–115: Freeman, Cooper.

Eliade, M. 1978. *A History of Religious Ideas*. Vol. 1, *From Stone Age to the Eleusinian Mysteries*. Originally published in French in 1976; English translation by Willard Track. Chicago: University of Chicago Press.

Eliade, M. 1982. *A History of Religious Ideas*. Vol. 2, *Gautama Buddha to the Triumph of Christianity*. Originally published in French in 1978; English translation by Willard Trask. Chicago: University of Chicago Press.

Eliade, M. 1985. *A History of Religious Ideas*. Vol. 3, *From Muhammad to the Age of Reforms*. Originally published in French in 1983; English translation by Alf Hiltebeitel and Diane Apostolos-Cappadona. Chicago: University of Chicago Press.

Elie, P. 2022. "The Long Holy War Behind Putin's Political War in Ukraine." *New Yorker*, April 21, 2022.

Eskildsen, S. 2004. *The Teachings and Practices of the Early Quanzhen Taoist Masters*. Albany: State University of New York Press.

Esposito, J., ed. 2003. *The Oxford Dictionary of Islam*. New York: Oxford University Press.

Esposito, J. 2011. *What Everyone Needs to Know About Islam*. 2nd ed. New York: Oxford University Press.

Evans, A. 1921. *The Palace of Minos: A Comparative Account of the Successive Stages of the Early Cretan Civilization as Illustrated by the Discoveries at Knossos*. London: Macmillan.

Evans, A. 1925. "The Early Nilotic, Libyan and Egyptian Relations with Minoan Crete." The Huxley Memorial Lecture for 1925. *Journal of the Royal Anthropological. Institute of Great Britain and Ireland* 55: 199–228.

Faisal, D. 2005. *Landscapes of the Jihad: Militancy, Morality, Modernity*. Ithaca, NY: Cornell University Press.

Fan, L., and N. Chen. 2014. "The Revival of Indigenous Religion in China" China Watch. Preprint from *The Oxford Handbook of Religious Conversion, 2014*.

Faust, A. 2006. *Israel's Ethnogenesis*. London: Equinox.

Feinman, G. 2000. "Cultural Evolutionary Approaches and Archaeology. Past, Present, and Future." In *Cultural Evolutionism. Contemporary Viewpoints*, ed. G. Feinman and L Manzanilla, 3–14. New York: Kluwer/Plenum.

Feinman, G. 2016. "Variation and Change in Archaic States. Ritual as a Mechanism of Sociopolitical Integration." In *Ritual and Archaic States*, ed. J. Murphy, 1–22. Gainesville: University of Florida Press.

Feinman, G. 2019. "The Prehispanic Mesoamerican World: Framing Interaction." In *Interregional Interaction in Ancient Mesoamerica*, ed. J. Englehardt and M. Carrasco. Louisville, 34–50: University Press of Colorado.

Fenech, L., and W. McLeod. 2014. *Historical Dictionary of Sikhism*. Lanham, MD: Rowman & Littlefield.

Feuerstein, G, S. Kak, and D. Frawley. 1995. *In Search of the Cradle of Civilization*. Madras, India: Quest.

Fieldhouse, D. 2006. *Western Imperialism in the Middle East 1914–1958*. Oxford: Oxford University Press.

Fitzgerald, T. 2007. *Discourse on Civility and Barbarity: A Critical History of Religion and Related Categories*. New York: Oxford University Press.

Flavius, J. 1800. *Complete work of Josephus*. New York: Bigelow, Brown.

Flood, G. 1996. *An Introduction to Hinduism*. Cambridge: Cambridge University Press.

Florescano, E. 1999 *The Myth of Quetzalcoatl*. Translation of 1995 edition by L. Hochroth. Baltimore, MD: Johns Hopkins University Press.

Flügel, P. 2005. "The Invention of Jainism: A Short History of Jaina Studies." *International Journal of Jain Studies* 1 (1): 1–14.

Flügel, P. 2006. "Demographic Trends in Jaina Monasticism." In *Studies in Jaina History and Culture: Disputes and Dialogs*, ed. P. Flügel, 312–398. London: Taylor and Francis.

Ford, E. 2017. *Cold War Monks. Buddhism and America's Secret Strategy in Southeast Asia*. New Haven, CT: Yale University Press.

Forsyth, J. 1994. *A History of the Peoples of Siberia*. Cambridge: Cambridge University Press.

Fortin, D., and J. Moon 2013. *The Ellen G. White Encyclopedia*. Hagerstown, MD: Review and Herald.

Foulston, L., and S. Abbott 2009. *Hindu Goddesses: Beliefs and Practices*. East Sussex, UK: Sussex Academic Press.

Fowler, M. 1999. *Buddhism: Beliefs and Practices*. East Sussex, UK: Sussex Academic Press.

Fowler, M. 2005. *Zen Buddhism: Beliefs and Practices*. East Sussex, UK: Sussex Academic Press.

Fox, R. 1986. *Pagans and Christians in the Mediterranean World from the Second Century CE to the Conversion of Constantine*. London: Penguin.

Fracchia, J., and R. Lewontin 2005. "The Price of Metaphor." *History and Theory* 44 (1): 14–29.

Frankfort, H. 2000. *Ancient Egyptian Religion: An Interpretation*. Slightly corrected republication of the 1948 first edition by Columbia University Press, New York. Mineola, NY: Dover Publications.

Freiberger, O. 2019. *Considering Comparison: A Method for Religious Studies*. New York: Oxford University Press.

Freidson, Y. 2016. "Ethiopian Judaism Nearly Identical to That Practiced During Second Temple Period." yNetNews. https://www.ynetnews.com/articles/0,7340,L-4891062,00.html.

Freud, S. 1939. *Moses and Monotheism*. London: Garden City.

Freud, S. 1922. *Group psychology and the analysis of ego*. New York: Boni and Liveright.

Frick, E., and B. Lautenschlager. 2007. *Auf Unendliches bezogen—Spirituelle Entdeckungen bei C. G. Jung*. Munich, Germany: Koesel.

Fudge, T. 1998. *The Magnificent Ride. The First Reformation in Hussite Bohemia*. New York: Routledge.

Futuyma, D. 2013. *Evolution*. 3rd. ed. Sunderland, MA: Sinauer.

Gandhi, M. 1949. *Communal Unity*. Ahmedabad, India: Navajivan.

Garland, R. 2013. *Ancient Greece. Everyday Life in the Birthplace of Western Civilization*. New York: Sterling.

Garsoian, N. 1967. *The Paulician Heresy*. Paris: Mouton.

Garza, M. 2002. "Sacred Forces of the Mayan Universe." In *Native Religions and Cultures of Central and South America*, ed. L. Sullivan. First English printing of 1997 Spanish edition. New York, 92–176: Continuum International.

Geertz, C. 1973. *The Interpretation of Cultures: Selected Essays*. New York: Basic Books.

George, A., trans. 1999. *The Epic of Gilgamesh*. London: Penguin.

Gerges, F. 2016. *ISIS: A History*. Princeton, NJ: Princeton University Press.

Gethin, R. 1998. *Foundations of Buddhism*. Oxford University Press.

Gimbutas, M. 1991. *The Civilization of the Goddess: The World of Old Europe*. San Francisco, CA: Harper.

Gnuse, K. 1997. *No Other Gods. Emergent Monotheism in Israel*. Sheffield, UK: Sheffield Academic Press.

Godse N. 1948. *Why I Assassinated Mahatma Gandhi*. Reprint, 1993. Delhi: Surya Bharti Parkashan.

Goelet, O., R. Faulkner, C. Andrews, J. Gunther, and J. Wasserman. 2015. *The Egyptian Book of the Dead. The Book of Going Forth by Day*. 3rd ed. San Francisco, CA: Chronicle.

Goldin, P. 2005. *After Confucius: Studies in Early Chinese Philosophy*. Honolulu: University of Hawaii Press.

Goldin, P. 2011. *Confucianism*. New York: Routledge.

Goleman, D. 1995. "Religious Faith and Social Activity Help to Heal, New Research Finds." *New York Times*, February 4, 1995.

Gontier, N. 2015. "Reticulate Evolution Everywhere." In *Reticulate Evolution*, ed. N. Gontier. Cham, Switzerland, 1–40: Springer.

Goodall, J. 1986. *The Chimpanzees of Gombe: Patterns of Behavior*. Cambridge, MA: Harvard University Press.

Goodman, J. 2022. *Who is Hadi Matar? NJ Man Charged With Attempting to Murder Salman Rushdie Attack*. NBC New York, last updated August 15, 2022, accessed May 10, 2023.

Gordon, B. 2009. *Calvin*. New Haven, CT: Yale University Press.

Gort, J. 2006. "Introduction." In *Religions View Religions in Pursuit of Understanding*, ed. J. Gort, H. Jansen, and H. Vroom, 1–8. New York: Rodopi.

Gostin, O., and A. Chong. 1994. "Living Wisdom: Aborigines and the Environment." In *Aboriginal Australia: An Introductory Reader in Aboriginal Studies*, ed. C. Bourke, E. Bourke, and B. Edwards, 147–167. St. Lucia: University of Queensland Press.

Gould, S. J. 1977. *Ontogeny and Phylogeny*. Cambridge, MA: Belknap Press of Harvard University Press.

Gould, S. J. 2010. *An Urchin in the Storm: Essays About Books and Ideas*. New York: W. W. Norton.

Gould, S. J. 2011. *Full House: The Spread of Excellence from Plato to Darwin*. Cambridge, MA: Harvard University Press.

Grabbe, L. 2010. *An Introduction to Second Temple Judaism*. London: A&C Black.

Graham, B. 2003. "General Apologizes for Talk of God, War." *Washington Post*, October 18, 2003.

Graham, B. n.d. "The Mysteries of God." Billy Graham Evangelistic Association. Accessed December 9, 2017. https://billygraham.org/devotion/the-mysteries-of-god/.

Grande, L. 1985. "The Use of Paleontology in Systematics and Biogeography, and a Time Control Refinement for Historical Biogeography." *Paleobiology* 11 (2): 1–11.

Grande, L. 1994. "Repeating Patterns in Nature, and 'Impact' in Science." In *Interpreting the Hierarchy of Nature: From Systematic Patterns to Evolutionary Process Theories*, ed. L. Grande and O. Rieppel, 61–84. San Diego: Academic Press.

Grande, L. 2017a. "Donn Rosen and the Perils of Paleontology." In *Assumptions Inhibiting Progress in Comparative Biology*, ed. B. Crowther and L. Parenti, 159–168. Boca Raton, FL: CRC Press.

Grande, L. 2017b. *Curators. Behind the Scenes of Natural History Museums*. University of Chicago Press.

Grande, L., and O. Rieppel, eds. 1994. *Interpreting the Hierarchy of Nature. From Systematic Patterns to Evolutionary Process Theories*. San Diego: Academic Press.

Grant, J. 2015. *For the Common Good: The Bohemian Land Law and the Beginning of the Hussite Revolution*. Leiden, Netherlands: Brill.

Grant, M. 1916. *The Passing of the Great Race; or, The Racial Basis of European History*. New York: Scribner's.

Grant, N., and D. Mansell. 2008. *A Guidebook to Religious and Spiritual Practices for People Who Work with People*. New York: Universe.

Grantham, T. 2004. "The Role of Fossils to Phylogeny Reconstruction: Why Is It So Difficult to Integrate Paleobiological and Neontological Evolutionary Data?" *Biology and Philosophy* 19: 687–720.

Graves, R. 1960. *Greek Gods and Heroes*. New York: Doubleday.

Gravrand, H. 1990. *Pangool: le genie religieux sereer*. Dakar: Dakari Nouvelles editions Africaines du Senegal.

Gray, P. 2001. "No, But I Read the Book." *Time*, June 24, 2001.

Gray, R., and Q. Atkinson. 2003. "Language-Tree Divergence Times Support the Anatolian Theory of Indo-European origin." *Nature* 426: 435–439.

Gray, R., and J. Watts. 2017. "Cultural Macroevolution Matters." *Proceedings of the National Academy of Sciences of the United States of America* 114: 7846–7852.

Green, M. 1992. *Animals in Celtic Life and Myth*. New York: Routledge.

Greenspahn, F. 2011. "Sadducees and Karites: the Rhetoric of Jewish Sectarianism." *Jewish Studies Quarterly* 18: 91–105.

Gresky, J., J. Haelm, and L. Clare. 2017. "Modified Human Crania from Göbekli Tepe Provide Evidence for a New Form of Neolithic Skull Cult." *Science Advances* 3 (6): e1700564.

Griffiths, G. 1996. *Triads and Trinities*. Cardiff: University of Wales Press.

Grigg, R. 1999. *The Tao of Zen*. Edison, NJ: Alva Press.

Gunn, G. 2003. *First Globalization: The Eurasian Exchange, 1500 to 1800*. Lanham, MD: Rowman & Littlefield.

Gustafson, A. 1961. A History of Swedish Literature. Minneapolis, MN: American-Scandinavian Foundation.

Guttmacher Institute. 2012. "Guttmacher Statistic on Catholic Women's Contraceptive Use." February 15, 2012. https://www.guttmacher.org/article/2012/02/guttmacher-statistic-catholic-womens-contraceptive-use.

Haak, W., et al. 2015. "Massive Migration from the Steppe Was a Source for Indo-European Languages in Europe." *Nature* 522: 207–211.

Haas, J., W. Creamer, and A. Ruiz. 2003. "Gourd Lord" (Peru Archaeology, Norte Chico). *Archaeology* 56: 9–9.

Haas, J, W. Creamer, and A. Ruiz. 2004. "Dating the Late Archaic Occupation of the Norte Chico Region in Peru." *Nature* 432: 1020–1023.

Hackett, J. A. 2001. "There Was No King in Israel: The Era of Judges." In *The Oxford History of the Biblical World*, ed. M. Coogan. Oxford: Oxford University Press.

Habito, R. 1986. "The Trikaya Doctrine in Buddhism." *Buddhist-Christian Studies* 6: 52–62.

Haeckel, E. 1883. *The History of Creation: or the Development of the Earth and its Inhabitants by the Action of Natural Causes*. Vol. II. English translation from German by E. R. Lankester. New York: Appleton and Company.

Haeckel, E. 1904. *The Wonders of Life*. London: Watts.

Hagan, P. 2021. "The Spirit Cave Mummy Is Over 10,000 Years Old." *The Archive*, August 6, 2021.

Haija, R. 2006. "The Armageddon Lobby: Dispensationalist Christian Zionism and the Shaping of U.S. Policy Toward Israel." *Holy Land Studies* 5 (1): 75–95.

Hall, A. 2013. "Time for Acknowledgement: Christian-Run Native American Boarding Schools Left Legacy of Destruction." *Sojourners*, December 16, 2013.

Hall, R. 1997. *An Archaeology of the Soul. North American Indian Belief and Ritual*. Chicago: University of Illinois Press.

Hallaq, W. 2005. *The Origins and Evolution of Islamic Law*. Cambridge: Cambridge University Press.

Hallaq, W. 2016. "Was al-Shafi'I the Master Architect of Islamic Jurisprudence?" Chapter 10 in *The Formation of Islamic Law*, ed. W. Hallaq. London: Routledge.

Halpin, M. 1981. *Totem Poles: An Illustrated Guide*. Vancouver: University of British Columbia Press.

Hamer, D. 2004. *The God Gene: How Faith Is Hardwired into Our Genes*. New York: Doubleday.

Hankins, B. 2002. *Uneasy in Babylon: Southern Baptist Conservatives and American Culture*. University of Alabama Press.

Hanson, R. 1988. *The Search for the Christian Doctrine of God: The Arian Controversy 318–381*. Edinburg: Clark.

Harari, Y. 2014 *Sapiens: A Brief History of Humankind*. English translation. McCelland and Stewart, division of Random House, Canada.

Hardacre, H. 2017. *Shinto: A History*. New York: Oxford University Press.

Harner, M. 1977. "The Enigma of Aztec Sacrifice." *Natural History* 86 (4): 46–51.

Harner, M. 1980. *The Way of the Shaman*. San Francisco, CA: Harper.

Harpur, T. 2004. *The Pagan Christ: Recovering the Lost Light*. New York: Walker.

Harris, M. 1968. *The Rise of Anthropological Theory*. New York: Crowell.

Harris, M. 1989. *Our Kind: Who We Are, Where We Came From, Where We Are Going*. New York: Harper & Row.

Harris, M. 2017. "Inside the First Church of Artificial Intelligence." *Wired*, November 15, 2017. https://www.wired.com/story/anthony-levandowski-artificial-intelligence-religion/.

Harris, S. 2004. *The End of Faith*. New York: W. W. Norton.

Harris, S. 2006. *Letter to a Christian Nation*. New York: Knopf.

Harrison, P., ed. 2018. *Setting Out on the Great Way: Essays on Early Mahayanna Buddhism*. Sheffield, UK: Equinox.

Harrison, W. 1992. "The Role of Women in Anabaptist Thought and Practice: The Hutterite Experience of the Sixteenth and Seventeenth Centuries." *Sixteenth Century Journal* 32 (1): 49–69.

Harrison, W. 2014. *In Praise of Mixed Religions: The Syncretism Solution in a Multifaithed World*. London: McGill-Queen's University Press.

Harrod, J. 2014. "The Case for Chimpanzee Religion." *Journal for the study of Religion, Nature, and Culture* 8 (1): 8–45.

Hart, G. 1997. *The Egyptian Myths*. Austin: University of Texas Press.

Hart, G. 2005. *The Routledge Dictionary of Egyptian Gods and Goddesses*. New York: Routledge.

Harvey, G. 2006. *Animism: Respecting the Living World*. New York: Columbia University Press.

Harvey, P. 2013. *An Introduction to Buddhism: Teachings, History and Practices*. 2nd ed. Cambridge: Cambridge University Press.

Hassig, R. 2003. "El Sacrificio y las Guerras Floridas."*Arquelogia Mexicana* XI (63): 46–51.

Hawks, J., K. Hunley, S. Lee, and M. Wolpoff. 2000. "Population Bottlenecks and Pleistocene Human Evolution." *Molecular Biology and Evolution* 17 (1): 2–22.

Hawley, J. and D. Wulff, eds. 1996. *Devi Goddesses of India*. Berkeley: University of California Press.

Hawting, G. 2000. *The First Dynasty of Islam: The Umayyad Caliphate AD 661–750*. 2nd ed. New York: Routledge.

Heggarty, P. 2006. "Interdisciplinary Indiscipline? Can Phylogenetic Methods Meaningfully Be Applied to Language Data—and to Dating Language?" In *Phylogenetic Methods and the Prehistory of Language*, ed. P. Foster and C. Renfew, 183–194. Cambridge, UK: McDonald Institute Monographs.

Hein, D., and G. Shattuck. 2004. *The Episcopalians*. New York: Church.

Hendrix, S. 2015. *Martin Luther. Visionary Reformer*. New Haven, CT: Yale University Press.

Henshilwood, C.S., F. d'Errico, K. van Niekerk, L. Dayet, A. Queffelec, and L. Pollarolo 2018. "An Abstract Drawing from the 73,000-Year-Old Levels at Blombos Cave, South Africa." *Nature* 562: 115–118.

Heydt-Coca, M. 1999. "When Worlds Collide: The Incorporation of the Andean World into the Emerging World Economy in the Colonial Period." *Dialectical Anthropology* 24: 1–43.

Heyes, C., and L. Huber, eds. 2000. *The Evolution of Cognition*. Cambridge, MA: MIT Press.

Heyrman, C. 1997. *Southern Cross: The Beginning of the Bible Belt*. New York: Alfred A. Knopf, Inc.

Hick, J. "Religious Pluralism." In, *The World's Religious Traditions. Current Perspectives in Religious Studies*, ed. F. Whaling, 147–164. New York: Bloomsbury.

Hidalgo, M. 2007. *Sexual Abuse and the Culture of Catholicism: How Priests and Nuns Become Perpetrators*. New York: Haworth.

Hildebrand, H., ed. 1996. *Oxford Encyclopedia of the Reformation*. Oxford: Oxford University Press.

Hillar, M. 2012. *From Logos to Trinity: The Evolution of Religious Beliefs from Pythagoras to Tertullian*. Cambridge: Cambridge University Press.

Hiltebeitel, A. 2013. "Hinduism." In *The Religious Traditions of Asia: Religion, History, and Culture*, ed. J. Kitagawa. New York: Routledge.

Hintze, A. 2014. "Monotheism the Zoroastrian Way." *Journal of the Royal Asiatic Society* 24: 225–249.

Hippolytus of Rome. 225 CE. *Philosophumena*. [*Refutation of all Heresies*]. 1921 translation, ed. S. Simpson. New York: MacMillan.

Hirakawa, A. 1990. *A History of Indian Buddhism: From Sakyamuni to Early Mahayana*. Translated and edited by Paul Groner. Honolulu: University of Hawaii Press.

Hitler, A. 1943 [1999 English translation by Ralph Manheim of 1943 German edition]. *Mein Kampf*. New York: Houghton Mifflin.

Hodgson, M. 1974. *The Venture of Islam: Conscience and History in a World Civilization*. 3 vols. Chicago: University of Chicago Press.

Hoffman, M. 1979. *Egypt Before the Pharaohs—The Prehistoric Foundation of Egyptian Civilization*. New York: Dorset.

Hoffman, M. 2007. *From Rebel to Rabbi: Reclaiming Jesus and the Making of Modern Jewish Culture*. Stanford, CA: Stanford University Press.

Hoffman, V. 2012. *The Essentials of Ibadi Islam*. Syracuse, NY: Syracuse University Press.

Hofstadter, R. 1955. *Social Darwinism in American Thought*. Boston, MA: Beacon.

Hofstede, G., and M. Bond. 1988. "The Confucius Connections: from Cultural Roots to Economic Growth." *Organizational Dynamics* 16 (4): 4–21.

Holden, C., and S. Shennan. 2005. "Introduction to Part 1: How Tree-Like Is Cultural Evolution?" In *The Evolution of Cultural Diversity*, ed. R. Mace, C. Holden, and S. Shennan. London: University College London.

Holmberg, P., B. Graslund, O. Sundqvist, and H. Williams 2020. "The Rok Runestone and the End of the World." *Futhark: International Journal of Runic Studies* 9–10: 7–38.

Holt, J. 2004. *The Buddhist Vishnu: Religious Transformation, Politics and Culture*. New York: Columbia University Press.

Hoppal, M. 1987. *Shamanism: An Archaic and/or Recent System of Beliefs.* Wheaton, IL: Theosophical Publishing.

Horsley, R. 2003. *Religion and Empire: People, Power, and the Life of the Spirit.* Minneapolis, MN: Fortress.

Horton, R. 1993. *Patterns of Thought in Africa and the West. Essays on Magic, Religion and Science.* Cambridge: Cambridge University Press.

Hostetler, J. 1997. *Hutterite Society.* 2nd ed. First edition, 1974. Baltimore, MD Johns Hopkins University Press.

Howorth, H. 1880. *History of the Mongols: From the Ninth to the Nineteenth Century.* New York: Burt Franklin.

Hudson, A., and A. Kenny. 2004. "Wyclif [Wycliffe], John [*called* Doctor Evangelicus]." *Oxford Dictionary of National Biography.* Oxford: Oxford University Press.

Hughey, J., P. Paschou, P. Drineas, D. Mastropaolo, D. Lotakis, P. Navas, M. Michalodimitrakis, J. A. Stamatoyannopoulos, and G. Stamatoyannopoulos. 2013. "A European Population in Minoan Bronze Age Crete." *Nature Communications* 4: 1861.

Hull, D. 1988. *Science as Progress: An Evolutionary Account of the Social and Conceptual Development of Science.* University of Chicago Press.

Hull, E. 1898. *The Cuchullin Saga in Irish Literature.* London: Nuff.

Human Rights Watch, 2008. *Invisible Civilians. The Challenge of Humanitarian Access in Yemen's Forgotten War.* New York: Human Rights Watch, accessed online May 10, 2023. www.hrw.org.

Huss, J. 1915. *De Ecclesia. The Church.* [Original 1413 Latin edition of *De Ecclesia* by J. Huss translated into 1915 English edition by David Schaffie for New York publisher Shribner's Sons.

Hutterian Brethren Website, n.d. "The Schmiedeleut Split." https://hutterites.org/the-leut/origins-leut/split/ Accessed January 6, 2021.

Hutton. R. 1991. *The Pagan Religions of the Ancient British Isles: Their Nature and Legacy.* Oxford: Oxford University Press.

ICZN (International Trust for Zoological Nomenclature). 1999. *International Code of Zoological Nomenclature.* 4th ed. London: ICZN.

Ingold, T. 2000. *The Perception of the Environment: Essays on Livelihood, Dwelling and Skill.* London: Routledge.

Insoll, T. 2004. *Archaeology, Ritual, Religion.* London: Routledge.

Irenaeus. 180. *Adversus haereses* [*Against Heresies*]. 5 vols.

Irwin, L. 2000. *Native American Spirituality: A Critical Review.* Lincoln: University of Nebraska Press.

Isbell, W. 2008. "Wari and Tiwanaku: International identities in the Central Andean Middle Horizon." In *The Handbook of South American Archaeology*, ed. H. Silverman and W. Isbell, 731–760. New York: Springer.

Isabella I and Ferdinand II [Queen of Castile and King of Aragon]. 1492. *Decreto de la Alhambra, Edicto de Granada* [Decree of the Alhambra].

Israel Central Bureau of Statistics. 2010. "Persons 20 and Over, by Religiosity and by Selected Characteristics." *Statistical Abstract of Israel* no. 61, *Society and Welfare* no. 7.3.

Jackson, S. 1901. *Huldreich Zwingli: The Reformer of German Switzerland. 1484–1531.* New York: Knickerbocker.

Jackson, J. and N. Weidman. 2004. *Race, Racism, and Science: Social Impact and Interaction*. New Brunswick, NJ: Rutgers University Press.

Jacobi, H. 1884. *Jaina Sutras*. Oxford: Oxford University Press.

Jacobs, S. 2010. *Hinduism Today*. London: Continuum International.

Jacobsen, K. 2012. *Yoga Powers. Extraordinary Capacities Attained Through Meditation and Concentration*. Boston: Brill.

Jacobson, T. 1976. *A History of Mesopotamian Religion*. New Haven, CT: Yale University Press.

Jain, U. 1975. *Jaina Sects and Schools*. New Delhi: Concept.

James, S. L., and S. Dillon, eds. 2012. *A Companion to Women in the Ancient World*. Malden, MA: Wiley-Blackwell.

Jansen, M., and G. Jimenez. 2011. *The Mixtec Pictorial Manuscripts*. Leiden, Netherlands: Brill.

Janusek, J. 2004. *Identity and Power in the Ancient Andes. Tiwanaku Cities Through Time*. New York: Routledge.

Jenkins, J. 2022. "How One Priest Turned Putin's Invasion into a Holy War." *Rolling Stone*, March 10, 2022.

Jewish Virtual Library. n.d. "Ancient Jewish History: The Birth and Evolution of Judaism." Accessed December 3, 2017. http://www.jewishvirtuallibrary.org/the-birth-and-evolution-of-judaism.

Jha, R. 2014. "Exploring Parallels Between the Philosophies of Upanishads and Daoism." In *Brahman and Dao: Comparative Studies of Indian and Chinese Philosophy and Religion*, ed. T. Ithmar and Z. Yao, 19–34. Lanham, MD: Lexington.

Johnson, M., and B. Grim. 2013. *The World's Religions in Figures: An Introduction to International Religious Demography*. Malden, MA: Wiley.

Johnson, W. 2009. *A Dictionary of Hinduism*. Oxford: Oxford University Press.

Jonas, H. 1958. *The Gnostic Religion*. Boston, MA: Beacon.

Jones, C., and J. Ryan. 2006. *Encyclopedia of Hinduism*. New York: InfoBase.

Jones, G., and G. Palffy, eds. 2013. *The Religions Book: Big Ideas Simply Explained*. New York: DK.

Jones, W. 1796. *Institutes of Hindu Law: or Ordinances of Menu, According to the Gloss of Culluca, Comprising the Indian System of Duties, Religious and Civil*. London: J. Sewell, Cornhill, and J. Debrett, Piccadilly.

Joordens, J., et al. 2015. "*Homo erectus* at Trinil on Java Used Shells for Tool Production and Engraving." *Nature* 518: 228–231.

Josephson-Storm, J. 2017. *The Myth of Disenchantment: Magic, Modernity, and the Birth of Human Sciences*. Chicago: University of Chicago Press.

Joyce, R., R. Edging, K. Lorenz, and S. Gollespie. 1991. "Olmec Bloodletting: An Iconographic Study." In *Sixth Palenque Round Table, 1986*, ed. M. Robertson. Norman, 143–159: University of Oklahoma Press.

Juergensmeyer, M. 2005. *Religion in Global Civil Society*. Oxford: Oxford University Press.

Kadir, S. 2017. *Islam Explained*. Singapore: Marshall Cavendish, International.

Kamath, M. 2007. *Gandhi, a Spiritual Journey*. Mumbai, India: Indus Source Books.

Kaplan, M. 1934. *Judaism as a Civilization: Toward Reconstruction of American-Jewish life*. New York: Thomas Yoseloff.

Kaplan, S. 1995. *The Beta Israel: Falasha in Ethiopia from Earliest Times to the Twentieth Century*. New York: New York University Press.

Keightley. D. 2004. "The Making of Ancestors: Late Shang Religion and its Legacy." In *Chinese Religion and Society: The Transformation of a Field*, ed. J. Lagerwey, 3–64. Hong Kong SAR: Chinese University Press.

Kelder, J. 2010. *The Kingdom of Mycenae: A Great Kingdom in the Late Bronze Age Aegean.* Bethesda, MD: CDL Press.

Kelly, J., and M. Walsh. 2010. *A Dictionary of Popes*. 3rd ed. Oxford: Oxford University Press.

Kennedy, H. 2016. *Caliphate: The History of an Idea*. New York: Basic Books.

Kenney, J. 2006. *Muslim Rebels: Kharijites and the Politics of Extremism on Egypt*. Oxford: Oxford University Press.

Kenoyer, J., J. Clark, J. Pal, and G. Sharma. 1983. "An Upper Paleolithic Shrine in India?" *Antiquity* 57 (220): 88–97.

Keys, D. 1999. *Catastrophe: An Investigation into the Origins of the Modern World*. New York: Ballantine.

Keyt, G. 1960. *Songs of Love, Gita Govinda: The Sanskrit Classic Translated and Illustrated by George Keyt*. Delhi: Hind Pocket.

Khan, Y. 2007. *The Great Partition: The Making of India and Pakistan*. New Haven, CT: Yale University Press.

Khodarkovsky, M. 2022. "Putin's Dream of Godliness: Holy Russia" *New York Times*, January 22, 2019.

Kibble, D. 2002. "The Attacks of 9/11: Evidence of a Clash of Religions?" *Parameters* 32 (3): 34–45.

Kind, C. N. Ebinger-Rist, S. Wolf, T. Beutelspacher, and K. Wehrberger. 2014. "The Smile of the Lion Man: Recent Excavations in Stadel Cave (Baden-Württemberg, Southwestern Germany) and Restoration of the Famous Upper Paleolithic Figurine." *Quartar* 61: 129–145.

King, K. 2003. *The Gospel of Mary Magdala: Jesus and the First Woman Apostle*. Santa Rosa, CA: Polebridge.

Kirkland, R. 2004. *Taoism: The Enduring Tradition*. New York: Routledge.

Knapp, H. 1998. "The Character of Puritan Missions: The Motivation, Methodology, and Effectiveness of the Puritan Evangelization of the Native Americans in New England." *Journal of Presbyterian History* 76 (2): 111–126.

Knowles-Berry, S. 1995. "Chontal of Tabasco." In *Encyclopedia of World Cultures*. Vol. 8, *Middle Americas and the Caribbean*, ed. J. Dow and R. Kemper. Boston, MA: G. K. Hall.

Kolata, A. 1993. *The Tiwanaku: Portrait of an Andean Civilization*. Cambridge, MA: Wiley-Blackwell.

Kot, G. 2017. "Ranking the Rolling Stones' Moments That Matter from the Navy Pier Show." *Chicago Tribune*, April 21, 2017.

Kramer, S. 1956. *History Begins at Sumer*. Philadelphia: University of Pennsylvania Press.

Kramer, S. 1963. *The Sumerians: Their History, Culture, and Character*. Chicago: University of Chicago Press.

Kramer, S. 2021. "Key Findings About the Religious Composition of India." Pew Research Center. https://www.pewresearch.org/short-reads/2021/09/21/key-findings-about-the-religious-composition-of-india/.

Krasinski, K., and D. Yesner. 2008. "Late Pleistocene/Early Holocene Site Structure in Beringia: A Case Study from the Broken Mammoth Site." *Alaska Journal of Anthropology* 6: 27–41.

Kraybill, D. 2010. *Concise Encyclopedia of Amish, Brethren, Hutterites and Mennonites.* Baltimore: Johns Hopkins University Press.

Krishan, Y. 1986. "Buddhism and the Caste System." *Journal of International Association of Buddhist Studies* 9: 71–84.

Kubler, G. 1945. "The Behavior of Atahualpa. 1531–1533." *Hispanic American Historical Review* 25: 413–427.

Lacey, M., and F. Oakley. 2011. *The Crisis of Authority in Catholic Modernity.* Oxford: Oxford University Press.

Lajimodiere, D. 2019. *Stringing Rosaries: The History, the Unforgivable, and the Healing of Northern Plains American Indian Boarding School Survivors.* Fargo: North Dakota State University Press.

Lambert, M. 1998. *The Cathars.* Oxford: Blackwell.

Lang, O. 2009. "Hindu Sacrifice of 250,000 Animals Begins." *Guardian,* November 24, 2009. https://www.theguardian.com/world/2009/nov/24/hindu-sacrifice-gadhimai-festival-nepal.

Lattimore, R. 1959. *[Translation of Hesiod's] The Works and Days. Theogany: The Shields of Herakles.* Ann Arbor: University of Michigan Press.

Lau, G. 2008. "Ancestor Images in the Andes." In *The Handbook of South American Archaeology,* ed. H. Silverman and W. Isbell, 1027–1046. New York: Springer.

Laumakis, S. 2008. *An Introduction to Buddhist Philosophy.* Cambridge: Cambridge University Press.

Lee, Ming-Huei. 2017. *Confucianism: Its Roots and Global Significance.* Honolulu: University of Hawaii Press.

Leglu, C., R. Rist, and C. Taylor, eds. 2014. *The Cathars and the Albigensian Crusade: A Sourcebook.* New York: Routledge.

Lehmann, A., and J. Myers. 1985. "Ghosts, Souls, and Ancestors: The Power of the Dead." In *Magic, Witchcraft, and Religion*: *an Anthropological study of the Supernatural,* ed. A. Lehmann and J. Myers, 283–286. Mayfield Publiching Co., CA.

Levenson, J. 2012. *Inheriting Abrahamism: The Legacy of the Patriarch in Judaism, Christianity, and Islam.* Princeton, NJ: Princeton University Press.

Levit, G., and U. Hossfeld. 2020. "Ernst Haeckel, Nikolai Miklucho-Maclay and the Racial Controversy Over the Papuans." *Frontiers in Zoology* 17: 16.

Levy, L. 1981. *Treason Against God: A History of the Offense of Blasphemy.* New York: Schocken.

Lewens, T. 2018. "Cultural Evolution." *The Stanford Encyclopedia of Philosophy,* ed. E. N. Zalta. https://plato.stanford.edu/archives/sum2020/entries/evolution-cultural.

Lewis, B. 1980. *The Sargon Legend: A Study of the Akkadian Text and the Tale of the Hero Who Was Exposed at Birth.* Dissertation Series 4. Cambridge, MA: American Schools of Oriental Research.

Lewis, B. 1998. "License to Kill. Usama bin Laden's Declaration of Jihad." *Foreign Affairs* 77 (6): 14–19.

Lewis, J. "Scientology: up stat, down stat." In, *The Cambridge Companion to New Religious Movements*. Cambridge: Cambridge University Press.

Lewis-Williams, D. 2002. *The Mind in the Cave*. London: Thames and Hudson.

Lewis-Williams, J. 1995. "Seeing and Construing. The Making and 'Meaning' of Southern African Rock Art Motif." *Cambridge Archaeological Journal* 5 (1): 3–32.

Li, C. 2008. "The Philosophy of Harmony in Classical Confucianism." *Philosophy Compass* 3 (3): 432–435.

Liang, Y. 2016. "The Anthropological Study of Religion in China: Contexts, Collaborations, Debates and Trends." *Asia Research Institute Working Paper Series* (250): 25.

Libbrecht, U. 2007. *Within the Four Seas . . . Introduction to Comparative Philosophy*. Paris: Peeters.

Libreria Editrice Vaticana, 2019 [Second Edition]. Paragraph 882 of the Catechism of the Catholic Church.

Lieberman, P. 1991. *Uniquely Human: The Evolution of Speech, Thought, and Selfless Behavior*. Cambridge, MA: Harvard University Press.

Lieu, J. 2015. *Marcion and the Making of a Heretic: God and Scripture in the Second Century Cambridge*. New York: Cambridge University Press.

Lieu, S. 1980. "Nestorians and Manichaeans on the South China Coast." *Vigiliae Christianae* 34 (1): 71–88.

Lieu, S. 1998. *Manichaeism in Central Asia and China*. Leiden, Netherlands: Brill.

Lincoln, B. 2018. *Apples and Oranges. In, On, and With Comparison*. Chicago: University of Chicago Press.

Lind, M. 2015. *Ancient Zapotec Religion: An Ethnohistorical and Archaeological Perspective*. Boulder: University Press of Colorado.

Lindner, E. 2011. *Yearbook of American and Canadian Churches*. Nashville, TN: Abingdon.

Lindow, J. 2001. *Norse Mythology: A Guide to the Gods, Heroes, Rituals, and Beliefs*. Oxford: Oxford University Press.

Linduff, K., and Y. Sun, eds. 2004. *Gender and Chinese Archaeology*. Walnut Creek, CA: AltaMira.

Linnaeus, C. 1735. *Systema naturae, sive regna tria naturae systematice proposita per classes, ordines, genera and species*. Leiden, Netherlands: Haak.

Lipner, J. 1994. *Hindus: Their Religious Beliefs and Practices*. New York: Routledge.

Littlejohn, R. 2010. *Confucianism: An Introduction*. London: Yaurus.

Litwa, M. D. 2016. *Refutation of all Heresies*. Atlanta, GA: Society of Biblical Literature.

Liu, X. 2010. *The Silk Road in World History*. Oxford: Oxford University Press.

Loads, A., ed. 1990. *Feminist Theology: A Reader*. Louisville, KY: Westminster John Knox.

Lorenzen, D. 2005. "Saivism: An Overview." In *Encyclopedia of Religion Online*, ed. L Jones. Gale Virtual Reference Library.

Lortsch, D. 1910. *Histoire de la Bible en France*. Geneva: Agency de la Societe Biblique Britannique et Etrangere.

Love, V. 2012. *Divining the Self: A Study in Yoruba Myth and Human Consciousness*. University Park: Pennsylvania State University Press.

Lovejoy, A. 1936. *The Great Chain of Being. A Study of the History of the Idea*. Massachusetts: Harvard University Press.

Lovejoy, P. 2012. *Transformation of Slavery: A History of Slavery in Africa*. New York: Cambridge University Press.

Lovgren, S. 2006. "Lost Gospel Revealed; Says Jesus Asked Judas to Betray Him." *National Geographic Online*. https://www.nationalgeographic.com/science/article/lost-gospel-judas-revealed-jesus-archaeology.

Lowenstein, S. 1994. *The Berlin Jewish Community. Enlightenment, Family and Crisis 1770–1830*. New York: Oxford University Press.

Lucas, S. 2006. "The Legal Principles of Muhammad B. Isma'il Al-Bukhari and Their Relationship to Classical Salafi Islam." *Islamic Law and Society* 13 (3): 289–324.

Lucas, S. 2011. "Sunnism, Sunni." *Encyclopedia of Christianity Online*. https://referenceworks.brillonline.com/entries/encyclopedia-of-christianity/sunnism-sunnis-SI.100?s.num=83&s.start=80.

Lurker, M. 1987. *Dictionary of Gods and Goddesses, Devils and Demons*. New York: Routledge.

Macdonell, A. 1995. *Vedic Mythology*. Delhi: Motilal Banarsidass.

MacCulloch, D. 2010. *Christianity. The First Three Thousand Years*. New York: Viking.

MacKenzie, M. 1979. "Mani's 'Sabuhragan.'" *Bulletin of the School of Oriental Studies, University of London* 42 (3): 500–534.

MacKillop, J. 2004. *A Dictionary of Celtic Mythology*. Oxford: Oxford University Press.

MacMillen, R. 1986. *Christianizing the Roman Empire, AD 100–400*. New Haven, CT: Yale University Press.

Maizland, L. 2020. "China's Repression of Uighurs in Xinjiang." Council on Foreign Relations. https://brewminate.com/chinas-repression-of-uighurs-in-xinjiang.

Mallinson, J., and M. Singleton. 2017. *Roots of Yoga*. London: Penguin Random House.

Mallory, F. 2020. "The Case Against Linguistic Paleontology." *Topoi* 40: 273–284. https://doi.org/10.1007/s11245-020-09691-5.

Mallory, J., and D. Adams. 2006. *The Oxford Introduction to Proto-Indo-European and the Proto-Indo-European World*. New York: Oxford University Press.

Mandair, A. 2013. *Sikhism: A Guide for the Perplexed*. London: Bloomsday.

Mann, C. 2005a. "Oldest Civilization in the Americas Revealed." *Science* 307 (5706): 34–35.

Mann, C. 2005b. *1491. New Revelations of the Americas Before Columbus*. New York: Random House.

Mann, C. 2013. "The Clovis Point and the Discovery of America's First Culture." *Smithsonian Magazine*, November 2013.

Manuila, A., ed. 1981. *Progress in Medical Terminology*. Basel, Switzerland: Karger.

March, J. 2014. *Dictionary of Classical Mythology*. Oxford: Oxbow.

Markham, I., and T. Ruparell. 2001. *Encountering Religion: An Introduction to Religions of the World*. Oxford: Blackwell.

Markman, R., and P. Markman. 1989. *Masks of the Spirit: Image and Metaphor in Mesoamerica*. Berkeley: University of California Press.

Marloes, J. 2016. "Unity Through Diversity: A Case Study of Chrislam in Lagos." *Africa: Journal of the International African Institute* 86 (4): 646–672.

Marshall, J. 1931. *Mohenjo Daro and the Indus Civilization*. London: Arthur Probsthain.

Marshall, P. 2017. *Heretics and Believers: A History of the English Reformation*. New Haven, CT: Yale University Press.

Martin, L. 2008. "Can Religion Really Evolve? (And What Is It Anyway?)." In *The Evolution of Religion: Studies, Theories, and Critiques*, ed. Bulbulia et al, 349–356. Santa Margarita, CA: Collins Foundation.

Martin, R., ed. 2016. *Encyclopedia of Islam and the Muslim World.* Farmington Hills, MI: Cengage Learning.

Martin, S. 2005. *The Cathars: The Most Successful Heresy of the Middle Ages.* New York: Thunder's Mouth.

Martinez, D. 1994. "Japanese Religion." In *Women in Religion*, ed. J. Holm and J Bowker, 168–177. Bloomsbury Academic.

Martinez, M., P. Ceballos, M. Coe, R. Diehl, and S. Houston. 2006. "Oldest Writing in the New World." *Science* 313: 1610–1614.

Marty, M. 1959. *A Short History of Christianity.* New York: New American Library.

Marty, M. 1988. "Conclusion: Proselytism in a Pluralistic World." In *Pushing the Faith*, ed. M. Marty and F. Greenspahn, 155–163. New York: Crossroad.

Massachusetts. 1648. *The Book of General Lauues and Libertyes Concerning the Inhabitants of Massachusetts.* Printed according to order of the General Court, and are to be solde at the shop of Hezekiah Usher. Early American Imprints Series 1, Cambridge, MA.

Mathorn, H. 2016. "Pope Francis Washes and Kisses Feet of Muslim, Hindu and Christian Refugees." *The Independent*, 25 March, 2016.

Mayer, D. B. Vandermeersch, and O. Bar-Yosef. 2009. "Shells and Ochre in Middle Paleolithic Qafzeh Cave, Israel: Indications for Modern Behavior." *Journal of Human Evolution* 56: 307–314.

Mayled, J. 2002. *Sikhism.* Oxford: Heinemann.

Maza, C. 2017. "Will Isis Attack the Vatican on Christmas?" *Newsweek*, November 16, 2017.

Mbiti, J. 1969. *African Religions and Philosophy.* Oxford: Heineman.

McBeth, L. 1987. *The Baptist Heritage. Four Centuries of Baptist Witness.* Academic, Nashville.

McEwan, G. 2006. *The Incas: New Perspectives.* New York: Norton.

McGivern, M. 2015. "The Call to Rescind the Doctrine of Discovery." *National Catholic Reporter*, July 9, 2015.

McGrath, A. 2005. *Dawkins' God: Genes, Memes, and the Meaning of Life.* Malden, MA: Blackwell.

McGrath, A., and J. McGrath. 2007. *The Dawkins Delusion: Atheist Fundamentalism and the Denial of the Divine.* Downers Grove, IL: InterVarsity.

McGrath, T. 2010. "Dominicans, Franciscans, and the Art of Political Rivalry: Two Drawings and a Fresco by Giovanni Battista della Rovere." *Renaissance Studies* 25 (2): 185–207.

McGuckin, J. 2020. *The Eastern Orthodox Church. A New History.* New Haven, CT: Yale University Press.

McLeod, W. 2009. *The A to Z of Sikhism.* Lanham, MD: Scarecrow.

McNutt, P. 1999. *Reconstructing the Society of Ancient Isra*el. Louisville, KY: Westminster John Knox.

McRae, J. 2003. *Seeing Through Zen.* Berkeley, CA: University of California Press.

Melchert, C. 2016. "The Formation of the Sunni Schools of Law." Chapter 13 in *The Formation of Islamic Law*, ed. W. Hallaq. London: Routledge.

Melton, J. 2014. *Faiths Across Time: An Exploration of Material Life and the Thought of People*. Santa Barbara, CA: ABC-CLIO.

Melton, J., and M. Baumann 2010. *Religions of the World: A Comprehensive Encyclopedia of Beliefs and Practices, Second Edition*. Santa Barbara, CA: ABC-CLIO.

Merriam-Webster Collegiate Dictionary. 2012.

Messier, R. 2010. *The Almoravids and the Meaning of Jihad*. Santa Barbara, CA: Praeger.

Meyendorff, J. 1989. *Imperial Unity and Christian Divisions*. Crestwood, NY: St. Vladimir's Seminary Press.

Miller, G., and C. Valente. 2016. *Population Policy: Abortion and Modern Contraceptives Are Substitutions. Demography* 53: 979–1009.

Miller, K. 2005. "Showdown at the O.K. Caral." *Discover* 26 (9). https://www.discovermagazine.com/planet-earth/showdown-at-the-ok-caral.

Min, O. 2012. "There Is No Need for Zhongguo Zhexue to Be Philosophy." *Asian Philosophy* 22: 199–223.

Mishler, D. 2021. *What, If Anything, Are Species?* Boca Raton, FL: CRC Press.

Modi, J. 1928. *The Funeral Ceremonies of the Parsis*. 4th ed. First edition, 1892. Bombay: Fort Printing Press.

Mohammed, R. and B. Kayode, 2021. "Present Day Terrorists are the Kharijites of the Time of the Prophet." *Journal of Islamic Studies and Culture*. 9 (2): 1–12.

Mohler, et al., 2018. "Report on slavery and Racism in the History of the Southern Baptist Theological Seminary." Southern Baptist Seminary, online. https://www.sbts.edu/southern-project/.

Moin, B. 1999. *Khomeini: Life of the Ayatollah*. London: Tauris.

Molloy, M. 2018. *Experiencing the World's Religions: Tradition, Challenge, and Change*. 6th ed. New York: McGraw-Hill Education.

Monaghan, P. 2004. *The Encyclopedia of Celtic Mythology and Folklore*. New York: Facts on File.

Monter, W. 1990. *Frontiers of Heresy. The Spanish Inquisition from the Basque Lands to Sicily*. Cambridge: Cambridge University Press.

Mooney, J. 1902. *Myths of the Cherokee*. Washington, DC: U.S. Government Publishing Office.

Moore, C. 1967. *The Chinese Mind: Essentials of Chinese Philosophy and Culture*. Honolulu: University of Hawaii Press.

Moore, J., and C. Mackey. 2008. "The Chimfi Empire." In *The Handbook of South American Archaeology*, eds. H. Silverman and W. Isbell, 783–808. New York: Springer.

Moore, R. 2012. *The War on Heresy*. Cambridge, MA: Harvard University Press.

Moosa, M. 1987. *Extremist Shiites: The Ghulat Sects*. Syracuse, NY: Syracuse University Press.

Mora, C., D. Tittensor, S. Adl, A. Simpson, and B. Worm. 2011. "How Many Species Are There on Earth and in the Ocean?" *PLOS Biology* e1001127. https://doi.org/10.1371/journal.pbio.1001127.

Moreno-Mayar, J., et al. 2018. "Terminal Pleistocene Alaskan Genome Revels First Founding Population of Native Americans." *Nature* 553: 203–207.

Morford, M., R. Lenardon, and M. Sham. 2011. *Classical Mythology*. 9th ed. Oxford: Oxford University Press.

Morgan, D. 2001. *The Best Guide to Eastern Philosophy and Religion*. New York: Renaissance.
Morris, B. 2006. *Religion and Anthropology: A Critical Introduction*. Cambridge: Cambridge University Press.
Morris, E. 2010. *The Pharaoh and Pharaonic Office*. In *A Companion to Ancient Egypt*, ed. A. Lloyd, 201–217. Vol. 1. Chichester, UK: Wiley-Blackwell.
Morritt, R. 2011. *Beringia: Archaic Migrations into North America*. Newcastle upon Tyne, UK: Cambridge Scholars Publishing.
Moseley, M. 2001. *The Incas and their Ancestors: The Archaeology of Peru*. 2nd ed. London: Thames and Hudson.
Motz, L. 1997. *The Faces of the Goddess*. New York: Oxford University Press.
Mounier, A., and M. Lahr 2019. "Deciphering African Late Middle Pleistocene Hominin Diversity and the Origin of Our Species." *Nature Communications* 10: 3406. https://doi:org.10.1038/s41467-019-11213-w.
Muller, F. Max 1867. *Chips from a German Workshop*. Vol. II. London: Longbottom.
Muller, F. Max. 1873. *Introduction to the Science of Religion*. London.
Muller, F. Max. 1889. *Natural Religion*. London: Longmans, Green.
Muller, G. 1902. *The Life and Letters of Right Honorable Friedrich Max Muller*. Vol. 1. London: Longman.
Mungazi, D. 2000. *In the Footsteps of the Masters: Desmond Tutu and Abel T. Muzorewa*. London: Praeger.
Muss-Arnolt, W. 1892. "The Names of the Assyro-Babylonian Months and Their Regents." *Journal of Biblical Literature* 11: 72–94, 160–176.
Nakamura, H. 1980. *Indian Buddhism: A Survey with Bibliographical Notes*. Delhi: Motilal Banarsidass.
National Geographic. 2003. "What Archaeology Is Telling Us About the Real Jesus." https://news.nationalgeographic.com/news/2003/06/0602_030602_untouchables.html.
Nelson, G., and N. Platnick 1981. *Systematics and Biogeography: Cladistics and Vicariance*. New York: Columbia University Press.
Nelson-Pallmeyer, J. 2003. *Is Religion Killing Us? Violence in the Bible and the Quran*. New York: Continuum International.
Neusner, J. 2004. *The Emergence of Judaism*. Louisville, KY: Westminster John Knox.
Neusner, J., and A. Avery-Peck. 2004. *The Routledge Dictionary of Judaism*. New York: Routledge.
Neville, R. 2018. *Defining Religion: Essays in Philosophy of Religion*. Albany: State University of New York Press.
Nicholson, A. 2013. "Is Yoga Hindu? On the Fuzziness of Religious Boundaries." *Common Knowledge* 19 (3): 490–505.
Nickalls, J. 1952. *The Journal of George Fox*. Cambridge: Cambridge University Press.
Nickerson, R. 1998. "Confirmation Bias: A Ubiquitous Phenomenon in Many Guises." *Review of General Psychology* 2 (2): 175–220.
Nigosian, S. 1993. *Zoroastrian Faith: Tradition and Modern Research*. Montreal: McGill Queens University Press.
Nikam, N., and R. McKeon, eds. and trans. 1959. *The Edicts of Ashoka*. Chicago: University of Chicago Press.

Nirbhai, S. 1990. *Philosophy of Sikhism: Reality and Its Manifestations*. New Delhi: Atlantic.

Nishida, T., T. Kano., J. Goodall, W. McGrew, and M. Nakamura. 1999. "Ethogram and Ethnography of Mahale Chimpanzees." *Anthropological Science* 107 (2): 141–188.

Nobutaka, I., ed., and I. Satoshi, E. Jun, and M. Mizue. 2003 [English translation by M. Teeuwen and J. Breen of original 1998 Tokyo edition]. *Shinto—A Short History*. London: RoutledgeCurzon.

Norenzayan, A. 2013. *Big Gods. How Religion Transformed Cooperation and Conflict*. Princeton, NJ: Princeton University Press.

Norenzayan, A., and W. Gervais. 2012. "The Cultural Evolution of Religion." In *Creating Consilience: Integrating the Sciences and the Humanities*, ed. E. Slingerland and M. Collard, 243–265. New York: Oxford University Press.

Norenzayan, A., and A. Sharif. 2008. "The Origin and Evolution of Religious Prosociality." *Science* 322: 58–62.

Oberoi, H. 1994. *The Construction of Religious Boundaries: Culture, Identity, and Diversity in the Sikh Tradition*. Chicago: University of Chicago Press.

O'Brien, M., et al. 2014. "On Thin Ice: Problems with Stanford and Bradley's Proposed Solutrean Colonization of North America." *Antiquity* 88 (340): 606–624.

Ohnuki-Tierney, E. 1991. "The Emperor of Japan as Deity (Kami)." *Ethnology* 30 (3): 199–215.

Olson, J. 1996. *The Peoples of Africa: An Ethnohistorical Dictionary*. London: Greenwood.

Omer, I. 2015. "Evidence Mounts of Ancient Jewish Roots of Beta Israel Ethiopian Jewry." Genetic Literacy Project. https://geneticliteracyproject.org/2015/06/16/evidence-mounts-of-ancient-jewish-roots-of-beta-israel-ethiopian-jewry.

O'Shea, S. 2000. *The Perfect Heresy: The Revolutionary Life and Death of the Medieval Cathars*. New York: Walker.

Osman, A. 2014. *The Zahiri Madhhab (3rd/9th–tenth/16th century): A Texturalist Theory of Islam*. Vol. 38, *Studies in Islamic Law and Society*, ed. B. Weiss. Leiden, Netherlands: Brill.

Paden, W. 2016. *New Patterns for Comparative Religion: Passages to an Evolutionary Perspective*. London: Bloomsbury.

Pagels, E. 1992. *The Gnostic Paul*. New York: Continuum International.

Parkinson, W. 2017. "Conclusions: Anthropology, Archaeology, and the Legacy of Franz Boas; or, 'Hello, My Name is Bill, I am an Anthropologist. No wait. I am an Archaeologist. No. Wait . . . .'" In *These "Thin Partitions" Bridging the Growing Divide Between Cultural Anthropology and Archaeology*, ed. J. Englehardt and I. Rieger. Boulder, 269–286: University Press of Colorado.

Parrinder, G. 1967. *African Mythology*. London: Paul Hamlyn.

Partin, H. 2005. "Classification of Religions." In *Encyclopedia of Religion*, 2nd ed, ed. L. Jones. Vol. 3, *1817–1821*. Farmington Hills, MI: Macmillan Reference.

Paul, G. 2002. *Against Race. Imagining Political Culture Beyond the Color Line*. Cambridge, MA: Harvard University Press.

"Paulicianism." 2020. Britannica Online Encyclopedia. https://www.britannica.com/topic/Paulicians.

Pavlinov, I. 2021. *Biological Systematics. History and Theory*. Boca Raton, FL: CRC Press.

Pearson, B. 2008. "Basilides the Gnostic." In *A Companion to Second-Century Christian Heretics*, ed. A. Marjanen and P. Luomanen, 1–31. Leiden, Netherlands: Brill.

Penn, L. 2004. *False Dawn: The United Religions Initiative, Globalism, and the Quest for a One-World Religion*. Hillsdale, NY: Sophia Perennis.

Peoples, H., P. Duda, and F. Marlowe. 2016. "Hunter-Gatherers and the Origins of Religion." *Human Nature* 27: 261–282.

Percy, N., prod. 2005. *Israel and the Arabs: Elusive Peace*. Aired on BBC, October, 2005.

Peters, E. 1980. *Heresy and Authority in Medieval Europe*. London: Secular Press.

Peters, E. 1998. *The First Crusade: The Chronicle of Fulcher of Chartres and Other Source Materials*. Philadelphia: University of Pennsylvania Press.

Peters, E. 2009. *The Quest: The Historians Search for Jesus and Muhammad*. Prince Frederick, MD: Recorded Books.

Peters, J., and K. Schmidt. 2004. "Animals in the Symbolic World of Pre-Pottery Neolithic Göbekli Tepe, South-Eastern Turkey: A Preliminary Assessment." *Anthropozoologica* 39: 179–218.

Peterson, A., I. Gilhus, L. Martin, J. Jensen, and J. Sorenson, eds. 2019. *Evolution, Cognition, and the History of Religion: A New Synthesis*. Leiden, Netherlands: Brill.

Petrovich, C. 2017. "More Than Forty Amish Affiliations? Charting the Fault Lines." *Journal of Amish and Plain Anabaptist Studies* 5 (1): 120–142.

Pettitt, P. 2011. *The Paleolithic Origins of Human Burial*. New York: Routledge.

Pew Research Center. 2012. "The World's Muslims: Unity and Diversity." Pew Research Center: Muslims Around the World, August 9, 2012. https://www.pewresearch.org/religion/2012/08/09/the-worlds-muslims-unity-and-diversity-executive-summary.

Pew Research Center. 2014. "Russians Return to Religion but Not to Church." Pew Research Center: Orthodox Christianity, February 10, 2014. https://www.pewresearch.org/religion/2014/02/10/russians-return-to-religion-but-not-to-church.

Phalet, K., M. Maliepaard, F. Fleischmann, and D. Gungor. 2013. "The Making and Unmaking of Religious Boundaries." *Comparative Migrational Studies* 1 (1): 123–145.

Philippi, D., trans. 1969. *Kojiki*. Princeton, NJ: Princeton University Press.

Phillips, M. 2015. "Mesoamerica: Toltec." In *Concise Encyclopedia of Mexico*, ed. M. S. Werner, 378–380. New York: Routledge.

Pitblado, B. L. 2011. "A Tale of Two Migrations: Reconciling Recent Biological and Archaeological Evidence for the Pleistocene Peopling of the Americas." *Journal of Archaeological Research* 19 (4): 327–375.

Pitts, W. 1995. "Davidians and the Branch Davidians." In, *Armageddon in Waco: Critical Perspectives on the Branch Davidian Conflict*, ed. S. Wright, 20–42. University of Chicago Press.

Pitulko, V., E. Pavlov, P. Nikolskiy, and V. Ivanova. 2012. "The Oldest Art of the Eurasian Arctic: Personal Ornaments and Symbolic Objects from Yana RHS, Arctic Siberia." *Antiquity* 86: 642–659.

Plato. 360 BCE. *Timaeus* [*A Dialog of Plato*].

Pluralism Project of Harvard University. 2022. "Orthodox Christian Churches." https://pluralism.org/orthodox-christian-churches.

Pool, C. 2007. *Olmec Archaeology and Early Mesoamerica*. Cambridge: Cambridge University Press.

Pope Alexander VI. 1493. *Inter Caetera*. Papal Encyclicals. Papal Encyclicals Online. https://www.papalencyclicals.net/alex06/alex06inter.htm.

Pope Francis. 2013. *The Joy of the Gospel*. Libreria Editrice Vaticana, 2013: par. 253.

Popper, K. 1945 [1995]. *The Open Society and Its Enemies*. 5th ed. London: Routledge.

Popper, K. 1959. *The Logic of Scientific Discovery*. New York: Basic Books.

Popson, C. 2002. "Grim Rites of the Moche." *Archaeology* 55 (2). https://archive.archaeology.org/0203/abstracts/moche.html.

Possehl, G. 2002. *The Indus Civilization: A Contemporary Perspective*. Walnut Creek, CA: AltaMira.

Poussin, V. 1917. *The Way to Nirvana: Six Lectures on Ancient Buddhism as a Discipline of Salvation*. Cambridge: Cambridge University Press.

Prebish, C. 1975. *Buddhism: A Modern Perspective*. University Park: Pennsylvania State University Press.

Pregadio, F., ed. 2008. *The Encyclopedia of Taoism*. London: Routledge.

Prentiss, K. 2014. *The Embodiment of Bhakti*. New York: Oxford University Press.

Prestigiacomo, A. 2019. "Pope Francis tells students how not to convert unbelievers." Daily Wire, December 24, 2019. www.dailywire.com.

Priest, K., and R. Priest. 2007. "Divergent Worship Practices in the Sunday Morning Hour: Analysis of an 'Interracial' Church Merger Attempt." In *This Side of Heaven: Race, Ethnicity, and Christian Faith*, ed. R. Priest and N. Alvaro, 275–292. New York: Oxford University Press.

Pritz, R. 1988. *Nazarene Jewish Christianity*. Jerusalem: Magnes Press.

Prochazka-Eisl, G. 2016. "The Alevis." Oxford Research Encyclopedia of Religion. https://oxfordre.com/religion/display/10.1093/acrefore/9780199340378.001.0001/acrefore-9780199340378-e-101;jsessionid=3CE049A980E6F5C28A7BE986AD81BF23?result=1&rskey=AKKfeK.

Prothero, S. 2011. *God Is Not One: The Eight Rival Religions That Run the World*. New York: Harper Collins.

Protzen, J.-P., and S. Nair. 2000. "On Reconstructing the Tiwanaku Architecture." *Journal of the Society of Architectural Historians* 59: 358–371.

Proulx, D. 2008. "Paracas and Nasca: Regional Cultures on the South Coast of Peru." In *The Handbook of South American Archaeology*, ed. H. Silverman and W. Isbell, 563–588. New York: Springer.

Puhvel, J. 1987. *Comparative Mythology*, Baltimore, MD: Johns Hopkins University Press.

Pummer, R. 2010. "Samaritanism: A Jewish Sect or an Independent form of Yahwehism?" In *Samaritans: Past and Present*, ed. M. Mor and F. Reiterer, 1–24. New York Walter de Gruyter.

Pummer, R. 2015. *The Samaritans: A Profile*. Michigan: William B. Eerdmans Publishing Co.

Puri, H. 2003. "Scheduled Castes in Sikh Community: A Historical Perspective." *Economic and Political Weekly* 38: 2693–2701.

Puri, S. 2014. *The Human Dharma*. Create Space Independent Publishing.

Quilter, J. 2011. "Moche Religion." In *The Oxford Handbook of the Archaeology of Ritual and Religion*, ed. T. Insoll, 585–599. Oxford: Oxford University Press.

Rabi, I. 1951. "Dianetics: The Modern Science of Mental Health, by L. Ron Hubbard." [Book review.] *Scientific American* 184 (January): 57–58.

Radhakrishnan, S., and C, Moore. 1957. *A Sourcebook in Indian Philosophy*. Princeton, NJ: Princeton University Press.

Raff, J., and D. Bolnick. 2014. "Paleogenomics: Genetic Roots Science of Mental Health of the First Americans." *Nature* 506: 162–163.

Raghavan, M. et al. (with 56 coauthors) 2014. "The Genetic Prehistory of the New World Arctic." *Science* 345, no. 6200. DOI:10.1126/science.1255832.

Randeria, J. 1993. *The Parsi Mind: A Zoroastrian Asset to Culture*. New Delhi: Munshiram Manoharlal.

Raof, M. 1990. *Cultural Atlas of Mesopotamia and the Ancient Near East*. Oxford, UK: Andromeda.

Rasimus, T. 2009. *Paradise Reconsidered in Gnostic Myth Making. Rethinking Sethianism in the Light of the Ophite Evidence*. Leiden, Netherlands: Brill.

Rasmussen, K. 1921. *Eskimo Folk Tales*. Edited and translated into English by W. Waster. London: Pinnacle.

Ratcliffe, S., ed. 2016. *Oxford Essential Quotations* (Fourth Edition). UK: Oxford University Press.

Rasmussen, M., et al. 2010. "Ancient Human Genome Sequence of an Extinct Paleo-Eskimo." *Nature* 463: 757–762.

Ravenstein, E. G. 1861. *The Russians on the Amur*. London: Trubner.

Razoux, P. 2015. *The Iran-Iraq War*. Cambridge, MA: Belknap Press of Harvard University Press.

Record, J. 2010. *Wanting War. Why the Bush Administration Invaded Iraq*. Washington, DC: Potomac.

Redden, E. 2016. "*The Religious War Against American Scholars of India*." *Inside Higher Ed*. https://www.insidehighered.com/news/2016/04/12/scholars-who-study-hinduism-and-india-face-hostile-climate.

Reddy, K. 1999. *Indian History: For the UPSC Civil Services Preliminary Examination*. New Delhi: Tata McGraw-Hill.

Redford, D., ed. 2001. *The Oxford Encyclopedia of Ancient Egypt*. New York: Oxford University Press.

Reeves, N. 2001. *Akhenaten: Egypt's False Prophet*. London: Thames and Hudson.

Renna, T. 2015. "The Holy Roman Empire was Neither Holy, nor Roman, nor and Empire." *Michigan Academician* 42 (1): 60–75.

Renzetti, C., and S. Yocum, eds. 2013. *Clergy Sexual Abuse: Social Science Perspectives*. Boston, MA: Northeastern University Press.

Restall, M. 2003. *Seven Myths of the Spanish Conquest*. New York: Oxford University Press.

Reston, J. 2001. *Warriors of God: Richard the Lionheart and Saladin in the Third Crusade*. New York: Doubleday.

Reynolds, A. 1993. *Jesus Versus Christianity*. London: Cambridge International.

Rhys, J. 1888. *Lectures on the Origin and Growth of Religion as Illustrated by Celtic Heathendom.* London: Williams and Norgate.

Rhodes, J. 1991. "An American Tradition: The Religious Persecution of Native Americans." *Montana Law Review* 52 (1). https://scholarworks.umt.edu/mlr/vol52/iss1/4.

Ricard, R. 1966. *The Spiritual Conquest of Mexico.* Translated by Lesley Byrd Simpson. Berkeley: University of California Press.

Richards, R. 2010. *The Species Problem: A Philosophical Analysis.* Cambridge: Cambridge University Press.

Richerson, P., and R. Boyde. 2005. *Not by Genes Alone. How Culture Transformed Human Evolution.* Chicago: University of Chicago Press.

Richerson, P., and J. Henrich. 2012. "Tribal Social Instincts and the Cultural Evolution of Institutions to Solve Collective Action Problems." *Cliodynamics* 3: 38–80.

Richter, P. 2003. "General Apologizes for Remarks on Islam, Says He's No Zealot." *Los Angeles Times,* October 18, 2003.

Rieppel, O. 2016. *Phylogenetic Systematics: Haeckel to Hennig.* Boca Raton, FL: CRC Press.

Riley-Smith, J. 1991. *The First Crusade and the Idea of Crusades.* Philadelphia: University of Pennsylvania Press.

Ringgren, H. 1974. *Religions of the Ancient Near East.* Translated by John Sturdy. Louisville, KY: Westminster John Knox.

Risse, G. 1995. *Mending Bodies, Saving Souls: A History of Hospitals.* New York: Oxford University Press.

Rives, J. 1995. "Human Sacrifice Among Pagans and Christians." *Journal of Roman Studies* 85: 65–85.

Roach, A. 2005. *The Devil's World: Heresy and Society.* New York: Routledge.

Robinet, I. 1997. *Taoism: Growth of a Religion.* Stanford, CA: Stanford University Press.

Rogers, H. 1896. *Mary Dyer of Rhode Island. The Quaker Martyr that was hanged on Boston Common, June 1, 1660.* Norwood, MA: Norwood Press..

Rogerson, A. 1969. *Millions Now Living Will Never Die: A Study of Jehovah's Witnesses.* London: Constable.

Rogerson, J. K. 2003. "Deuteronomy." In *Eerdmans Commentary on the Bible,* ed. J. Dunn and J. Rogerson. Grand Rapids, MI: Eerdmans.

Romey, K. 2018. "Ancient Mass Child Sacrifice May Be World's Largest." *National Geographic,* April 26, 2018. https://www.nationalgeographic.com/science/article/mass-child-human-animal-sacrifice-peru-chimu-science.

Roney, J., and R. Hard. 2009. "The Beginnings of Maize Agriculture." *Archaeology Southwest* 23 (1): 4–5.

Roscoe, P. 2006. "The Comparative Method." In *The Blackwell Companion to the Study of Religion,* ed. R. Segal, 25–46. Malden, MA: Blackwell.

Rose, J. 2011. *Zoroastrianism: A Guide for the Perplexed.* London: Continuum International.

Ross, A. 1974. *Pagan Celtic Britain: Studies in Iconography and Tradition.* London: Sphere.

Rossano, M. 2006. "The Religious Mind and the Evolution of Religion." *Review of General Psychology* 10: 346–364.

Roukema, R. 1998. *Gnosis and Faith in Early Christianity.* Harrisburg, PA: Trinity.

Rountree, K. 2001. "The Past Is a Foreigners' Country: Goddess Feminists, Archaeologists, and the Appropriation of Prehistory." *Journal of Contemporary Religion* 16: 5–27.

Rowe, J. 1948. "The Kingdom of Chimor." *Aus Acta Americana* 6: 27–59.

Rudy, W. 1984. *The Universities of Europe, 1100–1914.* London: Fairleigh Dickinson University.

Rüpke, J., ed. 2007. *A Companion to Roman Religion.* Malden, MA: Wiley-Blackwell.

Russell, F. 1975. *The Just War in the Middle Ages.* Cambridge: Cambridge University Press.

Ryan, J. 2009. *Imaginary Friends: Representing Quakers in American Culture, 1650–1950.* Madison: University of Wisconsin Press.

Ryder, J. and G. Nicholls. 2011. "Missing Data in a Stochastic Dollo Model for Binary Trait Data, and Its Application to the Dating of Proto-Indo-European." *Journal of the Royal Statistical Society, Series C* 60 (1): 71–92.

Sahagun, B. 1540–1585. *Florentine Codex: General History of the Things of New Spain.* Translation by C. Dibble and A. Anderson, 1950–1982. Salt Lake City: University of Utah Press.

Sahgal, N., and B. Grimm. 2013. "Egypt's Restrictions on Religion Coincide with Lack of Religious Tolerance. Pew Research Center: International Religious Freedom & Restrictions, July 2, 2013. https://www.pewresearch.org/short-reads/2013/07/02/egypts-restrictions-on-religion-coincide-with-lack-of-religious-tolerance.

Saler, B. 2001. "Comparativism: Some Suggestions for Improving the Inevitable." *Numen* 48: 267–275.

Salguero, C. P., ed. 2017. *Buddhism and Medicine: An Anthology of Premodern Sources.* New York: Columbia University Press.

Salguero, C. P., ed. 2020. *Buddhism and Medicine: An Anthology of Modern and Contemporary Sources.* New York: Columbia University Press.

Salguero, C. P. 2022. *A Global History of Buddhism and Medicine.* New York: Columbia University Press.

Saliba, J. A. 1976. *"Homo Religiousus" in Micea Eliade: an Anthropological Evaluation.* Leiden, Netherlands: Brill.

Samaritan Update. n.d. "Mount Gerizim, All the Days of Our Lives." Accessed August 24, 2018. http://www.thesamaritanupdate.com.

Sangave, V. 1980. *Jaina Community: A Social Survey.* 2nd ed. Mumbai: Popular Prakashan.

Sangave, V. 2001. *Facets of Jainology: Selected Research Papers on Jain Society, Religion, and Culture.* Mumbai: Popular Prakashan.

Sanmark, A. 2003. "Power and Conversion: A Comparative Study of Christianization in Scandinavia." *Occasional Papers in Archaeology* 34.

Saunders, N. 2017. "The Evolving Theology of the Nation of Islam." In *New Perceptions of The Nation of Islam,* ed. D. Gibson and H. Berg. New York: Routledge.

Schaferdiek, K. 2014. "Ulfila and the So-Called 'Gothic' Arianism-English Summary." In *Arianism: Roman Heresy and Barbarian Creed,* ed. G. Berndt and R. Steinacher, 45–48. New York: Routledge.

Schmidt, K. 2006. *Sie bauten die ersten Tempel. Das ratselhafte Heiligtum der Steinzeitjager.* Munich, Germany: Verlag.

Schmidt, K. 2010. "Gobekli Tepe—the Stone Age Sanctuaries: New Results of Ongoing Excavations with a Special Focus on Sculptures and High Reliefs." *Documenta Prehistorica* 37: 239–256.

Schomer, K., and W. McLeod, eds. 1987. *The Sants: Studies in a Devotional Tradition of India*. Delhi: Motilal Banarsidass.

Schopf, J. 2002. *Life's Origins: The Beginning of Biological Evolution*. Berkeley: University of California Press.

Schreiber, K. 2005. "Sacred Landscapes and Imperial Ideologies: The Wari Empire in Sondondo, Peru." *Archaeological Papers of the American Anthropological Association* 14: 131–150.

Schultz, H. 2000. *The Germanic Realms in Pre-Carolingian Central Europe*. New York: Peter Lang.

Seager, R. 1995. *The World's Parliament of Religions: The East/West Encounter, Chicago, 1993*. Bloomington: Indiana University Press.

Segal, R. 1999. *Theorizing About Myth*. Amherst: University of Massachusetts Press.

Sennis, A., ed. 2016. *Cathars in Question*. UK, York Medieval Press.

Serratelli, A. 2018. "Pope Francis: Proselytism vs. Evangelization." *Catholic News Agency*, May 11, 2018.

Shady, R. 2014. "Living Conditions, Social Systems, and Cultural Expressions of the Caral and Chinchorro Populations During the Archaic Period." In *The Chinchorro Culture: A Comparative Perspective, the Archaeology of the Earliest Human Mummification*, ed. N. Sanz, B. Arriaza, and V. Standen, 71–106. Paris, France: UNESCO.

Shady Solis, R., J. Haas, and W. Creamer. 2001. "Dating Caral, a Preceramic Site in the Supe Valley on the Central Coast of Peru." *Science* 292: 723–726.

Sharer, R. and L. Traxler. 2006. *The Ancient Maya*. 6th ed. Stanford, CA: Stanford University Press.

Sharma, V. 2001. *History of Jainism with special reference to Mathura. Reconstructing Indian History and Culture no. 23*. New Delhi: D.K. Printworld.

Sharot, S. 2001. *A Comparative Sociology of World Religions: Virtuosos, Priests, and Popular Religion*. New York: New York University Press.

Shirer, W. 1960. *The Rise and Fall of the Third Reich*. London: Secker & Warburg.

Shizuka, S. 1997. "A study on the Origin of Mahayana Buddhism." *Eastern Buddhist Society* 30 (1): 70–113.

Shoemaker, M. 1912. *Indian Pages and Pictures: Rajputana, Sikkim, the Punjab, and Kashmir*. New York: Knickerbocker.

Shulamit, S. 1994. "The Head of the Rabbinite, Karaite and Samaritan Jews: On the History of a Title." *Bulletin of the School of Oriental and African Studies* 57 (2): 255–267.

Sibly, W., and M. Sibly. 2003. *The Chronicle of William of Puylaurens: The Albigensian Crusade and its Aftermath*. Woodbridge, UK: Boyden.

Silverman, H., and D. Proulx. 2002. *The Nasca*. Malden, MA: Blackwell.

Singh, B. 2012. "Concepts of Species and Modes of Speciation." *Current Science* 103: 784–790.

Singh, B. n.d. "The Sikhs Are Neither Hindus nor Muslims." Accessed January 21, 2018. http://www.sikhmarg.com/english/The-Sikhs.html.

Singh, G. 2004. *Gandhi: Behind the Mask of Divinity*. Amherst, NY: Prometheus.

Singh, P. and L. Fenech, eds. 2014. *The Oxford Handbook of Sikh Studies*. Oxford University Press.

Singh, U. 2008. *History of Ancient and Early Medieval India: From the Stone Age to the Twelfth Century.* Delhi: Pearson Education India.

Singh Khalsa, H. 2015. "Yoga in Sikhism." SikhNet. https://www.sikhnet.com/news/yoga-sikhism.
Skjaervo, P. 2011. *The Spirit of Zoroastrianism*. New Haven, CT: Yale University Press.
Slade, S. 2019. "The New Theocrats are Neither Conservative Nor Christian." *Reason Magazine*, June 3, 2019. https://reason.com/2019/06/03/the-new-theocrats-are-neither-conservative-nor-christian.
Smita, N. 1999. *Broken People: Caste Violence Against India's "Untouchables."* New York: Human Rights Watch.
Smith, A. 2009. *Indigenous Peoples and Boarding Schools: A Comparative Study*. Report to the Secretariat of the United Nations Permanent Forum on Indigenous Issues.
Smith, J. 1971. "Adde Parvum Parvo Magnus acervus Erit." *History of Religions* 11 (1): 67–90.
Smith, J. 1982. *Imagining Religion. From Babylon to Jonestown*. Chicago: University of Chicago Press.
Smith, J. 1990. *Drudgery Divine. On the Comparison of Early Christianities and the Religions of Late Antiquity*. Chicago: University of Chicago Press.
Smith, J. 1998. "Religion, Religions, Religious." In *Critical Terms for Religious Studies*, edited by M. Taylor, 269–284. Chicago: University of Chicago Press.
Smith, J. 2004. *Relating Religions. Essays in the Study of Religion*. Chicago: University of Chicago Press.
Smith, M. 2001. *The Origins of Biblical Monotheism: Israel's Polytheistic Background and the Ugaritic Texts*. New York: Oxford University Press.
Smith, M. 2002. *The Early History of God: Yahweh and the Other Deities in Ancient Israel*. Grand Rapids, MI: Eerdmans.
Smith, M. 2010. *God in Translation: Deities in Cross-Cultural Discourse in the biblical World*. Grand Rapids, MI: Eerdmans.
Smith, M., J. Wharton, and J. Olson. 2003. "Aztec Feasts, Rituals and Markets." In *The Archaeology and Politics of Food and Feasting in Early States and Empires*, ed. T. L. Bray. New York, 235–270: Kluwer Academic/Plenum.
Smith, W. C. 1962. *The Meaning and End of Religion*. New York: MacMillan.
Sober, E. 1988. *Reconstructing the Past: Parsimony, Evolution, and Inference*. Cambridge, MA: MIT Press.
Sokal, C. 2019. "Washington State lawmaker, Matt Shea, defends advocacy for 'Holy Army' as Spokane sheriff refers his writings to FBI." Seattle Times, November 1, 2018, updated December 20, 2019.
Sonbol, A., ed. 2005. *Beyond the Exotic: Women's Histories in Islamic Societies*. Syracuse, NY: Syracuse University Press.
Sosis, R. 2019. "Why Cultural Evolutionary Models of Religion Need a Systematic Approach." In *Evolution, Cognition, and the History of Religion: A New Synthesis*, ed. A. Peterson, I. Gilhus, L. Martin, J. Jensen, and J. Sorenson, 45–61. Leiden, Netherlands: Brill.
Spencer, H. 1864. *Principles of Biology*. Vol. 1. London: Williams and Norgate.
Spencer, H. 1898. *Principles of Sociology*. Vols. 1 and 2. D. New York: Appleton.
Spilsbery, J. 1652. *A Treatise Concerning the Lawful Subject of Baptism*. London: Henry Hills in Fleet-Yard.
Spiro, J. 2009. *Defending the Master Race. Conservation, Eugenics, and the Legacy of Madison Grant*. University of Vermont Press.

Spong, J. 2007. *Jesus for the Non-Religious: Recovering the Divine at the Heart of the Human.* New York: HarperCollins.

Spores, R., and A. Balkansky. 2013. *The Mixtecs of Oaxaca: Ancient Times to the Present.* Norman: University of Oklahoma Press.

Spretnak, C. 2011. "Anatomy of Backlash: Concerning the Work of Marija Gimbutas." *Journal of Archaeomythology* 7: 25–51.

Srinivasan, A. 2011. *Hinduism for Dummies.* Hoboken, NJ: Wiley.

Srinivasan, D. 1975. "The So-Called Proto-Siva Seal from Mohenjo-Daro: An Iconological Assessment." *Archives of Asian Art* 29: 47–58.

Srivastava, S. 2001. "Parsis Turn to Solar Power." BBC News South Asia, July 18, 2001. http://news.bbc.co.uk/2/hi/south_asia/1443789.stm.

Stanglin, K., and T. McCall 2012. *Jacob Arminius. Theologian of Grace.* New York: Oxford University Press.

Stanish, C. 2003. *Ancient Titicaca: The Evolution of Complex Society in Southern Peru and Northern Bolivia.* Berkeley: University of California Press.

Stanish, C. 2017. *The Evolution of Human Co-operation. Ritual and Social Complexity in Stateless Societies.* Cambridge: Cambridge University Press.

Stannard, D. 1992. *American Holocaust: The Conquest of the New World.* New York: Oxford University Press.

Stark, R. 1997. *The Rise of Christianity: How the Obscure Marginal Jesus Movement Became the Dominate Religious Force in the Western World in a Few Centuries.* San Francisco, CA: HarperSanFrancisco.

Steadman, L., C. Palmer, and C. Tilley. 1996. "The Universality of Ancestor Worship." *Ethnology* 35: 63–76.

Stearns, P. 2005. *World History in Brief: Major Patterns of Change and Continuity.* Fifth Edition. New York: Pearson Longman.

Steele, R. 2004. *Handbook of Inca Mythology.* Santa Barbara, CA: ABC-CLIO.

Steinkerchner, S. 2020. "Introduction: Dominicans and Jesuits, Through the Centuries." *Journal of Jesuit Studies* 7 (3): 357–376.

Stern, P. 2013. *Historical Dictionary of the Inuit.* 2nd ed. Lanham, MD: Scarecrow.

Stevens, A. 2003. *Archetype Revisited: An Updated Natural History of Self.* Toronto, Canada: Inner City.

Stewart, D. 1998. *Islamic Legal Orthodoxy: Twelver Shiite Response to the Sunni Legal System.* Salt Lake City: University of Utah Press.

Stoddart, S. 2009. *Historical Dictionary of the Etruscans.* Lanham, MD: Scarecrow.

Stone, A. 2019. "The Lineages of the First Humans to Reach Northeastern Siberia and the Americas." *Nature* 570: 170–172.

Sturluson, S. 1220. *Edda.* Translation of A. Faulkes from 1987. London: J. M. Dent.

Sujato, B. 2006. *Sects and Sectarianism. The Origins of Buddhist Schools.* Taiwan: Buddha Education Foundation.

Sullivan, L. 2002. "Introduction." In *Native Religions and Cultures of Central and South America,* ed. L. Sullivan, 1–6. New York: Continuum International.

Sumption, J. 1999. *The Albigensian Crusade.* London: Faber and Faber.

Sutton, J., and M. Chaves. 2004. "Explaining Schism in American Protestant Denominations, 1890–1990." *Journal for the Scientific Study of Religion* 43 (2): 171–189.

Sweeny, K. 2008. "Early American Religious Traditions: Native Visionary and Christian Providence." *OAH Magazine of History* 22 (1): 8–13.

Sybesma, R., ed. 2015. *Encyclopedia of Chinese Language and Linguistics Online*. Leiden, Netherlands: Brill.

Taylor, B. 2009. *Dark Green Religion: Native Spirituality and the Planetary Future*. Berkeley: University of California Press.

Taylor, D. 2014. *Like a Mighty Army? The Salvation Army, the Church, and the Churches*. Cambridge: James Clark.

Taylor, J. 2001. *Death and the Afterlife in Ancient Egypt*. Chicago: University of Chicago Press.

Taylor, R., and O. Bar-Yosef. 2016. *Radiocarbon Dating: An Archaeological Perspective*. 2nd ed. New York: Routledge.

Taylor R. and H. Chow. 2005. *The Illustrated Encyclopedia of Confucianism*. Vol. 2. New York: Rosen.

Taysom, S. 2011. *Shakers, Mormons and Religious Worlds: Conflicting Visions, Contested Boundaries*. Indiana University Press.

Teasley, D. 1999. *The Branch Davidian Siege at Ranch Apocalypse Near Waco, Texas: A Chronology. CRS Report for Congress*. Congressional Research Service, Library of Congress.

Teeter, E. 2011. *Religion and Ritual in Ancient Egypt*. New York: Cambridge University Press.

Terberger, T., M. Zhilin, and S. Savchenko. 2021. "The Shigir Idol in the Context of Early Art in Eurasia." *Quaternary International* 573: 14–29.

Terraciano, K. 2001. *The Mixtecs of Colonial Oaxaca*. Stanford, CA: Stanford University Press.

Terrell, J. and J. Hart. 2002. "Introduction." In *Darwin and Archaeology: A Handbook of Key Concepts*, ed. J. Hart and J. Terrell, 1–14. Westport, CT: Bergin and Garvey.

Tesch, N. 2018. "Wahhabi." Britannica Online Encyclopedia. https://www.britannica.com/topic/Wahhabi/additional-info#history.

Thames, K. 2022. "Putin's Unholy War." *Atlantic Council*, April 22, 2022.

Theissen, G. 1985. *Biblical Faith: An Evolutionary Approach*. Philadelphia, PA: Fortress.

Thompson, J. 1970. *Maya History and Religion*. Norman: University of Oklahoma Press.

Tigay, J. 1982. *The Evolution of the Gilgamesh Epic*. Philadelphia: University of Pennsylvania Press.

Time. 1977. "Religion: A Sci-Fi Faith." April 5, 1976.

Toohey, J. 2013. "Feeding the Mountains: Sacred Landscapes, Mountain Worship, and Sacrifice in the Maya and Inca Worlds." *Reviews in Anthropology* 42: 161–178.

Toshio, K. 1981. "Shinto in the History of Japanese Religion." Trans. J. Dobbins and S. Gray. *Journal of Japanese Studies* 7 (1): 1–21.

Towsend, R. 1985. "Deciphering the Nazca World: Ceramic Images from Ancient Peru." *Art Institute of Chicago Museum Studies* 11 (2): 116–139.

Trinkaus, E., A. Buzhilova, M. Mednikova, and M. Dobrovolskaya 2014. *The People of Sunghir: Burials, Bodies, and Behavior in the Earlier Upper Paleolithic*. New York: Oxford University Press.

Tsu, L. 1972. *Tao Te Ching. A New Translation by Gia-Fu Feng and Jane English*. New York: Vintage.

Turchin, P. 2009. "A Theory for the Formation of Large Empires." *Journal of Global History* 4: 191–217.

Turchin, P., et al. 2015. "Seshat: The Global History Databank." *Cliodynamics* 6: 77–107.

Turchin, P., et al. 2022. "Big Gods and Big Science: Further Reflections on Theory, Data, and Analysis." *Religion, Brain, and Behavior* 13 (2): 218–231.

Turchin, P., and S. Garrilets. 2009. "Evolution of Complex Hierarchical Societies." *Social Evolution and History* 8: 167–198.

Turku, H. 2018. *The Destruction of Cultural Property as a Weapon of War: ISIS in Syria and Iraq*. Cham, Switzerland: Springer International.

Turner, J. 2001. *Sethian Gnosticism and the Platonic Tradition*. Paris: Peeters.

Turville-Petre, E. 1975. *Myth and Religion of the North: The Religion of Ancient Scandinavia*. Westport, CT: Greenwood.

Tutu, D., and J. Allen, ed. 2011. *God Is Not a Christian And Other Provocations*. New York; Harper One.

Twain, M. 1935. *Mark Twain's Notebook*. New York: Harper Brothers.

Tylor, E. 1871. *Primitive Culture: Researches into the Development of Mythology, Philosophy, Religion, Art, and Custom*. Vols. 1 and 2. London: John Murray.

United Nations News. 2021. "In Iraq, Pope Spreads Message of Peace, Religious Tolerance, and Humanities Resilience." *United Nations News*, March 7, 2021. https://news.un.org/en/story/2021/03/1086652.

University of Cambridge Darwin Correspondence Project. n.d. Accessed March 1, 2023. https://www.darwinproject.ac.uk.

URI (United Religions Initiative). 2000. URI Charter. https://www.uri.org/what-we-do/charter.

Urton, G. 1999. *Incan Myths*. Austin: University of Texas Press.

Uttal, W. 2004. *Dualism: The Original Sin of Cognitivism*. New York: Psychology Press.

Utter, G. and J. True 2004. *Conservative Christians and Political Participation*. California: ABC-CLIO. Inc.

Van Dijk, J. 2003. "Hell." *The Oxford Essential Guide to Egyptian Mythology*, ed. D. Redford, 161–162. New York: Berkley.

Van Norden, B., and N. Jones. 2019. "Huayan Buddhism." In *The Stanford Encyclopedia of Philosophy*, ed. E. N. Zalta. https://plato.stanford.edu/archives/win2019/entries/buddhism-huayan.

Vainio, O.-P. 2008. *Justification and Participation in Christ. The Development of the Lutheran Doctrine from Luther to the Formula of Concord*. Leiden, Netherlands: Brill.

Vatican Archives. 1993. "Catechism of the Catholic Church." http://www.vatican.va/archive/ENG0015/_INDEX.HTM.

Vatican, 2007. General Decree regarding the derelict of attempted sacred ordination of a woman. Cannon 30, Common Law, in the Ordinary Session of December 19, 2007. https://www.vatican.va/roman_curia/congregations/cfaith/documents/rc_con_cfaith_doc_20071219_attentata-ord-donna_en.html.

Verano, J. 2008. "Trophy Head-Taking and Human Sacrifice in Andean South America." In *The Handbook of South American Archaeology*, ed. H. Silverman and W. Isbell, 1047–1061. New York: Springer.

von Glasenapp, H. 1925. *Jainism: An Indian Religion of Salvation* [*Der Jainismus: Eine Indische Erlosungsreligion*] Translated by Shridhar B. Shrotri. Delhi: Motilal Banarsidass.

Wade, N. 2009. *The Faith Instinct. How Religion Evolved and Why It Endures*. New York: Penguin.

Wade, L. 2018. "Feeding the Gods." *Science* 360: 1288–1292.

Wakefield, M. 2009. *Understanding Scientology: The Demon Cult*. Morrisville, NC: Lulu Press, Inc.

Walbank, F. 1982. *The Hellenistic World*. Cambridge, MA: Harvard University Press.

Waldron, A. 1998. "Religious Revivals in Communist China." *Orbis* 42 (2): 325–334.

Walker, C. 1987. *Reading the Past: Cuneiform*. London: British Museum Press.

Wallace, A. 1966. *Religion: An Anthropological View*. New York: Random House.

Wallace, D. 2009. "All That." *New Yorker*, December 14, 2009.

Wallace, M. 2018. *When God Was a Bird: Christianity, Animism, and the Reenchantment of the World*. New York: Fordham University Press.

Waller, M. 2006. *Sovereign Ladies: The Six Reigning Queens of England*. New York: St. Martin's.

Wang, R. 2012. *Yinyang: The Way of Heaven and Earth in Chinese Thought*. Cambridge: Cambridge University Press.

Warrant, T. 2018. "Washington State Lawmaker, Matt Shea, Defends Advocacy for "Holy Army" as Spokane Sheriff Refers His Writings to the FBI." *Seattle Times*, November 1, 2018.

Webb, T. 2013. *Transatlantic Methodists: British Wesleyanism and the Formation of an Evangelical Culture in Nineteenth-Century Ontario and Quebec*. London: McGill-Queens, University Press.

Weber, T. 2004. *On the Road to Armageddon: How Evangelicals Became Israel's Best Friend*. Grand Rapids, MI: Baker Scientific.

Werblowsky, R., and G. Wigoder. 1997. *The Oxford Dictionary of Jewish Religion*. Oxford: Oxford University Press.

Werner, K. 1977. "Yoga and the Ṛig-Veda: An Interpretation of the Keśin Hymn (RV 10, 136)." *Religious Studies* 13: 289–302.

Werbach, A. 2010. "Climate Change Denier Aims to Chair Energy Committee." *The Atlantic*, November 10, 2010.

Wesler, K. 2012. *An Archaeology of Religion*. Lanham, MD: University Press of America.

Westenholz, J. 1997. *Legends of the Kings of Akkad*. Winona Lake, IN: Eisenbrauns.

Whitmarsh, T. 2021. "Ancient Greece." In *The Cambridge History of Atheism*, eds. S. Bullivant and M. Ruse, 84–99. Cambridge University Press.

Wigoder, G., F. Skolnik, and S. Himelstein, eds. 2002. *The New Encyclopedia of Judaism*. 2nd ed. New York: New York University Press.

Wilbur, E. 1952. *A History of Unitarianism in Transylvania, England and America*. Cambridge, MA: Harvard University Press.

Wiley, G., et al. 2016. "Pre-Columbian Civilizations." *Encyclopedia Britannica*. https://www.britannica.com/topic/pre-Columbian-civilizations.

Wilkinson, R. 2003. *The Complete Gods and Goddesses of Ancient Egypt*. London: Thames and Hudson.

Willerslev, R. 2007. *Soul Hunters: Hunting, Animism, and Personhood Among the Siberian Yukaghirs*. Berkeley: University of California Press.

Willerslev, R., and O. Ulturgasheva 2012. "Revisiting the Animism Versus Totemism Debate: Fabricating Persons Among the Envy and Chukchi of North-eastern Siberia." In *Animism in Rainforests and Tundra: Personhood, Animals, Plants, the Things in Contemporary Amazonia and Siberia*, ed. M. Brightman, V. Grotti, and O. Ulturgasheve, 48–68. New York: Berghahn.

Williams, A. 2021. "The Former Uber Exec Who Was Pardoned by Trump Has Closed His Church That Worshipped AI, Donating Its Funds to the NAACP." Business Insider, February 19, 2021. https://www.businessinsider.com/uber-google-ai-anthony-levandowski-trump-pardon-church-naacp-2021-2.

Williams, M. 1996. *Rethinking Gnosticism: An Argument for Dismantling a Dubious Category*. Princeton, NJ: Princeton University Press.

Williams, P. 2008. *Mahayana Buddhism: The Doctrinal Foundations*. 2nd ed. London: Routledge.

Williams, R., and D. Nash. 2016. "Religious Ritual and Wari State Expansion." In *Ritual and Archaic States*, ed J. Murphy, 131–156. Gainesville, University of Florida Press.

Wilson, D. S. 2003. *Darwin's Cathedral: Evolution, Religion and the Nature of Society*. Chicago: University of Chicago Press.

Wilson, D. S. 2007. "Atheism as a Stealth Religion." *Huffington Post*. December 14, 2007.

Wilson, E. O. 1998. *Consilience: The Unity of Knowledge*. New York: Vintage House.

Wilson, E. O. 2012. *The Social Conquest of Earth*. New York: Liveright.

Wilson, R. 1958. *The Gnostic Problem*. London: A. R. Mowbray.

Wine, S. 1995. *Judaism Beyond God*. Society for Humanistic Judaism. Hoboken, NJ: Milan Press.

Winfield, N., and P. Smith. 2022. "Pope Apologizes for 'Catastrophic' School Policy in Canada." *AP News*, July 26, 2022.

Winroth, A. 2014. *The Age of the Vikings*. Princeton, NJ: Princeton University Press.

Winzeler, R. 2008. *Anthropology and Religion*. Lanham, MD: AltaMira.

Wood, J. 2000. "Introduction." In *The Mythology of the British Islands: An Introduction to Celtic Myth, Legend, and Romance*, ed. C. Squire. Ware: Wordsworth Editions.

Wooden, C. 2015. "The devil knows Christians are one, says Pope Francis." *The Catholic Herald*, May 25, 2017.

Woods, C., G. Emberling, and E. Teeter, eds. 2010. *Visible Language: Invention of Writing in Ancient Middle East and Beyond*. Oriental Institute Museum Publication no. 2. Chicago: University of Chicago.

Woodard, R. 2010. *Indo-European Sacred Space: Vedic and Roman Cult*. Urbana: University of Illinois Press.

Wright, R. 2009. *The Evolution of God*. New York: Back Bay.

Wright, R. 2013. *Mysteries of the Jaguar Shamans of the Northwest Amazon*. Lincoln: University of Nebraska Press.

Wybraniec, J., and R. Finke. 2004. "Religious Regulations in the Courts." In *Regulating Religion: Case Studies from Around the Globe*, ed. J. Richardson. Boston, MA, 535–554: Springer.

Wypustek, A. 2013. *Images of Eternal Beauty in Funerary Verse. Inscriptions of the Hellenistic and Greco Roman Periods*. Leiden, The Netherlands: Brill.

Xing, G. 2004. *The Concept of Buddha: Its Evolution from Early Buddhism to Trikaya Theory*. New York: RoutledgeCurzon.

Yamakage, M., M. Gillespie, S. Mineko, G. Gillespie, Y. Komuro, P. Leeuw, and A. Rankin. 2007. *The Essence of Shinto: Japan's Spiritual Heart*. Tokyo: Kodansha International.

Yang, C. 1961. *Religion in Chinese Society; a Study of Contemporary Social Functions of Religion and Some of Their Historical Factors*. Berkeley: University of California Press.

Yang, L., D. An, and J. Turner. 2005. *Handbook of Chinese Mythology*. Oxford: Oxford University Press.

Yao, X. 2000. *An Introduction to Confucianism*. New York: Cambridge University Press.

Young, E. 2002. "Mass Human Sacrifice Unearthed in Peru." *New Scientist*, October 2002.

Zakaria, F. 2003. "And He's Head of Intelligence?" *Newsweek*, October 26, 2003. https://www.newsweek.com/and-hes-head-intelligence-139059.

Zeller, B. 2014. *Heaven's Gate. America's UFO Religion*. NY: New Youk University Press.

Zerin, E. 1998. "Karl Popper on God: The Lost Interview." *Skeptic* 6 (2): 46–49.

Zhilin, M., S. Savchenko, S. Hansen, K. Heussner, and T. Terberger. 2018. "Early Art in the Urals: New Research on the Wooden Sculpture from Shigir." *Antiquity* 92 (362): 334–350.

Zilhao, J. 2007. "The Emergence of Ornaments and Art: An Archaeological Perspective on the Origins of 'Behavioral Modernity.'" *Journal of Archaeological Research* 15: 1–54.

Zimmer, H. 1969. *Philosophies of India*. Ed. J. Campbell. Princeton, NJ: Princeton University Press.

Zsuzsanna, G. 2015. *Mani's Pictures: The Didactic Images of the Manichaeans from Sasanian Mesopotamia to Uygur Central Asia and Tang-Ming China*. Leiden, Netherlands: Brill.

Zuidema, R. 2002. "Inca Religion: Its Foundations in the Central American Context." In *Native Religions and Cultures of Central and South America*, ed. L. Sullivan, 236–253. New York: Continuum International.

Zuo, J. 1991. "Political Religion: The Case of the Cultural Revolution in China." *Sociological Analysis* 52: 99–110.

Zydenbos, R. 2006. *Jainism Today and Its Future*. Munich, Germany: Manya Verlag.

FIGURE CREDITS

PREFACE

Page x: Modified by the author. Annatsach, CC BY-SA 4.0, Wikimedia Commons. https://commons.wikimedia.org/wiki/File:Lone_tree_in_winter.jpg.

CHAPTER 1

All figures by the author.

CHAPTER 2

All figures and photography by the author.

CHAPTER 3

All figures by the author unless otherwise noted.

Page 49: Modified by the author. Chimpanzee Congo, Public domain, Wikimedia Commons. https://commons.wikimedia.org/wiki/File:Chimpanzee_congo_painting.jpg.

Page 50:

(*Top*) Modified by the author. Chris S. Henshilwood, CC BY-SA 4.0, Wikimedia Commons. https://commons.wikimedia.org/wiki/File:Blombo.jpg.

(*Bottom left*) Modified by the author. Basran Burhan, CC BY-SA 4.0, Wikimedia Commons. https://commons.wikimedia.org/wiki/File:Leang_Tedongnge_rock_art_panel_credit_Basran_Burhan.jpg.

(*Bottom right*) Modified by the author. @ photo Luc-Henri Fage, www.fage.fr., Public domain, Wikimedia Commons. https://commons.wikimedia.org/wiki/File:Lubang_Jeriji_Sal%C3%A9h_cave_painting_of_Bull.jpg.

Page 52: Modified by the author. José-Manuel Benito Álvarez, Public domain, Wikimedia Commons. https://commons.wikimedia.org/wiki/File:Sunghir-tumba_paleol%C3%ADtica.jpg.

Page 57: Specimen in the collection of the Ulm Museum, Ulm, Germany. Photo by Yvonne Muhleis, courtesy of Landesamt fur Denkmalpfledge im RP Stuttgart und Museum Ulm, Baden-Wurttemberg, Germany.

Page 60:

(*Top*) Image courtesy of the Institut für Ur- und Frühgeschichte und Archäologie des Mittelalters, Abteilung für Ältere Urgeschichte und Quartärökologie, Universität Tübingen.

(*Bottom left*) Original piece in the Moravske Zemske Museum, Czech Republic. Petr Novák, CC BY-SA 2.5, Wikimedia Commons. https://commons.wikimedia.org/wiki/File:Vestonicka_venuse.jpg.

(*Bottom center*) Original piece in the Naturhistorisches Museum, Vienna. Don Hitchcock, CC BY-SA 4.0, Wikimedia Commons. https://commons.wikimedia.org/wiki/File:Willendorf-Venus-1468.jpg.

(*Bottom right*) Modified by the author. Original piece in the Hermitage Museum, Saint Petersburg, Russia. Andreas Franzkowiak, Archäologisches Museum Hamburg und Stadtmuseum Harburg, CC BY-SA 3.0, Wikimedia Commons. https://commons.wikimedia.org/wiki/File:Venus_4_from_Kostenki_RU_front_left.jpg.

Page 64:

(*Top*) Modified with shadow figure of human for scale. Dick Osseman, CC BY-SA 4.0, Wikimedia Commons. https://commons.wikimedia.org/wiki/File:G%C3%B6beklitepe_Building_C_5299.jpg.

(*Bottom left*) Zhengan, CC BY-SA 4.0, Wikimedia Commons. https://commons.wikimedia.org/wiki/File:G%C3%B6bekli_Tepe_Pillar.JPG.

(*Bottom middle*) Cobija, CC BY-SA 4.0, Wikimedia Commons. https://commons.wikimedia.org/wiki/File:%C5%9Eanl%C4%B1urfa_M%C3%BCzesi_Neotilik_%C3%87a%C4%9F_totem.jpg.

(*Bottom right*) Sue Fleckney, CC BY-SA 2.0, Wikimedia Commons. https://commons.wikimedia.org/wiki/File:Vulture_Stone,_Gobekli_Tepe,_Sanliurfa,_South-east_Anatolia,_Turkey.jpg.

CHAPTER 4

All figures by the author unless otherwise noted.

Page 76: Images courtesy of Dr. Jonathan Mark Kenoyer.

Page 77:

(*Top Left*) Original specimen in the collection of the National Museum of India, New Delhi, India. Unknown, Indus Valley Civilization sealmaker from Mohenjodaro archaeological site, Public domain, Wikimedia Commons. https://commons.wikimedia.org/wiki/File:Shiva_Pashupati.jpg.

(*Top Right*) Sanyam Bahga, CC BY-SA 3.0, Wikimedia Commons. https://commons.wikimedia.org/wiki/File:BijapurShivaMonument.jpg.

(*Bottom Left*) Original specimen in the CSMVS Museum in Mumbai, India. Jean-Pierre Dalbéra, Paris, France, CC BY 2.0, Wikimedia Commons.https://commons.wikimedia.org/wiki/File:A_sculpture_of_Brahma_Hindu_deity_at_Museum_CSMVS_Mumbai.jpg.

(*Bottom right*) Original in the Walters Art Museum, Baltimore, Maryland. Anonymous, Walters Art Museum, Public domain, Wikimedia Commons. https://commons.wikimedia.org/wiki/File:Cambodian_-_The_Hindu_God_Brahma_-_Walters_542734.jpg.

Page 92: Mostly after Srinivisan (2011) and Bowker (2006).

Page 96:

(*Left*) Raja Ravi Varma, Public domain, Wikimedia Commons. https://commons.wikimedia.org/wiki/File:Raja_Ravi_Varma,_Goddess_Lakshmi,_1896.jpg.

(*Right*) Raja Ravi Varma, Public domain, Wikimedia Commons.https://en.wikipedia.org/wiki/Kali#/media/File:Kali_by_Raja_Ravi_Varma.jpg.

Page 107: Mostly after Flügel (2006).

Page 110: Mostly after Buswell and Lopez (2014), Bowker (2006), and Sujato (2006).

Page 112:

(*Left*) Tevaprapas Makklay, Public domain, Wikimedia Commons. https://commons.wikimedia.org/wiki/File:Buddha_in_Sarnath_Museum_(Dhammajak_Mutra).jpg.

(*Right*) Jean-Pierre Dalbéra, Paris, France, CC BY 2.0, Wikimedia Commons. https://commons.wikimedia.org/wiki/File:Le_temple_de_Chennakesava_(Somanathapura,_Inde)_(14487936463).jpg.

Page 127: Mostly after Buswell and Lopez (2014) and Bowker (2006).

Page 129: Mostly after Buswell and Lopez (2014).

Page 126: Modified by the author. Drew Heath, CC BY-SA 4.0, Wikimedia Commons. https://commons.wikimedia.org/wiki/File:%E5%96%84%E7%BC%98%E5%AF%BA-2.jpg.

Page 135: Modified by the author. Nyx Ning, CC BY-SA 3.0, Wikimedia Commons. https://commons.wikimedia.org/wiki/File:%E6%B2%B3%E5%8D%97_%E9%B2%81%E5%B1%B1%E5%8E%BF_%E4%B8%96%E7%95%8C%E6%9C%80%E9%AB%98%E7%9A%84%E4%BD%9B%E6%95%99%E9%80%A0%E5%83%8F_-_panoramio.jpg.

Page 138: (*Top*) Photographer Oleg Yunakov. Cropped and slightly modified with photoshop. Used under a Creative Commons license. File:Hamandir Sahib (Golden Temple).jpg (*Bottom*) Author Daniel Romanson. Cropped and slightly modified using photoshop. Used under a Creative Commons license. File: https://commons.wikimedia.org/wiki/File:Golden_temple_aerial_shot_in_Amritsar.jpg.

Page 141: Mostly after Molloy, 2018; Bowker, 2006; Nobutako et al., 2003, and the British Taoist Association website (accessed February 16, 1951).

Page 143: (*Left*) Original statue in the collection of the Asian Art Museum in San Francisco, California. Modified by the author. CC BY-SA 4.0, Wikimedia Commons. https://en.wikipedia.org/wiki/Doumu#/media/File:The_Taoist_deity_Doumu._Dehua_porcelain._Qing,_1700-1800._Asian_Art_Museum.jpg (*Right*) Author unknown. Wikimedia; used under a Creative Commons license. File:Mawangdui LaoTsu Ms2.JPG.

Page 146: (*Left*) Original carving in the Guimet Museum in Paris, France. Modified by the author. © Marie-Lan Nguyen, CC-BY 4.0, Wikimedia Commons. https://commons

.wikimedia.org/wiki/File:Votive_stele_Three_Pure_Ones_Guimet_EO1324.jpg(*Right*)Original in the Hunan Province Puseum, China. Wikimedia Commons File: Mawangdui Lao Tsu Ms2.JPG.

Page 155. Tsukioka Yoshitoshi (1839–1892), Public domain, Wikimedia Commons. https://commons.wikimedia.org/wiki/File:Emperor_Jimmu.jpg.

CHAPTER 5

All figures by the author unless otherwise noted.

Page 174: Nationalmuseet, Roberto Fortuna of Kira Ursem, CC BY-SA 3.0, Wikimedia Commons. https://commons.wikimedia.org/wiki/File:Gundestrupkedlen-_00054_(cropped).jpg.

Page 178:

(*Left*) Berig, CC BY-SA 4.0, Wikimedia Commons. https://en.wikipedia.org/wiki/Runestone#/media/File:S%C3%B6_111,_Stenkvista.jpg.

(*Right*) Public domain, Wikimedia Commons. https://en.wikipedia.org/wiki/Thor%27s_Fight_with_the_Giants.

Page 184: Original in the British Museum, London, UK. Osama Shukir Muhammad Amin FRCP (Glasg), CC BY-SA 4.0, Wikimedia Commons.

https://commons.wikimedia.org/w/index.php?search=ashur&title=Special:MediaSearch&go=Go&type=image.

Page 185: M. Chohan, Public domain, Wikimedia Commons. https://commons.wikimedia.org/wiki/File:Iraq;_Nimrud_-_Assyria,_Lamassu%27s_Guarding_Palace_Entrance.jpg.

Page 191: Modified by the author. Rémih, CC BY-SA 3.0, Wikimedia Commons. https://commons.wikimedia.org/wiki/File:Medinet_Habu_Ramses_III3.JPG.

Page 193 top figure: Specimen in the Field Museum Anthropology Collection. Photo by John Weinstein, Field Museum, Chicago, Illinois.

Page 193 bottom figure: Photo by the author of diorama in the Field Museum, Chicago, Illinois.

Page 206:

(*Left*) Modified by the author. Quatremère de Quincy, Public domain, Wikimedia Commons. https://commons.wikimedia.org/wiki/File:Le_Jupiter_Olympien_ou_l%27art_de_la_sculpture_antique.jpg.

(*Right*) Modified by the author. Andrew Bossi, CC BY-SA 2.0, Wikimedia Commons. https://commons.wikimedia.org/wiki/File:8646_-_St_Petersburg_-_Hermitage_-_Jupiter.jpg.

Page 209: Redrawn by the author. NOAA, Public domain, Wikimedia Commons. https://commons.wikimedia.org/wiki/File:Beringia_land_bridge-noaagov.gif.

Page 212:

(*Left* color image) Modified by the author. Vyacheslav Bukharov, CC BY-SA 4.0, Wikimedia Commons. https://commons.wikimedia.org/wiki/File:UOLE_museum-19.JPG.

(*Left* black-and-white image on gray background) Modified by the author. Tolmachev V. Y., Public domain, Wikimedia Commons. https://commons.wikimedia.org/wiki/File:Shigir_idol.jpg.

(*Center*) Drawing by Ravenstein, from Ravenstein, 1861.

(*Right*) Photo by John Weinstein, Field Museum, Chicago, Illinois.

Page 222:

(*Top left*) Audrey and George Delange, Attribution, Wikimedia Commons. https://commons.wikimedia.org/wiki/File:La_Venta_Stele_19_(Delange).jpg.

(*Top right*) Modified by author. Unknown, Public domain, Wikimedia Commons. https://commons.wikimedia.org/wiki/File:Quetzalcoatl_telleriano.jpg.

(*Bottom*) Modified by author. Diego Delso, CC BY-SA 3.0, Wikimedia Commons. https://commons.wikimedia.org/wiki/File:Teotihuac%C3%A1n,_M%C3%A9xico,_2013-10-13,_DD_80.JPG.

Page 242:

(*Left*) Redrawn from the Norte Chico gourd image taken from the photo archives of the Field Museum, Chicago, Illinois.

(*Middle*) Antonio Raimondi (1824–1890), Public domain, Wikimedia Commons. Originally drawn from carving on what became known as the Raimondi Stele monolith. https://commons.wikimedia.org/wiki/File:Raimondi_Stela_(Chavin_de_Huantar).png.

(*Right*) Redrawn from a photo of a Wari pottery fragment in the collection of Museo Historico Regional Hipolito, Ayacucho, Peru; based on photo by William Isbell illustrated on the Cotsen Institute of Archaeology website for DIG 15.44.

Page 247:

(*Left*) Mhwater, Public domain, Wikimedia Commons. https://en.wikipedia.org/wiki/Gate_of_the_Sun#/media/File:Zonnepoort_tiwanaku.jpg.

(*Right*) Modified by author. Charles Wiener (1880), Public domain, Wikimedia Commons https://commons.wikimedia.org/wiki/File:2Wiener-Tintin-Dieu_Soleil.jpg.

Page 251: Modified by author. Rubyk, CC BY-SA 2.5, Wikimedia Commons. https://commons.wikimedia.org/wiki/File:MachuPichuSacredValley_fir000202_edit.jpg.

CHAPTER 6

All figures by the author unless otherwise noted.

Page 264: With permission of SMB Agyptisches Museum und Papyrussammlung, SMB. Photo by Margarete Busing. Original specimen in the Neues Museum, Egyptian Museum of Berlin, Germany. Neoclassicism Enthusiast, CC BY-SA 4.0, Wikimedia Commons. https://commons.wikimedia.org/wiki/File:Relief_depicting_Akhenaton_and_Nefertiti_with_three_of_their_daughters_under_the_rays_of_Aton_01_(cropped).jpg.

Page 268:

(*Left*) Redrawn by the author.

(*Right*) Figure flipped horizontally. Furfur, CC BY-SA 4.0, Wikimedia Commons. https://commons.wikimedia.org/wiki/File:Faravahar_neu.svg.

Page 271: Modified by the author. From Shoemaker (1912), Public domain, Wikimedia Commons. https://commons.wikimedia.org/wiki/File:Tower_of_silence_interior.jpg.

Page 279: Original in the collection of the Hermitage Museum, Saint Petersburg, Russia. Rembrandt, Public domain, Wikimedia Commons. https://commons.wikimedia.org/wiki/File:Rembrandt_Abraham%27s_Sacrifice_(Hermitage).jpg.

CHAPTER 7

All figures by the author unless otherwise noted.

Page 290: Modified by the author. Oldtidens_Israel_&_Judea.svg: FinnWikiNoderivative work, Richardprins, CC BY-SA 3.0, Wikimedia Commons. https://commons.wikimedia.org/wiki/File:Kingdoms_of_Israel_and_Judah_map_830.svg.

Page 297: (*Top*) Photo by Ariely for Wikimedia. Cropped here. Used under Creative Commons liscense. (*Bottom*) Modified by the author. Gary Bembridge, London, UK, CC BY 2.0, Wikimedia Commons. https://commons.wikimedia.org/wiki/File:Western_Wall_and_Dome_of_the_Rock_Jerusalem_-10_(33400391921).jpg.

Page 313: *Top row, left to right.*

(*Top far left*) Yale University Art Gallery, Public domain, Wikimedia Commons. https://commons.wikimedia.org/wiki/File:Christ_Healing_the_Paralytic_-_Dura-Europos_circa_232.jpg.

(*Top left center*) Modified by the author. Unknown, Public domain, Wikimedia Commons. https://commons.wikimedia.org/wiki/File:Christ_with_beard.jpg.

(*Top right center*) Modified by the author. Unknown artist (sixth century), Public domain, Wikimedia Commons. https://commons.wikimedia.org/wiki/File:Spas_vsederzhitel_sinay.jpg.

(*Top far right*) Modified by the author. Original in the Riverside Church, New York. Heinrich Hofmann (1889), Public domain, Wikimedia Commons. https://commons.wikimedia.org/wiki/File:Christ,_by_Heinrich_Hofmann.jpg.

(*Bottom*) Modified by the author. Photograph by Vanoostjoost. Pierre-Jules Jollivet (1794–1871), CC BY-SA 3.0, Wikimedia Commons. https://commons.wikimedia.org/wiki/File:Saint-Vincent-de-Paul_Trinit%C3%A9_de_Jules_Jollivet.jpg.

CHAPTER 8

All figures by the author unless otherwise noted.

Page 336: Photograph by Marie-Lan Nguyen. BnF Museum, Public domain, Wikimedia Commons. https://commons.wikimedia.org/wiki/File:Constantine_multiple_CdM_Beistegui_233.jpg.

Page 340: Redrawn after the earliest known illustration of the Shield of the Trinity, from around 1210 CE (Cotton Faustina manuscript B. VII, folio 42v). That illustration was by an unknown scribe for *Compendium Historiae in Genealogia Christi* written by French Theologian Peter of Poitiers (1130–1215 CE). AnonMoos, Public domain, Wikimedia Commons. https://commons.wikimedia.org/wiki/File:Shield-Trinity-Scutum-Fidei-English.png.

Page 343: Modified by the author. Geuiwogbil, CC BY-SA 3.0, Wikimedia Commons. https://commons.wikimedia.org/wiki/File:Theodosius_I%27s_empire.png.

Page 358: Photos by the author. Published with permission of Techny Towers Retreat and Conference Center, Techny, Illinois.

Page 359:

(*Top*) Redrawn by the author. Central Intelligence Agency, Washington, DC (2008), Public domain, Wikimedia Commons. https://commons.wikimedia.org/wiki/File:Map_of_Vatican_City.jpg.

(*Bottom*) Modified by the author. Photo by Francois Malan. Dfmalan, CC BY-SA 3.0, Wikimedia Commons. https://commons.wikimedia.org/wiki/File:Vatican_StPeter_Square.jpg.

CHAPTER 9

All figures by the author unless otherwise noted.

Page 371: Ferdinand Pauwels (1872), Public domain, Wikimedia Commons. https://commons.wikimedia.org/wiki/File:Luther95theses.jpg.

Page 396:

(*Left*) Original portrait in the Thyssen-Bornemisza Museum, Madrid, Spain. Hans Holbein the Younger, Public domain, Wikimedia Commons. https://commons.wikimedia.org/wiki/File:Hans_Holbein,_the_Younger,_Around_1497-1543_-_Portrait_of_Henry_VIII_of_England_-_Google_Art_Project.jpg.

(*Right*) Unknown source, Public domain, Wikimedia Commons. https://commons.wikimedia.org/wiki/File:Book_of_common_prayer_1549.jpg.

CHAPTER 10

All figures by the author unless otherwise noted.

Page 419: Drawing from a Gnostic gem in Bernard de Montfaicon's *L'antiquite expliquee et representee en figures* (1722), Public domain, Wikimedia Commons. https://commons.wikimedia.org/wiki/File:Lion-faced_deity.jpg.

Page 430:

(*Left*) Modified by the author. http://www.csct.ugent.be/file/62, Public domain, Wikimedia Commons. https://commons.wikimedia.org/wiki/File:Manichaean_Diagram_of_the_Universe_(Detail_12).jpg.

(*Right*) Modified by the author. Zsuzsanna Gulácsi, *Mani's Pictures: The Didactic Images of the Manichaeans from Sasanian Mesopotamia to Uygur Central Asia and Tang-Ming China*, p. 274, Public domain, Wikimedia Commons. https://commons.wikimedia.org/wiki/File:Jesus_Image_on_a_Manichaean_Temple_Banner.jpg.

CHAPTER 11

All figures by the author unless otherwise noted.

Page 443: Modified by the author. Original painting in the British Library, London, UK. Public domain, Wikimedia Commons. https://commons.wikimedia.org/wiki/File:Miraj_by_Sultan_Muhammad.jpg.

Page 447: Modified by the author. Tahir mq, CC BY-SA 4.0, Wikimedia Commons. https://commons.wikimedia.org/wiki/File:Kaaba_111.jpg.

Page 455. Modified by the author. DieBuche, Public domain, Wikimedia Commons. https://commons.wikimedia.org/wiki/File:Map_of_expansion_of_Caliphate.svg.

Page 478: Michael D. Photos (2018), CC BY-SA 4.0, Wikimedia Commons. https://commons.wikimedia.org/wiki/File:Baha%27i_Temple_at_Dusk.jpg.

CHAPTER 12

All figures by the author based on data from the following sources:

Page 488: Based on Johnson and Grim (2013) and Pew Research Center (2012b).

Page 490: Based on Johnson and Grim (2013).

SUBJECT INDEX

GPSR Authorized Representative: Easy Access System Europe, Mustamäe tee 50, 10621 Tallinn, Estonia, gpsr.requests@easproject.com

www.ingramcontent.com/pod-product-compliance
Ingram Content Group UK Ltd.
Pitfield, Milton Keynes, MK11 3LW, UK
UKHW050644171025
463847UK00024B/1

* 9 7 8 0 2 3 1 2 1 2 3 0 4 *